HANDBOOK OF PLATELET PHYSIOLOGY AND PHARMACOLOGY

HANDBOOK OF PLATELET PHYSIOLOGY AND PHARMACOLOGY

Edited by

Gundu H. R. Rao

University of Minnesota

KLUWER ACADEMIC PUBLISHERS
Boston / Dordrecht / London

Distributors for North, Central and South America:
Kluwer Academic Publishers
101 Philip Drive
Assinippi Park
Norwell, Massachusetts 02061 USA
Telephone (781) 871-6600
Fax (781) 871-6528
E-Mail <kluwer@wkap.com>

Distributors for all other countries:
Kluwer Academic Publishers Group
Distribution Centre
Post Office Box 322
3300 AH Dordrecht, THE NETHERLANDS
Telephone 31 78 6392 392
Fax 31 78 6546 474
E-Mail <services@wkap.nl>

Electronic Services <http://www.wkap.nl>

Library of Congress Cataloging-in-Publication Data

Handbook of platelet physiology and pharmacology / edited by Gundu H.R. Rao.
p. cm.
Includes bibliographical references and index.
ISBN 0-7923-8538-1 (alk. paper)
1. Blood platelets Handbooks, manuals, etc. I. Rao, Gundu H.R.. 1938- .
QP97.H36 1999
612.1'17--dc21 99-27962
CIP

Printed on acid-free paper.

Contents

List of Contributors

1. Kailash C. Agarwal, Ph. D.
Department of
Molecular Pharmacology,
Brown University
Providence, RI
02912, USA

2. Colin N. Baigent, Ph. D.
ATT Collaboration
CTSU Harkness Building
Radcliffe Infirmary
Woodstock Road
Oxford OX2 6HE
United Kingdom

3. Rodger L. Bick, M. D., Ph. D.
Departments of
Pathology & Pharmacology
Loyola University Med. Center
21260 South First Ave.
Maywood, IL
60153, USA

4. David C. Calverley, M. D.
USC School of Medicine
1441 Eastlake Ave
NOR MS 34
Los Angeles, CA
90033, USA

5. Thomas Chandy, Ph. D.
Chemical Engineering and
Material Sciences
University of Minnesota
Minneapolis, MN
55455, USA

6. Kenneth J. Clemetson, Ph. D.
Theodor Kocher Institut
Der Universitat Bern
Freiestrasse 1, Ch-3012 Berne
Switzerland

7. Robert W. Colman, M. D.
Thrombosis Research Center
Temple Univ. Sch. of Medicine
3400 N. Broad Street
Philadelphia, PA
19140, USA

8. Maribel Diaz-Ricart, Ph. D.
Servicio de Hemoterapia
Hospital Clinico Provincial
Villarroel 170, Barcelona
08036, Spain

9. Gines Escolar, M. D. Ph. D.
Servicio de Hemoterapia
Hospital Clinico Provincial
Villarroel 170, Barcelona
08036, Spain.

10. Daniel Fareed, B.Sc.
Departments of
Pathology & Pharmacology
Loyola University Med. Center
21260 South First Ave.
Maywood, IL
60153, USA

11. Jawed Fareed, Ph. D.
Departments of
Pathology & Pharmacology
Loyola University Med. Center
21260 South First Ave.
Maywood, IL
60153, USA

12. Deborah French, M. D.
Department of Medicine
Mount Sinai Hospital &
Medical School
One Gustave L. Levy Place
New York, NY
10029-6574, USA

13. Mony M. Frojmovic, Ph. D.
McIntyre Medical
Science Building
McGill University
3655 Drummond Street
Montreal, QB
Canada, H3G IY6

14. Nicholas J.Greco, Ph. D.
Platelet Biology Laboratory
American Red cross
15601 Crabbs Branch Way
Rockville, MD
20855, USA

15. Holm Holmsen, Ph. D.
Department of Biochemistry and
Molecular Biology
University of Bergen
Astradveien 19, Bergen
N5009, Norway

16. Debra Hoppensteadt, Ph. D.
Departments of Pathology &
Pharmacology
Loyola University Med. Center
21260 South First Ave.
Maywood, IL
60153, USA

17. Huzoor-Akbar, Ph. D.
Molecular and Cellular Biology
Department of Biological
Sciences, Irvine Hall
Athens, OH
45701, USA

18. G.A. Jamieson, Ph. D, D. Sc.
Platelet Biology Laboratory
American Red Cross
15601 Crabbs Branch Way
Rockville, MD
20855, USA

19. Gerhard J. Johnson, M. D.
Veterans Affairs Medical Center
One Veterans Way
Minneapolis, MN
55417, USA

20. Beate Kehrel, Ph. D.
Experimental and Clinical
Haemostaseology
Department of Anaesthesiology
and Intensive Care Medicine
University of Muenster
D-48149 Muenster, Germany

21. Bruce R.Lester, Ph. D,.
Knowledge Frontiers
3989 Central Ave, N. E., # 625
Minneapolis, MN
55421, USA

22. Mahadev Murthy, Ph. D.
Division Endocrinolgy,
Metabolism & Nutrition
Department of Medicine
Hennepin County Medical Center
914 South Eighth Street, D-3
Minneapolis, MN
55404. USA

23. Ellinor I. Peerschke, Ph. D.
Cornell Medical Center
New York University
525 E 68th Street, Rm F511J
New York, NY
10021, USA

24. Anna S. Radomski
Division of R and D
Lacer, S.A.
08025 Barcelona
Spain

25. Marek W. Radomski, M.D, D.Sc.
Division of R and D
Lacer, S.A., 08025 Barcelona
Spain

26. Gundu H. R. Rao, Ph. D.
Departments of Lab. Med. &
Pathol. and Biomed. Engineering
P.B. 609 UMHC
Academic Health Center
University of Minnesota
Minneapolis, MN
55455, USA

27. A. Koneti Rao, M. D.
Department of Medicine
Temple University School of
Medicine
3400 N. Broad St Rm 300-OMS
Philadelphia, PA
19140, USA

28. Gerald J. Roth, M. D.
Division of Hematology
V.A. Medical Center
1660 South Columbian Way
Seattle, WA
98108, USA

29. Anita Ryningen, Ph. D.
Department of Biochemistry
and Molecular Biology
University of Bergen
Astradveien 19
Bergen N-5009
Norway

30. Shivendra D. Shukla, Ph. D.
University of Missouri-
Columbia
517B Medical Science Building
One Hospital Drive
Columbia, MO
65212, USA

31. Cathie Sudlow, Ph. D.
ATT Collaboration
CTSU Harkness Building
Radcliffe Infirmary
Woodstock Road
Oxford OX2 6HE
United Kingdom

32. Narendra N. Tandon, Ph. D.
Thrombosis & Vascular Biology
Otsuka America Pharmaceutical
9900 Medical Center Drive
Rockville, Maryland
20850, USA

33. Jeanine M. Walenga, Ph. D.
Departments of Pathology &
Pharmacology
Loyola University
Medical Center
21260 South First Ave.
Maywood, IL
60153, USA

34. Douglas J. Weiss, D.V.M., Ph. D.
Department of Pathobiology
and Veterinary Sciences
University of Minnesota
St. Paul, MN
55108, USA

35. Helmut Wolf, M. D, Ph. D.
Departments of Pathology &
Pharmacology
Loyola University
Medical Center
21260 South First Ave.
Maywood, IL
60153, USA

PREFACE

Despite my many years of research and teaching in platelet physiology and pharmacology at the University of Minnesota, I am often confronted with conflicting opinions as to the relevance of nonnucleated platelets in human health and disease. It is fascinating to think that how cells with no apparent nucleus, have such a towering impact on concepts, dealing with often overlapping physiological (i.e. hemostasis, wound healing, etc.) and pathophysiological (i.e. thrombosis, stroke, atherosclerosis, wound healing, diabetes, inflammation and cancer) components. Although the idea of compiling new frontiers of platelet research in the form of a book was quite simple at the beginning, the project turned out to be a major undertaking from my part. At the end, I am elated that the contributors to this book were gracious enough to write chapters in their area of research expertise despite their pressing and highly valuable time. For me, it has been an humbling experience as the chapters that I have compiled, are written by people with incredible recognition for their relentless contributions over the years to strengthen the understanding of platelet physiology and pharmacology. In my opinion, this has added an immense value to the book. I am proud to have been involved in this undertaking despite several unexpected problems and delays during this project. I am confident that this book would be highly useful to the community of scientists, including graduate students, researchers, academicians, physicians and other health care professionals, and pharmaceutical industry scientists.

Circulating platelets which lack nucleus neither adhere to the vessel wall nor aggregate unless they encounter a zone of injury. Upon encountering such a zone of injury, they become almost instantly activated, which leads to their adhesion and aggregation, both reactions are of fundamental importance to hemostasis and thrombosis. Because of this reason, platelet research has clearly led the way in the continuing development of new strategies and drugs that can help prevent and treat arterial thrombosis, stroke and atherosclerosis. Unquestionably, platelet research has also impacted concepts dealing with many other diseases. Nevertheless, considerable progress has been made in the development of new antiplatelet agents in recent years. These newer agents are aimed at interrupting specific sites and pathways of platelet activation. Inhibitors of specific platelet agonist-receptor interactions include antithrombins, thromboxane A_2 receptor antagonists, and adenosine diphosphate receptor blockers (i.e. ticlopidine, clopidogrel). In addition, inhibitors of arachidonic acid metabolism and thromboxane A_2 include aspirin, newer COX-2 inhibitors, other NSAIDs, thromboxane A_2 synthase inhibitors and ω-3 fatty acids. Moreover, long awaiting drugs that block ligand binding to the platelet glycoprotein IIb/IIIa complex (i.e. tirofiban) have now entered the market.

In this book, the chapters are organized into six major sections, including *Introduction, Receptor Biology, Platelet Biochemistry, Experimental Physiology, Platelet Pathology and Platelet Pharmacology*. Authoritative chapters in each section have provided a collective strength to our initial philosophy of accomplishing a comprehensive review of current concepts in each discipline. Although every attempt has been made to provide an interdisciplinary discussion on the subject of platelets in this book, there may still be some gaps and lapses for which readers are urged to consult other articles and reviews. I have deliberately avoided going into any specific comments on reviews in order to let the imagination of the readers flow freely. I believe that the readers are intelligent enough to judge and form their own critical opinion.

I must humbly express my deep gratitude to thirty five scientists in the field for their invaluable contributions. I now honestly believe that this publication would not have been possible without their meritorious contributions.

I am deeply indebted to my dear friend and close research collaborator, Mahadev Murthy, Ph. D., Director of Research, Division of Endocrinology, Metabolism and Nutrition, Department of Medicine, Hennepin County Medical Center, Minneapolis, MN, USA, for his commitment and contribution to this project. He has spent countless hours during this project in reviewing and preparing camera ready manuscripts for final submission to the Kluwer Academic Publisher. In addition, he has written two excellent chapters for the book. I must confess that this publication would not have been completed without his generous and truly dedicated efforts.

I would like to take this opportunity to thank Charles W. Schmieg, Jr., Acquisitions Editor, Kluwer Academic Publishers, 101 Philip Drive, Assinippi Park, Norwell, MA, 02061, USA, for facilitating the publication of this book. I am specially thankful for his cooperation and patience even though this project was delayed by about four months.

Finally, I would not be in this field today without my mentor, James G. White, M. D., Regents' Professor & Associate Dean, Academic Health Center, University of Minnesota, Minneapolis, MN, USA. I humbly dedicate this publication to James G. White, M. D., who has been my mentor, teacher, associate and dear friend, during my long career in platelet research. In the end, my academic success and accomplishments over the years, would not have been possible without the support of my wife Yashoda, my daughter Aupama and my son Prashanth. I sincerely acknowledge and appreciate their patience and support throughout my career.

Gundu H. R. Rao
Professor

University of Minnesota
Minneapolis, MN
55455

1

PLATELET PHYSIOLOGY AND PHARMACOLOGY: AN OVERVIEW

Gundu H. R. Rao, Ph. D.
Departments of Laboratory Medicine & Pathology
and Biomedical Engineering Institute
Academic Health Center
University of Minnesota, Minneapolis, MN
55455, USA

1.1 INTRODUCTION

Dr. Gundu Rao has spent the last 25 years studying and teaching platelet physiology and pharmacology at the University of Minnesota. Over the years, he has made significant contributions to the understanding of platelet function and its impact on the pathogenesis of atherosclerosis and thrombosis. In addition, he has been vocal in the national and international activities that advocate prevention, early diagnosis and treatment of Coronary Artery Disease (CAD). More recently, he led the establishment of South Asian Society for Atherosclerosis and Thrombosis (SASAT). He continues to champion activities on the prevention of CAD, in the International arena.

Clinical, experimental and epidemiological studies have demonstrated a ubiquitous role for platelets in the pathogenesis of thromboembolic events and hemorrhagic disorders (1-19). Therefore, there is great need for developing specific and effective drugs capable of modulating platelet function. A thorough understanding of the mechanisms involved in regulating platelet function will facilitate the development of better antiplatelet drugs. Agonists interact at specific sites on the plasma membrane of platelets and initiate a series of signaling events (20-37). Activation signals are communicated to intracellular effector enzymes through transmembrane receptors that are coupled to GTP-binding proteins (29).

Stimulation of intracellular enzymes facilitate the formation of second messengers capable of mobilizing cytosolic free calcium from internal membrane stores. Ionized calcium is the primary bioregulator of various platelet responses (31, 32). Agonists as well as antagonists are capable of modulating cytosolic calcium levels. Elevation of cytosolic calcium stimulates phospholipase A_2, liberates arachidonic acid from membrane phospholipids, promotes assembly of filamentous actin, granule mobilization and secretion of granule contents (31). The mechanisms involved in platelet activation facilitate the expression of binding sites for fibrinogen on platelet glycoprotein (GP) 11b/111a receptor (38-45). Activated GP 11b/111a receptor binds fibrinogen and this binding seems to contribute to the formation of platelet aggregates and growth of the thrombus. Major pharmacological approaches have focused on developing inhibitors capable of preventing one or more signaling mechanisms (33). Currently available antiplatelet drugs are quite effective in preventing platelet activation, including aggregation and secretion. However, they are less effective in preventing platelet interaction with biomaterials and cell matrix components. Intense research is in progress to develop newer antiplatelet drugs. A recent review of antiplatelet clinical trials concluded that use of any antiplatelet drug significantly reduces the development of cardiovascular events relating to coronary artery disease. The articles in this book which are written by experts, provide a comprehensive review on various aspects of platelet physiology and pharmacology.

1.2 Role of platelets in hemostasis and thrombosis

Platelets contribute significantly to the normal hemostatic process (1-6). They play a critical role in the recognition of injured vasculature, formation of hemostatic plugs, prevention of bleeding, retraction of clots and wound healing (6-9). When they develop severe dysfunction they contribute to the pathogenesis of hemorrhagic disorders. Whereas, when they are hyperactive, they can initiate events leading to clinical complications associated with cardiovascular disorders (10-16). Platelets circulate in the blood as nonadhesive disc shaped cells. When activated they undergo a series of discrete transformations. The degree of activation depends upon the strength of the activating stimuli and the information available on the interactive domains of the cell matrix components. For instance, interaction with laminin will result in minimum activation and formation of focal adhesion. However, fibronectin will promote complete spreading of platelets. Whereas, collagen will induce formation of aggregates as well as secretion of granule contents. Platelet responses that are well recognized include development of stickiness, adhesion, change in shape, irreversible aggregation and secretion of granule contents. Platelet activation is a prerequisite for the formation of a hemostatic plug and arrest of bleeding at the injured site. Although formation of a hemostatic plug is a natural response to injury, the role of platelets in atherosclerosis, thrombosis and stroke are pathological manifestations (1-19).

1.3 Platelet morphology and biochemistry

Blood platelets have a discoid form in their resting state. A circumferential band of microtubules support this shape. This characteristic shape facilitates their movement in the circulating blood at the periphery of the flow close to the vessel wall (Fig. 1).

In order to relate cell structure to the functional responses, the work by White has divided anatomy of the platelet into three distinct zones, the peripheral zone, the sol-gel zone and the organelle zone (20). The peripheral zone consists of membranes and closely associated structures including lipid bilayer, cytoskeletal proteins, transmembrane proteins, various

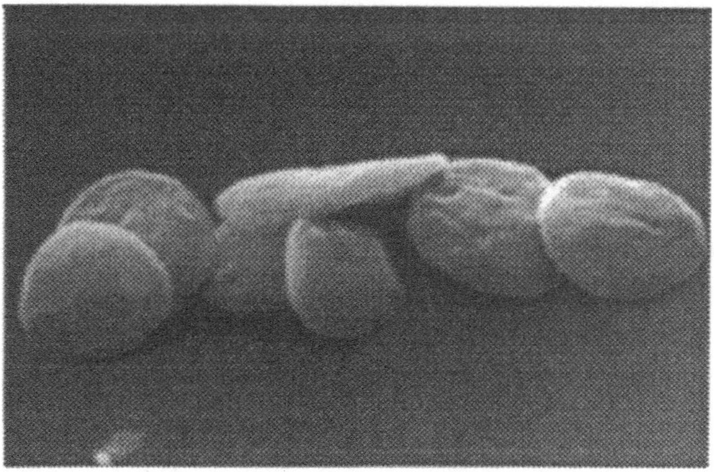

Figure 1. Scanning electronmicrograph of resting platelets The characteristic shape of circulating discoid platelets is critical for their passage in the vessel wall with smooth blood flow (Courtesy: James G. White, MD).

glycoprotein-rich interactive domains, transmembrane receptors for soluble agonists such as epinephrine, adenosine diphosphate (ADP), thrombin, thromboxane and platelet activating factor (PAF), GTP-binding proteins, integrin (GP 11b/111a, GP1a/11a, GP1c/11a) and non-integrin (GP1V, GPV, GP1b/1X) receptors and ion channels (reviewed in subsequent chapters).

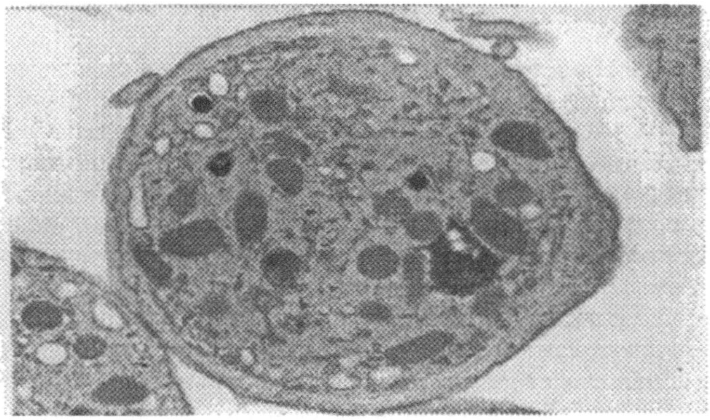

Figure 2. Electronmicrograph of a discoid platelet (sectioned in the equatorial plane) Microtubules that support the discoid form are seen beneath the plasma membrane; Organelles such as dense bodies, granules (alpha, lysosomal) and mitochondria can also be seen (Courtesy: James G. White, MD).

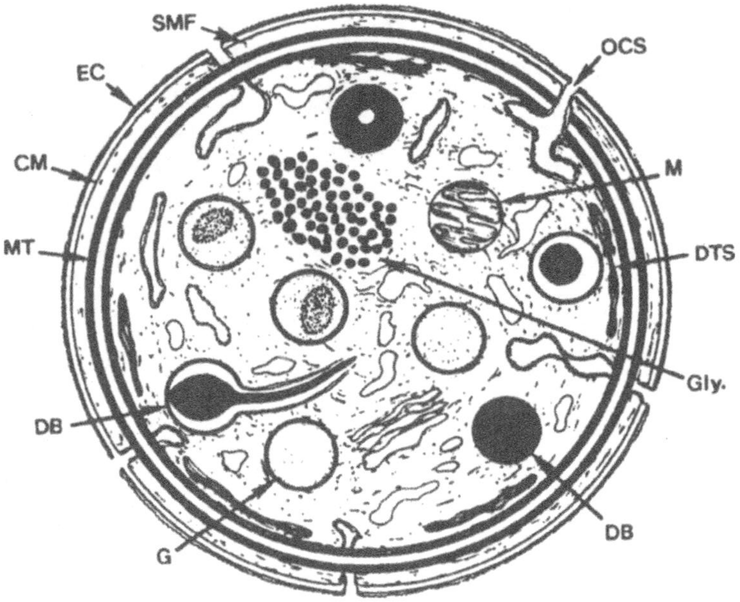

Figure 3. Schematic representation of discoid platelets: Features of a discoid platelet: EC, exterior coat; CM, cell membrane; CS, channels of the surface connected canalicular system; SMF, submembrane filaments; Sol-gel zone contains actin microfilaments, microtubule(MT), and glycogen (Gly), formed elements; α-granules (G), dense bodies (DB) and mitochondria.(M); DTS and OCS are part of the membrane system (Courtesy: James G. White, MD)

The morphological features described above support the discoid shape in resting platelets and thus, providing a contractile system that facilitates shape change, pseudopod formation, contraction and secretion. The major portion of the contractile system involves actin. Other proteins of the platelet contractile system include, myosin, tropomyosin, acting binding protein, α-actinin, gelsolin, profilin, vinculin and spectrin. The organelle zone consists of granules, dense bodies, peroxisomes, lysosomes, mitochondria and glycogen. This zone serves as the storage site for various enzymes, non-metabolic adenine nucleotides, serotonin, a variety of proteins, calcium, and antioxidants such as ascorbic acid, taurine, and glutathione. The unique complex membrane system (DTS & OCS) plays a very critical role in platelet pathophysiology. The dense dense bodies(DB) are the site for calcium, adenine nucleotides and serotonin sequestration. The DTS is the site where enzymes of arachidonic acid metabolism and prostaglandin synthesis are localized. The sol-gel zone constitutes components of the cytoplasm. It contains fibers, filaments and proteins in various states of polymerization.(Fig. 2)

The surface-connected open canalicular system (OCS) provides access to the interior for plasma-borne substances and it serves as a conduit for products secreted during the release reaction (Fig. 3).

1.4 Platelet physiology

Physiological agonists cannot penetrate the plasma membrane barrier and therefore, they must first couple to specific interactive domains on the platelet surface membrane in order to trigger the sequence of activation signals (21-29, reviewed by Ryningen and Holmsen in this book). The physiological agonists that are known to activate platelets include adenosine diphosphate (ADP), epinephrine (adrenaline), thromboxane A_2 (TxA_2), thrombin, and platelet activation factor (PAF). In addition, cell matrix components such as laminin, fibronectin, collagen and von Willebrand factor also trigger platelet activation. Platelet adherence to surfaces and resulting activation leads to the development of stickiness, change in shape, formation of pseudopods, adhesion, followed by aggregation and secretion of granule contents.(Fig. 4).

Receptors for ADP, epinephrine, thromboxane, thrombin and PAF have been well characterized (27, 31, reviewed in other chapters). Membrane spanning receptors of epinephrine, thrombin and thromboxane are coupled to the ubiquitous GTP-binding proteins. Platelets contain monomeric, low molecular weight G proteins as well as heterotrimeric membrane associated G-proteins. GTP binding to the α-subunit of G-proteins facilitates the interaction with effector enzymes, resulting in the hydrolysis of GTP to GDP, which terminates its stimulatory role (29).

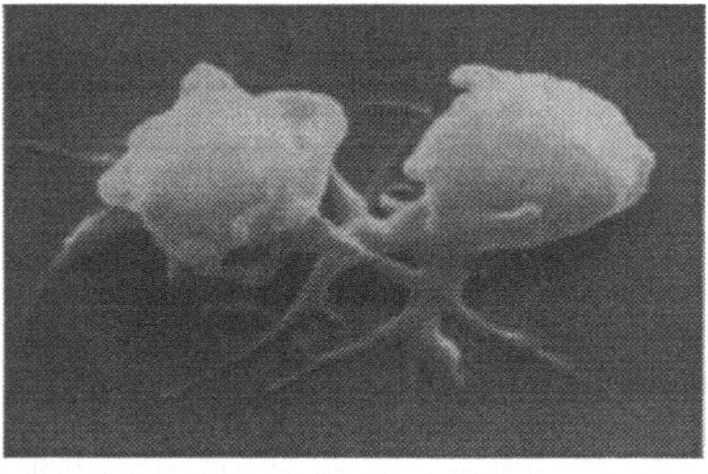

Figure 4. Scanning electronmicrograph of activated platelets Platelet activation resulting in shape change and formation of pseudopods (PS) (Courtesy: James G. White, MD).

Agonist-mediated activation of platelets stimulates phospholipase C (PLC) and it then triggers the hydrolysis of phosphatidyl inositol 4,5-bisphosphate (PIP_2), the formation of second messengers such as 1, 2-diacylglycerol (DAG) and inositol 1 ,4, 5-trisphosphate (IP_3) (30-38, also reviewed in this book). Diacylglycerol (DAG) is a substrate for protein kinase C (PKC), which is recognized as a multifunctional enzyme (25). This lipid intermediate is also a substrate for phosphatidic acid. Diacylglycerol (DAG) induces translocation of cytosolic PKC to membranes, which acts as a trigger mechanism for its activation. On the other hand, IP_3 is known to mobilize ionized calcium from internal membrane stores. However, thrombin and PAF can also mobilize calcium from the external milieu. Ionized calcium plays a central role in all platelet functional responses and it is therefore considered as an important bioregulator in platelet pathophysiology (31-36, also reviewed elsewhere in this book). Elevated cytosolic calcium is essential for the assembly of filamentous actin. Furthermore, it is also considered essential for the activation of phospholipase A_2, a key enzyme to mobilize arachidonic acid (AA) from membrane phospholipids for further metabolism (reviewed elsewhere in this book). It is well documented that free AA then is converted to cyclic endoperoxides such as PGG_2 and PGH_2 in the presence of cyclooxygenase (27, 11, 23). These transient metabolites are further transformed into novel thromboxanes by thromboxane synthetase. Thromboxane A_2 is the major metabolite of AA metabolism in platelets (31). It is a vasoconstrictor and potent platelet agonist. In endothelial cells, AA is converted to cyclic endoperoxides by cyclooxygenase, and these metabolites are further transformed to prostacylcin (PGI_2) by prostacyclin synthetase. Prostacyclin is a vasodilator and a potent platelet antagonist (31). Thus both phosphoinositol pathway and AA metabolism, contribute significantly to the activation of platelets by soluble agonists (27, 2, 10, 11, 14). These events promote the expression of an activation dependent epitope on the platelet glycoprotein GP11b/111a receptor (39-46). Activation of this receptor promotes fibrinogen binding and facilitates platelet adhesion, aggregation and growth of the thrombus. However, activation of this receptor is not essential for its interaction with surface bound fibrinogen. Apart from the agonists mentioned above, shear force also can induce platelet activation (discussed elsewhere in this book). It is believed that fibrinogen plays a role in the adhesion and aggregation of platelets under low shear rates. Furthermore, at high shear forces von Willebrand factor interaction with GP1b/1X seems to be important (31, also discussed in other chapters of this book). Circulating adhesive proteins such as fibrinogen, cell matrix components, bacterial membrane proteins, certain tumor cells and biomaterial surfaces also interact with the platelet plasma membrane at discrete domains. Binding of ligands to integrin and non-integrin receptors induce activation signaling mechanisms (27, 31, 62, also discussed in other chapters of this book). Binding results in the activation and stimulation of various effector enzymes and formation of second messengers, leading to aggregation and secretion of granule contents. Specific mechanisms involved in the process of centralization of granules and release of their contents are poorly understood(26, 27, 31).

1.5 Altered physiology and function

Many investigators have attempted to correlate the in vitro functional response of platelets to clinical manifestations of thrombotic episodes or bleeding diathesis (8, 10, 12, 15, 30, 46, 47, also discussed in other chapters). Yet, it remains a difficult task to establish a clear relationship between specific functional responses and their role in normal hemostasis. However, the presence of functional glycoprotein GP1b/1X and GP11b/111a receptors and the ability of platelets to undergo shape change, spread, become sticky, irreversibly aggregate, or release granule contents, are considered essential for normal hemostatic function. Drug-induced impairment of signal transduction and biochemical lesions, resulting from procedures such as surgery, dialysis, angioplasty, may also lead to platelet dysfunction. Platelets may also develop biochemical lesions during storage and thus, may carry dysfunctional characteristics making them less suitable for transfusion. Although intracellular calcium elevation is considered important for eliciting platelet responses such as contraction and secretion, its role remains questionable in platelet shape change and other processes such as the development of stickiness, fibrinogen binding, the adhesion or interaction with the vascular subendothelium (48-49, also discussed in other chapters).

Action of weak agonists for mediating secretion of granule contents depends upon availability of TxA_2. However, bovine and equine platelets which do not aggregate in response to the action of the arachidonic acid metabolite, thromboxane, support normal in vivo hemostasis in these animals. Similarly, the majority of dogs have platelets that do not respond to AA metabolites (50). On the other hand, epinephrine exposure restores the sensitivity of these refractory platelets to the action of AA (51). The observations in our laboratory have also led to the development of a novel concept of membrane modulation (17). The concept seems to offer a reasonable explanation for observed functional response of platelets with no detectable cyclooxygenase activity, which can still support normal hemostasis in adults. These findings raise the possibility that cyclooxygenase may not be obligatory for normal hemostasis, but may likely play an important role in thrombosis. Other laboratories have also shown such phenomenon, particularly for adrenaline and noradrenaline. Some reports also suggest that adrenaline stimulation of α-receptors may in fact amplify or restore signal transduction mechanisms in platelets.

Studies by Scrutton et. al. and Stormorken and his associates, have shown compromised functional response to epinephrine in apparently normal individuals (52, 53). Weiss et. al. have described secretion defects from patients with bleeding disorders (54). White et. al. have followed functional response of platelets of patients with diabetes and Hermasky-Pudlak Syndrome (HPS) whose platelets lack dense bodies. Platelets of patients with HPS exhibit compromised response to the action of agonists (55). Hardisty et. al., and Ware and associates, have provided further evidence for altered signal transduction mechanisms. These and other studies seem to suggest that an impaired intracellular calcium flux may be the chief cause of platelet dysfunction (56-58, discussed in other chapters).

A brief review of the literature on platelet disorders indicates that there is still a great deal of confusion in establishing a clear correlation between specific signal-driven responses and platelet function despite voluminous information on the biochemistry of signal transduction, second messengers and calcium pools and fluxes. It is further complicated by discrepancies between human and animal models. For instance, the dogma that cows, horses, sheep and goat, could still maintain normal hemostasis despite their platelets inability to respond to AA metabolites or epinephrine (17, also discussed in other chapters in this book). On the contrary, platelets obtained after the cardiopulmonary by-pass surgery (CPBS), not only have diminished number of glycoprotein 1b/1X receptors but also are highly refractory to thrombin. This functional defect seems to offer a reasonable explanation for reported post-CPBS bleeding. Therefore, a general hypothesis can be readily advanced that a compromise of one or more signaling events may clearly lead to a defective or diminished response, which then could impact platelet-related pathophysiology. For instance, there is evidence that altered membrane phospholipid composition may lead to the inactivation of α-adrenoreceptor function (59). Studies from our laboratory have shown that exogenously added phospholipase A_2 indeed adversely affect the platelet response, particularly to epinephrine (60). Recent studies have shown that in many malignant and inflammatory conditions plasma phospholipase A_2 activity may be elevated (61). It is therefore reasonable to rationalize that membrane phospholipid and fatty acid changes brought about by disease processes, oxidative stress and aging process could indeed contribute to the altered signal-driven cellular mechanisms and thus, impacting the overall pathophysiology. For example, the α-adrenergic receptor which mediates catecholamine responses, may be compromised by membrane modulation (17). It is not unreasonable to assume that lipid changes which could readily affect the stereospecific lipid bilayer assembly that may be important for pathophysiological responses. However, further studies may be required to document such relationships more precisely, linking to signal transducing mechanisms, second messenger systems, calcium fluxes in both human and animal models. Furthermore, a careful correlation between signal-driven responses and platelet function will lead to a better understanding of platelet pathophysiology (27, 31, also discussed in other chapters).

1.6 Platelet pharmacology

Earlier studies on platelet biochemistry, physiology and function suggest that release of granule contents is essential for recruitment of platelets, irreversible aggregation and formation of thrombi (31, 62). Discovery of novel vasoactive thromboxanes and prostacyclins from AA, established a crucial link between these AA metabolites and platelet-endothelial interactions (63, 64). As discussed earlier, agonist-mediated activation leads to the stimulation of phospholipases and formation of second messengers that facilitate mobilization of cytosolic free calcium, centralization of granules, assembly of filamentous actin, expression of binding sites for adhesion molecules, contraction, secretion of granules and irreversible aggregation (27, 31, 62, 12, 13, 14, also discussed in other chapters). These observations clearly support that the initial receptor-coupling triggers a complex series of biochemical events that eventually transforms into a functional response. Based on these new developments and understanding on specific platelet activation pathways, many antiplatelet drugs that prevent discrete steps have been developed (62). Early pharmacological approaches have focused on three specific areas: 1) the development of drugs that prevent release of AA, and its conversion to various prostanoids, including receptor antagonists; 2) the development of drugs that stimulate adenylyl and guanylyl cyclase enzymes and that inhibit phosphodiesterase; 3) the development of calcium antagonists (62, also discussed in other chapters). Studies from our laboratory have demonstrated that the inhibition of AA metabolism alone is insufficient to block platelet function (62-79). Furthermore, these studies demonstrate that irreversible platelet aggregation can be accomplished independent of prostanoids (PGG_2, PGH_2 and TxA_2) or released granule contents (ADP, serotonin) (62-75). In separate studies it has been shown that epinephrine induced membrane modulation can restore the sensitivity of drug-induced refractory platelets to the action of agonists (17). Studies with calcium specific chelators suggest that platelet responses such as shape change, stickiness, irreversible aggregation and activation caused by extracellular matrix components do not require elevation of cytosolic calcium (49).

Some of the clinical complications attributed to increased platelet activity include acute myocardial infarction, stroke (hemorrhagic and thrombotic), unstable angina, reocclusion following coronary thrombolysis, occlusion during thromboplasty and coronary restenosis and tumor metastasis (8, 10, 12-15, 30, 47). In addition to these clinical situations, platelets are known to contribute to atrial fibrillation, pulmonary embolism, and left ventricular dysfunction in a highly significant manner. In this regard, inhibitors of platelet function have proven effective in the secondary prevention of clinical vascular complications in patients with cardiovascular and cerebrovascular disease (80-87). The encouraging results obtained with low to moderate doses of aspirin, a known inhibitor of cyclooxygenase, in patients with coronary artery disease have prompted the use of antiplatelet drugs in the secondary prevention of cardiovascular events associated with coronary artery disease. A recent review of over 400 clinical trials revealed that use of any antiplatelet drug significantly reduced the risk for developing cardiac events associated with coronary artery disease (86, also discussed in other chapters). These clinical observations have prompted a greater need and urgency to develop new effective antiplatelet drugs.

1.7 Platelet function inhibitory drugs

Screening for drugs that prevent platelet activation

Promising new drugs that could affect/prevent one or more platelet responses (i.e. adhesion, spreading, aggregation and secretion of granule contents) can be readily screened. One of the extensively used methods involves measurement of platelet aggregation by aggregometry. Furthermore, one can monitor the secretion of adenosine triphosphate from the granules concurrent with platelet aggregation response, with the luciferin-luciferase system. Alternately, one can use flow cytometry to follow the expression of granule associated protein (GMP-140; P-selectin; CD 62) on activated platelets with fluorescent coupled monoclonal antibodies (88). By using fluorescent coupled monoclonal antibodies for human fibrinogen one can detect bound fibrinogen on activated platelets. The presence of fibrinogen and the absence of P-selectin would indicate platelet activation independent of secretion. One can also follow platelet micro particle formation using flow cytometry. This newly emerging technique could be used for rapid drug screening very effectively.

In addition to these techniques developed and used for measuring platelet activation in suspension, one can follow platelet adhesion, spreading and aggregation on various natural and synthetic surfaces. Various test materials can be exposed to platelets in plasma, buffer or in whole blood. If the use of anticoagulants are not preferred, one can draw blood directly on to the surface to be tested. After the appropriate incubation period, the degree of activation can be measured by using standard microscopical techniques. One can also use platelet-specific FITC-coupled monoclonal antibodies or phalloidin rhodamine to stain platelets on surfaces. Such preparations can be imaged using fluorescence microscopy.

To some extent, platelet interaction on surfaces is mediated by shear force. Therefore it is essential to screen antiplatelet drugs under flow conditions. Two different techniques are used to monitor platelet interaction on surfaces. The classical Baumgartner technique employs denuded rabbit aorta to evaluate platelet interaction with basement membrane components (89). The "flat chamber" technique uses a chamber in which cover slips coated with various test materials can be exposed to flowing blood (90). One can get a layer of cell matrix components for testing, by growing endothelial cells on cover slips and stripping them off the glass surface after they reach confluency.

Once the drugs are screened using appropriate methods, promising new compounds need to be tested in animal models for their safety, dose and efficacy. It is very important to remember that platelets of different species vary widely in their responses to agonists (91, also discussed in other chapters). For instance, majority of dogs have platelets that do not respond to AA with aggregation (50). However, when these AA refractory platelets are exposed to epinephrine first they regain their sensitivity to the action of AA and aggregate irreversibly (51). This phenomenon of membrane modulation observed in dog and human platelets does not seem to work in many other species. Platelets of sheep, goat, cow and horse do not respond with aggregation when exposed to epinephrine or AA. The large variations observed in platelet responses to aggregating agents that have been reported for different animal species warrant serious consideration in the choice of animal model for in vivo drug testing.

Platelet activation inhibitory drugs

Inhibitors of platelet activation are commonly described as antiplatelet or antithrombotic drugs. They can be grouped under the following headings: 1) Cyclooxygenase inhibitors, 2) Thromboxane synthetase inhibitors, 3) Receptor antagonists, 4) Adenylyl cyclase/ guanylyl cyclase stimulators, 5) Phosphodiesterase inhibitors, 6) Serine protease inhibitors, 7) Calcium antagonists, 8) Miscellaneous drugs. Some of the known antiplatelet drugs are listed in Table 1.

Most commonly used antiplatelet/antithrombotic drugs are aspirin, nitric oxide donors, dipyridamole, ticlopidine, adenylyl/guanylyl cyclase stimulators, phosphodiesterase inhibitors, calcium antagonists, heparin and coumadin (62. 80-86, also discussed in other chapters). Aspirin is the most widely used drug for the prevention, treatment and prophylaxis of ischemic heart disease (80-86). Several major clinical trials have been conducted with various doses of aspirin, ranging from 80 mg to several grams per day (62, 86). The outcome of these studies seem to indicate conclusively that antiplatelet drugs provide significant protection against clinical complications. The results with antiplatelet drugs are highly significant in clinical terms even though the protection is not 100 %. However, complex drug-disease interactions could easily explain the effectiveness of antiplatelet drugs on clinical outcomes.

Studies from our laboratory have consistently demonstrated the existence of an intrinsic mechanism that can restore the sensitivity of drug-induced refractory platelets to the action of agonists. We have termed this mechanism as "membrane modulation." This unique mechanism allows platelets to aggregate irreversibly independent of prostanoids, PAF, or dense granule contents (17, 49, 65-69). It is mediated by α-adrenergic receptor stimulation and it is calcium dependent. It facilitates calcium uptake and fibrinogen binding. Epinephrine-induced restoration of sensitivity seems to be independent of phosphatidyl inositol metabolism, formation of second messengers and elevation of cytosolic calcium. Studies with calcium chelators (Quin 2 free acid, chlortetracycline, Quin 2 AM) have shown that elevation of cytosolic calcium is not critical for adhesion, spreading or aggregation of platelets (49).

It is apparent that platelets have multiple activation signaling mechanisms. Interestingly, currently available antiplatelet drugs do not inhibit platelet activation on cell matrix components or biomaterial surfaces (27, 3, 62). Because, activation of GPIIb\IIIa is not required for its interaction with surface bound fibrinogen. Furthermore, platelets can interact with other lignads such as laminin, fibronectin, and von Willebrand factor. Therefore, platelet activation inhibitory drugs will only prevent aggreation, secretion of granule contents and growth of thrombus formation and not platelet-surface interaction.

Table 1. Antiplatelet/Antithrombotic drugs

Cyclooxygenase inhibitors:	Acetyl salicylic acid (aspirin), Nonsteroidal antiinflammatory drugs (NSAID), Phenylbutazone, Indomethacin, Ibuprofen, Fenoprofen, Fluriboprofen, Neproxen, Sulfinpyrazone, antioxidants (Butylated Hydroxytoluene, BHT; Butylated Hydroxyanisole, BHA; Dipheylamine, DPA).
Thromboxane synthetase inhibitors:	Benzydamine, Imidazole congeners, 9, 11-azo-13 oxa-15 hydroxy prostanoic acid, 9, 11-azoprosta-5-13 dienoic acid (U-51605), 9, 11(epoxmethano) prostanoic acid, 1(isopropyl-2-indloy (1)-3 pyridyl-3-ketone (L-8027).
Thromboxane receptor antagonists:	(n-penylamino)carbonyl 2-Oxazolyl-7-Oxa bicyclo Hept-2-yl methyl benzenpropanoic acid (BMS 180,291), Vapiprost (SN 309), 13 azaprostanoic acid, Isoprostane, (3 pyridinyl) alkanoic acids, (arylsulfonylamino) alkanoic acid.
Stimulators of adenylyl/guanylyl	Prostacyclin, PGE_1, PGD_2, adenosine, forskolin, coleonol; endothelium derived relaxing factor (EDRF, nitric oxide), cyclases: nitroglycerine, nitroprusside, Sin-1, nitrosoglutathione.
Phosphodiesterase inhibitors:	Dipyridamole and related compounds, RA233, RA433, VK744, VK774, caffeine, papaverine, aminophylline, theophylline, methylxanthines.
Serine protease inhibitors:	Heparin, hirudin, recombinant hirudin, hirudin analogues, peptide antagonists of thrombin receptor. Peptide aldehydes (D-phe-pro-arg-H; D-phe-pro-arg-CH2Cl).
Calcium antagonists:	cyclic adenine nucleotides (cAMP.cGMP), simulators of adenylyl/ guanylyl cyclases, inhibitors of cAMP/cGMP hydrolysis, Verapamil, Nifedipine, Diltiazem, Quin-2AM, BAPTA-AM.
Miscellaneous drugs:	Antibiotics, immune suppressive agents, antibodies for specific agonist receptors, synthetic peptide mimetics of interactive domains of cell matrix components, such as collagen, fibronectin; fibrinogen, ticlopidine, disintegrins, proteins from venoms, proteins from saliva of blood sucking animals and insects.

The knowledge gained over the years on platelet biochemistry, physiology and pharmacology has clearly led the development of antiplatelet drugs that can be effective on specific key biochemical events and thus, in the treatment of thrombotic and other platelet-related disorders. Earlier pharmacological approach seems to have focused on compounds that are effective, specifically on platelet aggregation and secretion (62). This approach was used because of the heavy emphasis on the role of secretion of granule contenets for achieving irreversible aggregation. In addition, not much was known about the mechanisms involved in platelet activation on surfaces.

In the absence of a clear understanding of the various biochemical mechanisms that underlie discrete platelet activation sequences, earlier studies concentrated on drugs that prevented agonist-mediated aggregation and secretion. However, there has been considerable progress made over the last few years in the understanding of the biochemistry of platelet activation (27, 31, also discussed in other chapters). This new knowledge on cell signaling pathways, molecular events associated with ligand receptor interactions, has given impetus for the development of specific inhibitory compounds and receptor antagonists. For instance, it has been shown that the n-terminal sequence of amino acids of thrombin receptor can mimic the action of thrombin (92, 93). Similarly, a synthetic "mini collagen" derived from Type IV collagen sequence has been shown to promote platelet activation (94). The amino acid sequence of the recognition site on various adhesive proteins for the integrin receptor contains arginine, glycine, and aspartic acid (RGD) (39-41, 95). Based on the known sequence of biologically active domains several inhibitory peptides have been synthesized and targeted to block, specifically the action of thrombin and collagen.

Antiplatelet drugs

Aspirin is the most thoroughly studied antiplatelet drug (80-86). It was used as an antiplatelet drug even before its inhibitory effect on cyclooxygenase was recognized (96, 97). Aspirin acetylates this enzyme and causes irreversible inhibition of its biological activity (98). Polyenoic acids such as eicospentaenoic acid (EPA) and docosahexaenoic acid (DHA) also impair AA metabolism by cyclooxygenase (99, 100). In addition to aspirin, there are also several other drugs that inhibit this enzyme in platelets. Apart from the inhibition of this enzyme, prostanoid synthesis could be prevented by blocking the release of the free fatty acid from the phospholipids. Since phospholipase A_2 liberates AA, inhibitors of this enzyme can be effective antiplatelet drugs. Other commonly used antiplatelet drugs include dipyridamole (persantin) and ticlopidine (thienopyridines: Ticlid, Clopidogrel), nitric oxide donors, and calcium antagonists (62, 101-110, also discussed in other chapters). Dipyridamole is a vasodilator and it has been used with aspirin in several clinical trials. This drug has also been used with warfarin (coumadin) in antithrombotic therapy. Advances in biotechnology and in biochemical separation techniques have led the way in understanding the nature of ligand-receptor interacting domains and cell activation signaling mechanisms. Undoubtedly, these new appreaches in drug designing have impacted the development of newer antiplatelet and antithrombotic drugs (110). The use of antiplatelet drugs has become an important therapeutic modality for the prevention of acute arterial thromboembolic occlusions (112). There is sufficient scientific base now to conclude that antiplatelet drugs are clinically effective in significantly lowering the risk of developing cardiovascular events associated with coronary artery disease (86,). Aspirin remains as the drug of choice for the treatment of coronary artery disease (CAD). There is sufficient evidence that low-to-moderate dose of aspirin is as effective as any other antiplatelet drug in reducing the risk and clinical complications associated with CAD (80-83, 109).

Antithrombotic drugs

Antithrombotic drugs have been in use clinically for over half a century (110-114, 116).

The first clinical trial was reported in 1948. In the same year, the American Heart Association recommended the use of anticoagulant therapy for the treatment of coronary thrombosis.(112). Heparin was one of the earliest antithrombotic drugs available for therapeutic use. Dicoumarol and warfarin became available for therapeutic use later. In the early years, there was considerable use of anticoagulants. However, there has been considerable shift towards the use of antiplatelet drugs in the treatment of CAD, in view of the large body of evidence available to implicate platelet hyperactivation in the pathophysiology of cardiovascular events. Coronary thrombus is by and large platele-rich. Earlier studies from our laboratory demonstrated that preformed aggregates can be dissociated by antiplatelet drugs(117). In these studies none of the thrombolytic drugs caused dissociation of platelet aggregates. After several clinical trials and the development of newer atniplatelet drugs(GP11b\111a antagonists) our earlier observation has been confirmed (118). It is important to note that platelet activation, thrombin generation and fibrin polymerization play a critical role in the development of thrombus. Therefore, antiplatelet drugs, antithrombotic compounds and thrombolytic agents are important in the treatment of coronary thrombosis (62, 109-114, 117).

New antithrombotic drugs represent a wide spectrum of natural, synthetic, semisynthetic compounds produced by new biotechnology techniques, with marked differences in their chemical composition, properties, biochemical and pharmacological actions (110-114). Hirudin is a single chain carbohydrate free polypeptide. Unlike heparin it can inactivate thrombus bound thrombin. It is a potent inhibitor of thrombin and binds with high affinity at the substrate binding site. Hirulog (bivalirudin) is a bifunctional 20-amino acid peptide developed to mimic the interactive domain of hirudin. Argatroban is an arginine-derivative which inhibits thrombin with great affinity and is considered a better inhibitor of thrombin action than heparin. Many more inhibitors of thrombin action are available including napasgatran (R046-6240), inogatran (pINN), efegatran sulfate (Ly294468), Du P714 and low molecular weight heparins (110-114). Badimon et. al. in a recent review concluded that the unique effects of specific thrombin inhibitors, including thrombin action on platelets and blood coagulation, demonstrated in experimental and preliminary clinical trials, suggest that specific thrombin inhibitors may represent a new generation of drugs in antithrombotic therapy of acute coronary syndromes (113).

Newer antiplatelet drugs include several receptor antagonists (62, 109). Platelet activation leads to the expression of binding sites for fibrinogen binding on platelet GPIIb/IIIa receptors(42-45). Activated GP11b/111a receptors bind fibrinogen and promote adhesion on surfaces as well as aggregation of platelets. This receptor is promiscuous and binds not only fibrinogen but also vonWillebrand factor, fibronectin, vitronectin. laminin and thrombospondin. These adhesion molecules have a common recognition site containing arginine-glycine-aspartic acid (RGD). This observation has prompted the development of a variety of antagonistic peptides. Monoclonal antibodies, murine or chimeric 7E3 (c7E3 Fab: abciximab, ReoPro) are the first receptor antagonists that have been studied in humans. A variety of RGD mimetics have been synthesized and tested in animals and humans. Blockade of the GP11b/111a receptor leads to effective inhibition of adhesion, platelet aggregation and thrombus formation promoted by fibrinogen (42, 43, 95). These drugs are specific for the GP11b/111 a receptor and do not interfere with the action of receptors for collagen, or von Willebrand factor.

Newer orally active non-peptide antiplatelet/antithrombotic drugs are currently being developed (109-116). Lamifiban (R044-9883) and tirofiban(MK-383) have been shown to inhibit platelet aggregation at micromolar concentrations. Tirofiban has been tested in human volunteers. The compound has been shown to be four orders of magnitude more selective for antagonizing GP11b/111a than the natural ligand, vitronectin. Lamifiban and tirofiban are the first non-antibody/non-peptide antagonists that have been subjected to large human clinical trials (109). When thrombin binds to fibrin or is associated with thrombus, it is relatively impervious to the action of heparin, the widely used anticoagulant. Direct thrombin inhibitors like D-Phe-L-Pro-L-Arg-CH2Cl(PPACK), hirudin, hirulog and others inhibit bound thrombin as effectively as free thrombin. Recent clinical trials suggest that direct thrombin inhibitors and specific platelet receptor antagonists may be the choice of drugs for antiplatelet and antithrombotic therapy (62, 109-116).

In conclusion, the concept of platelet activation and its relevance to health and disease, has evolved methodically over the last 2-3 decades, with greater understanding of biochemical pathways. These developments have clearly led to the defining of two major biochemical pathways, which are now impacting the CAD prevention and treatment strategies through new drug discoveries in a highly significant way. The first being the phosphoinositol metabolism leading to the formation of second messengers, DAG, IP3, and elevation of cytosolic calcium. The second pathway is mediated by the increase in ionized calcium, causing phospholipase A_2 activation and liberation of AA. Free AA is then converted to cyclic endoperoxides and thromboxanes. The second messengers produced by these two biochemical pathways appear to be involved in the modulation of a number of functional responses, including assembly of filamentous actin, centralization of granules, secretion of granule contents and expression of binding sites for fibrinogen. Based on the knowledge and understanding, many antiplatelet drugs that inhibit prostanoid synthesis, platelet aggregation and secretion of granule contents, have been developed over the last few years. However, the relationship between signal-driven biochemical events and platelet interaction with cell matrix components, injured vascular surface and biomaterial surfaces, remains a highly challenging field for further investigations. In any event, studies on newer antiplatelet and antithrombotic drugs have provided a great deal of excitement and have led to the discovery of many specific antagonists. Although the biochemistry of platelets remains highly complex, we now have well defined pathways that could impact platelet functional responses. It is very likely that further understanding of the role of these pathways and the specific mechanisms that underlie platelet activation on surfaces will lead to the development of newer and more specific anti-platelet drugs with greater efficacy and safety. The future for newer antiplatelet and antithrombotic drugs looks bright. In this regard, the combined therapeutic approaches should prove to be highly valuable in the treatment of coronary thrombosis, which is still a number one killer in the United States and in other industrialized countries. The compiled chapters in this book are comprehensive reviews written by authorities in their own area and therefore, the book should serve as an invaluable source of information for readers, including biochemists, pharmacologists, pharmacists, drug companies, academicians and many other groups.

1.8 Acknowledgements

The author is thankful to thank Mrs. Anupama R. Tate and Dr. Mahadev Murthy for their help in the preparation of this manuscript. This work was supported by grants from NIH

HL-18880, Laboratory Medicine & Pathology and Biomedical Engineering Institute.

References

1. Marcus AJ, Zucker MB. The physiology of blood platelets. New York: Crune and Scrutton, 1965.

2. Kowalski E, Niewiarowski S. Biochemistry of blood platelets. New York: Academic Press, 1966.

3. Caen J. Platelet aggregation. Paris: Masson and Cie, 1971.

4. Brinkhous KM, Sherman RW, Mostof FK. The Platelet. Baltimore: Williams and Wilkson Co, 1971.

5. Johnson SA. The circulating platelets. New York: Academic Press, 1971.

6. Weiss HJ. Platelets and their role in hemostasis. Ann Rev New York Acad Sci 1972; 201: 1-450

7. Ciba Foundation. Biochemistry and pharmacology of platelets. New York: Elsvier, 1975.

8. Caen JP, Cronberg S, Kubisz P. Platelet physiology and pathology. New York: Straton Intercontinental Medcial Book Co, 1977.

9. deGaetano G, Garattini S. Platelets: A multidisciplinary approach. New York: Raven Press, 1977.

10. Kwan HC, Bowie EJW. Thrombosis. Philadelphia: W. B. Saunders, 1982.

11. Phillips DR, Shuman MA. Biochemistry of platelets. New York: Academic Press, 1986.

12. Colman RW, Hirsch J, Maders VJ, Salzman EW. Hemostasis and thrombosis: basic principles and clinical practice. Philadelphia: J.B.Lippincott Co, 1987

13. Bloom Al, Thomas DP. Haemostasis and thrombosis. London: Churchill Livingstone, 1987.

14. Marguerie G, Zwaal RFA. Biochemistry and pathophysiolgy of platelet membrane. London: John Libbey, 1988.

15. Corriveau DM. Fristma GA. Haemostasis and thrombosis in the clinical laboratory. Philadelphia: J.B.Lippincott Co, 1988.

16. Mehta JL, RC. Conti. Thrombosis and Platelets in Myocardial Ischemia. Philadelphia: F.A.Davis Company, 1987.

17. Rao GHR, Whitte JG. "Epinephrine-induced platelet membrane modulation". In The platelet amine storage granule. Meyer KM, Barnes CD, eds, Boca Raton: CRC Press, 1992.

18. Authi KS, Watson SP, Kakkar VV. Mechanisms of platelet activation. New York: Academic Press, 1993.

19. von Bruchhausen F, Walter U. Platelets and their factors. Berlin: Springer-Verlag, 1997.

20. White JG. Platelet structural physiology: The ultrastructure of adhesion, secretion and aggregation in arterial thrombosis. Cardiovasc Clin. 1987; 18: 13-23

21. Hallam TJ, Sanchej A, Rink TJ. Stimulus-response coupling in human platelets. Biochem J 1984; 218: 819-827

22. Zucker MB Nachmias VT. Platelet activation. Arteriosclerosis 1985; 5: 2-18

23. Berridge MJ. The molecular basis of communication within the cell. Sci Am 1985; 253: 142-152

24. Holmsen H. Platelet metabolism and activation. Semin Hematol 1985; 22: 219-240

25. Nishizuka Y. Studies and perspectives of protein kinase C. Sci 1986; 233: 305-312

26. Knight DE. Calcium and diacylglycerol control secretion. Bio Sci Reports 1987; 7: 355-367

27. Seiss W. Molecular mechanisms of platelet activation. Physiol Rev 1989;69: 59- 178

28. Rana RS, Hokin LE. Role of phosphoinositides in transmembrane signalling. Physiol Rev 1990; 70: 115- 164

29. KaziroY, Itoh H, Kozasz T, Masato N, Satoh T. Structure and function of signal-transducing GTP-binding proteins. Ann Rev Biochem 1991; 60:349-400

30. Nachman RL. Thrombosis and atherogenesis: molecular connections. Blood 1992; 79: 1897-1906

31. Rao GHR. Physiology of blood platelet activation. Ind J Physiol Pharmacol 1993; 37: 263-275

32. Rao GHR. Signal transduction, second messengers and platelet function. J Lab Clin Med 1993; 121: 18-21

33. Rao GHR. Signal transduction, second messengers and platelet pharmacology. Pharmacol 1994; 13: 39-44

34. Massini P, Luscher EF. On the significance of the influx of calcium ion in stimulated human blood platelets. Biochim Biophys Acta1976;436: 652-663

35. Kroll MH, Schafer AI. Biochemical mechanisms of platelet activation. Blood 1989; 74: 1181-1195

36. Rink TJ, Hallam TJ. What turns platelets on? Trend in Biochm Sci 1984; 9: 53-56

37. Rittenhouse SE. Inositol lipid metabolism in the responses of stimulated platelets.Cell Calcium 1982; 3: 311-322

38. Rink TJ, Sage SO. Calcium signaling in human platelets. Ann Rv Physiol 1990; 52: 431-449.

39. Hynes RO. Integrins, a family of cell surface receptors. Cell 1987; 48: 229-242

40. Ruoslahti E. Integrins. J Clin Invest 1991; 87: 1-4

41. Ruoslahti E, Pierchbacher MD. New perspectives in cell adhesion. RGD and integrins. Sci 1987; 491-497.

42. Peerschke EIB. The platelet fibrinogen receptor. Semin Hematol 1985; 22: 241-259

43. Morinelli TA, Niewiarowski S Kornecki E, Figures WR, Wachtfoget Y, Colman RW. Platelet aggregation and exposure of fibrinogen receptors by prostaglandin endoperoxide analogues. Blood 1983; 61: 41-49

44. Coller BS. A new murine monoclonal antibody reports on activation-dependent change in the conformation/or microenvironments of the platelet glycoprotein 11b/111a complex. J Clin Invest 1985; 76:101-108

45. Shattil SJ, Brass LF. Induction of the fibrinogen receptor on human platelets by intracellular mediators. J Biol Chem 1987; 262: 992-1000

46. Packham MA. Role of platelets in thrombosis and hemostasis. Can J Physiol Pharmacol 1993; 72: 278-284

47. Rao GHR. Clinical relevance of platelet research in thrombosis and hemostasis. Int J Progr Cardiovasc Sci 1996; 3: 21-24

48. Rao GHR, Gerrard JM, Cohen I, Witkop CJ, White JG. "Origin and role of calcium in platelet activation-contraction-secretion coupling". In Cell Calcium, Fiskum G, ed. New York, NY: Plenum Publications, 1988.

49. Rao GHR, Gerrard JM, Mahadevappa M, White JG. Possible mechanisms of epinephrine actions in Quin-2-loaded platelets refractory to arachidonic acid. Biochem Med Met Biol 1993; 50: 322-337

50. Johnson GJ, Leis LA, Rao GHR, White JG. Arachidonate-induced platelet aggregation in the dog.Thromb Res 1979; 14: 147-154

51. Johnson GJ, Rao GHR, Leis LA, White JG. Effects of agents which alter cyclic AMP on arachidonate-induced platelet aggregation in the dog. Blood 1980; 55: 722-729

52. Scrutton MC, Clare KA, Hutton RA, Bruckdorfer RR. Depressed responsiveness to adrenalin in patients from apparently normal human donors: a familial trait. Br J Haematol 1981; 49: 303-314

53. Stormorken H, Gogstad G, Solum NO. A new bleeding disorder: lack of platelet aggregatory response to adrenalin and lack of secondary aggregation to ADP and platelet activating factor (PAF). Thromb Res. 1982; 29: 391-402

54. Weiss HJ, Platelet physiology and abnormalities of platelet function. N Eng J Med 1975; 293:531-541

55. White JG, Rao GHR, Gerrard JM. "Platelet stimulus activation-contraction-secretion coupling: a frequently fractured chain of events". In Mann Taylor eds, The regulation of coagulation, New York, NY: Elsevier Inc, 1980

56. Hardisty RM. Disorders of platelet function. Br. Med Bull 1977; 33: 207-212

57. HardistyRM, Machin SJ, Nokes TJC, Rink TJ, Smith SW. A new congenital defect of platelet secretion: impaired responsiveness of the platelets to cytoplasmic free calcium. Br J. Haematol 1983; 53: 543-557

58. Ware JA, Clark BA, Smith M, Salzman EW. Abnormalities of cytoplasmic Ca2+ in platelets from patients with uremia. Blood 1986; 226: 29-35

59. Shreeve SM, Valliere JE. The Alpha1 adrenoreceptor is inactivated by alterations in membrane phospholipids. Eur J Pharmacol 1992; 226: 29-35

60. Rao GHR White JG. Influence of phospholipase A2 on human blood platelet alpha adrenergic receptor function. Trhomb. Res. 1989;53:427-34

61. Lukas R. Trif I, Cucianu M. Plasma phospholipase A2 in children with malignant disorders. Eur J Clin Chem Clin Biochem 1992; 30: 127-130

62. Rao GHR, Rao AT. Pharmacology of platelet inhibitory drugs. Ind J Physiol 1994; 38: 69-84

63. Needleman P, Whitaker MO, Wyche K Spracher H, Raz A. Manipulation of platelet aggregation by prostaglandins and their fatty acid precursors : pharmacological basis for a therapeutic approach. Prostaglandins 1980; 19: 165-181

64. Hamberg M, Svensson J, Samuelson B. Thromboxanes: a new group of biologically active compounds derived from prostaglandin endoperoxides. Proc Natl Acad Sci USA 1975; 72: 2994-2998

65. Rao GHR Johnson GJ, White JG. Influence of epinephrine on the aggregation response of aspirin-treated platelets. Prost Med 1980; 5: 45-58

66. Rao GHR Reddy KR, White JG. Modification of human platelet response to sodium arachidonate by membrane modulation. Prost Med 1981; 6: 75-90

67. Rao GHR, White JG. Epinephrine potentiation of arachidonate induced aggregation of cyclooxygenase deficient platelets. Amer J Hematol 1981; 11: 355-366

68. Rao GHR, Gerrard JM, Witkop CJ, White JG, Platelet aggregation independent of ADP release or prostaglandin synthesis in patients with the Hermansky-Pudlak syndrome. Prost Med 1981; 6: 459-472

69. Rao GHR, White JG. Role of arachidonic acid metabolism in human platelet activation and irreversible aggregation. Amer J Hematol 1985; 19: 339-347

70. Rao GHR, White JG. Comparative pharmacology of cyclooxygenase inhibitors. Prost Leuko Med 1985; 18: 119-131

71. Rao GHR, Reddy KR, White JG. Low dose aspirin, platelet function and prostaglandin synthesis: Influence of adrenergic blockade. Prost Med 1981; 6: 485-494

72 Rao GHR, Radha E, White JG. Enteric coated aspirin, platelet cyclooxygenase activity and function. Prost Leukot Med 1984; 13: 3-12

73. Rao GHR, Radha E, White JG. Irreversible aggregation does not depend on lipoxygenase metabolites. Biochem Biophys Res Comm 1985; 132: 652-657

74. Rao GHR, Escolar G, White JG. Epinephrine reverses the inhibitory influence of aspirin on vessel-wall interaction. Throm Res 1986; 47: 625-637

75. Folts JD, Rowe GG, Rao GHR. Problem with aspirin as antithrombotic agent in coronary artery disease. Lancet 1988;1: 937

76. Rao GHR, White JG. Influence of various doses of aspirin (in vivo) on platelet arachidonic acid metabolism (ex vivo) and function. Prost Leukot Essential Fatty Acids Med 1994; 51: 63-69

77 Rao GHR, Johnson JG, White JG. Ibuprofen protects platelet cyclooxygenase from irreversible inhibition by aspirin. Arteriosclerosis 1983; 3: 384-388

78. Rao GHR, Escolar JG, Zavrol J, White JG. Influence of adrenergic receptor blockade on aspirin-nduced inhibition of platelet function. Platelets1990; 1: 145-150

79. Rao GHR, Ericson DG, Weiss DJ, Parks PG, White JG. Influence of aspirin and carbacyclin on bovine platelet function. ASAIO J 1992; 38: 830-834

80. Fuster V, Dyken Ml, Vokones PS, Hennekens C. Aspirin as a therapeutic agent in cardiovascular disease. Circ 1993; 87: 659-675.

81. Rao GHR, Rao ASC, White JG. Aspirin in ischemic heart disease-an overview. Ind Heart J 1993; 45: 73-79

82. Rao GHR. "Aspirin and coronary artery disease" In Coronary artery disease in South Asians: Epidemiology, Risk Factors and Prevention" Rao GHR, Kakkar VV eds. 1998.

83. Patrono C. Aspirin and human platelets: from clinical trials to acetylation of cyclooxygenase and back. Trends Pharmacol Sci 1989; 10: 45-458

84. Roth GJ, Calverley DC. Aspirin, platelets, and thrombosis: Theory and practice. Blood 1994; 83: 885-898

85. Patrono C. Aspirin as an antiplatelet drug. New Engl J Med 1994; 330: 1282-1294

86. Antiplatelet Trialists Collaboration: The aspirin papers. Brit Md J 1994; 308: 71- 72, 81-106.

87. Rao GHR. Role of platelet adhesion and aggregation in thrombus formation. Thromb Haemost 1998; 79: 454

88. Michelson AD. Flow cytometry: A clinical test of platelet function. Blood 1996; 87: 4925-4936

89. Baumgartner HR. The role of blood flow in platelet adhesion, fibrin deposition and formation of mural thrombi. Microvasc Res. 1973; 5: 167-179

90 deGraf JC, Banga JD, Moncada S, Palmer RMJ, deGroot PG, Sixma JJ. Nitric oxide functions as an inhibitor of platelet function under flow conditions. Circ 1992; 85: 2284-2290

91. Dodds WJ. "Platelet function in animals: species specificities". In Platelet: a multidisciplinary approach. deGaetano G, Garattini S eds., New York, NY:Raven Press, 1978.

92. Vu T-KH, Hung DT, Wheaton VI, Coughlin SR. Molecular cloning of a functional thrombin receptor reveals a novel proteolytic mechanism of receptor activation. Cell 1991; 288:123-128

93. Natarajan S, Riexniger D, Peluso M, Seiler SM. "Tethered ligand" derived penta peptide agonist of thrombin receptor: a study of side chain requirements for human platelet activtion and GTPase stimulation. Int J Peptide Protein Res 1994; 6: 1-7

94. Rao GHR, Fields CM, White JG, Fields G. Promotion of human platelet adhesion and aggregation by a synthetic, triple-helical "minicollagen". J Biol Chem 1994; 269: 13899-13903

95. Harfenist EJ, Packham MA, Mustard JF. Effects of the cell adhesion peptide, Arg-Gly-Asp-Ser, on responses of washed platelets from humans, rabbits and rats. Blood 1988; 71:132-136

96. O'Brien JR. Effects of salicylates on human platelets. Lancet 1968; 1: 779-783

97. Zucker MB, Peterson J. Inhibition of adenosine diphosphate-induced secondary aggregation and other platelet functions by acetylsalcicylic acid ingestion. Proc Soc. Exp Biol Med 1968, 127: 547-551

98. Vane JR. Inhibition of prostaglandins as a mechanism of action for aspirin-like drugs. Nature 1971; 231: 232-235

99. Rao GHR, Radha E, White JG. Effect of docosahexaenoic acid on arachidonic acid metabolism and platelet function. Biochem Biphys Res Comm 1983; 117: 549-556

100. Rao GHR, Kishore NP, Peller JD, White JG. "Influence of polyenoic acids on arachidonic acid metabolism and platelet function." In Cardiovascular Disease, Gallo L ed, New York: Plenum Publications 1988.

101. Rao GHR, Krishnamurthi S, Raij L, White JG. Influence of nitric oxide on agonist mediated calcium mobilization in platelets. Biochem Med Met Biol 1990; 43: 271-275

102. Krishnamurthi S, Wheeler-Jones CPD, Patel Y, Sadowska K, Kakkar VV, Rao GHR. Nitroprusside inhibits platelet function primarily by inhibiting Ca2+ mobilization. Biochem Soc Trans 1990; 18: 468-470

103. Rao GHR, Raij L, Lester B, White JG. "Inhibition of agonist- induced human platelet activation by nitric oxide." In Nitric oxide from L-arginine: a bioregulatory system. Moncada S, Higgs EA eds, London: Elsevier Publishing Co, 1990.

104. Rao GHR, Doni MG. Influence of nitrovasodilators on platelet function. Trombosi & Aterosclerosi 1993; 38: 69-84

105. Rao GHR, Reddy KR, White JG. Influence of epinephrine on prostacyclin induced dissociation of ADP aggregated platelets. Prost Med 1980; 4: 385-397

106. Rao GHR, Smith CM II, White JG. Influence of calcium antagonists on thrombin induced calcium mobilization and platelet-vessel interactions. Biochem Med Met Biol 1992; 17: 226-231

107. Rao GHR. Influence of anti-platelet drugs on platelet-vessel wall interactions. Prost Leukot Med 1987; 30: 133-145

108 Rao GHR, Peller JD, Semba CP, White JG. Influence of the calcium sensitive flurophore Quin 2 on platelet function. Blood 1986; 67: 354-361

109. Schror K. Antiplatelet drugs. A comparitive review. Drugs 1995; 50: 7-28

110. Fareed J. Current trends in antithrombotic drug and device development. Sem in Thromb & Hamestas 1996; 1: 3-8

111. Verstraete M. New developments in antiplatelet and antithrombotic therapy. Eur Heart J 1995; 16: 16-23

112. Cairns JA, Hirsh J, Lewis HD, Resnekov L, Theroux P. Antithrombotic agents in coronary artery disease. Chest 1992; 102: 456S-481S

113. Badimon L, Meyer BJ, Badimon JJ. Thrombin in arterial thrombosis. Haemost 1994; 24: 69-80

114. Fitzgerald D. Specific thrombin inhibitors in vivo. Ann N.Y. Acad Sci 1994; 714: 41-52

115. Becker RC. Antiplatelet therapy in coronary artey disease. Emerging strategies for the treatment and prevention of actute myocardial infarction. Arch Pathol Lab Med 1993; 117: 89-96

116. Verstraete M, Zoldhelyi P. Novel antithrombotic drugs in development. Drugs 1995; 49: 956-87

117. Rao GHR, Wilson RF, White CW, White JG. Influence of thrombolytic agents on human platelet function. Throm Res 1991; 62: 319-334

118. Topol EJ. Towaard a new frontier in myocardial reperfusion therpay. Emerging platelet preeminance. Cric 1998; 97: 211-218

2

HUMAN PLATELET THROMBIN RECEPTORS AND THE TWO RECEPTOR MODEL FOR PLATELET ACTIVATION

G. A. Jamieson, Narendra N. Tandon* and
Nicholas J. Greco
American Red Cross
Otsuka America Pharmaceutical, Inc.*

2.1 INTRODUCTION

The interaction of proteolytically-active α-thrombin with platelets and other cells of the vasculature, such as endothelial cells and smooth muscle cells, plays a major role in both normal hemostasis and atherosclerosis.[1,2] Despite extensive studies in numerous laboratories extending back over thirty years, major questions regarding the mechanism of these interactions remain unresolved. Furthermore, since thrombin can also induce chemotaxis and adhesion of inflammatory cells, and fibroblast mitogenesis, the importance of elucidating the nature of its receptor, or receptors, extends far beyond its role in platelet activation. However, this review will be restricted mainly to considerations of thrombin receptors in human platelets.

> The authors have collaborated for many years in studies on platelet activating receptors, including not only receptors for α-thrombin but also those for collagen (CD36 and GPVI, laminin (67kDA), and adenosine diphosphate.

The interaction of thrombin with platelets causes increases in cytoplasmic Ca^{2+}, shape change and the conversion of prothrombin to thrombin via the prothrombinase complex leading to further platelet activation, aggregation and secretion. The concentration of free α-thrombin in plasma following physiological activation has been determined to be in the range of 0.5-2nM due to the effects of thrombomodulin, antithrombin III and other factors.[3,4] *In vitro*, α-thrombin is one of the most potent known platelet agonists and can cause platelet activation at thrombin concentrations in the picomolar range. Thus, the high α-thrombin concentrations used in some studies, sometimes as high as 10U/ml or 100nM, could introduce artifacts unrelated to the activation process. Moreover, the α-thrombin used should have a specific activity of at least 3000U/mg since thrombin degradation products such as β- and γ-thrombins in preparations of lower purity would not be detectable since they would not clot fibrinogen but would still be capable of causing platelet activation.

2.2 Binding studies

Quantitation of the interaction of platelets with thrombin has been best defined by steady state binding studies using ^{125}I-labeled preparations of α-thrombin or inactivated thrombins such as PPACK-thrombin (α-thrombin treated with D-phenyl-L-prolyl-L-arginine chloromethyl ketone) followed by computer-assisted analysis of the binding data, most commonly by the use of the LIGAND program.[5] Binding isotherms are usually developed at room temperature by adding washed platelets to a constant tracer concentration (~10^{-10} M) of ^{125}I-labeled α-thrombin in the presence of increasing concentrations up to ~10^{-6} M of unlabeled a-thrombin and then separating bound and free ligands by centrifugation or filtration techniques. For adequate analysis of multisite binding models, it is essential that the isotherm be based on an adequate number of binding points: for example, effective resolution of a three site model requires a minimum of 20-24 triplicate data points. Under these optimal conditions, two different types of binding sites have been found within physiologically-relevant ranges of thrombin concentrations for both α-thrombin and PPACK-thrombin. About 50 high affinity binding sites with a dissociation constant of about 0.5nM have been identified together with ~2000 moderate affinity binding sites with a dissociation constant of ~10nM: thrombin binding data with various platelet substrates are summarized in Table 1.

A very low affinity thrombin binding component can be expressed either as non-specific binding or as a third site (~500,000sites/plt;Kd ~ 2500nM) but does not appear to be of physiological significance. The possible existence of a very low affinity site may explain the difficulty of demonstrating complete competition of the binding of ^{125}I-labeled thrombin at 0.1nM by 1μM unlabeled thrombin since complete competition at a site with a dissociation constant of 2.5μM would require an α-thrombin concentration of ~25μM.

Under steady state conditions and at infinite dilution, the theoretical binding activity of a given site is the product of the apparent dissociation constant and the molar concentration of these sites. Thus, under these conditions the relative contribution of these three sites to the total thrombin binding activity of platelets is approximately equal.

Further characterization of these binding sites on intact platelets and isolated membranes was carried out by radiation inactivation and target analysis.[6] When high energy radiation causes ionization in a molecule its functional capacity is destroyed. Since the target size is proportional to the mass of the functional unit, the radiation-induced loss of binding activity is a measure of the mass of the binding component. The apparent molecular sizes obtained by this procedure relate to the sum of the structural components required to elicit the measured activity. If this activity resides in a single polypeptide chain, radiation inactivation will indicate a functional size equivalent to its molecular weight. However, if an activity such as binding involves a complex with contributions from several components, the functional size determined by radiation inactivation will be related to the sum of the molecular weights of the interacting subunits.

Table 1. Thrombin Binding Parameters of Human Platelets

		High Affinity		Moderate Affinity	
		Kd (nM)	Sites/plt	Kd (nM)	Sites/plt
1.	α-T	0.60 ± 0.17	66 ± 20	25 ± 7	2800 ± 1570
2.	TM60	nd	Nd	3.3	190
3.	anti TNA	0.10 ± 0.04	10 ± 5	nd	Nd
4.	PPACK-T	0.83 ± 0.29	80 ± 30	30 ± 21	502 ± 340
5.	anti TNA	0.40 ± 0.17	60 ± 33	nd	Nd
6.	α-T	0.58 ± 0.19	36 ± 19	5.8 ± 4.0	260 ± 155
7.	SMR_x	nd	Nd	2.6 ± 0.8	180 ± 64
8.	α-T	0.25-1.3	105-344	30-77	3400-8200
9.	LJIb10	nd	nd	1.3-4.9	234-714
10.	B-S pts	nd	nd	3.4	1800

nd = not detectable. Abbreviations are α-T, α-thrombin; PPACK-T, PPACK-thrombin; SMRx pts, treated with Serratia protease; B-S pts, Bernard-Soulier. (Lines 1-3, 4&5, 6&7)[27,34] (Lines 8,9 & 10)[3].

Using radiation inactivation, we determined that the functional molecular size of the highest affinity component (Kd ~0.5nM) was almost one million (880,000±140,000) suggesting that a multimolecular complex was involved in the binding of α-thrombin at this site.[7] The moderate affinity site (Kd 10nM) had a functional molecular size of 30,000±9,000 while the lowest affinity site (Kd 2.5μM), which has generally been considered to be equivalent to nonspecific binding, has a functional molecular size of 3800±1800. The functional implications of these results will be considered later in this review.

Whether these results identified true receptors or were merely binding sites which did not play a functional role in platelet activation remained a matter of dispute. Several investigators have reported that catalytically-inactivated thrombin derivatives bind to the same sites on platelets as does α-thrombin but do not compete in platelet activation by α-thrombin. These results were then interpreted as indicating that the binding studies did not identify functional receptors coupled to signal transduction and that the observed binding was, therefore, unrelated to platelet activation.[3,8-11]

We have, however, been able to resolve this paradox in order to clarify the relationship between binding and activation.[12,13] We showed that the failure of previous investigators to detect inhibition by catalytically-inactivated thrombins may have been due to the fact that some of them, such as α-thrombin treated with tosyl-lysyl-chloromethyl ketone (TLCK-thrombin), were equilibrium mixtures containing about 4% active α-thrombin and that these mixtures would therefore activate platelets at the high concentrations (~1μM) used to evaluate inhibitory effects. In contrast, PPACK-thrombin has no detectable residual proteolytic activity, it competes with α-thrombin in binding to these two sites and it completely abrogates thrombin-induced platelet responses including shape change, aggregation, secretion and the corresponding increases in $[Ca^{2+}]_i$. These results demonstrate that these high and moderate affinity binding sites on platelets are true receptors occupancy of which is required for platelet activation.

2.3 Membrane microviscosity

Elevated plasma cholesterol is a major risk factor for atherosclerosis and platelets from patients with homozygous familial hypercholesterolemia (Type II hypercholesterolemia) have been known to have increased sensitivity to a variety of aggregating agents. In a number of *in vitro* studies, designed to provide a model system for elucidation of the basis for this hyperaggregability, platelet membrane microviscosity was modified by incubating platelets with cholesterol-rich liposomes and it was shown that the resulting cholesterol-modified platelets exhibited the expected hypersensitivity to aggregating agents.

In our own studies, we modified membrane microviscosity by incubating normal platelets overnight as 22° with cholesterol-rich, cholesterol-normal and cholesterol-poor liposomes.[14-16] Under these conditions, and considering only high affinity sites since moderate and low affinity sites were combined in this analysis, the *number* of receptors was proportional to membrane microviscosity while the *affinity* at those receptors was inversely proportional to membrane microviscosity (Table 2).

Table 2: High Affinity Thrombin Binding to Cholesterol-Modified Platelets

	C:PL	Kd (nM)	Sites/plt
chol-enriched	1.07	0.22	260
chol-normal	0.57	0.12	150
chol-depleted	0.38	0.05	80

Abbreviation: chol, cholesterol.

As compared with modified cholesterol-normal platelets, the dose response curves for both aggregation and serotonin secretion in response to α-thrombin underwent a leftward shift using cholesterol-enriched platelets and a rightward shift using cholesterol-depleted platelets: the concentrations required for half-maximal aggregation response were 0.17, 0.35 and 0.53nM, respectively, while the corresponding values for serotonin secretion were 0.50, 0.70 and 1.0nM. These results indicate that the maximum aggregation response in cholesterol-enriched platelets occurred with 30% occupancy of high affinity receptor sites, in cholesterol-normal platelets with 50% occupancy and in cholesterol-depleted platelets with 70% occupancy of the high affinity receptor sites. Thus the change in receptor number appeared to be the determinant for platelet responsiveness in these experiments. In each case, the occupancy of 40-50 high affinity receptors was sufficient to induce half-maximal aggregation. Thus changes in membrane microviscosity alter the number of high affinity thrombin receptors which in turn alters the responsiveness of the platelet but these changes in membrane microviscosity do not appear to alter the stimulus-response coupling between the occupied high affinity receptor sites and the bioresponse mechanism.

When these parameters were evaluated in platelets from patients with Type II hypercholesterolemia it was found that the patients could be divided into two groups: Platelets from Type II patients *without* overt cardiovascular disease bound lower amounts of thrombin than did normal controls but aggregation was within the normal range and the thrombin concentrations required for half maximal activation were identical, 0.24±0.05nM and 0.23±0.07nM respectively. However, in patients *with* overt cardiovascular disease, as indicated by femoral bruits and confirmed by cardiac catherization, the thrombin binding isotherm fell within the normal range but these platelets were hyperaggregable as shown by a leftward shift in the concentration dependence data and the fact that the thrombin concentration for half-maximal aggregation was decreased to 0.16±0.03nM. Since all Type II patients eventually develop cardiovascular disease, the phenotype characteristic of platelet-thrombin interaction is probably that seen in the group prior to the onset of cardiovascular disease: namely, a decreased binding of thrombin to the platelets but an increased aggregation response.

Several conclusions can be drawn from these studies: 1) changes in platelet membrane microviscosity affect expression of high and moderate affinity thrombin receptors; 2) these changes in receptor expression are not accompanied by changes in stimulus-response coupling in platelets that have had their cholesterol/phospholipid ratio changed by manipulation in vitro; 3) changes in receptor expression and responsiveness also occur in platelets from patients with Type II hypercholesterolemia but these changes are associated with enhanced stimulus-response coupling; 4) most importantly, these studies show that platelets whose membrane microviscosity has been changed by *in vitro* manipulation are not an appropriate model for platelet function in familial hypercholesterolemia.

Not only do these studies support the presence in blood platelets of receptors with two distinct binding affinities but this concept is also supported by earlier functional evidence that thrombin-induced platelet activation is mediated by two distinct pathways which differ in their sensitivity to proteolysis, their requirement for sodium ions, in the need for receptor occupancy and in the role of G proteins (reviewed in [17,18]). Furthermore, kinetic studies demonstrated that both ligand-receptor and proteolytic interactions occur in thrombin-induced platelet activation. Thus, the preponderance of evidence supports the view that two different types of receptor are involved in the interaction of thrombin with platelets. The major question is, however, what is the nature of these two receptors?

2.4 Candidate Receptors

The two platelet membrane glycoproteins for which the strongest evidence exists at the present time that they are thrombin receptors are a specific form of the GPIb-IX-V complex and PAR1, the protease-activated thrombin receptor: these will be discussed next in this review. Based on their apparent ability to form complexes with α-thrombin, several other proteins had previously been proposed as thrombin receptors but in only a few cases were the necessary further studies carried out to test these hypotheses. Protease nexin 1 was proposed as a thrombin receptor based on the similar time courses of complex formation and platelet activation but it was subsequently found that protease nexin 1 cannot be a receptor since it is an internal component of platelets that is expressed on the surface only after activation. Glycoprotein V has also been proposed as a thrombin receptor based on the fact that it can be cleaved by low concentrations of α-thrombin but subsequent studies (reviewed in[17]) showed that there was no consistent relationship between the rate or extent of GPV hydrolysis and the extent of platelet activation induced by α-thrombin. Another thrombin-activatable receptor, PAR3, has been identified in mouse platelets[19] but not in human platelets using a reverse transcriptase/ polymerase chain reaction approach (unpublished data and S.Coughlin, personal communication). PAR3 has, however, been reported cloned from a human platelet cDNA library.[20]

2.4.1 Protease Activated Receptor 1 (PAR1)

The mechanism of activation of PAR1, the moderate affinity platelet thrombin receptor, has been clearly defined. Expression cloning studies in Xenopus oocytes identified a thrombin receptor that is a member of the G protein-coupled, seven transmembrane domain family.[8,21] The binding of α-thrombin to this receptor results in proteolytic cleavage of the R^{41}/S^{42} bond leading to a new amino terminus which acts as a tethered ligand interacting with the receptor itself to effect activation which appears to be mediated through the second cytoplasmic loop.[22]. The synthetic tethered ligand peptide (TLP) with the sequence SFLLRN corresponding to the new amino terminal sequence can activate platelets and other cells in the absence of α-thrombin. Addition of TLP to platelets leads to the activation of both adenylate cyclase and phospholipase C,[23,24] and phosphoinositide formation has been found to be proportional to PAR1 cleavage.[25] The kinetics of the interaction of thrombin and other relevant proteases with the PAR1-derived peptide L^{38}-E^{60} have been determined.[26]

PAR1 has been confirmed as corresponding to the moderate affinity receptor based on inhibition of the moderate affinity component of thrombin binding by anti PAR1 antibodies[27] and by the TLP homolog FLLRN.[28] Furthermore, despite the fundamentally different basis of the two techniques, the functional molecular size of the moderate affinity binding site determined by radiation inactivation and target analysis (30,000±9,000[7]) is in reasonable agreement with the peptide molecular weight of PAR1 as deduced from the cloned cDNA (44,000). Finally, the number of platelet binding sites for the PAR1 monoclonal antibody ATAP138 (~1800)[29] is identical with the number of moderate affinity binding sites previously determined for α-thrombin.

Despite the extensive binding and functional evidence in the literature for the existence of two different types of thrombin receptors, PAR1 was initially termed "the" thrombin receptor and it was asserted that it alone was both necessary and sufficient to explain all of the cellular effects of α-thrombin. However, increasing evidence has shown that the tethered ligand peptide is only a partial agonist: for example, as compared with α-thrombin, TLP and its homologs induced less expression of the activated, binding-competent form of GPIIb/IIIa, failed to cause association of GPIIb and $pp60^{src}$ with the cytoskeleton, failed to sustain activation of $p42/44^{mapk}$, failed to induce platelet prothrombinase activity and produced weaker and more attenuated responses in terms of Ca^{2+} mobilization, arachidonate production, serotonin release, protein phosphorylation and phosphatidic acid production.[30-33] Similar differences have been found in other cell types: for example, unlike α-thrombin, TLP failed to induce the expression of ICAM-1 in neutrophils,[34] the translocation of protein kinase C in endothelial cells[35] or of MAP kinases in endothelial cells and Chinese hamster ovary fibroblasts,[36,37] and the two agonists differ in Ca^{2+} signalling in osteoblasts.[38]

Whether the tethered ligand peptide alone can induce fibroblast mitogenesis has not so far been satisfactorily resolved. PAR1-related peptides have been reported to be unable to induce mitogenesis in CCL39 hamster fibroblasts in the absence of added growth factors possessing tyrosine kinase activity in some studies[37,39] but not in others.[40]

The most compelling evidence that activation of PAR1 is not obligatory for thrombin-induced activation of human platelets is, however, the recent findings from two different laboratories that platelets from mice lacking the PAR1 gene exhibit no detectable bleeding diathesis and that their platelets give normal responses to α-thrombin as measured by aggregation and serotonin secretion.[41,42] These results clearly establish that the activation of PAR1 is not obligatory for the activation of platelets by α-thrombin.

2.5 The GPIb-IX-V complex

The detailed structure of GPIb-IX-V complex is discussed elsewhere in this monograph and in a recent excellent review.[43] In brief, this complex comprises disulfide-linked α- (130kDa) and β- (25kDa) subunits, the latter being tightly but non-covalently bound to GPIX (22kDa), and this GPIb-IX complex is more loosely associated with GPV(82kDa) in a ratio of 2:2:2:1.[44] The question of whether the GPIb-IX-V complex occurs in cells other than those of megakaryocytic origin has not been satisfactorily resolved.

The evidence that the GPIb-IX-V complex functions as the high affinity platelet receptor for α-thrombin may be summarized as follows:

(i) GPIbα contains a high affinity thrombin binding site in its N-terminal domain, specifically in the sequence D^{269}-D^{287}.[45]

(ii) The monoclonal antibodies TM60 and LJIb10 directed against this thrombin binding domain in GPIbα have been shown to inhibit both the high affinity binding of α-thrombin to platelets and platelet activation induced by low concentrations of α-thrombin (Fig.1; Table I).[3,27,46]

(iii) Snake venom proteins such as echicetin and crotalin, which bind to the thrombin binding site on GPIbα, have also been shown to block the binding of α-thrombin and thrombin-induced platelet activation.[47,48]

(iv) Kininogens, and kininogen-derived peptides, bind to the GPIb-IX-V complex and reduce both the amount of high affinity thrombin binding and thrombin-induced platelet aggregation.[49]

(v) Prothrombinase expression, arguably the most important consequence of platelet (iii) Snake venom proteins such as echicetin and crotalin, which bind to the thrombin binding site on GPIbα, have also been shown to block the binding of α-thrombin and thrombin-induced platelet activation.[47,48]

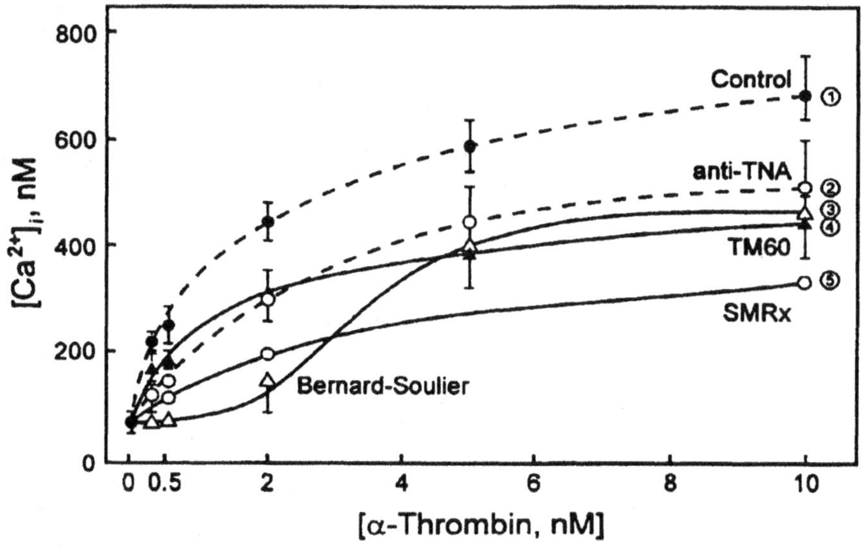

FIGURE 1: Peak $[Ca^{2+}]$ responses to increasing concentrations of α-thrombin. Mean ± SD values given are from 3-6 different donors except for Bernard-Soulier platelets which were two samples from a single patient. Curves 1 and 2 are shown as dashed lines to improve clarity. Curve 1, control platelets. Curve 2, platelets preincubated with polyclonal antiTNA antibody to PAR1. Curve 3, Bernard Soulier platelets. Curve 4, platelets preincubated with TM60 monoclonal antibody to GPIb. Curve 5, plkatelets pretreated with Serratia protease (100ng/ml). Modified from [27], with permission.

(vi) Bernard-Soulier platelets, which constitutively lack the GPIb-IX-V complex, show no high affinity thrombin binding and do not respond to low concentrations of α-thrombin.[3,27,51]

(vii) Selective proteolysis of GPIbα on intact platelets with Serratia marcescens metalloprotease essentially abrogates thrombin binding at the high affinity site and reduces thrombin-induced platelet activation while having little effect on platelet interaction with vWF (Fig.2; Table 1).[52-54]

(viii) Similarly, in vitro storage of platelet concentrates for transfusion results in a concomitant loss of the glycocalicin moiety of GPIbα, high affinity thrombin binding and thrombin responsiveness.[55]

(ix) The combined molecular weights of the individual components comprising the monomer form of the GPIb-IX-V complex is 460kDa. Since the high affinity thrombin binding site on platelets has a functional molecular size of ~900kDa, as determined by radiation inactivation and target analysis,[7] this suggests that the functional receptor most probably exists as a dimer of the GPIb-IX-V complex.

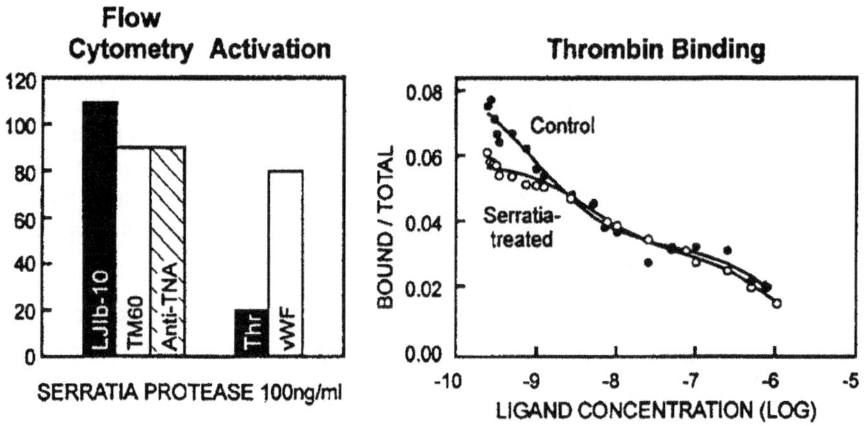

FIGURE 2: Reactions of Serratia protease-treated platelets. The right hand portion of the figure shows thrombin binding isotherms for control platelets and for platelets treated with 100ng/ml of Serratia protease. Note the absence of high affinity binding and the requirement for an adequate number of binding points to accurately reflect changes in ligand binding. The left hand portion of the figure demonstrates that there is little change in surface expression of GPIbα or in vWF/ristocetin reactivity but that ~80% of the Ca^{2+} response to 0.5nM α-thrombin is lost. Modified from [54], with permission.

It has been frequently commented upon that the total number of copies of the GPIb-IX-V complex present on platelets (~25,000) vastly exceeds the number of high affinity thrombin binding sites (~50). This apparent discrepancy probably reflects the fact that these high affinity sites are part of this "supercomplex" comprising a dimeric form of the GPIb-IX-V complex and, perhaps, other components. This supercomplex may be a stable structure of all of its individual components or it may reflect an equilibrium binding of monomeric and dimeric forms of the GPIb-IX-V complex.

Recent studies using immobilized glycocalicin, some carried out at 4°C, have provided further confirmation of the role of GPIb as a thrombin receptor but have posed questions regarding the affinity of its interaction with α-thrombin.[45,56,57] The reduced temperature used in these studies is of concern since this could affect binding affinity through changes in membrane microviscosity, as discussed above. Moreover, immobilized glycocalicin exhibits only a single thrombin binding site (Kd ~10^{-8}M) whereas binding to purified soluble GPIb-IX and glycocalicin demonstrates both high (Kd ~0.3nM) and moderate affinity (Kd ~50nM) sites.[58] Although the moderate affinity receptor is now known to be PAR1, these results suggest that immobilization of glycocalicin may prevent the self-association that may be required for expression of a high affinity binding component.

Both the GPIb-IX-V complex and PAR1 are required to ensure the maximal rate and extent of thrombin-induced platelet activation as measured by the increase in $[Ca^{2+}]_i$ and appear to act additively and independently (Fig. 2).[27] The thrombin-induced increase in $[Ca^{2+}]_i$ was inhibited by 50% in the presence of antibodies to either GPIb or PAR1 and was almost completely inhibited by the simultaneous addition of antibodies to both receptors: these results demonstrate that receptors other than GPIb and PAR1 are unlikely to be involved in thrombin-induced platelet activation. Specifically, the anti GPIb antibodies TM60 and LJIb10 maximally reduced the thrombin-induced increase in $[Ca^{2+}]_i$ to 57 ± 13% and 51 ± 4%, respectively, of control values while the anti PAR1 antibody anti TNA reduced it to 35 ± 13%. In the presence of both TM60 and anti TNA, $[Ca^{2+}]_i$ was reduced to 10 ± 5% of the control. This difference from the values obtained with each antibody separately is significant at $p<0.05$. The fact that Ca^{2+} mobilization decreases by 50% in the presence of these anti GPIb MoAbs is of importance since, if the role of GPIb were only to facilitate activation through PAR1, as has been suggested by several investigators, it would be expected to affect the rate but not the extent of activation as measured by changes in cytoplasmic Ca^{2+} concentration.

Reports that have questioned the relevance of the binding of α-thrombin to the GPIb-IX-V complex in mediating platelet activation have generally reached this conclusion as a result of the use of inappropriate assays. For example, ~22 point binding isotherms are required for the resolution of high, moderate and low affinity thrombin binding sites in a three site model.[5] However, 10 point binding isotherms, which would be sufficient for only a one site model, have been the basis for a proposal that GPIb acts as a negative regulator by sequestering small amounts of α-thrombin on the platelet surface.[9,10] This model would, in fact, imply that Bernard-Soulier platelets, which cannot sequester thrombin in this manner, should be more sensitive to α-thrombin than control platelets whereas the reverse is known to be the case.

Another study, which concluded that GPIb plays no role in either thrombin binding or thrombin-induced platelet activation, used flow cytometry to measure thrombin binding[59] although this technique cannot be used to assay less than about 500 binding sites per cell and is far too insensitive to detect the 50 high affinity thrombin binding sites that occur on platelets.

The cDNA of each of the components of the GPIb-IX-V complex has been cloned and the proteins have been expressed in fibroblasts and other cell types lacking these structures but the mechanism by which the binding of α-thrombin to GPIbα initiates platelet activation through the high affinity pathway remains to be determined.

It should be emphasized that, while proteolytically-active thrombin is required for activation through the GPIb-IX-V complex, there is at this time no *a priori* evidence that a proteolytic event is involved. Conformational changes in solution structure are known to occur in thrombin analogues following derivatization at its active site (PPACK-thrombin) or mutation at the anion binding exosite. These changes include differences in intrinsic fluorescence, interaction with recombinant hirudin peptides and susceptibility to proteolysis.[60,61] These changes could abrogate PPACK-thrombin's ability to effect activation although still capable of binding to the receptor.

There are several structural features of GPIb which may play a role in its receptor function and its capacity to induce signal transduction. It was early recognized that glycocalicin, the proteolytic fragment comprising essentially the whole extracellular domain of GPIbα, contains binding sites for both α-thrombin and vWF.[62] However, the two types of sites have been distinguished by the fact that the high affinity binding form of the receptor can be totally degraded with Serratia marcescens metalloprotease under conditions where vWF responsiveness is essentially unaffected.[54] The proximity of the thrombin and vWF binding sites on GPIbα may also play a role in the ability of fibrin monomer to facilitate the interaction of GPIbα with vWF.[63]

Further studies[45] have localized the thrombin binding domain of GPIbα to sequences in the region D^{269}-D^{287} which includes a heptapeptide sequence Y^{276} DYYPEE that is also important for the binding of vWF.[64] Desulfation or mutation of any one of these three tyrosine residues so that it cannot be sulfated has recently been shown to result in a loss of thrombin binding ability.[57,65] These results strongly support a significant role for these sulfation/desulfation processes in initiating thrombin-induced signalling through the GPIb-dependent high affinity pathway.

The possible involvement of phosphorylation/dephosphorylation events at the GPIb-IX-V complex in thrombin-induced platelet activation has not been experimentally established but cannot be unequivocally ruled out. Thrombin-induced platelet activation does not cause measurable changes in the level of phosphorylation at S^{166} of GPIbβ, the only phosphorylated residue in the GPIb-IX complex,[66] but this could reflect the fact that there are only ~50 copies of the high affinity, high molecular weight receptor complex so that changes that might occur in the complexed form of GPIbβ could be obscured by the absence of phosphorylation changes in the ~30,000 copies of the uncomplexed form.

2.6 Two receptor model

In our earlier model,[18] we had inferred from the literature that the high affinity GPIb-dependent receptor was activated by α-thrombin and was coupled to phospholipase A_2 while PAR1 was coupled to phospholipase C and could be activated by the tethered ligand peptide and by both α- and γ-thrombins: This hypothesis has been largely confirmed by recent work. The PAR1-mediated moderate affinity pathway has been shown to be coupled to both PLC and adenylate cyclase[23,67] and this coupling occurs through the G_q and G_i forms since their activities are stimulated and inhibited, respectively, by TLP.[24] Gαq knockout mice fail to aggregate or secrete ATP in response to α-thrombin.[68] They do, however, manifest a shape change response indicating that a Gαq-independent mechanism of thrombin-induced activation remains operational presumably through the high affinity pathway. Since TLP activates $p38^{mapk}$ this kinase may be assigned to the moderate affinity pathway.

The high affinity receptor form of the GPIb-IX complex probably exists as a dimer associated with GPV in the ratio of 2:1 and requires sulfation of specific tyrosine residues in the sequence Y^{276}DYY of GPIbα. This receptor mediates the inhibition of thrombin-induced platelet activation by kininogens[49] and, in part, thrombin-induced prothrombinase assembly.[50] The association of PLA_2 with the high affinity pathway mediated by the GPIb-IX-V complex has been supported by the demonstration that the PLA_2 zeta isoform (14-3-

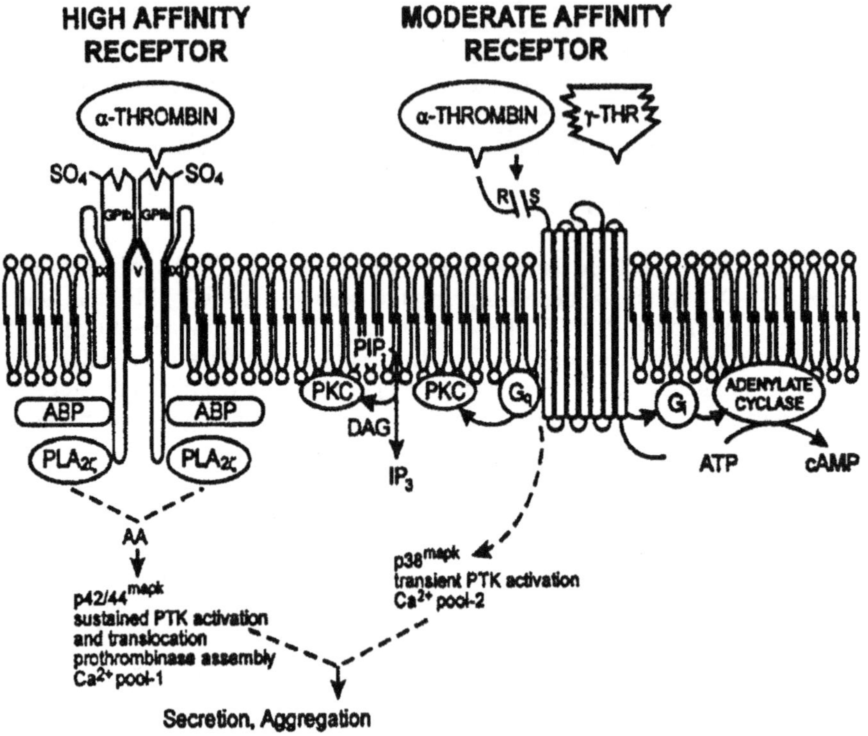

FIGURE 3: The two receptor model of thrombin-induced platelet activation. See text for details. Modified from [18], with permission.

3 protein) can be co-immunoprecipitated with GPIb[69] but the role of this interaction in the effects of α-thrombin on platelets, as compared with those of vWF, remains to be elucidated. However, the TLP-independent events described earlier, such as the sustained activation of p42/44^{mapk} and other kinases and their translocation to the cytoskeleton may be provisionally assigned to this high affinity GPIb-dependent pathway. The Ca^{2+} pool mobilized by this receptor may be different from that mobilized by the moderate affinity, PAR1-dependent pathway since immunoinhibition reaches a maximum of 50% in each case.

Taking the two receptor model as a whole, it may be noted that those activities which have been shown to require continuous receptor occupancy, such as arachidonate mobilization and the formation of phosphatidic acid, are mediated through the high affinity pathway and represent a receptor-ligand mediated mechanism while those receptor activities that require only transient receptor occupancy, such as the synthesis and breakdown of phosphoinositides, correspond to the moderate affinity proteolytic PAR1 pathway. The fact that immunoinhibitory effects at either receptor can be overcome at elevated thrombin levels merely indicates the relatively high affinities of thrombin binding at both sites (~0.5nM and 10nM, respectively) as compared with the relatively low affinities of the binding antibodies; for example the Kd of the binding of ATAP138 to PAR1 is ~80nM.[29]

In summary, the available evidence strongly supports the two receptor model for thrombin-induced platelet activation. In this model a supercomplexed form of GPIb acts as the high affinity receptor while PAR1 acts as the moderate affinity receptor. The applicability of this model in other cell systems, and the relative contributions of the two receptor pathways in the normal or pathophysiological activation of platelets, remain to be determined.

References

1. Fuster V, Badimon L, Badimon JJ, Chesebro JH: The pathogenesis of coronary artery disease and the acute coronary syndromes. Part 1. New Eng J of Med 326:242, 1992

2. Fuster V, Badimon L, Badimon JJ, Chesebro JH: The pathogenesis of coronary artery disease and the acute coronary syndromes. Part 2. New Eng J of Med 326:310, 1992

3. De Marco L, Mazzucato M, Masotti A, Fenton JW, Ruggeri ZM: Function of glycoprotein Ib alpha in platelet activation induced by alpha-thrombin. J Biol Chem 266:23776, 1991

4. Shuman MA, Levine SP: Thrombin generation and secretion of platelet factor 4 during blood clotting. J Clin Invest 61:1102, 1978

5. Munson PJ: Ligand: A computerized analysis of ligand binding data. Methods Enzymol 92:543, 1983

6. Kempner ES, Schlegel W: Size determination of enzymes by radiation inactivation. Anal Biochem 92:2, 1979

7. Harmon JT, Jamieson GA: Thrombin binds to a high-affinity ~900,000 dalton site on human platelets. Biochemistry 24:58, 1985

8. Vu T-KH, Hung DT, Wheaton VI, Coughlin SR: Molecular cloning of a functional thrombin receptor reveals a novel proteolytic mechanism of receptor activation. Cell 64:1057, 1991

9. Leong L, Henriksen RA, Kermode JC, Rittenhouse SE, Tracy PB: The thrombin high-affinity binding site on platelets is a negative regulator of thrombin-induced platelet activation. Structure-function studies using two mutant thrombins, Quick I and Quick II. Biochemistry 31:2567, 1992

10. Hayes KL, Leong L, Henriksen RA, Bouchard BA, Ouellette L, Church WR, Tracy B: α-Thrombin-induced human platelet activation results solely from formation of a specific enzyme-substrate complex. J Biol Chem 269:28606, 1994

11. Hourani SMO, Cusack NJ: Pharmacological receptors on blood platelets. Pharma Rev 43:243, 1991

12. Greco NJ, Tenner Jr,T.E., Tandon NN, Jamieson GA: PPACK-thrombin inhibits thrombin-induced platelet aggregation and cytoplasmic acidification but does not inhibit platelet shape change. Blood 75:1983, 1990

13. Greco NJ, Tandon NN, Jamieson GA: Thrombin binding to platelets defines functional receptors: Inhibition of thrombin-induced platelet activation by catalytically-inactivated thrombin. Platelets 6:270, 1995

14. Tandon NN, Harmon JT, Rodbard D, Jamieson GA: Thrombin receptors define responsiveness of cholesterol-modified platelets. J Biol Chem 258:11840, 1983

15. Harmon JT, Tandon NN, Hoeg JM, Jamieson GA: Thrombin binding and response in platelets from patients with dyslipoproteinemias: increased stimulus-response coupling in type II hyperlipoproteinemia. Blood 68:498, 1986

16. Tandon NN, Hoeg JM, Jamieson GA: Perfusion studies on the formation of mural thrombi with cholesterol-modified and hypercholesterolemic platelets. J Lab Clin Med 105:157, 1985

17. Jamieson GA: The activation of platelets by thrombin: A model for activation by high and moderate affinity receptor pathways, in Jamieson GA: (ed): Platelet Membrane Receptors: Molecular Biology, Immunology, Biochemistry, and Pathology, New York, Alan R. Liss, Inc., 1988, p 137

18. Greco NJ, Jamieson GA: High and moderate affinity pathways for α-thrombin-induced platelet activation. Proc Soc Exp Biol Med 198:792, 1991

19. Ishihara H, Connolly AJ, Zeng D, Kahn ML, Zheng YW, Timmons C, Tram T, Coughlin SR: Protease-activated receptor 3 is a second thrombin receptor in humans. Nature 386:502, 1997

20. Scase TJ, Heath MF, Evans RJ: Cloning of PAR3 cDNA from human platelets and human erthroleukemic and human promonocytic cell lines. Blood 90:2113, 1997

21. Rasmussen UB, Vouret-Craviari V, Jallat S, Schlesinger Y, Pages G, Pavirani A, Lecocq JP, Pouyssegur J, Van Obberghen-Schilling E: cDNA cloning and expression of a hamster α-thrombin receptor coupled to Ca^{2+} mobilization. FEBS Lett 288:123, 1991

22. Verrall S, Ishii M, Chen M, Wang L, Tram T, Coughlin SR: The thrombin receptor second cytoplasmic loop confers coupling to G_q-like G proteins in chimeric receptors. J Biol Chem 272:6898, 1997

23. Seiler SM, Michel IM, Fenton JW: Involvement of the "tethered-ligand" receptor in thrombin inhibition of platelet adenylate cyclase. Biochem Biophys Res Commun 182:1296, 1992

24. Baffy G, Yang L, Raj S, Manning DR, Williamson JR: G protein coupling to the thrombin receptor in chinese hamster lung fibroblasts. J Biol Chem 269:8483, 1994

25. Ishii K, Hein L, Kobilka B, Coughlin SR: Kinetics of thrombin receptor cleavage on intact cells. J Biol Chem 268:9780, 1993

26. Parry MAA, Myles t, Tschopp J, Stone SR: Cleavage of the thrombin receptor: dentification of potential activators and inactivators. Biochem J 320:335, 1996

27. Greco NJ, Tandon NN, Jones GD, Kornhauser R, Jackson B, Yamamoto N, Tanoue K, Jamieson GA: Contributions of glycoprotein Ib and the seven transmembrane domain receptor to increases in platelet cytoplasmic $[Ca^{2+}]_i$ induced by α-thrombin. Biochemistry 35:906, 1996

28. Gralnick HR, Williams S, McKeown LP, Hansmann K, Fenton JW,III, Krutzsch H: High-affinity α-thrombin binding to platelet glycoprotein Ibα: Identification of two binding domains. Proc Natl Acad Sci USA 91:6334, 1994

29. Brass LF, Vassallo RR,Jr., Belmonte E, Ahuja M, Cichowski K, Hoxie JA: Structure and function of the human platelet thrombin receptor. J Biol Chem 267:13795, 1992

30. Lau L-F, Pumiglia K, Cote YP, Feinstein MB: Thrombin-receptor agonist peptides, in contrast to thrombin itself, are not full agonists for activation and signal transduction in human platelets in the absence of platelet-derived secondary mediators. Biochem J 303:391, 1994

31. Kramer RM, Roberts EF, Hyslop PA, Utterback BG, Hui KY, Jakubowski JA: Differential activation of cytosolic phospholipase A_2 ($cPLA_2$) by thrombin and thrombin receptor agonist peptide in human platelets. J Biol Chem 270:14816, 1995

32. Goodwin CA, Wheeler-Jones CPD, Kakkar VV, Deadman JJ, Authi KS, Scully MF: Thrombin receptor activating peptide does not stimulate platelet procoagulant activity. Biochem Biophys Res Commun 202:321, 1994

33. Lasne D, Donato J, Falet H, Rendu F: Different abilities of thrombin receptor activating peptide and thrombin to induce platelet calcium rise and full release reaction. Thromb Haemost 74:1323, 1995

34. Sugama Y, Tiruppathi CM, Janakidevi K, Andersen TT, Fenton II,J.W., Main AB: Thrombin-induced expression of endothelial P-Selectin and intercellular adhesion molecule-1: A mechanism for stabilizing neutrophil adhesion. J Cell Biol 119:935, 1992

35. Lum H, Andersen TT, Siflinger-Birnboim A, Tiruppathi C, Goligorsky MS, Fenton II,J.W., Main AB: Thrombin receptor peptide inhibits thrombin-induced increase in endothelial permeability by receptor desensitization. J Cell Biol 120:1491, 1993

36. Lenormand P, Sardet C, Pages G, L'Alleman G, Brunet A, Pouyssegur J: Growth factors induce nuclear translocation of MAP kinases ($p42^{mapk}$ and $p44^{mapk}$) but not of their activator MAP kinase kinase ($p45^{mapkk}$) in fibroblasts. J Cell Biol 122:1079, 1993

37. Vouret-Craviari V, Van Obberghen-Schilling E, Scimeca JC, Van Obberghen E, Pouyssegur J: Differential activation of $p44^{mapk}$ (ERK1) by α-thrombin and thrombin-receptor peptide agonist. Biochem J 289:209, 1993

38. Jenkins AL, Bootman MD, Derridge MJ, Stone SR: Differences in intracellular calcium signaling after activation of the thrombin receptor by thrombin and agonist peptide in osteoblast-like cells. J Biol Chem 269:17104, 1994

39. Vouret-Craviari V, Van Obberghen-Schilling E, Rasmussen UB, Pavirani A, Lecocq J-P, Pouyssegur J: Synthetic α-thrombin receptor peptides activate G protein-coupled signaling pathways but are unable to induce mitogenesis. Mol Biol Cell 3:95, 1992

40. Trejo J, Connolly AJ, Coughlin SR: The cloned thrombin receptor is necessary and sufficient for activation of mitogen-activated protein kinase and mitogenesis in mouse lung fibroblasts. Loss of responses in fibroblasts from receptor knockout mice. J Biol Chem 271:21536, 1996

41. Connolly AJ, Ishihara H, Kahn ML, Farese Jr,R.V., Coughlin SR: Role of the thrombin receptor in development and evidence for a second receptor. Nature 381:516, 1996

42. Darrow AL, Fung-Leung W-P, Ye RD, Santulli RJ, Cheung W-M, Derian CK, Burns CL, Damiano BP, Zhou L, Keenan CM, Peterson PA, Andrade-Gordon P: Biological consequences of thrombin receptor deficiency in mice. Thromb Haemost 76:860, 1996

43. Ware J: Molecular analyses of the platelet glycoprotein Ib-IX-V receptor. Thromb Haemost 79:466, 1998

44. Lopez JA: The platelet glycoprotein Ib-IX complex. Blood Coagul Fibrinolysis 5:97, 1994

45. De Marco L, Mazzucato M, Masotti A, Ruggeri ZM: Localization and characterization of an α-thrombin-binding site on platelet glycoprotein Ibα. J Biol Chem 269:6478, 1994

46. Mazurov AV, Vinogradov DV, Vlasik TN, Repin VS, Booth WJ, Berndt MC: Characterization of an antiglycoprotein Ib monoclonal antibody that specifically inhibits platelet-thrombin interaction. Thromb Res 62:673, 1991

47. Peng M, Emig FA, Mao A, Lu W, Kirby EP, Niewiarowski S, Kowalska MA: Interaction of echicetin with a high affinity thrombin binding site on platelet glycoprotein GPIb. Thromb Haemost 74:954, 1995

48. Chang M-C, Lin H-K, Peng H-C, Huang T-F: Antithrombotic effect of crotalin, a platelet membrane glycoprotein Ib antagonist from venom of *crotalus atrox*. Blood 91:1582, 1998

49. Bradford HN, Dela Cadena RA, Kunapuli SP, Dong J-F, Lopez JA, Colman RW: Human kininogens regulate thrombin binding to platelets through the glycoprotein Ib-IX-V complex. Blood 90:1508, 1997

50. Pedicord DL, Greco NJ, Jamieson GA, Dicker IB: GPIb mediates thrombin-induced platelet procoagulant activity. Thromb Haemost 475, 1997(abstr)

51. McNicol A, Sutherland M, Zou r, Drouin J: Defective thrombin-induced calcium changes and aggregation of Bernard-Soulier platelets are not associated with deficient moderate-affinity receptors. Arterioscler Thromb & Vascular Biol 16:628, 1996

52. Harmon JT, Jamieson GA: Activation of platelets by alpha-thrombin is a receptor-mediated event. J Biol Chem 261:15928, 1986

53. Brower MS, Levin RI, Garry K: Human neutrophil elastase modulates platelet function by limited proteolysis of membrane glycoproteins. J Clin Invest 75:657, 1985

54. Greco NJ, Jones GD, Tandon NN, Kornhauser R, Jackson B, Jamieson GA: Differentiation of the two forms of GPIb functioning as receptors for α-thrombin and von Willebrand factor: Ca^{2+} responses of protease-treated human platelets activated with α-thrombin and the tethered ligand peptide. Biochemistry 35:915, 1996

55. Lozano ML, Rivera J, Gonzalez-Conejero R, Moraleda JM, Vicente V: Loss of high-affinity thrombin receptors during platelet concentrate storage impairs the reactivity of platelets to thrombin. Transfusion 37:368, 1997

56. Mazzucato M, DeMarco L, Masotti A, Pradella P, Bahou WF, Ruggeri ZM: Characterization of the initial α-thrombin interaction with glycoprotein Ibα in relation to platelet activation. J Biol Chem 273:1880, 1998

57. Marchese P, Murata M, Mazzucato M, Pradella P, DeMarco L, Ware J, Ruggeri ZM: Identification of three tyrosine residues of glycoprotein Ibα with distinct roles in von Willebrand factor and α-thrombin binding. J Biol Chem 16:9571, 1995

58. Harmon JT, Jamieson GA: The glycocalicin portion of platelet glycoprotein Ib expresses both high and moderate affinity receptor sites in thrombin. J Biol Chem 261:13224, 1986

59. Liu L, Freedman J, Hornstein A, Fenton JW,II, Song Y, Ofosu FA: Binding of thrombin to the G-protein-linked receptor, and not to glycoprotein Ib, precedes thrombin-mediated platelet activation. J Biol Chem 272:1997, 1997

60. Parry MAA, Stone SR, Hofsteenge J, Jackman MP: Evidence for common structural changes in thrombin induced by active-site or exosite binding. Biochem J 290:665, 1993

61. Wu Q, Sheehan JP, Tsiang M, Lentz SR, Birktoft JJ, Sadler JE: Single amino acid substitutions dissociate fibrinogen-clotting and thrombomodulin-binding activities of human thrombin. Proc Natl Acad Sci USA 88:6775, 1991

62. Okumura T, Jamieson GA: A single receptor for platelet aggregation induced by thrombin or ristocetin. Thromb Res 8:701, 1976

63. Parker RI, Gralnick HR: Fibrin monomer induces binding of endogenous platelet von Willebrand Factor to the glycocalicin portion of platelet glycoprotein IB. Blood 70:1589, 1987

64. Ward CM, Andrews RK, Smith AI, Berndt MC: Mocarhagin, a novel cobra venom metalloproteinase, cleaves the platelet von Willebrand factor receptor glycoprotein Ibα. Identification of the sulfated tyrosine/anionic sequence Tyr-276 --Glu-282 of glycoprotein Ibα as a binding site for von Willebrand factor and α-thrombin Biochemistry 35: 4929, 1996

65. Dong J-F, Li CQ, Lopez JA: Tyrosine sulfation of the glycoprotein Ib-IX complex: Identification of sulfated residues and effect on ligand binding. Biochemistry 33:13946, 1994

66. Wyler B, Bienz D, Clemetson KJ, Luscher EF: Glycoprotein Ibα is the only phosphorylated major membrane glycoprotein in human platelets. Biochem J 234:373, 1986

67. Huang R, Sorisky A, Church WR, Simons ER, Rittenhouse SE: "Thrombin" receptor-directed ligand accounts for activation by thrombin of platelet phospholipase C and accumulation of 3-phosphorylated phosphoinositides. J Biol Chem 266:18435, 1991

68. Offermanns S, Toombs CF, Hu Y-H, Simon MI: Defective platelet activation in $G\alpha_q$-deficient mice. Nature 389:183, 1997

69. Du X, Harris SJ, Tetaz TJ, Ginsberg MH, Berndt MC: Association of a phospholipase A_2 (14-3-3 protein) with the platelet glycoprotein Ib-IX complex. J Biol Chem 269:18287, 1994

3

PLATELET THROMBOXANE RECEPTORS: BIOLOGY AND FUNCTION

Gerhard J. Johnson, M.D.
University of Minnesota Medical School
Staff Physician, Hematology/Oncology Section
Veterans Affairs Medical Center
Minneapolis, MN
USA

3.1 INTRODUCTION

> Gerhard J. Johnson, M.D. is Professor of Medicine, University of Minnesota Medical School, and Staff Physician, Hematology/Oncology Section, Medical Service, Veterans Affairs Medical Center, Minneapolis, Minnesota. His platelet research program, funded by VA Merit Review and NIH grants since 1972, has focused on the study of platelet signal transduction via thromboxane receptors, with emphasis on the study of dog platelets. He has served as Chairman of the VA Merit Review Board for Hematology and the Subcommittee on Animal Models of the Scientific and Standardization Committee, ISTH.

Thromboxane A_2 (TXA_2), is a product of the cyclooxygenase pathway of arachidonic acid metabolism with platelet activation, vascular and bronchial smooth muscle constrictive and smooth muscle proliferative properties. TXA_2 plays an important role in normal hemostasis and in vascular and bronchial smooth muscle physiology, it contributes to the pathophysiology of vascular and pulmonary diseases and it also appears to play a role in some forms of renal disease. TXA_2

binds to specific receptors on platelets and blood vessels that stimulate signaling pathways that mediate its functional effects. Considerable progress has been made in recent years in definition of the structure and function of the receptor to which TXA_2_binds, and of post-receptor signaling pathways in platelets and other cells. This chapter will review this progress and summarize current understanding of the platelet TXA_2 receptor. It will also point out the limitations of the present state of knowledge. Since many studies of TXA_2 receptor structure and function have been performed on cells other than platelets, and because differences in pharmacological properties have been observed among cells of different origins, reference will be made to investigation of TXA_2 receptors in other cell types where appropriate. Receptors that bind TXA_2 have been referred to by several designations. These receptors bind TXA_2 and prostaglandin endoperoxides; thus, they have been termed TXA_2/PGH_2 receptors (1). However, the International Union of Pharmacology Classification of Prostanoid Receptors (2) nomenclature-TP receptors-will be used in this chapter.

3.2 BIOLOGICAL EFFECTS OF TP RECEPTOR ACTIVATION

The biological action of TXA_2 is mediated via specific, high affinity TP receptors on platelets, smooth muscle cells and other tissues (See 1-5 for reviews). TXA_2 binding to TP receptors results in multiple biological consequences. Among the best defined are platelet aggregation and secretion, smooth muscle contraction, proliferation and hypertrophy, CNS-mediated vascular pressor responses, and thymocyte proliferation.

3.2.1 PLATELETS

Platelet aggregation and secretion make important contributions to primary hemostasis. Although platelet aggregation and secretion in vivo are dependent upon platelet activation by multiple agonists, TXA_2 plays an amplification role in this process. Studies performed in the absence and presence of platelet cyclooxygenase inhibitors indicate that TXA_2 contributes substantially to the development of the hemostatic plug that is responsible for the cessation of bleeding following injury to small blood vessels (6-8). TXA_2 induced-platelet aggregation makes a critical contribution to the development of acute occlusion of human atherosclerotic arteries, resulting in unstable angina, myocardial infarction, cerebral transient ischemic attacks and infarction and peripheral vascular occlusion as indicated by the effectiveness of aspirin inhibition of platelet cyclooxygenase in their prevention (9). Platelet aggregation stimulated by multiple agonists is greatly potentiated by secretion of adenine nucleotides from dense granules. Dense granule secretion stimulated by weak agonists is dependent on TXA_2 formation .

3.3 SMOOTH MUSCLE CONTRACTION

3.3.1BLOOD VESSELS

TP receptors are widely distributed on arterial and venous vascular smooth muscle cells. TXA_2-induced smooth muscle contraction mediates vasoconstriction (10,11) that is accentuated in the absence of endothelium (12). TXA_2 contributes to closure of umbilical vessels at birth (2). TXA_2 and TXA_2 analogs can induce or contribute to pulmonary hypertension (13-15) and systemic hypertension (16,17). TXA_2 synthesis has been observed to be increased in several forms of experimental hypertension in animals (17), and in some human subjects with essential hypertension (18), but a direct role for TXA_2 is uncertain. Short-term treatment with a combined thromboxane synthase/TP receptor antagonist inhibited platelet aggregation induced by the TP receptor agonist U46619 (See IV. TP RECEPTOR FUNCTION, A. LIGAND BINDING below for structures of TP receptor agonists and antagonists) and decreased renal and extrarenal TXA_2 synthesis in patients with mild essential hypertension, but it had no effect on arterial blood pressure (19). However, longer duration treatment with thromboxane synthase or TP receptor inhibitors has been effective in lowering the blood pressure of rodents with genetic hypertension (16,20). Data from animal and human studies indicate that intrarenal generation of TXA_2 contributes to vasoconstriction that occurs in inflammatory and lupus-associated glomerular injury (21). The observation that TP receptor gene expression may be upregulated by interleukin 6, that is excreted in excess quantities in the urine of patients with glomerulonephritis, may be relevant to the pathophysiology of glomerular renal disease (22).

3.3.2 BRONCHIAL SMOOTH MUSCLE

TXA_2 facilitates bronchial smooth muscle contraction resulting in broncho-constriction that may contribute to obstructive pulmonary disease (23-25) and allergic airway reactions (26).

C. SMOOTH MUSCLE PROLIFERATION

Smooth muscle cell protein synthesis and hypertrophy (27) and mitogenesis (28) have been observed to be stimulated by TXA_2. Whether or not TXA_2 has direct mitogenic properties is debated (See 29 for review), but evidence of mitogenic effects of TXA_2 on smooth muscle has been presented (28,30-33). The observations of Hanasaki, et al (34) and of Grosser, et al (35) suggest that TXA_2 acts as a mitogen only in the presence of insulin or growth factors such as platelet-derived growth factor. Futher evidence of the potentiating effects of TXA_2 on mitogenesis induced by 5-HT (33) and by thrombin (36) supports this concept. The role of TXA_2-stimulated mitogenesis in arterial smooth muscle proliferation in human disease is uncertain, but it has been speculated that TXA_2 contributes to the pathophysiology of hypertension (20) and to coronary artery smooth muscle proliferation following revascularization (29,36).

D. CNS PRESSOR RESPONSES

Brain TP receptors, shown to be present in both animal (37,38) and human (38,39) brain tissues, were found to mediate calcium mobilization (38) and a hypertensive response in rats that was partially attributed to the release of arginine vasopressin (40).

E. THYMOCYTE PROLIFERATION

TP receptors were found to be highly expressed on thymocytes (41,42), and TP receptor agonists stimulated apoptosis of these cells (42). Since phagocytic cells of the thymic reticulum produced TXA_2, along with PGE_2 and 6-keto-$PGF_{1\alpha}$, and they were shown to regulate thymocyte proliferation in vitro (43), TXA_2 may exert a negative control on thymocyte proliferation.

III. TP RECEPTOR STRUCTURE

TP receptors are members of the broader class of prostanoid receptors that appear to constitute a subfamily within the rhodopsin-like, G protein-coupled receptor superfamily (3). These G protein-coupled receptors (GPCR) have as a common structural characteristic seven transmembrane alpha_helices connected by alternating extracellular and intracellular loops (44,45). **Figure 1** illustrates the deduced amino acid sequence and the trans-membrane topology of the human TP receptor.

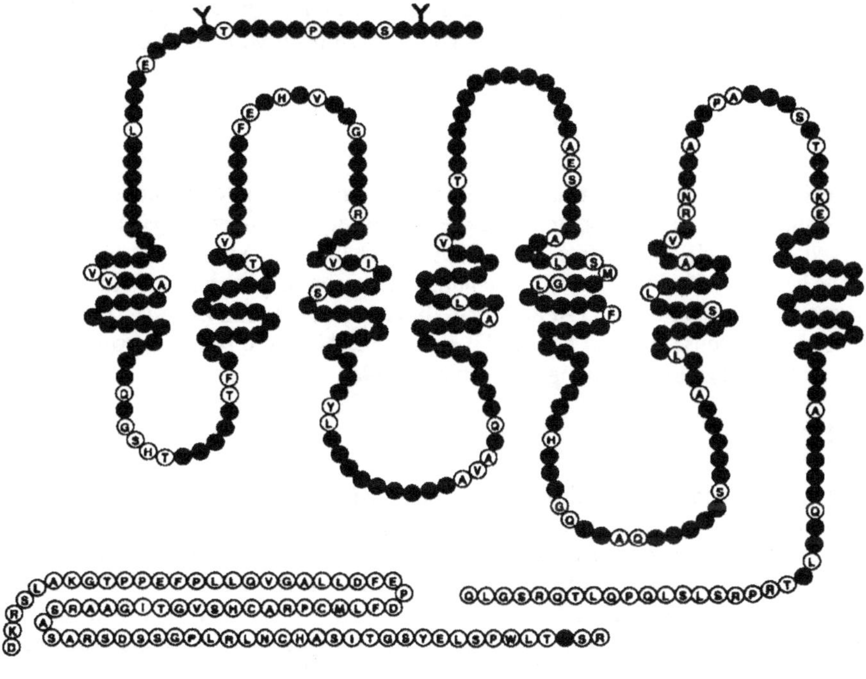

Figure 1. Y= glycosylation sites; Shaded circles=homology with mouse TP receptor. The sequence of the alternatively spliced endothelial-derived TP receptor (TP-β) carboxyl terminus is shown at the lower left. Figure (ref 242) reproduced with permission of Cayman Chemical, Ann Arbor, MI.

A. TP RECEPTOR ISOLATION

Binding sites for the TP receptor antagonist, $[^{125}I]$-PTA-OH, were detergent solubilized from human platelets by Burch, et al (46). The ligand binding and displacement characteristics of the solubilized protein were very similar to those of TP receptors present on intact human platelets. Subsequent isolation and purification of TP receptors from human platelet membranes was carried out by Ushikubi, et al (47) by the use of S-145, conjugated to Affi-Gel 102, as an affinity ligand. TP receptors were obtained from detergent-solubilized platelet membranes, and were isolated and purified by a series of chromatographic steps that resulted in ~8700 fold purification. The purified protein of M_r

57,000 had relatively low binding affinity that was corrected in the presence of asolectin. Later Knezevic, et al (48) and Borg, et al (49) purified TP receptors from rat brain and rabbit aorta, as well as human platelets, by the use of ligand-affinity and immunoaffinity columns prepared fom anti-peptide receptor antibodies. The molecular weight of the human platelet TP receptor was determined to be 55 kDa (49). Purification of TP receptors by ligand-affinity chromatography resulted in elution of receptor binding and GTPase activity in the same fraction. GTPase activity was stimulated by U46619. Immunoaffinity chromatography permitted the copurification of TP receptor-coupled G proteins, including $G\alpha_q/G\alpha_{11}$ and additional 42 kDa and 85 kDa proteins that immunoreacted with a $G\alpha_{,common}$ antiserum that did not recognize $G\alpha_q/G\alpha_{11}$ (48). These studies demonstrated that TP receptors could be isolated with functionally coupled G proteins.

B. MOLECULAR BIOLOGY

TP receptor cDNA has been isolated from numerous human cells including placenta, endothelial cells and monocytes, and the human erythroleukemia (HEL) and human-derived K562 cell culture lines, and from mouse lung, rat kidney, rat astrocytes and bovine heart (41, 50, 51-57). Only one human TP receptor gene has been identified (50), but two receptor isoforms (TP-α and TP-β) have been described (50,56). The full length sequences of mouse, rat and bovine TP receptor cDNA have been determined. A high level of homology exists among the species (**Table**).

1. CLONING

Working from partial amino acid sequences obtained from proteolytic fragments of the TP receptor purified by Ushikubi, et al (47), Hirata, et al (50) designed an oligo-nucleotide probe that allowed them to identify a partial cDNA clone in a human mega-karyocytic cell line (MEG-01). A 586 bp fragment of the MEG-01-derived partial clone was used to screen a human placenta cDNA library. A positive clone with a 2.9 kb insert that contained the entire coding region was identified. The open reading frame (1029 bp) was found to code for a protein of 343 amino acids (Mr 37,429).

2. MOLECULAR STRUCTURE

Three important characteristics common to GPCR that were identified in TP receptors (**Figure 1**) were 1) N-glycosylation sites on the extracellular amino-terminus, 2) two conserved cysteine residues, one in the first and the other in the second extracellular loop, that form an intramolecular disulfide bridge, and 3) an aspartic acid residue in the second transmembrane domain (44). Glycosylation appeared to influence the level of GPCR expression (45). The difference between the molecular weight of the isolated TP receptor (55-57 kDa) and its calculated molecular weight (37 kDa) was accounted for by glycosylation (58). In vitro mutagenesis studies of TP receptor cysteines demonstrated the importance of the disulfide bond to receptor-ligand interactions (59). Based on study of other receptors, the aspartic acid residue in the second transmembrane domain was presumed to play an important role in signal transduction (44). Hydrophobicity analysis revealed seven hydrophobic segments that corresponded to the transmembrane domains typical of GPCR (50). Significant homology was identified between TP receptors and other GPCR, particularly in the transmembrane domains (50). COS cells transfected with TP receptor cDNA expressed TP receptors as determined by specific binding of antagonist, [^{3}H]S-145. The K_d for [^{3}H]S-145 binding was equal to that of human platelets (50). The TP receptor was also shown to couple to PLC (50). The protein expressed by this TP receptor cDNA, (TP-α), was later identified in platelets (60).

3. CHROMOSOMAL LOCALIZATION

The human TP receptor gene is located on chromosome 19p13.3 (61), the mouse gene is on chromosome 10 (62) and the rat gene is on chromosome 7q11 (63).

4. GENE STRUCTURE

Nüsing, et al (61) described the organization of the human placenta TP (TP-α) receptor gene. The gene spanned more than 15kb. It contained three exons and two introns. This exon-intron relationship is conserved in other prostanoid receptors (44). The first intron (6.3 kb) was found to be located in the 5' noncoding region, 83 bp upstream from the ATG start site. Intron 2 (4.3 kb) was found to be located between the end of the sixth transmembrane region and the seventh transmembrane region and the 3'-untranslated region (61). Transcription initiation sites were determined by the use of 5'-RACE methodology applied to the study of poly(A)$^+$ RNA obtained from both placenta and MEG-01 cells. Approximately 90% of the transcription sites for both placental and MEG-01 transcripts were located within -102 to -190 bp in exon 1. Additional transcripts obtained from MEG-01 RNA were also identified. By restriction mapping these were found to be located in intron 1, 2.2 kb upstream of exon 2. Sequencing of the 5'-flanking regions of the major and minor transcripts revealed evidence of a promoter region I, upstream of exon 1, that contained the majority of the transcription initiation sites. Promoter region I had a high GC content, but no TATA or CAAT box, and it contained four consensus sequences for the RNA polymerase II transcription factor, Sp1 (61). Consensus sequences for AP-2 and phorbol response elements were also identified in promoter I. Promoter region II, also located upstream of exon 1, contained consensus sequences for Sp1, AP-2, the acute phase reactant regulatory element and a homolog of the glucocorticoid response element. Subsequent study (64) of the 5' untranslated region of TP receptor cDNA obtained from human uterus, lung and K562 cells, studied by PCR methodology rather than 5'-RACE, revealed substantial differences from those reported by Nüsing, et al (61). D'Angelo, et al (64) found the major transcription initiation site to be located ~380 bp 5' of that previously identified. This site was located ~470 bp 5' of the first exon-intron splice site (~560 bp 5' of the translation initiation codon) and, in contrast to the report of Nüsing et al (61), no evidence for transcription initiation within the first intron was found (64). A third transcription initiation site was found in the rat TP receptor gene, by use of PCR and primer extension methodology, to be located 63 bp upstream of the 5' end of the cloned TP receptor cDNA (22). These data indicate that agreement exists regarding several aspects of TP receptor structure, but the location of the promoter region is controversial.

5. GENE EXPRESSION

TP receptor gene expression was stimulated by phorbol ester in human mega-karyoblastic leukemia (CHRF-288) cells (65) and human HEL cells (66), but neither gluco-corticoids

TABLE

	1					60
MUSTXA2R	·····t···a	····v····q	···a······	·····al··g	··········	···p·ag*p·
RATTA2R	··l·st···a	····v····q	···a······	·····al··g	··········	···p·ag*p·
BOVINE	····a·····	····m·····	··········	·····l····	·········m	·····s·qs·
HUMHTAR	··········	····t·····	··········	·····v····	··········	·····g··t·
HSU11271	··········	····t·····	··········	·····v····	··········	·····g··t·
CONSENSUS	MWPNGSSLGP	CFRP-NITLE	ERRLIASPWF	AASFC-VGLA	SNLLALSVLA	GARQG-SH-R

	61					120
MUSTXA2R	····al····	··········	·····a····	··ld·r·t··	s····y····	a·········
RATTA2R	····al····	··········	···v·a····	··ld·r·t··	·····h···a	a·········
BOVINE	··········	·····m····	······t··f	v····q····	··s··h····	i·········
HUMHTAR	··········	··········	··t·······	·····h····	·····r····	v·i····s··
HSU11271	··········	··········	··t·······	·····h····	·····r····	v·i····s··
CONSENSUS	SSFLTFLCGL	VLTDFLGLLV	TGAIVVSQHA	ALFEW-AVDP	GCRLC-FMGV	-MVFFGLCPL

	121					180
MUSTXA2R	··········	·v········	·tat·*····	·······v··	g·········	l···s·····
RATTA2R	·······a··	·v········	··at·*····	·······vg·	gt········	l···s·····
BOVINE	··········	··········	··t·······	t·······s·	··········	··h·······
HUMHTAR	··········	y·········	··v·······	··········	··········	··········
HSU11271	··········	y·········	··v·······	··········	··········	··········
CONSENSUS	LLGAAMASER	FLGITRPFSR	PA-ASQRRAW	ATVGLVWAAA	LALGLLPLLG	VGRYTVQYPG

	181					240
MUSTXA2R	········tq	r···v···i·	····a·····	·l········	···r···tr·	·t*·······
RATTA2R	··········	r·······m·	····v·····	·l········	···r···ar·	·t*·······
BOVINE	········td	p·········	·····i···m	······i···	·········a	t·········
HUMHTAR	··········	s·········	sm··gl····	··········	··········	········s·
HSU11271	··········	s·········	sm··gl····	··········	··········	········s·
CONSENSUS	SWCFLTLGAE	-GDVAFGLLF	ALLGS-SVGL	SFLLNTVSVA	TLCHVYHGQE	AAQQRPRDCE

	241					300
MUSTXA2R	·······v··	····t·····	·····m··l·	·t··v··fs·	··l·a··h··	··········
RATTA2R	·······v··	····t·····	·····l··l·	·tl·v···s·	··l····r··	··········
BOVINE	·······m··	·····i····	··········	·s······t·	····l··r··	··········
HUMHTAR	····a··l··	········l·	··········	rn······a·	·······ke·	··········
HSU11271	····a··l··	········l·	··········	rn······a·	·······ke·	··········
CONSENSUS	VEMMVQL-GI	MVVASVCWNP	LLVFIAQTVL	Q-PPAMSP-G	QLSRTTE-QL	LIYLRVATWN

	301			
MUSTXA2R	··········	···s·····h	··f·sqlqav	··rrppaqamlsgp
RATTA2R	··········	···s·····h	··ftsqlqav	··happtqamlsgp
BOVINE	··········	·····iq·fy	······s·sl	··qpqltrrstih
HUMHTAR	··········	·········q	······p·sl	··qpqltqrsglq
HSU11271	··········	·········q	······p·r*	··tlwpsleysgtisahcnlrlpgssdsrasasr
				aagitqvshcarpcmlfdpsfd
				llagvqllpfepptgkalsrkd
CONSENSUS	QILDPWVYIL	FRRAVLRRL-	PRLSTR-R--	SL---- + tails

TXA_2 Receptor Amino Acid Sequences

Alignment of the TXA_2 receptor amino acid sequences of mouse (MUSTXA2R), rat (RATTA2R), bovine (BTU53485 & BTU53484), human placental (HUMHTAR), human endothelium (HSU11271). The amino acid sequences are available from the Genbank file server under accession numbers: D10849 (mouse lung), D21158 (rat kidney), U53485 (bovine liver), U53484 (bovine heart), U11271 (human endothelium) and D38081 (human placenta). Dots indicate identity with the consensus sequence, asterisks represent gaps, dashes represent the lack of a consensus and blank areas indicate sequence has not been completed.

nor the acute-phase response stimulators, IL-1, IL-6, lipopolysacharide, C-reactive protein or TNF, induced expression of TP receptor protein (66). These results indicated that TP receptor gene transcription in HEL cells was controlled primarily by promoter region I (66). The TP receptor gene promoter region was characterized by D'Angelo, et al (64). They determined that the PMA/protein kinase C (PKC) responsiveness of the TP receptor gene was attributable to AP-2-like binding to sites between 1.85 and 1.95 kb 5' of the major transcription initiation site (64). This site was futher defined as a 14-nucleotide C-rich sequence, flanked by an octanucleotide inverted repeat, located -1.938 to -1.925 kb 5' of the transcription start site (67). The PMA-responsive enhancer factor that bound to the C-rich sequence was identified as the transcription factor, Sp1 (67). However, Sp1 binding alone was not sufficient to facilitate phorbol enhancement of TP gene expression. Additional, unidentified down-stream elements, located between -1.91 and -1.84 kb, were also required (67).

6. TP RECEPTOR ISOTYPES

Only one TP receptor protein was found to be coded by the TP receptor gene identified by Nüsing et al. (61). Although these investigators found no evidence for alternative splicing of transcripts when RT-PCR of poly(A)$^+$ RNA obtained from human placenta, MEG-01 cells or mesangial cells was performed using primers flanking intron 2, a proposed second form of the human TP receptor (TP-β), considered to be a product of alternative splicing, was later cloned from human endothelial cells by Raychowdhury, et al (56,68). The nucleotide sequence of the full length TP-β_receptor cDNA obtained from endothelial cells was originally reported to be 1504 nucleotides with a single open reading frame, and a predicted sequence of 369 amino acids (56). However, the accurate nucleotide sequence, published as a correction of the original work (68), predicted 407 amino acids. The sequence of the first 328 amino acids was exactly the same as that of the TP receptor cDNA derived from placenta (TP-α) (50), but the sequences diverged substantially thereafter. The final 15 amino acids of the carboxyl terminus of the TP-α_receptor were found to be replaced by a 79 amino acid segment in the TP-β_receptor (**Figure 1, TABLE**). The spliced segment was located entirely in the cytoplasmic tail distal to the seventh transmembrane domain. The nucleotides that encoded the divergent portion of the cytoplasmic tail, plus the 3' untranslated region of TP-β, were found to match a sequence contained entirely within the 3' untranslated region of TP-α_(56). Although Raychowdhury, et al found only TP-β mRNA in endothelial cells, and both TP-α_and TP-β_mRNA in placenta, they were unable to demonstrate either TP-α_or TP-β mRNA in platelets (56). In contrast, Hirata, et al (69) identified mRNAs for both TP-α and TP-β variants of the human TP receptor in human platelets.

7. TP RECEPTOR MUTATION

Despite the description of several examples of impaired functional and biochemical responses to TP receptor agonists in human and animal platelets (70-76), only one TP receptor mutation has been described (77). A single amino acid substitution ($R^{60}\rightarrow L$) was found in human TXR-α_cDNA obtained by RT-PCR from platelets of patients with a dominantly inherited bleeding disorder (See V. ALTERED TP RECEPTOR FUNCTION, A. TP RECEPTOR MUTATION below).

C. STRUCTURE-FUNCTION RELATIONSHIPS

1. LIGAND BINDING

Complete definition of the ligand binding domains of the TP receptor has not been accomplished as yet; however, regions that appear to be important for ligand binding have been identified. Approachs to definition of these binding sites include the following:

a. Photoaffinity Labeling Photoaffinity labeling carried out with [^{125}I]-SAP-N_3, combined with proteolytic cleavage studies, suggested that ligand binding occurred primarily to the third and fourth transmembrane regions of TP receptors, with minor contributions of the amino terminus and the seventh transmembrane region plus the carboxy terminus (78). The significance of these observations for understanding agonist binding to intact receptors is uncertain because, a) these studies were carried out on isolated TP receptors independent of a lipid membrane environment, and b) studies of mutated TP receptors expressed in intact cells implicated different agonist binding domains (See III. TP RECEPTOR STRUCTURE, C. STRUCTURE-FUNCTION RELATIONSHIPS, 1. LIGAND BINDING, d. Mutagenesis below).

b. Fluorescence Labeling A biotinylated TP receptor probe, SQB, synthe-sized by Komiotis, et al (79) provided direct evidence of the existence of TP receptors on platelet plasma membranes. These studies also demonstrated that ligand binding domains of TP receptors were located at or near the external surface of platelet plasma membranes, since the maximum permissible depth of insertion of the probe (< 20.5 Å) was less than 30% of the width of the membrane.

c. Structural Modeling Three dimensional modeling of the TP receptor, based on an amino acid model of the β_2-adrenergic receptor complexed to isoproterenol, revealed a large hydrophobic pocket flanked by S^{201} and R^{295} (80). TP receptor interaction with TXA_2 was predicted, on the basis of this model, to occur by the carboxyl group of TXA_2 interacting with R^{295}, the hydroxyl group at position 15 interacting with S^{201} and the ligand hydrophobic moiety interacting with the receptor hydrophobic pocket. The accuracy of this model has not been tested by crystallographic analysis, but some of these predictions were confirmed by mutagenesis studies.

d. Mutagenesis Pharmacological studies of TP receptors suggested that cysteine residues played a role in ligand binding. Mutagenic analysis of the human K562 TP receptor identified cysteines at amino acid positions 102 and 105 (first extracellular loop) and 183 (second extracellular loop) to be critical for agonist ([^{125}I]-BOP) binding (59,81). A disulfide bond between C^{105} and C^{183}, cysteine residues conserved in all prostanoid receptors, was found to be important for receptor stability since substitution of serine for cysteine at these positions ($C^{105}\rightarrow S$ and $C^{183}\rightarrow S$) abolished TP receptor ligand binding (59,81). Substitution $C^{102}\rightarrow S$ also resulted in markedly reduced binding capacity and affinity for both agonist and antagonist ligands (59,81). Funk, et al (82) found amino acid residues within the seventh transmembrane domain that play an important role in ligand binding to TP receptors. A single nucleotide substitution ($T^{895}\rightarrow C$), that coded for a single amino acid substitution ($W^{299}\rightarrow R$), abolished antagonist ([^{3}H]-SQ29548) and agonist ([^{125}I]-BOP) binding to TP receptors expressed in HEK 293 cells (82). While mutant $W^{299}\rightarrow R$ failed to bind [^{125}I]-BOP, another mutation of the same residue ($W^{299}\rightarrow L$) bound [^{125}I]-BOP with the same affinity to an equal number of sites as did transfected wild-type TP receptor (82). An additional mutant ($R^{295} \rightarrow Q$) reduced antagonist and agonist binding. This suggested that R^{295} played an important role in ligand binding to TP receptors as predicted by Yamamoto, et al (80) from molecular modeling. Further support for the model came from the work of Chaing, et al (81) who found that mutation of a conserved fifth transmembrane domain serine ($S^{201} \rightarrow A$) resulted in decreased receptor affinity for an agonist containing a 15 hydroxyl group ([^{125}I]-BOP) but not for the non-hydroxylated antagonist ([^{3}H]-SQ29548). A sixth transmembrane domain serine also appeared to be important for agonist binding since a mutation in this domain ($S^{255} \rightarrow A$) increased the K_d for [^{125}I]-BOP from 1.63 to 5.41 while it had no effect on the K_d for [^{3}H]-SQ29548 (81).

Other mutagenesis studies employing rat/human TP receptor chimeras defined the extracellular domain and part of the first transmembrane domain as important for high affinity agonist binding (83). The higher affinity agonist binding observed in rat TP receptors, compared to human TP receptors, was found to require cooperative interactions between L^{37} and either A^{36} or G^{40} (83). The substitution $V^{37} \rightarrow L$ in the human TP receptor resulted in lengthening of the side chain of amino acid 37 that extended the isopropyl group further into the putative binding pocket (83). These studies yielded a very important insight regarding TP receptor structure-function relationships; namely, that rat and human TP receptors are species variants of one pharmacological TP receptor subtype, rather than distinct receptor subtypes (83). The cytoplasmic carboxyl tail of TP receptors was found not to play a role in agonist binding. These mutagenesis studies indicated that the binding site for TXA_2 and endoperoxides resided within the hydrophobic trans-membrane domains of TP receptors, particularly domains 1 and, as demonstrated by Funk, et al (82), domain 7.

In summary, the data obtained from mutagenesis studies indicates that agonist binding to TP receptors is dependent primarily on binding sites in the first and seventh transmembrane domains, but S^{255}, in the sixth transmembrane domain, may also play a role. The intracellular carboxyl terminus does not play a role in agonist binding. In the absence of crystallographic analysis, mutagenesis studies are probably the most revealing studies of structure-function relationships, but it should be kept in mind that these studies have limitations. As pointed out by Dorn, et al (83), a weakness of mutagenesis studies of receptor binding properties is that it is difficult to differentiate functional changes conferred by characteristics of individual amino acids from more general alterations in protein folding and tertiary structure.

e. Biochemical Modifications Conserved histidines in the second extra-cellular loop and/or the third intracellular loop were implicated in ligand binding to TP receptors (84). The effects of pH on TP receptor ligand affinity (See IV TP RECEPTOR FUNCTION, A. LIGAND BINDING, 3. EQUILIBRIUM BINDING below) were reversed by diethylpyrocarbonate and N-bromosuccinimide, reagents used to modify histidine residues. These observations were interpreted as evidence that, at acidic pH, histidine residues play a role in regulating TP receptor affinity (85). However, the role of histidines in TP receptor ligand binding is uncertain, because, a) no direct evaluation of the effect of diethyl-pyrocarbonate on cysteine residues was performed in the studies of Mayeux, et al (85) and b) histidine mutations ($H^{224} \rightarrow R$ and $H^{227} \rightarrow R$) resulted in no changes in TP receptor ligand binding or in calcium signaling (59).

Phosphorylation of the third extracellular loop also appeared to play a role in TP receptor function (66,86) (See IV TP RECEPTOR FUNCTION, E. TP RECEPTOR PHOSPHORYLATION, 2.ROLE OF PHOSPHORYLATION IN TP RECEPTOR ACTIVATION below).

2. EFFECTOR ACTIVATION

Mutation of C^{223} (third intracellular loop) resulted in TP receptors with normal ligand equilibrium binding characteristics but impaired calcium signaling (59). This suggested that this region of the receptor was important for G protein/effector (PLC-β) interaction. Signal transduction from transfected mutant mouse TP receptors to PLC-β was also shown to be dependent upon the carboxyl terminus (87). The $R^{60}\rightarrow L$ mutation identified in platelets of patients with an hereditary bleeding disorder (77) indicated that the first intracellular loop of the TP receptor is important for agonist-induced conformational change of Gq, the G protein that mediates PLC-β activation (See IV. TP RECEPTOR FUNCTION, C.SIGNAL TRANSDUCTION, 4. ROLE OF G PROTEINS below).

The two forms of TP receptors (TP-α and TP-β) expressed in transfected CHO cells bound antagonist ligands equivalently, and they did not differ in their ability to activate β-PLC (69). However, expression of TP-α and TP-β receptor cDNAs in CHO cells revealed opposite coupling to adenylyl cyclase. I-BOP increased the level of cAMP in cells expressing TP-α but not in cells expressing TP-β.__A mutation ($R^{60}\rightarrow L$) of TP-α receptor cDNA abolished agonist-induced elevation of cAMP, as well as activation of PLC-β. The effects of I-BOP stimulation on cAMP levels were interpreted as evidence that the carboxyl terminal tails of TP receptor isoforms determined the specificity of G protein coupling, and that the R^{60} residue was important for the interaction of TP receptors with G_q and possibly G_s but not G_i (69). However, alternative interpretations are possible.

$G\alpha_q$ has been shown to have either stimulatory or inhibitory effects on various isoforms of adenylyl cyclase (88). Since the particular isoforms with which $G\alpha_q$ might have interacted in CHO cells with overexpressed TP receptors are unknown, it is not certain that the elevated cAMP levels observed only at high I-BOP concentrations (10-30 nM) (69) have any relevance to TP receptor function in intact platelets. However, type VII adenylyl cyclase mRNA was found to be present, along with mRNA for types III and IV, in human erythroleukemia (HEL) cells, studied as a model for platelets (89), and platelets may contain type VII (90). Since type VII adenylyl cyclase, like type II to which it is closely related, was shown to be activated by PKC (91), and since human platelet adenylyl cyclase was less inhibited by $G\alpha_i$ following treatment with phorbol esters (92), the increase in cAMP, observed in CHO cells transfected with TP receptor subtypes could have been contributed to by $G\alpha_q$ stimulation of adenylyl cyclase via PKC. However, this stimulation probably would have required activation of the cyclase by $G\alpha_s$ (88), and no obvious stimulator of G_s was present in the studies of Hirata, et al (69). Obviously, further definition of the type(s) of adenylyl cyclase present in platelets and the role of $G\alpha_q$ in its control will be required to define the mechanism by which TP receptor activation might modulate platelet adenylyl cyclase. Although the transfection studies of Hirata, et al (69) suggested that TP-α and TP-β receptors had differing roles in signal transduction in transfected cells, it is important to note that spontaneously expressed TP-β receptor *protein* has not been demonstrated in any cell type. To date, the functional role, if any, of the TP-β receptor isoform in platelets or other cells is unknown (1,53,60).

IV. TP RECEPTOR FUNCTION

A. LIGAND BINDING

Studies of TP receptor binding have been carried out with several radioligands that have structural similarity to TXA_2 and PGH_2.

1. AGONISTS AND ANTAGONISTS

Some of the more commonly used agonist ligands include, **U44069** (9α,11α-epoxymethano-15S-hydroxy-prosta-5Z,13E-dienoic acid), **U46619** (11α,9α-epoxymethano-15S-hydroxy-prosta-5Z,13E-dienoic acid), **STA_2** (9α,11α-thia-11α-carba-prosta-5Z,13E-dienoic acid), **PTA-OH** (9,11-dimethyl-methano-11,12-methano-16-(3-iodo-4-hydroxyl)-13-aza-15αβ-ω-tetranor-TXA_2), **9-11azoPGH$_2$** [(5Z,9α,11α,13E,15S)-9,11-azo-15-hydroxyprosta-5,13-dienoic acid], **CTA_2** (2β(Z),3α(1E,3R*)-3-aza-15αβ-ω-tetranor-TXA_2 and **I-BOP** ([1S-(1α,2β(5Z),3α_(1E,3S*) ,4α)-7-[3-(3-hydroxy-4-(4'-iodophen-oxy)-1-butenyl)-7-oxabicyclo-[2.2.1]heptan-2-yl]-5-heptenoic acid). I-BOP (93) is the most specific, high affinity agonist ligand. Antagonist ligands include **SQ29548** ([15-[1α,2β(Z5),3β,4α]-7-[3-[2-(phenlyamino)-carbonly]hydrazino]methyl]-7-oxobicyclo-[2,2,1]-hept-2-yl]-5-heptanoic acid), **GR 32191** ([1R-[1α(Z)2β,3α,5α]]-(+)-7-[5-[[(1,1'-biphenyl)-4-yl]methoxy]-3-hydroxy-2-(1-piperidinyl)cyclopentyl]-4-heptenoic acid, **S-145** (1S,2R,3R,4R)-(5Z)-7-(3-[phenyl,sulphonyl-amino-bicyclo[2.2.1]hept-2-yl)hept-5-enoic acid, **SAP-N$_3$** 7-[(1R,2S,3S,5R)-6,6-dimethyl-3-(4-azido-3-iodobenzesulfonylamino)bicyclo[3.1.1]hept-2-yl]-5(Z)-heptanoic acid, and others (See 2,4,94 for reviews).

2. SPECIFICITY

TXA_2 and PGH_2 both stimulate platelet aggregation by binding with high affinity to platelet TP receptors (11,12,95). In plasma, TXA_2 was found to be 10-30 times more potent than PGH_2 at inducing platelet aggregation (12,95), but in washed platelets PGH_2 was a more potent platelet aggregating agent (EC_{50} 45 nM vs 163 nM) (95). TP receptors are quite specific for TXA_2 and PGH_2. Other prostanoids, such as PGD_2, $PGF_{2\alpha}$ and 8-epi-$PGF_{2\alpha}$, have been found to bind to these receptors, but their agonist potency is significantly less than that of highly selective TP receptor analogs and their effects on TP receptors have varied depending on the tissue studied (2,96-99). 8-epi-$PGF_{2\alpha}$ was found to be a weak agonist (96,97) or partial agonist (100) of human platelet TP receptors, but it did not displace [^{125}I]-BOP binding to either TP-α_or TP-β_receptor isoforms expressed in HEK 293 cells and it was much less potent in displacing [^{125}I]-BOP or [^{3}H]-SQ29548 from human platelet membranes than U46619 or I-BOP (97,100). $PGF_{2\alpha}$ was even less effective in competing for I-BOP binding sites on platelets. These data suggested that 8-epi-$PGF_{2\alpha}$ activated platelets via a closely related but distinct receptor (97). Subsequent studies identified a separate binding site for 8-epi-$PGF_{2\alpha}$ on rat aortic smooth muscle cells (99). Krauss, et al (101) reported the results of a study of the smooth muscle contracting and platelet aggregating properties of a series of 9,11-cyclic carbonate derivatives of $PGF_{2\alpha}$. Although two of these prostanoids had potent agonist properties in smooth muscle, only one was a potent platelet agonist (EC_{50} 4 nM in washed platelets) whose effects were antagonized by SQ29548. The differential effects of these structurally similar prostanoids on smooth muscle and platelets was interpreted as further evidence for TP receptor subtypes, but the role of the 8-epi-$PGF_{2\alpha}$ binding site in the functional effects of these compounds was not assessed. Since the biological profile of the platelet receptor to which GR 32191 bound reversibly (See IV. TP RECEPTOR FUNCTION, A. LIGAND BINDING, 3. EQUILIBRIUM BINDING below) was similar to that of the platelet 8-epi-$PGF_{2\alpha}$ receptor (97), it is possible that the 9,11-cyclic $PGF_{2\alpha}$ derivatives mediated their effects by binding to this site rather than to TP receptors.

3. EQUILIBRIUM BINDING

The binding studies performed with radioligands have been predominantly equilibrium binding studies. Studies performed with agonist ligands have yielded data that have been analyzed as either one or two binding sites or affinities. Armstrong, et al (102) observed two platelet binding sites for U44069, while Kattelman, et al (103) identified only a single platelet binding site for U46619. Two binding sites observed with agonists have been correlated with platelet function. Armstrong, et al (102) found the smaller (~15%) of two displaceable U46619 binding components, observed at 1-20 nM, was functionally related to shape change, while the larger component was correlated with aggregation. Morinelli, et al (104) found the higher affinity platelet [^{3}H]-U46619 binding site (Kd 0.04 μM, 1,166 sites/platelet) to be related to shape change, while the lower affinity site (Kd 1.5μM, 15,163 sites/platelet) correlated with secretion. Dorn and DeJesus (105) identified two platelet binding sites for [^{125}I]-BOP. The high affinity site (K_d 270 pM, 222 sites/ platelet) was functionally related to shape change, while the low affinity site (K_d 3.9 nM, 818 sites/platelet) was coupled to aggregation. In one study, analyses performed by two computer programs yielded different results (104), while in another (106) two different programs yielded the same interpretation that the data best described two classes of binding sites. In other studies, binding of [^{125}I]-BOP to human platelets at pH 7.4 revealed only a single class of binding sites, but at pH 6.0-6.5, high (K_d 0.2-0.3nM) and low (K_d 4-11 nM) affinity classes of binding sites were observed (85,107). Receptor site numbers were also increased at pH 6.0-6.5 compared to pH 7.4. The K_d for the high affinity site correlated well with the EC_{50} value for I-BOP-induced shape change, and the K_d for the low affinity site correlated with the EC_{50} for I-BOP-induced calcium flux and aggregation (85,107). Other investigators have also observed only a single class of [^{125}I]-BOP binding sites to platelets (53,93,108).

Two platelet TP receptor binding sites have been observed with antagonists as well. Takahara, et al (109) found that the antagonist ligand, GR 32191, appeared to discriminate two TP receptor binding sites on human platelets. GR 32191 bound irreversibly to a site linked to PLC-β activation and to U46619-induced aggregation, and it bound reversibly to another site that mediated shape change and the majority of agonist-induced rise in cytosolic calcium. As noted above (See IV. TP RECEPTOR FUNCTION, A. LIGAND BINDING, 2. SPECIFICITY), the reversible GR 32191binding site may be the $PGF_{2\alpha}$ receptor; therefore, the significance of binding to this site is uncertain. GR 32191 binding data cannot be interpreted as unequivocal evidence of two TP receptor antagonist binding sites.

Studies of TP receptor agonist binding performed in other cells and tissues have revealed similar variability in regard to the presence of one or two binding affinities. Nonlinear analysis of [^{125}I]-BOP binding to placental tissue indicated a superior fit to two affinity sites (52), while in mouse kidney [^{125}I]-BOP bound to one site on cells in the medulla and to two sites on cells in the cortex (110). Others observed single site binding to smooth muscle cells (111) and to human peripheral blood monocytes (51). Evaluation of GR 32191 binding revealed evidence of different TP receptors on rat aortic smooth muscle compared to human platelets, since only the reversible GR 32191 binding site was identified on vascular smooth muscle cells (112).

It has been suggested that the low affinity class of binding sites identified by computer analysis of radioligand binding data may represent non-specific binding (104). Since studies performed with [^{3}H]-U46619 revealed evidence of ligand uptake into a platelet compartment that was competitive but nondisplaceable (103), platelet incorporation of ligands, such as U46619, might be interpreted as specific binding (104). However this appears to be unlikely since Morinelli, et al (104) demonstrated that an equal amount of [^{3}H]-U46619 was displaced by the antagonist, PTA-OH, after 40 minutes incubation as after 5 minutes. Since equilibrium binding was reached at 5 minutes and the incubations were carried out for 15 minutes, incorporation of isotope into the platelets could not explain the low affinity binding. Our laboratory has confirmed the observation of Kattleman, et al (103) that a portion of binding is competitive but nondisplaceable. However, prolonged incubation (>60 min) is required for significant [^{3}H]-U46619 incorporation into platelets (113). Therefore, the results of our studies agree with those of Morinelli, et al (104) that in equilibrium binding studies carried out with incubation periods of 10 minutes (106) no significant platelet uptake of [^{3}H]-U46619 is evident.

Another potential cause for differences in one and two site binding is the use of agents, such as prostacyclin, that block platelet activation. Kattelman, et al (103) reported a single binding site for U46619 to prostacyclin-incubated platelets, while others studies (76,102, 104,106) in which prostacyclin was not used found evidence of two site or two affinity binding. Morinelli, et al (104) found that secretion did not affect [^{3}H]-U46619 binding to platelets, but the specific effect of inhibitory prostaglandins on agonist binding was not evaluated.

The issue of whether or not platelets have one or two binding sites or affinities remains unresolved; however the evidence derived from equilibrium binding studies favors two binding sites or affinities. Additional support for two binding affinities comes from the fact that kinetic agonist binding to intact platelets also revealed two binding sites on both human and dog platelets (114). Although the evidence for two binding sites has suggested the presence of two TP receptors, it is, in light of the molecular evidence discussed above, more likely that one TP receptor has two affinity states.

B. TP RECEPTOR SUBTYPES

1. EVIDENCE FOR SUBTYPES

The observation of two TP receptor binding sites on platelets, that correlated with function, (See IV. TP RECEPTOR FUNCTION, A. LIGAND BINDING, 3. EQUILIBRIUM BINDING above) suggested that receptor subtypes might be responsible. Another reason to consider the possibility that TP receptor subtypes existed was the fact that subtypes of prostaglandin E (EP) receptors were recognized (2). Support for the existence of receptor subtypes derived primarily from data obtained from studies of the pharmacological characteristics of TP receptors on platelets compared to those on blood vessels (See 1 for review). Equilibrium binding studies performed with a large series of 13 azapinane TXA_2 analogs demonstrated differences between TP receptors on platelets and saphenous veins (1). CTA_2, 13-azapinane analogs and difluorinated TXA_2 derivatives acted as antagonists to platelets receptors, but they were agonists to blood vessel TP receptors (115-117). Although some of the differences between platelets and vascular tissues could be attributed to species differences, studies of the same ligands revealed significant differences in rank order potency between platelets and vascular tissues within the same species (108,118). Other studies observed differences between platelets and vascular tissues when antagonist but not agonist binding was studied (119).

2. EVIDENCE AGAINST SUBTYPES

Although some investigators found differences in the pharmacological character-istics of platelet TP receptors compared to those on blood vessels, others did not (119-122). As noted above, multiple EP receptor subtypes exist. However, the occurrence of EP receptor subtypes does not necessarily imply that TP receptors subtypes exist, since EP receptor subtypes are products of separate genes, while only one TP receptor gene has been identified. Another possible explanation for variable TP receptor function is tissue-specific differential expression of receptor isoforms. Isoforms of EP3 receptors that are derived from alternative splicing have been characterized (123,124). Evidence for TP receptor isoforms (TP-α_and TP-β) attributable to alternative splicing has been presented (See III. TP RECEPTOR STRUCTURE, B. MOLECULAR BIOLOGY, 6. TP RECEPTOR ISOTYPES above). If the ligand binding characteristics of these isotypes differed, and expression of one of these splice variants predominated in a tissue-specific manner, it would be theoretically possible to explain tissue-specific functional receptor differences on the basis of receptor isotypes. However, no differences in ligand binding to TP-α_and TP-β or displacement by agonists or antagonists were observed when their respective cDNAs were expressed in COS cells (60,69). In addition, no evidence of alternatively spliced mRNA was found in mouse kidney even though [^{125}I]-BOP bound to two sites on cells obtained from mouse kidney cortex and to only one site on cells obtained from mouse kidney medulla (110). Furthermore, no differences were noted in ligand binding specificities among the four EP3 isoforms (124).

3. SPECULATION REGARDING CAUSE OF TP RECEPTOR VARIABILITY

Some of the tissue variability in TP receptor ligand binding can be attributed to species differences, but other factors are probably responsible. Among factors known to influence ligand binding parameters are G proteins. When G protein availability is limiting, total receptor density measured by agonists may differ from that measured by antagonists (125). This phenomenon is observed in recombinant systems where the natural stoichi-ometry in distorted. Allan, et al (53) demonstrated that binding to transfected TP receptors was influenced by cotransfection of G protein constituents, $G\alpha_q$ and $G\alpha_{13}$. Thus variable cell-specific expression of G proteins that couple to TP receptors could influence their ligand binding characteristics. This is particularly likely when platelets are compared to other tissues, since platelets lack $G\alpha_{11}$, a G protein subunit nearly universally expressed in other cells (126). Another G protein that could influence TP receptor binding in platelets, but not in vascular cells, is $G\alpha_{16}$, since this subunit is expressed only in hematopoietic cells (127). Van der Vuurst, et al (128) have reported the presence of $G\alpha_{16}$ in megakar-yocytes and platelets, but it is unlikely to be present in vascular tissues. Differences in local membrane environments could also influence ligand access to TP receptors that might be responsible for tissue differences in binding parameters. The demonstration that enrichment of platelet phospholipids with eicosapentaenoic acid or docosahexaenoic acid decreased TP receptor affinity for [^{3}H]-U46619 (See V. ALTERED TP RECEPTOR FUNCTION, B. TP RECEPTOR SIGNAL TRANSDUCTION DEFECTS, 3. DRUG EFFECTS, b. OMEGA-3 FATTY ACIDS below) supports this contention. Differences in G proteins or in membrane constituents could also explain why TP receptors transfected into cells that do not naturally express these receptors have routinely yielded Scatchard plots indicative of one site binding, whereas native cells often yielded two affinity state binding profiles (See IV. TP RECEPTOR FUNCTION, A. LIGAND BINDING, 3. EQUILIBRIUM BINDING above).

C. SIGNAL TRANSDUCTION

Agonist binding to TP receptors results in platelet shape change, aggregation and secretion. TXA_2 stimulates fibrinogen binding that is responsible for primary aggregation, and signal transduction via G proteins that results in platelet secretion.

1. SHAPE CHANGE

In human platelets, shape change stimulated by 9,11-azo-PGH_2, STA_2 and U46619 (104,129,130), as well as ADP and collagen, was determined to be a consequence of myosin light chain phosphorylation mediated via activation of myosin light chain kinase (129). Shape change followed specific agonist binding to TP receptors at lower concentrations of agonist than aggregation or secretion. Shape change induced by U46619 had an EC_{50} of 0.035 μM, while the EC_{50} was 0.53 μM for fibrinogen binding, 0.54 μM for 5HT secretion, and 1.3 μM for aggregation (104). The EC_{50} for shape change stimulated by I-BOP was ~0.2 nM, and the relationship of I-BOP occupancy of the high affinity receptor binding site to shape change was linear (105). However, in rabbit platelets shape change induced by U46619 was not inhibited by myosin light chain kinase inhibitors or calmodulin, and it occurred without a rise in cytosolic calcium concentration (131). This suggested that TP receptor-mediated shape change in rabbit platelets occurred by an alternative pathway.

2. AGGREGATION

Fibrinogen binding to integrin $\alpha_{IIb}\beta_3$ occurs following a conformational change in the integrin mediated by inside-out signaling in activated platelets (132). The studies of Morinelli, et al (133) indicated that fibrinogen binding stimulated by U46619 was dependent upon the presence of ADP, but the results of other studies suggested that fibrinogen binding could be directly stimulated by TP receptor agonists as well (134,135). Regardless of whether TXA_2 stimulated fibrinogen binding independent of ADP, TXA_2 analogs and ADP interacted to optimally stimulate fibrinogen binding (133), and it was suggested that their interaction was synergistic (104). Platelet aggregation was stimulated by TXA_2 in the absence of platelet secretion (136,137). A biochemical event that appeared to be required for platelet aggregation induced by minimal concentrations of U46619 was activation of p38 MAPK (138). In contrast to shape change, the relationship of I-BOP occupancy of the low affinity TP receptor binding site was found to be hyperbolic with half-maximal aggregation occurring after 25% receptor occupancy (105).

3. SECRETION

Secreted constituents of platelet storage granules, particularly ADP, amplify primary aggregation stimulated by weak agonists. An important event necessary for TP receptor-mediated platelet secretion is the activation of PLC-β (139,140). As in other cells, activation of PLC-β results in the formation of two important intracellular mediators, inositol 1,4,5 trisphosphate (IP_3) and 1,2 diacylglycerol (DG). IP_3 releases calcium from intracellular stores, resulting in a rise in cytosolic calcium concentration, and DG activates protein kinase C. TP receptors are coupled to PLC-β by guanine nucleotide binding proteins commonly refered to as G proteins.

4. ROLE OF G PROTEINS

G proteins are composed of α subunits (Gα) and βγ subunits (Gβγ) (See 141 for review). Agonist bound receptors promote dissociation of GDP from Gα. GTP binding to Gα results in separation of Gα from Gβγ. Hydrolysis of bound GTP by an intrinsic Gα GTPase activity terminates the Gα-effector interaction, and GDP-bound Gα rebinds to Gβγ. Therefore, ligand binding to receptors results in effector activation, such as stimulation of PLC-β by Gq family G proteins, and increased GTP hydrolysis. The specificity of G proteins for effectors has been defined primarily by Gα, but Gβγ also play important roles in signal transduction. TP receptors interact with members of the G_q family ($G\alpha_q$), the G_{12} family ($G\alpha_{12}$ and $G\alpha_{13}$) and the G_i family ($G\alpha_{i2}$, $G\alpha_{i3}$, $G\alpha_z$) (139,140). G_q family G proteins link receptors to PLC-β, and G_i family G proteins mediate signal transduction from receptors to adenylyl cyclase. $G\alpha_z$ can inhibit cyclic AMP accumulation (142), and $G\alpha_{12}$ and $G\alpha_{13}$ function in Na^+-H^+ exchange (143) and in regulation of the actin cytoskeleton (144) that may play a role in platelet shape change (140), but their precise roles in cellular function have not been defined. $G\alpha_{13}$ was recently shown to activate the phospholipase D1 isozyme by a pathway that required Rho family GTPases (145). Phospholipase D1 was also activated by a constitutively activated mutant $G\alpha_q$. Since $G\alpha_{13}$ and $G\alpha_q$ are abundantly expressed in platelets, activation of phospholipase D via TP receptors may have biological significance but, the implications of this observation for TP receptor function in platelets or other cells is currently unknown.

Platelets have at least two G protein-mediated pathways that link receptors to PLC-β (**Figure 2**). Thrombin and platelet activating factor (PAF) receptors activate high affinity GTPase and PLC-β via pertussis toxin-sensitive and toxin-insensitive G proteins (139, 146). In contrast, activation of PLC-β via TP receptors is pertussis toxin-insensitive (139). The pertussis toxin-sensitive pathway is mediated by G_i. Since $G\alpha_i$ has not been demonstrated to reconstitute PLC-β stimulation in vitro (147), it is likely that activation of PLC-β via G_i is mediated by Gβγ rather than Gα. The toxin-sensitive pathway is mediated via a Gq family G protein now known to be $G\alpha_q$ (140,148,149).

Earlier studies indicated that the G protein responsible for PLC-β activation was G_q, G_{11} or both G_q and G_{11} (150). It was not possible to determine whether one or both of these G proteins was the precise mediator, since nearly all cells contain both $G\alpha_q$ and $G\alpha_{11}$ (126), and because it has been difficult to distinguish $G\alpha_q$ from $G\alpha_{11}$ by antisera, due to their high degree of homology. Subsequent observations of differences in electrophoretic mobility of $G\alpha_q$ and $G\alpha_{11}$ first indicated that human platelets lacked $G\alpha_{11}$ protein (148). Further investigation revealed that human and dog platelets contained $G\alpha_q$ mRNA but they lacked $G\alpha_{11}$ mRNA (149). Offermanns, et al (140) confirmed the deficiency of $G\alpha_{11}$ in platelets obtained from mice. Since platelets lacked $G\alpha_{11}$, it was apparent that signal transduction from TP receptors to PLC-β was mediated via $G\alpha_q$ (149). This observation was recently elegantly confirmed in $G\alpha_q$ knockout mice whose platelets had markedly impaired PLC-β activation, aggregation and secretion (140). Since PLC-β activation was deficient in thrombin as well as U46619-stimulated $G\alpha_q$-deficient platelets, the stimulatory regulation of PLC-β appeared to be mediated primarily through Gq for both receptors (140).

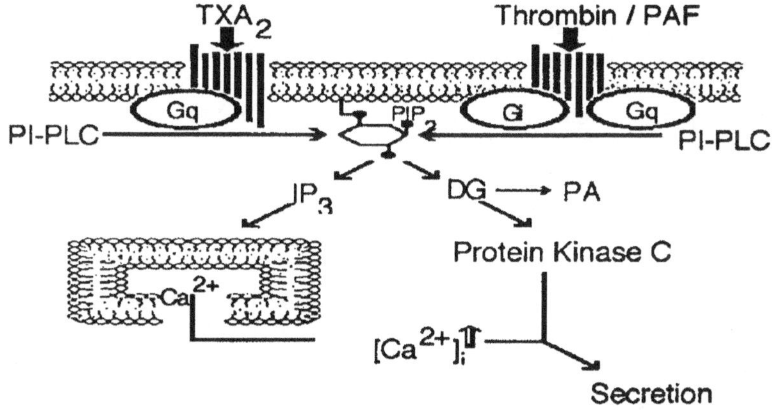

Figure 2.* Signaling Pathways for TXA_2 and Thrombin Receptors

* Adapted from Brass, et al, (139)

A comprehensive model for the role of G proteins in agonist binding to TP receptors, similar to those developed for other GPCR (See 125, 151 for reviews), has not been presented. However, the allosteric ternary complex model (also called extended ternary complex model), formulated on the basis of data obtained from the study of β-adrenergic receptors (152, 153), may be applicable. The allosteric ternary complex model predicts that these receptors exist in active (R*) and inactive (R) equilibrium states governed by an isomerization constant. In this model G proteins bind only to R* resulting in R*G, but agonist binding 1) shifts the R to R* equilibrium by an allosteric constant and 2) affects R*G stability by a second allosteric constant. The allosteric constants affect agonist efficacy and the apparent dissociation constants of any ligand. Therefore binding analysis of receptor function necessitates an evaluation of receptors in terms of their spontaneous isomerization, ligand dependent allosteric constant and ligand G protein allosteric constant. This type of evaluation has not been performed for TP receptors.

D. PLATELET TP RECEPTOR DESENSITIZATION

Desensitization of TP receptor-mediated platelet activation has been studied extensively. Liel, et al (154) and Murray and FitzGerald (155) carried out the first studies of homologous desensitization of normal TP receptors. They observed that exposure of platelets to TP receptor agonists resulted in attenuation of subsequent agonist-induced platelet aggregation and a reduction in receptor sites identified by radioligand binding. Okwu, et al (107) found that shape change, as well as aggregation, induced by I-BOP was inhibited following incubation of human platelets with U46619. Binding of both agonist and antagonist radioligands was decreased following homologous desensitization (107,154,155). Murray and FitzGerald (155) also provided evidence that TP receptor desensitization resulted in receptor-G protein uncoupling. Studies carried out in human astrocytoma cells indicated that homologous desensitization of TP receptors occurred in two phases (156). The early phase of desensitization to STA_2 involved uncoupling of receptors from G proteins, while a reduction of binding sites occurred later. Spurney (157) observed that desensitization of transfected mouse TP receptors was dependent upon the level of receptor expression achieved. Prior exposure (10 min.) to U46619 inhibited subsequent U46619-induced IP_3 formation and calcium flux ~50% in mouse mesangial cells expressing low levels of mouse TP receptors, while these indicators of PLC-β_activation were very little affected in cells with high levels of receptor expresssion.

Dorn and Davis (158) found that PKC activation was necessary for TP receptor desensitization. Subsequent studies of the mechanism of desensitization revealed evidence of TP receptor phosphorylation (See IV. TP RECEPTOR FUNCTION, E. TP RECEPTOR PHOSPHORYLATION below) that was in part mediated by PKC. Studies of CHRF-288 cells indicated that TP receptor internalization also occurred following agonist exposure (159). Desensitization of receptor-effector coupling, that occurred early (T1/2 ~1 hour) following agonist exposure, was separable from receptor internalization, that occurred later. [^{125}I]-BOP binding sites decreased 57%, without a change in binding affinity, after 24 hour incubation with U 46619. Both internalization and degradation were found to be necessary for maximal desenstization of TP receptors (159).

Transfected TP-α and TP-β receptors in CHO cells demonstrated different patterns of homologous desensitization. TP-α receptors, overexpressed in fibroblasts, decreased in number, as did native platelet receptors, following 24 hour exposure to I-BOP, but TP-β receptors increased following similar long-term exposure to I-BOP (160). Activation of PKC uncoupled calcium mobilization from TP-β receptors to a greater extent than from TP-α_receptors (160). These differences cannot be attributed to differences in phosphorylation, since both isoforms transfected into HEK cells were phosphorylated similarly on agonist stimulation (60).

E. TP RECEPTOR PHOSPHORYLATION

Both basal and agonist-induced TP receptor phosphorylation has been observed. A basal level of phosphorylation exists in TP receptors, and phosphorylation increases following agonist binding (60,66,161). Agonist-stimulated phosphorylation appears to contribute significantly to receptor desensitization (161,162), and the level of basal phosphorylation may influence TP receptor activation.

1. ROLE OF PHOSPHORYLATION IN TP RECEPTOR DESENSITIZATION

Sites of agonist-induced TP receptor phosphorylation on serine-threonine residues in the third extracellular loop and the carboxyl-terminal tail of TP receptor fusion proteins have been identified by the use of competitive synthetic peptides, and roles for PKC and protein kinase A (PKA) have been defined (66). Other kinases may contribute to phosphorylation of TP receptors following activation (60). Antagonist ([^{3}H]-SQ29548) binding to TP receptors expressed in transfected HEL cells was inhibited following incubation with PMA, presumably as a result of TP receptor phosphorylation (66). More recent studies of agonist-dependent TP receptor isoform phosphorylation demonstrated a 2-3 fold increase in basal phosphorylation of transfected TP receptors in response to PMA, and a reduction in basal phosphorylation of both TP-α and TP-β receptors in response to PKC inhibition (60). The role of PKC and PKA in agonist-induced TP phosphorylation is uncertain (60,66). Therefore, other kinases, such as GPCR-dependent kinases, may play a role in vivo (60); however, a TP receptor-dependent kinase has not been described.

2. ROLE OF PHOSPHORYLATION IN TP RECEPTOR ACTIVATION

The basal level of TP receptor phosphorylation may influence its activation. Since the previously discussed isomerization states of GPCRs influence ligand binding (See IV. TP RECEPTOR FUNCTION, C. SIGNAL TRANSDUCTION, 4. ROLE OF G PROTEINS above), receptor agonist binding affinity may be influenced by the basal state of phosphorylation. Evidence in favor of this hypothesis comes from the studies of Kinsella, et al and Habib, et al (60,66). Inhibition of PKC by staurosporin resulted in a modest increase in antagonist binding to TP receptors (66), and PKC inhibitors reduced TP receptor basal phosphorylation (60). A PKA inhibitor also increased antagonist binding to TP receptors (66).

PKA and PKC ecto-kinases could contribute to control of TP receptor activation by phosphorylation of the third extracellular loop (66, 86, 163). Evidence in support of these kinases having an effect on TP receptor function includes the total inhibition of I-BOP-induced platelet aggregation and of arachidonic acid, EP171 (a stable endoperoxide analog) and U46619-induced platelet aggregation and secretion by alkaline phosphatase that was reversed by pretreatment with inorganic phosphate or ATP plus creatine phosphate/creatine phosphokinase (86,164). Also, studies of a chimeric β_2 adrenergic receptor indicated that the third extracellular loop of this GPCR modulated receptor-G protein affinity (153).

Other studies suggested a role for platelet ecto-protein kinase C (ecto-PKC) in inhibition of platelet aggregation via phosphorylation of surface membrane proteins (165). Inhibition of ecto-PKC activity by antibody or inhibitory peptides resulted in spontaneous human platelet aggregation. Platelet aggregation induced by these agents could be blocked by phosphatase inhibitors. Therefore, inhibition of phosphorylation combined with continuous dephosphorylation, that resulted in a decrease in the phosphorylation state of the membrane, appeared to be responsible for platelet aggregation. No evidence of effects on receptor phosphorylation was presented, but the anti-ecto-PKC antibody inhibited phosphorylation of an ~50 kDa membrane protein. The third extracellular loop of the TP receptor is in a position where it could be phosphorylated by ecto-PKC. However, evidence against an effect of phosphorylation on TP receptor binding is the observation that neither the affinity nor the number of [^{125}I]-BOP platelet binding sites was altered by pretreatment with alkaline phosphatase (164). Alkaline phosphatase significantly decreased I-BOP-stimulated GTPase activity, and addition of inorganic phosphate to platelets exposed to alkaline phosphatase and I-BOP resulted in aggregation within seconds, plus restoration of increased I-BOP-induced GTPase activity. These studies indicated that the phosphor-ylation state of TP receptors, or closely associated proteins, influenced TP receptor-mediated platelet activation, but the mechanism responsible was not defined.

In addition to the serine-threonine phosphorylation sites modulated by PKC and PKA, a basal state of tyrosine phosphorylation exists in platelet TP receptors, as in bradykinin receptors, and this phosphorylation increases on agonist stimulation (166). The tyrosine kinase responsible for this phosphorylation, as well as that of the parallel phosphorylation of phosphatidylinositol 3-kinase (PI_3-kinase), that occurs on agonist stimulation, is unknown, but $p27^{syk}$ known to be activated by TP receptor agonists (167), is a candidate (166). It is also possible that PI_3 kinase itself might phosphorylate TP receptors. The functional role of tyrosine kinase-mediated phosphorylation of TP receptors is unknown.

3. DEPHOSPHORYLATION

If basal phosphorylation is a mechanism of control of TP receptor function, then phosphatases would be necessary to effect dephosphorylation. Phosphatases that dephosphorylate TP receptors appear to exist, since Kinsella, et al (66) observed that inhibition of PKC or PKA resulted in increased ligand binding to TP receptors. This study did not specifically demonstrate a decrease in TP receptor phosphorylation, but as noted previously (See IV. TP RECEPTOR FUNCTION, E. TP RECEPTOR PHOSPHORYLATION, 2. ROLE OF PHOSPHORYLATION IN TP RECEPTOR ACTIVATION above), stimulation of both PKC and PKA resulted in decreased [^{3}H]-SQ29548 binding that was associated with increased receptor phosphorylation (66). The phosphatases responsible for dephosphorylation of TP receptors are unknown. Several novel serine/threonine phosphatases have recently been identified (168). Protein phosphatase 5 (PP5), a member of the PPP family, has interesting properties that suggest it as a candidate for TP receptor dephosphorylation. PP5 interacts with glucocorticoid receptors and with atrial natriuretic peptide (ANP)/guanylate cyclase receptors that are phosphorylated on many serine and threonine residues in the absence of ligand. Following agonist binding, ANP receptors are desensitized by dephosphorylation that may be mediated by PP5 (168). Another characteristic of PP5 that suggests a possible association with TP receptors is its activation by polyunsaturated fatty acids such as arachidonic acid (169). A candidate for a tyrosine phosphatase that may mediate TP receptor dephosphorylation is SHP-1. SHP-1 has been shown to down-regulate phosphorylation and mitogenesis in hematopoietic cells and to be associated with the somatostatin GPCR (170).

V. ALTERED TP RECEPTOR FUNCTION

A. TP RECEPTOR MUTATION

The one human TP receptor mutation ($R^{60}\rightarrow L$), described by Hirata, et al (77) was manifested by defective signal transduction from receptors to PLC-β and a mild bleeding disorder. The proband's platelets had absent platelet aggregation induced by the TP receptor agonist, STA_2, and impaired secondary aggregation induced by ADP and by epinephrine, but a normal aggregation response to thrombin. A previously reported investigation of this patient revealed normal equilibrium binding of TP agonist and antagonist to his platelets, but defective activation of PLC-β, as manifested by impaired inositol phosphate formation and calcium flux (70). A male sibling and one male and one female offspring had similar platelet aggregation abnormalities, suggesting autosomal dominant inheritance. The proband was found to be homozygous for the (R^{60}→L) mutation, while his brother and two children were heterozygous (77). A second, unrelated subject with the same platelet aggregation abnormality was also found to have the same TP receptor mutation. These data indicated that the (R^{60}→L) mutation was responsible for impaired TP receptor function in these subjects.

B. TP RECEPTOR SIGNAL TRANSDUCTION DEFECTS

1. HUMAN

Other human subjects with defective TP receptor agonist-stimulated platelet aggre-gation have been described (70-73), but a specific signal transduction abnormality has been defined in only one individual (73).

a. $G\alpha_q$ deficiency

Gabbeta, et al (73) described a patient with life-long mucocutaneous bleeding whose platelets demonstrated abnormal aggregation and secretion in response to multiple agonists including TXA_2 analogs. Stimulation of GTPase activity by thrombin, PAF and U46619 was significantly impaired in this patient's platelet membranes, and immuno-blotting with specific antisera revealed a >50% decrease in $G\alpha_q$. TP receptor structure or function was not evaluated, but this patient's impaired platelet response to TXA_2 was likely to be a consequence of a deficiency of $G\alpha_q$ (73).

b. Preterm Infants

Cord blood platelets obtained from premature infants were found to have impaired aggregation responses to several agonists, including U46619, and decreased secretion in response to collagen and U46619 (171). Although these platelets were deficient in TXB_2 production stimulated by collagen, they bound [^{3}H]-U46619 to an equal number of TP receptor binding sites with equal affinity compared to platelets obtained from full term infant and adult controls. Thus, their impaired function was not attributable to defective TP receptor binding, but dysfunction of TP receptor signaling could not be excluded.

c. Deficient Inositol 1,4,5-Triphosphate Formation

Impaired platelet aggregation responses to arachidonic acid and U46619 have been observed in an 11 year old girl with a mild life-long bleeding disorder (72). TP receptor binding of [^{3}H]-U46619 was normal and an appropriate increase in GTPase activity was observed following binding of agonist to TP receptors. This implied that the TP receptor was intact and that it coupled to a G protein, but G protein activation failed to activate

PLC-β. The cause of this defect was not determined, but a mutation in the TP gene was excluded. A defect in PLC-β or in $G\alpha_q$ was postulated (72). Since the patient's platelets also demonstrated impaired aggregation in response to ADP it is possible that she had mild PLC-β deficiency, a defect found to be a cause of impaired platelet aggregation and secretion stimulated by multiple agonists (73). However, this is less likely since a normal response to thrombin was observed.

2. ANIMAL

Compared to human platelets, the functional responses of animal platelets to TP receptor agonists are quite varied. Platelets from some species, such as cats, guinea pigs and rabbits, were reported to aggregate irreversibly when stirred with sodium arachidonate or 9,11-azo-PGH_2, while sheep and dog platelets failed to aggregate under the same conditions (74, 172). Horse platelets aggregated reversibly in response to arachidonate (172). Also, cat and rabbit platelets demonstrated partial agonist responses to the TXA_2 analogs, PTA_2 and CTA_2, that were completely inhibitory to human and guinea pig platelet TP receptors (172). Cow, pig and mink platelets were observed to undergo shape change when stirred with arachidonate, but they did not aggregate (172). [^{125}I]-PTA-OH bound similarly to TP receptors on dog and human platelets, but it did not bind to rabbit platelets (173). More extensive studies have been carried out of the impaired TXA_2 responses observed in dogs, rabbits and rats.

a. Dogs

Most dogs, mongrel or purpose-bred, have platelets that form TXA_2, but fail to aggregate irreversibly or to secrete granular contents in response to the TXA_2 formed (76). These TXA_2 insensitive (TXA_2-) platelets also fail to aggregate irreversibly or to secrete in response to TXA_2 analogs such as U46619 or I-BOP (76). However, some mongrel dogs, and a few purpose-bred strains of dogs, have TXA_2 sensitive (TXA_2+) platelets that aggregate and secrete in response to the TXA_2 formed as do human platelets (75,76, 174,175). Each dog's platelet response to TXA_2 is consistent and genetically defined. The uncommon TXA_2+ phenotype is inherited as an autosomal recessive trait (176). TXA_2- platelets fail to aggregate or secrete when stirred with very low concentrations of thrombin that are dependent upon TXA_2 formation to stimulate secretion; however, they aggregate irreversibly and secrete if exposed to higher concentrations of thrombin or to epinephrine in addition to TXA_2 or TXA_2 analogs (74-76,174,175). Dogs with TXA_2- platelets do not have a bleeding diathesis; however, they have significantly less platelet deposition on prosthetic vascular grafts and significantly higher rates of graft patency than dogs with TXA_2+ platelets (177). Dog TXA_2 - platelets share many biochemical and functional characteristics with TXA_2 insensitive human platelets that have the $R^{60}\rightarrow L$ mutation (77). However, they differ significantly in that dogs with TXA_2- platelets do not have abnormal bleeding, and our laboratory has found that their TP receptor gene does not have the $R^{60}\rightarrow L$ mutation.

The biochemical basis for TXA_2- dog platelets has been extensively studied and partially defined. Both TXA_2- and TXA_2+ dog platelets were found to bind TP receptor agonists and antagonists with equal affinity to a comparable number of high affinity (nM Kd) receptors in equilibrium binding assays; however, agonist binding to TP receptors on TXA_2- platelets resulted in markedly impaired GTPase and PLC-β activation compared to TXA_2+ or human platelets (76). These observations suggested that $G\alpha_q$ was either deficient, as in the patient described by Gabbeta, et al (73), or mutated. However, $G\alpha_q$ was not deficient in TXA_2- platelets, and no differences in the sequence of $G\alpha_q$ were observed between TXA_2- and TXA_2+ platelets (149). Despite these findings, however, additional data suggest that the signal transduction defect of dog TXA_2- platelets exists at the receptor-G protein interface. Impaired activation of PLC-β is corrected by prior exposure of platelets to epinephrine by a mechanism that may involve Gβγ subunits (114). The mechanism responsible for the signaling defect in TXA_2- platelets is currently under active investigation.

Soslau, et al (178) made the observation that dog TXA_2- platelets aggregated in response to U46619 alone after platelet-rich plasma was stored at room temperature for 3.5-5 hours, while human platelets stored under the same conditions failed to aggregate when stirred with U46619 alone. These investigators speculated that dog platelets had two TP receptor populations, one of which was cryptic and that presumably became exposed after several hours at room temperture. However, no direct evidence in support of this hypothesis was presented. This phenomenon has not been observed in our laboratory.

b. Rabbits

Endothelium-dependent pulmonary artery contractions in response to arachidonic acid or methacholine appeared to be mediated by TXA_2 in rabbits (179). However, approximately 25% of normal New Zealand White rabbits failed to demonstrate pulmonary artery contraction in response to arachidonic acid, U46619 or I-BOP. These "unresponsive" rabbits were found to have only 23% of the TP receptors present on pulmonary and aortic crude membranes of "responsive" rabbits (179). Of interest was the fact that platelets from "unresponsive" rabbits aggregated as well or better than platelets from "responsive" animals when stirred with U46619, and no deficiency of TP receptors nor binding affinity was found. These data suggested that TP receptor expression may be tissue specific. This conclusion was supported by the observation that inhibition of testosterone 5α-reductase in rats decreased TP receptor density on aortic membranes but not on platelets (180) (See V. ALTERED TP RECEPTOR FUNCTION, 3. DRUG EFFECTS, a. HORMONES below).

c. Rats

U46619 was observed to induce shape change in rat platelets, but it did not stimulate phosphoinositide hydrolysis, aggregation or secretion (181). As in other species, however, TXA_2 and TXA_2 mimetics were found to synergize with agonists for other receptors to stimulate platelet activation (181).

3. DRUG EFFECTS

a. Hormones

Androgens have been found to influence TP receptor expression . Testosterone and androstenedione increased the in vitro expression of TP receptors on human erythroleukemia cells (182,183), and the administration of testosterone to rats increased TP receptor density of both platelets and aortic tissue (184). Although TP receptor expression in both platelets and vascular tissue was stimulated by androgens, tissue-specific regulation was demonstrated (180). TP receptors were up-regulated in vascular tissues, but not in platelets, by the active metabolite of testosterone, dihydrotestosterone. In contrast, admin-istration of testosterone to healthy men increased platelet TP receptor density greater than two fold without a change in K_d (185).

I-BOP-induced aortic contractions were decreased ~50% in dexamethasone-treated rabbits, and this functional effect of corticosteroid administration was accompanied by an ~25% reduction in aortic cell membrane TP receptor number (186). Although the functional and receptor changes were corrrelated, the magnitude of the reduction in receptor density was small. It is possible that the effects of corticosteroids on aortic contraction were due to an influence on post-receptor events as well as on receptors (186).

b. Omega 3 Fatty Acids

Dietary intake of large quantities of ω-3 fatty acids have been associated with a decreased incidence of cardiovasular disease and with impaired platelet aggregation. The inhibitory effects of these fatty acids on platelet function have been attributed to decreased TXA_2 formation by inhibition of phospholipase A_2 or cyclooxygenase by one or more mechanisms. These include competition with arachidonic acid for cyclooxygenase, the production of thromboxane A_3, a biologically less active prostanoid, and by perturbation of membrane fluidity (187). In addition to possible effects of ω-3 fatty acids on TXA_2 production, the incorporation of eicosapentaenoic acid (20:5n-3) and docosahexaenoic acid (22:6n-3) into platelet phospholipids in vitro (188,189), or in vivo following oral ingestion (190), inhibited aggregation stimulated by U46619. The functional effects of these ω-3 fatty acids were specific for TP receptors, and appeared to be due to inhibition of [^{3}H]-U46619-receptor binding (188). Antagonist binding of [^{3}H]-SQ29548 was also inhibited, but α_2 adrenergic receptor binding was not affected (187,188). The inhibitory effect of 22:6n-3 on platelet ligand binding to TP receptors was found to be dependent upon its esterification in specific phospholipids, but not on alterations in membrane fluidity (191). It has been postulated that 20:5n-3 and 22:6n-3 inhibit TP receptors by interacting with the receptor at its binding site as a function of structural similarity to TXA_2 (187, 191), but direct confirmation is lacking.

c. Dihydropyridines

The dihydropyridine calcium channel antagonists and agonists competitively antagonize ligand binding to TP receptors. The calcium channel agonist, BAY 8644, was found to be a stereoslective competitive antagonist of TP receptor agonist binding to human platelets (106). In this study BAY 8644 inhibited [^{3}H]-U46619 binding with a K_i of 1.47 μM and it blocked U46619-induced platelet aggregation, [^{14}C]-serotonin secretion, calcium flux and pleckstrin phosphorylation. Its inhibitory effect on TP receptors was attributable primarily to the (+)-(4*R*) enantiomer. The calcium channel antagonist, nifedipine, had similar inhibitory effects on platelelts at concentrations of 10-20 μM (106). These findings suggested that dihydropyridines, that were previously shown to inhibit human platelet function when administered in therapeutic doses, may function by receptor inhibition, since platelets appear not to have voltage-dependent calcium channels. Mayeux, et al (192) also observed similar effects of three dihydropyridines on platelet [^{125}I]-BOP and

$[^{125}I]$-PTA-OH binding and calcium flux stimulated by I-BOP. They found stereoselective inhibition of specific photoaffinity labeling of TP receptors.

d. Inhaled Anesthetics

Inhaled anesthetics, particularly halothane, have been shown to inhibit human platelet function in vitro at clinically attainable concentrations (193-196). Secondary aggregation induced by ADP and epinephrine and collagen-induced aggregation ex vivo were found to be inhibited (197) and bleeding times were frequently prolonged (194,198) in patients anesthetized with these agents. The secondary wave of aggregation induced by weak agonists was eliminated or inhibited by halothane and isoflurane, and platelet aggre-gation induced by the TP receptor agonist, STA_2, was significantly suppressed by halo-thane, isoflurane and enflurane (195,196). Hirakata, et al (196) demonstrated that halo-thane (14 mM) inhibited binding of $[^3H]$-S-145 to human platelets. The Kd increased from 0.53 nM to 14.3 nM, but B_{max} did not change. However, neither enflurane (20mM) nor isoflurane (20 mM) inhibited $[^3H]$-S-145 binding. Since these concentrations substantially exceed those achieved in vivo, the significance of these findings in regard to the mechanism by which halothane inhibits platelet aggregation and prolongs the bleeding time is uncertain. Other mechanisms, such as inhibition of TXA_2 formation (195), or a rise in platelet cyclic AMP (193), may also contribute . Halothane and isoflurane may have different mechanisms of action since they were observed to have differential effects on prostenoid-induced vasoconstriction of dog coronary arteries (195).

e. Sulfonylureas

Prostanoid-induced vasoconstriction, including venous and arterial constriction stimulated by U46619, has been shown to be inhibited by sulfonylureas (199-201). McPherson, et al (202) found a significant correlation between sulfonylurea-derivative induced inhibition of vascular relaxation, mediated via ATP-dependent potassium channels, and their ability to inhibit U46619-mediated vascular contraction of pig coronary arteries. Subsequent studies by Stanke, et al (201) found that the second generation sulfonylurea, glibenclamide, also inhibited U46619-induced contractions of human coronary arteries and saphenous veins. These results suggested that sulfonylureas might antagonize TP recep-tors. The results of animal studies indicated that the sulfonylureas were competitive antag-onists of U46619-induced vascular contraction (199,202); however, the data derived from other animal and human studies (201) indicated a noncompetitive interaction at high drug concentrations. Since both glibenclamide and tolbutamide inhibited rat aortic contractions stimulated by G protein activation, it is possible that the sulfonylureas inhibit TP receptor-mediated events by inhibiting signal transduction via G proteins (200). It has been suggested that the effects of these drugs on TXA_2-induced vasoconstriction may have clinical relevance (201), but no evidence of clinical benefit has been presented. The effect of sulfonylureas on platelet TP receptors is unknown.

f. TP Receptor Antagonists

Numerous TP receptor antagonists have been synthesized with the intent that they would be useful agents to block the adverse effects of TXA_2 in human diseases (See 94, 222 for reviews). Some of these agents are specific TP receptor antagonists, while others have combined TP receptor antagonist and thromboxane synthase inhibitory properties.

Extensive experimental studies with these inhibitors have been performed in vitro and in vivo, and a limited number of clinical trials have been carried out. As yet none of these agents have a defined role in clinical medicine. Although they have several theoretical advantages over cyclooxygenase inhibitors (203,204), definite superiority of these agents over aspirin in human therapy remains to be demonstrated (205).

1. Cardiovascular Disease A clinical trial of ridogrel (206), a combined thromboxane synthase/TP receptor inhibitor (207) as an adjunct to thrombolysis in patients with acute myocardial infarction demonstrated no advantage over aspirin in enhancing the fibrinolytic efficacy of streptokinase, but a lower incidence of new ischemic events was observed in patients treated with ridogrel. TP receptor antagonists, GR32191B (208) and sulotroban have shown no benefit in prevention of restenosis following angioplasty of human coronary arteries (209,210). Another specific TP receptor antagonist, BAY U3405 (211), also demonstrated no benefit in patients with peripheral arterial occlusive disease (212), but picotamide, a dual thromboxane synthase/TP receptor antagonist (213) had a beneficial effect on claudication in similar patients (214).

2. Bronchoconstriction In contrast to the minimal beneficial effects seen in cardiovascular disease clinical trials, studies of TP receptor antagonists, BAY U3405 and ICI 192605, have indicated beneficial effects on drug-induced bronchoconstriction in humans (23,24).

Other TP receptor antagonists have been developed (see 215-219 for examples), but their clinical utility remains to be evaluated.

4. DISEASE-RELATED ALTERATIONS

a. Myocardial Ischemia

Studies of platelets obtained from patients with angina or myocardial infarction have revealed increased aggregability to several agents, including TP receptor agonists (5,220, 221). Dorn, et al (221) observed an increase in TP receptors, assessed by analysis of binding of the TP receptor antagonist [^{125}I]-PTA-OH, without a change in affinity, on platelets obtained from patients who had experienced a recent myocardial infaction. These observations were confirmed by Modesti, et al (222). Since the number of TP receptors decreased to normal with time following the infarction (221), the increase appeared to be a consequence of the infarct rather than a contributory cause. The pathogenesis of the increase in receptor number may be linked to thrombin formation, because increased platelet receptor density was correlated with increased plasma concentrations of fibrinopeptide A, and because heparin therapy for 48 hours lowered the elevated fibrinopeptide A levels and decreased the number of [^{125}I]-PTA-OH binding sites (222). In vitro exposure of platelets to thrombin also increased TP receptor density without a change in affinity (222). The results of these studies suggested that thrombin formation in vivo could be responsible for increased TP receptor exposure on platelets that could contribute to increased platelet activation during thrombosis. An alternative interpretation is that the increased TP receptor sites observed could have been a consequence of larger platelets appearing in the circulation following utilization of platelets in thrombus formation (5).

b. Diabetes

Increased platelet TXA_2 synthesis occurs in diabetes mellitus (223,224). Halushka, et al (223) found that platelets from diabetic subjects were significantly less sensitive to the antiaggretory efffects of the TP receptor antagonist, 13-azaprostenoic acid. This suggested the possibility that platelet TP receptors were increased in diabetic subjects. Subsequent studies observed variable numbers of TP receptors to be present on platelets of patients with diabetes, some of whom were found to have increased senstitvity to TP receptor agonists (225). Modesti, et al (226) found platelet TP receptor binding of [^{125}I]-PTA-OH to be decreased ~50%. Jaschonek, et al (227) observed no significant differences in platelet [^{3}H]-U46619 binding between diabetic and control subjects, but they found that diabetic subjects with retinopathy had 35% fewer receptors than control subjects. Similarly, Morinelli, et al (228) found no change in platelet TP receptor number or affinity of rats made diabetic by streptozotocin, although these investigators found insulin therapy to significantly increase aortic membrane TP receptor affinity and to decrease receptor number. The data available do not indicate that TP receptors on platelets of diabetics are uniformly altered; however, it is possible that some patients have platelets that are down-regulated by exposure to the increased TXA_2 produced by diabetic platelets activated in vivo (226,229).

c. Pregnancy-Induced Hypertension

Enhanced platelet activation in vivo, as evidenced by increased plasma beta-thromboglobulin, has been observed in normal pregnancy (230), and increased platelet thromboxane production, measured by malondialdehyde production, and decreased platelet survival were found in both normotensive and hypertensive pregnancies that yielded small-for-gestational age infants (231). Horn, et al (232) observed that platelets from healthy pregnant females displayed increased aggregation in both platelet-rich plasma and whole blood and increased [^{14}C]-5HT secretion in response to U46619, without an increase in TXB_2 production. Liel, et al (233) observed an increased number of TP receptors, assayed with [^{125}I]-PTA-OH, on platelets obtained from patients with pregnancy-induced hyper-tension compared to non-hypertensive pregnant women . They found the greatest increases in subjects with the highest blood pressure. Since preeclampsia, and especially eclampsia, are often preceded by evidence of platelet activation (234), and accompanied by consumptive coagulopathy, it is possible that increased TP receptors in this context, as in myocardial ischemia, were also due to increased platelet size secondary to increased platelet turnover. The study of Liel, et al (233) did not include a non-pregnant hypertensive control group. In both these clinical contexts an increase in TP receptors is paradoxical (5) since a decrease would be expected in situations, like preeclampsia/eclampsia (235-237), or intracoronary thrombosis, where increased platelet exposure to TXA_2 occurs. Therefore, an increase in TP receptors on platelets could account for the increased reactivity to TP receptor agonists observed in hypertensive pregnant subjects, but it does not explain the increased reactivity observed in healthy pregnant women.

In contrast to the results of studies of human platelets, aggregation induced by U46619 was decreased in platelets obtained from pregnant rabbits compared to platelets from non-pregnant female rabbits (238). [^{125}I]-I-BOP binding studies revealed no differences in TP receptor affinity or density between pregnant and non-pregnant rabbits, so the differences in platelet aggregation could not be attributed to TP receptor differences.

d. Androgen Toxicity

The abuse of anabolic steroids has been associated with stroke and myocardial infarction in young male athletes (239). These events may be linked to increased platelet reactivity (240). Since androgens increase TP receptor density on rat platelets and arteries (184), and because androgens increased TP receptor gene expression in HEL cells (182,183), it was postulated that the increased platelet reactivity observed in humans ingesting anabolic steroids was attributable to increased responsiveness to TXA_2. The observations of Ajayi, et al (185) supported this contention. They found that the administration of testosterone to normal human males resulted in a doubling of platelet TP receptor density and increased platelet responses to TP receptor agonists.

e. Nephrotic Syndrome

Children with nephrotic syndrome were found to have elevated excretion of TXB_2, and the TP receptor agonist, STA_2, stimulated a greater calcium flux in their platelets than in platelets from control subjects or from patients with nephrotic syndome who were in remission (241). TP receptor binding was not studied, and the mechanism responsible for the accentuated response to STA_2 was not defined.

Acknowledgment: The author is indebted to his colleagues, Patricia C Dunlop, Ph.D. and Linda A Leis, B.S., for invaluable contributions to the work of his laboratory reported in this chapter, and to the preparation of this manuscript.

References

1. Halushka PV, Allan CJ, Davis-Bruno KL. Thromboxane A_2 receptors. J Lipid Mediat Cell Signalling 1995;12:361-78

2. Coleman RA, Smith WL, Narumiya S. VIII. International union of pharmacology classification of prostanoid receptors: properties, distribution, and structure of the receptors and their subtypes. Pharmacol Rev 1994;46:205-29

3. Ushibuki F, Hirata M, Narumiya S. Molecular biology of prostanoid receptors; an overview. J Lipid Mediat Cell Signalling 1995;12:343-59

4. Armstrong RA. Platelet prostanoid receptors. Pharmacol Ther 1996;72:171-91

5. Halushka PV, Pawate S, Martin ML. "Thromboxane A_2 and Other Eicosanoids" In *Platelets and Their Factors, Handbook of Experimental Pharmacology*, vol 126. F. von Bruchhausen and U. Walter, eds. Berlin, Heidelberg: Springer-Verlag, 1997, pp 459-82

6. Sixma JJ, Wester J. The hemostatic plug. Semin Hematol 1977;14:265-88

7. Schwartz BS, Leis LA, Johnson GJ. In vivo platelet retention in human bleeding-time wounds II. Effect of aspirin ingestion. J Lab Clin Med 1979;94:574-84

8. Thorngren M, Shafi S, Born GVR. Thromboxane A_2 in skin-bleeding-time blood and in clotted venous blood before and after administration of acetylsaliclylic acid. Lancet 1983;1:1075-8

9. Roth GJ, Calverley DC. Aspirin, platelets, and thrombosis: theory and practice. Blood 1994;83:885-98

10. Hamberg M, Svensson J, Samuelsson B. Thromboxanes: a group of biologically active compounds derived from prostaglandin endoperoxides. Proc Nat Acad Sci USA 1975;72:2994-8

11. Bhagwat SS, Hamann PR, Still WC, Bunting S, Fitzpatrick FA. Synthesis and structure of the platelet aggregating factor thromboxane A_2. Nature 1985;315:511-3

12. Houston DS, Shepherd JT, Vanhoutte PM. Aggregrating human platelets cause direct contraction and endothelium-dependent relaxation of isolated canine coronary arteries: role of serotonin, thromboxane A_2, and adenine nucleotides. J Clin Invest 1986;78:539-44

13. Bertolino F, Valentin J-P, Maffre M, Jover B, Bessac A-M, John GW. Prevention of thromboxane A_2 receptor-mediated pulmonary hypertension by a nonpeptide angiotensin II type 1 receptor antagonist. J Pharmacol Exp Ther 1994;268:747-52

14. Fuse S, Kamiya T. Plasma thromboxane B_2 concentration in pulmonary hypertension associated with congenital heart disease. Circulation 1994;90:2952-5

15. Jino H, Kurahashi K, Usui H, Nakata Y, Shimizu Y, Temma S. Pharmacological nature of TP receptor mediated contraction in human intrapulmonary artery. Life Sci 1996;59:2059-65

16. Boussairi EH, Sacquet J, Sassard J, Benzoni D. Thromboxane A_2-prostaglandin H_2 and renovascular hypertension in rats. Am J Physiol 1994;267:R1190-7

17. Gao H, Welch WJ, Dibona GF, Wilcox CS. Sympathetic nervous system and hypertension during prolonged TxA_2/PGH_2 receptor activation in rats. Am J Physiol 1997;273:H734-9

18. Hornych A, Safar M, Bariéty J, Simon A, London G, Levenson J. Thromboxane B_2 in borderline and essential hypertensive patients. Prostaglandins Leukotrienes and Medicine 1983;10:145-55

19. Ritter JM, Barrow SE, Doktor HS, Stratton PD, Edwards JS, Henry JA, Gould S. Thromboxane A_2 receptor antagonism and synthase inhibition in essential hypertension. Hypertension 1993;22:197-203

20. Geoffroy J, Benzoni D, Vincent M, Sassard J. Thromboxane A_2 and development of genetic hypertension in the Lyon rat strain. Hypertension 1990;16:655-61

21. Breyer, Matthew D. and Badr, Kamal F. "Arachidonic Acid Metabolites and the Kidney.", In *The Kidney*, Barry M. Brenner, ed. Philadelphia, PA: W.B. Saunders, 1996, pp 754-88

22. Takahashi N, Takeuchi K, Sugawara A, Taniyama Y, Kato T, Wilcox CS, Abe K, Ito S. Structure and transcriptional function of the 5'-flanking region of rat thromboxane receptor gene. Biochem Biophys Res Commun 1998;244:489-93

23. Al Jarad N, Hui KP, Barnes N. Effects of a thromboxane receptor antagonist on prostaglandin D_2 and histamine induced bronchoconstriction in man. Br J Clin Pharmacol 1994;37:97-100

24. Aizawa H, Shigyo M, Nogami H, Hirose T, Hara N. BAY u3405, a thromboxane A_2 antagonist, reduces bronchial hyperresponsiveness in asthmatics. Chest 1996;109:338-42

25. Devillier P, Bessard G. Thromboxane A_2 and related prostaglandins in airways. Fundam Clin Pharmacol 1997;11:2-18

26. Nagai H, Takeda H, Yamaguchi S, Tanaka H, Matsuo A, Inagaki N. The effect of a thromboxane A_2 receptor antagonist BAY-u-3405 on experimental allergic reactions. Prostaglandins 1995;50:75-87

27. Dorn II GW, Becker MW, Davis MG. Dissociation of the contractile and hypertrophic effects of vasoconstrictor prostanoids in vascular smooth muscle. J Biol Chem 1992;267:24897-905

28. Morinelli TA, Zhang L-M, Newman WH, Meier KE. Thromboxane A_2/prostaglandin H_2-stimulated mitogenesis of coronary artery smooth muscle cells involves activation of mitogen-activated protein kinase and S6 kinase. J Biol Chem 1994;269:5693-8

29. Dorn II GW. Role of thromboxane A_2 in mitogenesis of vascular smooth muscle cells. Agents Actions 1997;48 (Suppl):42-62

30. Sachinidis A, Flesch M, Ko Y, Schrör K, Böhm M, Düsing R, Vetter H. Thromboxane A_2 and vascular smooth muscle cell proliferation. Hypertension 1995;26:771-80

31. Ko FN. Low-affinity thromboxane receptor mediates proliferation in cultured vascular smooth muscle cells of rats. Arterioscler Thromb Vasc Biol 1997;17:1274-82

32. Zhao Y, Foegh ML. Angiopeptin, a somatostatin analogue, inhibits rat coronary artery and aorta smooth muscle cell proliferation induced by the thromboxane A_2 mimetic U46619. Prostaglandins 1997;54:781-93

33. Pakala R, Willerson JT, Benedict CR. Effect of serotonin, thromboxane A_2, and specific receptor antagonists on vascular smooth muscle cell proliferation. Circulation 1997;96:2280-6

34. Hanasaki K, Nakano T, Arita H. Receptor-mediated mitogenic effect of thromboxane A_2 in vascular smooth muscle cells. Biochem Pharmacol 1990;40:2535-42

35. Grosser, T, Zucker T-P, Weber A-A, Schulte K, Sachinidis A, Vetter H, Schrör K. Thromboxane A_2 induces cell signaling but requires platelet-derived growth factor to act as a mitogen. Eur J Pharmacol 1997;319:327-32

36. Zucker T-Ph, Bönisch D, Muck S, Weber A-A, Bretschneider E, Glusa E, Schrör K. Thrombin-induced mitogenesis in coronary artery smooth muscle cells is potentiated by thromboxane A_2 and involves upregulation of thromboxane receptor mRNA. Circulation 1998;97:589-95

37. Gao H, Peng B,Welch WJ, Wilcox CS. Central thromboxane receptors: mRNA expression and mediation of pressor responses. Am J Physiol 1997;272:R1493-500

38. Blackman SC, Dawson G, Antonakis K, Le Breton GC. The identification and characterization of oligodendrocyte thromboxane A_2 receptors. J Biol Chem 1998;273:475-83

39. Sakai K, Nakahata N, Ono H, Yamamoto T, Ohizumi Y. Homologous desensitization of thromboxane A_2 receptor in 1321N1 human astrocytoma cells. J Pharmacol Exp Ther 1996;276:829-36

40. Wilcox CS, Gao H, Verbalis JG, Welch WJ. Role of AVP in pressor responses during activation of central TxA_2/PGH_2 receptors. Am J Physiol 1997;273:H1927-32

41. Namba T, Sugimoto Y, Hirata M, Hayashi Y, Honda A, Watabe A, Negishi M, Ichikawa A, Narumiya S. Mouse thromboxane A_2 receptor: cDNA cloning, expression and nothern blot analysis. Biochem Biophys Res Commun 1992;184:1197-203

42. Ushikubi F, Aiba Y-I, Nakamura K-I, Namba T, Hirata M, Mazda O, Katsura Y, Narumiya S. Thromboxane A_2 receptor is highly expressed in mouse immature thym-ocytes and mediates DNA fragmentation and apoptosis. J Exp Med 1993;178:1825-30

43. Homo-Delarche F, Duval D, Papiernik M. Prostaglandin production by phagocytic cells of the mouse thymic reticulum in culture and its modulation by indomethacin and corticosteroids. J Immunol 1985;135:506-12

44. Ushikubi F, Hirata M, Narumiya S. "Platelet Prostaglandin Receptors." In *Platelets and Their Factors, Handbook of Experimental Pharmacology,* vol 126, F. von Bruchhausen and U. Walter, eds. Berlin, Heidelberg: Springer-Verlag, 1997, pp 135-54

45. Gudermann T, Schöneberg T, Schulz G. Functional and structural complexity of signal transduction via G-protein-coupled receptors. Annu Rev Neurosci 1997;20:399-427

46. Burch RM, Mais DE, Saussy DL Jr, Halushka PV. Solubilization of a thromboxane A_2/prostaglanding H_2 antagonist binding site from human platelets. Proc Natl Acad Sci USA 1985;82:7434-8

47. Ushikubi F, Nakajima M, Hirata M, Okuma M, Fujiwara M, Narumiya S. Purification of the thromboxane A_2/prostaglandin H_2 receptor from human blood platelets. J Biol Chem 1989;264:16496-501

48. Knezevic I, Borg C, LeBreton GC. Identification of G_q as one of the G-proteins which copurify with human platelet thromboxane A_2/prostaglandin H_2 receptors. J Biol Chem 1993; 268:26011-7

49. Borg C, Lim CT, Yeomans DC, Dieter JP, Komiotis D, Anderson EG, LeBreton GC. Purification of rat brain, rabbit aorta, and human platelet thromboxane A_2/ prostaglandin H_2 receptors by immunoaffinity chromatography employing anti-peptide and anti-receptor antibodies. J Biol Chem 1994;269:6109-16

50. Hirata M, Hayashi Y, Ushikubi F, Yokota Y, Kageyama R, Nakanishi S, Narumiya S. Cloning and expression of cDNA for a human thromboxane A_2 receptor. Nature 1991;349:617-20

51. Allan CJ, Halushka PV. Characterization of human peripheral blood monocyte thromboxane A_2 receptors. J Pharmacol Exp Ther 1994;270:446-52

52. D'Angelo DD, Davis MG, Ali S, Dorn II, GW. Cloning and pharmacologic characterization of a thromboxane A_2 receptor from K562 (human chronic myelogenous leukemia) cells. J Pharmacol Exp Ther 1994;271:1034-41

53. Allan CJ, Higashiura K, Martin M, Morinelli TA, Kurtz DT, Geoffroy O, Meier GP, Gettys TW, Halushka PV. Characterization of the cloned HEL cell thromboxane A_2 receptor: evidence that the affinity state can be altered by $G\alpha_{13}$ and $G\alpha_q$. J Pharmacol Exp Ther 1996;277:1132-9

54. Abe T, Takeuchi K, Takahashi N, Tsutsuni E, Taniyama Y, Abe K. Rat kidney thromboxane receptor: molecular cloning, signal transduction, and intrarenal expression localization. J Clin Invest 1995;96:657-64

55. Kitanaka J-i, Hashimoto H, Sugimoto Y, Sawada M, Negishi M, Suzumura A, Marunouchi T, Ichikawa A, Baba A. cDNA cloning of a thromboxane A_2 receptor from rat astrocytes. Biochim Biophys Acta 1995;1265:220-3

56. Raychowdhury MK, Yukawa M, Collins LJ, McGrail SH, Kent KC, Ware JA. Alternative splicing produces a divergent cytoplasmic tail in the human endothelial thromboxane A_2 receptor. J Biol Chem 1994; 269:19256-61

57. Muck S, Weber A-A, Meyer-Kirchrath J, Schrör K. The bovine thromboxane A_2 receptor: molecular cloning, expression, and functional characterization. Arch Pharmacol 1998;357:10-6

58. Mais DE, True TA, Martinelli MJ. Characterization by photoaffinity labelling of the human platelet thromboxane A_2/prostaglandin H_2 receptor: evidence for N-linked glycosylation. Eur J Pharmacol 1992;227:267-74

59. D'Angelo DD, Eubank JJ, Davis MG, Dorn II GW. Mutagenic analysis of platelet thromboxane receptor cysteines. Roles in ligand binding and receptor-effector coupling. J Biol Chem 1996;271:6233-40

60. Habib A, Vezza R, Créminon C, Maclouf J, FitzGerald GA. Rapid, agonist-dependent phosphorylation *in vivo* of human thromboxane receptor isoforms. J Biol Chem 1997;272:7191-200

61. Nüsing RM, Hirata M, Kakizuka A, Eki T, Ozawa K, Narumiya S. Characterization and chromosomal mapping of the human thromboxane A_2 receptor gene. J Biol Chem 1993;268:25253-9

62. Taketo M, Rochelle JM, Sugimoto Y, Namba T, Honda A, Negishi M, Ichikawa A, Narumiya S, Seldin MF. Mapping of the genes encoding mouse thromboxane A_2 receptor and prostaglandin E receptor subtypes EP_2 and EP_3. Genomics 1994;19:585-8

63. Takeuchi K, Nakagawara K, Mori M, Abe T, Takahashi N, Tsutsumi E, Taniyama Y, Kato T, Takeuchi T, Abe K. Assignment of the gene for rat thromboxane receptor (Tbxa2r) to chromosome 7q11 by fluorescence in situ hybridization. Cytogenet Cell Genet 1996;73:79-80

64. D'Angelo DD, Davis MG, Houser WA, Eubank JJ, Ritchie ME, Dorn II GW. Characterization of 5' end of human thromboxane receptor gene. Circ Res 1995;77:466-74

65. Dorn II GW, Davis MG, D'Angelo DD. Gene expression during phorbol ester-induced differentiation of cultured human megakaryoblastic cells. Am J Physiol 1994;266:C1231-9

66. Kinsella BT, O'Mahony DJ, FitzGerald GA. Phosphorylation and regulated expression of the human thromboxane A_2 receptor. J Biol Chem 1994;269:29914-9

67. D'Angelo DD, Oliver BG,Davis MG, McCluskey TS, Dorn II GW. Novel role for Sp1 in phorbol ester enhancement of human platelet receptor gene expression. J Biol Chem 1996;271:19696-704

68. Raychowdhury MK, Yukawa M, Collins LJ, McGrail SH, Kent KC, Ware JA. Alternative splicing produces a divergent cytoplasmic tail in the human endothelial thromboxane A_2 receptor. Correction. J Biol Chem 1995;270:7011

69. Hirata T, Ushikubi F, Kakizuka A, Okuma M, Narumiya S. Two thromboxane A_2 receptor isoforms in human platelets. J Clin Invest 1996;97:949-56

70. Fuse I, Mito M, Hattori A, Higuchi W, Shibata A, Ushikubi F, Okuma M, Yahata K. Defective signal transduction induced by thromboxane A_2 in a patient with mild bleeding disorder: impaired phospholipase C activation despite normal phospholipase A_2 activation. Blood 1993;81:994-1000

71. Fuse I, Hattori A, Mito M, Higuchi W, Yahata K, Shibata A, Aizawa Y. Pathogenetic analysis of five cases with a platelet disorder characterized by the absence of thromboxane A_2 (TXA_2)-induced platelet aggregation in spite of normal TXA_2 binding activity. Thromb Haemost 1996;76:1080-5

72. Mitsui T, Yokoyama S, Shimizu Y, Katzuura M, Akiba K, Hayasaka K. Defective signal transduction through the thromboxane A_2 receptor in a patient with a mild bleeding disorder: deficiency of the inositol 1,4,5-triphosphate formation despite normal G-protein activation. Thromb Haemost 1997;77:991-5

73. Gabbeta J, Yang X, Kowalska MA, Sun L, Dhanasekaran N, Rao AK. Platelet signal transduction defect with Gα_subunit dysfunction and dimished $G\alpha_q$ in a patient with abnormal platelet responses. Proc Natl Acad Sci USA 1997;94:8750-5

74. Burke SE, Lefer AM, Nicolaou KC, Smith GM, Smith JB. Responsiveness of platelets and coronary arteries from different species to synthetic thromboxane and prostaglandin endoperoxide analogues. Br J Pharmacol 1983;78:287-92

75. Clemmons RM, Meyers KM. Acquisition and aggregation of canine blood platelets: basic mechanisms of function and differences because of breed origin. Am J Vet Res 1984;45:137-44

76. Johnson GJ, Leis LA, Dunlop PC. Thromboxane-insensitive dog platelets have impaired activation of phospholipase C due to receptor-linked G protein dysfunction. J Clin Invest 1993;92:2469-79

77. Hirata T, Kakizuka A, Ushikubi F, Fuse I, Okuma M, Narumiya S. Arg^{60} to leu mutation of the human thromboxane A_2 receptor in a dominantly inherited bleeding disorder. J Clin Invest 1994; 94:1662-7

78. True TA, Mais DE. Localization of [^{125}I]SAP-N_3 binding in the human platelet thromboxane A_2/prostaglandin H_2 receptor by proteolytic cleavage analysis. Eur J Pharmacol 1994;266:51-5

79. Komiotis D, Wencel-Drake JD, Dieter JP, Lim CT, LeBreton GC. Labeling of human platelet plasma membrane thromboxane A_2/prostaglandin H_2 receptors using SQB, a novel biotinylated receptor probe. Biochem Pharmacol 1996;52:763-70

80. Yamamoto Y, Kamiya K, Terao S. Modeling of human thromboxane A_2 receptor and analysis of the receptor-ligand interaction. J Med Chem 1993;36:820-5

81. Chiang N, Kan WM, Tai H-H. Site-directed mutagenesis of cysteinyl and serine residues of human thromboxane A_2 receptor in insect cells. Arch Biochem Biophys 1996;334:9-17

82. Funk CD, Furci L, Moran N, Fitzgerald GA. Point mutation in the seventh hydrophobic domain of the human thromboxane A_2 receptor allows discrimination between agonist and antagonist binding sites. Mol Pharmacol 1993;44:934-9

83. Dorn II GW, Davis MG, D'Angelo DD. Structural determinants for agonist binding affinity to thromboxane/prostaglandin endoperoxide (TP) receptors. J Biol Chem 1997;272:12399-405

84. Schrör K, Davis-Bruno K, Halushka PV. Inhibition of ligand binding to thromboxane A_2/prostaglandin H_2 receptors by diethylpyrocarbonate. Protection by receptor ligands and reversal by hydroxylamine. Biochem Pharmacol 1995;49:921-7

85. Mayeux PR, Morinelli TA, Williams TC, Hazard ES, Mais DE, Oatis JE, Baron DA, Halushka PV. Differential effect of pH on thromboxane A_2/prostaglandin H_2 receptor agonist and antagonist binding in human platelets. J Biol Chem 1991;266:13752-8

86. Hatmi M, Haye B, Gavaret JM, Vargaftig BB, Jacquemin C. Alkaline phosphatase prevents platelet stimulation by thromboxane-mimetics. Br J Pharmacol 1991;104:554-8

87. Spurney RF, Coffman TM. The C-terminus of the thromboxane receptor contributes to coupling and desensitization in a mouse mesangial cell line. J Pharmacol Exp Ther 1997;283:207-15

88. Taussig R, Zimmermann G. "Type-specific Regulation of Mammalian Adenylyl Cyclases by G Protein Pathways." In *Advances in Second Messenger and Phosphoprotein Research*, vol 32, Dermot M. F. Cooper ed. Philadelphia, PA, Lippincott-Raven Publishers, 1998, pp 81-98

89. Hellevuo K, Yoshimura M, Kao M, Hoffman PL, Cooper DMF, Tabakoff B. A novel adenylyl cyclase sequence cloned from the human erythroleukemia cell line.Biochem Biophys Res Commun 1993;192:311-8

90. Iyengar R. Molecular and functional diversity of mammalian G_s-stimulated adenylyl cyclases. FASEB J 1993;7:768-75

91. Hellevuo K, Yoshimura M, Mons N, Hoffman PL, Cooper DMF, Tabakoff B. The characterization of a novel human adenylyl cyclase which is present in brain and other tissues. J Biol Chem 1995;270:11581-9

92. Simmoteit R, Schulzki H-D, Palm D, Mollner S, Pfeuffer T. Chemical and functional analysis of components of adenylyl cyclase from human platelets treated with phorbolesters. FEBS Lett 1991;285:99-103

93. Morinelli TA, Oatis Jr. JE, Okwu AK, Mais DE, Mayeux PR, Masuda A, Knapp DR, Halushka PV. Characterization of an ^{125}I-labeled thromboxane A_2/prostaglandin H_2 receptor agonist. J Pharmacol Exp Ther 1989;251:557-62

94. Hall SE. Thromboxane A_2 receptor antagonists. Med Res Rev 1991;11:503-79

95. Mayeux PR, Morton HE, Gillard J, Lord A, Morinelli TA, Boehm A, Mais DE, Halushka PV. The affinities of prostaglandin H_2 and thromboxane A_2 for their receptor are similar in washed human platelets. Biochem Biophys Res Commun 1988;157:733-9

96. Morrow JD, Minton TA, Roberts II LJ. The F_2-isoprostane, 8-epi-prostaglandin $F_{2\alpha}$, a potent agonist of the vascular thromboxane/endoperoxide receptor, is a platelet thromboxane/endoperoxide receptor antagonist. Prostaglandins 1992;44:155-63

97. Praticò D, Smyth EM, Violi F, FitzGerald GA. Local amplification of platelet function by 8-epi prostaglandin $F_{2\alpha}$ is not mediated by thromboxane receptor isoforms. J Biol Chem 1996:271:14916-24

98. Mohler ER, Franklin MT, Adam LP. Intracellular signaling by 8-epi-prostaglandin $F_{2\alpha}$ is mediated by thromboxane A_2/prostaglandin endoperoxide receptors in porcine carotid arteries. Biochem Biophys Res Commun 1996;225:915-23

99. Fukunaga M, Yura T, Grygorczyk R, Badr KF. Evidence for distinct nature of F_2-isoprostane receptors from those of thromboxane A_2. Am J Physiol 1997;272:F477-83

100. Yin K, Halushka PV, Yan Y-T, Wong PY-K. Antiaggregatory activity of 8-epi-prostaglandin $F_{2\alpha}$ and other F-series prostanoids and their binding to thromboxane A_2/ prostaglandin H_2 receptors in human platelets. J Pharmacol Exp Ther 1994;270:1192-6

101. Krauss AH, Woodward DF, Gibson LL, Protzman CE, Williams LS, Burk RM, Gac TS, Roof MB, Abbas F, Marshall K, Senior J. Evidence for human thromboxane receptor heterogeneity using a novel series of 9,11-cyclic carbonate derivatives of prostaglandin F_2 alpha. Br J Pharmacol 1996;117:1171-80

102. Armstrong RA, Jones RL, Wilson NH. Ligand binding to thromboxane receptors on human platelets: correlation with biological activity. Br J Pharmacol 1983;79:953-64

103. Kattelman EJ, Venton DL, Le Breton GC. Characterization of U46619 binding in unactivated, intact human platelets and determination of binding site affinities of four TXA_2/PGH_2 receptor antagonists. Thromb Res 1986; 41:471-81

104. Morinelli TA, Niewiarowski S, Daniel JL, Smith JB. Receptor-mediated effects of a PGH_2 analogue (U46619) on human platelets. Am J Physiol 1987;253:H1035-43

105. Dorn II GW, DeJesus A. Human platelet aggregation and shape change are coupled to separate thromboxane A_2-prostaglandin H_2 receptors. Am J Physiol 1991;260:H327-34

106. Johnson GJ, Dunlop PC, Leis LA, From AHL. Dihydropyridine agonist bay k 8644 inhibits platelet activation by competitive antagonism of thromboxane A_2 -prostaglandin H_2 receptor. Circ Res 1988;62:494-505

107. Okwu AK, Ullian ME, Halushka PV. Homologous desensitization of human thromboxane A_2/prostaglandin H_2 receptors. J Pharmacol Exp Ther 1992;262:238-45

108. Masuda A, Mais DE, Oatis Jr JE, Halushka PV. Platelet and vascular thromboxane A_2/prostaglandin H_2 receptors. Biochem Pharmacol 1991; 42:537-44

109. Takahara K, Murray R, FitzGerald GA, Fitzgerald DJ. The response to thromboxane A_2 analogues in human platelets. Discrimination of two binding sites linked to distinct effector systems. J Biol Chem 1990;265:6836-44

110. Mannon RB, Coffman TM, Mannon PJ. Distribution of binding sites for thromboxane A_2 in the mouse kidney. Am J Physiol 1996;271:F1131-8

111. Morinelli TA, Mais DE, Oatis Jr JE, Crumbley III AJ, Halushka PV. Characterization of thromboxane A_2/Prostaglandin H_2 receptors in human vascular smooth muscle cells. Life Sci 1990;46:1765-72

112. Furci L, Fitzgerald DJ, Fitzgerald GA. Heterogeneity of prostaglandin H_2/ thromboxane A_2 receptors: distinct subtypes mediate vascular smooth muscle contraction and platelet aggregation. J Pharmacol Exp Ther 1991;258:74-81

113. Dunlop PC, Johnson GJ Unpublished observations

114. Johnson GJ, Leis LA, Dunlop PC. Epinephrine potentiation of platelet activation: the role of G protein beta/gamma subunits. Thromb Haemost (suppl) 1997;603

115. Lefer AM, Smith III EF, Araki H, Smith JB, Aharony D, Claremon DA, Magolda RL, Nicolaou KC. Dissociation of vasoconstrictor and platelet aggregatory activities of thromboxane by carbocyclic thromboxane A_2, a stable analog of thromboxane A_2. Proc Natl Acad Sci USA 1980;77:1706-10

116. Mais DE, DeHoll D, Sightler H, Halushka PV. Different pharmacologic activities for 13-azapinane thromboxane A_2 analogs in platelet and blood vessels. Eur J Pharmacol 1988;148:309-15

117. Morinelli TA, Okwu AK, Mais DE, Halushka PV, John V, Chen C-K, Fried J. Difluorothromboxane A_2 and stereoisomers: stable derivatives of TXA_2 with differential effects on platelets and blood vessels. Proc Natl Acad Sci USA 1989;86:5600-4

118. Mais DE, Saussy Jr DL, Chaikhouni A, Kochel PJ, Knapp DR, Hamanaka N, Halushka PV. Pharmacologic characterization of human and canine thromboxane A_2/prostaglandin H_2 receptors in platelets and blood vessels: evidence for different receptors. J Pharmacol Exp Ther 1985;233:418-24

119. Tymkewycz PM, Jones RL, Wilson NH, Marr CG. Heterogeneity of thromboxane A_2 (TP-) receptors: evidence from antagonist but not agonist potency measurements. Br J Pharmacol 1991;102:607-14

120. Armstrong RA, Jones RL, Peesapati V, Will SG, Wilson NH. Competitive antagonism at thromboxane receptors in human platelets. Br J Pharmacol 1985;84:595-607

121. Huzoor-Akbar, Mukhopadhyay A, Anderson KS, Navran SS, Romstedt K, Miller DD, Feller DR. Antagonism of prostaglandin-mediated responses in platelets and vascular smooth muscle by 13-azaprostanoic acid analogs. Evidence for selective blockade of thromboxane A_2 responses. Biochem Pharmacol 1985;34:641-7

122. DiGrande A, Malatino LS, Boura ALA, Read MA, Walters WAW. Modulation by adenosine of thromboxane A_2 receptor-mediated constriction in the human umbilical artery. Int J Clin Pharmacol Ther Toxicol 1994;32:344-6

123. Sugimoto Y, Negishi M, Hayashi Y, Namba T, Honda A, Watabe A, Hirata M, Narumiya S, Ichikawa A. Two isoforms of the EP_3 receptor with different carboxyl-terminal domains. Identical ligand binding properties and different coupling properties with Gi proteins. J Biol Chem 1993;268:2712-8

124. Namba T, Sugimoto Y, Negishi M, Irie A, Ushikubi F, Kakizuka A, Ito S, Ichikawa A, Narumiya S. Alternative splicing of C-terminal tail of prostaglandin E receptor subtype EP_3 determines G-protein specificity. Nature 1993;365:166-70

125. Kenakin T. Differences between natural and recombinant G protein-coupled receptor systems with varying receptor/G protein stoichiometry. Trends Pharmacol Sci 1997;18:456-64

126. Strathmann M, Simon MI. G protein diversity: a distinct class of α_subunits is present in vertebrates and invertebrates. Proc Natl Acad Sci USA 1990;87:9113-7

127. Amatruda III TT, Steele DA, Slepak VZ, Simon MI. $G_{\alpha 16}$, a G protein α_subunit specifically expressed in hematopoietic cells. Proc Natl Acad Sci USA 1991;88:5587-91

128. Van der Vuurst H, van Willigan G, van Spronsen A, Hendriks M, Donath J, Akkerman J-WN. Signal transduction through trimeric G proteins in megakaryoblastic cell lines. Arterioscler Thromb Vasc Biol 1997;17:1830-6

129. Daniel JL, Molish IR, Rigmaiden M, Stewart G. Evidence for a role of myosin phosphorylation in the initiation of the platelet shape change response. J Biol Chem 1984;259:9826-31

130. Kawahara Y, Fukuzaki H, Kaibuchi K, Tsuda T, Hoshijima M, Takai Y. Activation of protein kinase C by the action of 9,11-epithio-11,12-methano-thromboxane A_2 (STA_2), a stable analogue of thromboxane A_2, in human platelets. Thromb Res 1986;41:811-8

131. Ohkubo S, Nakahata N, Ohizumi Y. Thromboxane A_2-mediated shape change: independent of G_q-phospholipase C--Ca^{2+} pathway in rabbit platelets. Br J Pharmacol 1996;117:1095-104

132. Shattil SJ, Kashiwagi H, Pampori N. Integrin signaling: the platelet paradigm. Blood 1998;91:2645-57

133. Morinelli TA, Niewiarowski S, Kornecki E, Figures WR, Wachtfogel Y, Colman RW. Platelet aggregation and exposure of fibrinogen receptors by prostaglandin endoperoxide analogues. Blood 1983;61:41-9

134. Bennett JS, Vilaire G, Burch JW. A role for prostaglandins and thromboxanes in the exposure of platelet fibrinogen receptors. J Clin Invest 1981;68:981-7

135. Di Minno G, Thiagarajan P, Perussia B, Martinez J, Shapiro S, Trinchieri G, Murphy S. Exposure of platelet fibrinogen-binding sites by collagen, arachidonic acid, and ADP: inhibition by a monoclonal antibody to the glycoprotein IIb-IIIa complex. Blood 1983;61:140-8

136. Kinlough-Rathbone RL, Reimers HJ, Mustard JF. Sodium arachidonate can induce platelet shape change and aggregation which are independent of the release reaction. Science 1976;192:1011-2

137. Charo IF, Feinman RD, Detwiler TC. Prostaglandin endoperoxides and thromboxane A_2 can induce platelet aggregation in the absence of secretion. Nature 1977;269:66-9

138. Saklatvala J, Rawlinson L, Waller RJ, Sarsfield S, Lee JC, Morton LF, Barnes MJ, Farndale RW. Role for p38 mitogen-activated protein kinase in platelet aggregation caused by collagen or a thromboxane analogue. J Biol Chem 1996;271:6586-9

139. Brass LF, Manning DR, Cichowski K, Abrams CS. Signaling through G proteins in platelets: to the integrins and beyond. Thromb Haemost 1997;78:581-9

140. Offermanns S, Toombs CF, Hu Y-H, Simon MI. Defective platelet activation in $G\alpha_q$-deficient mice. Nature 1997; 389:183-6

141. Offermanns S, Simon MI. Organization of transmembrane signalling by heterotrimeric G proteins. Cancer Sur 1996;27:177-98

142. WongYH, Conklin BR, Bourne HR. G_z-mediated hormonal inhibition of cyclic AMP accumulation. Science 1992;255:339-42

143. Hooley R, Yu C-Y, Symons M, Barber DL. $G\alpha_{13}$ stimulates Na^+-H^+ exchange through distinct Cdc42-dependent and RhoA-dependent pathways. J Biol Chem 1996;271:6152-8

144. Buhl AM, Johnson NL, Dhanasekaran N, Johnson GL. $G\alpha_{12}$ and $G\alpha_{13}$ stimulate rho-dependent stress fiber formation and focal adhesion assembly. J Biol Chem 1995;270:24631-4

145. Plonk SG, Park S-K, Exton JH. The α-subunit of the heterotrimetric G protein G_{13} activates a phospholipase D isozyme by a pathway requiring rho family GTPases. J Biol Chem 1998;273:4823-6

146. Hung DT, Wong YH, Vu T-KH, Coughlin SR. The cloned platelet thrombin receptor couples to at least two distinct effectors to stimulate phosphoinositide hydrolysis and inhibit adenylyl cyclase. J Biol Chem 1992;267:20831-4

147. Smrcka AV, Sternweis PC. Regulation of purified subtypes of phosphatidylinositol-specific phospholipase Cβ by G protein α and βγ subunits. J Biol Chem 1993;268:9667-74

148. Milligan G, Mullaney I, McCallum JF. Distribution and relative levels of expression of the phosphoinositidase-C-linked G-proteins $G_q\alpha$_and G_{11} α: absence of G_{11} α in human platelets amd haemopoietically derived cell lines. Biochim Biophys Acta 1993;1179:208-12

149. Johnson GJ, Leis LA, Dunlop PC. Specificity of $G\alpha_q$ and $G\alpha_{11}$ gene expression in platelets and erythrocytes. Biochem J 1996;318:1023-31

150. Shenker A, Goldsmith P, Unson CA, Spiegel AM. The G protein coupled to the thromboxane A_2 receptor in human platelets is a member of the novel G_q family. J Biol Chem 1991; 266:9309-13

151. Leff, P. The two-state model of receptor activation. Trends Pharmacol Sci 1995;16:89-97

152. Samama P, Cotecchia S, Costa T, Lefkowitz RJ. A mutation-induced activated state of the β_2-adrenergic receptor. Extending the ternary complex model. J Biol Chem 1993;268:4625-36

153. Zhao M-M, Gaivin RJ, Perez DM. The third extracellular loop of the_β_2-adrenergic receptor can modulate receptor/G protein affinity. Mol Pharmacol 1998;53:524-9

154. Liel N, Mais DE, Halushka PV. Desensitization of platelet thromboxane A_2/prostaglandin H_2 receptors by the mimetic U46619. J Pharmacol Exp Ther 1988;247:1133-8

155. Murray R, FitzGerald GA. Regulation of thromboxane receptor activation in human platelets. Proc Natl Acad Sci USA 1989;86:124-8

156. Sakai K, Nakahata N, Ono H, Yamamoto T, Ohizumi Y. Homologous desensitization of thromboxane A_2 receptor in 1321N1 human astrocytoma cells. J Pharmacol Exp Ther 1996;276:829-36

157. Spurney RF. Effect of receptor number on desensitization of the mouse thromboxane receptor. Biochem Pharmacol 1998;55:1271-81

158. Dorn II GW, Davis MG. Differential megakaryocytic desensitization to platelet agonists. Am J Physiol 1992;263:C864-72

159. Dorn II GW. Regulation of response to thromboxane A_2 in CHRF-288 megakaryocytic cells. Am J Physiol 1992;262:C991-9

160. Yukawa M, Yokota R, Eberhardt RT, von Andrian L, Ware JA. Differential desensitization of thromboxane A_2 receptor subtypes. Circ Res 1997;80:551-6

161 Okwu AK, Mais DE, Halushka PV. Agonist-induced phosphorylation of human platelet TXA_2/PGH_2 receptors. Biochim Biophys Acta 1994;1221:83-8

162. Houser WA, D'Angelo DD, Eubank JJ, Dorn II GW. Distinct phosphorylation sites in carboxyl terminus of thromboxane receptors mediate homologous and heterologous desensitization. Circulation (suppl I) 1995;92:I803-4

163. Hatmi M, Gavaret J-M, Elalamy I, Vargaftig BB, Jacquemin C. Evidence for cAMP-dependent platelet ectoprotein kinase activity that phosphorylates platelet glycoprotein IV (CD36). J Biol Chem 1996;271:24776-80
164. Margolin N, True TA, Saussy Jr DL, Mais DE. Effect of alkaline phosphatase on thromboxane mimetic induced platelet activation. Prostaglandins 1994; 48:235-46

165. Babinska A, Ehrlich YH, Kornecki E. Activation of human platelets by protein kinase C antibody: role for surface phosphorylation in homeostasis. Am J Physiol 1996;271:H2134-44

166. Morinelli TA, Finley EL, Jaffa AA, Kurtz DT, Ullian ME. Tyrosine phosphorylation of phosphatidylinositiol 3-kinase and of the thromboxane A_2 (TXA_2) receptor by TXA_2 mimetic I-BOP in A7r5 cells. Biochem Pharmacol 1997;53:1823-32.

167. Maeda H, Taniguchi T, Inazu T, Yang C, Nakagawara G, Yamamura H. Protein-tyrosine kinase p72^{syk} is activated by thromboxane A_2 mimetic U44069 in platelets. Biochem Biophys Res Commun 1993;197:62-7

168. Cohen PTW. Novel protein serine/threonine phosphatases: variety is the spice of life. Trends Biochem Sci 1997;22:245-51

169. Skinner J, Sinclair C, Romeo C, Armstrong D, Charbonneau H, Rossie S. Purification of a fatty acid-stimulated protein-serine/threonine phosphatase from bovine brain and its identification as a homolog of protein phosphatase 5. J Biol Chem 1997;272:22464-71

170. Lopez F, Esteve J-P, Buscail L, Delesque N, Saint-Laurent N, Théveniau M, Nahmias C, Vaysse N, Susini C. The tyrosine phosphatase SHP-1 associates with the sst2 somatostatin receptor and is an essential component of sst2-mediated inhibitory growth signaling. J Biol Chem 1997; 272:24448-54

171. Israels SJ, Odaibo FS, Robertson C, McMillan EM, McNicol A. Deficient thromboxane synthesis and response in platelets from premature infants. Pediatr Res 1997;41:218-23

172. Meyers KM, Katz JB, Clemmons RM, Smith JB, Holmsen H. An evaluation of the arachidonate pathway of platelets from companion and food-producing animals, mink and man. Thromb Res 1980;20:13-24

173. Narumiya S, Okuma M, Ushikubi F. Binding of a radioiodinated 13-azapinane thromboxane antagonist to platelets: correlation with antiaggregatory activity in different species. Br J Pharmacol 1986;88:323-31

174. Johnson GJ, Leis LA, Rao GHR, White JG. Arachidonate-induced platelet aggregation in the dog. Thromb Res 1979;14:147-54

175. Johnson, GJ, Rao GHR, Leis LA, White JG. Effect of agents that alter cyclic AMP on arachidonate-induced platelet aggregation in the dog. Blood 1980;55:722-9

176. Johnson GJ, Leis LA, King RA. Thromboxane responsiveness of dog platelets is inherited as an autosomal recessive trait. Thromb Haemost 1991;65:578-80

177. Mc Goff MA, Allen BT, Der T, Sicard GA, Santoro SA. Mechanisms of vascular graft thrombosis: role of altered canine platelet sensitivity to thromboxane. Thromb Res 1989; 55:695-707

178. Soslau G, Arabe L, Parker J, Pelleg A. Aggregation of human and canine platelets: modulation by purine nucleotides. Thromb Res 1993;72:127-37

179. Buzzard CJ, Pfister SL, Halushka PV, Campbell WB. Decrease in vascular TxA_2 receptors in a subgroup of rabbits unresponsive to a TxA_2 mimetic. Am J Physiol 1994;266:H2320-6

180. Higashiura K, Blaney B, Morgan E, Mathur RS, Halushka PV. Inhibition of testosterone 5α-reductase: evidence for tissue-specific regulation of thromboxane A_2 receptors. J Pharmacol Exp Ther 1996;279:1386-91

181. Nakano T, Terawaki A, Arita H. Signal transduction in collagen-stimulated rat platelets is composed of three stages. J Biochem 1987;101:1169-80

182. Matsuda K, Mathur RS, Duzic E, Halushka PV. Androgen regulation of thromboxane A_2/prostagladin H_2 receptor expression in human erythroleukemia cells. Am J Physiol 1993;265:E928-34

183. Zucker T-P, Higashiura K, Mathur RS, Halushka PV. Androstenedione increases thromboxane A_2 receptors in human erythroleukemia cells. Life Sci 1996;58:683-90

184. Matsuda K, Ruff A, Morinelli TA, Mathur RS, Halushka PV. Testosterone increases thromboxane A_2 receptor density and responsiveness in rat aortas and platelets. Am J Physiol 1994;267:H887-93

185. Ajayi AAL, Mathur R, Halushka PV. Testosterone increases human platelet thromboxane A_2 receptor density and aggregation responses. Circulation 1995;91:2742-7

186. Sessa WC, Halushka PV, Okwu A, Nasjletti A. Characterization of the vascular thromboxane A_2/prostaglandin endoperoxide receptor in rabbit aorta. Circ Res 1990;67:1562-9

187. Parent CA, Lagarde M, Venton DL, Le Breton GC. Selective modulation of the human platelet thromboxane A_2/prostaglandin H_2 receptor by eicosapentaenoic and docosahexaenoic acids in intact platelets and solubilized platelet membranes. J Biol Chem 1992;267:6541-7

188. Swann PG, Parent CA, Croset M, Fonlupt P, Lagarde M, Venton DL, Le Breton GC. Enrichment of platelet phospholipids with eicosapentaenoic acid and docosahexaenoic acid inhibits thromboxane A_2/prostaglandin H_2 receptor binding and function. J Biol Chem 1990;265:21692-7

189. Kramer HJ, Stevens J, Grimminger F, Seeger W. Fish oil fatty acids and human platelets: dose-dependent decrease in dienoic and increase in trienoic thromboxane generation. Biochem Pharmacol 1996;52:1211-7

190. Scheurlen M, Kirchner M, Clemens MR, Jaschonek K. Fish oil preparations rich in docosahexaenoic acid modify platelet responsiveness to prostaglandin-endoperoxide/ thromboxane A_2 receptor agonists. Biochem Pharmacol 1993;46:245-9

191. Bayon Y, Croset M, Daveloose D, Guerbette F, Chirouze V, Viret J, Kader J-C, Lagarde M. Effect of specific phospholipid molecular species incorporated in human platelet membranes on thromboxane A_2/prostaglandin H_2 receptors. J Lipid Res 1995;36:47-56

192. Mayeux PR, Mais DE, Halushka PV. Interactions of dihydropyridine Ca^{2+} channel agonists with the human platelet thromboxane A_2/prostaglandin H_2 receptor. Eur J Pharmacol 1991;206:15-21

193. Walter F, Vulliemoz Y, Verosky M, Triner L. Effects of halothane on the cyclic 3', 5'-adenosine monophosphate enzyme system in human platelets. Anesth Analg 1980;59:856-61

194. Dalsgaard-Nielsen J, Risbo A, Simmelkjaer P, Gormsen J. Impaired platelet aggregation and increased bleeding time during general anaethesia with halothane. Br J Anaesth 1981;53:1039-42

195. Blaise GA, Parent M, Laurin S, Omri A, Reader TA, Moutquin JM. Platelet-induced vasomotion of isolated canine coronary artery in the presence of halothane or isoflurane. J Cardiothorac Vasc Anesth 1994;8:175-81

196. Hirakata H, Ushikubi F, Narumiya S, Hatano Y, Nakamura K, Mori K. The effect of inhaled anesthetics on the platelet aggregation and the ligand-binding affinity of the platelet thromboxane A_2 receptor. Anesth Analg 1995;81:114-8

197. Sweeney D, Williams V. The effect of halothane general anaesthesia on platelet function. Anaesth Intensive Care 1987;15:278-81

198. Fyman PN, Triner L, Schranz H, Hartung J, Casthely PA, Abrams LM, Keaney AE, Cottrell JE. Effect of volatile anaesthetics and nitrous oxide-fentanyl anaesthesia on bleeding time. Br J Anaesth 1984;56:1197-200

199. Cocks TM, King SJ, Angus JA. Glibenclamide is a competitive antagonist of the thromboxane A_2 receptor in dog coronary artery *in vitro*. Br J Pharmacol 1990;100:375-8.

200. Delaey C, Van de Voorde J. Prostanoid-induced contractions are blocked by sulfonylureas. Eur J Pharmacol 1995;280:179-84

201. Stanke F, Cracowski J-L, Chavanon O, Magne J-L, Blin D, Bessard G, Devillier P. Glibenclamide inhibits thromboxane A_2-induced contraction in human internal mammary artery and saphenous vein. Eur J Pharmacol 1998;341:65-71

202. McPherson GA, Choi RT, Kong DCM, Iskander MN. The thromboxane A_2 and K_{ATP} channel antagonist actions of a series of sulphonylurea derivatives in the pig coronery artery. Eur J Pharmacol 1997;324:193-200

203. Golino P, Ambrosio G, Gresele P, Pascucci I, Ragni M, Russolillo E, Leproux GB, Chiariello M. The *in vivo* antiplatelet effects of thromboxane A_2 synthase inhibitors are potentiated by simultaneous thromboxane A_2/prostaglandin H_2 receptor blockade. J Pharmacol Exp Ther 1993;266:511-7

204. Salvati P, Dho L, Ukmar G, Vaga, L, Rimoldi O, Patrono C. A comparative evaluation of thromboxane receptor blockade, thromboxane synthase inhibition and both in animal models of arterial thrombosis. J Pharmacol Exp Ther 1994;269:238-45.

205. Schafer AI. Antiplatelet therapy. Am J Med 1996;101:199-209

206. The RAPT Investigators. Randomized trial of ridogrel, a combined thromboxane A_2 synthase inhibitor and thromboxane A_2/prostaglandin endoperoxide receptor antagonist, versus aspirin as adjunct to thrombolysis in patients with acute myocardial infarction. The ridogrel versus aspirin patency trial (RAPT). Circulation 1994;89:588-95

207. Hoet B, Arnout J, Van Geet C, Deckmyn H, Verhaeghe R, Vermylen J. Ridogrel, a combined thromboxane synthase inhibitor and receptor blocker, decreases elevated plasma β-thromboglobulin levels in patients with documented peripheral arterial disease. Thromb Haemost 1990;64:87-90

208. Hornby EJ, Foster MR, McCabe PJ, Stratton LE. The inhibitory effect of GR32191, a thromboxane receptor blocking drug, on human platelet aggregation, adhesion and secretion. Thromb Haemost 1989;61:429-36

209. Serruys PW, Rutsch W, Heyndrickx GR, Danchin N, Mast EG, Wijns W, Rensing BJ, Vos J, Stibbe J. Prevention of restenosis after percutaneous transluminal coronary angioplasty with thromboxane A_2-receptor blockade. A randomized, double-blind, placebo-controlled trial. Circulation 1991;84:1568-80

210. Savage MP, Goldberg S, Bove AA, Deutsch E, Vetrovec G, Macdonald RG, Bass T, Margolis JR, Whitworth HB, Taussig A, Hirshfeld JW, Cowley M, Hill JA, Marks RG, Fischman DL, Handberg E, Herrmann H, Pepine CJ; for the M-HEART study group. Effect of thromboxane A_2 blockade on clinical outcome and restenosis after successful coronary angioplasty. Multi-hospital eastern atlantic restenosis trial (m-heart II). Circulation 1995;92:3194-200

211. Theis JGW, Dellweg H, Perzborn E, Gross R. Binding characteristics of the new thromboxane A_2/prostaglandin H_2 receptor antagonist [^{3}H]BAY U 3405 to washed human platelets and platelet membranes. Biochem Pharmacol 1992;44:495-503

212. Bellucci S, Kedra W, Groussin H, Jaillet N, Molho-Sabatier P, Andreassian B, Tobelem G. Ex vivo inhibition of platelet aggregation in patients with severe limb arteriopathy treated with BAY U3405, a specific TXA_2 receptor antagonist. Thromb Haemost 1994;72:659-62

213. Gresele P, Deckmyn H, Arnout J, Nenci GG, Vermylen J. Characterization of N,N'-bis(3-picolyl)-4-methoxy-isophthalamide (picotamide) as a dual thromboxane synthase inhibitor/thromboxane A_2 receptor antagonist in human platelets. Thromb Haemost 1989;61:479-84

214. Coto V, Cocozza M, Oliviero U, Lucariello A, Picano T, Coto F, Cacciatore L. Clinical efficacy of picotamide in long-term treatment of intermittent claudication. Angiology 1989;40:880-5

215. Chang T-S, Kim H-M, Lee K-S, Khil L-Y, Mar W-C, Ryu C-K, Moon C-K. Thromboxane A_2 synthase inhibition and thromboxane A_2 receptor blockade by 2-[(4-cyanophenyl)amino]-3-chloro-1,4-naphthalenedione (NQ-Y15) in rat platelets. Biochem Pharmacol 1997;54:259-68

216. Bertolino F, Valentin J-P, Maffre M, Grelac F, Bessac A-M, Maclouf J, Delhon A, Levy-Toledano S, Patoiseau J-F, Colpaert FC, John GW. Intrinsic activity of the non-prostanoid thromboxane A_2 receptor antagonist, daltroban (BM 13,505), in human platelets *in vitro* and in the rat vasculature *in vivo*. Br J Pharmacol 1995;115:210-6

217. Romstedt KJ, Shin Y, Shams G, Doyle K, Tantishaiyakul V, Clark MT, Adejare A, Hamada A, Miller DD, Feller DR. Halogen-substituted trimetoquinol analogs as thromboxane A_2 receptor antagonist in platelets and aorta. Biochem Pharmacol 1993;46:2051-9

218. Roald HE, Barstad RM, Engen A, Kierulf P, Skjorten F, Sakariassen KS. HN-11500-a novel thromboxane A_2 receptor antagonist with antithrombotic activity in humans at arterial blood flow conditions. Thromb Haemost 1994;71:103-9

219. Ogletree ML, Harris DN, Schumacher WA, Webb ML, Misra RN. Pharmocological profile of BMS 180,291: a potent, long-acting, orally active thromboxane A_2/prostaglandin endoperoxide receptor antagonist. J Pharmacol Exp Ther 1993;264:570-8

220. Mehta J, Mehta P, Conti CR. Platelet function studies in coronary heart disease. IX increased platelet prostaglandin generation and abnormal platelet sensitivity to prostacyclin and endoperoxide analog in angina pectoris. Am J Cardiol 1980;46:943-7

221. Dorn II GW, Liel N, Trask JL, Mais DE, Assey ME, Halushka PV. Increased platelet thromboxane A_2/prostaglandin H_2 receptors in patients with acute myocardial infarction. Circulation 1990;81:212-8

222. Modesti PA, Colella A, Cecioni I, Costoli A, Biagini D, Migliorini A, Serneri GGN. Increased number of thromboxane A_2-prostaglandin H_2 platelet receptors in active unstable angina and causative role of enhanced thrombin formation. Am Heart J 1995;129:873-9

223. Halushka PV, Rogers RC, Loadholt CB, Colwell JA Increased platelet thromboxane synthesis in diabetes mellitus. J Lab Clin Med 1981;97:87-96

224. Winocour PD. Platelets, vascular disease, and diabetes mellitus. Can J Physiol Pharmacol 1994;72:295-303

225. Collier A, Tymkewycz P, Armstrong R, Young RJ, Jones RL, Clarke BF. Increased platelet thromboxane receptor sensitivity in diabetic patients with proliferative retinopathy. Diabetologia 1986;29:471-4

226. Modesti PA, Abbate R, Gensini GF, Colella A, Serneri GGN. Platelet thromboxane A_2 receptors in type I diabetes. Clin Sci 1991;80:101-5

227. Jaschonek K, Faul C, Weisenberger H, Krönert K, Schröder H, Renn W. Platelet thromboxane A_2/endoperoxide (TXA_2/PGH_2) receptors in type I diabetes mellitus. Thromb Haemost 1989;61:535-6

228. Morinelli TA, Tempel GE, Jaffa AA, Silva RH, Naka M, Folger W, Halushka PV. Thromboxane A_2/prostaglandin H_2 receptors in streptozotocin-induced diabetes: effects of insulin therapy in the rat. Prostaglandins 1993;45:427-38

229. Davi G, Gresele P, Violi F, Basili S, Catalano M, Giammarresi C, Volpato R, Nenci GG, Ciabattoni G, Patrono C. Diabetes mellitus, hypercholesterolemia, and hypertension but not vascular disease per se are associated with persistent platelet activation *in vivo*. Evidence derived from the study of peripheral arterial disease. Circulation 1997;96:69-75

230. Pekonen F, Rasi V, Ammälä M, Viinikka L, Ylikorkala O. Platelet function and coagulation in normal and preeclamptic pregnancy. Thromb Res 1986;43:553-60

231. Wallenburg HCS, Rotmans N. Enhanced reactivity of the platelet thromboxane pathway in normotensive and hypertensive pregnancies with insufficient fetal growth. Am J Obstet Gynecol 1982;144:523-8

232. Horn EH, Hardy E, Cooper J, Heptinstall S, Rubin PC. Platelet reactivity in vitro in relation to thromboxane in healthy pregnancy. Thromb Haemost 1996;75:346-51

233. Liel N, Nathan I, Yermiyahu T, Zolotov Z, Lieberman JR, Dvilansky A, Halushka PV. Increased platelet thromboxane A_2/prostaglandin H_2 receptors in patients with pregnancy induced hypertension. Thromb Res 1993;70:205-10

234. Janes SL, Kyle PM, Redman C, Goodall AH. Flow cytometric detection of activated platelets in pregnant women prior to the development of pre-eclampsia. Thromb Haemost 1995;74:1059-63

235. Fitzgerald DJ, Mayo G, Catella F, Entman SS, FitzGerald GA. Increased thromboxane biosynthesis in normal pregnancy is mainly derived from platelets. Am J Obstet Gynecol 1987;157:325-30

236. Fitzgerald DJ, Rocki W, Murray R, Mayo G, FitzGerald GA. Thromboxane A_2 synthesis in pregnancy-induced hypertension. Lancet 1990;335:751-4

237. Walsh SW. Preeclampsia: an imbalance in placental prostacyclin and thromboxane production. Am J Obstet Gynecol 1985;152:335-40

238. Losonczy G, Mucha I, DiPirro J, Sweeney J, Brown G, Brentjens J, Venuto R. The effect of pregnancy on the response to the TxA_2/PGH_2 analogue U-46619 in rabbits. Am J Physiol 1993;265:R772-80

239. Rockhold RW. Cardiovascular toxicity of anabolic steroids. Annu Rev Pharmacol Toxicol 1993;33:497-520

240. Ferenchick G, Schwartz D, Ball M, Schwartz K. Androgenic-anabolic steroid abuse and platelet aggregation: a pilot study in weight lifters. Am J Med Sci 1992;303:78-82

241. Kobayashi T, Suzuki J, Watanabe M, Suzuki S, Yoshida K, Kume K, Suzuki H. Changes in platelet calcium concentration by thromboxane A_2 stimulation in patients with nephrotic syndrome of childhood. Nephron 1997;77:309-14

242. Maxey KM. Eicosanoid receptors. The classical receptors. Cayman Currents 1997;5:1-10

4

COLLAGEN RECEPTORS: BIOLOGY AND FUNCTIONS

Beate Kehrel, Ph.D.
Experimental and Clinical Haemostaseology
Department of Anaesthesiology
and Intensive Care Medicine
University of Muenster
D-48149 Muenster, Germany

4.1 INTRODUCTION

When a blood vessel wall is damaged, platelets are exposed to a variety of activating agents including collagen. The adhesion of platelets to vascular subendothelium is a critical initial step in haemostasis and thrombosis. Collagens of the subendothelium are major determinants of the thrombogenicity of the blood vessel wall. The interaction of platelets with collagens is a complex process since collagen is not only a potent platelet agonist but also an adhesive protein[1]. When platelets encounter fibrillar collagen they do not just adhere but, following adhesion, undergo a complex series of intracellular reactions which result in platelet activation, secretion of the contents of the granules and the induction of surface receptor sites

Beate Kehrel studied biology and chemistry at the University of Muenster. After her Ph. D., she studied medicine and started her scientific career at the Institute for Arteriosclerosis Research in Muenster., where she came in-now 15 years ago-into initial contact with her main field of research., the interaction of platelets and extracellular matrix, especially collagen. After working for 12 years on thrombocytopathias in the Department of Internal Medicine at the University of Muenster, she is now Head of the Laboratoriesfor Experimental and Clinical Haemostaseology in the Department of Anaesthesiology.

for adhesive proteins, which reinforce the initial mechanisms of adhesion and mediate platelet aggregation[2].

Platelet/collagen adhesion can be classified into two categories: primary adhesion and secondary reinforcing adhesion[2]. Primary adhesion can be categorised into divalent cation-independent and Mg^{2+}-dependent adhesion.

Monomeric and fibrillar collagens effectively support platelet adhesion, whereas the native, triple-helical structure of collagen and the polymerisation of the monomeric collagen are required for collagen-induced platelet aggregation and secretion[3-5].

Various mechanisms and receptor populations may be involved in both these processes[6,7](Figure 1). The relationship between the various receptors and the mechanisms is still not fully understood. The initial interaction between the subendothelium and the quiescent platelets in flowing blood is thought to occur via von-Willebrand-factor (vWf) bound to collagen on the subendothelium and GPIb on the platelet (for detailed information see chapter 7).

Over the years a large number of molecules have been proposed as direct platelet/collagen receptors but recently these have been narrowed down to a few:

$_2\beta_1$ integrin[8,9]
c.CD36[10]
c.GPVI[11]
c.P65[12]

and, indeed, may involve a complex containing these components.

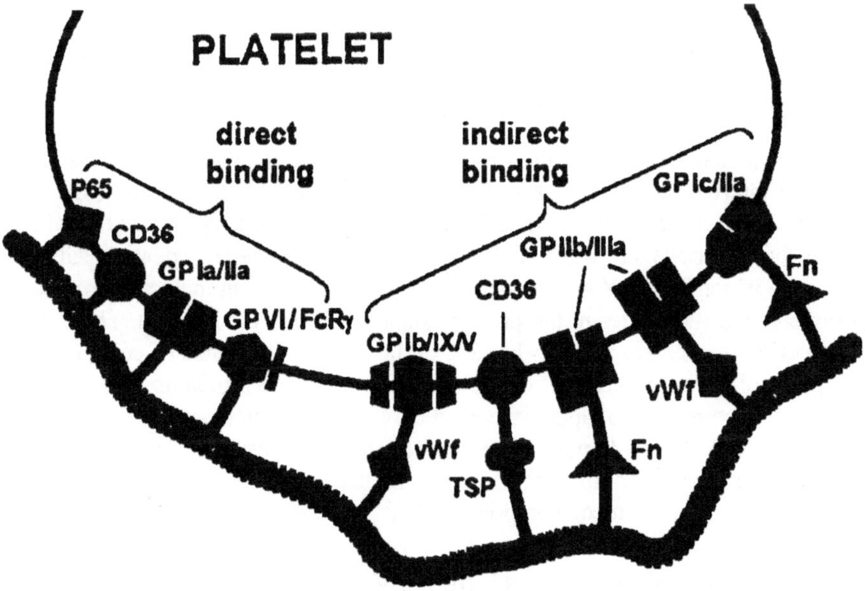

Figure 1: Model for collagen binding to platelets and the various binding proteins involved.

4.2. Collagens

Collagens, the most abundant components of the extracellular matrix, are a family of characteristic fibrous proteins found in all multicellular animals[13].

Thus far more than 20 collagens have been characterised at the protein and/or nucleic acid levels.

Based on their supramolecular structures the collagens are categorised as shown in Table 1.

Table 1: Collagens

	Fibril-forming[14]		Non fibril-forming[14]
Collagen types	Fibril-associated collagens with interrupted triple helices (FACIT)[15]	Sheet-forming[16]	Others
I C.II C.III C.V C.XI	C.IX C.XII C.XIV C.XVI	C.IV C.VIII C.X	VI: micro-fibrils [17,18] VII: anchoring fibres for keratinocytes [19,20]

All collagen molecules have parts that are built up of three chains in a triple-helical conformation. The individual polypeptide chains are called α-chains and are numbered in arabic numerals for a given collagen type, which is itself indicated by roman numerals in parentheses.

Within the molecule each α-chain forms a left-handed polyproline II helix and three helices are intertwined to form a right-handed superhelix, the collagen triple helix. Molecules then associate very specifically to yield the highly ordered quaternary structure of the collagen fibre. All collagen α-chains have repeating Gly-X-Y-sequences in which X and Y are frequently represented by the amino acids Pro and Hyp respectively. But they differ in the precise amino acids and their length. The Gly-Pro-Hyp triplet represents approximately 12% of the primary sequence of collagen type I. Large parts of some collagens consist of non-collagenous domains. An interesting example of a complex collagen is collagen type VI, as it shares non-collagenous domains with the vWf A-domain, platelet glycoprotein Ib, fibronectin type III repeats, and Kunitz type protease inhibitor[21,22]. Along with these collagens, a number of secreted proteins, such as complement component C1q[23], contain collagenous sequences and short triple helical sections.

At least nine of the collagens - types I, III, IV, V, VI, VIII, XII, XIII, and XIV - have been found in the vessel wall[24]. The ratio of the various collagen molecules in a normal and in an arteriosclerotic vessel wall is different[24,25]. The fibril-forming collagens type I,

III and V constitute the highest proportion of vessel wall collagens[25]. The collagen fibrils often aggregate into large bundles which can be seen by light microscopy as collagen fibres. Immunohistochemical studies of adhesive proteins showed enrichment of collagen types I, III, V, and VI, vitronectin, fibronectin, fibrinogen/fibrin, and thrombospondin in the arteriosclerotic plaque. The pattern of increased platelet deposition in serial cross sections corresponded best with areas in which collagen types I and III were enriched[26].

4.3. Von-Willebrand-Factor

The von-Willebrand-factor (vWf) is an essential intermediary in the adhesion of platelets to collagen under conditions of arterial flow[27]. The vWf-GPIb is critical for stopping rapidly moving platelets under high shear so that they can interact effectively with subendothelium components, as demonstrated by the bleeding problems associated with Bernard-Soulier syndrome, where GPIb/V/IX is deficient or defective[28] and von-Willebrand disease, where vWf is affected[29].

The vWf mediates not only platelet adhesion but also thrombus formation at intermediate and high shear rates[30].

vWf is a large multimeric glycoprotein that circulates in blood and is present inside platelets as well as in endothelial cells and subendothelial matrix of the vessel wall[31,32]. Mature vWf has a multimeric structure and exists as a series of oligomers containing a variable number of subunits, from a minimum of two to a maximum of fifty to one hundred. This molecular organisation provides the potential for multiple contact sites with platelets and collagen. vWf has a domain structure consisting of repeated A, B, C and D domains. Two collagen type I or III binding sites have been identified on the A1 and A3 domain of the vWf subunit[33,34]. The A1 domain is involved in binding to the platelet receptor glycoprotein (GP) Ib (6chapter 7) and the protease sensitive site of vWf is located in the A2 domain[35]. For the vWf-binding to collagens type I and III the A3 domain contains the major binding site. The major binding site on the A3 domain and the minor site present on A1 bind to different sites of collagen[36].

The crystal structure of the A3 domain suggests that adhesion to collagen types I and III is achieved primarily through interactions between negatively charged residues on A3 and positively charged residues on collagen[37]. The A3 domain does not contain a metal ion-dependent adhesion site motif. The collagen recognition by vWf A3 therefore seems to occur by a mechanism differing from that of the integrin $\alpha_2\beta_1$[38]. In contrast to the interaction with collagen types I and III, vWf binds to collagen type VI primarily via its A1 domain[39].

Several inhibitors of the vWf/collagen interaction show clearly its importance for platelet adhesion to damaged vessel wall or to disrupted arteriosclerotic plaque. The salivary glands of the leech *Haementeria officinalis* contain a protein, leech anti-platelet protein (LAPP). This protein, which binds to collagen and works by inhibiting binding of vWf to collagen, is a potent platelet adhesion inhibitor at high shear rate[40]. Calin from the saliva of the medicinal leech *Hirudo medicinalis* is a potent inhibitor of collagen-mediated platelet adhesion and activation as well. Calin interferes with both the direct platelet/collagen interaction and the vWf/collagen binding[41].

In contrast to platelets adhering to a collagen surface platelets adhering to the vWf surface are mainly single platelets[42]. Small numbers of vWf molecules are sufficient to attract and slow down platelets flowing near the surface. The subsequent processes of adhesion and activation are mediated by the interaction between direct platelet receptors with collagen (see chapter 4.4).

4.4. P65

In the early Eighties Chiang and Kang isolated a platelet membrane collagen-binding protein by affinity chromatography on chicken skin collagen type I[43-46]. The protein had a molecular weight of 65kD under reducing and nonreducing conditions, showing that it did not contain disulphide bonds. A monoclonal antibody raised against this protein inhibited collagen type I-induced platelet aggregation and ATP secretion of dense bodies but had no effect on collagen type III-induced platelet activation.

The protein was cloned[46] and expressed in a prokaryotic and a eukaryotic expression system. In a eukaryotic expression vector, a 65kD protein was identified that was recognised by the monoclonal antibody against purified P65 from platelet membranes. Endoglycosidase H treatment suggests that P65 is a glycoprotein with a 54kD protein core. The glycosylation of P65 does not seem to affect the function of the protein, for both recombinant proteins expressed either in the prokaryotic or in the eukaryotic system inhibited collagen type I-induced platelet adhesion and aggregation. Expression of P65 in naturally nonadherent T cells rendered them adherent to collagen type I[46]. The recombinant protein inhibits collagen type I-induced platelet aggregation and the adhesion of [^{14}C]serotonine-labelled platelets to collagen type I. Interestingly, the recombinant protein neither binds to collagen type III nor inhibits collagen type III-induced aggregation. Therefore it seems to be highly specific for collagen type I[46]. Deduced from the hydrophilicity diagram, the protein contains three transmembrane domains (residues 1-21, 35-55, and 213-233), an extracellular domain (residue 55-212) and a short extracellular NH_2-terminus. An 18 amino acid residue peptide derived from the extracellular domain (residue 84-101) contents the recognition sequence for collagen type I[47].

4.5. CD36

CD36, a multifunctional membrane glycoprotein also known as GPIIIb and GPIV, is a member of a recently discovered family of scavenger receptors, members of which so far recorded are LIMPII[48], CLA-1[49], FAT[50], and SR-BI[51].

The FAT protein of adipocyte membranes is 58% homologous with that of CD36 identified in human platelets and is involved in binding and transportation of long-chain fatty acids[50].

CD36 is involved in binding to long-chain fatty acids as well[50]. Long-chain fatty acids are one of the major cardiac energy substrates. It is proposed that CD36 deficiency type I is an aetiology of hereditary hypertrophic cardiomyopathy[52,53]. CD36 is a high affinity receptor for the native lipoproteins HDL, LDL, and VLDL[54] as well as for oxidised LDL[55]. CD36 is thought to have a role in platelet aggregation mediating the phagocytosis of apoptotic cells[56,57] and binds to thrombospondin-I[58]. Thrombospondin induced dimerisation of membrane-bound but not soluble CD36[59], and it has been reported that in endothelial cells thrombin-stimulated calcium mobilisation is inhibited by thrombospondin via CD36[60]. On the other hand, platelets deficient in CD36 expressed endogenous thrombospondin-1 as control platelets and their binding capacity for exogenous thrombospondin-1 was the same[61]. Cytoadherence of erythrocytes infected with *Plasmodium falciparum* to capillary endothelial cells is thought to be mediated by CD36[62,63].

CD36 has been shown to be closely associated with the pp60$^{c\text{-}src}$-related protein tyrosine kinases fyn, lyn, and yes in non-activated platelets but not after their activation[64] and to be involved in signal transduction[60,65]. The translocation of the csk homologous kinase (Chk/Hyl) controls the activity of CD36-anchored lyn tyrosine kinase in thrombin-stimulated platelets[66]. On resting platelets CD36 is spatially associated with the glycoprotein IIb/IIIa complex[67] in cholesterol-rich membrane microdomains[68].

The function of CD36 as a collagen receptor is discussed in detail at a later stage.

In addition to platelets, CD36 is expressed on various other cells such as megakaryocytes[69], monocytes[70], fetal erythrocytes[71], reticulocytes[72], microvascular endothelial cells[73], and several tumour cells[74].

CD36 is an integral membrane glycoprotein with a molecular weight of 88kD (reduced and nonreduced)[75]. It has a topology with two transmembrane domains and is palmitoylated on both N- and C-terminal cytoplasmic tails[76]. The proposed topology of CD36 predicts that, of ten cysteine residues, six lie in the extracellular domains, whereas four are equally distributed in the two short terminal tails flanking the N-terminal and C-terminal hydrophobic stretches[77].

A recently publication reports that CD36 forms dimers and multimers in platelets and transfected COS-7-cells by intermolecular cysteine-bridging[78]. It is likely that these dimers are more effective in CD36-mediated signal transduction. CD36 is highly glycosylated, very hydrophobic, and extremely resistant to proteolysis when it is inserted into the platelet membrane[75,79].

The CD36 gene was assigned to chromosome 7 and localised on q11.2[80]. The human CD36 gene is encoded by 15 exons that extend more than 32kilobases: the mRNA 5-untranslated region is encoded by three exons, and cytoplasmic terminal ends are encoded by single exons, whereas the extracellular domain is encoded by 11 exons[81].

In Japanese and other East Asian populations as well as in African populations a distinct proportion (1-4%) of individuals lack CD36 on the platelets with no negative effects on haemostasis[82]. In type I CD36 deficiency, neither platelets nor monocytes/macrophages express CD36; in type II, monocytes/macrophages express CD36 but platelets do not[83,84]. Three known mutations cause CD36 deficiency, i.e. a 478C6T substitution in codon 90 (proline 906serine)[85], a dinucleotide deletion at nucleotide 539 in codon 110[86] and a single nucleotide insertion at nucleotide 1159 in codon 317 leading to a frameshift[87]. Individuals with type I CD36 deficiency risk developing anti-CD36 isoantibodies upon blood transfusions or during pregnancy[88].

CD36 has been implicated as a collagen receptor. Purified CD36 binds to collagen type I fibrils, and F(ab) fragments of polyclonal antibodies raised against CD36 inhibit collagen-induced platelet aggregation[89,90]. Monoclonal antibodies were described that inhibited platelet adhesion to collagen in static systems under Mg^{2+}-independent conditions but had little effect in the presence of Mg^{2+} [91]. These monoclonal antibodies inhibited platelet adhesion under flow conditions (shear rates $800s^{-1}$) to arterial subendothelium and endothelial cell extracellular matrix by 40% after 2 minutes of perfusion and by 30% after 5 minutes. However,after 10 minutes no significant inhibition was seen[92].

Furthermore, soluble CD36 competes with membrane-bound CD36 and inhibits platelet activation by type I collagen. McGregor et al[90], described the inhibition of platelet aggregation by anti CD36 $F(ab)_2$ in response to low doses of type I collagen. But the antibodies used also inhibited the aggregation induced by low doses of ADP, epinephrine or thrombin. This indicates that the inhibitory effect is of a more general nature and not confined to collagen. In the presence of Mg^{2+}, anti-CD36 and anti-GPIa/IIa or anti-GPIV antibodies in combinations were more effective in inhibiting adhesion than anti-GPIa/IIa or anti-GPVI alone[93].

In comparison with control platelets, CD36-deficient platelets showed at shear rate $800s^{-1}$ in the absence of Mg^{2+} reduced platelet adhesion to collagen type I at early time points[94,95]. In order to work under more physiologically divalent cation conditions in a co-operative study between our group and the group headed by Sixma (Utrecht) we determined the adhesion of CD36-deficient platelets to collagen types I, III, IV, and V, using heparinised blood. No differences in platelet adhesion and subsequent aggregate formation on collagens type I, III, and IV were observed under static or flow ($300s^{-1}$, $1600s^{-1}$) conditions, whereas the adhesion of CD36-deficient platelets to collagen type V was strongly reduced under static conditions. Collagen type V is nonadhesive under flow conditions. Only contact CD36-deficient platelets were observed on collagen type V. These platelets did not spread or aggregate[96]. McKeown et al[97] also compared the adhesion and aggregate formation on collagen types I, III, and IV surfaces in static and to collagen type I in flow systems at high shear force and found no difference between CD36-deficient and control platelets. CD36-deficient platelets aggregate normally in response to collagen type I and III but fail to aggregate in response to collagen type V[98,99]. The collagen type I-induced intracellular mobilisation of Ca^{2+} in the CD36-deficient platelets was of the same magnitude as that seen in platelets from normal donors. In addition, serotonine secretion, βTg secretion and PF4 secretion appeared not to be impaired[98,100]. Tyrosine phosphorylation was also similar between the CD36-deficient and normal platelets[100].

Thus it is unlikely that CD36 plays a major role in platelet/collagen interactions.

Since CD36 appears not to be involved in primary adhesion but possibly rather in amplification of the signal from collagen adhesion, it cannot be excluded that changes in the stoichiometry of signalling molecules can compensate for the absence of CD36 itself[101].

4.6. a_2b_1-integrin (GPIa/IIa, VLA_2, ECMRII)

The protein complex that is universally accepted as a collagen receptor is the $\alpha_2\beta_1$ integrin[102]. It is also known as platelet glycoprotein Ia/IIa complex or as very late activation antigen (VLA) 2 on activated T-cells or human fibroblast class II extracellular matrix receptor (ECMRII)[103]. It has also been designated CD49b/CD29.

The strongest evidence that $\alpha_2\beta_1$ is involved in platelet adhesion to collagen and in platelet aggregation comes from clinical studies. Nieuwenhuis et al[8] described a patient with a mild bleeding disorder attributable to an 80% reduction in the expression of the α_2 integrin (GPIa) subunit. The platelets of this patient showed impaired collagen-induced aggregation but normal responsiveness to all other platelet agonists. In addition, impaired adhesion of the patient's platelets to extracellular matrix of vascular endothelium and to different types of purified collagens was demonstrated[104]. In contrast to the adhesion effect in von Willebrand's disease, which is most pronounced under conditions of high shear forces, the patient's platelets showed impaired adhesion at both low and high shear forces. The few GPIa-deficient platelets adhering to collagen type I fibres remained in the contact phase and did not spread.

We described a patient with a bleeding disorder, whose platelets were completely deficient in GPIa and intact thrombospondin and did not aggregate in response to collagen[105]. Acquired deficiencies of the platelet GPIa/IIa complex with absence of responsiveness of the platelets to collagens due to myeloproliferative disorders have been described[106].

On platelets about 1400 to 2000 copies of GPIa/IIa were detected[107]. Kritzik et al[108] found that three allelic differences in the α_2 gene are associated with expression levels of the $\alpha_2\beta_1$ integrin on the platelet surface. 807T/873A individuals have on average a 1.6 fold higher level of $\alpha_2\beta_1$ expression than 807C/873G individuals. The variation in receptor expression correlated with the ability of nonactivated washed platelets to adhere to collagen types I and III[109]. It has been speculated that increased $\alpha_2\beta_1$ expression might be a risk factor for stroke or myocardial infarction.

Human autoantibodies[110] and mice monoclonal antibodies[107,111] that bind to GPIa and block platelet adhesion to collagen types I and III and collagen-induced platelet aggregation have been described.

Jararhagin, a snake venom from *Bothrops jararaca*, contains a metalloproteinase that cleaves the β_1 subunit (GPIIa) of the receptor complex, resulting in an inhibition of platelet adhesion to collagen[112] as well as an inhibition of early platelet signalling events in response to collagen, as demonstrated in the loss of pp72(syk) phosphorylation.

Santoro et al[113] found that the binding of GPIa/IIa complex to collagen type I is Mg^{2+}-dependent and is inhibited by Ca^{2+}, a finding which matches the earlier finding that platelet adhesion to collagen is mainly Mg^{2+}-dependent[113]. Staatz et al[?] described that the DGEA sequence of the α_1(I)CB3 fragment of collagen type I binds to GPIa/IIa in an Mg^{2+}-dependent manner. Grab et al reported, however, that a triple-helical peptide containing DGEA was unable to support integrin(β_1)-mediated fibroblast adhesion to collagen[114]. Staatz et al recently reported that $\alpha_2\beta_1$ integrin also binds to the CB4 peptide of the α_2(I) collagen chain[115].

Knight et al identified the $\alpha_2\beta_1$ integrin binding site of $_1\alpha$ (I)CB3 using seven overlapping α_1(I)CB3 fragment peptides that spontaneously assembled into triple-helices after cross-linking[116]. In this study the Barnes group identified a 15-residue sequence, GFP*GERGVEGPP*GPA, corresponding to residues 502-516 of the parent α_1(I) chain, as the $\alpha_2\beta_1$ recognition sequence of collagen type I. Earlier the same group identified the collagen type III sequence GGPP*GPR residues 522-528 of the collagen α_1(III) chain as the minimum structure required for the recognition of $\alpha\ \beta$ using the same elegant technique[117].

The $\alpha_2\beta_1$ integrin is expressed in a variety of cell types. In cells of haematopoietic lineage, expression is restricted to megakaryocytes and platelets. On endothelial cells and fibroblasts and many epithelial cell types, $\alpha_2\beta_1$ additionally binds to laminin[118]. The echovirus 1, a non-enveloped RNA-virus implicated in febrile illness and viral meningitis, binds to $\alpha_2\beta_1$[119,120]. $\alpha_2\beta_1$ bears the clinically important Br(a) and Br(b) alloantigenic determinants, which can result in antibody-mediated platelet destruction. A single base change results in a substitution of Lys in Br(a) to Glu in Br(b) at amino acid residue 505[121].

The primary structure of GPIa was established by Takada and Hemler[122]. An overall sequence homology of 18-25% with other integrin α-subunits has been observed. The GPIa sequence contained a similar distribution of cysteine residues, divalent cation metal binding domains and a transmembrane domain[102].

Like the other integrins, the $\alpha_2\beta_1$ integrin has a noncovalently associated heterodimeric structure. The α subunit of GPIa/IIa has an M_r of 165,000 and 150,000 on reduced and nonreduced SDS-PAGE respectively, whereas the β_1 subunit exhibits 130,000 and 110,000 respectively.

The α_2 subunit is composed of one single polypeptide chain with three metal-binding domains followed by an I-domain insert (residues 140-359) on its extracellular side. The β_1 subunit, GPIIa, has a short cytoplasmic domain, a single transmembrane domain and a large extracellular domain with conserved cysteine residues, forming four internally folded cystine repeat units. The amino terminal 45-50kDa regions in association with the α subunit may contribute to ligand-binding activity[102]. Some anti-β_1 antibodies activate the integrin and enhance the ligand-binding activity. The epitope recognised by these activating antibodies has been mapped to the same small region of the β_1 subunit, residue 207-218, recognised by monoclonal antibodies that inhibit ligand binding[123]. The authors have suggested that activating antibodies might fix the flexible turn in a conformation that facilitates ligand binding. $\alpha_2\beta_1$ binds to collagen by a 200 amino acid inserted domain (I-domain) present in the NH_2-terminus of the α_2 subunit. This I-domain has a certain homology with the A-domain of von Willebrand factor (vWf), the I-domain of CR3, and cartilage matrix protein[124].

The binding site of several inhibitory anti-α_2 antibodies, including the famous 6F1, has been localised within the α_2-I-domain[125]. The crystal structure of the I-domain from $\alpha_2\beta_1$ integrin identified a Mg^{2+}-binding metal-ion dependent adhesion site (MIDAS). The authors additionally described a structural feature designated the C helix[126]. This C helix exhibits a groove which is able to accommodate a collagen triple helix[127]. Depraetere et al[128] prepared a recombinant α_2-I-subunit and confirmed that the $_2\alpha$ -

I-domain has all the necessary information to mediate the binding of $\alpha_2\beta_1$ to collagen. The recombinant I-domain prevented platelet adhesion to collagen under static conditions as well as collagen-induced platelet aggregation.

A hotly disputed topic is whether $\alpha_2\beta_1$ is constitutively active with the binding site for collagen always available or, like other integrins, needs to be activated. Very recently Jung and Moroi[129] showed that platelets needs to be activated to bind to soluble collagen but not to fibrillar collagen, which is able to activate platelets itself via GPVI. The binding to soluble collagen is mediated via $\alpha_2\beta_1$ integrin. Lahav and colleagues found that collagen binding induces the formation of a newly discovered disulfide bond in a fraction of $\alpha_2\beta_1$ and that this change increases the affinity of the $\alpha\ \beta$ to its ligand collagen[130].

The role of $\alpha_2\beta_1$ in collagen-induced signal transduction is discussed in 4.8.

4.7. GPVI/FcRg

Another so far uncharacterised 62kDa (reduced conditions) membrane protein, named GPVI, meets several criteria as a collagen receptor.

Four GPVI-deficient patients have been described whose platelets exhibit impaired activation by collagen but not by other stimuli[11,131,132,133,]. All patients have mild bleeding problems. Sugiyama et al[131] reported on a patient with ITP, whose platelets were defective in collagen-induced aggregation. The patient's plasma contained an antibody that recognised GPVI on control platelets but not on her own. This antibody induced aggregation of control platelets, and F(ab) fragments of the antibody inhibited collagen-induced platelet aggregation as well as platelet adhesion to collagen under static conditions[134]. Takahashi et al[135] described another patient with an auto-antibody against GPVI and a mild bleeding tendency with a slightly prolonged bleeding time.

The second line of evidence that GPVI is a collagen receptor is provided by comparison of the signalling events mediated by collagen and by cross-linking GPVI (for details see 4.8). Collagen-induced platelet activation requires the expression by collagen of both tertiary (triple-helical) and quaternary (polymeric) structure[3,4,5]. Recently, Morton et al[136] have described simple collagen-related peptides (CPPs) comprising a repeating Gly-Pro-Hyp motif, that mimic the collagen tertiary (triple-helical) structure. These authors have shown that, when cross-linked to impact quaternary (polymeric) structure, these peptides (CRP-XL) are extremely active platelet agonists, being even more reactive than collagen fibres. This activity is not mediated by $\alpha_2\beta_1$ integrin[136,137]. In contrast, non-cross-linked (non-polymeric) CRP antagonises platelet activation stimulated by either native collagen or CRP-XL[136]. Using platelets of patients with deficiencies in defined platelet glycoproteins we excluded CD36, the von Willebrand factor and GPIIb/IIa as the primary signalling receptor that recognises the quaternary structure of collagen[7]. The results confirmed the view that CD36, GPIIb/IIIa, and the vWf are not essential for platelet activation by collagen.

In contrast, GPVI-deficient platelets were unresponsive to CRP-XL, showing that GPVI is a signalling receptor for the polymeric cross-linked triple-helical Gly-Pro-Hyp motif[7].

The Gly-Pro-Hyp sequence is a highly specific platelet recognition sequence and cannot be substituted by Gly-Pro-Pro, Gly-Pro-Ala or Gly-Pro-Arg, showing that the platelet-reactivity of fibrillar collagen is not just an intrinsic property of the triple-helical conformation[138]. The venom from the tropical rattlesnake ***Crotalus durissus terrificus*** contains a powerful platelet activator, convulxin, that specifically recognises GPVI[139]. Convulxin is a heterodimeric C-type lectin that activates platelets via clustering of GPVI molecules. Convulxin subunits do not induce platelet aggregation or signal transduction but inhibit platelet aggregation induced by collagen while having no effect on platelet aggregation induced by thrombin or vWf-ristocetin[139]. ^{125}J-labelled convulxin bound to GPVI immunoprecipitated by the patient's anti-GPVI antibody mentioned above. In part, convulxin induces signal transduction like collagen[139](see 4.8).

Jandrot-Perrus et al[140] reported that the anti-$\alpha_2\beta_1$ antibody 6F1 inhibited the platelet activation by low concentrations of convulxin. This finding leads to the still open question of how $\alpha_2\beta_1$ integrin and GPVI interact corresponding with collagen.

Takayama and colleagues[141] found that, in GPVI-deficient platelets, the FcRγ chain was absent as well. In a patient with 10% GPVI, both FcRγ and GPVI were proportionally absent. These findings suggest that FcRγ chain and GPVI are co-expressed on the platelet surface to jointly build the collagen receptor.

4.8. Collagen-induced signal transduction

Platelet aggregation induced by collagen is mediated by collagen itself, by thromboxane A_2 synthesised by the platelets, and by ADP secreted from the dense bodies. Gαq-deficient mice are protected from collagen-induced thromboembolism probably because platelet activation by thromboxane A_2 and by ADP is inhibited in these animals[142].

Direct platelet activation by collagen is mediated through a tyrosine kinase-dependent pathway. Collagen stimulation leads to the phosphorylation of numerous proteins. This in turn leads to the tyrosine phosphorylation and activation of phospholipase Cγ2, which cleaves phosphatidylinositol 4,5 bis-phosphate into diacylglycerol and inositol 1,4,5 triphosphate. These second messengers in turn cause the activation of protein kinase C and the Ca^{2+} influx in human platelets. The release of intracellular calcium is necessary for granule secretion, activation of GPIIb/IIIa, actin polymerisation and aggregation[143].

Collagen signals through the same pathway as that of immune receptors[144]. It induces phosphorylation on tyrosine on Fc receptor γ-chain (FcRγ), the non-receptor tyrosine kinase Syk and phospholipase Cγ2 (PLCγ2)[145,146,147].

Syk assembles into signalling complex via interaction between its tandem Src homology 2 (SH2) domains and the tyrosine-phosphorylated immuno receptor tyrosine-based activation motif (ITAM) present in the FcRγ. The FcRγ and Syk are essential for activation of mouse platelets by collagen[148].

Convulxin, anti-GPVI $F(ab)_2$ and CRP-XL while cross-linking GPVI induce signal transduction in part like collagen[139,140,149,150,151].

The role of $\alpha_2\beta_1$ in signalling is a subject of controversial discussion. Whereas some authors think that "the role of $\alpha_2\beta_1$ is in primary attachment rather than in subsequent activation"[27] or that "recognition of $\alpha_2\beta_1$ is insufficient to cause activation"[127], we and others believe in a role of $\alpha_2\beta_1$ in signalling. Keely and Parise described $\alpha_2\beta_1$ integrin as a necessary co-receptor for collagen-induced activation of Syk[152].

Ichinohe et al studied collagen-induced signal transduction in platelets of GPVI-deficient patients. Severe defects have been observed in collagen-stimulated Syk activation and tyrosine phosphorylation of PLCγ2, Vav and FAK125, implicating a role of GPVI in recruiting these proteins into the signalling cascade[153]. GPVI-deficient platelets exhibit $\alpha_2\beta_1$-dependent c-Src activation accompanied by tyrosine phosphorylation of several substrates including cortactin. GPVI-deficient platelets showed fibrinogen binding in response to collagen but failed to aggregate and to express CD62 and CD63[7]. It is likely that collagen induces some signalling via $\alpha_2\beta_1$, leading to activation of GPIIb/IIIa. I would like to speculate that the GPIIb/IIIa molecules are not able to perform capping because a post-fibrinogen binding signal is missing in GPVI-deficient platelets, preventing these platelets from aggregating in response to collagen. Collagen-induced signal transduction is shown in Figure 2.

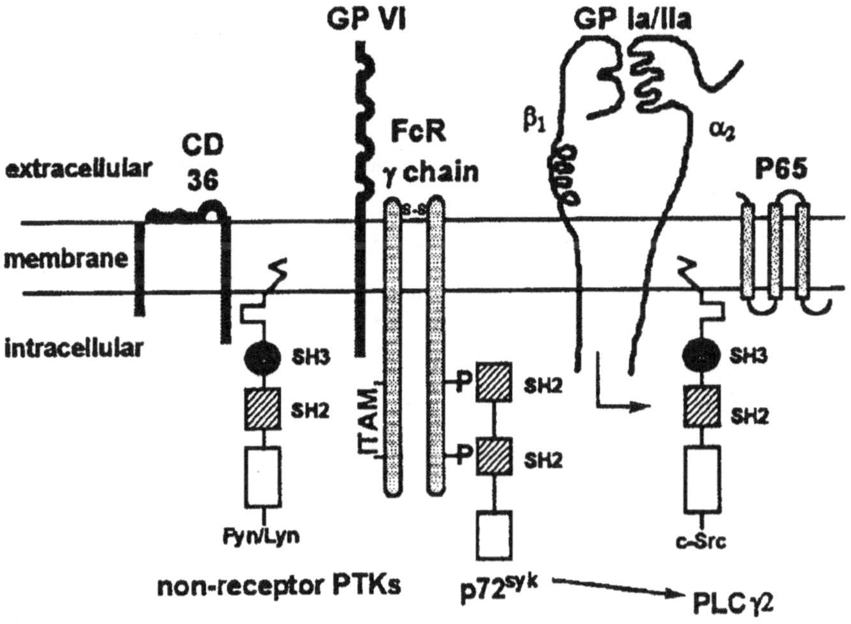

Figure2: Model of collagen-induced signal transduction via the collagen receptors α2β1 integrin and GPVI/FcRγ

The questions of whether $\alpha_2\beta_1$ integrin has to be activated to bind to collagen and how $\alpha_2\beta_1$, GPVI/FcRγ and possible other collagen receptors cross-talk and interact have still to be clarified.

References

1. Sixma JJ, van Zanten GH, Huizinga EG, van der Plas RM, Verkley M, Wu YP, Gros P, de Groot PG: Platelet adhesion to collagen: an update. Thromb Haemost 78:434-438, 1997

2. Santoro SA: Molecular basis of platelet adhesion to collagen. In: Jamieson GA (ed): Platelet membrane receptors: molecular biology, immunology, biochemistry, and pathology. Alan R. Liss, New York; 291-314, 1988

3. Muggli R, Baumgartner HR: Collagen-induced platelet aggregation: Requirement for tropocollagen multimers. Thromb Res 3:715-728, 1973

4. Jaffe R, Deykin D: Evidence for a structural requirement for the aggregation of platelets by collagen. J Clin Invest 53:875-883, 1974

5. Brass LF, Bensusan HB: The role of quaternary structure in platelet-collagen interaction. J Clin Invest 54:1480-1487, 1974

6. Morton LF, Peachey AR, Barnes MJ: Platelet-reactive sites in collagens type I and type III. Evidence for separate adhesion and aggregatory sites. Biochem J 258:157-163, 1989

7 Kehrel B, Wierwille S, Clemetson KJ, Anders O, Steiner M, Knight CG, Farndale RW, Okuma M, Barnes MJ: Glycoprotein VI is a major collagen receptor for platelet activation: It recognizes the platelet-activating quaternary structure of collagen, whereas CD36, GPIIb/IIIa and vWf do not. Blood 91:491-499, 1998

8. Nieuwenhuis HK, Akkermann JWN, Houdijk WPM, Sixma JJ: Human blood platelets showing no response to collagen fail to express surface glycoprotein Ia. Nature 318:470-472, 1985

9. Staatz WD, Walsh JJ, Pexton T, Santoro SA: The $\alpha2\beta1$ integrin cell surface collagen receptor binds to the $\alpha1$(I)-CB3 peptide of collagen. J Biol Chem 265:4778-4781, 1990

10. Tandon NN, Kralisz U, Jamieson GA: Identification of glycoprotein IV (CD36) as a primary receptor for platelet-collagen adhesion. J Biol Chem 264:7576-7583, 1989

11. Moroi M, Jung SM, Okuma M, Shinmyozu K: A patient with platelets deficient in glycoprotein VI that lack both collagen-induced aggregation and adhesion. J Clin Invest 84:1440-1445, 1989

12. Chiang TM, Kang AH: Isolation and purification of collagen $\alpha1$(I) receptor from human platelet membrane. J Biol Chem 257:7581-7586, 1982

13. Kielty CM, Hopkinson I, Grant ME: Collagen: The collagen familiy: Structure, assembly, and organization in the extracellular matrix. In: Royce PM, Steinman B (eds): Conncective tissue and its heritable disorders. Wiley-Liss, New York; 103-147, 1993

14. Tanzer ML: Collagens and elastin: Structure and interactions. Curr Opin Cell Biol 1:968-973, 1989

15. Shaw LM, Olsen BR: FACIT collagens: Diverse molecular bridges in extracellular matrices. Trends Biochem Sci 16:191-134, 1991

16 .van der Rest M, Garrone R: Collagen family of proteins. FASEB J 5:2814-2823, 1991

17. Rauterberg J, Jander R, Troyer D: Type VI collagen - a structural glycoprotein with a collagenous domain. Front Matrix Biol 11:90-109, 1986

18. Engvall E, Hessle H, Klier G: Molecular assembly, secretion, and matrix deposition of type VI collagen. J Cell Biol 102:703-710, 1986

19. Burgeson RE, Lunstrum GP, Rokosova B, Rimberg CS, Rosenbaum LM, Keene DR: The structure and function of type VII collagen. Ann N Y Acad Sci 580:32-43, 1990

20. Bruckner Tuderman L, Mitsuhashi Y, Schnyder UW, Bruckner P: Anchoring fibrils and type VII collagen are absent from skin in severe recessive dystrophic epidermolysis bullosa. J Invest Dermatol 93:3-9, 1989

21. Bonaldo P, Colombatti A: The carboxyl terminus of the chicken α3 chain of collagen VI is a unique mosaic structure with glycoprotein Ib-like, fibronectin type III and Kunitz modules. J Biol Chem 264:20235-20239, 1989

22. Colombatti A, Bonaldo P: The superfamily of proteins with von Willebrand factor type A-like domains: one theme common to components of extracellular matrix, hemostasis, cellular adhesion, and defense mechanisms. Blood 77:2305-2315, 1991

23. Reid KBM: Complete amino acid sequences of the three collagen-like regions present in subcomponent C1q of the first component of human complement. Biochem J 179:367-371, 1979

24. Kehrel B: Platelet-collagen interactions. Semin Thromb Hemost 21:123-129, 1995

25. Rauterberg J, Jaeger E: Collagen and collagen synthesis in the atherosclerotic vessel wall. In: Robenek N, Severs NJ (eds): Cell interactions in atherosclerosis. CRC Press, Boca Raton; 102-135, 1992

26. van Zanten GH, de Graaf S, Slootweg PJ, Heijnen HF, Connolly TM, de Groot PG, Sixma JJ: Increased platelet deposition on atherosclerotic coronary arteries. J Clin Invest 93:615-632, 1994

27. Sixma JJ, van Zanten GH, Saelman EU, Verkleij M, Lankhof H, Nieuwenhuis-HH, de Groot PG: Platelet adhesion to collagen. Thromb Haemost 74:454-459, 1995

28. Clemetson KJ, McGregor JL, James E, Dechavanne M, Lüscher EF: Characterization of the platelet membrane glycoprotein abnormalities in Bernard-Soulier syndrome and comparison with normal by surface-labeling techniques and high-resolution two-dimensional gel electrophoresis. J Clin Invest 70:304-311, 1982

29. Tschopp TB, Weiss HJ, Baumgartner HR: Decreased adhesion of platelets to subendothelium in von Willebrand's disease. J Lab Clin Med 83:296-300, 1974

30. Fressinaud E, Sakariassen KS, Rothschild C, Baumgartner HR, Meyer D: Shear rate-dependent impairment of thrombus growth on collagen in nonanticoagulated blood from patients with von Willebrand disease and hemophilia A. Blood 80:988-994, 1992

31. Ruggeri ZM, Ware J: von Willebrand factor. FASEB J 7:308-316, 1993

32. Sixma JJ, de Groot PG: von Willebrand factor and the blood vessel wall. Mayo Clin Proc 66:628-633, 1991

33. Cruz MA, Yuan H, Lee JR, Wise RJ, Handin RI: Interaction of the von Willebrand factor (vWF) with collagen. Localization of the primary collagen-binding site by analysis of recombinant vWF A domain polypeptides. J Biol Chem 270:10822-10827, 1995

34. Sixma JJ, Schiphorst ME, Verweij CL, Pannekoek H: Effect of deletion of the A1 domain of von Willebrand factor on its binding to heparin, collagen and platelets in the presence of ristocetin. Eur J Biochem 196:369-375, 1991

35. Lankhof H, Damas C, Schiphorst ME, Ijsseldijk MJ, Bracke M, Furlan M, Tsai HM, de Groot PG, Sixma JJ, Vink T: Von Willebrand factor without the A2 domain is resistant to proteolysis. Thromb Haemost 77:1008-1013, 1997

36. Lankhof H, van Hoeij M, Schiphorst ME, Bracke M, Wu YP, Ijsseldijk MJ, Vink T, de Groot PG, Sixma JJ: A3 domain is essential for interaction of von Willebrand factor with collagen type III. Thromb Haemost 75:950-958, 1996

37. Huizinga EG, Martijn van der Plas R, Kroon J, Sixma JJ, Gros P: Crystal structure of the A3 domain of human von Willebrand factor: implications for collagen binding. Structure 5:1147-1156, 1997

38. Bienkowska J, Cruz M, Atiemo A, Handin R, Liddington R: The von willebrand factor A3 domain does not contain a metal ion-dependent adhesion site motif. J Biol Chem 272:25162-25167, 1997

39. Hoylaerts MF, Yamamoto H, Nuyts K, Vreys I, Deckmyn H, Vermylen J: Von Willebrand factor binds to native collagen VI primarily via its A1 domain. Biochem J 324:185-191, 1997

40. van Zanten GH, Connolly TM, Schiphorst ME, de Graaf S, Slootweg PJ, Sixma JJ: Recombinant leech antiplatelet protein specifically blocks platelet deposition on collagen surfaces under flow conditions. Arterioscler Thromb Vasc Biol 15:1424-1431, 1995

41. Harsfalvi J, Stassen JM, Hoylaerts MF, Van Houtte E, Sawyer RT, Vermylen J, Deckmyn H: Calin from Hirudo medicinalis, an inhibitor of von Willebrand factor binding to collagen under static and flow conditions. Blood 85:705-711, 1995

42. Moroi M, Jung SM, Nomura S, Sekiguchi S, Ordinas A, Diaz-Ricart M: Analysis of the involvement of the von Willebrand factor-glycoprotein Ib interaction in platelet adhesion to a collagen-coated surface under flow conditions. Blood 90:4413-4424, 1997

43. Chiang TM, Kang AH, Dale JB, Beachey EH: Immunochemical studies of the purified human platelet receptor for the alpha 1(I)-chain of chick skin collagen. J Immunol 133:872-876, 1984

44. Chiang TM, Jin A, Kang AH: Platelet-collagen interaction. Inhibition by a monoclonal antibody raised against collagen receptor. J Immunol 139:887-892, 1987

45. Chiang TM, Seyer JM, Kang AH: Collagen-platelet interaction: separate receptor sites for types I and III collagen. Thromb Res 71:443-456, 1993

46. Chiang TM, Rinaldy A, Kang AH: Cloning, characterization, and functional studies of a nonintegrin platelet receptor for type I collagen. J Clin Invest 100:514-521, 1997

47. Chiang TM, Kang AH: A synthetic peptide derived from the sequence of a type I collagen receptor inhibits type I collagen-mediated platelet aggregation. J Clin Invest 100:2079-2084, 1997

48. Vega MA, Segui-Real B, Garcia JA, Cales C, Rodriguez F, Vanderkerckhove J, Sandoval IV: Cloning, sequencing, and expression of a cDNA encoding rat LIMP II, a novel 74-kDa lysosomal membrane protein related to the surface adhesion protein CD36. J Biol Chem 266:16818-16824, 1991

49. Calvo D, Vega MA: Identification, primary structure, and distribution of CLA-1, a novel member of the CD36/LIMPII gene family. J Biol Chem 268:18929-18935, 1993

50. Abumrad NA, el Maghrabi MR, Amri EZ, Lopez E, Grimaldi PA: Cloning of a rat adipocyte membrane protein implicated in binding or transport of long-chain fatty acids that is induced during preadipocyte differentiation. Homology with human CD36. J Biol Chem 268:17665-17668, 1993

51. Acton SL, Scherer PE, Lodish HF, Krieger M: Expression cloning of SR-B1, a CD36-related class B scavenger receptor. J Biol Chem 269:21003-21009, 1994

52. Tanaka T, Sohmiya K, Kawamura K: Is CD36 deficiency an etiology of hereditary hypertrophic cardiomyopathy?. J Mol Cell Cardiol 29:121-127, 1997

53. Watanabe K, Toba K, Ogawa Y, Aizawa Y, Tanabe N, Miyajima S, Kusano Y, Nagatomo T, Hirokawa Y: Different patterns of 123I-BMIPP myocardial accumulation in patients with type I and II CD36 deficiency. Kaku Igaku 34:1125-1130, 1997

54. Calvo D, Gomez-Coronado D, Suarez Y, Lasuncion MA, Vega MA: Human CD36 is a high affinity receptor for the native lipoproteins HDL, LDL, and VLDL. J Lipid Res 39:777-788, 1998

55. Endemann G, Stanton LW, Madden KS, Bryant CM, White RT, Protter AA: CD36 is a receptor for oxidized low density lipoprotein. J Biol Chem 268:11811-11816, 1993

56. Savill J: Recognition and phagocytosis of cells undergoing apoptosis. Br Med Bull 53:491-508, 1997

57. Navazo MD, Daviet L, Savill J, Ren Y, Leung LL, McGregor JL: Identification of a domain (155-183) on CD36 implicated in the phagocytosis of apoptotic neutrophils. J Biol Chem 271:15381-15385, 1996

58. Asch AS, Silbiger S, Heimer E, Nachman RL: Thrombospondin sequence motif (CSVTCG) is responsible for CD36 binding. Biochem Biophys Res Commun 182:1208-1217, 1992

59. Daviet L, Malvoisin E, Wild TF, McGregor JL: Thrombospondin induces dimerization of membrane-bound, but not soluble CD36. Thromb Haemost 78:897-901, 1997

60. Enenstein J, Gupta K, Vercellotti GM, Hebbel RP: Thrombin-stimulated calcium mobilization is inhibited by thrombospondin via CD36. Exp Cell Res 238:465-471, 1998

61. Kehrel B, Kronenberg A, Schwippert B, Niesing-Bresch D, Niehues U, Tschöpe D, van de Loo J, Clemetson KJ: Thrombospondin binds normally to glycoprotein IIIb deficient platelets. Biochem Biophys Res Commun 179:985-991, 1991

62. Ockenhouse CF, Tandon NN, Magowan C, Jamieson GA, Chulay J: Identification of a platelet membrane glycoprotein as a falciparum malaria sequestration receptor. Science 243:1469-1471, 1989

63. McCormick CJ, Craig A, Roberts D, Newbold CI, Berendt AR: Intercellular adhesion molecule-1 and CD36 synergize to mediate adherence of Plasmodium falciparum-infected erythrocytes to cultured human microvascular endothelial cells. J Clin Invest 100:2521-2529, 1997

64. Huang MM, Bolen JB, Barnwell JW, Shattil SJ, Brugge JS: Membrane glycoprotein IV (CD36) is physically associated with fyn, lyn and yes protein-tyrosine kinases in human platelets. Proc Natl Acad Sci U S A 88:7844-7848, 1991

65. Schüepp BJ, Pfister H, Clemetson KJ, Silverstein RL, Jungi TW: CD36-mediated signal transduction in human monocytes by anti-CD36 antibodies but not by anti-thrombospondin antibodies recognizing cell membrane-bound thrombospondin. Biochem Biophys Res Commun 175:263-270, 1991

66. Hirao A, Hamaguchi I, Suda T, Yamaguchi N: Translocation of the Csk homologous kinase (Chk/Hyl) controls activity of CD36-anchored Lyn tyrosine kinase in thrombin-stimulated platelets. EMBO J 16:2342-2351, 1997

67. Dorahy J, Berndt MC, Shafren DR, Burns GF: CD36 is spatially associated with glycoprotein IIb-IIIa (alpha IIb beta 3) on the surface of resting platelets. Biochem Biophys Res Commun 218:575-581, 1996

68. Dorahy DJ, Lincz LF, Meldrum CJ, Burns GF: Biochemical isolation of a membrane microdomain from resting platelets highly enriched in the plasma membrane glycoprotein CD36. Biochem J 319:67-72, 1996

69. Berger G, Caen JP, Berndt MC, Cramer EM: Ultrastructural demonstration of CD36 in the α-granule membrane of human platelets and megakaryocytes. Blood 82:3034-3044, 1993

70. Asch AS, Barnwell J, Silverstein RL, Nachman RL: Isolation of the thrombospondin membrane receptor. J Clin Invest 79:1054-1061, 1987

71. Kieffer N, Bettaieb A, Legrand C, Coulombel L, Vainchenker W, Edelman L, Breton-Gorius J: Developmentally regulated expression of a 78kDa erythroblast membrane glycoprotein immunologically related to the platelet thrombospondin receptor. Biochem J 262:835-842, 1989

72. Joneckis CC, Ackley RL, Orringer EP, Wayner EA, Parise LV: Integrin α4β1 and glycoprotein IV (CD36) are expressed on circulating reticulocytes in sickle cell anemia. Blood 82:3548-3555, 1993

73. Greenwalt DE, Watt KW, Hasler T, Howard RJ, Patel S: Structural, functional, and antigenic differences between bovine heart endothelial CD36 and human platelet CD36. J Biol Chem 265:16296-16299, 1990

74. Clezardin P, Frappart L, Clerget M, Pechoux C, Delmas PD: Expression of thrombospondin (TSP1) and its receptors (CD36 and CD51) in normal, hyperplastic, and neoplastic human breast.. Cancer Res 53:1421-1430, 1993

75. Tandon NN, Lipsky RH, Burgess WH, Jamieson GA: Isolation and characterization of platelet glycoprotein IV (CD36). J Biol Chem 264:7570-7575, 1989

76. Tao N, Wagner SJ, Lublin DM: CD36 is palmitoylated on both N- and C-terminal cytoplasmic tails. J Biol Chem 271:22315-22320, 1996

77. Gruarin P, Sitia R, Alessio M: Formation of one or more intrachain disulphide bonds is required for the intracellular processing and transport of CD36. Biochem J 328:635-642, 1997

78. Thorne RF, Meldrum CJ, Harris SJ, Dorahy DJ, Shafren DR, Berndt MC, Burns GF, Gibson PG: CD36 forms covalently associated dimers and multimers in platelets and transfected COS-7 cells. Biochem Biophys Res Commun 240:812-818, 1997

79. Daviet L, McGregor JL: Functional roles fo membrane glycoprotein CD36. Platelets 7:117-124, 1996

80. Fernandez Ruiz E, Armesilla AL, Sanchez Madrid F, Vega MA: Gene encoding the collagen type I and thrombospondin receptor CD36 is located on chromosome 7q11.2. Genomics 17:759-761, 1993

81. Armesilla AL, Vega MA: Structural organization of the gene for human CD36 glycoprotein. J Biol Chem 269:18985-18991, 1994

82. Greenwalt DE, Lipsky RH, Ockenhouse CF, Ikeda H, Tandon NN, Jamieson GA: Membrane glycoprotein CD36: A Review of its role in adherence, signal transduction, and transfusion medicine. Blood 80:1105-1115, 1992

83. Take H, Kashiwagi H, Tomiyama Y, Honda S, Honda Y, Mizutani H, Furubayashi T, Karasuno T, Nishiura T, Kanayama Y, et al: Expression of GPIV and N(aka) antigen on monocytes in N(aka)-negative subjects whose platelets lack GPIV. Br J Haematol 84:387-391, 1993

84. Yamamoto N, Akamatsu N, Sakuraba H, Yamazaki H, Tanoue K: Platelet glycoprotein IV (CD36) deficiency is associated with the absence (type I) or the presence (type II) of glycoprotein IV on monocytes. Blood 83:392-397, 1994

85. Kashiwagi H, Tomiyama Y, Honda S, Kosugi S, Shiraga M, Nagao N, Sekiguchi S, Kanayama Y, Kurata Y, Matsuzawa Y: Molecular basis of CD36 deficiency. Evidence that a 478C-->T substitution (proline90-->serine) in CD36 cDNA accounts for CD36 deficiency. J Clin Invest 95:1040-1046, 1995

86. Kashiwagi H, Tomiyama Y, Kosugi S, Shiraga M, Lipsky RH, Nagao N, Kanakura Y, Kurata Y, Matsuzawa Y: Family studies of type II CD36 deficient subjects: linkage of a CD36 allele to a platelet-specific mRNA expression defect(s) causing type II CD36 deficiency. Thromb Haemost 74:758-763, 1995

87. Kashiwagi H, Tomiyama Y, Nozaki S, Honda S, Kosugi S, Shiraga M, Nakagawa T, Nagao N, Kanakura Y, Kurata Y, Matsuzawa Y: A single nucleotide insertion in codon 317 of the CD36 gene leads to CD36 deficiency. Arterioscler Thromb Vasc Biol 16:1026-1032, 1996

88. Ikeda H, Mitani T, Okuma M, Haga H, Ohtzuka S, Kato T, Nakase T, Sekiguchi S: A new platelet specific antigen Nak(a) involved in the refractoriness of HLA-matched platelet transfusion. Vox Sang 57:213-217, 1989

89. Tandon NN, Jamieson GA: Role of GPIV as a platelet receptor for collagen adhesion. Prog Clin Biol Res 283:315-318, 1988

90. McGregor JL, Catimel B, Parmentier S, Clezardin P, Dechavanne M, Leung LLK: Rapid purification and partial characterization of human platelet glycoprotein IIIb - Interaction with thrombospondin and its role in platelet aggregation. J Biol Chem 264:501-506, 1989

91. Matsuno K, Diaz-Ricart M, Montgomery RR, Aster RH, Jamieson GA, Tandon NN: Inhibition of platelet adhesion to collagen by monoclonal anti-CD36 antibodies. Br J Haematol 92:960-967, 1996

92. Diaz-Ricart M, Tandon NN, Gomez Ortiz G, Carretero M, Escolar G, Ordinas A, Jamieson GA: Antibodies to CD36 (GPIV) inhibit platelet adhesion to subendothelial surfaces under flow conditions. Arterioscler Thromb Vasc Biol 16:883-888, 1996

93. Nakamura T, Jamieson GA, Okuma M, Kambayashi J, Tandon NN: Platelet adhesion to native type I collagen fibrils. Role of GPVI in divalent cation-dependent and -independent adhesion and hromboxane A2 generation. J Biol Chem 273:4338-4344, 1998

94. Tandon NN, Ockenhouse CF, Greco NJ, Jamieson GA: Adhesive functions of platelets lacking glycoprotein IV (CD36). Blood 78:2809-2813, 1991

95. Diaz-Ricart M, Tandon NN, Carretero M, Ordinas A, Bastida E, Jamieson GA: Platelets lacking functional CD36 (glycoprotein IV) show reduced adhesion to collagen in flowing whole blood. Blood 82:491-496, 1993

96. Saelman EUM, Kehrel B, Hese KM, de Groot PG, Sixma JJ, Nieuwenhuis HK: Platelet adhesion to collagen and endothelial cell matrix under flow conditions is not dependent on platelet glycoprotein IV. Blood 83:3240-3244, 1994

97. McKeown L, Vail M, Williams S, Kramer W, Hansmann K, Gralnick H: Platelet adhesion to collagen in individuals lacking glycoprotein IV. Blood 83:2866-2871, 1994

98. Kehrel B, Kronenberg A, Rauterberg J, Niesing-Bresch D, Niehues U, Kardoeus J, Schwippert B, Tschöpe D, van de Loo J, Clemetson KJ: Platelets deficient in glycoprotein IIIb aggregate normally to collagens type I and III but not to collagen type V. Blood 82:3364-3370, 1993

99. Yamamoto N, Akamatsu N, Yamazaki H, Tanoue K: Normal aggregations of glycoprotein IV (CD36)-deficient platelets from seven healthy Japanese donors. Br J Haematol 81:86-92, 1992

100. Daniel JL, Dangelmaier C, Strouse R, Smith JB: Collagen induces normal signal transduction in platelets deficient in CD36 (platelet glycoprotein IV). Thromb Haemost 71:353-356, 1994

101. Clemetson KJ: Platelet activation: signal transduction via membrane receptors. Thromb Haemost 74:111-116, 1995

102. Santoro SA, Zutter MM: The α2β1 integrin: a collagen receptor on platelets and other cells. Thromb Haemost 74:813-821, 1995

103. Kunicki TJ, Nugent DF, Staatz S, Orchekowski RP, Wayner EA, Carter WG: The human fibroblast class II extracellular matrix receptor mediates platelet adhesion to collagen and is identical to the platelet glycoprotein Ia-IIa complex. J Biol Chem 263:4516-4519, 1988

104. Nieuwenhuis HK, Sakariassen KJ, Houdyk WPM, Nievelstein PFEM, Sixma JJ: Deficiency of platelet membrane glycoprotein Ia is associated with a decreased platelet adhesion to subendothelium: A defect in platelet spreading. Blood 68:692-695, 1986

105. Kehrel B, Balleisen L, Kokott R, Mesters R, Stenzinger W, Clemetson KJ, van de Loo J: Deficiency of thrombospondin and membrane glycoprotein Ia in platelets with defective collagen-induced aggregation and spontaneous loss of disorder. Blood 71:1074-1078, 1988

106. Handa M, Watanabe K, Kawai Y, Kamata T, Koyama T, Nagai H, Ikeda Y: Platelet unresponsiveness to collagen: involvement of glycoprotein Ia-IIa (alpha 2 beta 1 integrin) deficiency associated with a myeloproliferative disorder. Thromb Haemost 73:521-528, 1995

107. Coller BS, Beer JH, Scudder LE, Steinberg MH: Collagen-platelet interactions: Evidence for a direct interaction of collagen with platelet GPIa/IIa and indirect interaction with platelet GPIIb/IIIa mediated by adhesive proteins. Blood 74:182-192, 1989

108. Kritzik M, Tarantino MD, Nugent DJ, Santoso S, Kunicki TJ: Nucleotide polymorphisms in the α2 gene define multiple alleles which are associated with differences in platelet α2β1. Blood 90a:27, 1997 (abstract)

109. Kunicki TJ, Orchekowski R, Annis D, Honda Y: Variability of integrin α2β1 activity on human platelets. Blood 82:2693-2703, 1993

110. Deckmyn H, Chews SL, Vermylen J: Lack of platelet response to collagen associated with an autoantibody against glycoprotein Ia: A novel cause of acquired platelet dysfunction. Thromb Haemost 64:74-79, 1990

111. Gralnick HR, McKeown LP, Williams SS, Lawrence JC, Shafer BC, Kramer WS, Hanemann KE, Vail MLI, Magruder LE: A murine monoclonal antibody that identifies a 157/130kDa platelet collagen receptor. Circulation 78:308, 1988 (abstract)

112. Kamiguti AS, Markland FS, Zhou Q, Laing GD, Theakston RD, Zuzel M: Proteolytic cleavage of the beta1 subunit of platelet α2β1 integrin by the metalloproteinase jararhagin compromises collagen-stimulated phosphorylation of pp72. J Biol Chem 272:32599-32605, 1997

113. Santoro SA: Identification of a 160 000 Dalton platelet membrane protein that mediates the initial divalent cation-dependent adhesion to collagen. Cell 46:913-920, 1986

114. Grab B, Miles AJ, Furcht LT, Fields GB: Promotion of fibroblast adhesion by triple-helical peptide models of type I collagen-derived sequences. J Biol Chem 271:12234-12240, 1996

115. Staatz WD, Walsh JJ, Santoro SA: The α2β1 integrin binds to the CB4 peptide of the α2(I) collagen chain. Biochem Mol Biol Int 42:577-582, 1997

116. Knight CG, Morton LF, Onley DJ, Peachey AR, Messent AJ, Smethurst PA, Tuckwell DS, Farndale RW, Barnes MJ: Identification in collagen type I in an integrin α2β1-binding site containing an essential GER sequence. J Biol Chem, submitted

117. Farndale RW, Barnes MJ: The platelet reactivity of synthetic peptides based on the collagen III fragment α1(III)CB4. Evidence for an integrin α2β1 recognition site involving residues 522-528 of the α1(III) collagen chain. J Biol Chem 272:11044-11048, 1997

118. Kirchhofer D, Languino LR, Ruoslahti E, Pierschbacher MD: α2β1 integrins from different cell types show different binding specificities. J Biol Chem 265:615-618, 1990

119. Bergelson JM, Shepley MP, Chan BM, Hemler ME, Finberg RW: Identification of the integrin VLA-2 as a receptor for echovirus. Science 255:1718-1720, 1992

120. Bergelson JM, Chan BM, Finberg RW, Hemler ME: The integrin VLA-2 binds echovirus 1 and extracellular matrix ligands by different mechanisms. J Clin Invest 92:232-239, 1993

121. Santoso S, Kalb R, Walka M, Kiefel V, Müller-Eckhardt C, Newman PJ: The human platelet alloantigens Br(a) and Brb are associated with a single amino acid polymorphism on glycoprotein Ia (integrin subunit α2). J Clin Invest 92:2427-2432, 1993

122. Takada Y, Hemler ME: The primary structure of the VLA-2 collagen receptor α2 subunit (platelet GPIa): homology to other integrins and the presence of a possible collagen-binding domain. J Cell Biol 109:397-407, 1989

123. Takada Y, Ylanne J, Mandelman D, Puzon W, Ginsberg MH: A point mutation of integrin β1 subunit blocks binding of α5β1 to fibronectin and invasin but not recruitment to adhesion plaques. J Cell Biol 119:913-921, 1992

124. Lee JO, Rieu P, Arnaout MA, Liddington R: Crystal structure of the A domain from the α subunit of integrin CR3 (CD11b/CD18). Cell 80:631-638, 1995

125. Kamata T, Takada Y: Direct binding of collagen to the I domain of integrin α2β1 (VLA-2, CD49b/CD29) in a divalent cation-dependent manner. J Biol Chem 269:26006-26010, 1994

126. Emsley J, King SL, Bergelson JM, Liddington RC: Crystal structure of the I domain from integrin α2β1. J Biol Chem 272:28512-28517, 1997

127. Barnes MJ, Knight CG, Farndale RW: The collagen-platelet interaction. Curr Opin Hematol, 1998 (in press)

128. Depraetere H, Wille C, Gansemans Y, Stanssens P, Lauwereys M, Baruch D, De Reys S, Deckmyn H: The integrin α2β1 (GPIa/IIa)-I-domain inhibits platelet-collagen interaction. Thromb Haemost 77:981-985, 1997

129. Jung SM, Moroi M: Platelets interact with soluble and insoluble collagens through characteristically different reactions. J Biol Chem 273:14827-14837, 1998

130. Gofer-Dadosh N, Klepfish A, Schmilowitz H, Shaklai M, Lahav J: Affinity modulation in platelet α2β1 following ligand binding. Biochem Biophys Res Commun 232:724-727, 1997

131. Sugiyama T, Okuma M, Ushikubi F, Sensaki S, Kanaji K, Uchino H: A novel platelet aggregating factor found in a patient with defective collagen-induced platelet aggregation and auto-immune thrombocytopenia. Blood 69:1712-1720, 1987

132. Arai M, Yamamoto N, Moroi M, Akamatsu N, Fukutake K, Tanoue K: Platelets with 10% of the normal amount of glycoprotein VI have an impaired response to collagen that results in a mild bleeding tendency. Br J Haematol 89:124-130, 1995

133. Ryo R, Yoshida A, Sugano W, Yasunaga M, Nakayama K, Saigo K, Adachi M, Yamaguchi N, Okuma M: Deficiency of P62 a putative collagen receptor in platelets from a patient with defective collagen-induced platelet aggregation. Am J Hematol 39:25-31, 1992

134. Moroi M, Okuma M, Jung SM: Platelet adhesion to collagen-coated wells: analysis of this complex process and a comparison with the adhesion to matrigel-coated wells. Biochim Biophys Acta 1137:1-9, 1992

135. Takahashi H, Hanano M, Moroi M, Shibata A: Platelet membrane glycoprotein VI (GPVI) is necessary for collagen-induced aggregation and adhesion and anti GPVI antibody induces platelet aggregation: An evidence obtained from a patient with systemic lupus erythematosus. Thromb Haemost 73:1197, 1995 (abstract)

136. Morton LF, Peachey AR, Zijenah LS, Goodall AH, Humphries MJ, Barnes MJ: Conformation-dependent platelet adhesion to collagen involving integrin α2β1-mediated and other mechanisms: multiple α2β1-recognition sites in collagen type I. Biochem J 299:791-797, 1994

137. Barnes MJ, Knight CG, Farndale RW: The use of collagen-based model peptides to investigate platelet-reactive sequences in collagen. Biopolymers 40:383-397, 1996

138. Knight CG, Morton LF, Onley DJ, Peachey AR, Ichinohe T, Okuma M, Farndale RW, Barnes MJ: Collagen-platelet interaction: Gly-Pro-Hyp is uniquely specific for platelet gpVI and mediates platelet activation by collagen. Cardiovasc Res (submitted)

139. Polgar J, Clemetson JM, Kehrel BE, Wiedemann M, Magnenat E, Wells TNC, Clemetson KJ: Platelet activation and signal transduction by convulxin, a C-type lectin from crotalus durissus terrificus venom via the p62/GPVI collagen receptor, phosphorylation of p72syk and Fcγ. J Biol Chem 272:13576-13583, 1997

140. Jandrot-Perrus M, Lagrue A-H, Okuma M, Bon C: Adhesion and activation of human platelets induced by convulxin involve glycoprotein VI and integrin α2β1. J Biol Chem 272:27035-27041,

141. Tsuji M, Ezumi Y, Arai M, Takayama H: A novel association of Fc receptor γ-chain with glycoprotein VI and their co-expression as a collagen receptor in human platelets. J Biol Chem 272:23528-23531, 1997

142. Offermanns S, Toombs CF, Hu YH, Simon MI: Defective platelet activation in G α(q)-deficient mice. Nature 389:183-186, 1997

143. Smith JB, Selak MA, Dangelmaier C, Daniel JL: Cytosolic calcium as a second messenger for collagen-induced platelet responses. Biochem J 288:925-929, 1992

144. Watson SP, Gibbins J: Collagen receptor signalling in platelets: extending the role of the ITAM. Immunol Today 19:260-264, 1998

145. Blake RA, Schieven GL, Watson SP: Collagen stimulates tyrosine phosphorylation of phospholipase Cγ2 but not phospholipase C-γ1 in human platelets. FEBS Lett 353:212-216, 1994

146. Yanaga F, Poole A, Asselin J, Blake R, Schieven G, Clark EA, Law CL, Watson SP: Syk interacts with tyrosine phosphorylated proteins in human platelets activated by collagen and crosslinking of the Fcγ-IIA receptor. Biochem J 311:471-478, 1995

147. Gibbins J, Asselin J, Farndale R, Barnes M, Law CL, Watson SP: Tyrosine phosphorylation of the Fc receptor γ-chain in collagen-stimulated platelets.. J Biol Chem 271:18095-18099, 1996

148. Poole A, Gibbins JM, Turner M, van Vugt MJ, van de Winkel JGJ, Saito T, Tybulewicz VLJ, Watson SP: The Fc receptor γ-chain and the tyrosine kinase Syk are essential for activation of mouse platelets by collagen. EMBO J 16:2333-2341, 1997

149. Achison M, Joel C, Hargreaves PG, Sage SO, Barnes MJ, Farndale RW: Signal elicited from human platelets by synthetic, triple helical, collagen-like peptides. Blood Coagul Fibrinolysis 7:149-152, 1996

150. Asselin J, Gibbins JM, Achison M, Lee YH, Morton LF, Farndale RW, Barnes MJ, Watson SP: A collagen-like peptide stimulates tyrosine phosphorylation of syk and phospholipase Cγ2 in platelets independent of the integrin α2β1. Blood 89:1235-1242, 1997

151 Ichinohe T, Takayama H, Ezumi Y, Yanagi S, Yamamura H, Okuma M: Cyclic AMP-insensitive activation of c-Src and Syk protein-tyrosine kinases through platelet membrane glycoprotein VI. J Biol Chem 270:28029-28036, 1995

152. Keely PJ, Parise LV: The α2β 1 integrin is a necessary co-receptor for collagen-induced activation of syk and the subsequent phosphorylation of phospholipase Cγ 2 in platelets. J Biol Chem 271:26668-26676, 1996

153 Ichinohe T, Takayama H, Ezumi Y, Arai M, Yamamoto N, Takahashi H, Okuma M: Collagen-stimulated activation of Syk but not c-Src is severely compromised in human platelets lacking membrane glycoprotein VI. J Biol Chem 272:63-68, 1997

5

ADENOSINE RECEPTORS: BIOLOGY AND FUNCTION

Kailash C. Agarwal, Ph.D.
Department of Molecular Pharmacology,
Physiology and Biotechnology
Brown University
Providence, RI 02912, USA

5.1. INTRODUCTION

Adenosine, a natural purine metabolite of adenine nucleotides, is a key regulator of many physiologic functions including vascular blood flow, platelet thrombotic activity, lipolysis and neurotransmission (1-3). It is continuously produced by many tissues of the body which include the vascular endothelium (4,5), heart (6,7), brain (6), skeleton muscle (6) and platelets (8). During the period of ischemia the adenosine concentrations in many tissues increase by several fold (5,6,9). Adenosine concentrations may also increase considerably in tissues under conditions of high energy demands (6,9). The common pathway for production of adenosine involves ATP Y ADP Y AMP Y Adenosine. Although, in cardiac tissue, the production of adenosine from *S*-adenosylhomocysteine (*SAH*, a product of *S*-adenosyl-methionine-dependent

Kailash C. Agarwal, Ph.D., Professor of Molecular and Biochemical Pharmacology. During the past 30 years at Brown University, he has contributed extensively in the development of antithrombotic and cardiovascular drugs both from synthetic sources and medicinal plants. Dr. Agarwal has given many invited lectures both nationally and internationally in his field of research, and he is an active member of several national and international scientific societies.

effects by binding to specific receptors coupled to membrane-bound adenylate cyclase, an enzyme that regulates intracellular cAMP levels. Depending on the cell or tissue type, adenosine may inhibit or stimulate adenylate cyclase thus decreasing or increasing intracellular cAMP levels (2,3). The adenosine A_1 receptor interacts with the inhibitory G_i protein and inhibits adenylate cyclase, whereas the$_2$A receptor interacts with the stimulatory G_s protein and stimulates the enzyme (10). Adenosine via its two receptors plays a key role in maintaining oxygen supply/demand balance: A_1 responses normally bring about a reduction in oxygen demand (e.g. reductions in the heart rate and contractility, inhibition of locomotion activity, hypothermia, lipolysis inhibition, platelet activation), whereas A_2 responses increase oxygen supply (e.g. vasodilatation, inhibition of platelet aggregation) (10-12). During the past two decades, many agonists and antagonists selective to A_1 and A_2 receptors have been chemically synthesized and developed as potential drugs.

5.2 Adenosine receptors

Adenosine released from the vascular endothelium and platelets is an important endogenous regulator of platelet function by acting via its specific adenosine receptors. Platelet adenosine receptors have been characterized with the help of adenosine and its analogs in studies on radioligand binding to platelet membrane and adenylate cyclase activity. 2-Choloroadenosine, N^6-phenylisopropyladenosine and 5=-*N*-ethylcarboxamidoadenosine (NECA), which strongly inhibit platelet aggregation by stimulating adenylate cyclase, were employed for adenosine A_2 receptor ligand binding studies (13,14). NECA is 5-10 fold more potent than adenosine in inhibiting platelet aggregation and stimulating membrane adenylate cyclase (15). It has been suggested that an intramolecular hydrogen bond between purine *N-3* and *C-5=* amide nitrogen of NECA may stabilize the nucleoside in a *syn* conformation and, therefore, favors binding. The binding of NECA is rapid and highest at 0°C but reversible with a $t_{1/2}$ of about 20 s (13). Two different binding sites for NECA have been identified with K_d of 0.16 and 2.9 μM, and B_{max} of 8.4 and 33.4 pmoles mg^{-1} protein (13). These studies have also demonstrated that the NECA binding sites are different from the adenosine-transport or ADP-receptor sites since dipyridamole, a potent adenosine transport inhibitor, or ADP did not affect NECA binding to human platelet membranes. Another 5=-*N*-carboxamide derivative, 5=-*N*-cyclopropylcarboxamido-adenosine (CPCA) is about 2-fold more potent than NECA in inhibiting platelet aggregation and stimulating adenylate cyclase (14). CPCA is also an excellent agonist for adenosine A_2 receptor binding studies. A newly synthesized A_2 selective agonist, 2-(*p*-(2-carboxyethyl)phenethylamino-5=-*N*-ethyl-carboxamidoadenosine (CGS 21680) has been employed for its binding characteristics on human platelet membranes (16). The saturation studies reveals a single class of binding sites with K_d and B_{max} of 1.4 μM and 5.9 pmoles mg^{-1} protein respectively. CGS 21680 is a potent inhibitor of platelet aggregation in human and rabbit blood (17). An adenosine binding protein, adenotin has been purified from human platelet membranes (18). NECA binds adenotin with a K_d of 155 nM and B_{max} of 1.85 ± 0.10 nmole mg^{-1} protein (18), although the affinity of adenosine for adenotin is relatively low (80 times lower than for NECA).

The adenosine A_2 receptors have been further characterized for high and low affinity subtypes, A_{2a} and A_{2b}, on the basis of biochemical and molecular cloning technics (19-21). The A_{2a} receptor, which is coupled to stimulation of adenylate cyclase, is a high affinity receptor found in the brain striatum, platelets, and vascular smooth muscle cells (19,20). The analog, CGS 21680 was examined to study binding characteristics with the purified platelet membrane preparation enriched with adenosine receptors (22). The findings revealed a single class of binding site with K_d and B_{max} values of 285 nM and 2.07 pmoles mg^{-1} protein respectively (22). Adenosine antagonists have also been employed to characterize platelet adenosine receptor. Recently, a new nonxanthine selective adenosine A_{2a} antagonist, 5-amo-7-(2-phenyl-ethyl)-2-(2-furyl)-pyrazolo(4,3-e)-1,2,4-triazolo(1,5-c)pyrimidine (SCH 58261) has been examined revealing a single class of recognition sites with high affinity, K_d of 0.85 nM and B_{max} of 85 fmoles mg^{-1} protein (23). The binding data are in good agreement with the results from functional studies such as A_{2a} agonist-induced stimulation of adenylate cyclase or platelet aggregation inhibition (23).

5.3 Antiplatelet action of adenosine

Adenosine was first identified as an inhibitor of platelet aggregation in 1962 by Born et al (24). Subsequently, many other investigators examined antiplatelet activity of adenosine and its derivatives (25-29). Most of these investigators examined adenosine antiaggregatory effects by adding exogenous adenosine (2-40 μM) to platelet-rich plasma (PRP). Optimal inhibition requires incubation of adenosine with PRP for 5-10 m prior to addition of an aggregating agent (30). Since adenosine can be rapidly metabolized by cellular adenosine deaminase and adenosine kinase, preincubation for a longer period of time may lead to loss of adenosine inhibitory effect (30). In whole blood, the extracellularly added adenosine rapidly disappears primarily due to erythrocytic uptake and metabolism, and therefore no significant inhibition by adenosine (10-50 μM) is seen on platelet aggregation (31). Experiments employing [^{14}C]adenosine (10 μM) show that it disappears with a $t_{1/2}$ of <15 s from the extracellular medium in human whole blood (32). In the presence of dipyridamole, the $t_{1/2}$ for adenosine is increased to about 5 m (32). Using a nucleoside transport inhibitor such as dipyridamole, dilazep, or p-nitrobenzyl-thioinosine, adenosine is strongly inhibitory (IC_{50}, 2 μM) of platelet aggregation in human whole blood (30,31). Also, if adenosine metabolism is blocked by the inhibitors of adenosine deaminase (e.g. 2=-deoxycoformicin) and adenosine kinase (e.g. 5-iodotubercidin), adenosine acts as a potent inhibitor (IC_{50}, 1.5 μM) of platelet aggregation in whole blood (31). Acadesine, an adenosine analog, has been examined in both *in vitro* and *in vivo* systems for its effects on platelet aggregation (33). Interestingly, acadesine which shows no effect on ADP-induced platelet aggregation or intracellular platelet cAMP levels in human PRP, causes inhibition of platelet aggregation in whole blood with 50% inhibition at 240 ± 60 μM (33). These investigators have reported that acadesine is first phosphorylated to acadesine monophosphate (ZMP) by erythrocytic adenosine kinase. ZMP blocks adenosine metabolism by inhibiting adenosine kinase causing an increase in extracellular adenosine concentrations (33). Acadesine is also shown to be inhibitory of adenosine deaminase (ADA) (33) which may further elevate extracellular adenosine levels. Other studies have previously shown that adenosine, in presence of 5-iodotubercidin 2=-deoxycoformycin, the inhibitors of adenosine kinase and ADA respectively, produces strong inhibition of platelet aggregation in human whole blood (8,31). In addition, the antiplatelet actions of adenosine can be potentiated by calcium channel blockers (e.g. nifedipine, verapamil and diltiazem (34).

5.3.1 Species differences

Wide variations in response to aggregating agents and their inhibition by adenosine have been reported in blood samples of various species (28,35,36). Adenosine (10-100 μM) inhibits ADP-induced platelet aggregation in PRP samples from human, rabbit, dog (37,38) and sheep (39), whereas adenosine is non-inhibitory of platelet aggregation in PRP from guinea-pig, horse, rat or cat blood (37,38). In mouse PRP, adenosine (10-1000 μM) inhibits only 33% of submaximal ADP (1 μM)-induced platelet aggregation (40). The data from these studies are summarized in Table 1.

Table 1. Species differences in the inhibitory effect of adenosine on ADP-induced platelet aggregation in PRP samples

Species (μM)	*Adenosine Aggregation (%)*	*Inhibition of*	*Ref. #*
Human	10, 100	85-100	28,37
Rabbit	10, 80	50-85	37,38
Dog	100	70	37
Sheep	1.2	<50	39
Mouse	10-1000	33	40
Guinea-pig	80, 500	<5	37
Horse	500	<5	37
Rat	80, 500	<5	37,43
Cat	80	<5	38

The mechanism for non-inhibitory action of adenosine in guinea-pig, horse, cat and rat blood is not well understood, although their platelets aggregate normally with ADP (28,35,36,41). Adenosine-mediated increases in intracellular cAMP levels were similar in rat and human platelets, however adenosine shows no inhibition of platelet aggregation in rat PRP (42,43). It has been suggested that the difference in the inhibitory effectiveness of adenosine on rat and human platelets is quantitative rather than qualitative and apparently depended on the inability of adenosine to increase intracellular cAMP sufficiently in rat platelets (42).

Interestingly, human platelets resuspended in rat plasma retain adenosine inhibition of ADP-induced aggregation, whereas rat platelets resuspended in human plasma behave like rat PRP, suggesting that the non-inhibitory effect of adenosine in rat PRP is an intrinsic property of the platelets and is not due to extrinsic plasma factors (43). Rat plasma contains much higher adenosine levels (586 ± 124 nM) as compared to human plasma (230 ± 90 nM) (8). ADP causes only reversible (primary phase) platelet aggregation in rat PRP. Perhaps the presence of high levels of adenosine in the plasma plays a role in desensitizing the rat platelets to granular release by ADP and thus producing only primary phase of aggregation. This suggests that adenosine is more inhibitory of the secondary phase of aggregation than the primary phase of aggregation. Also, 5-10 fold greater amounts of collagen are required to induce platelet aggregation in rat PRP as compared to human PRP, perhaps due to higher levels of plasma adenosine (44,45).

5.4 Adenosine production and platelet inactivation

Human platelets contain large amounts of adenine nucleotides (9-11 mM) distributed between the metabolic (cytoplasmic) and storage (dense granules) pools (26,30). Aggregating agents such as collagen and thrombin cause the release of the storage pool nucleotides which consist of about 60% of the total adenine nucleotides (30,46). The experiments of Figure 1 demonstrate that platelets continuously produce adenosine in human PRP, causing large increases in plasma adenosine levels in the presence of the inhibitors of adenosine deaminase (ADA), 2=-deoxycoformycin (dCF) and nucleoside transport system, dilazep. Plasma adenosine levels of 1806 ± 702 nM are seen after 3 h of PRP storage at room temperature (22° C), demonstrating that adenosine is formed at the rate of about 500 pmoles/h/ml. This suggests that the platelet storage and/or metabolic pools of adenine nucleotides contribute in the adenosine production. To examine whether the release of adenosine is from the metabolic and/or storage pools,

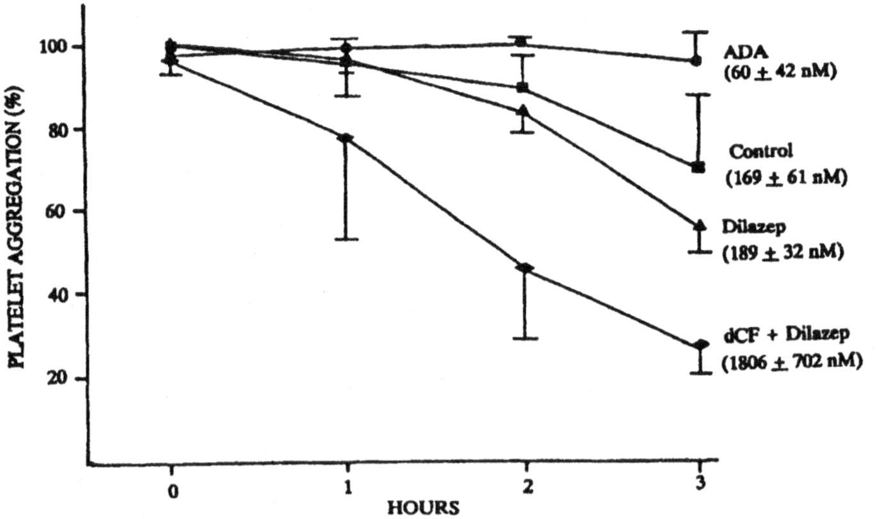

Figure 1. Adenosine production in human PRP and platelet inactivation. Fresh human PRP was incubated at 22° C with NaCl (0.9%) (Control); ADA (2 units/ml); dilazep (2 µM); or dCF (5 µM) + dilazep (2 µM). After incubation (0, 1, 2 or 3 h), ADP (4 µM) was added to induce platelet aggregation. Plasma adenosine levels were determined by HPLC with fluorescent detector. (Reproduced with permission: publication of Agarwal, KC; Ref. 8).

the platelet pools were labeled with [^{14}C]guanosine and [^{14}C]5-HT. The PRP containing the labeled platelet metabolic pool after 3 h of incubation caused a slight release (1%) of [^{14}C] radioactivity as guanosine and its metabolites, whereas the released radioactivity of [^{14}C]5-HT from the storage pool was about 8% (Figure 2). In presence of indomethacin, the granular release of 5-HT was decreased to 4% demonstrating that indomethacin blocks the 5-HT release only by 50%. Similarly, indomethacin was only partially inhibitory (25-45%) of adenosine production. In contrast, indomethacin inhibits 100% 5-HT release induced by ADP (8). These findings suggest that the 5-HT release and adenosine production occur from another storage pool which is not blocked by indomethacin. No other studies have been carried out to support these findings.

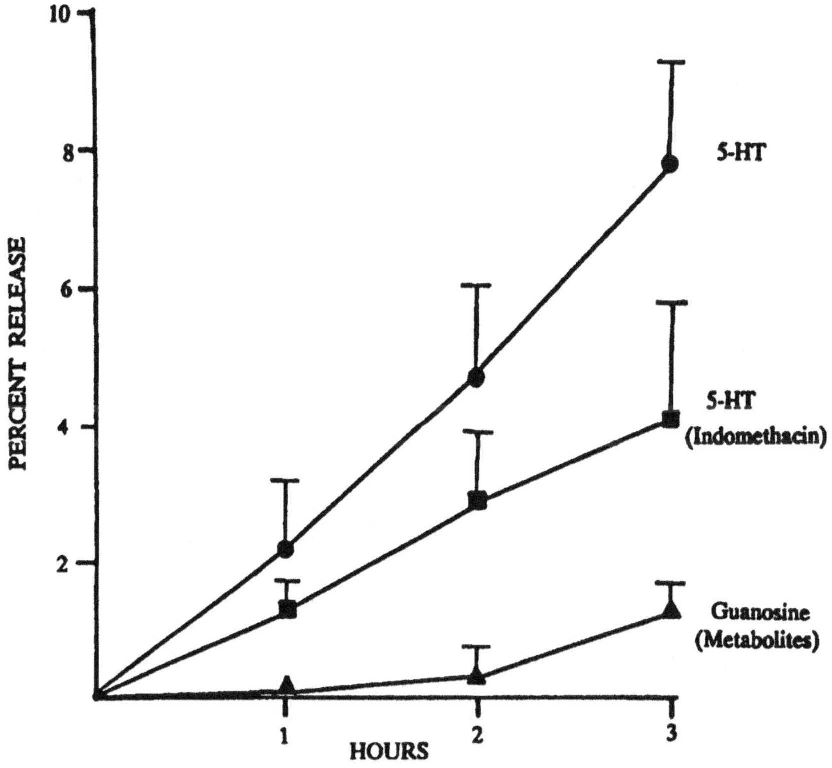

Figure 2. Release of 5-HT and guanosine (plus metabolites) from [^{14}C]5-HT and [^{14}C]-guanosine labeled storage and metabolic pools of platelets respectively in human PRP. (Reproduced with permission: publication of Agarwal, KC; Ref. 8).

The studies of Figure 1 further demonstrate that the increase in plasma adenosine levels during the PRP incubation causes platelets to poorly respond (30%) to aggregation. In contrast, if ADA is added to PRP prior to its incubation, ADP-induced aggregation remains excellent (>95%) after 3 h, demonstrating that adenosine produced from platelets contribute significantly in the inactivation of platelet function. Furthermore, another study has shown that platelet-derived adenosine protects isolated perfused rat heart from injury and dysfunction after ischemia and reperfusion in isolated rat heart (47).

Table 2. Role of plasma adenosine (Ado) in platelet aggregation and 5-HT release induced by ADP, epinephrine, PAF, or vasopressin.

Agonists	*EC_{50} values (µM)* Platelet aggregation (%)	*5-HT release*	*Ref. #*
ADP			
Ado-depleted PRP	1.7	2.5	(8)
Normal PRP	3.2	12	
PAF			
Ado-depleted PRP	0.07	0.07	(50)
Normal PRP	>1.2	>1.2	
Vasopressin*			
Ado-depleted PRP	0.006	0.007	(49)
Normal PRP	0.058	0.077	

Ado-depleted PRP: PRP was treated with adenosine deaminase (ADA, 2 units/ml) to degrade plasma adenosine.

Normal PRP: Blood was collected in presence of 2=-deoxycoformycin (adenosine deaminase inhibitor) and dilazep (adenosine transport inhibitor) to maintain steady-state levels of plasma adenosine.

*Blood from female donors.

Normally human plasma contains low levels of adenosine, 100-300 nM (48,49). The role of plasma adenosine in platelet aggregating and 5-HT release activities of several agonists has been recently examined. The data in Table 2 compare the platelet aggregating and 5-HT release activities in adenosine-depleted PRP and PRP with normal adenosine levels. The EC_{50} values for various agonists of platelet aggregation are much lower in adenosine-depleted PRP than the PRP with normal adenosine levels. PAF- or vasopressin-induced effects on platelet aggregation and 5-HT release are most affected after depletion of plasma adenosine causing >10-fold decrease in the EC_{50} values. Gender differences in the platelet aggregating activity of vasopressin have been reported (49). Platelets from most female subjects respond to platelet aggregation and 5-HT release at much lower levels of vasopressin as compared to most male subjects (49). The mechanism of action underlying these differences are not known. One may speculate that it may be due to differences in the platelet vasopressin receptor density and/or vasopressin-induced intracellular Ca^{2+} mobilization.

5.5 Agents affecting adenosine actions

5.5.1. Forskolin and cAMP PDE inhibitors

Studies from this laboratory have previously shown that plasma adenosine contributes greatly in the antiplatelet activities of forskolin (45) and the inhibitors of cAMP PDE (44,51,52). Although adenosine is non-inhibitory of platelet aggregation in rat PRP, it plays an important role in the antiplatelet effects of cAMP PDE inhibitors and forskolin (44,52). The data in Table 3 compare the antiplatelet effects of forskolin and HL-725, an extremely potent inhibitor of cAMP PDE (IC_{50}, 2.5 x 10^{-10} M) (53) in adenosine-depleted PRP, and PRP with normal adenosine levels of human and rat blood. Both forskolin and HL-725 were much more inhibitory in rat PRP than in human PRP, perhaps due to higher levels of adenosine in the rat plasma (586 $\pm$124 nM) than in the human plasma (230 $\pm$ 90 nM). The IC_{50} values for forskolin and HL-725 to inhibit collagen-induced platelet aggregation were lower (2-4 fold) in PRP with normal adenosine levels than in adenosine-depleted PRP. Furthermore, both forskolin and HL-725 produced 10-20 fold more inhibition of platelet aggregation in human whole blood (44,45). These studies demonstrate that plasma adenosine plays a key role in the antiplatelet actions of forskolin and cAMP PDE inhibitors. Furthermore, the antiplatelet effects can be greatly potentiated if combined with agents that cause an increase in plasma adenosine levels (e.g. dilazep and dipyridamole).

Table 3: Antiplatelet activity of forskolin and HL-725 in human and rat PRP

Agents	*Adenosine Deaminase*	*IC_{50} (mean $\pm$ SD, μM)* *Human PRP*	*Rat PRP*
Forskolin[a]	-	2.4 $\pm$ 1.2	0.6 $\pm$ 0.2
	+	4.7 $\pm$ 1.3	2.5 $\pm$ 1.1
HL-725[b]	-	0.09 $\pm$ 0.03	0.05 $\pm$ 0.01
	+	0.26 $\pm$ 0.08	0.18 $\pm$ 0.03

Data from publications of Agarwal et al., [a] *Ref.* 45; and [b] *Ref.* 44.

The data of Figure 3 further demonstrate the role of adenosine in the antiplatelet effect of forskolin. Forskolin is weakly inhibitory (IC_{50}, 4.2 μM) of ADP-induced platelet aggregation in adenosine-depleted PRP. However, if the PRP is replenished with adenosine (50 or 300 nM), forskolin becomes much more inhibitory with IC_{50} of 0.23 and 0.53 μM respectively. These findings suggest that forskolin may prove to be much more effective to block platelet thrombotic activity in the vascular system where adenosine is continuously produced by the endothelial cells.

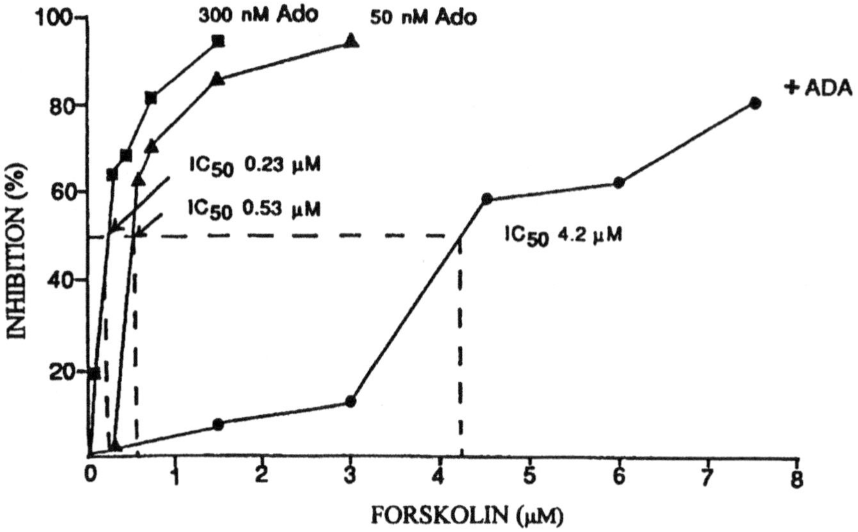

Figure 3. Antiplatelet effects of forskolin in adenosine-depleted (ADA-treated) PRP, and PRP replenished with adenosine (50 and 300 nM). Before replenishing the ADA-treated PRP with adenosine, the PRP was treated with the inhibitors of ADA (2=-deoxycoformycin, 5 μM) and adenosine uptake (dilazep, 2 μM). Platelet aggregation was induced by ADP (4 μM). (Reproduced with permission: publication of Agarwal et al., Ref. 45).

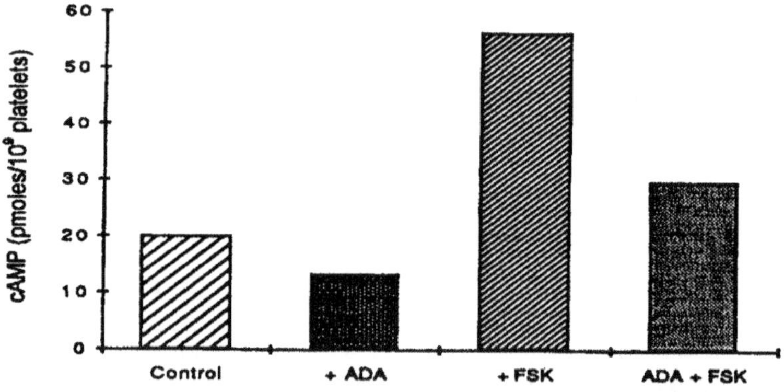

Figure 4. Effect of forskolin (3 μM) on platelet cAMP levels in adenosine-depleted PRP, and PRP with normal adenosine levels. (Reproduced with permission: publication of Agarwal, KC; Ref. 8).

The role of plasma adenosine in the forskolin-stimulating activity on platelet cAMP has been studied (8) (Figure 4). Platelet cAMP levels decrease by about 30% after depletion of plasma adenosine. Forskolin (3 μM) induces a 3-fold increase in platelet cAMP in PRP containing normal levels of adenosine, whereas only a 1.5 fold increase in platelet cAMP levels is seen in adenosine-depleted PRP. Similar effects of plasma adenosine have been reported in potentiating the actions of HL-725 on platelet cAMP levels (8). These studies demonstrate that the synergistic antiplatelet effects of forskolin or HL-725 with plasma adenosine are due to their potentiated effects on platelet cAMP levels.

5.5.2. Dipyridamole

Dipyridamole (persantine), a vasodilatory and antiplatelet agent, has been widely used clinically for many years (54). Dipyridamole is a weak inhibitor of cAMP PDE (K_i, 20 μM) (55), whereas, it is a potent inhibitor of the nucleoside transport system in a variety of cells (K_i, 0.05 μM; 400-fold lower compared to the K_i for cAMP PDE) (56). Dipyridamole binds extensively (>99%) to plasma proteins (alpha-acid glycoprotein and albumin) (57). Peak therapeutic levels of dipyridamole are usually in the range of 1-4 μM (total, bound plus free) (58). Therefore, it appears that clinically relevant blood levels are too low to inhibit cAMP PDE. Since dipyridamole is a potent nucleoside transport inhibitor, it causes an elevation in plasma adenosine levels (59). Adenosine, which is produced by many tissues in the body (4-8), is rapidly transported into cells (erythrocytes and other cells) and metabolized primarily by adenosine deaminase and adenosine kinase (30,31), resulting in low levels of plasma adenosine (100 - 300 nM) (8,48). Studies from this laboratory have previously shown that on addition of adenosine (200 nM or 10 μM) to human whole blood, it disappears with a $t_{1/2}$ of <15 s (31,51). However, in the presence of dipyridamole (10 μM) the $t_{1/2}$ of adenosine is increased to about 5 m (31). Clinically dipyridamole is shown to elevate plasma adenosine levels from 0.15 μM to 0.29 μM in 30 m after its administration (0.4 mg/kg i.v.) (59). Platelet aggregation is inhibited in whole blood samples collected 2 h after an oral intake of dipyridamole (200 mg) (60). Furthermore, dipyridamole is shown to inhibit platelet aggregation in whole blood by blocking the re-uptake of adenosine formed as a result of erythrocytic microtrauma, and the inhibitory effect of dipyridamole can be reversed by adenosine deaminase or adenosine receptor antagonists, theophylline and 5=-methythioadenosine (60). Similarly, studies from our laboratory have shown that adenosine, which is non-inhibitory in human whole blood, strongly inhibits platelet aggregation in the presence of dipyridamole (10 μM) with IC_{50} of 1.7 μM (51). Vasopressin-induced platelet aggregation in whole blood can be strongly inhibited if the blood is collected in presence of dipyridamole (10 μM) (61). Dipyridamole is shown to inhibit platelet-derived growth factor (PDGF) release from platelet alpha granules due to its inhibitory effect on the nucleoside transport system (62). Dipyridamole is also shown to inhibit cGMP PDE (IC_{50}, 0.3 μM) (63) and potentiates nitric oxide (NO)-mediated inhibition of platelet aggregation (64).

3.5.3 Methylxanthines (Theophyline and caffeine)

The methylxanthines, theophylline and caffeine, act primarily as antagonists for both adenosine A_1 and A_2 receptors (65-68). Caffeine is a relatively weak adenosine antagonist as compared to theophylline (69). Studies to examine these methylxanthines as adenosine antagonists on platelet function have been limited. Both theophylline and caffeine are shown to stimulate platelet reactivity *in vitro* and *in vivo* studies (49,50,67,69,70). Platelets from subjects after chronic consumption of caffeine show decreased antiplatelet activity of the adenosine analog NECA on thrombin-induced platelet aggregation, which may be caused by up-regulation of the adenosine receptors (67). A clinical study from our laboratory employing 22 stable outpatients, 12 patients who were receiving chronic treatment of theophylline (250-750 mg/day) with plasma theophylline levels of 9.2 ± 1.1 μg/ml, showed significant greater responses of PAF on platelet aggregation in comparison with the other 10 patients of the control-group (50).

As discussed in the above studies (Table 2), plasma adenosine is strongly inhibitory of platelet aggregation induced by vasopressin or PAF. However, if the blood contains therapeutic levels of theophylline, the aggregating responses of both vasopressin and PAF are greatly potentiated (49,50). As seen in Figure 5, theophylline and 3,7-dimethyl-1-propargylxanthine (DMPX), a relatively selective adenosine A_2 receptor antagonist, strongly potentiate vasopressin-induced platelet aggregation in PRP of blood samples containing normal adenosine levels (blood collected in presence of dCF and dilazep to prevent adenosine metabolism). Vasopressin response was also examined in healthy female subjects before and one hour after oral administration of theophylline (300 mg) (Figure 6). Vasopressin induces only reversible platelet aggregation in blood samples collected before theophylline administration, whereas vasopressin causes potentiated platelet aggregation after theophylline ingestion. These findings suggest that such methyl-xanthines ingestion may produce greater platelet thrombotic activity, which may further induce complications of coronary artery disease.

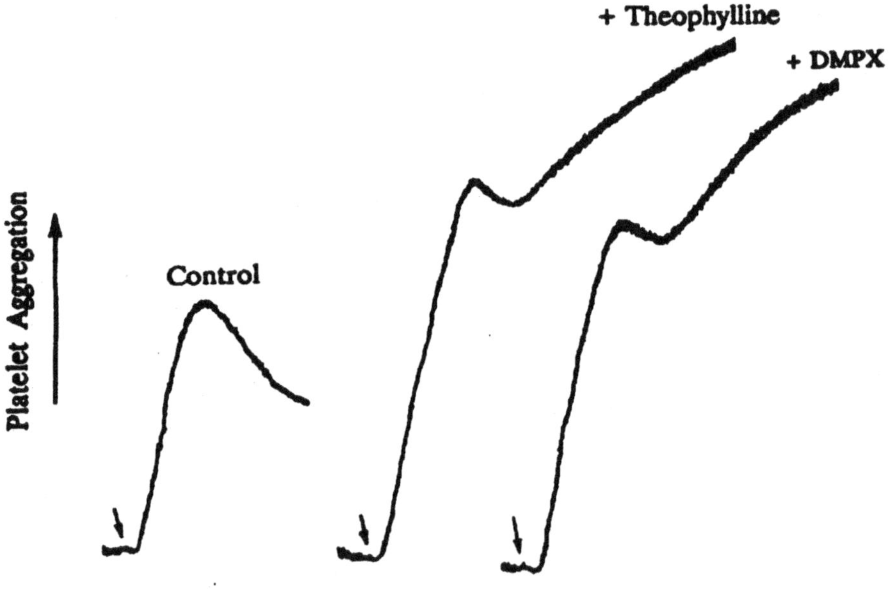

Figure 5. Effects of adenosine antagonists (theophylline, 50 μM and 3,7-dimethyl-1-propargylxanthine, DMPX, 20 μM) on vasopressin-induced platelet aggregation in PRP with normal amounts of adenosine (blood collected in presence of dCF and dilazep from a female donor). (Reproduced with permission: poblication of Agarwal, KC; Ref. 49).

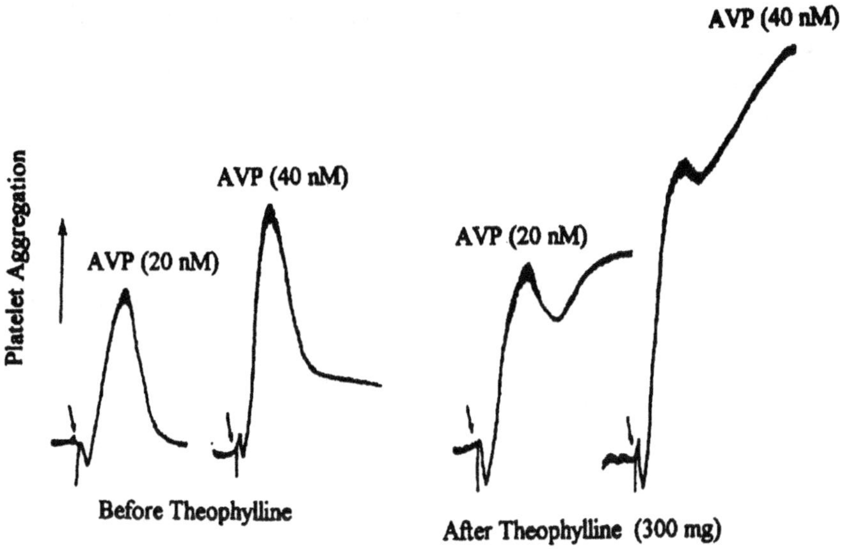

Figure 6. Potentiation of vasopressin-induced platelet aggregation after theophylline administration. Blood was collected from a female donor before and after 1 h of theophylline oral administration (300 mg)

5.6 Adenosine effects on intracellular Ca^{2+} mobilization

Platelet agonists induce an increase in intracellular free Ca^{2+} comprising 10-20% mobilization from intracellular Ca^{2+} stores and 80-90% from influx of extracellular Ca^{2+} (71,72). Adenosine is shown to inhibit thrombin-induced Ca^{2+} influx and the release of Ca^{2+} from intracellular stores in human platelets with an IC_{50} of about 0.5 μM (73). Both caffeine and the inhibitor of adenylate cyclase, 2=,5=-didoexy-adenosine antagonize the inhibitory effect of adenosine on intracellular Ca^{2+} mobilization (73). In contrast, the inhibitory effect of adenosine on intracellular Ca^{2+} mobilization is potentiated by the cAMP PDE inhibitor, papaverine (73). Recently studies from our laboratory reported that adenosine acts as a potent inhibitor of PAF-induced Ca^{2+} mobilization in human platelets with an IC_{50} value of 0.3 μM, demonstrating that adenosine is a key modulator of PAF-induced Ca^{2+} mobilization, a process that is essential for platelet aggregation (49,50,71,72).

Adenosine effects on vasopressin-induced intracellular Ca^{2+} mobilization in human platelets have been examined (Figure 7). Washed platelet suspension (1.5×10^8 platelets/ml) in Tyrode buffer (pH 7.4) containing $MgCl_2$ (1 mM), $CaCl_2$ (2 mM), and Fura 2 (5 μM), was employed for Ca^{2+} measurements. The platelet suspension was treated with ADA (2 units/ml) to degrade any adenosine that may be present. Vasopressin causes a rapid elevation in Ca^{2+} levels (250% in 15 s), followed by a decrease in the Ca^{2+} (25% after 120 s). The platelet suspension was then treated with 2=-deoxycoformycin (5 μM) and dilazep (2 μM) to block the ADA activity and adenosine uptake by platelets respectively. Various amounts of adenosine (0.1 - 5.0 μM) were added to the platelet suspension and the effects of vasopressin-induced on Ca^{2+} mobilization were measured. The data show strong inhibition of Ca^{2+} mobilization (IC_{50}, 0.5 μM). These studies suggest that plasma adenosine acts as a calcium antagonist blocking the vasopressin-induced intracellular Ca^{2+} mobilization, a process that is essential for vasopressin V_1 receptor-induced platelet aggregation.

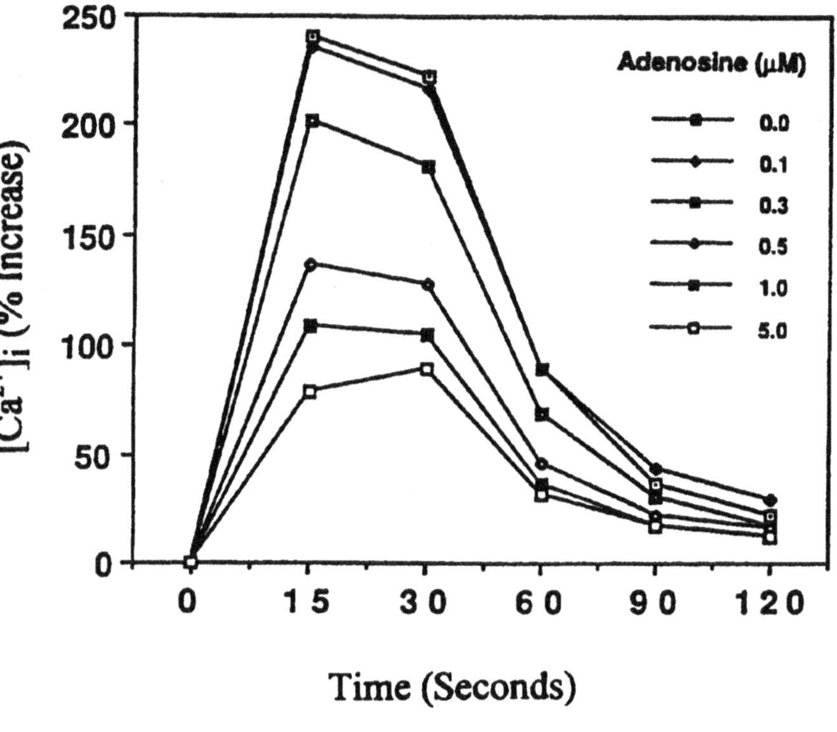

Figure 7. Adenosine effect on vasopressin-induced intracellular Ca^{2+} mobilization in washed platelet suspension. (Reproduced from publication of Agarwal, et al., Ref. 74)

5.7 Conclusions

Both the vascular endothelium and platelet, which contain large amounts of adenine nucleotides (10 - 14 mM), contribute significantly to the plasma adenosine levels. However, the plasma adenosine levels remain low (100 - 300 nM) due to its rapid cellular uptake and metabolism. Even these low plasma adenosine levels play a key role in the regulation of intracellular cAMP levels and platelet function. On depletion or lowering of plasma adenosine levels platelet cAMP concentrations decrease and the platelet response to various agonists is greatly increased. Several studies demonstrate that plasma adenosine is much more inhibitory of platelet granular release than the platelet aggregation induced by various agonists. This suggests that adenosine is a key negative modulator of platelet agonist activity which causes the granular release of vasoactive and platelet-activating agents (e.g. 5-HT, ADP, Ca^{2+} etc.). Studies from our laboratory have shown that plasma adenosine is strongly inhibitory of vasopressin- and platelet-activating factor-induced platelet aggregation and 5-HT release from the dense granules. Adenosine causes platelet inactivation by acting on its specific membrane receptor A_{2a}, which is coupled to the G_s stimulatory site of adenylate cyclase. Several studies suggest that adenosine acts as a potent calcium antagonist blocking the intracellular Ca^{2+} mobilization, an important process that plays a key role in the platelet activation. Adenosine receptor antagonists, methylxanthines (e.g. theophylline, caffeine) block the inhibitory effect of adenosine causing platelet activation as seen by depletion of plasma adenosine levels. In contrast, nucleoside transport inhibitors (e.g. dipyridamole, dilazep), which cause an increase in plasma adenosine levels, produce significant effects in the inactivation of platelet function. Furthermore, cAMP PDE inhibitors and forskolin which act synergistically with plasma adenosine may prove to be clinically important in thromboembolic disorders. The human platelet serves as a suitable model to study both *in vitro* and *in vivo* biochemical and pharmacological actions of adenosine, adenosine antagonists, and agents that affect adenosine activity, in tissues (e.g. brain striatum and vascular smooth muscle) which possess adenosine A_{2a} receptors.

References

1. Arch, JRS, Newsholme, EA, The control of the metabolism and the hormonal role of adenosine, *Essays Biochem*, 1978, 14; 82-123.

2. Gerlach, E, Becker, BF, *Topics and Perspectives in Adenosine Research*. New York: Springer-Verlag, 1987.

3. Imai, S, Nakazawa, M, *Role of Adenosine and Adenine Nucleotides in the Biological System*. New York: Elsevier Science Publishers BV, 1991.

4. Nees, S and Gerlach, E, Adenine nucleotide and adenosine metabolism in cultured coronary endothelial cells, In: *Regulatory functions of adenosine*, (eds. Berne, RM, Rall, TW and Rubio, R), Martinus Nijhoff Publishers, The Hague, 1983, 347-360.

5. Lowenstein, JM, Yu, MK, Naito, Y, Regulation of adenosine metabolism by 5=-nucleotidase, In: *Regulatory functions of adenosine*, (eds. Berne, RM., Rall, TW and Rubio, R), Martinus Nijhoff Publishers, The Hague, 1983, 117-131.

6. Berne, RM, Winn, HR, Knabb, RM, Ely, SW and Rubio, R, Blood flow regulation by adenosine in heart, brain and skeletal muscle, In: *Regulatory functions of adenosine*, (eds. Berne, RM, Rall, TW and Rubio, R), Martinus Nijhoff Publishers, The Hague, 1983, 293-317.

7. Nees, S, Dendorfer, A, New perspectives of myocardial adenine nucleotide metabolism, In: *Role of adenosine and adenine nucleotides in the biological system*, (eds: Imai, S and Nakazawa, M), Elsevier Science Publishers, Amsterdam, 1991, 273-288.

8. Agarwal, KC, Modulation of platelet functions by plasma adenosine, In: *Role of adenosine and adenine nucleotides in the biological system* (eds: Imai, S and Nakazawa, M), Elsevier Science Publishers, Amsterdam, 1991, 457-468.

9. Schrader, J, Borst, M, Kelm, M, Smolenski, T, Deussen, A, Intra- and extracellular formation of adenosine by cardiac tissue, In: *Role of adenosine and adenine nucleotides in the biological system*, (eds: Imai, S and Nakazawa, M), Elsevier Science Publishers, Amsterdam, 1991, 261-271.

10 Burnstock, G, Overview (Purinergic receptors), In: *Role of adenosine and adenine nucleotides in the biological system*, (eds: Imai, S and Nakazawa, M), Elsevier Science Publishers, Amsterdam, 1991, 3-16.

11. Daly, JW, Ukena, D, Jacobson, KA, Analogues of adenosine, theophylline, and caffeine: Selective interactions with A_1 and A_2 adenosine receptors, In: *Topics and perspectives in adenosine research* (eds: Gerlach, E and Backer, BF), Springer-Verlag, Heidelberg, 1987, 23-36.

12. Belardinelli, L, Adenosine system in the heart. *Drug Dev Res* 1993, 28; 263-267.

13. Huttemann, E, Ukena, D, Lenschow, V, Schwabe, U, Adenosine receptors in human platelets: Characterization by 5=-*N*-ethylcarboxamido[^{3}H]adenosine binding in relation to adenylate cyclase activity. *Nauyn-Schmiedeberg=s Arch Pharmacol* 1984, 325;226-285.

14. Ukena, D, Bohme, E, Schwabe, U, Effects of several 5=-carboxamide derivatives of adenosine on adenosine receptors of human platelets and rat fat cells. *Nauyn-Schmiedeberg=s Arch Pharmacol* 1984, 327, 36-42.

15. Cusack, NJ, Hourani, SMO, 5=-*N*-ethylcarboxamidoadenosine: A potent inhibitor of human platelet aggregation. *Br J Pharmacol* 1981, 72; 443-447.

16. Varani, K, Borea, PA, Guerra, L, Dionisotti, S, Zocchi, C, Ongini, E, Binding characteristics of the adenosine A_2 receptor ligand [^{3}H]CGS 21680 to human platelet membranes., *Biochem Pharmacol* 1994, 48; 1658-1661.

17. Dionisotti, S, Zocchi, C, Varani, K, Borea, PA, Ongini, E, Effects of adenosine derivatives on human and rabbit platelet aggregation: Correlation of adenosine receptor affinities and antiaggregatory activity. *Nauyn-Schmiedeberg=s Arch Pharmacol* 1992, 346; 673-676.

18. Fein, T, Schulze, E, Bar, J, Schwabe, U, Purifictaion and characterization of adenotin-like adenosine binding protein from human platelets. *Nauyn-Schmiedeberg=s Arch Pharmacol* 1994, 349;374-380.

19 Abbracchio, MP, Cattabeni, F, Fredholm, BB, Williams, M, Purinoreceptor nomenclature: A status report. *Drug Dev Res* 1993, 28; 207-213.

20 Fredholm, BB, Abbracchio, MP, Burnstock, G, Daly, JW, Harden, TK, Jacobson, KA, Leff, P, Williams, M, Nomenclature and classification of purinoceptors. *Pharmacol Rev.* 1994, 46; 143-156.

21. Linden, J, Jacobson, KA, Hutchins, C, Williams, M, Adenosine receptors, In: *Handbook of receptors and Ca^{2+} channels* (ed: Peroutka, SJ), Boca Raton, FL, 1994, 29-44.

22. Varani, K, Gessi, S, Dalpiaz, A, Borea, PA, Pharmacological and biochemical characterization of purified A_{2a} adenosine receptors in human platelet membranes by [^{3}H]CGS 21680, *J Pharmacol* 1996; 117, 1693-1701.

23. Dionisotti, S, Ferrara, S, Molta, C, Zocchi, C, Ongini, E, Labeling of A_{2a} adenosine receptors in human platelets by use of the new nonxanthine antagonist radioligand [^{3}H]SCH 58261, *J Pharmacol Exp Therapeu* 1996, 278; 1209-1214.

24. Born, GVR, Aggregation of blood platelets by adenosine diphosphate and its reversal. *Nature* 1962, 194; 927-929.

25. Born, GVR, Haslam, RJ, Goldman, M, Lowe, RD, Comparative effectiveness of adenosine analogues as inhibitors of blood platelet aggregation and as vasodilators in man. *Nature* 1965, 205; 678-680.

26. Agarwal, KC, Parks, RE, Jr, Townsend, L, Adenosine analogs and human platelets-II. Inhibition of ADP-induced aggregation by carbocyclic adenosine and imidazole-ring modified analogs. Significance of alterations in the nucleotide pools, *Biochem. Pharmacol.*, 1979, 28;501- 510.

27. Haslam, R.J, Cusack, NJ, Blood platelet receptors for ADP and adenosine, In: *Purinergic receptors*, (ed. Burnstock, G), Chapman and Hall, London, 1981, 223-285.

28. Agarwal, KC, Adenosine and platelet function, In: *Role of adenosine in cerebral metabolism and blood flow*, (eds. Stefanovich V, Okyayuz- Baklouti I), VNU Science Press BV, Germany, 1988, 107-124.

29. Mills, DCB, Smith, JB, The influence on platelet aggregation of drugs, *Biochem. J.*, 1971, 121; 185-196.

30. Agarwal, KC, Parks, RE, Jr, Adenosine analogs and human platelets: effects on nucleotide pools and the aggregation phenomenon, *Biochem. Pharmacol.*, 1975, 24; 2239-2248.

31. Dawicki DD, Agarwal KC, Parks, RE, Jr, Role of adenosine uptake and metabolism by blood cells in the antiplatelet actions of dipyridamole, dilazep and nitrobenzylthioinosine, *Biochem. Pharmacol.*, 1985, 34; 3965-3972.

32. Dawicki DD, Agarwal KC, Parks RE, Jr, Adenosine metabolism in human whole blood: Effects of nucleoside transport inhibitors and phosphate concentration, *Biochem. Pharmacol.*, 1988, 37; 621-626.

33. Bullough, DA, Zhang, C, Montag, A, Mullane, KM, Young, MA, Adenosine-mediated inhibition of platelet aggregation by acadesine: A novel antithrombotic mechanism *in vitro* and *in vivo*, *J Clin Invest*, 1994, 94; 1524-1532.

34. Johnson, GJ, Leis, LA, Francis, GS, Adenosine potentiates the inhibitory effects of calcium channel antagonists on human platelet aggregation induced by thromboxane A_2 or U46619. *Thromb Res* 1990, 5; 139-148.

35. Dodds, WJ, Platelet functions in animals: Species specificities, In: Platelets: A multidisciplinary approach (eds: de Gaetano, G, Garattini, S), Raven Press, NY, 1978, 45-59.

36. Hwang, DH, Species variation in platelet aggregation, In: The platelets: Physiology and pharmacology (ed: Longenecker, GS), Academic Press, NY, 1985, 289-305.

37. Sinakos, Z, Caen, JP, Platelet aggregation in mammalians (human, rat, rabbit, guinea-pig, horse, dog): A comparative study. *Thromb Diath Haemorrh* 1967,17; 99-111.

38. Philp RB, Bishop, B, Species-dependent inhibition of platelet aggregation by adenosine and dipyridamole: Paradoxical potentiation in the rat. *Thromb Diath Haemorrh* 1970, 23; 129-139.

39. Michal, F, Maguire, MH, Gough, G, 2-Methylthioadenosine-5=-phosphate, a specific inhibitor of platelet aggregation. *Nature* 1969, 222; 1073-1074.

40. Nunn, B, Some characteristics of mouse platelet aggregation and a comparison of the activity of a range of compounds in mouse and human platelet-rich plasma *in vitro*. *Thromb Haemost* 1981, 45; 1-5.

41. Agarwal, KC, Platelet aggregation and inhibitors, In: *Principles of cell adhesion* (eds. Richardson, PD, Steiner, M), CRC Press, Boca Raton, FL, 1995, 287-305.

42. Michel, H, Caen, JP, Born, GVR, Miller, R, d=Auriac, GA, Meyer, P, Relationship between the inhibition of aggregation and the concentration of cAMP in human and rat platelets. *Br J Haematol* 1976, 33; 27-38.

43. Agarwal, KC, Haskel, EJ, Parks, RE, Jr, Effects of adenosine analogs and adenine nucleotides on adenosine 5'-diphosphate-induced rat platelet aggregation, *Biochem Pharmacol.*, 1980, 29; 1799-1805.

44. Agarwal, KC, Buckley, RS, Parks, RE, Jr, Role of plasma adenosine in the antiplatelet action of HL-725, a potent inhibitor of cAMP phosphodiesterase: species differences, *Thromb. Res.*, 1987, 47; 191-200.

45. Agarwal, KC, Zielinski, BA, Maitra, RS, Significance of plasma adenosine in the antiplatelet activity of forskolin: Potentiation by dipyridamole and dilazep, *Thromb. Haemost.*, 1989, 61; 106-110.

46. Reimers, HJ, adenine nucleotides in blood platelets, In: *The platelets: Physiology and pharmacology* (ed: Longenecker, GS), Academic Press, NY, 1985, 85-112.

47. Yang, BC, Mehta, JL, Platelet-derived adenosine contributes to the cardioprotective effects of platelets against ischemia-reperfusion injury in isolated rat heart, *J Cardiovasc Pharmacol*, 1994, 24; 779-785.

48. Sollevi, A, Torssell, L, Owall, A, Edlund, A, Lagerkranser, M, Levels and cardiovascular effects of adenosine in humans, In: *Topics and perspectives in adenosine research* (eds: Gerlach, E, Backer, BF), Springer-Verlag, Heidelberg, 1987, 559-613.

49. Agarwal, KC, Modulation of vasopressin actions on human platelets by plasma adenosine and theophylline: gender differences, *J. Cardiovasc. Pharmacol.*, 1993, 21; 1012-1018.

50. Agarwal, KC, Clarke, E, Rounds, S, Parks, RE, Jr, Huzoor- Akbar, Platelet-activating factor (PAF)-induced platelet aggregation: modulation by plasma adenosine and methylxanthines, *Biochem. Pharmacol.* 1994, 48; 1909-1916.

51. Dawicki, DD, Agarwal, KC, Parks, RE, Jr, Potentiation of the antiplatelet action of adenosine in whole blood by dipyridamole or dilazep and the cAMP phosphodiesterase inhibitor, RA-233, *Thromb. Res,* 1986, 43; 161-175.

52. Agarwal, KC, Dieroff, CC, Buckley, RS, Parks, RE, Jr, Weisenberger, H, Antiplatelet activity of cyclic AMP phosphodiesterase (PDE) inhibitors: Role of plasma adenosine and species differences. *Thromb Haemost* 1989, 62; 406.

53. Ruppert, D, Weithmann, KU, HL-725, an extremely potent inhibitor of platelet phosphodiesterase and induced platelet aggregation *in vitro*. *Life Sci* 1982, 31; 2037-2043.

54. Stafford, A, Potentiation of adenosine and adenine nucleotides by dipyridamole, *Br. J. Pharmacol.,* 1966, 28; 218-227.

55. Asano, T, Ochiai, Y, Hidaka, H, Selective inhibition of separated forms of human platelet cyclic nucleotide phosphodiestease by platelet aggregation inhibitors, *Mol. Pharmacol.,* 1977, 13; 400-406.

56. Paterson, ARP, Jakobs, BS, Harley, ER, Fu, NU, Robins, MJ, Cass, CE, Inhibition of nucleoside transport, In: *Regulatory functions of adenosine,* (Eds. Berne, R.M., Rall, T.W. and Rubio, R.), Martinus Nijhoff Publishers, The Hague, 1983, 203-220.

57. El-Gamel, S, Wollert, U, Muller, WE, Optical studies on the specific interaction of dipyridamole with alpha1-acid glycoprotein (orosomucoid)., *J. Pharm. Pharmacol.,* 1981, 34; 152-157.

58. Klabunde, RE, Dipyridamole inhibition of adenosine metabolism in human blood, *Eur. J. Pharmacol.,* 1983, 93; 21-26.

59. Sollevi, A, Ostergren, J, Fagrell, B, Hjemdahl, P, Theophylline antagonizes cardiovascular responses to dipyridamole in man without affecting increases in plasma adenosine, *Acta Physiol. Scand.,* 1984, 121; 165-171.

60. Gresele, P, Arnout, J, Deckmyn, H, Vermylen, J, Mechanism of the antiplatelet action of the dipyridamole in whole blood: Modulation of adenosine concentration and activity, *Thomb. Haemost.,* 1986, 55; 12-18.

61. Agarwal, KC, Role of vasopressin in coronary artery disease of female subjects. *Indian Pharmacological Soc* (XXIX National Conference), 1996, 26.

62. Takehara K, Igarashi A, Ishibashi Y, Dipyridamole specifically decreases platelet-derived growth factor release from platelets, *Pharmacol, 1990, 40; 150-156.*

63. Lugnier, C, Schoeffler, P, Bec, AL, Strouthou, E, Stoclet, JC, Selective inhibition of cyclic nucleotide phosphodiesterases of human, bovine and rat aorta., *Biochem. Pharmacol.,* 1986, 35; 1743-1751.

64. Bult, H, Fret, HRL, Jordaens, FH, Herman, AG, Dipyridamole potentiates platelet inhibition by nitric oxide, *Thromb. Haemost.,* 1991, 66; 343-349.

65. Fredholm, BB, Are methylxanthine effects due to antagonism of endogenous adenosine ? *Trends Pharmacol. Sci.,* 1982, 1; 129-132.

66. Ukena, D, Shamim, MT, Padgett, W, Daly, JW, Analogs of caffeine: antagonists with selectivity for A_2 adenosine receptors, *Life Sci.,* 1986, 39; 743-750.

67. Biaggioni, I, Paul, S, Puckett, A, Arzubiaga, C, Caffeine and theophylline as adenosine receptor antagonists in humans, *J. Pharmacol. Exp. Therap.,* 1991, 258; 588-593.

68. Daly, JW, Analogs of caffeine and theophylline: activity as antagonists at adenosine receptors, In: *Role of adenosine and adenine nucleotides in the biological system,* (eds. Imai, S, Nakazawa, M), Elsevier science BV, Amsterdam, 1991, 119-129.

69. Ammaturo, V, Perricome, C, Cananio, A, Ripaldi, M, Ruggiano, A, Zuccarelli, B, Monti, M, Caffeine stimulates *in vivo* platelet reactivity, *Acta Med Scand,* 1988, 224; 245-247.

70. Agarwal, KC, Purines and regulation of platelet activation, In: *Purines and Myocardial Protection* (eds: Abd-Elfattah, AS, Wechsler, AS), Kluwer Academics, 1996, 409-418.

71. Hallam, TJ, Thompson, NT, Scrutton, MC, Rink, TJ, The role of cytoplasmic free calcium in the responses of quin2-loaded human platelets to vasopressin. *Biochem J.* 1984, 221; 897-901.

72. Rink TJ, Hallam TJ, What turns platelets on ?. *Trends Biochem Sci* 1984, 9; 215-218.

73. Paul SI, Feoktistov I, Hollister AS, Robertson D, Biaggioni, I, Adenosine inhibits the rise in intracellular calcium and platelet aggregation produced by thrombin: Evidence that both effects are coupled to adenylate cyclase. *Mol Pharmacol* 1990, 37; 870-875.

74. Agarwal, KC, Huzoor-Akbar, Parks, RE, Jr, Vasopressin-induced intracellular calcium mobilization in human platelets: Blockade by adenosine, *Thromb Haemost.*, 1993, 69; 1243.

6

PLATELET ACTIVATING FACTOR AND PLATELETS

Shivendra D. Shukla, Ph. D.
Department of Pharmacology
School of Medicine
University of Missouri Columbia
Columbia, MO
65212, USA

6.1 PAF DISCOVERY, STRUCTURE AND HETEROGENEITY

For the past two decades Dr. Shivendra D. Shukla's major interest has been phosphoinositide turnover, platelet activating factor receptor signaling and platelet pharmacology. He has made original contributions in this field. His research has been supported by American Heart Association, National Institute of Health and NIH Research Career Development Award.

Among agents which activate platelets PAF is unique since it is the only phospholipid known to date which stimulates platelets at nanomolar range. The first indications of such a mediator dates back to 1966 when it was observed that a mixture of rabbit platelets and stimulated leukocytes released histamine on specific antigen challenge (Barbaro and Zvaifler, 1966). In 1971 it was observed that a A fluid phase mediator was released from leukocytes of Ig E sensitized rabbits. This mediator activated platelets releasing vasoactive amines (Henson, 1970). Subsequently, two groups independently confirmed these observations (Sirganian and Osler, 1971; Benveniste et. al. 1972) and the soluble factor was named 'platelet activating factor, PAF (Benveniste et. al. 1972). These initial observations led to research on chemical identification of PAF. In 1979 three groups of investigators independently contributed towards the structural identification of PAF (Demopoulos et. al. 1979; Benveniste et. al., 1979; Blank et. al.

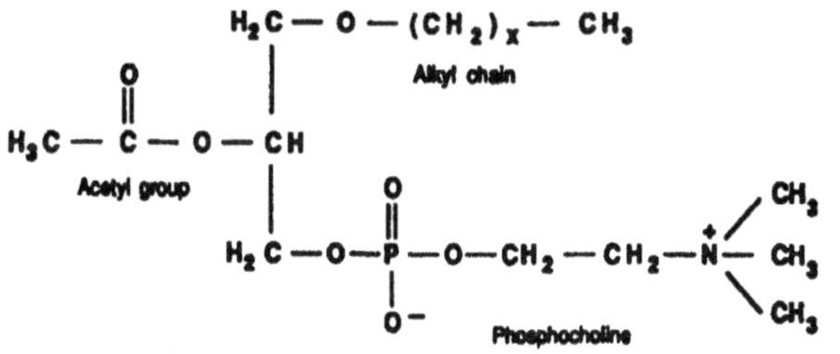

PLATELET ACTIVATING FACTOR

1979). PAF is an ether linked phosphoglyceride. Chemical structure of this lipid mediator was determined to be 1-0-alkyl-2-acetyl-sn-glycero-3-phosphorylcholine (Fig. 1). In literature PAF has also been termed as PAF-acether, and AGEPC (alkyl acetyl glycero ether phosphoryl choline). Excellent reviews have documented these historical developments (Hanahan, 1986, Snyder, 1985). In this article, research developments on PAF are summarized for platelets and covers its biochemical pharmacology and patho-physiological implications.

Removal of the acetyl group from sn-2 position results in the formation of biologically inactive derivative 1-alkyl-2-lyso-sn-3-glycerylphosphorylcholine (lyso-PAF). Step wise removal of the methyl group from the choline head group of PAF gives rise to dimethyl-, monomethyl- and ethanolamine derivatives. These head group analogos exhibit a graded loss of biological activity with removal of methyl group (Satouchi et. al. 1981). In addition, increasing the carbon chain length at Sn-2 position to propionyl, butyryl, hexanoyl groups causes increasing loss in the PAF biological activity (Hanahan, 1986). Similarly, changes in the alkyl chain length at C-1 position of the glycerol backbone of the phosphoglyceride also influenced the activity with C-16 and C-18 alkyl chains showing the highest activity. These observations established the specific structural requirement for the PAF activity. It may also be noted that differences in alkyl chains at C1 or short chain acyl groups at position C2 can give rise to derivatives of PAF with varying degrees of biological activity. Since all of these can be produced endogenously, therefore, PAF can exist in heterogeneous forms in vivo (Pinckard, et. al., 1994). However, PAF molecule with C16 chain, i.e. 1-0-hexadecyl-sn-glycero-3-phosphorylcholine is the most widely used (commercially available) PAF molecule. PAF has been shown to affect the functions of a variety of cells and organs including platelets, neutrophils, endothelial cells, macrophages, monocytes, kidneys, heart, lung, liver, skin, brain, smooth muscles, eyes, etc. (Koltai et. al., 1991a). It is a potent hypotensive agent and has been implicated in e.g. inflammation, allergy, asthma, reproduction, etc. Thus, PAF has diverse effects besides being a platelet agonist.

6.2 PAF Biosynthesis in Platelets

Two enzymatic pathways, de novo and remodeling have been described for the biosynthesis of PAF . The de novo route consists of several stepwise sequence of reactions. The ether bond in glycerolipids originates from a reaction between long chain fatty alcohols and acyldihydroxy-acetone phosphate (acyl DHAP) which gives rise to alkyl-DHAP. Subsequently, enzymatic conversions are as follows: alkyl DHAP → alkylglycero → P → alkylacetylglycero → P →alkylacetylglycerol → alkylacetyl GPC (PAF). The final step is mediated by CDP-choline-alkyl-acetyl-glycerol cholinephosphotransferase enzyme (See Snyder, 1995). The remodeling pathway is rapid and involves an endogenous lipid precursor (Fig. 2). Platelets contain and metabolize 1-alkyl-2-acyl-glycerophosphoryl-choline (alkyl PC), a major precursor for PAF (Mueller et. al. 1983; Natarajan et. al. 1983; Alam and Smith, 1987). A deacylation and reacetylation

Major Pathways for PAF Metabolism in Platelets

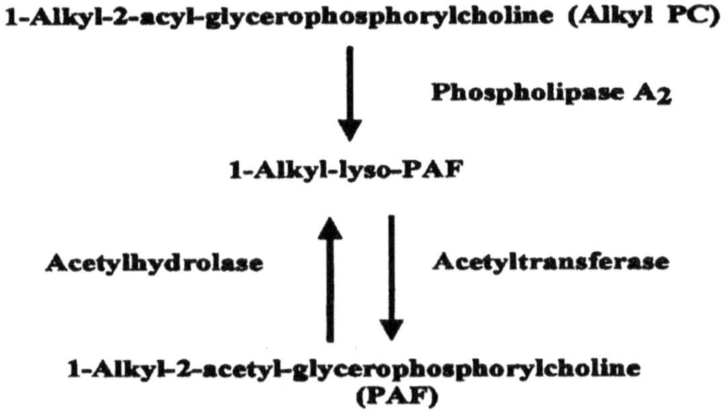

pathway converts alkyl PC to PAF. The deacylation of alkyl PC is mediated by the enzyme phospholipase A_2 (EC 3.1.1.4) giving rise to lyso-PAF. This is next acetylated by an acetyl-CoA: 1-alkyl-2-lyso-GPC acetyltransferase (EC 2.3.1.67) to PAF (Snyder 1985; 1995). Platelets do not contain or store preformed PAF. However in a report lipoprotein associated PAF was found in washed human platelets (Korth et. al. 1994) and its significance in platelet adherence to endothelium was raised. It has been demonstrated that platelets synthesize PAF (range: 0.25 to 3 pmole / 10^8 platelets) when challenged with Ca^{2+} + A23187 (Chignard et. al. 1979; Chap et. al. 1981; Marcus et. al. 1981), thrombin (Chignard et. al. 1979; Benveniste et. al. 1982) and collagen but not with ADP and arachidonic acid (Chignard et. al. 1980). PAF is hydrolysed by PAF acetylhydrolase to lyso PAF in platelets. PAF acetylhydrolase is almost equally distributed in membranes and cytosol of mouse platelets.

Blood bound PAF is located mainly in plasma (67±4%), erythroytes (18±1%) and platelets (5±1%) (Kelefiotis and Vakirtzi - Lemonias, 1997). Addition of [^{3}H] PAF to freshly drawn, rabbit or dog blood showed that [^{3}H] PAF was approximately equally distributed in plasma and blood cells fractions (unpublished). Serum albumin and lipoproteins have been considered as endogenous vehicles for PAF movement (Benveniste et. al. 1988; Clay et. al. 1990). A PAF releasing factor may also function as a PAF carrier in blood (Okamoto et. al. 1993).

6.3 Responses of Platelets to PAF

PAF stimulates platelet aggregation and serotonin secretion in a structurally specific manner. For example, lyso-PAF, the inactive structural analog has no platelet activating property even at as high as 1μM concentration. Critical micellar concentration (CMC) for PAF is near 1 μM concentration (Snyder, 1985). Modifications in the alkyl chain, acetyl group or at the head group also diminish its platelet stimulating activity (Snyder, 1985). Although platelets from human (Chignard et. al. 1979, McManus et. al. 1981), rabbit (Keraly and Benveniste, 1982), pig (Duronio, 1990; Rezaul, K., et. al., 1994), cow, (da Silva, et. al., 1996, Cheryk, et. al., 1996), sheep (Samiei, et. al., 1993), horse (Jarvis & Evans, 1994, Dillon & Heath, 1995, Lapetina, 1982), killer whale (Patterson, et. al., 1993) dog (Golino et. al. 1993), and cat respond to PAF, their sensitivity varies remarkably. Rabbit platelets respond to PAF at subnanomolar concentrations, whereas human platelets require submicromolar PAF for the activation (Rao and White, 1982, Rao et. al. 1984). Rat and mice platelets exhibit little response to PAF. Thus, the sensitivity of platelets from different species can be arbitrarily described as follows in increasing order (weak6strong), rat or mice<human<sheep<horse<dog>cow<pig≤rabbit. PAF also increases glycolytic metabolism in human platelets (Kim et. al. 1996).

6.4 PAF Receptor and Signal Transduction Pathways in Platelets

PAF activation of platelets is via its interaction with PAF-receptor in platelets. Specific PAF binding sites have been demonstrated in platelets through the use of either radioactive ligand [^{3}H] PAF or radioactive antagonist e.g. [^{3}H] WEB 2086, [^{3}H] L-659-989 (Hwang, 1993, see also Table 1). In both rabbit and human platelets high and low affinity binding sites exist (Hwang, 1990). However, Kd values (equilibrium dissociation contact) varies from 0.06 to 37 nM with B max (maximum binding capacity) values from 50 to 1400 binding sites per platelets (Chignard et. al., 1987, Hwang et. al. 1990). Rat platelets do not contain [^{3}H] PAF binding sites. Human platelets possess about 300 receptors for PAF with a Kd of about 0.2 nM. Inhibition of PKC or increase in cyclic-AMP did not influence [^{3}H] PAF binding. W7, a Ca^{2+}-calmodulin inhibitor and sphingosine decreased the binding by 50 - 80%. It was concluded that human platelet PAF receptors are likely to be under close intracellular control via a Ca^{2+}-calmodulin-dependent phosphorylation/dephosphorylation process (Burgers & Akkerman, 1993). PAF binding also is influenced by various ions. For example, Li^+ and Na^+ inhibit the specific [^{3}H] PAF binding whereas, Ca^{2+}, Mg^{2+}, Mn^{2+} and K^+ increase the binding. These ionic effects may, in part, be due to changes in receptor affinity (Hwang et. al., 1990).

It must be noted that PAF being a lipid exhibits a high non-specific binding when [^{3}H] PAF is used as a ligand. On the other hand, PAF antagonist [^{3}H] WEB2086 has a lower (less than 10%) non-specific binding and therefore is a preferred ligand for binding studies. PAF is also internalized (Homma et. al., 1987) and metabolized by platelets to long chains 1-alkyl-2-acyl PC (Pieroni and Hanahan, 1983, Tokumura, 1995). Therefore, in binding studies, internalization/ metabolism should be accounted for. In general the metabolism of PAF by platelets within the time frame used for binding studies (15-30 min.) is relatively low. In this context a non-metabolizable analog of PAF (1-0-hexadecyl-1'-2'-[^{3}H]-N-methylcarbamyl glycero-phosphocholine) is also available (O'Flaherty et. al. 1987; Travers et. al. 1992).

Table 1. Some known characteristics of platelet PAF receptors

Characteristics		Reference
PAF Receptor Protein	Mol. Wt.	
Rabbit Platelet	220 KDa Complex	Chau and Jii 1988
	52 Kda	Chau et. al. 1989
	350 Kda	Rogers et. al. 1991
Human Platelet	180 Kda	Valone, 1984
	160 Kda	Nishihara et. al 1985
	65 Kda	Shen et. al. 1989
PAF Receptor Number Binding Sites	KD (nM)	
100	2.5 ± 1.4	Rogers et. al. 1991
150 - 300	1.36 ± 0.05	Hwang et. al. 1993
160	0.83	Winslow 1984
242 ± 64	18.86 ± 4.82	Kloprogge &Akkerman 1984
245 ± 30	0.259 ± 0.03	Tuffin et. al. 1985
300	0.2	Burgers & Akkerman 1993
320 ± 38	0.1 ± 0.02	Chesney et. al. 1985
1399 ± 498	37.0 ± 13.0	Valone et. al. 1982
1577 ± 461	20.1 ± 6.3	Bussolini et. al. 1984
983 ± 391	1.58 ± 0.36	Inarrea et. al. 1984
19386 ± 6588	0.9 ± 0.5	Inarrea et. al. 1984

6.5 Antagonist

A large number of PAF receptor antagonists have been developed and used in PAF research (Koltai et. al. 1991 a&b, Shukla, 1996; Table II). Broadly, these antagonists can be classified in 5 categories: (i) nonconstrained PAF structure related (e.g. CV 3988, CV6209) (ii) constrained PAF structure-related (e.g. SRI63441, BN52111) (iii) synthetic compounds unrelated to PAF structure (e.g. WEB 2086) (iv) naturally occurring compounds (e.g. Ginkgolides, BN52020) and (v) dual antagonists (e.g. SCH 37370) (Piwinski et. al., 1991). Triazolobenzodiazepines inhibit platelet aggregation by PAF (Korneckie et. al., 1984). Dual antagonists are of interest. For example, rupatadine is an orally active dual antagonist of histamine and PAF with Ki (app) values of 0.1 and 0.55 MM respectively. It is a competitive PAF receptor antagonist (Merlos et. al. 1997). SR 27388 is a potent and orally active dual PAF receptor antagonist and antioxidant (Herbert et. al., 1993). The PAF receptor antagonists block PAF induced platelet aggregation, secretion and signaling responses. (Shukla, 1996). Several agents from natural sources also inhibit PAF induced platelet response. Leech saliva contains PAF antagonist (Eldor et. al., 1996). Extract of acacia nilotica blocked PAF induced aggregations of platelets (Shah et.al. 1997). Clove oil inhibited human platelet aggregation by PAF (Saeed, and Gilani, 1994). In ayurvedic medicine rubia cordifolia linn has been used for treatment of blood disorders. A partially purified fraction of this plant inhibited [^{3}H] PAF binding and aggregation in rabbit platelets (Tripathi et. al., 1993). Obviously some of these effects of natural substances may be relevant to non-traditional or alternate medical approaches involving PAF.

6.6 PAF Receptor

PAF receptor from platelets remains to be characterized. It must be noted here that PAF receptor has been cloned from guinea pig lung, human leukocytes and HL60 cells (see below). It is expected that platelet PAF receptor would be similar to the cloned receptor. However, there is no direct evidence for it yet. In this regard, it is interesting to note that almost all the efforts for isolation and purification of PAF receptor have been made using platelets (see Table 1). In 1984, a protein of 180 KDa was isolated from human platelet plasma membranes and was suggested to be a constituent of PAF receptor (Valone, 1984). Another protein of 160 KDa was isolated from human platelet membranes labeled with [^{3}H] PAF. This protein contained [^{3}H] PAF and was considered to represent PAF receptor (Nishihara, 1985). A protein of 65KDa from human platelets was concluded to be PAF receptor in a study by Shen et. al. 1989. From rabbit platelets a protein with apparent mol.wt. 220KDa was solubilized (Chau & Jii, 1988). GTP promoted the dissociation of [^{3}H] PAF from this protein suggesting the coupling of a GTP binding protein to the solubilized PAF receptor complex. These investigators later identified a protein of mol.wt. 52KDa and suggested that the binding site for PAF resides in this protein (Chau et. al. 1989). In another study, a protein complex of 350 KDa was solubilized from rabbit platelets and exhibited PAF binding. Approximately 100 receptors per cell with Kd value of 2.5 ± 1.4nM were reported (Rogers et. al. 1991).

Foregoing observations demonstrate the presence of PAF specific receptors/proteins in platelets. At present, the molecular nature of the PAF receptor in platelets remains unknown. The differences in the mol.wt for platelet PAF receptor reported in literature vary from 52 to 350 KDa. Other chemical and regulatory characteristics of the platelet PAF receptor also remain to be studied. The amino acid sequences, hydrophobicity, transmembrane domains, regulatory sites on the protein (e. g. G-protein binding, protein kinase phosphorylatin sites) are not known. It is obvious that compared to platelet thrombin or TXA_2 receptors the information on PAF receptor protein is lacking. In this regard it is relevant here to review the cloning of the PAF receptor from other sources. It can be assumed that platelet PAF receptor may be similar to them.

In 1991 PAF receptor was first cloned from guinea pig lungs (Honda et. al. 1991) using expression cloning strategy. The 3020 nucleotide sequence of the cloned PAF receptor encoded a 342 amino acid peptide and had a deduced mol.wt. of 38,982. The hydropathy profile indicated seven transmembrane spanning domains in the receptor. This was immediately followed by reports of cloning of PAF receptors from human leukocytes (Kunz et. al., 1992; Nakamura et. al. 1991) and from HL 60 cells (Ye et. al. 1991). The overall sequence similarity between guinea pig lungs and leukocyte receptors was 83% and similarly in transmembrane domain was 70%. The N-linked glycosylation site at the NH_2 terminus was absent in the leukocyte PAF receptor compared to guinea pig lung PAF receptor. Cloned PAF receptors from HL 60 cells showed 29% homology to N-formyl peptide receptor sequence. Based on the noticeable differences in the binding and cloning data available on PAF receptors, it can be speculated that heterogeneity in PAF receptors may exist. Structure and regulation of PAF receptor gene is also under active investigation (Shimizu & Mutoh, 1997).

6.6.1 PAF Receptor Signaling

In platelets, PAF triggers multiple signaling cascades (Fig. 3). Thus, activation of phospholipases, protein kinases, Ca^{2+} mobilization, MAP kinases mediated pathways have

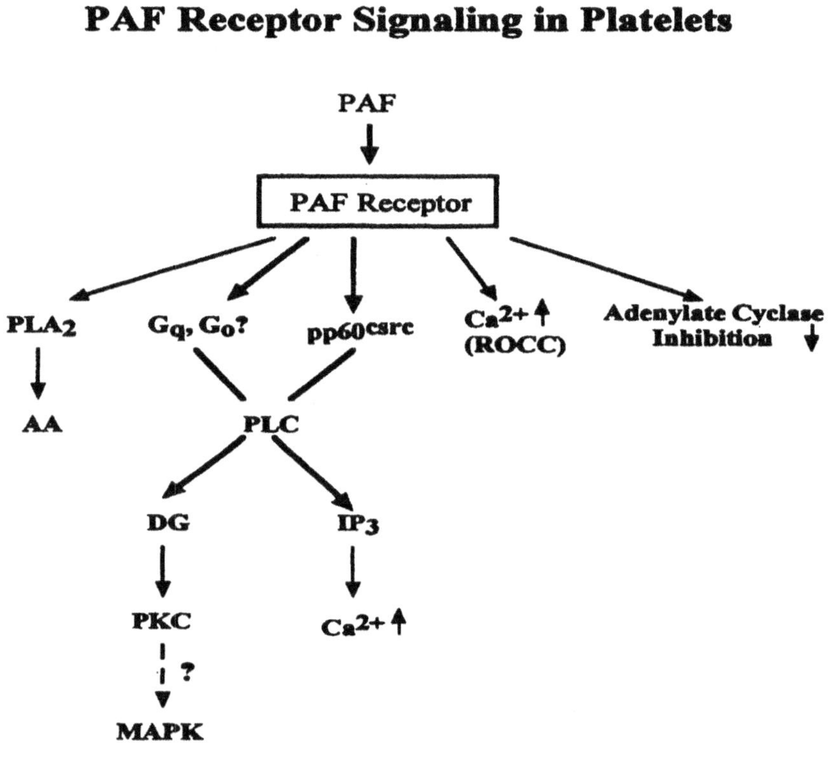

been documented in platelets (Shukla, 1992, Izumi and Shimizu, 1995). It must be noted that all the signaling pathways activated by PAF are receptor dependent since antagonists block these responses. The following section summarizes the PAF receptor coupled components involved in signaling pathways.

6.7 Phospholipases

Initially it was demonstrated in rabbit platelets that PAF activated turnover of phosphoinositides monitored by [^{32}P] labeling of PI, PIP and PIP_2 (Shukla and Hanahan, 1982). Subsequently, it was shown using [^{3}H] inositol labeled rabbit platelet that PAF activated phospholipase C pathway resulting in the generation of [^{3}H] inositol phoshates

(IP_3, IP_2, and IP) (Shukla, 1985). The activation of phosphoinositide turnover was not dependent on extracellular Ca^{2+} but the aggregation and serotonin secretion were dependent. Thus, early signaling events are considered external Ca^{2+} independent whereas the subsequent aggregation and secretion responses are dependent. Phosphatidic acid (PA), a metabolic product of PLC pathway, is also increased in platelets exposed to PAF and its level correlated with Ca^{2+} mobilization (Shukla and Hanahan, 1984). These increases in PA have been monitored by PAF stimulated [^{32}P] labeled platelets and also by mass (phosphorus) determinations of PA. Increases in PA due to phospholipase D activation by PAF have not been observed (Halenda et. al. 1996). PAF causes phosphoinositide turnover in platelets from different species including human, rabbit, horse (Lapetina, 1982; MacIntyre and Pollock, 1983) and thus suggests that this signaling pathway is common in PAF responses of platelets (Shukla, 1991). Lyso PAF is inactive in eliciting these responses and head group analogs of PAF, as discussed above, show the expected structure activity relationships. Activation of human platelets by PAF released EET (epoxyeicosatrieonic acid) and a role for PAF-receptor mediated hydrolysis of phospholipids containing EETs have been purposed (Zhu et. al., 1995). DG kinase inhibitor R59022 inhibited [^3H] PAF binding to human and rabbit platelets (Pakham et. al. 1996). Exogenously added diacylgylcerrol analog 1,2 - dihexanoyl - sn - glycerol (DHG), which is cell permeant, inhibited PAF induced aggregation of rabbit platelets but paradoxically potentiated human platelet aggregation. These results perhaps reflect upon species differences in the sensitivity of PAF signaling pathways (Packham et. al. 1996).

6.7.1 G-Proteins

Initial evidence for the G-protein involvement in the platelet PAF receptor came from the observations where (a) GTP was shown to shift the binding of PAF and (b) it stimulated GTPase activity (Avdonin et. al. 1985, Houslay et. al. 1986, Hwang, 1990). It was also shown that pertussis toxin inhibited the PAF responses in many cells (but not all). The PAF responses affected included e.g. IP_3 production, Ca^{2+} mobilization. Among known G proteins platelets possess Gs, Gi1 Gi2, Gi3, Gq, and Gz (Brass et.al., 1993). The fact that the cloned PAF receptor exhibits homology to G-protein coupled receptors, further supported G-protein involvement. The nature of the type of G-protein involved in platelet PAF receptor is not established. It is relevant to note that a protein distinct from Gs or Gi has been proposed to be involved in PAF mediated phosphoinositide turnover in platelets (Houslay et. al. 1986). In a reconstituted system it was recently suggested that α-subunit of Go may be coupled to PAF receptor to activate a novel PKC dependent mitogenic signaling pathway (Biesen et. al., 1996).

6.7.2 Protein Kinase C & Tyrosine Kinase Activation

PAF receptor is coupled to the activation of both protein kinase C and tyrosine kinase (Shukla, 1992, Izumi and Shimizu, 1995). PAF receptor antagonists block both pathways. The activation of PKC is through the known PLC generated diglyceride + Ca^{2+} pathway (Ieyasu et. al. 1982). Activation of PKC in platelets (eg. by phrobol myristate acetates) inhibited PAF stimulation of platelets. On the other hand PKC inhibitors (eg. staurosporine) enhanced PAF stimulated IP_3 production. In rabbit platelets PKC activation by PAF was suggested to be independent of enzyme translocation (Pelech et. al. 1990). Human platelets have 6 PKC isozymes (alpha, β, δ, zeta, eta and theta). PAF stimulated 200% and 175% increase in the levels of membrane - bound PKC eta and theta respectively with a concomitant decrease in these isoforms in the cytoplasm (Wang et. al., 1993). Activation of PKC by PMA or PAF is not affected by N-ethylmaleimide, an agent which uncouples Gi (Nieuwland et. al., 1994). PAF increases association of PKC delta and PI - 3 Kinase and this does not appear to be mediated by tyrosine phosphorylation of PKC delta (Ettinger et. al., 1996). Inhibitors of tyrosine kinase (eg. genistein, tyrphostin, 3, 5 dihydroxy cinnamate) inhibited PAF induced aggregation of platelets, IP_3 production (Dhar et. al., 1990; Zhu and Shukla, 1993; Murphy et. al. 1993). A role for $pp60^{c\text{-}src}$ tyrosine kinase in the activation of platelet PLC by PAF has been proposed (Dhar and Shukla, 1993). Electro-transjection of $pp60^{v\text{-}src}$ antibodies in rabbit platelets inhibited PAF stimulation of IP_3 production (Dhar and Shukla, 1994) suggesting a possible relationship between $pp60^{c\text{-}src}$ and PLC in platelets. PAF activates phosphorylation of $pp60^{c\text{-}src}$ increases its intrinsic tyrosine kinase activity and causes its translocation from cytsol to membranes (Dhar and Shukla, 1991, Zhu and Shukla, 1993, Murphy and Westwick, 1995). PAF also stimulated tyrosine phosphorylation of horse platelet proteins (Dillon and Heath, 1995). In porcine platelets, PAF activated tyrosine kinase activity of $p72^{syk}$ and its tyrosine phosphorylation. PAF receptor antagonist CV3988, tyrosine kinase inhibitor genistein inhibited this response but aspirin and apyrase were without effect. It was suggested that p72syk may play a role in PAF induced aggregation and Ca^{2+} mobilization (Rezaul et. al. 1994).

Role of endogenously activated type 1, protein phosphatese 1 (PP1) and type 2A (PP2A) in the signal transduction pathway of PAF stimulated rabbit platelets has been studied. Calyculin A, an inhibitor of PP1 and PP2A, caused inhibition of Ca^{2+} influx by PAF and correlated with inhibition of Ins (1,4,5) P_3 formation (Murphy and Westwick, 1994). It is therefore abundantly clear that both PKC and tyrosine kinase are integral components of PAF receptor signaling pathways. PAF is one of the first G-protein coupled receptors reported to be coupled to tyrosine kinase (Dhar et.al. 1990).

PAF activates mitogen activated protein kinase (MAP Kinase) in platelets. PAF stimulated P42 MAP Kinase (MAPK) in sheep platelets and to a lesser extent P44 MAPK. S6 kinases were also stimulated by PAF (Samiei et. al., 1993). It is likely that PKC is involved in this process. However, any precise role for platelet MAPK in PAF responses awaits future investigation.

6.7.3 Ca^{2+} and PAF Responses

Platelet intracellular Ca^{2+} is elevated by PAF (Lee et. al. 1981; Valone and Johnson, 1985, Sage et. al. 1992) in a PAF receptor coupled manner. The increase in Ca^{2+} can occur both by IP_3-mediated release of stored intracellular Ca^{2+} or by influx (Lee et. al., 1981, Shukla and Hanahan 1984, Sage et. al. 1992). It has been proposed that Ca^{2+} influx can also occur by PAF receptor operated Ca^{2+} channels (Avdonin et. al. 1987, Avdonin et. al. 1991, James-Kracke et. al. 1994). The rise in Ca^{2+} is obligatory for physiological responses of PAF. Recent results also tend to support the notion that IP_3 independent pathways for Ca^{2+} influx may also be involved in PAF activation of platelets (James- Kracke et. al 1994). In this study PAF induced increases in intracellular Ca^{2+} and aggregation of rabbit platelets were simultaneously monitored in a flourophotometer. It was concluded that PKC plays a role in PAF induced aggregation but not Ca^{2+} mobilization and shape change since staurosporine (PKC inhibitor) inhibited aggregation but not Ca^{2+} influx by PAF. In another study the effect of mibefradil, a novel nondihydropyridine Ca^{2+} antagonist, on increases in intracellular Ca^{2+} and aggregation of human platelets induced by PAF were examined. In order to differentiate effects of Ca^{2+} mobilization from intracellular stores from those on Ca^{2+} influx through receptor operated Ca $^{2+}$channels, intracellular Ca $^{2+}$ concentrations were measured either in fura-2 loaded platelets or loaded both with BAPTA and fura-2. It was concluded that mibefradil, which blocks L-and T-type Ca^{2+} channels, also blocked PAF receptor operated Ca^{2+} channels (Hahn, et. al., 1995). It is therefore likely that PAF stimulated Ca^{2+} mobilization in platelets occurs through more than one mechanism.

6.7.4 PAF Responses of Platelets in Relation to Other Agents

Various endogenous and exogenous agents modulate PAF responses of platelets (Table 2). The effect of drugs and inhibitors has been described earlier. In this section the interactions with endogenous substances will be discussed.

Platelets exhibit homologous desensitization to PAF. However, exposure of rabbit platelets to thrombin causes heterologous desensitization to PAF as monitored by IP_3 production and protein phosphorylation (Morrison and Shukla, 1988, Shukla et. al. 1989). Like other platelet agonists the responses of PAF are also inhibited by increases in cyclic-AMP and cyclic-GMP. In this regard, PAF was demonstrated to inhibit PGE1 stimulated adenylate cyclase (Haslam and Vanderwel, 1982). It had no effect on basal adenylate cyclase activity. In the presence of aspirin and CP/CPK (creatine phosphate and creatine phosphokinase) collagen causes whole blood aggregation monitored using impedance aggregometry. A role for PAF in collagen induced platelet aggregation was proposed (Oura, 1994). Agents that can increase plasma adenosine levels (e.g. inhibitors of adenosine uptake and adenosine metabolism) or methylxanthines reduce the PAF activation of human platelets. Patients receiving theophylline showed significantly greater platelet aggregation induced by PAF (Agarwal et. al., 1994). Addition of c-reactive protein to bovine platelets, suspended in homologous plasma, consistently reduced PAF induced aggregation of platelets (Cheryk et.al., 1996). Activation of equine platelets by PAF caused release of, among other proteins, thrombospondin (Lipscomb et. al. 1997). PAF induced release of ATP and histamine from platelets may be involved in the allergy reactions (Nakahodo et. al. 1994).

IgG monoclonal antibody NNKY5-5, directed against the Von Willebrand factor binding domain of glycoprotein Iba, potentiated PAF induced aggregation (Yanabu et. al., 1997). A possible mechanism involving p72Syk tyrosine kinase and a 64Kda protein was proposed. It is of interest to mention that DMP754, a GP II b/III a antagonist, inhibited human platelet aggregation induced by PAF (Mousa et.al., 1996). Threshold concentrations of PAF increased epinephrine induced Na^+/H^+ exchange and a possible role of PKC was postulated (Nieuwland et. al., 1993).

6.8 Platelet and PAF in Pathophysiological and Disease States

Half life of PAF in blood ranges seconds to a few minutes (Pickard et. al. 1979) and it is inactivated by plasma acetyl hydrolase (Farr et. al. 1980; Cabot et. al., 1984). Endogenous levels of PAF rise in a variety of pathological state e.g. allergy, asthma, sepsis, ischemia (Braquet et. al. 1987). It is therefore logical to expect that in vivo, under elevated blood PAF levels, desensitization of platelets to PAF may occur.

Table 2. Pharmacological modulation of PAF responses of platelets

Inhibitory Agents	
Agent / Action	**Reference**
PAF Recptor Antagonists[1]	
CV-3988, CV-6209, SRI 63,441, BN-52111, BN52115,	Braquet et. al. 1987;
BN-52020,	Koltai et. al. 1991;
WEB-2086, ONO-6240, L-652-731	Triazolam
	Shukla, 1996;
ABT 491	Albert et. al. 1997
PKC Activators	
Phorbol myristate acetate	Morrison et. al. 1989
Diacylglycerol (rabbit platelets)[2]	Packham et. al. 1996
Oleoyl acetylglycerol (OAG)[2]	Packham et. al. 1996
PLC Inhibitors	
Neomycin	Tysnes et. al. 1987
Quinacrine	Jarvis & Evans, 1996
Compound 48/80	Bronner et. al. 1987
Tyrosine Kinase Inhibitors	
Genistein	Dhar et. al. 1990
Erbstatin	Salari et. al. 1990
Tyrphostin, 3.5 dihydroxycinnamate	Zhu & Shukla, 1993
DG Kinase Inhibitor	
R59022	Packham et. al. 1996
Serine Protease Inhibitors	
Diethyl p-nitrophenyl phosphate (DNPP)	Sugatani et. al. 1987
N-acetyl-1-tryptophan ethyl ester (NATrpEE)	
PKC Inhibitors	
Staurosporine[3]	James-Kracke et. al. 1994
K252a	Yamada et. al. 1988
Ca^{2+} Channel Blockers	
Verapamil	Khan et. al. 1985
Mibefradil	Hahn et. al. 1995
Others	

Table 2 (Contd). Pharmacological modulation of PAF responses of platelets

Agent/Action	Reference
Forskolin	Shukla, 1985
PGI_2	Lapetina, 1982
Pertussis toxin	Brass et. al. 1988
	Shukla et. al. 1993
EGTA, EDTA	Shukla, 1985
Cardiolipin	Tsoukatos et. al. 1993
Trifluoperazine (Calmodulin antagonist)	Lapetina 1982
Zinc	Nunez et. al. 1989
Lanthanum	Khan et. al. 1985
TMB8 (intracellular Ca^{2+} antagonist)	Khan et. al. 1985
Oleic acid	Nunez et. al. 1990
Non-inhibitory or potentiating agents	
NSAID	Shukla, 1985
Aspirin	
Indomethacin	

1. Selected list; please see referred reviews.
2. In human platelets DE and OAG, paradoxically potentiated PAF induced aggregation.
3. Staurosporine inhibited aggregation but either potentiated or had no effect on IP_3 production and Ca^{2+} influx stimulated by PAF.

It appears to be the case. Platelets are desensitized (56% reduction in aggregation) to PAF caused by exposure of platelets to PAF released in vivo in acute asthma provoked by allergens (Beer, et. al., 1995). Endotoxin induced platelet aggregation in horse is mediated via PAF demonstrating an endogenous role of this lipid mediator (Jarvis and Evans 1994). Computerized thromboelastography assay has also been used to monitor PAF induced whole blood coagulation (Dambisya et. al., 1995) and significant role for endogenous PAF was postulated. Platelets of asthmatics show increased responses to PAF (Block et. al. 1988).

Role of PAF in the pathogenesis of microvascular vaso-occlusion in sickle cell disease (SCD) has been studied and it was demonstrated that patients with SCD are susceptible to a higher circulating level of PAF in vivo during steady state conditions. It was speculated that higher levels of PAF may be a contributing factor to the persistent stress and inflammatory state of the microcirculation of patients with SCD (Oh et. al., 1997). Impaired responses of platelets of a patient to PAF and other agonists may be due to defect in human platelet G-protein alpha subunit (Gabbeta et. al., 1997). In resting human platelets the basal level of hydrogen peroxide was significantly higher in diabetic subjects (non-insulin dependent, NIDDM). Moreover, after stimulation with thrombin, collagen and PAF platelets of diabetic subjects generated significantly higher amounts of hydrogen peroxide than controls (Leoncini et. al., 1997). Diabetic human platelets exhibit hypersensitivity to PAF as monitored by aggregation and PLC mediated phosphatidic acid production. The specific [^{3}H] PAF binding was not altered in diabetic platelets (compared to control) and therefore, a defect in signaling cascade may underlie the mechanism for hypersensitivity (Shukla et. al. 1992).

Although various platelet stimulating agents (eg. ADP, thrombin, collagen, PAF etc.) have same functional end point (eg. aggregation), they still have important distinguishing features and mechanistic pathways. PAF possesses such uniqueness. It has been proposed, for example, that PAF acts via a third pathway, the other two being ADP and arachidonate mediated (Cazenave et. al. 1979; Vargaftig et. al. 1981). In contrast to thrombin and several other platelet agonists, PAF stimulated aggregation is independent of cyclooxygenase pathway. Aspirin and indomethacin do not affect PAF responses (Vargaftig et. al. 1981, Shukla et. al. 1993). However, in some studies with human platelets cyclo oxygenase dependence has been shown at low doses of PAF and this minor discrepancy remains (Miller et. al. 1982; Macconi et. al. 1985; Cattaneo et. al. 1985) unresolved. There is overwhelming evidence indicating the cyclooxygenase-independent nature of PAF actions. It is likely that blockade of cyclooxygenase by drugs like aspirin or indomethacin will still render platelets responsive to PAF. Any clinical ramifications of this distinction has not been investigated in cardiovascular diseases eg. in atherosclerosis, stroke.

Theophylline administration in human subjects reduces the sensitivity of platelets to PAF (Agarwal et. al. 1994). Platelets from ploycythemia vera (PV) patients exhibited reduced aggregatory response to PAF but not to ADP and PMA. It was further shown that platelet aggregation induced by PAF (or ADP and PMA) was not significantly reduced in patients with chronic myeloid leukemia, essential thrombocythemia and multiple mycloma (LeBlanc et. al., 1994). A defect in the stimulus response coupling for PAF has been suggested in PV platelets. In chronic myeloproliferative disorder a defective platelet aggregation response to PAF was observed and suggested to be due to defficiency of an aspirin-independent mechanism (Viero et. al. 1986). Several studies have indicated platelet activation in migraine. PAF induced release of serotonin and PF4 (Platelet factor 4) was monitored in platelets from patients with migraine with aura (MA) and without aura (MWA). There was no significant difference in aggregation by PAF between migraine vs. healthy controls. However, PF4 secretion by PAF was increased in MWA and MA and there was increased serotonin secretion by PAF. These results indicate an abnormal platelet "-granule secretion in those with migraine and focus attention on PAF as possible factor contributing to the platelet activation associated with migraine (D'Andrea et. al., 1994).

6.8.1 PAF in Platelet and Vascular Endothelium Interactions

Both platelets and vascular endothelium are known to synthesize PAF (Chignard et. al. 1980; Lorant et. al. 1995). Endothelial cells form PAF upon challenge with thrombin and other agents. It has been suggested that PAF and platelets are involved in adhesion of neutrophils and platelets to endothelium. Platelets synthesize PAF and may release it associated with platelet microparticles (Iwamoto et. al., 1996). PAF mediates procoagulant activity on the surface of endothelial cells by promoting leukocyte adhesion via P-selectin (Lorant et. al. 1995). This also leads to activation of platelet at the site. Platelet mediated mechanisms have been implicated in intimal lesion formation following vascular injury. Mitogenic role of platelet factors released from platelets was investigated using [^{3} H] thymidine incorporation in smooth muscle cells and in this study no appreciable effect of PAF on proliferation was observed (Crowley et. al., 1994).

Cyclic GMP may be involved in the PAF and leukocyte dependent platelet adhesion to endothelial cells (Hirafuji et. al. 1996). Using isolated, perfused segment of rabbit ear central artery and antologous platelets, it was shown that PAF acts on blood vessel contractility through platelets (Michibayashi, 1996). Lipoprotein associated PAF (LA-PAF) was found in washed human platelets and it was suggested that it may modulate platelet adherence to vascular endothelial cells (Korth et. al., 1994). Enhancement of anoxic contraction of rat aortic rings by platelets is mediated via a PAF dependent mechanism (Yang et. al., 1993). PAF plays an important role as a mediator of platelet aggregation in vivo in rabbits and dogs (Golino et. al., 1993).

In conclusion, PAF is a unique and potent endogenous lipid mediator which activates platelets. Increases or decreases in PAF levels in vivo can profoundly influence vascular hemostasis. PAF activates multiple signaling pathways in platelets but the molecular properties of platelet PAF receptor and its regulation are poorly understood. Similarly the expression of PAF receptor in megakaryocyte development in normal and disease states would also be a topic of interest in the future.

6.9 Acknowledgement

The author is grateful to Ms. Pam Burgess for expert word processing, and to Mr. Daniel Weng for help in drawing figures.

References

Agarwal KC, Clarke E, Rounds S, Parks RE Jr, Huzoor-Akbar. Platelet-activating factor (PAF)-induced platelet aggregation. Modulation by plasma adenosine and methylxanthines. Biochem Pharmacol 1994; 48: 1909-1916

Albert DH, Magoc TJ, Tapang P, Luo G, Morgan DW, Curtin M, Sheppard GS, Xu L, Heyman HR, Davidsen SK, Summers JB, Carter GW. Pharmacology of ABT-491, a highly potent platelet-activating factor receptor antagonist. Eur J Pharmacol 1997; 325: 69-80

Alam I, Silver MJ. Metabolism of 1-alkyl-2-acyl-GPC in human platelets in response to stimulation by thrombin. Thromb Res 1987; 45: 311-322

Avdonin PV, Cheglakov IB, Boogry EM, Svitina-Ulitina IV, Mazaev AV, Tkachuk VA. Evidence for the receptor-operated calcium channels in human platelet plasma membrane. Thromb Res 1987; 46: 29-37

Avdonin PV, Cheglakov IB, Tkachuk VA. Stimulation of non-selective cation channels providing Ca^{2+} influx into platelets by platelet-activating factor and other aggregation inducers. Eur J Biochem 1991; 198: 267-273

Avdonin PV, Svitina-Ulitina IV, Kulikov VI. Stimulation of high-affinity hormone-sensitive GTPase of human platelets by 1-0-alkyl-2-0-acetyl-sn-glyceryl-3-phosphocholine (Platelet activating factor). Biochem Biophys Res Commun 1985; 131: 307-313

Barbaro JF, Zvaifler NJ. Antigen induced histamine release from platelets of rabbits producing homologous PCA antibody. Proc Soc Exp Biol (N.Y.) 1966; 122: 1245-1247

Beer JH, Wuthrich B, von Felten A. Allergen exposure in acute asthma causes the release of platelet-activating factor. Int Arch Aller Immun 1995; 106: 291-296

Benveniste J, Henson PM, Cochrane CG. Leukocyte dependent histamine release from rabbit platelets. The role of IgE, Basophils and a platelet activating factor. J Exp Med 1972; 136: 1356-1377

Benveniste J, Tence M, Varene P, Bidault J, Boullet C, Polonsky J. Semisynthese et structur proposes du facteur activant leg plaquettes (PAF): PAF-acether, un alkyl ether analogue de la lysophosphatidylcholine. C R Acad Sci (Paris) 1979; 289: 1037-1040

Benveniste J, Chignard M, Le Couedic JP, Vargaftig BB. Biosynthesis of PAF-acether II. Involvement of phospholipase A2 in the formation of PAF-acether and lyso PAF from rabbit platelets. Thrombosis Res 1982; 25: 375-385

Benveniste J, Nunez D, Duriez P, Korth R, Bidault J, Fruchart JC. Preformed PAF-acether and lyso PAF-acether are bound to blood lipoproteins. FEBS Lett 1988; 226: 371-376

Biesen T, Hawes BE, Raymond JR, Luttrell LM, Koch WJ, Lefkowitz RJ. Go-protein ∝ subunits activate mitogen-activated protein kinase via a novel protein kinase C dependent mechanism. J Biol Chem 1996; 271: 1266-1269

Blank ML, Snyder F, Byers LW, Brooks W, Muirhead EE. Antihypertensive activity of an alkyl ether analogue of phosphatidylcholine. Biochem Biophys Res Commun 1979; 90: 1194-2000

Block LH, Imhof E, Perruchoud AP. Platelets of asthmatics show increased phosphatidylinositol turnover in response to PAF. Amer Rev Resp Dis 1988; 137: 235

Braquet P, Touqui L, Shen Y, Vargaftig B. Perspectives in platelet activating factor research. Pharmacological Reviews 1987; 39: 97-145

Brass LF, Woolkalis MJ, Manning DR. Interactions in platelets between G proteins and the agonists that stimulate phospholipase C and inhibit adenylyl cyclase. J Biol Chem 1988; 263: 5348-5355

Brass LF, Hoxie JA, Kieber-Emmons T, Manning DR, Poncz M, Woolkalis M. Agonist receptors and G proteins as mediators of platelet activation. Adv Exper Med Biol 1993; 344: 17-36

Burgers JA, Akkerman JW. Regulation of the receptor for platelet-activating factor on human platelets. Biochem J 1993; 291: 157-161

Bussolini F, Tetta C, Camussi G. Specific binding of ^{3}H-PAF by human PMN. Agents and Action 1984; 15: 15-17

Cabot MC, Faulkner LA, Lackey RJ, Snyder F. Vertebrate class distribution of 1-alkyl-2-acetyl-*sn*-glycero-3-phosphocholine acetylhydrolase in serum. Comp Biochem Physiol 1984; 78B: 37-40

Cattaneo M, Canciani MT, Mannucci PM. Human platelet aggregation and release reaction induced by platelet activating factor (PAF-acether) - effects of acetylsalicylic acid and external ionized calcium. Thromb Haemostas 1985: 221-224

Cazenave, JP, Benveniste J, Mustard JF. Aggregation of rabbit platelets by platelet-activating factor is independent of the release reaction and the arachidonate pathway and inhibited by membrane-active drugs. Lab Invest 1979; 41: 275-285

Chap M, Manco G, Simon MF, Beneveniste J, Douste-Blazy L. Biosynthetic labelling of PAF from radioactive acetate by stimulated platelets. Nature 1981; 289: 312-314

Chau LY, Jii YJ. Characterization of ^{3}H-labelled platelet activating factor receptor complex solubilized from rabbit platelet membranes. Biochim BiophysActa 1988; 970: 103-112

Chau LY, Tsai YM, Cheng JR. Photoaffinity labeling of platelet activating factor binding sites in rabbit platelet membranes. Biochem Biophys Res Commun 1989; 161: 1070-1076

Cheryk LA, Hayes MA, Gentry PA. Modulation of bovine platelet function by C-reactive protein. Veterin Immunol Immunopath 1996; 52: 27-36

Chesney CM, Pifer DD, Huch KM. Desensitization of human platelets by platelet activating factor. Biochem Biophys Res Commun 1985; 127: 24-30

Chignard M, Le Coudic JP, Tence M, Vargaftig BB, Benveniste J. The role of PAF in platelet aggregation. Nature 1979; 799-800

Chignard M, Le Couedic JP, Vargaftig BB, Benveniste J. Platelet-activating factor (PAF-acether) secretion from platelets: effect of aggregating agents. Brit J Haematol 1980; 46: 455-464

Chignard M, Keraly CL, Nunez D, Coeffier E, Benveniste J. PAF-acether and platelets. Platelets in Biol and Pathol III Ed. D.E. MacIntyre and J.L. Gordon. Elsevier 1987; 279: 289-315

Clay KL, Johnson C, Henson P. Binding of platelet activating factor to albumin. Biochim Biophys Acta 1990; 1046: 309-314

Crowley ST, Dempsey EC, Horwitz KB, Horwitz LD. Platelet-induced vascular smooth muscle cell proliferation is modulated by the growth amplification factors serotonin and adenosine diphosphate. Circ 1994; 90: 1908-1918

D'Andrea G, Hasselmark L, Alecci M, Cazanzi A, Perini F, Welch KM. Platelet secretion from dense and alpha-granules in vitro in migraine with or without aura. J Neurol Neurosur Psychi 1994; 57: 557-561

da Silva MB, Herion F, Raskinet R, David JL, Gustin P, Dessy C, Lekeux P. Inhibition of PAF-induced platelet aggregation by WEB 2086 'in-vitro', an antagonist to the receptor for platelet-activating factor, in bovine. Zentralblatt Fur Veterinarmedizin-Reihe A. 1996; 43: 399-413

Dambisya YM, Lee TL, Hussaini IM. Effects of the platelet-activating factor receptor antagonist WEB 2086 on whole blood coagulation and fibrinolysis in a thromboelastography assay. Blood Coagul Fibrinol 1995; 6: 733-737

Demopoulos CA, Pinckard RN, Hanahan DJ. Platelet activating factor: evidence for 1-0-alkyl-2-acetyl-sn-glycerol-3-phosphoryl-choline as the active component (a new class of lipid chemical mediator). J Biol Chem 1979; 254: 9355-9358

Dhar A, Paul A, Shukla SD. Involvement of tyrosine kinase in PAF stimulation of phospholipase C in rabbit platelets: Studies with genistein and monocional antibody of phosphotyrosine. Mol Pharmacol 1990; 37: 519-525

Dhar A, Shukla SD. Involvement of $pp60^{c-src}$ in PAF stimulated platelets: Evidence for translocation from cytosol to membranes. J Biol Chem1991; 266: 18797-18801

Dhar A, Shukla SD. Tyrosine kinases in platelet signaling systems. Brit J Hematol 1993; 84: 1-7

Dhar A, Shukla SD. Electrotransjection of pp60v-src monoclonal antibody inhibits activation of phospholipase C in platelets. A new mechanism for platelet-activating factor responses. J Biol Chem 1994; 269: 9123-9127

Dillon AM, Heath MF. Protein tyrosine phosphorylation in equine platelets: the effect of stimulation by thrombin and platelet-activating factor. Equine Veterin J 1995; 27: 448-458

Duronio V. Characterization of platelet-activating factor receptors in porcine platelets. Can J Physiol Pharmacol 1990; 68: 1514-1519

Eldor A, Orevi M, Rigbi M. The role of the leech in medical therapeutics. Blood Reviews 1996; 10: 201-209

Ettinger SL, Lauener RW, Duronio V. Protein kinase C delta specifically associates with phosphatidylinositol 3-kinase following cytokine stimulation. J Biol Chem 1996; 271: 14514-14518

Farr RS, Cox CP, Wardlow ML, Jorgnsen R. Preliminary studies of an acid-labile factor (ALF) in human sera that inactivates PAF. Clin Immun Immpathol 1980; 15: 318-330

Gabbeta J, Yang X, Kowalska MA, Sun L, Dhanasekaran N, Rao AK. Platelet signal transduction defect with galpha subunit dysfunction and diminished galphaq in a patient with abnormal platelet responses. Proceed Nat Acad Sci USA 1997; 94: 8750-8755

Golino P, Ambrosio G, Ragni M, Pascucci I, Triggiani M, Oriente A, McNatt J, Buja LM, Condorelli M, Chiariello M, et. al. Short-term and long-term role of platelet activating factor as a mediator of in vivo platelet aggregation. Circulation 1993; 88: 1205-1214

Hahn M, Tkachuk VA, Bochkov VN, Cheglakov IB, Clozel JP. Blockade of receptor-operated calcium channels by mibefradil (Ro 40-5967): effects on intracellular calcium and platelet aggregation. Cardiovas Drugs Therapy 1995; 9: 815-821

Halenda SP, Wu H Jones AW, Shukla SD. Phospholipase D in platelets and megakaryocytic cells. Chem Phys Lipids 1996; 8: 21-26

Hanahan DJ. Platelet activating factor: a biologically active phosphoglyceride. Ann Rev Biochem 1986; 55: 483-509

Haslam RJ, Vanderwel M. Inhibition of platelete adenylate cyclase by 1-0-alkyl-2-0-acetyl-sn-glyceryl-3-phosphorylcholine (platelet-activating factor). J Biol Chem 1982; 257: 6879-6885
Henson PM. Release of vasoactive amine from rabbit platelets induced by sensitized mononuclear leukocytes and antigens. J Exp Med 1970; 131: 287-304

Herbert JM, Fraisse L, Bachy A, Valette G, Savi P, Laplace MC, Lasalle J, Roche B, Lale A, Keane PE, et. al. Biochemical and pharmacological properties of SR 27388, a dual antioxidant and PAF receptor antagonist. J Lipid Mediators 1993; 8: 31-51

Hirafuji M, Nezu A, Shinoda H, Minami M. Involvement of platelet cyclic GMP but not cyclic AMP suppression in leukocyte-dependent platelet adhesion to endothelial cells induced by platelet-activating factor in vitro. Brit J Pharmacol 1996; 117: 299-304

Homma H, Tokumura A, Hanahan DJ. Binding and internalization of platelet-activating factor 1-0-alkyl-2-acetyl-*sn*-glycero-3-phosphocholine in washed rabbit platelets. J Biol Chem 1987; 262: 10582-10587

Honda Z, Nakamura M, Miki I, Minami M, Wantanabe T, Seyama Y, Okado H, Toh, H, Ito K, Miyamoto T, Shimizu T. Cloning functional expression of platelet activating factor receptor from guinea pig lung. Nature 1991; 349: 342-346

Houslay MD, Bojanic D, Wilson A. Platelet activating factor and U44069 stimulate a GTPase activity in human platelets which is distinct from the guanine nucleotide regulatory proteins, N_s and N_i. Biochem J 1986; 234: 737-740

Hwang SB. Specific receptors for PAF, receptor heterogeneity and signal transduction mechanisms. J Lipid Mediators 1990; 2: 123-158

Hwang SB. Characterization of PAF receptors using radioligand binding studies. In PAF receptor: signal mechanisms of molecular biology, ed. S. D. Shukla. CRC Press 1993; 7-17

Ieyasu H, Takai Y, Kaibuchi K, Sawamura M, Nishizuka Y. A role of calcium-activated, phospholipid-dependent protein kinase in platelet-activating factor-induced serotonin release from rabbit platelets. Biochem Biophys Res Commun 1982; 108: 1701-1708

Inarrea P, Gomez-Cambronero J, Nieto M, Sanchez-Crespo M. Characteristics of the binding of PAF to platelets of different animal species. Eur J Pharmacol 1984; 105: 309-315

Iwamoto S, Kawasaki T, Kambayashi J, Ariyoshi H, Monden M. Platelet microparticles: a carrier of platelet-activating factor? Biochem Biophys Res Commun 1996; 218: 940-944

Izumi T, Shimizu T. Platelet activating factor receptor-gene expression and signal transduction. Biochim Biophys Acta 1995; 1259: 317-333

James-Kracke MR, Sexe RB, Shukla SD. Picomolar platelet-activating factor mobilizes Ca to change platelet shape without activating phospholipase C or protein kinase C; simultaneous fluorometric measurement of intracellular free Ca concentration and aggregation. J Pharmacol Exp Ther 1994; 271:824-831

Jarvis GE, Evans RJ. Endotoxin-induced platelet aggregation in heparinised equine whole blood in vitro. Res Vet Sci 1994; 57: 317-324

Jarvis GE, Evans RJ. PAF and not TXA_2 is an important mediator of endotoxin induced platelet aggregation in equine heparinised whole blood in vitro. Blood Coagulation and Fibrinolysis 1996; 7: 194-198

Kelefiotis D, Vakirtzi-Lemonias C. Kinetics of PAF transfer, distribution and metabolism in human blood in vitro. Life Sci 1997; 60: 2099-2109

Keraly CL, Benveniste J. Specific desensitization of rabbit platelets by platelet-activating factor (PAF-acether) and derivatives. Brit J Haematol 1982; 51: 313-322

Khan SN, Lane PA, Smith AD. Desaggregation of PAF-acether-aggregated platelets by verapamil and TMB-8 with reversal of phosphorylation of 40K and 20K proteins. Eur J Pharmacol 1985; 107:189-198

Kim HD, Burnett J, Shukla SD. Stimulation of glycolysis in human platelets by platelet activating factor. Cellular Pharmacol 1996; 3: 57-62

Kloprogge E, Akkerman JWN. Binding kinetics of PAF acether to human platelets. Biochem J 1984; 223: 901-909

Koltai M, Hosford D, Guinot P, Esanu A, Braquet P. Platelet activating factor: a review of its effects, antagonists and possible future clinical implications (Part I). Drugs 1991a; 42: 9-29, 1991b; 42: 174-204

Kornecki E, Ehrlich YH, Lenox RH. Platelet-activating factor-induced aggregation of human platelets specifically inhibited by triazolobenzodiazepines. Science 1984; 226: 1454-1456

Korth R, Zimmermann K. Richter WO. Lipoprotein-associated paf (LA-paf) was found in washed human platelets and monocyte/macrophage-like U937 cells. Chem Physics Lipids 1994; 70: 109-119

Kunz D, Gerard NP, Gerard C. The human leukocyte platelet-activating factor receptor, cDNA cloning, cell surface expression, and construction of a novel epitope-bearing analog. J Biol Chem 1992; 267: 9101-9106

Lapetina E. Platelet-activating factor stimulates the phosphatidylinositol cycle. J Biol Chem 1982; 257: 7314-7317

Le Blanc K, Berg A, Palmblad J, Samuelsson J. Stimulus-specific defect in platelet aggregation in polycythemia vera. Euro J Haematol 1994; 53: 145-149

Lee TC, Malone B, Blank ML, Snyder F. 1-alkyl-2-acetyl-sn-glycero-3-phosphocholine (platelet-activating factor) stimulates calcium influx in rabbit platelets. Biochem Biophys Res Commun 1981; 102: 1262-1268

Leoncini G, Signorello MG, Piana A, Carrubba M, Armani U. Hyperactivity and increased hydrogen peroxide formation in platelets of NIDDM patients. Thromb Res 1997; 86: 153-160

Lipscomb DL, Boudreaux MK, Paxton R, Spano J, Welles EG, Schumacher J. Identification of thrombospondin as a high molecular mass protein released from activated equine platelets. Amer J Vet Res 1997; 58: 954-960

Lorant DE, Zimmerman GA, McIntyre TM, Prescott SM. Platelet-activating factor mediates procoagulant activity on the surface of endothelial cells by promoting leukocyte adhesion. Sem Cell Biol 1995; 6: 295-303

Macconi D, Morzenti G, Livio M, Morelli C, Cassina G, Remuzzi G. Acetyl glycerylphosphorylcholine aggregates human platelets through two distinct pathways, both dependent on arachidonic acid metabolism. Lab Invest 1985; 52: 159-168

MacIntyre DE, Pollock WK. Platelet-activating factor stimulates phosphatidylinositol turnover in human platelets. Biochem J 1983; 212: 433-437

Marcus AJ, Safier LB, Ullman HL, Wong KTH, Broekman MJ, Weskler BB, Kaplan KL. Effects of acetyl glycerylether phosphorylcholine on human platelet function in vitro. Blood 1981; 58: 1027-1031

McManus LM, Hanahan DJ, Pinckard RN. Human platelet stimulation by acetyl glyceryl ether phosphorylcholine. J Clin Invest 1981; 67: 903-906

Merlos M, Giral M, Balsa D, Ferrando R, Queralt M, Puigdemont A, Garcia-Rafanell J, Forn J. Rupatadine, a new potent, orally active dual antagonist of histamine and platelet-activating factor. J Pharmacol Exper Therapeu 1997; 280: 114-121

Michibayashi T. Vascular responsiveness of platelet activating factor—with special reference to its action in the presence of platelets. J Smooth Musc Res 1996; 32: 195-202

Miller OV, Ayer DE, Gorman RR. Acetyl glycerylphosphorylcholine inhibition of prostaglandin I_2-stimulated adenosine 3', 5'-cyclic monophosphate levels in human platelets: evidence for thromboxane A_2 dependence. Biochem Biophys Acta 1982; 711: 445-451

Morrison WJ, Shukla SD. Desensitization of receptor coupled activation of phosphoinositide specific phospholipase C in platelets: Evidence for distinct mechanisms for platelet activating factor and thrombin. Mol Pharmacol 1988; 33: 58-63

Morrison WJ, Dhar A, Shukla SD. Staurosporine potentiates platelet activating factor stimulated phospholipase C activity in rabbit platelets but does not block its desensitization. Life Sci 1989; 45: 333-339

Mousa SA, Forsyth M, Lorelli W, Bozarth J, Xue CB, Wityak J, Sielecki TM, Olson RE, DeGrado W, Kapil R, Hussain M, Wexler R, Thoolen MJ, Reilly TM. Novel nonpeptide anatiplatelet glycoprotein IIb/IIIa receptor antagonist, DMP754: receptor binding affinity and specificity. Coronary Artery Disease 1996; 7: 767-774.

Mueller HW, Purdon AD, Smith JB, Wykle RL. 1-*0*-alkyl-linked phosphoglycerides of human platelets: distribution of arachidonate and other acyl residues in the ether-linked and diacyl species. Lipids 1983; 18: 814-819

Murphy CT, Kellie S, Westwick J. Tyrosine-kinase activity in rabbit platelets stimulated with platelet-activating factor. The effect of inhibiting tyrosine kinase with genistein on platelet-signal-molecule elevation and functional responses. Euro J Biochem 1993; 216: 639-651

Murphy CT, Westwick J. Role of type 1 and type 2A phosphatases in signal transduction of platelet-activating-factor-stimulated rabbit platelets. Biochem J 1994; 301: 531-537

Murphy CT, Westwick J. Stimulation of platelets with platelet-activating factor induces changes in the subcellular distribution and activity of the tyrosine kinase pp60c-src. Biochem Soc Transact 1995; 23: 194S

Nakahodo K, Saitoh S, Nakamura M, Kosugi T. Histamine release from rabbit platelets by platelet-activating factor. Arerugi-Japanese J Allergology 1994; 43:501-510

Nakamura M, Honda Z, Izumi T, Sakanaka C, Mutoh H, Minami M, Bito H, Seyama Y, Matsumoto T, Noma M, Shimizu T. Molecular cloning and expression of platelet-activating factor receptor from human leukocytes. J Biol Chem 1991; 266: 20400-20405

Natarajan V, Zuzarte-Augstin, Schmid HHO, Graff G. 1-alkylacyl and alkenylacyl glycerophospholipids of human platelets. Thromb Res 1983; 30: 119-125

Nieuwland R, Siffert W, Akkerman JW. Effect of epinephrine on the regulation of Na+/H+ exchange in human platelets. Biochim Biophys Acta 1993; 1148: 185-190

Nishihara J, Ishibashi T, Imai Y, Muramatsu T. Purification and characterization of the specific binding protein for platelet activating factor from human platelets. Tohuku J Exp Med 1985; 147: 145-152

Nunez D, Kumar R, Hanahan DJ. Inhibition of [^{3}H] platelet activating factor (PAF) binding by Zn^{2+}: a possible explanation for its specific PAF antiaggregating effects in human platelets. Arch Biochem Biophys 1989; 272: 466-475

Nunez D, Randon J, Gandhi C, Siafa-Kapadia A, Olson MS, Hanahan DJ. The inhibition of PAF induced platelet activation by oleic acid is associated with a decrease in polyphosphoinositide metabolism. J Biol Chem 1990; 265: 18330-18338

O'Flaherty JT, Redman JF, Schmitt JD, Ellis JM, Surles JR, Marx MH, Piantadosi C, Wykle RL. 1- O - alkyl - 2 - N - methylcarbamyl-glycerophosphocholine: a biologically potent, non-metabolizable analog of platelet-activating factor. Biochem Biophys Res Commun 1987; 147: 18-24

Oh SO, Ibe BO, Johnson D, Kurantsin-Mills J, Raj JU. Platelet-activating factor in plasma of patients with sickle cell disease in steady state. J Lab Clin Med 1997; 130: 191-196

Okamoto Y, Yoshida H, Ino M, Nakamura S, Okamoto G, Mizoguchi K, Suzuki Y, Sugatani J, Miwa M. PAF-releasing factor in human serum and inflammatory exudate. J Lipid Mediators 1993; 8: 151-156

Oura Y. The effects of platelet-activating factor (PAF) on the collagen-induced whole blood aggregation. Rinsho Byori-Japanese J Clin Path 1994; 42: 63-69

Packham MA, Ruben DH, Rand ML. Exogenous diacylglycerols synergize with PAF with human platelets, but inhibit PAF-induced responses of rabbit platelets. J Lipid Mediators Cell Signal 1996; 15: 69-87

Patterson WR, Dalton LM, McGlasson DL, Cissik JH. Aggregation of killer whale platelets. Thromb Res 1993 70: 225-231

Pelech SL, Charest DL, Howard SL, Paddon HB, Salari H. Protein kinase C activation by platelet-activating factor is independent of enzyme translocation. Biochim Biophys Acta 1990; 1051: 100-107

Pieroni G, Hanahan DJ. Metabolic behavior of acetyl glyceryl ether phosphorylcholine on interaction with rabbit platelets. Arch Biochem Biophys 1983; 224: 485-493

Pinckard RN, Farr RS, Hanahan DJ. Physicochemical and functional identity of rabbit PAF released in vivo during IgE sensitized basophils. J Immunol 1979; 123: 1847-1857

Pinckard RN, Woodard DS, Showell HJ, Conklyn MJ, Novak MJ, McManus LM. Structure and (patho) physiological diversity of PAF. Clin Rev in Allergy 1994; 12: 329-359

Piwinski JJ, Wong JK, Green MJ, Ganguli AK, Billah MM, West Jr, RE, Kreutner W. Dual antagonists of platelet activating factor and histamine. J Med Chem 1991; 34: 457-461

Rao AK, Willis J, Hassell B, Dangelmaier C, Holmsen H, Smith JB. Platelet-activating factor is a weak platelet agonist: evidence from normal human platelets and platelets with congenital secretion defects. Amer J Hematol 1984; 17: 153-165

Rao GHR, White JG. Platelet activating factor (PAF) causes human platelet aggregation through the mechanism of membrane modulation. Prostaglandins Leukotrienes Med 1982; 9: 459-472

Rezaul K, Yanagi S, Sada K, Taniguchi T, Yamamura H. Protein-tyrosine kinase p72syk is activated by platelet activating factor in platelets. Thromb Haemos 1994; 72: 937-941

Rogers JE, Duronio V, Wong SI, McNeil M, Salari H. Solubilization of a functionally active platelet-activating factor receptor from rabbit platelets. Biochem J 1991; 278: 405-410

Saeed SA, Gilani AH Antithrombotic activity of clove oil. JPMA - J Pakistan Med Assoc 1994; 44: 112-115

Sage SO, Mahaut-Smith MP, Rink TJ. Calcium entry in nonexcitable cells: lessons from human platelets. NIPS 1992; 7: 108-113

Salari H, Duronio, Howard S, Demos M, Pelech SL. Translocation-independent activation of protein kinase C by platelet-activating factor, thrombin and prostacyclin. Biochem J 1990; 267: 689-696

Samiei M, Sanghera JS, Pelech SL. Activation of myelin basic protein and S6 peptide kinases in phorbol ester- and PAF-treated sheep platelets. Biochim Biophys Acta 1993; 1176: 287-298

Satouchi K, Pinckard RN, McManus LM, Hanahan DJ. Modification of the polar head group of acetyl glyceryl ether phosphorylcholine and subsequent effects upon platelet activation. J Biol Chem 1981; 256: 4425-4432

Shah BH, Safdar B, Virani SS, Nawaz Z, Saeed SA, Gilani AH. The antiplatelet aggregatory activity of acacia nilotica is due to blockade of calcium influx through membrane calcium channels. Gen Pharmacol 1997; 29: 251-255

Shen TY, Hussaini I, Hwang SB, Chang MN. Recent developments of platelet activating factor antagonists. Adv Prostaglandin Thromboxane Leukotriene Res 1989; 19: 359-362

Shimizu T, Mutoh H. Structure and regulation of PAF receptor gene. Adv Exp Med and Biol 1997; 407: 197-204

Shukla SD, Hanahan DJ. AGEPC (platelet activating factor) induced stimulation of rabbit platelets: Effects on phosphatidylinositol, di- and tri-phosphoinositides and phosphatidic acid metabolism. Biochem Biophys Res Commun 1982; 106: 697-703

Shukla SD, Hanahan DJ. AGEPC induced stimulation of rabbit platelets: correlation between PA level, $^{45}Ca^{2+}$ uptake ^{3}[H] serotonin secretion. Arch Biochem Biophys 1984; 259: 1468-1471

Shukla SD. Platelet activating factor stimulated formation of inositol triphosphate in platelets and its modulation by various agents including Ca^{2+}, indomethacin, CV3988 and forskolin. Arch Biochem Biophys 1985; 240: 674-681

Shukla SD,Morrison WJ, Dhar A. Desensitization of platelet activating factor stimulated protein phosphorylation in platelets. Mol Pharmacol 1989; 29: 528-533

Shukla SD, Inositol phospholipid turnover in PAF transmembrane signaling. Lipids 1991; 26: 1028-1033

Shukla SD, Paul A, Klachko D. Hypersensitivity of diabetic human platelets to platelet activating factor. Thromb Res 1992; 66: 239-246

Shukla SD. Platelet activating factor receptor and signal transduction mechanisms. FASEB 1992; 6: 2296-2301

Shukla SD, Thurston AW, Zhu CY, Dhar A. Platelet activating factor receptor functions via phosphoinositide turnover and tyrosine kinase in "Platelet activating factor receptor: Signal mechanisms and molecular biology." Edited by S. D. Shukla, CRC Press 1993; 41-59

Shukla SD. Platelet activating factor receptor. In Biomembranes. Edited by A.G. Lee 1996; 2B: 219-235

Sirganian RP, Olser AG. Destruction of rabbit platelets in the allergic response of sensitized leukocytes: demonstration of a fluid phase intermediate. J Immunol 1971; 106: 1244-1251

Snyder F. Chemical and biochemical aspects of PAF: a novel class of acetylated ether-linked choline phospholipids. Medicinal Res Reviews 1985; 5: 107-140

Snyder F. Platelet activating factor, the biosynthetic and catabolic enzymes. Biochem J 1995; 305: 689-705

Sugatani J, Miwa M, Hanahan DJ. Platelet-activating factor stimulation of rabbit platelets is blocked by serine protease inhibitor (chymotryptic protease inhibitor). J Biol Chem 1987; 262: 5740-5747

Tokumura A. A family of phospholipid antocoids, occurrence, metabolism and bioactions. Progress in Lipid Research 1995; 34: 151-184

Travers JB, Li Q, Sprecher H, Fentel RH. Binding of carbamyl PAF to Raji lymphoblast PAF receptor. Int J Immunopharmacol 1992; 14: 515-523

Tripathi YB, Pandey S, Shukla SD. Anti-platelet activating factor property of Rubia cordifolia Linn. Indian J Exper Biol 1993; 31: 533-535

Tsoukatos D, Demopoulos CA, Tselepis AD, Moschidis MC, Donos A, Evangelou A, Benveniste J. Inhibition by cardiolipins of platelet-activating factor -induced rabbit platelet activation. Lipids 1993; 28: 1119-1124

Tuffin DP, Davey P, Dyer RL, Lunt DO, Wade PJ. Specific binding of [^{3}H] -1-0-octadecyl PAF-acether to washed human platelets. Adv Exp Med Biol. Ed Westwick, J. et. al. Plenum Press 1985; 83-96

Tysnes OB, Verhoeven AJ, Holmsen H. Neomycin inhibits agonist-stimulated polyphosphoinositide metabolism and responses in human platelets. Biochem Biophys Res Commun 1987; 144: 454-462

Valone FH, Coles E, Reinhold VR, Goetzl EF. Specific binding of phospholipid platelet activating factor by human platelets. J Immunol 1982; 129: 1637-1641

Valone FH. Isolation of a platelet membrane protein which binds the platelet-activating factor 1-0- hexadecyl-2-acetyl-SN-glycero-3-phosphorylcholine. Immunology 1984; 52: 169-174

Valone FH, Johnson B. Decay of the activating signal after platelet stimulation with 1-0-alkyl-2-acetyl-sn-glycero-3-phosphorylcholine: changes in calcium permeability. Thrombosis Res 1985; 40: 385-392

Vargaftig BB, Chignard M, Benveniste J. Present concepts on the mechanisms of platelet aggregation. Biochem Pharmacol 1981; 30: 263-271

Viero P, Cortelazzo S, Barbui T, Gaetano G. Defective platelet aggregation induced by platelet activating factor in myeloproliferative disorders: deficiency of an aspirin-independent mechanism. Haemostasis 1986; 16: 27-33

Wang F, Naik UP, Ehrlich YH, Freyberg Z, Osada S, Ohno S, Kuroki T, Suzuki K, Kornecki E. A new protein kinase C, nPKC eta', and nPKC theta are expressed in human platelets: involvement of nPKC eta' and nPKC theta in signal transduction stimulated by PAF. Biochem Biophys Res Commun 1993; 191: 240-246

Winslow CM. Effects of known pharmacologic receptor antagonists and platelet aggregation inhibitors on specific binding of [^{3}H] PAF to human platelets. Fed Proc 1984; 43: 1941

Yamada K, Iwahashi K, Kase H. Parallel inhibition of platelet-activating factor-induced protein phosphorylation and serotonin release by K-252a, a new inhibitor of protein kinases, in rabbit platelets. Biochem Pharmacol 1988; 37: 1161-1166

Yanabu M, Ozaki Y, Nomura S, Mayake T, Miyazaki Y, Kagawa H, Yamanaka Y, Asazuma N, Satoh K, Kume S, Komiyama Y, Fukuhara S. Tyrosine phosphorylation and p72syk activation by an anti-glycoprotein Ib monoclonal antibody. Blood 1997; 89: 1590-1598

Yang BC, Chandna VK, Lawson DL, Mehta JL. Platelets enhance anoxic contraction of rat aortic rings through platelet-activating factor-dependent mechanism. J Cardio Pharmacol 1993; 21: 249-255

Ye, RD, Prossnitz, ER, Zou AH, Cochrane CG. Characterization of a human cDNA that encodes a functional receptor for platelet activating factor. Biochem Biophys Res Commun 1991; 180: 105-111

Zhu CY and Shukla, S.D.. Increased tyrosine kinase activity in pp60c-src immunoprecipitate from platelet activating factor stimulated human platelets: in vitro phosphorylation of a synthetic peptide. Life Sci 1993 ; 53: 175-183

Zhu Y, Schieber EB, McGiff JC, Balazy M. Identification of arachidonate P-450 metabolites in human platelet phospholipids. Hyperten 1995; 25: 854-859

7

PLATELET GLYCOPROTEIN Ib-V-IX: BIOLOGY AND FUNCTION

K. J. Clemetson, Ph. D.
Theodor Kocher Institut
Der Universitat Bern
Freiestrasse 1
Ch-3012 Berne
Switzerland

7.1 INTRODUCTION

Dr. K. J. Clemetson has spent the last 25 years of his teaching and research career on platelet receptor biohcmeistry. Molecular biology and signal transduction mechanisms at the Throdor Kocher Institute of the University of Berne. He has made major contributions to the isolation of platelet receptor glycoproteins as well as to the identification of receptor function and to the analysis of genetic defects in the bleeding disorder Bernard-Soulier Syndrome. He has been active in national and international societies, including the International Society on Thrombosis and Haemostasis and was preseident of the European Thrombosis Research Organization from 1993-1997. He was awarded the Mogens-Ellerman Prize of the Swiss Haematology Society in 1990.

Platelet adhesion to damaged vessel wall in order to stop bleeding and start repair mechanisms is a complex process involving a number of platelet receptors as well as adhesive molecules and extracellular matrix proteins. On top of this are feed-back mechanisms both positive and negative that control the level of platelet response to the injury. The whole process is carefully regulated and it is this regulation that distinguishes normal physiological haemostasis from its pathological counterpart thrombosis. The GPIb-V-IX complex (CD42) is a major platelet receptor involved in both the early stages of adhesion to damaged vessel wall under high stress conditions as well as the platelet response to low levels of thrombin. The complex consists of four subunits GPIba, GPIbb, GPIX and GPV which all belong to the leucine-rich repeat protein family. The importance of these molecules in haemostasis can be judged by the bleeding problems resulting when they are defective or absent such as in Bernard-Soulier syndrome or when the counterpart adhesive ligand von Willebrand factor (vWf) is defective or absent in von Willebrand's disease. The GPIb-V-IX complex remains an interesting alternative target for control of platelet activation from those at present in clinical use or being tested. There are many aspects of this receptor and how it works which are still obscure although the last few years have improved knowledge in this area. This chapter deals with current knowledge of the structure and function of this receptor complex as well as some aspects of its molecular biology and regulation of its tissue specific expression and how signals are transduced to the platelet interior and lead to aggregation.

7.2 Structure

The GPIb-V-IX complex (CD42) consists of four subunits which are produced by genes present on three different chromosomes. GPIb is composed of two subunits a- (CD42b) (1) and b- (CD42c) (2) connected by a disulfide bond. The non-reduced form has a mass of 170 kDa while the reduced subunits have masses of 140 kDa (a-) and 27 kDa (b-) respectively. The gene for GPIba is on chromosome 17p12ter while that for GPIbb is on chromosome 22q11.2 (3). GPIX (CD42a) is a non-covalently associated subunit with a mass of 22 kDa (4), the gene is on chromosome 3q29 (5-6), while GPV (CD42d) is a still weaker non-covalently associated subunit with a mass of 82 kDa (7), the gene is on chromosome 3q21 (6). Whereas the other members are present at a 1:1 stoichiometry, GPV forms a 1:2 relationship with the rest of the complex (8). All these subunits belong to the same general family of proteins, the leucine-rich repeat family, and share several features of this protein class. The prototype of this family, leucine-rich a2 glycoprotein, was discovered in 1985 (9). It is a minor plasma protein and its function remains unknown. A characteristic feature of this family is the presence of repetitive sequence motifs containing many leucines and an asparagine in a typical distribution. All of the GPIb-V-IX subunits as well as leucine-rich glycoprotein belong to the 24 amino acid repeat family but repeats containing different sized units - often 28 amino acids - are known. GPIba contains six complete leucine-rich repeats plus two flanking domains (or half repeats) at each end. Both GPIbb and GPIX contain one complete repeat repeats plus two flanking domains (or half repeats) at each end and GPV contains 15 complete repeat repeats plus two flanking domains (or half repeats) at each end. Each 24 amino acid repeat consists of a part containing the leucines and asparagine which folds as a b-sheet followed by the remainder of the motif which forms an a-helix. Based upon X-ray crystallography of ribonuclease inhibitor containing 28 amino acid leucine-rich repeats (10) it is possible to predict that, in structures such as GPIba and GPV containing several repeats, they fit together as wedge-shaped units to form arc-like or horse shoe-like structures. The a-helical part forms a convex, and the b-sheet a concave surface. In GPIb-V-IX the function of these domains remains obscure but in related structures such as the small collagen-binding proteins the concave surfaces formed by the b-sheets are physiological important and engage in the binding (11, 12). On each side of the leucine-rich repeat domains are conserved disulfide bonds which form loops relevant to the stability and function of the overall structure. The loops formed by those on the N-terminal side are either simple or double. When they are double they most likely form loops one within another similar to other members of this family (1-4, 2-3 pattern). The loops on the C-terminal side are more complex and a 1-3,

2-4 pattern seems to be conserved (13). In GPIba it is the combination of the double loop with the leucine-rich domain that forms the major part of the physiological binding sites. From this point on, the structures of the various subunits diverge more. GPIba, for example, contains a highly negatively charged segment including three sulfated tyrosine residues (14, 15), which is also rich in cleavage sites for major physiological proteases (16). Much of the rest of the extracellular domain of GPIba is rich in serine and threonine residues and is highly O-glycosylated. This highly O-glycosylated region contains a

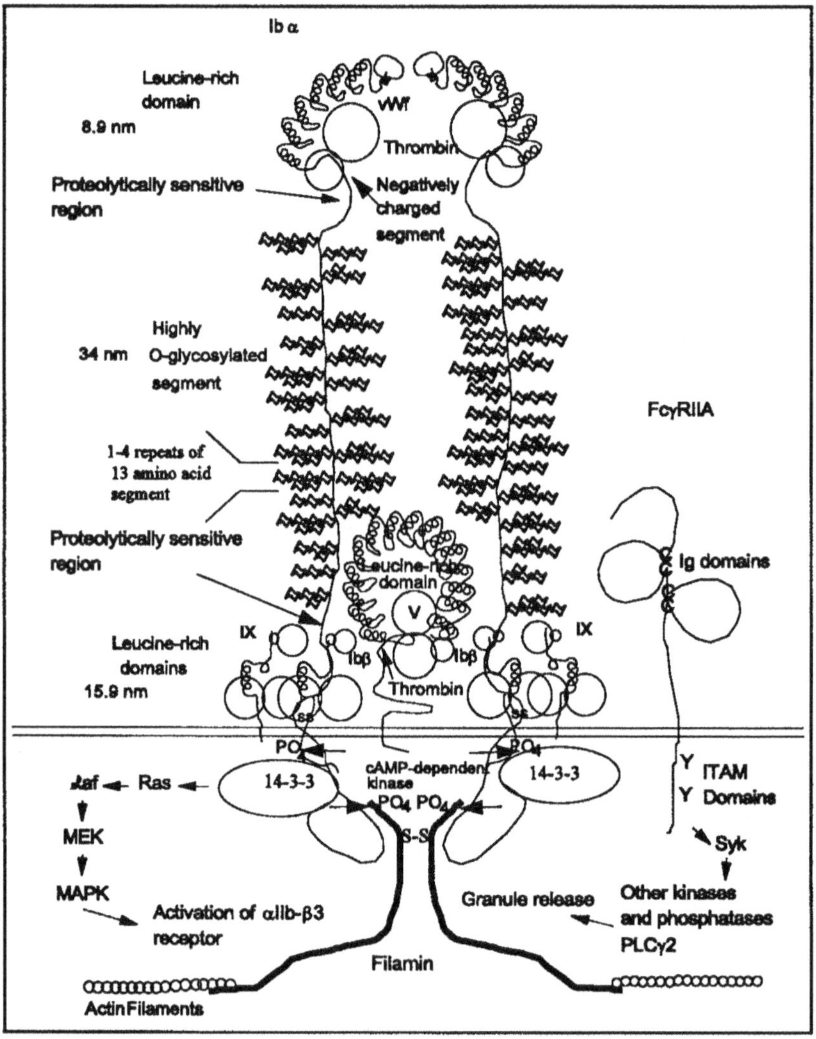

Figure 1. Schematic drawing of the GPIb-V-IX complex. Domain structure, putative binding sites and interactions between subunits and with the cytoskeleton as well as signalling pathways are indicated.

polymorphic 13 amino acid stretch which is present either in single, double, triple or quadruple forms called D, C, B and A, respectively, (17-19) that will be dealt with in more detail later. Just above the membrane is an unglycosylated region which again contains cleavage sites for several proteases. At the surface of the membrane is the cysteine which is linked to the b-subunit. After the transmembrane domain comes a cytoplasmic section of 130 amino acids. The other two small subunits as well as GPV have relatively short cytoplasmic sections. Fig. 1 shows the overall structure of the GPIb-V-IX structure with the main features indicated.

7.3 Post-translational modification of GPIb-V-IX

N- and O-glycosylation

Several forms of post-translational modification have been detected. The most important is glycosylation, both N- and O-. The type of glycosylation and the structure of the major oligosaccharides have been determined for GPIba (20-23). One of the most significant features of this subunit is the large number of O-glycosylation sites that form a mucin-like domain contributing to the stiffness of the overall structure and holding the receptor domains out from the membrane surface. Both the GPIbb and GPIX subunits have single biantennary N-linked oligosaccharides and no O-glycosylation (24). The glycosylation of GPV is less well studied. There are 8 putative N-glycosylation sites of which 4 have been definitively shown to be occupied. It is still unclear whether there is any O-glycosylation but there is some evidence for a few such sites in the domain just above the membrane, which contains Ser or Thr residues, based on the DNA sequence but gave blanks at these positions in direct peptide sequencing (25).

7.3.1 Phosphorylation

Another important modification is phosphorylation of Ser, Thr or Tyr residues within cytoplasmic domains as a signalling mechanism. In the GPIb-V-IX complex only GPIbb has been found to be modified in this way (26) and Ser166 is phosphorylated by cyclic AMP dependent kinase (27). This phosphorylation seems to play a role in stabilizing resting platelets. Ser166 is dephosphorylated when platelets are activated by a range of agonists and may be involved in the interactions of GPIb-V-IX with cytosolic components such as 14-3-3z.

7.3.2 Sulfatation

Although the modification of the hydroxyl group of tyrosine residues by a sulfate group was known for soluble proteins such as hirudin, such modifications of membrane proteins are fairly recent discoveries, largely due to the observation of discrepancies between recombinant and natural molecules. In the GPIb-V-IX complex there are three tyrosine residues Tyr276, Tyr 278 and Tyr 279, lying within the highly negatively charged domain between the double loop and the mucin-like domain, which are sulfated and which play a role in the thrombin and vWf binding sites (14, 15).

7.3.3 Acylation

Post-translational modification of Cys residues within the transmembrane and cytoplasmic domains of membrane-bound proteins are thought to play a role in modification of the membrane anchoring. Myristoylation can also occur by co-translational mechanisms and may have a role in protein-protein interactions. In the GPIb-V-IX complex there is evidence that both GPIb and GPIX are modified in these ways. While it was shown earlier that both GPIbb and GPIX are palmitoylated (28) other studies have shown that GPIX is also a major myristoylated protein, probably via an amide bond to an e-amino group of a lysine residue (29).

7.4 Basic Functions

7.4.1 Receptor properties

The major established roles of GPIb-V-IX are as a platelet receptor for vWf and as a receptor for thrombin but it also has ancillary functions in stabilising membrane-cytoskeleton interactions. These functions were first recognised because of their absence in Bernard-Soulier syndrome where this receptor complex is either not expressed or is functionally defective (30). Normal platelets, with intact GPIb-V-IX, agglutinate when treated with either ristocetin or botrocetin in the presence of human vWf or directly with either bovine or porcine vWf or asialo or Type IIB human vWf. These reagents somehow mimic the physiological interactions between GPIb and vWf that occur *in vivo*. However the *in vivo* conditions require the association of the vWf with collagen as well as adequate shear rates in order to induce binding between these structures. The differences between these animal vWf or modified human vWf and normal human vWf is thought to lead to an increased binding to human GPIb. Similarly, both ristocetin and botrocetin (31) induce interactions between normal vWf and GPIb. The aminoglycopeptide antibiotic ristocetin forms dimers (32) which bind to proline-containing peptides in both GPIb and vWf and therefore stabilises primary interactions. Botrocetin, a snake C-type lectin from various species of Bothrops, interacts with vWf and is thought to induce and stabilise the change in conformation which is responsible for the physiological interaction with GPIb. The A1 domain of vWf contains the binding site for all of these interactions with GPIb. This domain belongs to the I-domain class and like the similar domain from the a_2 integrin subunit it has been crystallised and its structure determined (33). It is a typical barrel-shaped structure with the putative binding site near the apex and is dependent on cations such as Ca2+ for a binding conformation. A number of mutations within the A1 domain give rise either to decreased interactions with GPIb in Type IIA von Willebrand's disease, or to enhanced interactions with GPIb in Type IIB von Willebrand's disease (34).

Many aspects of the structures within GPIba which interact with vWf still remain obscure. Two recurrent mutations within the larger loop of the major double loop system in GPIba have been described (Figure 2) which lead to increased binding to normal vWf (35, 36). In the same way that the mutations in Type IIB von Willebrand's disease are thought to mimic the conformation of the A1 domain of vWf binding to GPIb these may mimic the conformation of GPIb which binds to vWf (37). There is evidence for other domains of GPIb playing a role in vWf interactions. The negatively charged, tyrosine sulfate containing sequence is one of these and may participate in the initial contacts between the GPIb and vWf necessary for adaptation in conformation to facilitate the binding between these structures. Although the role of the leucine-rich domains in these interactions is unknown it would be surprising if they did not participate because they provide an ideal protein-protein interface.

The role of GPIb in platelet thrombin receptor function has remained controversial, largely because it is most evident at low thrombin concentrations (38), and these conditions have not always been used (39). Nevertheless there are a long series of reports, starting with Bernard-Soulier syndrome platelets (40) and extending to studies with antibodies to GPIb (41, 42) as well as with specific proteolytic cleavage of GPIb (43) and the use of various specific inhibitors that point to a role for GPIb as a thrombin receptor. Indeed it was demonstrated at an early stage that GPIb does contain a thrombin binding site and that the soluble fragment of GPIb, glycocalicin, is able to inhibit platelet reponses to thrombin (44). Since then it has become clear that, as thrombin can bind to both GPIb as well as the seven transmembrane (PAR-1) thrombin receptor, this provides a possible mechanism for enhancing the platelet response to thrombin by GPIb (45). It has also been suggested that the site for GPIb binding (via the negatively charged domain) to thrombin is the same as that used by heparin, so that one of the effects of heparin may be to reduce the platelet response to low doses of thrombin (46). GPV contains a thrombin cleavage site just under its double loop structure but all the evidence is against a major role for this in platelet activation by thrombin (47). The function of this cleavage site, as the role of GPV itself, still remains obscure. Current efforts to prepare GPV "knock-out" mice may cast some light on its function.

7.4.2 Proteolytic cleavage sites

Both GPIba and GPV are easily cleaved on the platelet surface by a wide range of proteases. This was a major cause of confusion whether or not GPIba was membrane-bound during the early days of research on platelet membrane receptors. The question of whether this proteolysis has an important role in physiology and pathology has still not been answered satisfactorily. In GPIba there are two areas which are particularly susceptible, one just above the membrane and the other between the double loop domain and the highly O-glycosylated domain. Cleavage just above the membrane gives the large extracellular fragment called glycocalicin because it is a major contributor to the glycocalix or sugar-rich domain surrounding the platelet (48). Endogenous platelet calpain is a major protease cleaving glycocalicin from the platelet but the cleavage sites are still not known precisely. Other proteases, both endogenous and from other cells may also be involved in removal of a glycocalicin-like molecule from the platelet but until the cleavage sites can be determined it is difficult to demonstrate these differences (49). Elastase and cathepsin G from leukocytes, as well as trypsin, also cleave GPIba in this region but it is not the primary cutting site for these enzymes and they release first a 45 kDa fragment which comes from cleavage at the other site between the double loop domain and the highly O-glycosylated domain. The cleavage sites in this area have been determined (50). A protease, mocarhagin, from snake venom was shown to be specific for the sulfated tyrosines domain and to cleave in the middle of this region (51). It should be noted that all of these cuts remove the vWf and thrombin binding sites from the platelet surface (Figure 2).

As mentioned above, GPV contains a thrombin cleavage site which liberates a 69 kDa fragment containing the leucine rich repeats and the double loop domain. GPV is also cut by other proteases, some of which cleave just above the membrane releasing a fragment that is only slightly smaller at 80 kDa than the complete membrane-bound molecule. Endogenous calpain is again thought to be the major enzyme cleaving at this site (47). There is some evidence for other granule or lysozomal enzymes cleaving in this region but they are poorly characterised. A better definition of the physiological cleavage sites on these receptors might be of considerable clinical interest as they could be used as longer term markers for platelet activation *in vivo*. Several other markers of platelet activation have been used and are currently in use, but all have the disadvantage that they disappear relatively rapidly and therefore say little about longer term or brief platelet activation events that may be indicative of dangerous states preceeding myocardial infarction or stroke. Thus, platelet specific substances released from a-granules or granule membrane proteins exposed at the surface following activation are indicative of recent platelet activation but within an hour are no longer detectable. Very little is known of the fate of the residual fragments left on the platelet surface after cleavage of GPIba and GPV. It is possible that because of their linkage with the membrane associated cytoskeleton these stumps are selectively endocytosed into the platelets. However, they may remain on the surface as useful markers of platelet exposure to proteases during circulation and would undoubtedly be increased in platelets which had been activated or exposed to proteases. Higher levels of glycocalicin in plasma are found associated with disease states involving increased platelet destruction (52).

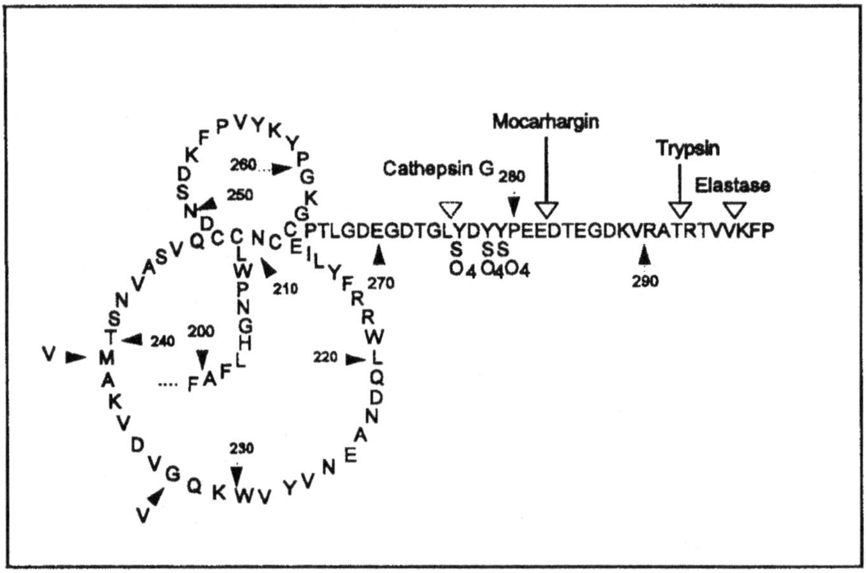

Figure 2. Diagram of the double loop and negatively charged domain of GPIba showing the mutations leading to platelet-type von Willebrand's disease, Gly233Val and Met239Val, as well as the major proteolytic cleavage sites.

7.5 Signal Transduction

VWf and GPIb interact under specific physiological conditions involving high shear stress conditions and this can be simulated under static conditions by changes in either of the partners leading to enhanced interactions. It was thought for a long time that these interactions are essentially passive and for this reason the platelet-platelet adhesion resulting was referred to as agglutination in contrast to the term aggregation used to describe the process involving activated platelets sticking together via fibrinogen.

However, evidence accumulated over the last decade has clearly shown that the initial interactions between GPIb and vWf lead to an activation of the platelets and to downstream signal transduction, probably involving activation of GPIIb-IIIa (53, 54). The initial stages of this process probably involve changes in the interaction of the cytoplasmic domain with other cytoplasmic proteins. A major component in this process linking GPIb-V-IX to downstream signalling is 14.3.3z which was found as a GPIb complex associated component (55). More recently two groups have shown that there are sequences, not only within the GPIba cytoplasmic domain but also in the cytoplasmic domains of GPIbb and possibly GPV that can interact with this protein (56). In particular studies with two-hybrid systems point to an important role for the cytoplasmic domain of GPIbb containing serine166 that is phosphorylated by cyclic AMP-dependent kinase (57). Thus, the interaction of 14.3.3z with GPIbb may therefore be stabilised by conditions that activate cyclic AMP-dependent kinase. How interactions between vWf and GPIb influence the state of the cytoplasm via 14.3.3z is still not clear. Perhaps this interaction induces dephosphorylation of GPIbb and therefore dissociates the 14.3.3z allowing it to activate downstream molecules such as Raf in a cascade reaction (58).

Another cytoplasmic molecule with which GPIb is intimately involved is actin binding protein (filamin). Part of the cytoplasmic domain of GPIba has been shown to interact with filamin (59, 60). Is this interaction necessary for platelet function? One of the features of Bernard-Soulier syndrome platelets where GPIb-V-IX is deficient is the abnormally large size of the platelets as well as an abnormal phospholipid distribution in these platelets (61). It has been suggested that this may related to the lack of linkage between GPIb and filamin. Recently, studies on the expression of filamin in cell lines on the interaction between GPIb-V-IX and the cell membrane have indicated that filamin has indirect effects on membrane incorporation of GPIb lacking filamin binding sequence as well as on that of the b_1 integrin subunit (62).

It is still unclear how signal transduction is induced by vWf binding to GPIb. One possible way could involve the FcgRIIA receptor in activation of tyrosine kinases (63). There have been recent studies suggesting that activation of GPIb associated tyrosine kinases (64, 65) may represent a plausible mechanism and it is known that several enzymes including PI3 kinase and p60src are associated with the cytoskeleton following platelet activation via GPIb and vWf (66).

Several different approaches to platelet activation via GPIb have been studied. The classical way has been via vWf itself using either ristocetin or botrocetin to induce interactions with GPIb. vWf is a complex molecule with many different interaction sites including an RGD sequence that can bind to activated GPIIb-IIIa. Thus, any interaction between vWf and GPIb probably automatically implicates more complex interactions between vWf and platelets. However, cross-linking with specific antibodies to GPIb has also been used and probably simulates more directly this first part of the process. Similarly, several snake C-type lectins with dimeric or tetrameric structures and binding sites specific for GPIb also activate platelets (67, 68).

7.6 GPIb-V-IX As a Target for Pharmacological Inhibition

Inhibition of platelets has become common as a method of controlling thrombosis under both chronic and acute situations. The major target structure for inhibition has been GPIIb-

IIIa since the eventual activation of this receptor is a common mechanism to which all pathways lead and therefore its blockage can have a systematic effect (69, 70). Although this presents some obvious advantages in terms of effectiveness, current studies tend to point to a number of disadvantages as well. One major problem is the steep dose/response curve precise dosing. A second associated problem which is often mentioned in connection with inhibition of GPIIb-IIIa is bleeding problems (71). There is therefore considerable interest to know if alternative strategies of inhibiting platelet activation by blocking other agonist or adhesion receptors may offer an alternative with fewer side effects. Other alternatives that are being extensively explored are blockage of the major feedback loops of platelet activation such as those involving thromboxanes and ADP by using combinations of aspirin and ticopidine or clopidogrel (72). However, strategies which reduce the activation rates of GPIIb-IIIa by blocking upstream adhesion receptors offer a further possibility. We have seen that GPIb-V-IX is probably the first receptor to be engaged during platelet adhesion to exposed subendothelium and there is recent evidence that is is also involved in the interactions between incoming quiescent platelets and a growing thrombus (73). Studies with several inhibitors of the GPIb-V-IX/vWf axis in animal thrombotic models have indicated that these can be effective blockers of the thrombotic process with surprising little effect on background haemostasis. Several peptides derived from the vWf A1 loop have been used which bind spontaneously to GPIb (74, 75). More recent studies have employed the subunits of snake C-type lectins as specific inhibitors (76-78). The use of the subunits avoids platelet activation due to cross-linking which can occur when the intact C-type lectins are used. An additional advantage in blocking receptors upstream of GPIIb-IIIa engagement is that the overall platelet activation is reduced. This means, for example, that release of all of the growth factors and hormones from the granules does not occur.

7.6.1 Polymorphisms Within the GPIb-V-IX Complex

There are two simple polymorphisms known within the GPIb-V-IX complex that give rise to immunological complications after blood transfusions. These are at Thr145/Met on GPIba (79) and at a location near the N-terminal cysteine loop on GPIbb giving rise to the *ly* alloantigen (80). A polymorphism was detected on GPIX at Ala139/Thr but no immune reactions have been reported (81). As well as these single amino acid substitutions GPIba also contains a 13 amino acid sequence from Ser399 to Thr411which is present as single, double, triple or quadruple copies (D,C, B and A) (Figure 3) leading to a molecule with different lengths (18). The longest version (A) is practically only found in populations in Japan and East Asia (19).The size polymorphism is linked to the Thr145/Met polymorphism so that the two and three copy polymorphisms are associated with the Met145 polymorphism and the one and two with the Thr145 polymorphism. At the moment there is considerable speculation about possible effects of these polymorphisms in sensitivity to activation of platelets or in binding to vWf which might influence susceptibility to cardiovascular disease in the same way that has been suggested for polymorphisms in GPIIb-IIIa (82-84).

7.6.2 Immunogenicity of GPIb-V-IX and role in immune thrombocytopenia

The large number of copies of GPIb-V-IX on the platelet surface is certainly a factor in the role of this complex in immunological disorders. As mentioned above various polymorphisms can be involved in the development of alloantibodies after blood or platelet transfusions. However, cases have been reported where autoantibodies were produced with epitopes on the GPIb-V-IX complex, in some cases giving rise to an acquired BSS state by blocking vWf binding to GPIb (85). Drug-induced thrombocytopenia involving antibody formation against neoantigens present in complexes of drugs with GPIb-V-IX as well as other platelet receptors has been often reported. The chinchona alkaloids, quinine and quinidine, used for prophylaxis and treatment of malaria, are particularly prone to cause this problem, perhaps because of a particular affinity for the platelet receptor (86).

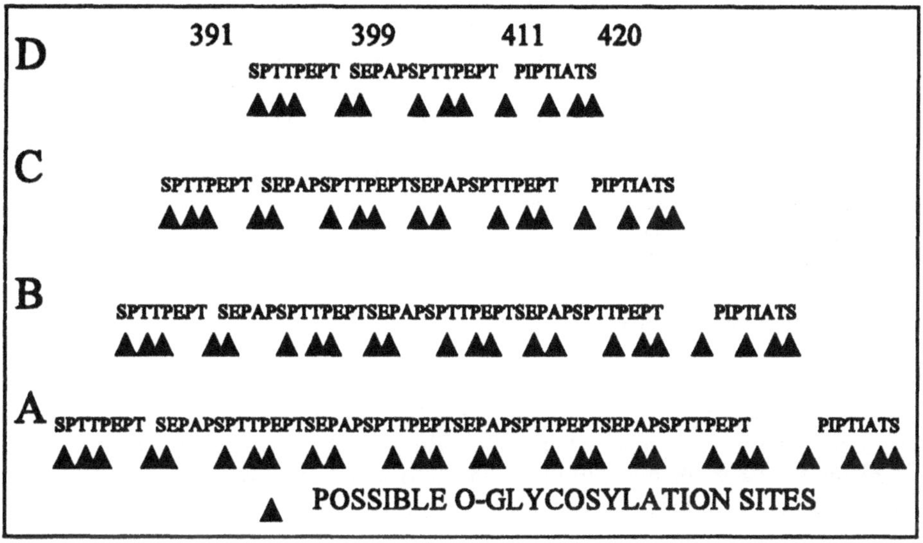

Figure 3. Diagram of size polymorphic domain of GPIba showing single, double, triple and quadruple forms. Putative O-glycosylation sites are indicated which may account for the differences in molecular mass found between these polymorphisms. Because of steric hindrance it is unlikely that all sites are glycosylated.

7.7 Genetic Disorders Affecting GPIb-V-IX

7.7.1 Bernard-Soulier syndrome

This disorder is caused by defects in one of the subunits leading to a molecule which does not integrate into the plasma membrane directly or which does not fold correctly and which is therefore degraded. Mutations leading to BSS have been detected in all of the subunits with the exception of GPV. Although this may merely mean that mutations in GPV have simply not been detected, another possible interpretation is, that, while the expression of the other subunits are necessary for the expression of GPV, GPV expression may not be necessary for the expression of the rest of the complex. The types of mutations that have been detected so far fall into a number of categories. Several mutations in GPIba are the type that lead to a frameshift and to a nonsense or stop codon before the transmembrane domain is translated, with the result that this subunit does not become embedded in the membrane and is expressed as a soluble protein (87, 88). Recently, patients with a variation on this were described with a mutation within the transmembrane domain of GPIba causing a frameshift so that this subunit was poorly incorporated in the membrane and the cytoplasmic domain was completely altered and, therefore, could not interact normally with cytoplasmic proteins (89, 90). Two categories of mutation that have been found in all three essential subunits affect either the leucine-rich repeats (91, 92) or else the cysteines of the loop systems (93, 94). Both these classes of mutation have major effects on folding and the stability of the subunit structure. Unlike Glanzmann's thrombasthenia there are hardly any examples of mutations leading to normal levels of expression of a malfunctioning complex. There are a few rare cases which have been described where the mutation in a single allelle is known and does not completely account for the phenotype. One of these has been described as showing a dominant inheritance which would imply that a mutation in a single allele would be adequate to cause the defect (95). So far there have been no models of the structure of the GPIb complex incorporating features that can explain such properties. One interesting mutation affecting one allele of the GPIbb gene (the other allele contained a deletion affecting this gene) was a point mutation present in a GATA transcription factor binding element, showing its importance for tissue specific expression (3). This is the first example of a mutation in a non-protein coding region causing BSS.

7.7.2 Bolin-Jamieson syndrome

There have been a few cases described of a dominantly inherited disorder affecting GPIb. These affect the length of the molecule and have been described as Bolin-Jamieson syndrome after the first reported case (96). The patients generally have mild bleeding problems and have one allele of the GPIba gene which expresses a molecule which is clearly longer than the longest polymorphic forms (41). The molecular explanation for this difference in size as well as the reason that it causes bleeding problems are still lacking. It had been suggested that perhaps there is a further duplication of the size polymorphism region leading to the larger form and that this larger size not only functions less well than the normal size range but that it somehow interferes with the function of the normal phenotype GPIb, however this remains speculation.

7.7.3 Platelet-type (Pseudo) von Willebrand's disease

As well as loss of function mutations there are also gain of function mutations which are classified as platelet-type von Willebrand's disease (vWd) because originally the symptoms showed some similarity to , in particular Type IIB. The reasons for this are clear. Because of the enhanced avidity between GPIb and vWf platelets aggregate and are removed from the circulation along with the larger vWf multimers. The patient therefore has bleeding complications because of both the thrombocytopenia and the loss of these higher, haemostatically more effective, multimers. Analysis of the vWf multimer composition of plasma shows a typical reduction in higher multimers as found also with Type IIB von Willebrand disease. In the latter case the disorder is caused by mutations in vWf leading to higher avidity for platelet GPIb. Two mutations have been detected in GPIb leading to platelet-type vWd (35, 36). Both of these lie in the larger of the loops making up the double loop structure (Figure 2) and in a domain which is implicated in binding to both vWf as well as thrombin.

7.8 Tissue Specific Expression of GPIb-V-IX Subunits

GPIb-V-IX subunits belong to the class of proteins that are thought to be normally specific for the megakaryocytic cell lineage. There are a few other platelet proteins that also fall in this category including GPIIb, probably also GPVI and, of course, the a-granule proteins PF4 and platelet basic protein. Despite this lineage specificity there have been several reports of expression of GPIb complex components on non-megakaryocytic cells. These include endothelial cells, tonsillar cells and tumour cells. All of the components of the GPIb-V-IX complex have been reported to be expressed on cultivated HUVECs as well as on adult endothelial cells and on tonsillar cells (97, 98) . The expression on endothelial cells may be related to inflammatory conditions (99, 100). It was also suggested that the GPIbb subunit expressed in endothelial cells may be larger than that expressed in platelets (101). However, an alternative explanation has been proposed for the expression of a larger form of GPIbb mRNA in a number of tissues which is based upon the observation that there is a gene (hCDC*rel*-1) upstream of that for GPIbb which has an inefficient polyadenylation signal so that the RNA polymerase tends to read through until it reaches the GPIbb gene stop signal and thus produce an abnormally long message (102). The protein coded by this abnormal message has not been identified but would not be expected to be physiologically relevant. This longer message has been found in several cell types not just in endothelial cells. The lack of expression of the normal length mRNA in endothelial cells would then argue against functional relevance. If expression of GPIb-V-IX on endothelial cells is confirmed it might be related to strengthening the attachment of these cells to subendothelium, via vWf under high shear stress conditions.

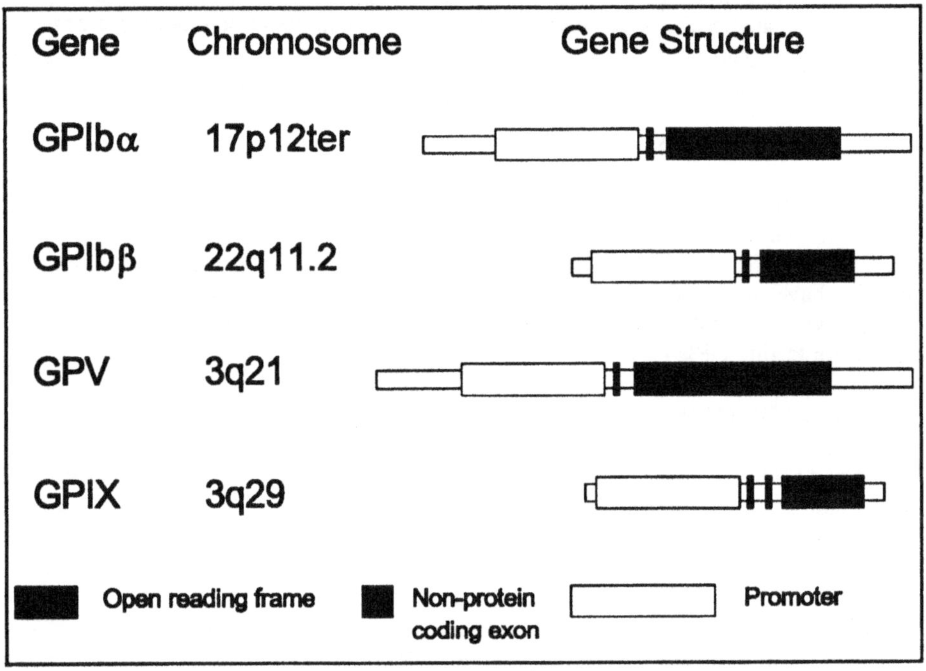

Figure 4. Gene structure and chromosomal location of the genes for the GPIb-V-IX complex.

There have also been several reports of GPIb complex proteins produced by neoplastic tissue implying that normal tissue specificity controls have failed to operate. A protein immunorelated to GPIba was detected on human breast carcinoma tissue but not on normal breast tissue (103). Aberrant expression of GPIb and vWf by splenic marginal zone lymphoma cells lead to acquired von Willebrand's disease (104). The mechanisms of this abnormal expression are of considerable interest because other megakaryocyte lineage specific markers are apparently not expressed at the same time.

7.8.1 Gene Structure And Tissue Specific Control of Gene Expression

The genes for the subunits of the GPIb-V-IX complex have simple structures compared to genes for other megakaryocytic specific proteins (Figure 4). There are only single copies of each gene and unlike PF4 and PBP no pseudogenes have been found. Three of the genes have a single exon containing all protein-encoding DNA separated by a small intron from a small 5' untranslated exon (105-107). In the case of the gene for GPIX there are two small 5' untranslated exons. These small exons are probably involved in the tissue specificity and efficiency of transcription (108). This structure is conserved in other species where the sequence of these genes is known. In normal tissue GPIb-V-IX may only be found in megakaryocytes and platelets and thus has a relatively restricted expression pattern. The jury is still out on normal expression in other cells although, as mentioned above, expression in endothelial cells has been reported by two groups. Analysis of promotor regions of the genes for these subunits as well as other known megakaryocyte specific genes such as PF4 and GPIIb, showed the presence of GATA and Ets binding sequences which have a positive effect on expression whereas all these genes appear to contain negative elements which prevent their expression in other types of tissue. The promoter region of the gene for GPV has not yet been analysed in much detail as the other but also contains conserved GATA and Ets binding sequences.

7.8.2 GPIb-V-IX in Other Species

Several members of the GPIb-V-IX family have been cloned and sequenced from species other than human. Those so far reported are mouse (109) and dog (GenBank U18489) GPIba, mouse GPIbb (110), and rat and mouse GPV (111). No reports of GPIX from other species have yet appeared. Because it is not yet possible to produce platelets from megakaryocytes in vitro, mouse models of Bernard-Soulier syndrome will be particularly useful for testing GPIb-V-IX function in vivo.

7.9 Future Developments

The present period is a very exciting one for those interested in GPIb-V-IX because in the near future it is expected that a number of important advances will be made. Although we have some ideas about the general structure of the complex we must admit that at the moment the detailed structure, and in particular the molecular details of the interactions between GPIb and vWf and how these two molecules adapt to and affect one another are totally unknown. The structure of the A1 domain of vWf was recently determined by X-ray crystallography and it must just be a question of time now until the structure of the outer domain of GPIba containing the vWf binding site, is determined, first on its own, and then complexed with the vWf domain, so that this riddle will be solved. Clearly this is not a trivial undertaking but structural techniques continue to improve rapidly. Another area where rapid developments are expected which should cast more light on GPIb-V-IX function, is the mechanism of signal transduction. The main driving force here is the identification and characterisation of the range of kinases, phosphatases and adaptor molecules involved. Still, it is a kind of four dimensional jigsaw puzzle and new techniques will need to be developed to simplify the analysis of all the pathways involved. The development of lines of mice with each of the subunits of the complex, as well as those with some of the essential signalling molecules "knocked out" will also provide modified platelets for a clearer study of the function of this complex. Last but not least, the development of practical inhibitors of GPIb-vWf interactions will allow the role of these molecules in the development of thrombosis and its control to be clarified.

7.9.1 Acknowledgements

Work performed at the Theodor Kocher Institute was supported by a grant 31-42336.94 from the Swiss National Science Foundation. We thank the Central Laboratory of the Swiss Red Cross Blood Transfusion Service, Berne, for the supply of buffy coats.

REFERENCES

1. Lopez JA, Chung DW, Fujikawa K, Hagen FS, Papayannopoulou T, Roth GJ. Cloning of the alpha chain of human platelet glycoprotein Ib: a transmembrane protein with homology to leucine-rich a-2-glycoprotein. Proc Natl Acad Sci U S A 1987;84:5615-9.

2. Lopez JA, Chung DW, Fujikawa K, Hagen FS, Davie EW, Roth GJ. The alpha and beta chains of human platelet glycoprotein Ib are both transmembrane proteins containing a leucine-rich amino acid sequence. Proc Natl Acad Sci U S A 1988;85:2135-9.

3. Ludlow LB, Schick BP, Budarf ML, Driscoll DA, Zackai EH, Cohen A, Konkle BA. Identification of a mutation in a GATA binding site of the platelet glycoprotein Ibb promoter resulting in the Bernard-Soulier syndrome. J Biol Chem 1996;271:22076-80.

4. Hickey MJ, Williams SA, Roth GJ. Human platelet glycoprotein IX: an adhesive prototype of leucine-rich glycoproteins with flank-center-flank structures. Proc Natl Acad Sci U S A 1989;86:6773-7.

5. Hickey MJ, Deaven LL, Roth GJ. Human platelet glycoprotein IX: Characterization of cDNA and localization of the gene to chromosome 3. FEBS Lett. 1990;274:189-192.

6. Yagi M, Edelhoff S, Disteche CM, Roth GJ. Human platelet glycoproteins V and IX: mapping of two leucine-rich glycoprotein genes to chromosome 3 and analysis of structures. Biochemistry 1995;34:16132-7.

7. Lanza F, Morales M, de La Salle C, Cazenave JP, Clemetson KJ, Shimomura T, Phillips DR. Cloning and characterization of the gene encoding the human platelet glycoprotein V. A member of the leucine-rich glycoprotein family cleaved during thrombin-induced platelet activation. J Biol Chem 1993;268:20801-7.

8. Modderman PW, Admiraal LG, Sonnenberg A, Von dem Borne AEGK. Glycoproteins V and Ib-IX form a noncovalent complex in the platelet membrane. J.Biol.Chem. 1992;267:364-369.

9. Takahashi N, Takahashi Y, Putnam FW. Periodicity of leucine and tandem repetition of a 24-amino acid segment in the primary structure of leucine-rich a2-glycoprotein of human serum. Proc Natl Acad Sci U S A 1985;82:1906-10.

10. Kobe B, Deisenhofer J. Crystal structure of porcine ribonuclease inhibitor, a protein with leucine-rich repeats. Nature 1993;366:751-756.

11. Svensson L, Heinegard D, Oldberg A. Decorin-binding sites for collagen type I are mainly located in leucine- rich repeats 4-5. J Biol Chem 1995;270:20712-6.

12. Kobe B, Deisenhofer J. A structural basis of the interactions between leucine-rich repeats and protein ligands. Nature 1995;374:183-186.

13. Hess D, Schaller J, Rickli EE, Clemetson KJ. Identification of the disulphide bonds in human platelet glycocalicin. Eur.J.Biochem. 1991;199:389-393.

14. Dong J, Li CQ, López JA. Tyrosine sulfation of the glycoprotein Ib-IX complex: Identification of sulfated residues and effect on ligand binding. Biochemistry 1994;33:13946-13953.

15. Marchese P, Murata M, Mazzucato M, Pradella P, De Marco L, Ware J, Ruggeri ZM. Identification of three tyrosine residues of glycoprotein Ib alpha with distinct roles in von Willebrand factor and alpha-thrombin binding. J Biol Chem 1995;270:9571-8.

16. Pidard D, Renesto P, Berndt MC, Rabhi S, Clemetson KJ, Chignard M. Neutrophil proteinase cathepsin G is proteolytically active on the human platelet glycoprotein Ib-IX receptor: characterization of the cleavage sites within the glycoprotein Iba subunit. Biochem J 1994;303:489-98.

17. Moroi M, Jung SM, Yoshida N. Genetic polymorphism of platelet glycoprotein Ib. Blood 1984;64:622-9.

18. Lopez JA, Ludwig EH, McCarthy BJ. Polymorphism of human glycoprotein Iba results from a variable number of tandem repeats of a 13-amino acid sequence in the mucin-like macroglycopeptide region. Structure/function implications. J Biol Chem 1992;267:10055-61.

19. Ishida F, Furihata K, Ishida K, Yan J, Kitano K, Kiyosawa K, Furuta S. The largest variant of platelet glycoprotein Iba has four tandem repeats of 13 amino acids in the macroglycopeptide region and a genetic linkage with methionine145. Blood 1995;86:1357-60.

20. Tsuji T, Tsunehisa S, Watanabe Y, Yamamoto K, Tohyama H, Osawa T. The carbohydrate moiety of human platelet glycocalicin. J Biol Chem 1983;258:6335-9.

21. Korrel SA, Clemetson KJ, Van Halbeek H, Kamerling JP, Sixma JJ, Vliegenthart JF. Structural studies on the O-linked carbohydrate chains of human platelet glycocalicin. Eur J Biochem 1984;140:571-6.

22. Tsuji T, Osawa T. The carbohydrate moiety of human platelet glycocalicin: the structures of the major Asn-linked sugar chains. J Biochem (Tokyo) 1987;101:241-9.

23. Korrel SA, Clemetson KJ, van Halbeek H, Kamerling JP, Sixma JJ, Vliegenthart JF. Identification of a tetrasialylated monofucosylated tetraantennary N- linked carbohydrate chain in human platelet glycocalicin. FEBS Lett 1988;228:321-6.

24. Wicki AN, Clemetson KJ. The glycoprotein Ib complex of human blood platelets. Eur J Biochem 1987;163:43-50.

25. Shimomura T, Fujimura K, Maehama S, Takemoto M, Oda K, Fujimoto T, Oyama R, Suzuki M, Ichihara-Tanaka K, Titani K, et al. Rapid purification and characterization of human platelet glycoprotein V: the amino acid sequence contains leucine-rich repetitive modules as in glycoprotein Ib. Blood 1990; 75: 2349-56.

26. Wyler B, Bienz D, Clemetson KJ, Luscher EF. Glycoprotein Ibb is the only phosphorylated major membrane glycoprotein in human platelets. Biochem J 1986;234:373-9.

27. Wardell MR, Reynolds CC, Berndt MC, Wallace RW, Fox JE. Platelet glycoprotein Ibb is phosphorylated on serine 166 by cyclic AMP-dependent protein kinase. J Biol Chem 1989;264:15656-61.

28. Muszbek L, Laposata M. Glycoprotein Ib and glycoprotein IX in human platelets are acylated with palmitic acid through thioester linkages. J Biol Chem 1989;264:9716-9.

29. Schick PK, Walker J. The acylation of megakaryocyte proteins: glycoprotein IX is primarily myristoylated while glycoprotein Ib is palmitoylated. Blood 1996;87:1377-84.

30. Clemetson KJ, Clemetson JM. Molecular abnormalities in Glanzmann's thrombasthenia, Bernard-Soulier syndrome, and platelet-type von Willebrand's disease. Curr Opin Hematol 1994;1:388-93.

31. Read MS, Smith SV, Lamb MA, Brinkhous KM. Role of botrocetin in platelet agglutination: formation of an activated complex of botrocetin and von Willebrand factor. Blood 1989;74:1031-5.

32. Hoylaerts MF, Nuyts K, Peerlinck K, Deckmyn H, Vermylen J. Promotion of binding of von Willebrand factor to platelet glycoprotein Ib by dimers of ristocetin. Biochem J 1995;306:453-63.

33. Emsley J, King SL, Bergelson JM, Liddington RC. Crystal structure of the I domain from integrin alpha2beta1. J Biol Chem 1997;272:28512-7.

34. Jenkins PV, Pasi KJ, Perkins SJ. Molecular Modeling of Ligand and Mutation Sites of the Type A Domains of Human von Willebrand Factor and Their Relevance to von Willebrand's Disease. Blood 1998;91:2032-44.

35. Miller JL, Cunningham D, Lyle VA, Finch CN. Mutation in the gene encoding the a chain of platelet glycoprotein Ib in platelet-type von Willebrand disease. Proc.Natl.Acad.Sci.USA 1991;88:4761-4765.

36. Russell SD, Roth GJ. Pseudo-von Willebrand disease: A mutation in the platelet glycoprotein Iba gene associated with a hyperactive surface receptor. Blood 1993;81:1787-1791.

37. Miller JL, Lyle VA. Mimotope/anti-mimotope probing of structural relationships in platelet glycoprotein Iba. Proc Natl Acad Sci U S A 1996;93:3565-9.

38. Yamamoto N, Greco NJ, Barnard MR, Tanoue K, Yamazaki H, Jamieson GA, Michelson AD. Glycoprotein Ib (GPIb)-dependent and GPIb-independent pathways of thrombin-induced platelet activation. Blood 1991;77:1740-1748.

39. Liu L, Freedman J, Hornstein A, Fenton JW, 2nd, Song Y, Ofosu FA. Binding of thrombin to the G-protein-linked receptor, and not to glycoprotein Ib, precedes thrombin-mediated platelet activation. J Biol Chem 1997;272:1997-2004.

40. Jamieson GA, Okumura T. Reduced thrombin binding and aggregation in Bernard-Soulier platelets. J Clin Invest 1978;61:861-4.

41. McGregor JL, Brochier J, Wild F, Follea G, Trzeciak MC, James E, Dechavanne M, McGregor L, Clemetson KJ. Monoclonal antibodies against platelet membrane glycoproteins. Characterization and effect on platelet function. Eur J Biochem 1983;131:427-36.

42. Yamamoto K, Yamamoto N, Kitagawa H, Tanoue K, Kosaki G, Yamazaki H. Localization of a thrombin-binding site on human platelet membrane glycoprotein Ib determined by a monoclonal antibody. Thromb Haemost 1986;55:162-7.

43. Wicki AN, Clemetson KJ. Structure and function of platelet membrane glycoproteins Ib and V. Effects of leukocyte elastase and other proteases on platelets response to von Willebrand factor and thrombin. Eur J Biochem 1985;153:1-11.

44. Okumura T, Hasitz M, Jamieson GA. Platelet glycocalicin. Interaction with thrombin and role as thrombin receptor of the platelet surface. J Biol Chem 1978;253:3435-43.

45. Greco NJ, Tandon NN, Jones GD, Kornhauser R, Jackson B, Yamamoto N, Tanoue K, Jamieson GA. Contributions of glycoprotein Ib and the seven transmembrane domain receptor to increases in platelet cytoplasmic [Ca2+] induced by alpha- thrombin. Biochemistry 1996;35:906-14.

46. De Candia E, De Cristofaro R, De Marco L, Mazzucato M, Picozzi M, Landolfi R. Thrombin interaction with platelet GPIb: role of the heparin binding domain. Thromb Haemost 1997;77:735-40.

47. Bienz D, Schnippering W, Clemetson KJ. Glycoprotein V is not the thrombin activation receptor on human blood platelets. Blood 1986;68:720-5.

48. Okumura J, Lombart C, Jamieson GA. Platelet glycocalicin. II. Purification and characterization. J Biol Chem 1976;251:5950-5.

49. Beer JH, Büchi L, Steiner B. Glycocalicin: A new assay--The normal plasma levels and its potential usefulness in selected diseases. Blood 1994;83:691-702.

50. Pidard D, Renesto P, Berndt MC, Rabhi S, Clemetson KJ, Chignard M. Neutrophil proteinase cathepsin G is proteolytically active on the human platelet glycoprotein Ib-IX receptor: Characterization of the cleavage sites within the glycoprotein Iba subunit. Biochem.J. 1994;303:489-498.

51. Ward CM, Andrews RK, Smith AI, Berndt MC. Mocarhagin, a novel cobra venom metalloproteinase, cleaves the platelet von Willebrand factor receptor glycoprotein Iba. Identification of the sulfated tyrosine/anionic sequence Tyr-276-Glu-282 of glycoprotein Ibalpha as a binding site for von Willebrand factor and alpha-thrombin. Biochemistry 1996;35:4929-38.

52. Beer JH, Clerici N, Baillod P, von Felten A, Schlappritzi E, Buchi L. Quantitative and qualitative analysis of platelet GPIb and von Willebrand factor in liver cirrhosis. Thromb Haemost 1995;73:601-9.

53. Ikeda Y, Handa M, Kamata T, Kawano K, Kawai Y, Watanabe K, Kawakami K, Sakai K, Fukuyama M, Itagaki I, Yoshioka A, Ruggeri ZM. Transmembrane calcium influx associated with von Willebrand factor binding to GP Ib in the initiation of shear-induced platelet aggregation. Thromb.Haemost. 1993;69:496-502.

54. Yuan Y, Dopheide SM, Ivanidis C, Salem HH, Jackson SP. Calpain regulation of cytoskeletal signaling complexes in von Willebrand factor-stimulated platelets. Distinct roles for glycoprotein Ib-V-IX and glycoprotein IIb-IIIa (integrin alphaIIbbeta3) in von Willebrand factor-induced signal transduction. J Biol Chem 1997;272:21847-54.

55. Du X, Harris SJ, Tetaz TJ, Ginsberg MH, Berndt MC. Association of a phospholipase A_2 (14-3-3 protein) with the platelet glycoprotein Ib-IX complex. J.Biol.Chem. 1994;269:18287-18290.

56. Andrews RK, Harris SJ, McNally T, Berndt MC. Binding of purified 14-3-3z signaling protein to discrete amino acid sequences within the cytoplasmic domain of the platelet membrane glycoprotein Ib-IX-V complex. Biochemistry 1998;37:638-47.

57. Calverley DC, Kavanagh TJ, Roth GJ. Human signaling protein 14-3-3z interacts with platelet glycoprotein Ib subunits Ibalpha and Ibbeta. Blood 1998;91:1295-303.

58. Fantl WJ, Muslin AJ, Kikuchi A, Martin JA, MacNicol AM, Gross RW, Williams LT. Activation of Raf-1 by 14-3-3 proteins. Nature 1994;371:612-614.

59. Solum NO, Olsen TM. Glycoprotein Ib in the Triton-insoluble (cytoskeletal) fraction of blood platelets. Biochim Biophys Acta 1984;799:209-20.

60. Okita JR, Pidard D, Newman PJ, Montgomery RR, Kunicki TJ. On the association of glycoprotein Ib and actin-binding protein in human platelets. J Cell Biol 1985;100:317-21.

61. Bevers EM, Rosing J, Zwaal RF. Development of procoagulant binding sites on the platelet surface. Adv Exp Med Biol 1985;192:359-71.

62. Meyer SC, Sanan DA, Fox JE. Role of actin-binding protein in insertion of adhesion receptors into the membrane. J Biol Chem 1998;273:3013-20.

63. Sullam PM, Hyun WC, Szollosi J, Dong J, Foss WM, Lopez JA. Physical proximity and functional interplay of the glycoprotein Ib-IX-V complex and the Fc receptor FcgRIIA on the platelet plasma membrane. J Biol Chem 1998;273:5331-6.

64. Razdan K, Hellums JD, Kroll MH. Shear-stress-induced von Willebrand factor binding to platelets causes the activation of tyrosine kinase(s). Biochem.J. 1994;302:681-686.

65. Asazuma N, Ozaki Y, Satoh K, Yatomi Y, Handa M, Fujimura Y, Miura S, Kume S. Glycoprotein Ib-von Willebrand factor interactions activate tyrosine kinases in human platelets. Blood 1997;90:4789-98.

66. Jackson SP, Schoenwaelder SM, Yuan Y, Rabinowitz I, Salem HH, Mitchell CA. Adhesion receptor activation of phosphatidylinositol 3-kinase. von Willebrand factor stimulates the cytoskeletal association and activation of phosphatidylinositol 3-kinase and pp60^{c-src} in human platelets. J.Biol.Chem. 1994;269:27093-27099.

67. Peng M, Lu W, Kirby EP. Alboaggregin-B: A new platelet agonist that binds to platelet membrane glycoprotein Ib. Biochemistry 1991;30:11529-11536.

68. Andrews RK, Kroll MH, Ward CM, Rose JW, Scarborough RM, Smith AI, Lopez JA, Berndt MC. Binding of a novel 50-kilodalton alboaggregin from Trimeresurus albolabris and related viper venom proteins to the platelet membrane glycoprotein Ib-IX-V complex. Effect on platelet aggregation and glycoprotein Ib-mediated platelet activation. Biochemistry 1996;35:12629-39.

69. Coller BS. Platelet GPIIb/IIIa antagonists: the first anti-integrin receptor therapeutics. J Clin Invest 1997;100:S57-60.

70. Harrington RA. Design and methodology of the PURSUIT trial: evaluating eptifibatide for acute ischemic coronary syndromes. Platelet Glycoprotein IIb-IIIa in Unstable Angina: Receptor Suppression Using Integrilin Therapy. Am J Cardiol 1997;80:34B-38B.

71. Theroux P. Antiplatelet therapy: do the new platelet inhibitors add significantly to the clinical benefits of aspirin? Am Heart J 1997;134:S62-70.

72. CAPRIE. A randomised, blinded, trial of clopidogrel versus aspirin in patients at risk of ischaemic events (CAPRIE). CAPRIE Steering Committee. Lancet 1996;348:1329-39.

73. Goto S, Ikeda Y, Saldivar E, Ruggeri ZM. Distinct mechanisms of platelet aggregation as a consequence of different shearing flow conditions. J Clin Invest 1998;101:479-86.

74. Sixma JJ, Ijsseldijk MJ, Hindriks G, van Zanten GH, Gorecki M, Panet A, Garfinkel LI, de Groot PG. Adhesion of blood platelets is inhibited by VCL, a recombinant fragment (leucine504 to lysine728) of von Willebrand factor. Arterioscler Thromb Vasc Biol 1996;16:64-71.

75. Andre P, Hainaud P, Bal dit Sollier C, Garfinkel LI, Caen JP, Drouet LO. Relative involvement of GPIb/IX-vWF axis and GPIIb/IIIa in thrombus growth at high shear rates in the guinea pig. Arterioscler Thromb Vasc Biol 1997;17:919-24.

76. Chang MC, Lin HK, Peng HC, Huang TF. Antithrombotic effect of crotalin, a platelet membrane glycoprotein Ib antagonist from venom of *Crotalus atrox*. Blood 1998;91:1582-9.

77. Peng M, Lu W, Beviglia L, Niewiarowski S, Kirby EP. Echicetin: A snake venom protein that inhibits binding of von Willebrand factor and alboaggregins to platelet glycoprotein Ib. Blood 1993;81:2321-2328.

78. Taniuchi Y, Kawasaki T, Fujimura Y, Suzuki M, Titani K, Sakai Y, Kaku S, Hisamichi N, Satoh N, Takenaka T, et al. Flavocetin-A and -B, two high molecular mass glycoprotein Ib binding proteins with high affinity purified from *Trimeresurus flavoviridis* venom, inhibit platelet aggregation at high shear stress. Biochim Biophys Acta 1995;1244:331-8.

79. Kuijpers RW, Faber NM, Cuypers HT, Ouwehand WH, von dem Borne AE. NH2-terminal globular domain of human platelet glycoprotein Ib alpha has a methionine 145/threonine145 amino acid polymorphism, which is associated with the HPA-2 (Ko) alloantigens. J Clin Invest 1992;89:381-4.

80. Kiefel V, Vicariot M, Giovangrandi Y, Kroll H, Bohringer M, Greinacher A, Breitfeld C, Santoso S, Mueller-Eckhardt C. Alloimmunization against Iy, a low-frequency antigen on platelet glycoprotein Ib/IX as a cause of severe neonatal alloimmune thrombocytopenic purpura. Vox Sang 1995;69:250-4.

81. Hayashi T, Suzuki K, Akiba J, Yahagi A, Tajima K, Satoh S, Sasaki H. G-->A transition at nucleotide 2110 in the human platelet glycoprotein (GP) IX gene resulting in Ala139(ACC)-->Thr(GCC) substitution. Jpn J Hum Genet 1997;42:369-71.

82. Murata M, Matsubara Y, Kawano K, Zama T, Aoki N, Yoshino H, Watanabe G, Ishikawa K, Ikeda Y. Coronary artery disease and polymorphisms in a receptor mediating shear stress-dependent platelet activation. Circulation 1997;96:3281-6.

83. Carlsson LE, Greinacher A, Spitzer C, Walther R, Kessler C. Polymorphisms of the human platelet antigens HPA-1, HPA-2, HPA-3, and HPA-5 on the platelet receptors for fibrinogen (GPIIb/IIIa), von Willebrand factor (GPIb/IX), and collagen (GPIa/IIa) are not correlated with an increased risk for stroke. Stroke 1997;28:1392-5.

84. Weiss EJ, Bray PF, Tayback M, Schulman SP, Kickler TS, Becker LC, Weiss JL,Gerstenblith G, Goldschmidt-Clermont PJ. A polymorphism of a platelet glycoprotein receptor as an inherited risk factor for coronary thrombosis. N Engl J Med 1996;334:1090-4.

85. Chong BH, Du X, Berndt MC, Horn S, Chesterman CN. Characterization of the binding domains on platelet glycoproteins Ib-IX and IIb/IIIa complexes for the quinine/quinidine-dependent antibodies. Blood 1991;77:2190-2199.

86. Kanaji T, Okamura T, Kurolwa M, Noda M, Fujimura K, Kuramoto A, Sano M, Nakano S, Niho Y. Molecular and genetic analysis of two patients with Bernard-Soulier syndrome--identification of new mutations in glycoprotein Iba gene. Thromb Haemost 1997;77:1055-61.

87. Kunishima S, Miura H, Fukutani H, Yoshida H, Osumi K, Kobayashi S, Ohno R, Naoe T. Bernard-Soulier syndrome Kagoshima: Ser 444 --> stop mutation of glycoprotein (GP) Iba resulting in circulating truncated GPIba and surface expression of GPIbb and GPIX. Blood 1994;84:3356-3362.

88. Kenny D, Newman PJ, Morateck PA. Montgomery RR. A dinucleotide deletion results in defective membrane anchoring and circulating soluble glycoprotein Iba in a novel form of Bernard- Soulier syndrome. Blood 1997;90:2626-33.

89. Afshar-Kharghan V, Lopez JA. Bernard-Soulier syndrome caused by a dinucleotide deletion and reading frameshift in the region encoding the glycoprotein Iba transmembrane domain. Blood 1997;90:2634-43.

90. Ware J, Russell SR, Marchese P, Murata M, Mazzucato M, De Marco L, Ruggeri ZM. Point mutation in a leucine-rich repeat of platelet glycoprotein Iba resulting in the Bernard-Soulier syndrome. J.Clin.Invest. 1993;92:1213-1220.

91. Wright SD, Michaelides K, Johnson DJD, West NC, Tuddenham EGD. Double heterozygosity for mutations in the platelet glycoprotein IX gene in three siblings with Bernard-Soulier syndrome. Blood 1993;81:2339-2347.

92. Simsek S, Noris P, Lozano M, Pico M, von dem Borne AE, Ribera A, Gallardo D. Cys209 Ser mutation in the platelet membrane glycoprotein Iba gene is associated with Bernard-Soulier syndrome. Br J Haematol 1994;88:839-44.

93. Noda M, Fujimura K, Takafuta T, Shimomura T, Fujii T, Katsutani S, Fujimoto T, Kuramoto A, Yamazaki T, Mochizuki T, Matsuzaki M, Sano M. A point mutation in glycoprotein IX coding sequence (Cys73 (TGT) to Tyr(TAT)) causes impaired surface expression of GPIb/IX/V complex in two families with Bernard-Soulier syndrome. Thromb Haemost 1996;76:874-8.

94. Miller JL, Lyle VA, Cunningham D. Mutation of leucine-57 to phenylalanine in a platelet glycoprotein Iba leucine tandem repeat occurring in patients with an autosomal dominant variant of Bernard-Soulier disease. Blood 1992;79:439-446.

95. Bolin RB, Okumra T, Jamieson GA. New polymorphism of platelet membrane glycoproteins. Nature 1977;269:69-70.

96. Beacham DA, Cruz MA, Handin RI. Glycoprotein Ib can mediate endothelial cell attachment to a von Willebrand factor substratum. Thromb Haemost 1995;73:309-17.

97. Wu G, Essex DW, Meloni FJ, Takafuta T, Fujimura K, Konkle BA, Shapiro SS. Human endothelial cells in culture and in vivo express on their surface all four components of the glycoprotein Ib/IX/V complex. Blood 1997;90:2660-9.

98. Konkle BA, Shapiro SS, Asch AS, Nachman RL. Cytokine-enhanced expression of glycoprotein Iba in human endothelium. J.Biol.Chem. 1990;265:19833-19838.

99. Beacham DA, Tran LP, Shapiro SS. Cytokine treatment of endothelial cells increases glycoprotein Iba-dependent adhesion to von Willebrand factor. Blood 1997;89:4071-7.

100. Kelly MD, Essex DW, Shapiro SS, Meloni FJ, Druck T, Huebner K, Konkle BA. Complementary DNA cloning of the alternatively expressed endothelial cell glycoprotein Ibb (GPIbb) and localization of the GPIbb gene to chromosome 22. J.Clin.Invest. 1994;93:2417-2424.

101. Zieger B, Hashimoto Y, Ware J. Alternative expression of platelet glycoprotein Ib(b) mRNA from an adjacent 5' gene with an imperfect polyadenylation signal sequence. J Clin Invest 1997;99:520-5.

102. Oleksowicz L, Dutcher JP, Deleon-Fernandez M, Paietta E, Etkind P. Human breast carcinoma cells synthesize a protein immunorelated to platelet glycoprotein-Iba with different functional properties. J Lab Clin Med 1997;129:337-46.

103. Tefferi A, Hanson CA, Kurtin PJ, Katzmann JA, Dalton RJ, Nichols WL. Acquired von Willebrand's disease due to aberrant expression of platelet glycoprotein Ib by marginal zone lymphoma cells. Br J Haematol 1997;96:850-3.

104. Wenger RH, Kieffer N, Wicki AN, Clemetson KJ. Structure of the human blood platelet membrane glycoprotein Iba gene. Biochem Biophys Res Commun 1988;156:389-95.

105. Yagi M, Edelhoff S, Disteche CM, Roth GJ. Structural characterization and chromosomal location of the gene encoding human platelet glycoprotein Ibb. J.Biol.Chem. 1994;269:17424-17427.

106. Lanza F, Morales M, De La Salle C, Cazenave J-P, Clemetson KJ, Shimomura T, Phillips DR. Cloning and characterization of the gene encoding the human platelet glycoprotein V. A member of the leucine-rich glycoprotein family cleaved during thrombin-induced platelet activation. J.Biol.Chem. 1993;268:20801-20807.

107. Hickey MJ, Roth GJ. Characterization of the gene encoding human platelet glycoprotein IX. J.Biol.Chem. 1993;268:3438-3443.

108. Ware J, Russell S, Ruggeri ZM. Cloning of the murine platelet glycoprotein Iba gene highlighting species-specific platelet adhesion. Blood Cells Mol Dis 1997;23:292-301.

109. Kitaguchi T, Murata M, Anbo H, Moriki T, Ikeda Y. Characterization of the gene encoding mouse platelet glycoprotein Ibb. Thromb Res 1997;87:235-44.

110. Ravanat C, Morales M, Azorsa DO, Moog S, Schuhler S, Grunert P, Loew D, Van Dorsselaer A, Cazenave JP, Lanza F. Gene cloning of rat and mouse platelet glycoprotein V: identification of megakaryocyte-specific promoters and demonstration of functional thrombin cleavage. Blood 1997;89:3253-62.

8

FIBRINOGEN RECEPTORS: BIOLOGY AND FUNCTION

Ellinor I. Peerschke, Ph. D.
Department of Pathology
Cornell Medical Center
New York,NY
10021, USA

8.1 INTRODUCTION

Platelets participate in hemostasis and thrombosis by accumulating at sites of blood vessel injury (1). Platelet deposition at these sites is important for the physiologic arrest of bleeding. Under pathologic conditions, however, platelet deposition can lead to vascular occlusion followed by ischemic tissue damage and thrombus embolization.

Dr. Peerschke is Professor of Pathology at Cornell University Medical College, and Director of Laboratories for Clinical Hematology at the New York Hospital. She has been interested in platelet physiology and biochemistry since 1980. Her laboratory has made significant contributions to understanding platelet-fibrinogen interactions and more recently platelet interactions with complement receptors for Clq and their roles in hemostasis, inflammation, and thrombosis. Dr. Peerschke received the Paul N. Yu Research Award from the American Heart Association in 1995 for her work towards understanding the role of platelets in cardiovascular disease and stroke.

Platelets can be activated by a variety of substances, including small molecules, enzymes, and macromolecular protein complexes (2). The plasma protein fibrinogen is both necessary and sufficient for platelet interactions leading to platelet cohesion or aggregation. Studies performed in the early 1960s identified fibrinogen as a cofactor for platelet aggregation (3-5). Patients with severe afibrinogenemia (6,7) were found to have long bleeding times, and their platelet aggregation in citrated platelet rich plasma was abnormal. Similar observations were made with platelets in plasma depleted of fibrinogen in vitro by heat denaturation, absorption with bentonite, or precipitation with tannic acid (8).The addition of purified fibrinogen to any of these deficient plasmas (4), including the platelet-rich plasma from patients with afibrinogenemia, restored the ability of platelets to aggregate. Since platelets from patients with the autosomal recessive disorder thrombasthenia are deficient in two surface membrane, carbohydrate rich glycoproteins, designated GPIIb and GPIIIa, and demonstrate similar functional abnormalities as platelets from patients with severe afibrinogenemia (9-11), GPIIb and GPIIIa were proposed to function in fibrinogen binding (12-15).

8.2 CHARACTERIZATION OF THE PLATELET FIBRINOGEN RECEPTOR

Identification of the Platelet Fibrinogen Receptor GPIIb and GPIIIa were tentatively identified as the platelet fibrinogen receptor based on studies performed with platelets from patients with thrombasthenia (9,12-15). Platelets from these patients do not aggregate in response to any agonists that require fibrinogen as a cofactor including ADP, epinephrine, thrombin, or collagen (9,10), fail to bind fibrinogen after appropriate stimulation (16-19), and either completely lack GPIIb and GPIIIa as assessed by SDS-polyacrylamide gel electrophoresis, or possess significantly reduced quantities of these glycoproteins (12-15). Biochemical evidence for the role of GPIIb and/or GPIIIa as the platelet fibrinogen receptor was derived from studies using heterobifunctional, light sensitive, crosslinking reagents (20,21). Additional, more specific evidence was provided using monoclonal antibodies directed against GPIIb and/or IIIa which inhibited fibrinogen binding and platelet aggregation, and whose reactivity with platelets from patients with thrombasthenia was markedly reduced or absent (22-25).

GPIIb-IIIa Structure

The structure of GPIIb and GPIIIa has been elucidated by biochemical and molecular methods (Fig. 1). Both glycoproteins are integral platelet membrane constituents. Together they account for approximately 18% of the platelet membrane protein mass (26-28).

Glycoproteins IIb and IIIa form a calcium-dependent heterodimer complex (29,30) that dissociates in the presence of divalent cation chelators. Although mM concentrations of either calcium or magnesium prevent complex dissociation, only calcium supports GPIIb-IIIa reassociation (31). Two high affinity (Kd 9 nM) and approximately six intermediate affinity (Kd 400 nM) calcium binding sites have been identified within the GPIIb-IIIa complex (32,33). In addition, GPIIb-IIIa contains a class of low affinity binding sites (Kd ~ mM) for divalent cations. The role of these low affinity sites in platelet aggregation and ligand binding is controversial (34-36).

Although dissociation of the GPIIb-IIIa complex by divalent cation chelators is reversible initially, progressive monomer self-association occurs at 37oC, producing nonfunctional, high molecular weight irreversible glycoprotein aggregates (31). The inability to reform heterodimer complexes is primarily associated with a calcium dependent transition in GPIIIa which results in subunit unfolding and subsequent GPIIIa aggregation (31). A similar structural transition is observed in GPIIb, but at tenfold-lower calcium concentrations.

GPIIb has a molecular weight of 136,000 daltons. It is composed of a larger alpha (125,000 daltons) and a smaller beta (23,000 daltons) subunit, linked by a single disulfide bond (37). The beta subunit serves to anchor GPIIb in the platelet membrane. The disulfide linked alpha subunit is entirely extracellular.

GPIIIa is also an integral membrane protein. It is a disulfide-bond-rich, single-chain protein with an apparent molecular weight of 90,000 daltons on nonreduced, SDS-polyacrylamide gel electrophoresis, and an apparent molecular weight of 110,000 daltons following disulfide bond reduction (37). GPIIIa contains 56 cysteine residues, each involved in an intrachain disulfide bond (38), and organized in 4 cysteine rich tandem repeats. A large, disulfide-bonded loop, composed of at least 325 and 384 amino acids is susceptible to proteolytic cleavage by chymotrypsin, trypsin, and plasmin (39,40). Because the cleaved membrane-bound GPIIIa no longer interacts with adhesive proteins, it is likely that a portion of the receptor's ligand-binding site resides in the proteolytic fragment or is affected by conformational changes resulting from proteolysis (41-43).

The GPIIb-IIIa complex is a member of the integrin family of heterodimeric adhesion receptors. It is composed of an alpha (GPIIb) and beta 3 (GPIIIa) subunit. GPIIb and GPIIIa are type I transmembrane glycoproteins containing 1008 and 762 amino acids, respectively (44), with a single membrane spanning domain and an extracellular amino terminus. GPIIb-IIIa distribution is limited to platelets, megakaryocytes, and some transformed cells (45). Normal platelets contain approximately 80,000 molecules of GPIIb-IIIa (46). Additional GPIIb-IIIa complexes are present in membranes of the platelet surface-connected canalicuar system (47) and alpha granules (48). These intracellular receptor pools can be expressed at the cell surface following platelet activation. Protein redistributions from one compartment to another during platelet activation reflects a change in platelet function correlating with an increase in platelet cohesion (49).

The genes for GPIIb and GPIIIa are each located in the proximal portion of the long arm of chromosome 17 at q 21-23 (50), and are in linkage disequilibrium. GPIIb contains two strongly hydrophobic regions, one representing a typical signal peptide at the amino terminus of the alpha chain, and the other located near the carboxyterminus of the beta chain, representing a putative transmembrane domain. The cytoplasmic tail consists of 20 hydrophilic amino acids. The extracellular domain of GPIIb beta contains a single, potential, N-linked glycosylation site. GPIIb alpha contains four possible, N-linked glycosylation sites and four stretches of 12 amino acids resembling calcium-binding regions.

The cDNA sequence of GPIIIa indicates the presence of an amino-terminal hydrophobic region representing a signal peptide, and a carboxy-terminal hydrophobic sequence constituting the transmembrane domain. The cytoplasmic tail consists of 41 amino acids and contains two tyrosine residues at positions 747 and 759 (50). Recent studies suggest that GPIIIa may be phosphorylated on tyrosine residues in response to thrombin induced platelet aggregation (51), and that this phosphorylation may be required for the binding of signaling proteins such as GRB2 and SHC. Phosphorylation of GPIIIa on serine and threonine residues has been reported also as a consequence of intracellular signaling events involving GPIIb-IIIa activation, and receptor occupancy and clustering (52).

The intracellular assembly and membrane expression of the GPIIb-IIIa complex has been studied in a variety of cultured cells (50). The data suggest that GPIIb is synthesized as an approximately 130,000 molecular weight single chain precursor which undergoes core glycosylation, disulfide bond formation, and folding in the endoplasmic reticulum (ER). Calcium binding domains of GPIIb play an important role in maintaining the overall conformation of GPIIb-IIIa, allowing further intracellular processing or transport of the complex to the cell surface (53,54). GPIIb posttranslational processing, consisting of cleavage into two subunits and final carbohydrate modification, does not go to completion unless the GPIIb-IIIa heterodimer has been assembled in the ER. In contrast, GPIIIa precursors are fully processed without GPIIb, and can complex with other alpha subunits such as alphav to form the vitronectin receptor (55,56).

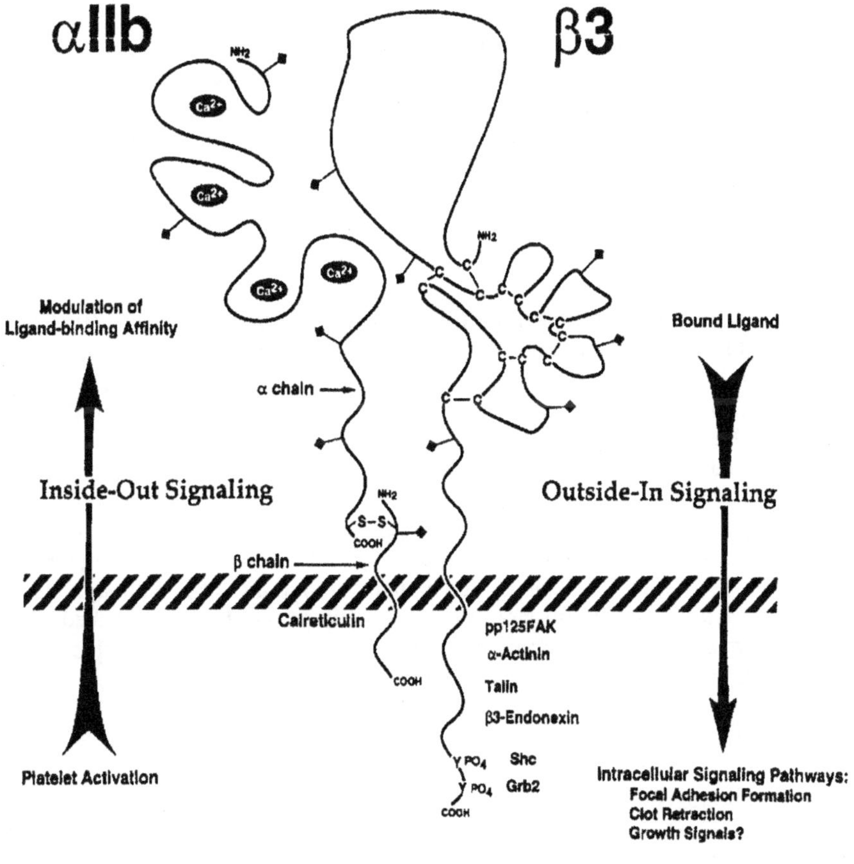

Figure 1. Structural and biochemical features of the GPIIb-IIIa fibrinogen receptor. (Solid diamonds) N-linked carbohydrate side changes; (Y) tyrosine phosphorylation sites; (S-S) disulfide bonds; (C) cysteine, (NH2) amino-terminus; (COOH) carboxy-terminus; (PO4) phosphorylation sites. Reprinted with permission from Peerschke EIB and Lopez JA, "Platelet Membranes and Receptors," In Thrombosis and Hemorrhage 2, J. Loscalzo and A.I. Schafer, eds., Williams and Wilkins, Baltimore, MD, 1998.

Neither GPIIb nor GPIIIa are expressed on the cell surface without first engaging in complex formation. Analysis of GPIIb from different patients with Glanzmann's thrombasthenia, also indicate, however, that mutations in the GPIIb alpha chain, including deletions of each calcium binding domain, result in lack of GPIIb-IIIa surface expression, despite GPIIb-IIIa heterodimer assembly (53-57). Thus, heterodimer formation alone is not sufficient for GPIIb-IIIa surface expression. Intact GPIIb calcium binding domains are also essential.

Recent studies indicate that a limited region of GPIIIa (V275GSDNH280) (58) expressed in an extended loop of the metal ion-dependent adhesion site appears to be involved in heterodimer assembly. In addition, the expression of isolated residues 1-233 of GPIIb and 111-318 of GPIIIa leads to the formation of soluble peptide heterodimers that recognize the RGD sequence (58).

Using a variety of biochemical and genetic techniques, the ligand specificity of GPIIb has been attributed to the first third of the N-terminal portion of the extracellular domain, which includes the first two calcium binding sites (59). A 21 amino acid N-terminal region of GPIIb (A294-M314) (60) participates in ligand binding but inhibits clot retraction and platelet aggregation (61). Specific mutations of Gly 184, Tyr 189, Tyr 190, Phe 191, and Gly 193 within the predicted turn structure of the third amino terminal repeat of GPIIb alpha significantly block the binding of soluble fibrinogen to the GPIIb-IIIa complex, but are not required for GPIIb-IIIa surface expression (62).

In addition to ligand binding sites in GPIIb, a highly conserved region in GPIIIa (D109-E171) appears to be involved in the recognition of RGD sequences in adhesive protein ligands. In addition, GPIIIa residues 274-368 appear to interact specifically with the C-terminal dodecapeptide (H12) of the fibrinogen gamma chain in a calcium independent manner (63). Antibodies directed against GPIIIa 274-368 inhibit fibrinogen binding as well as platelet adhesion and aggregation, supporting a role for this sequence as a binding domain for fibrinogen, distinct from the RGD binding site. Critical residues in GPIIIa for ligand binding include residues D217 and E220 (64). Furthermore, the cell adhesion regulatory domain of GPIIIa has been mapped to residues 747-762 in the cytoplasmic domain, with Y747 and Y759 being functionally important (65). Although the fine structure of the binding site for fibrinogen on GPIIb-IIIa has not been established, current evidence favors a model in which several discontinuous regions of both glycoproteins form a three-dimensional, ligand-binding pocket (66).

8.3 FUNCTION

In addition to fibrinogen, the GPIIb-IIIa complex can bind von Willebrand Factor (vWF), fibronectin, and vitronectin (67). Recognition of these ligands in part involves an RGD tripeptide sequence (55,67), which function as a common recognition sequence for receptors of the integrin family of adhesion receptors. In the case of fibrinogen, RGD sequences are located at positions 95-97 and 572-574 of each of its two A alpha chains. The fibrinogen molecule, however, also contains GPIIb-IIIa recognition domains encompassing residues 400-411 (H12) in the C-terminal of each of its two gamma chains (68). Synthetic peptides representing both RGD and H12 sequences completely inhibit fibrinogen as well as vWF and fibronectin binding, despite the absence of H12 sequences in the latter ligands (55,67). Studies with natural variants of fibrinogen, mutant forms of recombinant fibrinogen, and proteolytically derived fibrinogen fragments have emphasized the importance of the H12 sequence in the carboxy terminus of the fibrinogen gamma chain for fibrinogen binding and platelet aggregation (69,70). Whereas fibrinogen is the most abundant natural GPIIb-IIIa ligand in plasma and supports platelet aggregation at low shear, von Willebrand factor may be the predominant crossbridging molecule mediating aggregation under moderate shear stress (71).

Ligand Binding

In 1978, Mustard and co-workers (72) described the first interaction of radio-labeled fibrinogen with washed human and rabbit platelets during ADP-induced aggregation. This observation was confirmed and rapidly extended by numerous other laboratories (summarized in 73).

Although there is general agreement that the binding of fibrinogen to activated platelets is specific and saturable (73), several issues remain controversial, including the presence of one or more classes of binding sites and cooperativity among receptors. Careful evaluation of fibrinogen binding isotherms suggest the presence of heterogeneous ligand receptor interactions, and interactions among receptors (74) compatible with negative cooperativity, ligand polymerization, and receptor clustering and/or crosslinking (74-76). More recent studies demonstrate not only that the GPIIb-IIIa complex exists in both a low and high affinity state (77), but also provide evidence for ligand polymerization and receptor clustering by electron microscopy (78,79) and confocal scanning laser microscopy (76,80,81).

Fibrinogen binding to GPIIb-IIIa requires divalent cations, calcium or magnesium (73). A detailed analysis of the role of these cations in ligand binding to beta 3 integrins (36), suggests the presence of one class of magnesium binding sites that supports ligand binding, and two classes of calcium binding sites, a high affinity site that promotes ligand binding, and a low affinity site that may be inhibitory and increases the rate of ligand dissociation.

Fibrinogen binding is modulated also by manganese (82), which has been shown to influence integrin mediated cell attachment and spreading (83-85), and alters the binding specificity of purified GPIIb-IIIa from its natural ligand to a non-selective linear RGD containing peptide (82).

Inhibition/Potentiation of Fibrinogen Binding

Fibrinogen binding to platelets stimulated with most agonists is inhibited by chelation of divalent cations with EDTA (ethylene diamine tetra acetic acid), an increase in H+ ion concentration, PGE1 (prostaglandin E1), dibutyryl cyclic AMP (db c-AMP), metabolic inhibitors, or high concentrations of cytochalasins B or D (73) (Table 1). Studies comparing the binding of fibrinogen to ADP-treated platelets and platelets incubated with chymotrypsin have helped elucidate the mechanism of action of these inhibitors. Chelation of divalent cations and decreases in pH appear to inhibit fibrinogen binding directly, whereas other agents appear to affect the exposure or activation of GPIIb-IIIa fibrinogen receptors (summarized in 73).

Table 1. Activation and inhibition of fibrinogen binding to human platelets

AGONISTS	INHIBITORS
Adenosine Diphosphate (ADP)	Prostaglandin E1 (PGE1)
Epinephrine	dibutyryl cyclic adenosine monophosphate (c-AMP)
Thrombin	EDTA (ethylene diamine tetraacetic acid)
Collagen	pH < 6.8
Arachidonic Acid	cytochalasin B (high doses)
A23187	metabolic inhibitors
Dithiothreitol (DTT)	anti GPIIb/IIIa antibodies
ZnCl2	RGD containing peptides
Phorbol myristate acetate (PMA)	ibrinogen gamma chain dodecapeptide (H12)
chymotrypsin	
plasmin	
Prostaglandin H2 (PGH2)	
Low Temperature (4oC)	

Fibrinogen binding to platelets can be enhanced by the release of platelet dense granule ADP and prostaglandin endoperoxide synthesis. The most dramatic increase in fibrinogen binding is achieved, however, when agonists are used in combination with epinephrine (86). Epinephrine can overcome the inhibition of aspirin on ADP-induced aggregation by increasing fibrinogen receptor exposure independently of prostaglandin endoperoxide and thromboxane A2 synthesis (87,88), and restores the ability of platelets from stored concentrates to bind fibrinogen and to aggregate (87,89).

Importantly, several regulatory mechanism have been suggested. Recent data indicate that regulation of platelet activation and fibrinogen binding may occur as a direct result of GPIIb-IIIa occupancy by fibrinogen triggering as yet undefined signals to prevent further activation of circulating platelets by low intensity stimuli such as ADP and inhibit intravascular aggregation (90). Moreover, fibrinogen receptors that are activated but fail to interact with fibrinogen return to an inactive state (91), and platelet bound fibrinogen that does not participate in platelet crosslinking or aggregation is cleared from the platelet surface (76,80,81).

GPIIb-IIIa Activation

On resting platelets, the GPIIb-IIIa complex is unable to bind large extracellular adhesive ligands (55,67,73,77). Short peptides containing the RGD tripeptide sequence or peptides encompassing the fibrinogen gamma chain H12 sequence can bind, however, perhaps due to lack of stearic hindrance (92). When platelets become activated (Table 1) by such agents as ADP, epinephrine, thrombin, collagen, prostaglandin endoperoxides and arachidonic acid (73), specific GPIIb-IIIa dependent binding is observed which is simultaneously associated with conformational changes in GPIIb-IIIa (93) (Fig. 1).

In addition to traditional agonists, platelets can be induced to bind fibrinogen by exposure to low temperatures (94), chymotrypsin (95,96), plasmin (97,98), dithiothreitol (99,100), or ZnCl2 (101). The mechanism whereby these agents induce fibrinogen binding is incompletely understood, but may involve a change in membrane fluidity (94), proteolysis of the environment around the GPIIb-IIIa complex or the complex itself (41,95), reduction of membrane disulfide bonds (99), or direct activation of protein kinase C (PKC) (101), respectively. Although attempts have been made to correlate GPIIb or GPIIIa proteolysis with activation, the "gain of function" cleavage has not been identified (102,103).

Activation of GPIIb-IIIa can also be achieved by artificial modulation of the integrin with specific antibodies or peptides (104). RGD peptides reversibly activate GPIIb-IIIa in intact cells and cell membranes (105,106). Nuclear magnetic resonance studies indicate that RGD peptides induce conformational transitions in purified GPIIb-IIIa SDS systems, yielding multiple binding states for fibrinogen gamma chain peptides and modify peptide binding affinity (107).

Studies using monoclonal or polyclonal antibodies have suggested that conformational changes in GPIIb-IIIa and/or its microenvironment are associated with GPIIb-IIIa activation (108) (Fig. 2). Direct evidence for activation induced conformational changes in GPIIb-IIIa in intact platelets has been obtained using fluorescence-resonance energy transfer techniques (109). Platelet activation induces a conformational change in GPIIb-IIIa, exposing a ligand binding site that is likely to be composed of several discontinuous regions in the amino terminal portions of both subunits (109,110). In addition, clustering of GPIIb-IIIa heterodimers into oligomers has been observed following ligand binding (111) and may affect the ultimate strength and reversibility of the interaction.

The conversion of GPIIb-IIIa to a ligand binding receptor is tightly regulated, and involves cytoplasmic or "inside- out" signaling which may require extracellular ionized calcium. Activation also involves at least two intracellular pathways (112,113) one of which is initiated by G-protein dependent phosphoinositide hydrolysis leading to a rise in cytosolic calcium and activation of serine-threonine kinases such as protein kinase C (PKC) and tyrosine kinases. Many but not all agonists activate GPIIb-IIIa via PKC mediated pathways (113). Moreover, phosphoinositide 3-kinase activation is increased by protein kinase C stimulation and by GPIIb-IIIa engagement (114). Inhibition of phosphoinositide 3-kinase inhibits maintenance of the platelet GPIIb-IIIa in its active state and results in reversal of platelet aggregation (115). Substances such as prostacyclin and nitric oxide which stimulate protein kinase A and protein kinase G, respectively inhibit or reverse ligand binding (116).

The cytoplasmic domains of both GPIIb and GPIIIa appear to participate in regulating ligand binding affinity. Deletion of GPIIIa Leu717 - Asp723, as well as disruption of the GPIIb 991-995 GFFKP sequence, results in energy independent, constitutive GPIIb-IIIa activation (109,117). The mechanism of GPIIb -IIIa affinity modulation by mutational analysis of its constitutive cytoplasmic tails, however, is incompletely understood. Whereas deletion of the GPIIb cytoplasmic domain results in the constitutive activation of GPIIb-IIIa expressed in chinese hamster ovary (CHO) cells (118), the same mutation has no effect on phorbol myristate acetate (PMA) induced fibrinogen binding to transformed B lymphocytes (119). Reasons for these differences are not understood but may reflect cell type specific differences in "inside-out" signalling.

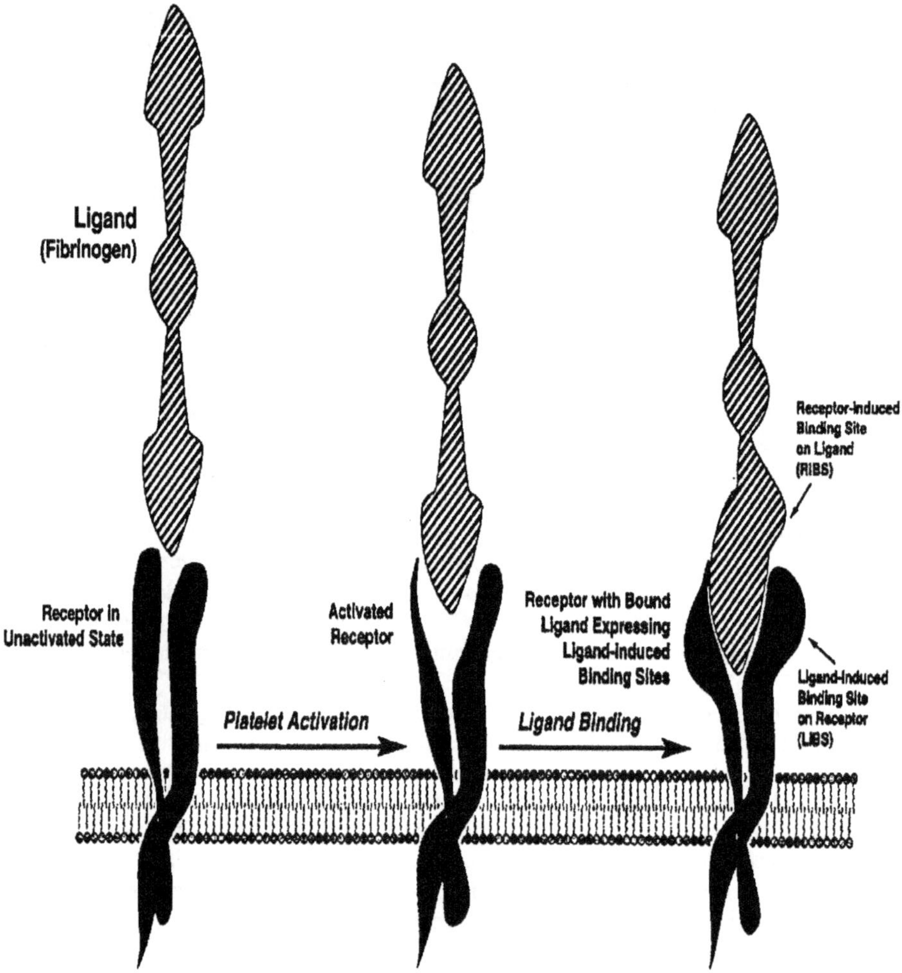

Figure 2. Model of GPIIb-IIIa activation and consequences of fibrinogen binding. Reprinted with permission from Peerschke EIB and Lopez JA, "Platelet Membranes and Receptors," In Thrombosis and Hemorrhage 2, J. Loscalzo and A.I. Schafer, eds., Williams and Wilkins, Baltimore, MD, 1998.

Further mutational analyses of the cytoplasmic domains of GPIIb-IIIa have identified specific cytoplasmic membrane proximal motifs of GPIIb (GFFKR) and GPIIIa (LLVIHDR) which may interact to form a hinge via a possible salt bridge between R995 in GPIIb and D723 in GPIIIa (120). This interaction is believed to maintain the integrin in a low affinity state until disrupted by platelet activation. A direct interaction between cytoplasmic domain peptides consisting of GPIIb (999-1008) and GPIIIa (721-740) has also been demonstrated (121). This peptide complex binds divalent cations, but how this correlates with regulating GPIIb-IIIa affinity is not understood. Interestingly, a naturally occurring point mutation (Ser752-Pro) in the GPIIIa cytoplasmic domain, disrupts "inside-out" signalling (54), implicating the GPIIIa cytoplasmic domain in inducing or maintaining the high affinity state.

The ability of integrin cytoplasmic domains to regulate integrin affinity may be due to their interaction with other cytosolic constituents. Consistent with this hypothesis is the observation (122) that coexpression of the GPIIIa and the beta 1 cytoplasmic and transmembrane domains in cells inhibits activation of a constitutively active mutant GPIIb-IIIa by competition between beta 1 and beta 3 (GPIIIa) cytoplasmic domains for critical intracellular factors. Similarly, cell-permeable GPIIIa cytoplasmic peptides inhibit the adhesive function of GPIIb-IIIa (66).

Using the yeast two-hybrid system, a 110 amino acid polypeptide, named beta 3 endonexin, has been identified recently (123). Beta 3 endonexin binds specifically to the GPIIIa cytoplasmic tail. The membrane distal NITY motif at GPIIIa 756-759 appears critical for interaction with beta 3 endonexin (120). Interestingly, overexpression of beta 3 endonexin in CHO cells leads to activation of wild type GPIIb-IIIa (124).

Using a similar experimental approach, a novel calcium binding protein of approximately 25,000 daltons termed CIB (calcium and integrin binding protein) has been cloned and sequenced (125). CIB specifically interacts with the GPIIb cytoplasmic domain. CIB has significant homology to two other regulatory proteins, calcineurin B and calmodulin, and may thus serve as a regulatory molecule docking specific enzymes to integrins.

Molecular studies have been useful also in identifying important GPIIb-IIIa ligand binding domains, including a 23 residue segment of the beta 3 chain extracellular domain (E312-S335) important for energy dependent activation. This sequence participates either in the transmission of a conformational change or in binding of a cofactor required for integrin activation (126). Furthermore, potential inhibitors of integrin activation have been identified and may include components of the Ras-linked MAP kinase pathway (127,128). Unlike H-Ras, which appears to inhibit GPIIb-IIIa activation in CHO cells, coexpression of constitutively active R-Ras activates the complex to bind large ligands (129).

Recently the existence of a marked gender difference in GPIIb-IIIa function was described (130). GPIIb-IIIa receptors on platelets from premenopausal women are more readily activated compared to receptors on platelets from young men. Enhanced activation is manifest by greater agonist induced ligand binding (increased Bmax) without changes in receptor affinity (Kd). Although the clinical implications of gender differences in platelet function remain unclear, it may have implications for customizing antithrombotic therapy.

8.4 POST-FIBRINOGEN BINDING EVENTS

Stabilization of fibrinogen binding

The interaction between fibrinogen and the GPIIb-IIIa complex is a multiphasic process (73,131). Initial binding occurs rapidly and is readily reversible, either by removing divalent cations with chelating agents, or by adding excess unlabeled fibrinogen as competitor (73). With time, this interaction becomes increasingly less reversible (73,132). These observations were recently confirmed by plasmon-resonance technology, which measures kinetic parameters of fibrinogen binding to purified GPIIb-IIIa (133). Given the number of potential fibrinogen recognition sequences on the GPIIb-IIIa complex, it is conceivable that stabilization of fibrinogen binding reflects the development of a higher affinity binding state involving multivalent ligand-receptor interactions. After fibrinogen binds to GPIIb-IIIa, both ligand and receptor undergo conformational changes referred to as RIBS (receptor induced binding sites) and LIBS (ligand induced binding sites)(134,135) (Fig. 2). Whether these changes contribute to stabilize fibrinogen binding or modulate platelet function is unclear.

Stabilization of fibrinogen binding has been described in platelet-rich plasma or washed platelets stimulated with ADP, thrombin, or epinephrine (summarized in 136) (Table 2). Unlike platelets stimulated with ADP or thrombin, however, platelets activated with chymotrypsin bind fibrinogen but fail to support irreversible binding unless costimulated with ADP (136). Similarly, fibrinogen receptor expression in response to platelet exposure to ZnCl2, PMA, or the calcium ionophore, A23187, supports only reversible fibrinogen binding (101). Thus, the mechanism of platelet activation influences the development of irreversible fibrinogen binding. Indeed, a strong association between the development of irreversible platelet fibrinogen interactions and activation of the platelet cytoskeleton is evolving (137). Moreover, post-ligand binding signaling events may play as yet undefined roles.

Mn+2 has been shown to influence integrin mediated cell attachment and spreading (83-85) and dramatically alters the binding specificity of purified GPIIb-IIIa, shifting receptor preference from its natural ligand to non-selective linear RGD containing peptides (82). Whereas Mn+2 has little effect on reversible fibrinogen binding induced by ADP or thrombin, two-fold decreases in the extent of irreversible fibrinogen binding have been observed (138). Because Mn+2 appears to modulate GPIIb-IIIa recognition of RGD sequences, these data may suggest the lack of a direct involvement of RGD sequences in stabilizing fibrinogen binding. This hypothesis is consistent with the observation that an early fibrinogen plasmic degradation product, lacking the fibrinogen alpha 572-574 RGD sequence, binds to activated platelets in both a reversible and irreversible manner (139), and a later degradation product, fibrinogen fragment D100, which lacks both alpha 572-574 and alpha 95-97 RGD sequences, continues to bind irreversibly to thrombin stimulated platelets (140). The stabilization of D100 binding to platelets further suggests that the dimeric structure of fibrinogen is not required.

Bound Fibrinogen Redistribution

The stabilization of platelet fibrinogen interactions is temporally correlated with the decreased accessibility of bound fibrinogen to antibody and enzymes (131). Additional evidence from low voltage, high resolution scanning electron microscopy with gold-conjugated fibrinogen (141) suggests that fibrinogen binding is associated with the formation of specific nonfibrillar fibrinogen aggregates. Studies using confocal scanning laser microscopy have demonstrated the clearance of bound fibrinogen from the surface of activated platelets, as well as the colocalization of GPIIb-IIIa and fibrinogen in internal areas, probably corresponding to the surface canalicular system (76,80). Other reports additionally suggest the incorporation of fibrinogen into alpha granules (81). These internal pools of ligand occupied GPIIb-IIIa are generally not accessible to exogenous antibody (142).

The relationship between fibrinogen redistribution on the surface of activated platelets and the progressive stabilization of fibrinogen binding is incompletely understood. Although the two phenomena are temporally correlated, they can be modulated independently (136). Platelets activated with ZnCl2 or PMA, for example, demonstrate progressive decreases in bound fibrinogen accessibility to antibody but fail to bind fibrinogen in and EDTA-resistant manner (101) (Table 2).

Moreover, EDTA resistant fibrinogen binding has been observed with purified, immobilized GPIIb-IIIa (143,144) indicating not only that fibrinogen redistribution and sequestration is not required for irreversible binding, but also that irreversible fibrinogen binding is an integral property of the GPIIb-IIIa complex which is independent of the platelet cellular and membrane microenvironment. Studies with purified GPIIb-IIIa further suggest that unlike GPIIb-IIIa on activated platelets, fibrinogen binding to immobilized GPIIb-IIIa fails to undergo the transition from reversible to irreversible fibrinogen binding observed with whole platelets (144). Thus, purified GPIIb-IIIa may exist in discrete active conformers, one binding fibrinogen reversibly and the other binding fibrinogen in an irreversible manner. In whole platelets, these GPIIb-IIIa conformers likely exist in equilibrium, modulated by intracellular signaling events, particularly activation of the actin cytoskeleton.

Table 2. Phases of fibrinogen binding induced by various agonists

AGONIST	PHASES OF FIBRINOGEN BINDING			
	Reversible Binding	Irreversible Binding	Fibrinogen Redistribution	Fibrinogen Association With Triton X-100 Insoluble Cytoskeleton
ADP	+	+	+	+
Thrombin	+	+	+	+
A23187	+	-	-	-
Chymotrypsin	+	-	-	-
Dithiothreitol	+	-	-	-
ZnCl2	+	-	+	-
PMA	+	-	+	-

(+) present
(-) absent or markedly decreased

The GPIIb-IIIa complex on platelets also binds other adhesive proteins including vWF and fibronectin (55,67). Interestingly, fibrinogen, fibronectin and vWF are not coordinately expressed on thrombin stimulated platelets (145). Rather, fibrinogen expression appears to be transient as described above, compared to the expression of fibronectin or vWF which remains relatively unchanged (80,145). Recent evidence (146) suggests that the fibrinogen gamma chain H12 sequence has a fluidizing effect on the platelet membrane lipid bilayer when bound to GPIIb-IIIa, whereas the interaction of RGDS peptides with platelet membrane integrins has the contrary effect. These observations further support the concept of specific platelet responses to ligation of GPIIb-IIIa, and are consistent with observed differences in fibrinogen, fibronectin, and vWF redistribution on stimulated platelets as fibrinogen binding is mediated largely by the gamma chain H12 sequence, and vWF and fibronectin bind via RGD containing sequences.The ability of platelets to organize adhesive protein on their surface is likely to have important implications for hemostasis and thrombosis.

GPIIb-IIIa Interactions With the Platelet Cytoskeleton

It has been suggested that GPIIb-IIIa is associated with the membrane skeleton of activated platelets (76). Membrane skeletal GPIIb-IIIa binds fibrinogen and triggers cytoskeletal reorganization, resulting in selective movement of occupied receptors. Following platelet activation and aggregation, the GPIIb-IIIa complex becomes associated with actin filaments in the platelet cytoskeleton (147). The association of GPIIb-IIIa with submembranous actin filaments is likely mediated by an intermediary protein such as talin (148). Indeed, the direct binding of talin to GPIIb-IIIa was recently demonstrated to involve both GPIIb and GPIIIa cytoplasmic domains (149). Other potential protein intermediates that form a link between the actin cytoskeleton and ligand bound integrins are vinculin, fibrin, alpha actinin, and tensin (148). These interactions may be controlled by small, GTP-binding proteins such as Rho. Presumably, the association of GPIIb-IIIa with the platelet cytoskeleton facilitates the contractile response, particularly clot retraction, which requires the interaction of fibrin with GPIIb-IIIa, GPIIb-IIIa association with actin, and the interaction of actin with myosin (150). Activation of the platelet cytoskeleton is implicated also in the exposure of fibrinogen receptors and the stabilization of platelet fibrinogen interactions (131).

GPIIb-IIIa Dependent Protein Signaling

The GPIIb-IIIa complex participates in the transfer of information from the extracellular to intracellular milieu ("outside-in signaling") by inducing tyrosine phosphorylation of cytoplasmic proteins (151,152). Ligand occupancy of GPIIb-GPIIIa is a prerequisite for induction of the full range of intracellular tyrosine phosphorylation events (152). These signaling events are distinct from the initial signals required for GPIIb-IIIa activation (152), and may involve GPIIb-IIIa clustering within the plasma of the membrane. Sequential activation of protein tyrosine kinases such as pp72 syk, pp60src, pp125FAK, as well as cortactin have been observed (51,151).

Within seconds, fibrinogen binding results in tyrosine phosphorylation of proteins of 50-72kDa and 140kDa (152). The protein tyrosine kinase pp72 syk is among the first proteins to become tyrosine phosphorylated and activated after fibrinogen binds to GPIIb-IIIa.This early pattern of phosphorylation is specific for ligand binding, as it is prevented by function blocking monoclonal antibodies. Early cytosolic protein phosphorylation does not require platelet aggregation or secretion and is not accompanied by increases in cytosolic calcium or protein kinase C activation (152).

Platelet aggregation not only results in increased phosphorylation of proteins such as Syk and FAK, but GPIIIa phosphorylation on residues of Tyr 747 and Tyr 759 has been described also following fibrinogen binding and platelet aggregation (51). SHC and GRB2, two proteins known to participate in the Ras signaling pathway (153) are also phosphorylated during platelet aggregation (51) and may associate with phosphorylated GPIIIa cytoplasmic domains.

Fibrinogen binding and platelet aggregation are accompanied by a dramatic reorganization of the actin cytoskeleton and redistribution of tyrosine kinases and other signaling elements to the detergent insoluble platelet cytoskeleton. Several protein tyrosine kinases (pp60 src, pp62yes, pp72syk) become activated and associated with the Triton insoluble GPIIb-IIIa rich cytoskeleton following thrombin induced platelet aggregation (summarized in 152). This cytoskeletal fraction is also enriched in other signaling molecules including phosphoinositide 3 kinase, protein kinase C, rap1b, rap2b, p21rasGAP and cytoskeletal proteins vinculin, alpha actinin, talin, and spectrin (152).

Platelet aggregation

The dimeric structure of fibrinogen allows it to form bridges between GPIIb-IIIa receptors on adjacent platelets, leading to platelet aggregation (Fig. 3). This model of platelet - platelet interactions is supported by experimental evidence demonstrating not only a direct relationship between platelet aggregation and fibrinogen binding, but also decreased platelet aggregation at high fibrinogen concentrations, when platelet receptors become saturated with fibrinogen and insufficient free receptors remain to link platelets into aggregates (summarized in 73). Moreover, expression of GPIIb-IIIa in CHO cells leads to aggregation upon appropriate GPIIb-IIIa activation and fibrinogen binding (124). Fibrinogen mediated platelet-platelet interactions may be reinforced by the release of alpha granule thrombospondin (154). In addition, conformational changes in fibrinogen following binding to activated GPIIb-IIIa, particularly the RIBS 1 (gamma 373-385) epitope, may participate in platelet aggregation (155).

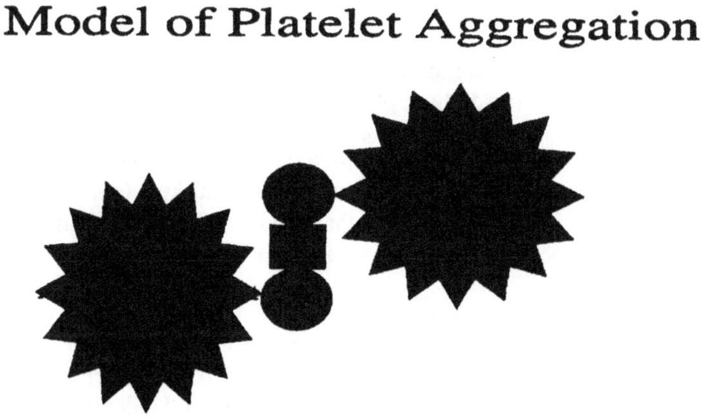

Figure 3. Model of fibrinogen mediated platelet aggregation. Fibrinogen crosslinks adjacent platelets into aggregates predominantly via its gamma chain, carboxyterminal, dodecapeptide sequence (H12) located in the fibrinogen D domain. (E) Amino terminal fibrinogen domain, (D) carboxyterminal fibrinogen domain.

The reversibility of platelet aggregation appears to be modulated in part by stabilization of fibrinogen binding. This concept is supported not only by the observed correlation between inhibition of irreversible platelet aggregation and irreversible fibrinogen binding by inhibition of actin polymerization (136,156) but also by observations that platelets which fail to bind fibrinogen irreversibly, form smaller more readily reversible aggregates than control platelets (100). Although the time to peak aggregation is significantly shorter than the time required for maximal stabilization of fibrinogen binding, the progressive stabilization of fibrinogen binding may occur more rapidly during platelet-platelet contact. Indeed, a number of intracellular events associated with stabilization of fibrinogen binding occur more rapidly in aggregated platelets than in stimulated but unaggregated platelets, including cytoskeleton reorganization (157), and tyrosine phosphorylation (152). Further evidence for the importance of post-fibrinogen binding events in GPIIb-IIIa mediated platelet aggregation comes from observations that human recombinant GPIIb-IIIa complexes, expressed on the surface of CHO cells bind fibrinogen reversibly and support cell-cell interactions at low but not high shear rates (124).

The described GPIIb-IIIa mediated, actin-dependent clearance of bound fibrinogen from the activated platelet surface (76,80,81) suggests a mechanism for preventing and limiting thrombus development. Indeed, a direct correlation between the surface accessibility of platelet-associated fibrinogen and platelet aggregation has been obtained (158). If platelets are stimulated with ADP or thrombin for example, and bind fibrinogen but are not aggregated, they become refractory to aggregation when stirred at a later time. Such platelets also fail to react with antifibrinogen antibodies and corresponding fibrinogen clearance from the external membrane to internal pools is observed by confocal microscopy (158). Moreover platelets activated with A23187 or chymotrypsin retain their ability to aggregate under similar conditions, and bound fibrinogen remains surface expressed (158). Furthermore, the clearance of bound fibrinogen from the activated platelet surface can be prevented by pretreatment of platelets with low concentrations of cytochalasin B or D, and this prevents the loss of platelet aggregability (158).

Platelet adhesion

Fibrinogen plays an important role in platelet adhesion to artificial surfaces (159) and the extracellular matrix (160). In addition, platelet thrombus formation on artificial surfaces is largely mediated by adsorbed fibrinogen (161), and fibrinogen may participate in activated platelet adhesion to damaged endothelial cells (162). Evidence is emerging that fibrinogen contains multiple functional domains which contribute to the expression of full platelet binding activity. In addition to alpha chain RGD and gamma chain H12 sequences (163), the beta 15-42 domain has been shown to support platelet spreading on fibrin (164).

Unlike soluble fibrinogen binding to platelets, platelet recognition of immobilized fibrinogen does not require GPIIb-IIIa activation (165). The adhesion of unstimulated platelets to fibrinogen under static conditions is supported exclusively by residues in the carboxyterminal of the A alpha chain RGD 572-574 (163). Under moderate wall shear stress, however, platelet adhesion is mediated by the fibrinogen gamma 400-411 sequence, and platelets fail to adhere to fibrinogen with mutated gamma chain domains but not RGD or RGE mutations (166).

The adhesion of stimulated platelets under static conditions is supported also by plasmic fibrinogen digests that contain no known platelet recognition sequence (167). Thus, plasmic degradation and immobilization of fibrinogen may expose neoepitopes required for platelet recognition. It is possible that distinct fibrinogen domains participate in the interaction between platelets and immobilized fibrinogen and become engaged as a function of platelet activation and changes in fibrinogen conformation.

It has been suggested that presentation of platelets with multivalent ligands, such as fibrinogen on a surface, contributes to platelet activation and accumulation. Indeed, platelet adhesion to fibrinogen is accompanied by phosphatidic acid production and a concomitant decrease in phosphatidylinositol bisphosphate,suggesting activation of phospholipase C (168). Increases in intracellular calcium concentrations measured using the intracellular fura-2 indicator have also been reported (169). Platelet adhesion to immobilized fibrinogen stimulates the induction of tyrosine phosphorylation of multiple proteins. Inhibitors of protein kinase C have been shown to prevent tyrosine phosphorylation of pp125FAK, pp101, and pp105, and abolish platelet spreading (170). Moreover, platelet spreading on immobilized fibrinogen requires activation of GPIIb-IIIa (171).

Platelet interactions with immobilized fibrinogen appear to be accompanied by the stabilization of the GPIIb-IIIa fibrinogen interaction. GPIIb-IIIa complexes can be recovered on immobilized fibrinogen after Triton X-100 lysis of adherent platelets (172). Moreover, this recovery of GPIIb-IIIa is resistant to chelating agents such as EDTA and enhanced by platelet stimulation with ADP or thrombin, but not chymotrypsin or the calcium ionophore A23187. These agonists demonstrate similar effects on stabilizing platelet fibrinogen interactions in solution (136).

Clot retraction

Ligand binding to GPIIb-IIIa induces conformational changes in the GPIIb-IIIa complex such that neoepitopes termed LIBS are expressed (173). LIBS may form on the extracellular and intracellular portions of GPIIb-IIIa, and may provide sites of interaction for additional plasma or cytoplasmic proteins. LIBS may facilitate "outside-in" signaling events leading to clot retraction. LIBS binding sites may also play an important role in second wave platelet aggregation, and adhesion to collagen (174,175).

Fibrinogen binding to GPIIb-IIIa and GPIIb-IIIa linkage to the cytoskeleton appears necessary for normal clot retraction. Interestingly, mutant fibrinogen molecules lacking the gamma chain H12 sequence, critical for platelet aggregation, are non-the-less able to support clot retraction (176,177). Perhaps RGD sequences or other as yet unidentified sequences in the altered conformation of fibrin become accessible for GPIIb-IIIa binding. Activation of calpain during platelet stimulation may retard clot retraction due to cleavage of critical intracellular molecules, particularly those affecting the association of GPIIb-IIIa with the cytoskeleton. A correlation between irreversible fibrinogen binding and clot retraction has also been proposed (136).

8.5 CONCLUSION

Platelet adhesion to the extracellular matrix exposed at sites of vascular damage, and subsequent platelet aggregation are integral parts of the hemostatic system designed to prevent hemorrhage. Under pathologic conditions, these same processes lead to thrombosis, vasoocclusion, ischemia, and infarction. The interaction of fibrinogen with GPIIb-IIIa plays an important role in both hemostasis and thrombosis. Recent advances in the cell and molecular biology of the platelet confirm that fibrinogen binding to GPIIb-IIIa is the final common pathway leading to platelet aggregation and platelet accumulation at sites of vascular injury. Novel inhibitors of fibrinogen binding to GPIIb-IIIa include specific monoclonal antibodies and synthetic peptides and peptidomimetics that act as direct competitive inhibitors (reviewed in 178). Platelet GPIIb-IIIa antagonists have been shown to be more potent than aspirin in inhibiting platelet function in vitro and in decreasing thrombotic vascular events in human and animal models in vivo (179). GPIIb-IIIa activation is tightly regulated by as yet incompletely understood signaling events, as is fibrinogen binding and the attendant platelet functions including platelet adhesion, spreading, aggregation and clot retraction. Better understanding of the molecular events regulating the GPIIb-IIIa-fibrinogen interaction and its functional sequelae will lead to the design of more potent, specific, and appropriate anti-platelet therapy.

References

1. Weiss HJ. Platelet Pathophysiology and Antiplatelet Drug Therapy. New York: Liss, 1982.

2. Zucker MB, Nachmias VT. Platelet activation. Arteriosclerosis 1985; 5:2-18.

3. Born GVR, Cross MJ. Effects of inorganic ions and plasma proteins on the aggregation of blood platelets by adenosine diphosphate. J Physiol Lond 1964; 170:397-412.

4. Cross MJ. Effect of fibrinogen on the aggregation of platelets by adenosine diphosphate. Thromb Diath Haemorrh 1964; 12:524-527.

5. McLean JR, Maxwell RE, Herther D. Fibrinogen and adenosine diphosphate induce aggregation of platelets. Nature 1965: 202:650-606.

6. Gugler E, Luescher EF. Platelet function in congenital afibrinogenemia. Thromb Diath Haemorrh 1965; 14:361-367.

7. Inceman S, Caen J, Bernard J. Aggregation, adhesion, and viscous metamorphosis of platelets in congenital fibrinogen deficiency. J Lab Clin Med 1966; 68: 21-32.

8. Solum NO, Stormorken H. Influence of fibrinogen on the aggregation of washed human blood platelets induced by adenosine diphosphate, thrombin, collagen, and adrenalin. Scand J Lab Invest 1965; 17:170-182.

9. Caen JP, Castaldi PA, Leclerc JC, et al. Congenital bleeding disorders with long bleeding times and normal platelet count: Glanzmann's thrombasthenia. Report of 15 patients. Am J Med 1966; 41: 4-26.

10. Hardisty RM, Dormandy KM, Hutton RA. Thrombasthenia. Brit J Haematol 1974; 10:371-387.

11. Mustard JF, Packham MA. Factors influencing platelet function: Adhesion, release and aggregation. Pharmacol Rev 1970; 22:97-187.

12. Hagen I, Nurden A, Bjerrum OJ, et al. Immunochemical evidence for protein abnormalities in platelets from patients with Glanzmann's thrombasthenia and Bernard Soulier Syndrome. J Clin Invest 1980; 65:722-731.

13. Nurden AT, Caen JP. The different glycoprotein abnormalities in thrombasthenic and Bernard Soulier platelets. Semin Hematol 1979; 16:234-250.

14. Phillips DR, Agin PP. Platelet membrane defects in Glanzmann's thrombasthenia. Evidence for decreased amounts of two major glycoproteins. J Clin Invest 1977; 60:535-545.

15. Shulman S, Karpatkin S. Crossed immunoelectrophoresis of human platelet membranes. Diminution of major antigens in Glanzmann's thrombasthenia and Bernard Soulier Syndrome. J Biol Chem 1980; 255:4320-4327.

16. Bennett JS, Vilaire G. Exposure of platelet fibrinogen receptors by ADP and epinephrine. J Clin Invest 1979; 64:1393-1400.

17. Kornecki E, Niewiarowski S, Morinelli TA, et al. Effect of chymotrypsin and adenosine diphosphate on the exposure of fibrinogen receptors on normal human and Glanzmann's thrombasthenia platelets. J Biol Chem 1981; 256:5696-5701.

18. Mustard JF, Kinlough-Rathbone RL, Packham MA, et al. Comparison of fibrinogen associated with normal and thrombasthenic platelets on exposure to ADP and chymotrypsin. Blood 1979; 545:987-993.

19. Peerschke EI, Zucker MB, Grant RA, et al. Correlation between fibrinogen binding to human platelets and platelet aggregability. Blood 1980; 55:841-847.

20. Bennett JS, Vilaire G, Cines DB. Identification of the fibrinogen receptor on human platelets by photoaffinity labeling. J Biol Chem 1982; 257:8049-8045.

21. Marguerie GA, Thomas-Maison N, Ginsberg MH, et al. The platelet-fibrinogen interaction. Evidence for proximity of the A alpha chain of fibrinogen to platelet membrane GPIIb/IIIa. Eur J Biochem 1984; 139:5-11.

22. Bennett JS, Hoxie JA, Leitman SF, et al. Inhibition of fibrinogen binding to stimulated human platelets by a monoclonal antibody. Proc Natl Acad Sci 1983; 80:2417-2421.

23. Coller BS, Peerschke EI, Scudder LE, et al. A murine monoclonal antibody that completely blocks the binding of fibrinogen to platelets produces a thrombasthenic like state in normal platelets and binds GPIIb and/or GPIIIa. J Clin Invest 1983; 72:325-338.

24. DiMinno G, Thiagarajan P, Perussia B, et al. Exposure of platelet fibrinogen binding sites by collagen, arachidonic acid and ADP: Inhibition by a monoclonal antibody to the GPIIb-IIIa complex. Blood 1983; 61:140-148.

25. McEver RP, Baenziger NL, Majerus PW. Isolation and quantitation of the platelet membrane glycoprotein deficiency in thrombasthenia using a monoclonal hybridoma antibody. J Clin Invest 1980; 66:1311-1318.

26. Kunicki TJ, Pidard D, Rosa JP, et al. The formation of a calcium dependent complex of platelet membrane GPIIb and IIIa in solution as detected by crossed immunoelectrophoresis. Blood 1981; 58:268-278.

27. Newman PJ, Knipp MA, Kahn RA. Extraction and identification of human platelet integral membrane proteins using Triton X-114. Thromb Res 1982; 27:221-224.

28. Phillips DR, Agin PP. Platelet plasma membrane glycoproteins. Evidence for the presence of nonequivalent disulfide bonds using nonreduced-reduced two dimensional gel electrophoresis. J Biol Chem 1977; 252:2121-2126.

29. Kunicki TJ, Pidard D, Rosa JP, Nurden AT. The formation of Ca++ dependent complexes of platelet membrane glycoproteins IIb and IIIa in solution as determined by crossed immunoelectrophoresis. Blood 1981; 58:268-278.

30. Jennings LK, Phillips DR. Purification of glycoproteins IIb and III from human platelet plasma membranes and characterization of a calcium dependent glycoprotein ib-III complex. J Biol Chem 1982; 257:10458-10463.

31. Steiner B, Parise LV, Leung B, Phillips DR. Ca+2 dependent structural transitions of the platelet glycoprotein IIb-IIIa complex. Preparation of stable glycoprotein IIb and IIIa monomers. J Biol Chem 1991; 266:14986-14991.

32. Brass LF, Shattil SJ. Identification and function of the high-affinity binding sites for Ca+2 on the surface of platelets. J Clin Invest 1984; 73:626-632.

33. Brass LF, Shattil SJ. Changes in surface bound and exchangeable calcium during platelet activation. J Biol Chem 1982: 257:14000-14005.

34. D'Souza SE, Haas TA, Piotrowicz TS, et al. Ligand and cation binding are dual functions of a discrete segment of the integrin beta3 subunit: cation displacement is involved in ligand binding. Cell 1994; 79:659-667.

35. Gulino D, Boudignon C, Shang L, et al. Ca(2+)-binding properties of the platelet glycoprotein IIb ligand interaction domain. J Biol Chem 1992; 267:1001-1007.

36. Hu DD, Barbes CE 3rd, Smith JW. An allosteric Ca2+ binding site of the b3-integrins that regulates the dissociation rate for RGD ligands. J Biol Chem 1996; 271:21745-21751.

37. Phillips DR, Charo IF, Parise LV, Fitzgerald LA. The platelet membrane glycoprotein IIb-IIIa complex. Blood 1988; 71:831-843.

38. Calvete JJ, Henschen A, Gonzalez-Rodriguez J. Assignment of disulfide bonds in human platelet GPIIIa. A disulfide pattern for the beta subunits of the integrin family. Biochem J 1991; 274:63-71.

39 .Kouns WC, Newman PJ, Puckett KJ, et al. Further characterization of the loop structure of platelet glycoprotein IIIa: partial mapping of functionally significant glycoprotein IIIa epitopes. Blood 1991; 78:3215-3233.

40. Calvette JJ, Mann K, Schafer W, et al. Proteolytic degradation of the RGD binding and non-RGD binding conformers of human platelet integrin glycoprotein IIb/IIIa: clues for identification of regions involved in the receptor's activation. Biochem J 1994; 298:1-7.

41. Peerschke EIB, Coller BS. A murine monoclonal antibody that blocks fibrinogen binding to normal platelets also inhibits fibrinogen interactions with chymotrypsin-treated platelets. Blood 1984; 64:59-63.

42. Niewiarowski S, Norton KJ, Eckardt A, et al. Structural and functional characterization of major platelet membrane components derived by limited proteolysis of glycoprotein IIIa. Biochim Biophys Acta 1989; 983:91-99.

43. Calvete JJ, Arias J, Alvarez MV, et al. Further studies on the topography of the N-terminal region of human platelet glycoprotein IIIa. Biochem J 1991; 274:457-463.

44. Abrams C, Shattil SJ. "The Platelet Integrin, GPIIb-IIIa (aIIbb3)." In The Platelet, E. Lapetina, ed, Greenwich, CT: JAI Press Inc, 1994.

45. Naik UP, Parise LV. Structure and function of platelet aIIbb3. Current Opin Hematol 1997; 4:317-322.

46. Wagner CL, Mascelli MA, Neblock DS, et al. Analysis of GPIIb/IIIa receptor number by quantification of 7E3 binding to human platelets. Blood 1996; 88:907-914.

47. Woods VL Jr, Wolff LE, Keller DM. Resting platelets contain a substantial centrally located pool of glycoprotein IIb-IIIa complexes which may be accessible to some but not other extracellular proteins. J Biol Chem 1986: 261:15242-15251.

48. Cramer EM, Savidge GF, Vainchenker W, et al. Alpha granule pool of glycoprotein Iib-IIIa in normal and pathologic platelets and megakaryocytes. Blood 1990; 75:1220-1227.

49. Cramer E, Berger G. Immunoelectron microscopic changes of the platelet plasma membrane after activation. Blood Coagul Fibrinolysis 1996; 7:172-177.

50. Bennett JS. The molecular biology of platelet membrane proteins. Semin Hematol 1990; 27:186-204.

51. Law DA, Nannizzi-Alaimo L, Phillips DR. Outside-in integrin signal transduction. Alpha IIb-Beta 3-(GPIIb-IIIa) tyrosine phosphorylation induced by platelet aggregation. J Biol Chem 1996; 171:10811-10815.

52. Shattil SJ. Function and regulation of the beta3 integrins in hemostasis and vascular biology. Thromb Haemost 1995; 74:149-155.

53. Basani RB, Vilaire G, Shattil SJ, et al. Glanzmann thrombasthenia due to a two amino acid deletion in the fourth calcium binding domain of alpha IIb: Demonstration of the importance of calcium binding domains in the conformation of alphaIIbbeta3. Blood 1996; 88:167-173.

54. Kahn MJ, Kieber-Emmons T, Vilaire G, et al. Effect of mutagenesis of GPIIb amino acid 273 on the expression and conformation of the platelet integrin GPIIb-IIIa. Biochemistry 1996; 35:14304-14311.

55. Hynes RO. The complexity of platelet adhesion to extracellular matrixes. Thromb Haemost 1991; 66:40-43.

56. Coller BS, Cheresh DA, Asch E, Seligsohn U. Platelet vitronectin receptor expression differentiates Iraqi-Jewish from Arab patients with Glanzmann thrombasthenia in Israel. Blood 1991; 77:75-83.

57. Vinciguerra C, Khelif A, Alemany M, et al. A nonsense mutation in the GPIIb heavy chain (Ser 870 stop) impairs platelet GPIIb-IIIa expression. Br J Hematol 1996; 95:399-407.

58. McKay BS, Annis DS, Honda S, et al. Molecular requirements for assembly and function of a minimized human integrin alphaIIbbeta3. J Biol Chem 1996; 271:30544-30547.

59. Loftus JC, Halloran CE, Ginsberg MH, et al. The amino-terminal one-third of alphaIIb defines the ligand recognition specificity of integrin alpha IIb beta 3. J Biol Chem 1996; 271:2033-2039.

60. D'Souza SE, Ginsberg MH, Matsueda GR, Plow EF. A discrete sequence in a platelet integrin is involved in ligand recognition. Nature 1991; 350:66-68.

61. Taylor DB, Gartner T. A peptide corresponding to GPIIb alpha 300-312, a presumptive fibrinogen gamma chain binding site on the platelet integrin GPIIb/IIIa, inhibits the adhesion of platelets to at least four adhesive ligands. J Biol Chem 1992; 267:11729-11733.

62. Kamata T, Irie A, Tokuhira M, Takada Y. Critical residues of integrin alpha IIb subunit for binding of alphaIIbbeta3 (glycoprotein IIb-IIIa) to fibrinogen and ligand mimetic antibodies (Pac-1,OP-G2, and LJ-CP3). J Biol Chem 1996; 271:18610-18615.

63. Alemany M, Concord E, Garin J, et al. Sequence 274-368 in the beta 3-subunit of the integrin alpha IIb beta 3 provides a ligand recognition and binding domain for the gamma chain of fibrinogen that is independent of platelet activation. Blood 1996; 87:592-601.

64. Tozer EC, Liddington RC, Sutcliffe MJ, et al. Ligand binding to integrin alphaIIb beta3 is dependent on a MIDAS-like domain in the beta 3 subunit. J Biol Chem 1996; 271:21978-21984.

65. Liu X-Y, Timmons S, Lin Y-Z, Hawiger J. Identification of a functionally important sequence in the cytoplasmic tail of integrin beta 3 by using cell-permeable peptide analogs. Proc Natl Acad Sci USA 1996; 93:11819-11824.

66. Plow EF, D'Souza SE, Ginsberg MH. Ligand binding to GPIIb-IIIa. A status report. Semin Thromb Hemost 1992; 18:324-332.

67. Ginsberg MH, Du X, O-Toole TE, Loftus JC. Platelet integrins. Thromb Haemostas 1995; 74:352-359.

68. Hawiger J. Adhesive interactions of platelets and their blockade. Ann NY Acad Sci 1991; 614:270-278.

69. Kirschbaum NE, Mosesson MW, Amrani DL. Characterization of the gamma chain platelet binding site on fibrinogen fragment D. Blood 1992; 79:2643-2648.

70. Farrell DH, Thiagarajan P, Chung DW, Davie EW. Role of fibrinogen alpha and gamma chain sites in platelet aggregation. Proc Natl Acad Sci USA 1993; 89:10729-10732.

71. Frojmovic MM, Kasirer-Friede A, Goldsmith HL, Brown EA. Surface secreted von Willebrand factor mediates aggregation of ADP-activated platelets at moderate shear stress; facilitated by GPIb but not controlled by GPIIb-IIIa. Thromb Haemost 1997; 77:568-576.

72. Mustard JF, Packham MA, Kinlough-Rathbone RL, et al. Fibrinogen and ADP-induced platelet aggregation. Blood 1978; 52:453-466.

73. Peerschke EIB. The platelet fibrinogen receptor. Semin Hematol 1985; 22:241-259.

74. Peerschke EIB. Evidence for interaction between fibrinogen receptors. Blood 1982; 60:973-977.

75. Loftus JC, Albrecht RM. Redistribution of the fibrinogen receptor on human platelets following surface activation. J Biol Chem 1984; 99:822-829.

76. Fox JEB, Shattil SJ, Kinlough-Rathbone RL, Richardson M, et al. The platelet cytoskeleton stabilizes the interaction between alpha IIb beta 3 and its ligand and induces selective movements of ligand-occupied integrin. J Biol Chem 1996; 271:7004-7011.

77. Shattil SJ. Function and regulation of the beta 3 integrins in hemostasis and vascular biology. Thromb Haemostas 1995; 74:149-155.

78. Olorundare OE, Simmons SR, Albrecht RM. Cytochalasin D and E: effects on fibrinogen receptor movements and cytoskeletal reorganization of fully spread, surface-activated platelets: a correlative light and electron microscopic investigation. Blood 1992; 79:99-109.

79. Olorundare OE, Simmons SR, Albrecht RM. Evidence for two mechanisms of ligand-receptor movement on surface activated platelets. Eur J Cell Biol 1993; 60:131-145.

80. Peerschke EIB. Bound fibrinogen distribution on stimulated platelets. Examination by confocal scanning laser microscopy. Am J Pathol 1995; 147:678-687.

81. Wencel-Drake JD, Boudignon-Proudhon C, Dieter MG, et al. Internalization of bound fibrinogen modulates platelet aggregation. Blood 1996; 87:602-612.

82. Kirchhofer D, Gailit J, Ruoslahti E, et al. Cation-dependent changes in the binding specificity of the platelet receptor GPIIb/IIIa. J Biol Chem 1990; 265:19525-19530.

83. Gailit J, Ruoslahti E. Regulation of the fibronectin receptor affinity by divalent cations. J Biol Chem 1988; 263:12927-12932.

84. Elice MJ, Urry LA, Hemler ME. Receptor functions for the integrin VLA-3: fibronectin, collagen, and laminin binding are differentially influenced by Arg-Gly-Asp peptide and by divalent cations. J Cell Biol 1991; 112:169-181.

85. Dransfiedl I, Cabanas C, Craig A, Hogg N. Divalent cation regulation of the function of the leukocyte integrin LFA-1. J Cell Biol 1992; 116:219-226.

86. Plow EF, Marguerie GA. Induction of the fibrinogen receptor on human platelets by epinephrine and the combination of epinephrine and ADP. J Biol Chem 1980; 255:10971-10977.

87. Peerschke EIB. Effect of epinephrine on fibrinogen receptor exposure by aspirin-treated platelets and platelets from concentrates in response to ADP and thrombin. Am J Haematol 1984; 16:335-345.

88. Rao GHR, Johnson GJ, White JG. Influence of epinephrine on the aggregation of aspirin-treated platelets. Prostaglandin Med 1980; 5:45-58.

89. DiMinno G, Capitana AM, Thiagarajan P, et al. Exposure of fibrinogen receptors on fresh and stored platelets by ADP and epinephrine as single agents and as a pair. Blood 1983; 61:1054-1059.

90. Padoin E, Alexandre A, Cavallini L, et al. Human platelet activation is inhibited by he occupancy of glycoprotein IIb/IIIa receptor. Arch Biochem Biophys 1996; 333:497-413.

91. Van Willigen G, Akkerman JWN. Protein kinase C and cyclic AMP regulate reversible exposure of binding sites for fibrinogen on the glycoprotein IIb-IIIa complex of human platelets. Biochem J 1991; 273:115-120.

92. Du XP, Plow EF, Frelinger AL 3d, et al. Platelet glycoprotein IIb-IIIa (alpha IIb beta 3 integrin) confers fibrinogen-and activation-dependent aggregation on heterologous cells. Blood 1991; 78:369-376.

93. Sims PJ, Ginsberg MH, Plow EF, Shattil SJ. Effect of platelet activation on the conformation of the plasma membrane glycoprotein IIb-IIIa complex. J Biol Chem 1991; 266:7345-7352.

94. Peerschke EI, Zucker MB. Fibrinogen receptor exposure and aggregation of human blood platelets produced by ADP and chilling. Blood 1981; 57:663-339.

95. Kornecki E, Niewiarowski S, Morinelli TA, et al. Effect of chymotrypsin and adenosine diphosphate on the exposure of fibrinogen receptors on normal human and Glanzmann's thrombasthenia platelets. J Biol Chem 1981; 256:5696-5701.

96. Niewiarowski S, Budzynski AZ, Morinelli TA, et al. Exposure of fibrinogen receptors on human platelets by proteolytic enzymes. J Biol Chem 1981; 256:917-925.

97. Loscalzo J, Pasche B, Ouimet H, Freedman JE. Platelets and plasminogen activation. Thromb Haemost 1995; 74:291-293.

98. Rabhi-Sabile S, Pidard D. Exposure of human platelets to plasmin results in the expression of irreversibly activated fibrinogen receptors. Thromb Haemost 1995; 73:693-701.

99. Zucker MB, Masiello NC. Platelet aggregation caused by dithiothreitol. Thromb Haemost 1984; 51:119-124.

100. Peerschke EIB. Regulation of platelet aggregation by post-fibrinogen binding events. Insights provided by dithiothreitol-treated platelets. Thromb Haemostas 1995; 73:862-867.

101. Peerschke EIB. Effect of stimulation on the stabilization of platelet-fibrinogen interactions. Thromb Haemostas 1992; 68:346-351.

102. Pidard D, Frelinger AL, Bouillot C, Nurden AT. Activation of the fibrinogen receptor on human platelets exposed to alpha chymotrypsin. Relationship with a major proteolytic cleavage at the carboxyterminus of the membrane glycoprotein IIb heavy chain. Eur J Biochem 1991; 200:437-447.

103. Pasche B, Ouimet H, Francis S, Loscalzo J. Structural changes in platelet glycoprotein IIb/IIIa by plasmin: determinants and functional consequences. Blood 1994; 83:404-414.

104. Derrick JM, Taylor DB, Loudon RG, Gartner TK. The peptide LSARLAF causes platelet secretion and aggregation by directly activating the integrin alphaIIb beta 3. Biochem J 1997; 325:309-313.

105. Du X, Plow EF, Frelinger AL, et al. Ligands " activate" integrin alpha IIb beta3 (platelet GPIIb-IIIa). Cell 1991; 65:408-416.

106. Smyth SS, Parise LV. Regulation of ligand binding to glycoprotein IIb-IIIa (integrin alpha IIb beta 3) in isolated platelet membranes. Biochem J 1993; 292:749-759.

107. Mayo KH, Fan F, Beavers MP, et al. RGD induces conformational transitions in purified platelet integrin GPIIb-IIIa SDS systems yielding multiple binding states for fibrinogen gamma chain C-terminal peptide. FEBS Lett 1996; 378:79-82.

108. Coller BS. "Activation Specific Platelet Antigens." In Platelet Immunobiology: Molecular and Clinical Aspects, TJ Kunicki, JN George, eds. New York: JB Lippincott, 1989, pp 166-192.

109. Sims PJ, Ginsberg MH, Plow EF, Shattil SJ. Effect of platelet activation on the conformation of the plasma membrane glycoprotein IIb-IIIa complex. J Biol Chem 1991; 266:7345-7352.

110. Plow EF, D'Souza SE, Ginsberg MH. Ligand binding to GPIIb-IIIa. A Status Report. Semin Thromb Haemost 1992; 18:324-332.

111. Isenberg WM, McEver RP, Phillips DR, et al. The platelet fibrinogen receptor: an immunogold-surface replica study of agonist-induced ligand binding and receptor clustering. J Cell Biol 1987; 104:1655-1663.

112. Shattil SJ. Function and regulation of the beta3 integrins in hemostasis and vascular biology. Thromb Haemost 1995; 74:149-155.

113. Blockman D, Deckmyn H, Vermylen J. Platelet activation. Blood Reviews 1995; 9:143-156.

114. King, WG, Kucera G, Sorisky A, et al. Protein kinase C regulates the stimulated accumulation of 3-phosphorylated phosphoinositedes in platelets. Biochem J 1991; 278:475-480.

115. Kovascovics TJ, Bachelot C, Toker A, Vlahos CJ, et al. Phosphoinositide 3-kinase inhibition spares actin assembly in activating platelets and reverses platelet aggregation. J Biol Chem 1995; 270:11358-11366.

116. Graber SE, Hawiger J. Evidence that changes in platelet cyclic AMP levels regulate the fibrinogen receptor on human platelets. J Biol Chem 1982; 257:14606-14609.

117. Hughes PE, O'Toole TE, Ylanne J, et al. The conserved membrane-proximal region of an integrin cytoplasmic domain specifies ligand binding affinity. J Biol Chem 1995; 270:12311-12417.

118. O'Toole TE, Mandelman D, Forsyth J, Shattil SJ, Plow EF, Ginsberg MH. Modulation of the affinity of integrin alpha IIb beta 3(GPIIb-IIIa) by the cytoplasmic domain of alpha IIb. Science 1991; 254:845-847.

119. Loh E, Qi W, Vilaire G, Bennett JS. Effect of cytoplasmic domain mutations on the agonist-stimulated ligand binding activity of the platelet integrin alpha IIb beta 3. J Biol Chem 1996; 271:30233-30241.

120. Hughes PE, Diaz-Gonzalez F, Leong L, et al. Breaking the integrin hinge: defined structural constraint regulates integrin signaling. J Biol Chem 1996; 271:6571-6574.

121. Haas TA, Plow EF. The cytoplasmic domain of alphaIIb beta 3: a ternary complex of the integrin alpha and beta subunits and a divalent cation. J Biol Chem 1996; 271:6017-6026.

122. Chen Y-P, O'Toole TE, Shiple YT, et al. "Inside-out" signal transduction inhibited by isolated integrin cytoplasmic domains. J Biol Chem 1994; 269:18307-18310.

123. Shattil SJ, O'Toole T, Eigenthaler M, et al. Beta 3-endonexin, a novel polypeptide that interacts specifically with the cytoplasmic tail of the integrin beta 3 subunit. J Cell Biol 1995; 131:807-816.

124. Kashiwagi H, Eigenthaler E, Schwartz MA, et al. Affinity modulation of platelet integrin alpha IIb beta 3 endonexin, a selective binding partner of the beta 3 integrin cytoplasmic tail. J Cell Biol 1997; 137:1433-1443.

125. Naij UP, Patel PM, Parise LV. Identification of a novel calcium-binding protein that interacts with the integrin alpha IIb cytoplasmic domain. J Biol Chem 1997; 272:4651-4654.

126. Baker EK, Tozer EC, Pfaff M, et al. A genetic analysis of integrin function: Glanzmann thrombasthenia in vitro. Proc Natl Acad Sci (USA) 1997; 94:1972-1978.

127. Hughes PE, Renshaw MW, Pfaff M, et al. Suppression of integrin activation: a novel function of a Ras/Raf-initiated MAP kinase pathway. Cell 1997; 88:521-530.

128. Shick DD, Wencel-Drake JD, Parise LV. Ras activation in platelets after stimulation of the thrombin receptor, thromboxane A2 receptor or protein kinase C. Biochem J 1997; 321:525-530.

129. Zhange Z, Vuori K, Wang H-G, et al. Integrin activation by R-ras. Cell 1996; 85:61-69.

130. Faraday N, Goldschmidt-Clermont PJ, Bray PF. Gender differences in platelet GPIIb-IIIa activation. Thromb Haemost 1997; 77:748-54.

131. Peerschke EIB. Events occurring after thrombin-induced fibrinogen binding to platelets. Semin Thromb Hemost 1992; 18:34-43.

132. Marguerie GA, Plow EF. The fibrinogen-dependent pathway of platelet aggregation. Ann NY Acad Sci 1983; 408:556-566.

133. Huber W, Hurst J, Schlatter D, et al. Determination of kinetic constants for the interaction between the platelet GPIIb-IIIa and fibrinogen by means of surface plasmon resonance. Eur J Biochem 1995; 227:647-756.

134. Ugarova TP, Budzynski AZ, Shattil SJ, et al. Conformational changes in fibrinogen elicited by its interaction with platelet membrane glycoprotein GPIIb-IIIa. J Biol Chem 1993; 268:21080-21087.

135. Du X, Gu M, Weisel J, Nagaswami C, et al. Long range propagation of conformational changes in integrin alpha IIb beta 3. J Biol Chem 1993; 268:23087- 23092.

136. Peerschke EIB. Reversible and irreversible binding of fibrinogen to platelets. Platelets 1997; 8:311-317.

137. Peerschke EIB. Time dependent association between platelet bound fibrinogen and the Triton X-100 insoluble cytoskeleton. Blood 1991; 77: 508-514.

138. Peerschke EIB. Stabilization of platelet fibrinogen interactions: modulation by divalent cations. J Lab Clin Med 1993; 121:135-141.

139. Peerschke EIB. Irreversible platelet-fibrinogen interactions occur independently of fibrinogen alpha chain degradation and are not mediated by intact platelet membrane glycoprotein IIb-IIIa complex. J Lab Clin Med 1988; 111:84-92.

140. Peerschke EIB. Reversibility of fibrinogen fragment D1 binding to human platelets: comparison with native fibrinogen. Thromb Res 1992; 68:259-267.

141. Simmons SR, Albrecht RM. Self-association of bound fibrinogen on platelet surfaces. J Lab Clin Med 1996; 128:39-50.

142. Woods VL, Wolf LE, Keller DM. Resting platelets contain a substantial centrally located pool of glycoprotein IIb-IIIa complex which may be accessible to some but not other extracellular proteins. J Biol Chem 1986; 261:15242-15251.

143. Muller V, Zerwes HG, Tangemann K, et al. Two-step binding mechanism of fibrinogen to alphaIIb/beta3 integrin reconstituted into planar lipid bilayers. J Biol Chem 1993; 268:6800-6808.

144. Peerschke EIB. Stabilization of platelet-fibrinogen interactions is an integral property of the glycoprotein Iib-IIIa complex. J Lab Clin Med 1994; 124:439-446.

145. Peerschke EIB. Adhesive protein expression on thrombin-stimulated platelets: time dependent modulation of anti-fibrinogen , -fibronectin, and -von Willebrand Factor antibody binding. Blood 1992; 79: 2028-2033.

146. Watala C, Gwozdinski K, Pluskota E, et al. Microenvironmental changes in platelet membranes induced by the interaction of fibrinogen derived peptide ligands with platelet integrins. Eur J Biochem 1996; 235:281-288.

147. Carrol RC, Butler RG, Morris PA, Gerrard JM. Separable assembly of platelet pseudopodal and contractile cytoskeletons. Cells 1982; 30:385-393.

148. Furman MI, Gardner TM, Goldschmidt-Clermont PJ. Mechanisms of cytoskeletal reorganization during platelet activation. Thromb Haemost 1993; 70:229-232.

149. Knezvic I, Leisner TM, Lam SC-T. Direct binding of the platelet integrin alpha IIb beta 3 (GPIIb-IIIa) to talin: evidence that interaction is mediated through the cytoplasmic domains of both alpha IIb and beta 3. J Biol Chem 1996; 271:16416-16421.

150. Cohen I. "The mechanism of clot retraction." In Platelet Membrane Glycoproteins, JN George, AT Nurden, DR Phillips, eds. New York: Plenum Press, 1985, pp 299-323.

151. Clark EA, Brugge JA. Integrins and signal transduction pathways: the road taken. Science 1995; 268:233-239.

152. Shattil SJ, Ginsberg MH, Brugge JS. Adhesive signaling in platelets. Curr Opin Cell Biol 1994; 6:695-704.

153. Egan SE, Giddings BW, Brooks MW, et al. Association of Sos Ras exchange protein with Grb 2 is implicated in tyrosine kinase signal transduction and transformation. Nature 1993; 363:45-51.

154. Asch AS, Nachman RL. "Thrombospondin: Phenomenology to Function." In Progress in Hemostasis and Thrombosis. Vol 9, BS Coller, ed. Philadelphia: WB Saunders, 1989; pp 157-175.

155. Liu O, Frojmovic M, Xia Z, et al. Role of RGDS/gamma 400-411 and RIBS domains on platelet bound fibrinogen in mediating platelet aggregation. Thromb Haemost 1995; 73:1059a.

156. Torti M, Testetics ET, Bertoni A, et al. Agonist-induced actin polymerization is required for the irreversibility of platelet aggregation. Thromb Haemostas 1996; 76: 444-449.

157. Kouns WC, Fix CF, Lamoreaux WJ, et al. The effect of glycoprotein IIb-IIIa receptor occupancy on the cytoskeleton of resting and activated platelets. J Biol Chem 1991; 266:13891-13900.

158. Peerschke EIB. Recognition of platelet associated fibrinogen by polyclonal antibodies. Correlation with platelet aggregation. Blood 1992; 79: 948-953.

159. Lindon JN, McManama GP, Kushner L, et al. Does the conformation of adsorbed fibrinogen dictate platelet interaction with artificial surfaces? Blood 1986; 68:355-362.

160. Godyna S, Diaz-Ricart M, Argraves WS. Fibulin-1 mediates platelet adhesion via a bridge of fibrinogen. Blood 1996; 88:2569-2577.

161. Ip JH, Stein B, Fuster V, Badimon L. Antithrombotic therapy in cardiovascular diseases. Future direction based on pathogenesis and risk. Ann NY Acad Sci 1991; 614:289-311.

162. Li JM, Podolsky RS, Rohrer MJ, et al. Adhesion of activated platelets to venous endothelial cells is mediated via GPIIb/IIIa. J Surg Res 1996; 61:543-548.

163. Savage B, Ruggeri ZM. Selective recognition of adhesive sites in surface-bound fibrinogen by glycoprotein IIb-IIIa on non activated platelets. J Biol Chem 1991; 266:11227-11233.

164. Hamaguchi M, Bunce LA, Sporn LA, Francis CW. Spreading of platelets on fibrin is mediated by the amino terminus of the beta chain including peptide beta 15-42. Blood 1993; 81:2348-2356.

165. Ruggeri ZM. The role of von Willebrand Factor and fibrinogen in the initiation of platelet adhesion to thrombogenic surfaces. Thromb Haemost 1995; 74:460-463.

166. Zaidi TN, McIntire LV, Farrell DH, Thiagarajan P. Adhesion of platelets to surface bound fibrinogen under flow. Blood 1996; 88:2967-2972.

167. Peerschke EIB, Galanakis DK. Platelet adhesion to late fibrinogen degradation products. Blood Coag Fibrinolys 1996; 7:353-360.

168. Gironcel D, Racaud-Sultan C, Payrastre B, et al. Alpha IIb beta 3 integrin mediated adhesion of human platelets to a fibrinogen matrix triggers phospholipase C activation and phosphatidylinositol 3',4'-biphosphate accumulation. FEBS Lett 1996; 389:253-256.

169. Jen CJ, Chen HI, Lai KC, Usami S. Changes in cytosolic calcium concentrations and cell morphology in single platelets adhered to fibrinogen-coated surfaces under flow. Blood 1996; 87:3775-3782.

170. Haimovich B, Kaneshiki N, Ji P. Protein kinase C regulates tyrosine phosphorylation of pp125FAK in platelets adherent to fibrinogen. Blood 1996; 87:152-161.

171. Nakatani S, Hato T, Minamoto Y, Fujita S. Differential inhibition of fibrinogen binding to agonist- and RGDS peptide-activated states of GPIIb-IIIa by an anti-GPIIIa monoclonal antibody, PMA 5. Thromb Haemost 1996; 76:1030-1037.

172. Peerschke EIB. Glycoprotein IIb and IIIa retention on fibrinogen-coated surfaces after lysis of adherent platelets. Blood 1995; 82:3358-3363.

173. Frelinger AL 3d, Du XP, Plow EF, Ginsberg MH. Monoclonal antibodies to ligand-occupied conformers of integrin alpha IIb beta 3 (glycoprotein IIb-IIIa) alter receptor affinity, specificity, and function. J Biol Chem 1991; 266:17106-17111.

174. Frelinger AL III, Cohen I, Plow EF, et al. Selective inhibition of integrin function by antibodies specific for ligand-occupied receptor conformers. J Biol Chem 1990; 265:6346-6352.

175. Mondoro TH, Wall CD, White MM, Jennings LK. Selective induction of a glycoprotein IIIa ligand-induced binding site by fibrinogen and von Willebrand Factor. Blood 1996; 88:3824-3830.

176. Rooney MM, Raise LV, Lord ST. Dissecting clot retraction and platelet aggregation: clot retraction does not require an intact fibrinogen gamma chain C-terminus. J Biol Chem 1996; 271:8553-8555.

177. Holmback K, Danton MJ, Suh TT, et al. Impaired platelet aggregation and sustained bleeding in mice lacking the fibrinogen motif bound by integrin alpha IIb beta 3. EMBO J 1996; 15:5760-5771.

178. Eccleston D, Topol EJ. Inhibitors of platelet glycoprotein IIb/IIIa as augmenters of thrombolysis. Coronary Artery Dis 1995; 6:947-955.

179. Coller BS. Augmentation of thrombolysis with antiplatelet drugs. Coronary Artery Disease 1995; 6:911-914.

9

BIOCHEMISTRY OF PLATELET ACTIVATION

Anita Ryningen and Holm Holmsen
Department of Biochemistry
and Molecular Biology
University of Bergen
NORWAY

9.1 FUNCTION

Platelets circulate in the blood stream as anucleate discoid cells, at a number of 150,000-350,000 per µl and have a life span of 10 days in man. They are involved in the early phase of the hemostatic process. When a blood vessel is damaged, platelets are exposed to subendothelial collagen. Together with von Willebrand factor (vWF) an activation process is started in which the platelets immediately adhere to the exposed collagen fibrils, change their shape from discs to spheres with long pseudopods and empty their granule contents (degranulation). Few seconds later more platelets are deposited on the collagen fibrils and the platelets are now also sticking to each other (platelet aggregation). A growing platelet plug is formed in the lesion of the vessel wall as more platelets aggregate. At a certain stage in this process, red blood cells are no

Anita Ryningen has done research since 1991 until the present on signaling mechanisms involved in activation and inhibition of platelet responses at the Department of Biochemistry and Molecular Biology, University of Bergen. She is currently finalizing her Ph. D. thesis. **Holm Holmsen** is Professor at the same institution and received his Dr. Phil. from the University of Oslo in 1969 on adenine nucleotide metabolism in platelets. He continued these studies at the Department of Medicine, Temple University Medical School, Philadelphia (1970-1982). His current research interest is lipid-mediated, cellular signal transduction.

longer capable of force their way through, only plasma continues to leak out, and primary hemostasis is achieved. Through this activation process the platelets expose different epitopes on their surface (they obtain procoagulant activity) which primes the coagulation cascade. The latter process culminates in the formation of thrombin which causes a network of fibrin to form between the platelets which strengthens the plug and makes it impermeable to both blood cells and plasma (secondary hemostasis).

The Cartoon: Taken from Derenleau (1987)

9.2 MORPHOLOGY AND SUBCELLULAR ORGANELLES

Platelets are small fragments of cytoplasm derived from megakaryocytes (Wright, 1906, 1910). They are about 3 µm in diameter and 1 µm thick and are therefore the smallest of the blood cells (White, 1987c). The plasma membrane consists of phospholipids where the negatively charged phosphatidylinositols and phosphatidylserines are located at the inner leaflet of the membrane and involved in active signaling. Glycoproteins (parts of various receptors) are extruding through the plasma membrane, with their cytoplasmic tails linked to the cytoskeleton via proteins like actin (Painter *et al.*, 1985) and actin-binding protein (Fox, 1985). The cytoskeleton in platelets consists primarily of actin filaments. It has been estimated that 20-30 % of total platelet protein is actin (Harris and Weeds, 1978; Rosenberg *et al.*, 1981; Fox *et al.*, 1984). For review of the organization of the platelet cytoskeleton, see Fox (1993b). Platelets have most of the subcellular organelles that are found in other cells except for nucleus. They are therefore deficient in DNA, except for mitochondrial DNA. There are numerous organelles in the cytoplasm of the platelet: some few and small, but active, mitochondria, storage granules, where α-granules and dense granules are platelet-specific and empty their contents to the extracellular milieu upon activation, peroxisomes, lysosomes and glycogen granules. The platelets contain two different intracellular membrane systems. Firstly, there is an internal system called "dense tubular system" (DTS) which is a closed system like the endoplasmatic reticulum in other cells. DTS is presumably the location for prostaglandin production (Gerrard *et al.*, 1978) and storage of calcium ions (Isenberg *et al.*, 1995). Secondly, there is a penetrating channel system, "open canalicular system" (OCS), which are invaginations of the plasma membrane and serves as an export system for secretory products from the storage granules. This channel system greatly increases the available surface area of the platelets. When the platelets are activated *in vitro*, they loose their discoid shape and become more spherical with long pseudopds. The organelles move to the center and become surrounded by a ring of newly made microtubules and microfilaments (Stenberg *et al.*, 1984; White, 1974, 1987c). Finally, the contents of the secretory organelles are extruded. At this stage in the activation process the platelets are not able to return to their discoid shape. They are sequestered and degraded by the liver and spleen and will be substituted with newly made platelets from the megakaryocytes.

9.3 PLATELET ACTIVATION AND RESPONSES

Platelets are multiresponding cells both with respect to the number of agonists and number of responses. The responses include mainly adhesion (to foreign surfaces, e.g., collagen or glass), shape change (from smooth discs to spheres with pseudopods), aggregation (platelets stick to each other) and secretion from three different storage granules (dense granules, α-granules and lysosomes), exposure of neoepitopes, shedding of microvesicles, formation of platelet procoagulant activity and retraction of fibrin clots.

9.3.1 Adhesion

Adhesion could be defined as an platelet activation process where insoluble high molecular weight proteins with repetitive structures (collagen and immune complexes) or surfaces normally not in contact with platelets, e.g. nonphysiological agonists like artificial surfaces, act as "agonists" and trigger platelet spreading on the surfaces themselves. Resting platelets tend to adhere to almost any foreign surface they are exposed to, and this process depends on the shear rate of the flowing blood. At low shear rates such as those found in large veins, the platelets make contact with and spread onto components of the subendothelial matrix such as collagen and fibronectin. See Table 9.5.1. for possible receptors involved. Under conditions of high shear, such as found in arterioles and in the microcirculation, the contact with and spreading on the subendothelium of platelets require vWF (Baumgartner *et al.*, 1980; Turitto *et al.*, 1983, 1985). Glycoprotein Ib appears to be the major vWF-binding site on the platelet surface that mediates platelet adhesion (Wagner and Marder, 1984; Ruggeri and Zimmerman, 1987; Weiss *et al.*, 1974; Ruan *et al.*, 1981). The adhesion of platelets to glass or other surfaces are probably also mediated by glycoproteins on the platelet surface (discussed further below in 9.5).

9.3.2 Shape change

In suspension the first response of platelets to stimulation is shape change which involves a change from the normal discoid shape to spiny spheres containing many long pseudopodia about 0.1 μm in diameter and more than one platelet diameter long. This process is mediated by the contractile microtubular system and is characterized morphologically by the extension of short and long dendritic pseudopodia (White, 1987b, 1987c). The response is very fast, with a half life of 6 seconds (Derenleau *et al.*, 1986), and is independent of extracellular Ca^{2+} (Holmsen *et al.*, 1974). The circumferential band of microtubules disappears and actin filaments and single microtubules appear in the newly formed pseudopods (Castle and Crawford, 1979). The mechanisms underlying shape change seem to involve changes in cytoskeletal proteins (Pribluda and Rotman, 1982; Kometani *et al.*, 1986; Yoshida *et al.*, 1986), phosphorylation of myosin light chain (Daniel *et al.*, 1984) and the response is inhibited by microtubule-active drugs like colchicine and vinca alkaloids (Menche *et al.*, 1980). Shape change is an energy-requiring process (Mills, 1973; Holmsen *et al.*, 1974) due to formation of pseudopods and recruitment of the membrane that would be required (the platelet volume remain constant, but the platelet surface area increases by at least 50 %). During shape change, cryptic receptors for fibrinogen are exposed on the platelet surface.

9.3.3 Aggregation

When spherical, activated platelets with exposed fibrinogen receptors are brought in close contact with each other in a medium containing Ca^{2+} and fibrinogen, they stick to each other and form aggregates in a process that takes 2-3 minutes to be completed. Fibrinogen is required as a cofactor for aggregation and because it is a large dimeric molecule, it is likely that it binds to receptors on adjacent activated platelets, cross-linking them into growing aggregates (McLean *et al.*, 1964; Cross, 1964; Blockmans *et al.*, 1995). After the initial shape change there is no further decrease in the adenylate energy charge (Holmsen *et al*, 1974), which suggests that aggregation may be a passive process which occurs after shape change and fibrinogen receptor exposure. Two types of aggregation responses can be distinguished *in vitro* depending on the potency and concentration of the stimulating agent: a primary aggregation that is reversible and occurs without secretion and a secondary aggregation that is irreversible and associated with secretion. A biphasic aggregation response can be seen by certain stimulating agents which induce a primary aggregation followed by secondary aggregation caused by autocrine stimulators (ADP, TXA_2, fibrinogen, serotonin, PAF) that are released from the platelets during the secretionprocess.

9.3.4 Secretion

Platelets contain three different types of secretory granules: dense granules, α-granules and acid hydrolase-containing granules. Following platelet stimulation, the contents of these granules are extruded from the platelet interior in a process involving exocytosis with fusion of the granules and the surface-connected open canacular system (White, 1970; Holme *et al.*, 1973; Holme *et al*, 1974). The granules differ from each other with the respect of the strength necessary to induce secretion. Lysosomal secretion occurs only after stimulation with strong agonists (i.e. thrombin, collagen), it is slow (several minutes) and is never complete (maximal 60 % of the total amounts in platelets). Dense- and α-granule secretion can be induced by weak agonists like ADP, TXA_2 and adrenaline (in synergism with one of the other), it is nearly 100 % within 1-2 minutes (Mills *et al*, 1968; Holmsen and Day, 1970; Kaplan *et al.*, 1979; Holmsen *et al.*, 1982; Verhoeven *et al.*, 1984). In contrast to other secretory cells, platelets are not able to restore the contents of their granules after secretion because of their lack of a nucleus. The different storing granules in platelets and their contents are shown in Table 9.3.

Table 9.3 The contents of the three storage granules in platelets

α-granules	Plasma proteins [PL]	Albumin
	Coagulation factors [PL]	Factor Va, VIII
		Fibrinogen
	Complement factors [PL]	C5-C9
		Vitronectin
		Clusterin
		C1-INH
		Factor D
		Factor H
	Platelet specific proteins	β-thromboglobulin
		Platelet factor 4-proteoglycan
	Growth factors	PDGF
		TGF-β
		EGF
		ECGF
	Glycoproteins [PL]	Thrombospondin
		Fibronectin
		vWF
Dense granules		ATP^{o}, GTP^{o}
		ADP^{o}, GDP^{o}
		PP_i^{o}, P_i^{o}
		Ca^{2+}, Mg
		Serotonin
Lysosomes	Acid hydrolases [PL]	β-hexosaminidase (low uptake form)
		β-glucuronidase
		β-galactosedase
		α-arabinosidase

[PL] denotes that the substance have its counterpart in plasma, while "o" indicates a substance that does not exchange with their cytoplasmic counterparts (metabolically inactive). Serotonin is the only granule-stored substance that is readily taken up from the extracellular phase.

9.3.5 Exposure of neoepitopes

Platelet adhesion and aggregation are associated with changes on the platelet surface. Stimulation with weak agonists such as ADP causes a conformational change in the platelet GPIIb/IIIa complex which exposes the receptor site for fibrinogen. When secretion occurs, granule membranes are translocated to the surface (OCS), carrying with them specific membrane glycoproteins (originally facing the interior of the granules) that appears as neo-antigenes on the surface (Abrams and Shattil, 1991). These can be detected by flow cytometry using antibodies against the new epitopes. These include activated fibrinogen-receptor detected with PAC-1 (Shattil *et al.*, 1985), P-selectin (also called GMP140, PADGEM and CD62), released from the α- and dense granules (McEver and Martin, 1984; Hsu-Lin *et al.*, 1984; Stenberg *et al.*, 1985; Berman *et al*, 1986; Bevilaqua *et al.*, 1991; Israels *et al*, 1992), GMP-33 from the α-granule membranes (Metzelaar *et al*, 1992), granulophysin from the dense granules (Gerrard *et al*, 1991; Israels *et al.*, 1992), glycoprotein 53 (CD63) and LAMP1 plus LAMP2 from the lysosomes (Chen *et al.*, 1985; Nieuwenhuis *et al.*, 1987; Fukuda *et al.*, 1988; Febbraio and Silverstein, 1990; Cox and Goodal, 1991).

Proteins released from the platelet granules may also bind to the platelet surface. Components of the α-granules such as thrombospondin (Aiken *et al.*, 1987), platelet factor 4 (Capitanio *et al.*, 1985), fibrinogen, fibronectin and PDGF bind to their respective receptors on the platelet surface. This concerns also ADP and serotonin from the dense granules (discussed further in 9.7). Phosphatidylserine (PS) which is normally located on the inner bilayer of the platelet membrane, becomes exposed on the outer surface of the membrane in activated platelets (Bevers *et al.*, 1983).

9.3.6 Microvesicles

Platelet activation produces formation of small microparticles or vesicles (50-800 nm in diameter) shed from the plasma membrane. The biological function of microparticles (MP) is still speculative. They contain most of the lipids and proteins found in platelets. MP formation is dependent on activated GPIIb-IIIa, and one or all of the GPIIb-IIIa-adhesive ligands. Platelets lacking GPIIb-IIIa (Glanzmann's thrombasthenia) do not vesiculate (Gemmel *et al.*, 1993). Cytoskeletal reorganization, which is regulated by serine and threonine phosphorylation, is also necessary for MP formation (Yano *et al.*, 1994). MPs are rich in exposed, negatively charged PS which can bind the procoagulant proteins Factor VIIIa (Gilbert *et al.*, 1991), Factor Va (Sims *et al.*, 1988) and Factor IXa (Hoffman *et al.*, 1992), suggesting a role of the MPs in modulating hemostasis and thrombosis. Platelet factor 3 (PF3), which may be identical to exposed PS (Koerner *et al.*, 1992; Dachary-Prigent *et al.*, 1996), is predominantly associated with MP and elevated in thrombotic conditions. The exposed PF3 is suggested to activate neutrophils and mediate platelet-neutrophil interaction (Jy *et al.*, 1995). Also C5b-9 is found to a higher degree on MP suggesting a mechanism to avoid cytolysis after formation of the complement pore (Wiedmer and Sims, 1985; Sims and Wiedmer, 1986). It is shown that the majority of the PAF released from platelets (87 %) are associated with MP and located on their outer surface (Iwamoto *et al.*, 1996).

9.3.7 Platelet procoagulant activity

Activated platelets provides a surface for assembly of the enzyme complexes that lead to the generation of thrombin. The prothrombinase reaction results in the enzymatic conversion of prothrombin to thrombin. Prothrombinase is a complex composed of activated factor V, activated Factor X, Ca^{2+} and prothrombin (Tracy and Mann, 1986; Kane and Davie, 1988). As described in 9.3.6, these sites for assembly are also present on MPs.

9.3.8 Retraction of fibrin clots

Fibrin clot formation in blood is followed by retraction of the clot in a process that requires activated platelets (Shaman and Greenberg, 1986). It is suggested that activated platelets interact with fibrin within the clot and that the contraction of the platelet cytoskeleton generates the force needed for the process. The platelet structure that is involved in the fibrin-binding is unclear. GPIIb-IIIa and Factor XIIIa have been shown to be necessary for generating the force of cytoskeletal contraction and to enzymatically cross-link fibrin within the clots, respectively (Kitchens and Newcomb, 1979; Cohen *et al.*, 1982; Gartner and Ogilvie, 1988).

9.4 SIGNAL TRANSDUCTION SYSTEMS

9.4.1 Heterotrimeric GTP-binding proteins

Many platelet receptors mediate their responses through heterotrimeric GTP-binding proteins (G proteins) which act as transducers between seven transmembrane domain receptors and a variety of enzymes and ion channels (Neer, 1995). In general, G proteins are made up of three polypeptides: an α subunit of about 39-52 kDa that binds and hydrolyzes GTP, and a β and γ subunit of 35-36 kDa and 5-8 kDa, respectively, that form a dimeric complex. The β and γ subunits are highly conserved while the α subunit is different for each G protein. In resting cells the α subunit contains bound GDP and form an inactive heterotrimer with the βγ subunits. When an agonist binds to its receptor, the receptor becomes activated and changes its conformation. The subunit responds with a conformational change leading to exchange of its bound GDP for GTP and dissociation of the α-GTP complex from the βγ complex. The free α-GTP and βγ subunits each activate effector systems. GTP is hydrolyzed to GDP by the intrinsic GTPase activity of the α subunit and reassociates with βγ (Gilman, 1987). The rate of GTP hydrolysis is a timing mechanism that controls the duration of both α and βγ subunit activation (Goody, 1994; Neer, 1995). One receptor can activate more than one type of G protein, and their α subunits can have different intrinsic rates of hydrolysis. Thus one signal can be quickly terminated while the other signal sustains, and give specificity of signal generation (Carty *et al.*, 1990; Neer, 1995). To date, over twenty different G protein α subunits have been identified and divided into four classes based on their amino acid sequences ($G\alpha_s$, $G\alpha_i$, $G\alpha_q$ and $G\alpha_{12}$) reviewed by Kaziro *et al.* (1991) and Simon *et al.* (1991). Mammals contain over 5 β and 12 γ subunits which form dimers, but not all combinations are possible. There seems to be some preferred combinations of isoforms that interacts maybe due to their three-dimensional structures (Pronin and Gautam, 1992; Ueda *et al.*, 1994; Ray *et al.*, 1995). The βγ subunits from G_i protein scavenge free $G_s\alpha$ subunits and block activation of adenylate cyclase (Gilman, 1987). The βγ subunits can further activate K^+-channels (Logothetis *et al.*, 1987), modulate adenylate cyclase, phospholipase Cβ, cytosolic phospholipase A_2, phosphoinositide 3-kinase and β-ARK (Clapham and Neer, 1993; Zhang *et al.*, 1995) and act via ras to activate mitogen-activated protein kinase pathways (Crespo *et al.*, 1994). Modification of the G proteins by lipids are necessary for efficient signal transduction (Casey, 1994; Ross, 1995). α_i, α_o and α_z subunits are myristylated (Jones *et al.*, 1990). Palmitoylation occurs on most of the α subunits (Linder *et al.*, 1993; Ross 1995) and secures the membrane localization and thus activation of G proteins. Upon activation α_s is depalmitoylated and released in the cytosol. In the cytosol it will be repalmitoylated and translocates to the membrane (Wedegaertner and Bourne,

1995; Milligan *et al.*, 1995). Since the α subunits are inactive in the cytosol, this process could also serve as a turn-off mechanism (Wedegaertner *et al.*, 1993). All γ subunits are prenylated and are necessary for membrane attachment to form interactions with Gα subunits or effectors (Casey, 1995).

Platelets contain the following types of heterotrimeric G proteins: G_i1, G_i2, G_i3, G_z, G_s, G_q, G_{11}, G_{12}, G_{13} and G_{16} proteins (Brass *et al.*, 1993; Brass, 1995; Carlson *et al.*, 1989; Gagnon *et al.*, 1991; Milligan *et al.*, 1993; Amatruda *et al.*, 1991; van Willigen *et al.*, 1995). Their specific connections to receptors and effectors will be discussed further below in "Platelet agonists and their signaling systems".

9.4.2 Monomeric G proteins

Monomeric G proteins are single peptide proteins with a molecular weight of about 20-30 kDa, and alternate between a GDP-bound resting state and a GTP-bound active state. They show 30-60 % sequence homology with proteins encoded by the *ras* oncogene. The members of the ras superfamily can be divided into ras, rho, rab, ran and arf families (Kahn *et al.*, 1992; Downward, 1990). The ras proteins have been implicated in cell proliferation and differentiation (Lowy and Willumsen, 1993), rho proteins in the control of actin cytoskeleton (Chardin *et al.*, 1988; Ridley, 1992a, 1992b; Zigmond, 1996), rab and arf proteins in sorting of intracellular vesicles (Novick and Brennwald, 1993; Serafini *et al.*, 1991) and ran proteins in nuclear functions, transition between S and M phases of the cell cycle (Bischoff and Ponstingl, 1991; Matsumoto and Beach, 1991) and protein import into the nucleus (Moore and Blobel, 1993). GTP-binding and hydrolysis are very low in the small G proteins, and they are dependent on regulatory factors for stimulation. GDP dissociation stimulator (GDS) induces the release of GDP and causes GTP to bind, and converts the protein to its active form. GTPase activating protein (GAP) stimulates the intrinsic GTPase activity of the protein, and leads to inactivation. GDP dissociation inhibitor (GDI) inhibits the dissociation of GDP and keeps the protein in the inactive state (Bokoch and Der, 1993).

Despite lacking the nucleus, platelets are surprisingly rich in several small G proteins e.g. ral, rac1, rap1A, rap1B, rap2A, rap2B, rab1, rab3B, rab4, rab6, rab8 and rhoA (Ohmori *et al.*, 1989; Polakis *et al.*, 1989; Bhullar *et al.*, 1990; Ohmstede *et al.*, 1990; Karniguian *et al.*, 1993). The physiological role of these proteins in platelets has not been clearly elucidated. However, rap1B has been postulated to be involved in translocation of PLCγ1 from the cytosol to the plasma membrane in thrombin-stimulated platelets (Torti and Lapetina, 1994); it has been shown to be phosphorylated by protein kinase A (PKA) and translocated from the plasma membrane to cytosol, something that could explain the inhibition of inositol phosphate formation after PKA activation (Lapetina *et al.*, 1989; Fisher *et al.*, 1990) and a role in epinephrine-enhanced signal transduction (Bencke-Marti and Lapetina, 1992). rab6 and rab8 are localized in the α-granules and phosphorylated upon platelet activation (Karniguian *et al.*, 1993). Rho activates phosphatidylinositide 3-kinase (PI 3-kinase) which binds to the platelet membrane cytoskeleton where it possibly causes cytoskeleton reorganization (Zhang *et al.*, 1993; Downward, 1997). Small G proteins have also been assigned a role in GPIIb/IIIa exposure (Torti *et al.*, 1991) and linkage to integrins (Brass *et al.*, 1997).

9.4.3 Phospholipase C and the polyphosphoinositide cycle

Platelets contain several isoforms of phospholipase C (PLC): PLCβ1, PLCβ2, PLCβ3, PLCγ1, PLCγ2 and PLCδ (Banno, 1990, 1992, 1995; Torti and Lapetina, 1992). PLCβ forms, in general, are regulated by heterotrimeric G proteins, i.e., PLCβ1 by Gα subunits and PLCβ2 by βγ subunits (Smrcka and Sternweis, 1993). PLCγ isoforms are regulated by tyrosine-phosphorylated receptors (e.g. EGF, PDGF, collagen) (Berrigde, 1993; Cockcroft and Thomas, 1992). The regulation of PLCδ is still unknown.

PLC is the signal-generating enzyme in the polyphosphoinositide cycle (Figure 9.4.3). Activated PLC mediates hydrolysis of membrane phosphatidylinositol-4,5-bisphosphate (PIP_2) to membrane-bound diacylglycerol (DAG) and cytosolic inositol-1,4,5-trisphosphate (IP_3), which both are signal molecules (see below). There are different PLCs in platelets, both in the plasma membrane and cytosol which differ from each other with respect to substrate affinity, catalysis and Ca^{2+} requirement (Banno *et al.*, 1988). PLC has also different affinity for molecular species of PIP_2, and prefers PIP_2 with stearic acid (C18:0) in the sn-1 position and arachidonic acid (C20:4) in the sn-2 position (Mauco *et al.*, 1984). The inositol phospholipids represent about 4 % of the total lipid content in platelets (Abdel-Latif, 1983), where PI comprises most of the mass, 1408 nmol/10^{11} platelets, while PIP and PIP_2 represents 168 and 132 nmol/10^{11} platelets, respectively (Tysnes *et al.*, 1987). Only 55 % of PI, PIP and PIP_2 join the metabolic pool of phospholipids. The non-metabolic pool is thought to be localized to internal membranes and bound to cytoskeletal proteins (Frølich *et al.*, 1992). Metabolic phosphatidylinositol (PI) is phosphorylated by PI 4-kinase to phosphatidylinositol 4-phosphate (PIP), which is further phosphorylated to PIP_2 by PIP 5-kinase. The phosphates in 4-position in PIP and 5-position of PIP_2 are removed by PIP phosphohydrolase and PIP_2 phosphohydrolase, respectively. Metabolic PIP and PIP_2 are thus continuously synthesized and broken down and their levels are kept in a steady state by the expenditure of 7 % of the total ATP production of resting platelets (Verhoeven *et al.*, 1987). The total ATP turnover in resting platelets is 3000 nmoles/10^{11} platelets x min^{-1}, which means that 2/3 of the metabolic PIP and PIP_2 pool is turning over every minute in resting platelets. It is postulated that this expenditure is linked to maintenance of responsivity (Holmsen *et al.*, 1992).

When the platelets are stimulated by most agonists, polyphosphoinositide(PPI) specific PLC is activated via G proteins, and splits PIP_2 to DAG and IP_3. There is only a transient increase in the level of DAG, before it is rapidly converted to PA by DAG kinase in an ATP-requiring reaction. PA is dephosphorylated by PA phosphohydrolase, converting PA to DAG. Together with PI ↔ PIP and PIP ↔ PIP_2, the DAG ↔ PA reaction represents a futile cycle which consumes ATP for unknown purposes. Net formation of PA is greater than the rate of its removal (Tysnes *et al.*, 1988) which causes PA to accumulate. PA is further converted to CDPDAG by CTP:PA cytidylyl transferase. CDPDAG reacts with free inositol to produce PI in a reaction catalyzed by CDPDAG:inositol 3-phosphatidyl transferase, and this reaction completes the PPI cycle. In stimulated platelets the metabolic turnover of ATP increases 2-3 times (Akkerman and Verhoeven, 1987). The phosphomonoesters (the P in the boxes in figure 9.4.3) in PIP and PIP_2 are still in equilibrium with the β- and γ-phosphorylgroups of metabolic ATP, so that the turnover between PI, PIP and PIP_2 could be above 6000 nmoles/10^{11} platelets x min^{-1} in stimulated platelets.

Stimulation of platelets also leads to the formation of polyphosphoinositides that are phosphorylated in the 3-position on the inositol ring, especially PI 3-phosphate, PI 3,4-bisphosphate and PI 3,4,5-trisphosphate (Kucera and Rittenhouse, 1990), due to a PI 3-kinase that becomes activated upon receptor stimulation. The 3-phosphorylated phosphoinositides are not hydrolyzed by PPI-PLC and are probably not taking part in the PPI-metabolism. It has been postulated that they contribute to the actin polymerization that takes place upon platelet stimulation (Downes and Carter, 1991).

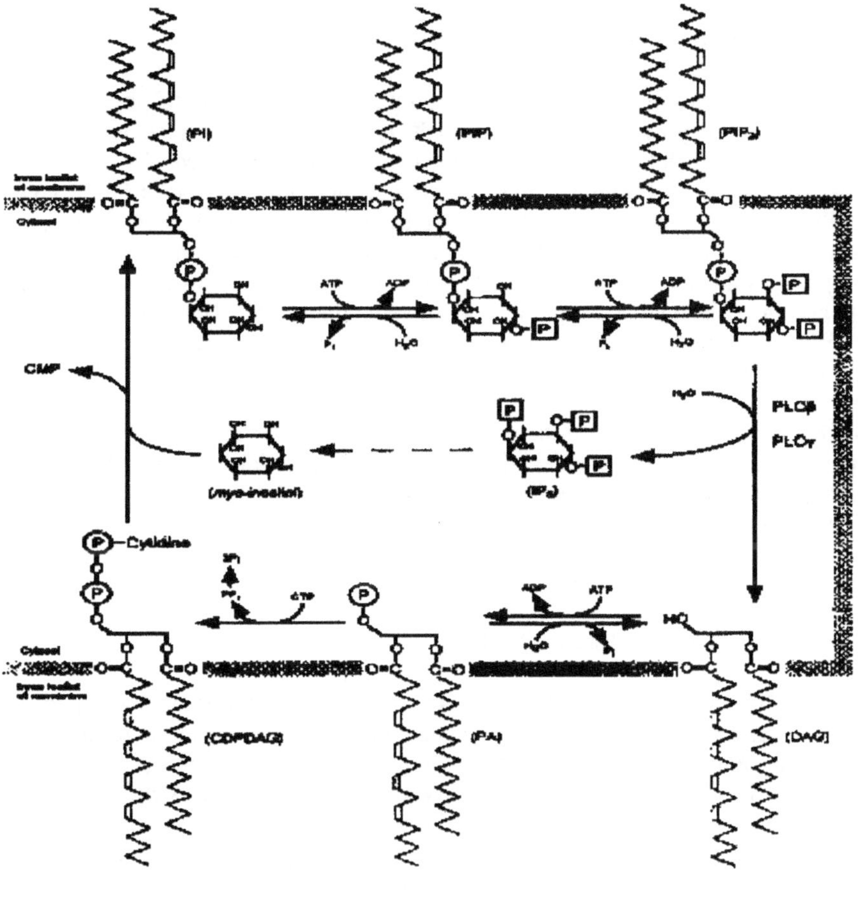

Figure 9.4.3. The polyphosphoinositide cycle: The enzymes involved are described in the text.

The phosphoinositides are postulated to be localized in the plasma membrane (Mauco *et al.*, 1987) because of the agonist-induced hydrolysis of PIP_2, but there are reason to believe that the phosphoinositides and their enzymes are compartmentalized (Vickers and Mustard, 1986; Nahas *et al.*, 1989; Kanoh *et al.*, 1990; Suzuki *et al.*, 1991; Grondin *et al.*, 1991) and, as already mentioned, only 55 % of the phosphoinositides are metabolically active (Frølich *et al.*, 1992) in human platelets. PIP and PIP_2 are in addition to be bound to PKC and plasmamembrane Ca^{2+}-transport ATPase, also found to be associated with actin-binding proteins in the cytoskeleton (profilin and gelsolin), and probably involved in the rapid actin polymerization that follows platelet activation (Lassing and Lindberg, 1985, 1988, 1990; Janmey and Stossel, 1987; Janmey and Matsudaira 1988; Goldschmidt-Clermont *et al.*, 1990). The synthesis of PI via CMP-PA from PA is postulated to take place in the ER in several cell types. In platelets a PI-transport protein is characterized that may transport the finished product to the plasma membrane (George and Helkamp, 1985). It is also suggested that the phospholipids are gathered in domains in the plasma membrane, which is suggested for PS (as a cofactor for PKC) and for PIP_2 which complexes with PPI-PLC (Monaco and Gershengorn, 1992).

9.4.4 Inositol trisphosphate (IP_3) and Calcium

The formation of cytosolic inositol 1,4,5-trisphosphate (IP_3) from the PIP_2-hydrolysis by PPI-PLC have two major functions: it binds to the type I IP_3 receptor in the dense tubular system (DTS) by which Ca^{2+} is released to the cytosol (O'Rourke *et al.*, 1995), and it is converted to inositol 1,3,4,5-tetrakisphosphate (IP_4) which stimulate Ca^{2+} influx from the extracellular space. Influx may also take place by receptor-operated mechanisms or by store regulation (capacitive Ca^{2+} entry) (Berridgde, 1993; Heemskerk and Sage, 1994). In resting platelets the level of cytosolic Ca^{2+} is 50-100 nM. Upon activation with a strong agonist, like thrombin, there is a rapid increase to more than 1 μM (Rink and Sage, 1990). The rise in the cytosolic Ca^{2+} causes activation of a variety of enzymes, such as PLC, PLA_2, myosin light chain kinase, calpain, phosphorylase kinase and calmodulin (Brass, 1995). The rise in $[Ca^{2+}]_i$ is returned to basal levels through active Ca^{2+} extrusion into the extracellular space and restoration of Ca^{2+} in the DTS. The increases in $[Ca\ ]_i$ are pulsative as they are seen as a series of repetitive spikes (Berridge, 1990), which can be explained by two models: (1) Ca^{2+} spiking is driven by cyclical changes in IP_3 production, depending on a Ca^{2+}-sensitive PLC, (2) repetitive Ca^{2+} release is due to a calcium-induced calcium release (CICR) (Berridge, 1993; Heemskerk and Sage, 1994; Bootman and Berridgde, 1995).

9.4.5 Activation of protein kinase C (PKC)

PKC is a Ser/Thr protein kinase and plays an important role in the signaling response in platelets. In general, the PKC family are divided into three classes on the basis of their structure: the classic PKCs (cPKCs) which comprise α-, β1 and β2 and γ-isoforms, the novel PKCs (nPKCs) which comprise the δ-, ϵ-, η-, θ- and κ-isoforms and the atypical group (aPKCs), comprising the ζ-, ι-, λ- and μ-isoforms. All three PKC classes require phosphatidylserine (PS) for activity. cPKCs require Ca^{2+} for catalytic activation, while the n- and aPKCs do not require Ca^{2+} but rather require DAG/PS or phorbolester/PS (Rasmussen *et al.*, 1995). The PKC isoforms so far identified do not contain SH2-domains or pleckstrin homology-domains. In unstimulated cells, most of the cPKCs are located in the cytosol, but are translocated to the cytoskeleton or membrane upon activation. Platelets contain at least the α, β, δ, ζ, η and θ-forms of PKC (Tsukuda *et al.*, 1988; Baldassare *et al.*, 1992; Grabarek *et al.*, 1992; Wang *et al.*, 1993). Activated PKC has different effects in platelets, e.g. modulation of inositol phosphate metabolism, GPIIb/IIIa receptor expression, secretion and protein phosphorylation. Substrates for PKC are pleckstrin, myosin light chain and the Na^+/H^+-exchanger which exchange intracellular protons for extracellular Na^+ in order to control cytoplasmic pH (Kaibuchi *et al.*, 1983; Lapetina *et al.*, 1985; Shattil and Brass, 1987; van Willigen and Akkerman, 1991). PKC activation has also been reported to affect signal transduction at G protein level (Lapetina *et al.*, 1986a;

Halenda *et al.*, 1986). PKC is believed to be activated indirectly by all platelet agonists through the DAG (Fukami and Holmsen, 1995) produced by phosphoinositide-specific PLC.

9.4.6 Pleckstrin and Myosin light chain

Pleckstrin (*p*latelet and *le*ukocyte *C k*inase sub*str*ate, phosphorylated at serine and threonine) or P47 is a 40 kDa protein present in platelets and leukocytes and was first described in platelets as a major substrate for PKC (Haslam *et al.*, 1979). Proteins that are important in signal transduction often contain specific domains that mediate protein-protein interaction. Examples are the abundant Src homology domains 2 and 3 (SH2 and SH3) which interacts with tyrosine-phosphorylated and proline-rich sequences, respectively (Cohen *et al.*, 1995). Many proteins also share a common domain called the pleckstrin homology (PH) domain, named after the platelet protein first recognized to contain this domain, and it has been proposed that this domain represent an important family of interaction domains (Haslam *et al.*, 1993; Mayer *et al.*, 1993). Although pleckstrin was cloned already in 1988 (Tyers *et al.*, 1988), structure and function of the PH domain was not reported until 1994 (Yoon *et al.*, 1994). The PH domain consists of a 100 amino acids structural motif, which form a β barrel with an amphiphatic α helix at the C-terminus. Even if the primary sequence varies, the secondary structure is highly preserved. PH domains are found in protein kinases, phospholipase Cs, GTPase activating proteins, GTP exchange factors, adapter proteins, cytoskeletal proteins and kinase substrates (Haslam *et al.*, 1993; Mayer *et al.*, 1993; Musacchio *et al.*, 1993; Ingley and Hemmings, 1994), and are thought to target proteins to lipid membranes or hydrophobic domains in proteins. Platelet pleckstrin contain two PH domains, one at the N-terminus (residues 6-99) and one at the C-terminus (residues 246-345) (Tyers *et al.*, 1988). PKC phosphorylates at ser^{113}, thr^{114} and ser^{117} adjacent to the N-terminal PH domain and is essential for the functional activity of pleckstrin in inhibiting phosphoinositide hydrolysis (Abrams *et al.*, 1995). C-terminal PH domain binds to the βγ subunits of G proteins and thereby presumably inhibits further signaling from an agonist-activated receptor (Harlan *et al.*, 1994; Touhara *et al.*, 1994; Inglese *et al.*, 1995), but the function of this interaction is still not clear. Pleckstrin phosphorylation seems also to be important in GPIIb/IIIa activation (Hartwig and Barkalow, 1997), inhibition of the Gβγ-activated PI-3 kinase (Abrams *et al.*, 1996) and activation of inositol polyphosphate 5-phosphatase I (Auethavekiat *et al.*, 1997), processes that could facilitate the catabolism of inositol triphosphate and terminate Ca^{2+}-dependent signaling events. PH domains bind to PIP_2, a process that targets phosphopleckstrin to the cytoplasmic surface of the plasma membrane, and inhibits polyphosphoinositide hydrolysis (Harlan *et al.*, 1994; Abrams *et al.*, 1995a; Abrams *et al.*, 1995b). The function of phosphorylated pleckstrin is still not elucidated, but it seems to regulate and control platelet activation.

Myosin light chain (MLC), a 20 kDa protein, is present in the dephosphorylated form in resting platelets and in the phosphorylated form in the stimulated platelet (Haslam *et al.*, 1979; Daniel *et al.*, 1981). MLC is the major Ca^{2+}/Calmodulin-dependent protein kinase substrate in platelets. Phosphorylation of MLC stimulates the actin-dependent ATPase of myosin (Lebowitz and Cooke, 1978) which result in formation of contracting actomyosin bundles that are involved in the rapid shape change process, a phosphorylation-process which precedes platelet secretion (Daniel *et al.*, 1984) The phosphorylation of myosin light chain is regulated by myosin light chain kinase (MLCK), which is regulated by Ca^{2+}

calmodulin and by a phosphomyosin phosphatase. MLCK is further regulated by a cAMP-dependent protein kinase which phosphorylates the kinase to an inactive form (Conti and Adelstein, 1980). Platelet myosin light chain is also phosphorylated by protein kinase C at another site than by the cyclic AMP-activated protein kinase (Naka *et al.*, 1983), which may modulate the secretory response (Inagaki *et al.*, 1984).

9.4.7 Intracellular alkalinization

A large number of cells respond to stimulation with an elevation of cytoplasmic pH (pH_i) (Madshus, 1988). The alkalinization, which is a immediate, overshooting response to intracellular acidification due to ATP hydrolase and lactate production (Akkerman *et al.*, 1979; Akkerman and Holmsen, 1981), is mediated through a Na^+/H^+ antiport present in the plasma membrane and exchanges Na^+ for intracellular H^+. Human platelets stimulated by receptor-mediated agonists undergo rapid alkalinization via the Na^+/H^+ exchanger (Horne *et al.*, 1981; Siffert *et al.*, 1984, 1987) and is suggested to be regulated by PKC (Siffert *et al.*, 1987). The exchanger activity in human blood platelets, which is dependent on the phosphorylation state, is regulated by the activities of protein kinases and protein phosphatases. The Na^+/H^+ exchanger regulates the cell volume and contributes to signal transduction (De-Caterina *et al.*, 1994; Siess, 1989). The exact mechanisms of the Na^+/H^+ exchanger in these processes are still not clear. Thrombin, PKC activators and calcium ionophores have been reported to activate the Na^+/H^+ exchanger, while the effects of ADP, PAF, adrenaline, collagen, angiotensin II and endothelin I are controversial, as reviewed by Siffert (1995). The Na^+/H^+ exchanger is also reported to be directly activated by the α-subunit of the G protein G_{13} (Voyno-Yasenetskaya *et al.*, 1994).

9.4.8 The phospholipase A_2 pathway

Cytosolic phospholipase A_2 ($cPLA_2$) is primarily located in the cytosol but associates with the dense tubular system and plasma membrane upon platelet activation. $cPLA_2$ is activated during agonist-induced platelet activation through G proteins, probably by their βγ-subunits (Kim *et al.*, 1989), and by an increase in cytosolic Ca^{2+} (Rittenhouse-Simmons, 1981; Rittenhouse and Horne, 1984; Borsch-Haubold *et al.*, 1995). There is also evidence that MAP kinases (ERKs) activate phospholipase A_2 ($cPLA_2$) in platelets (Kramer *et al.*, 1995). Activated $cPLA_2$ releases arachidonic acid (AA) from phosphatidyl choline, phosphatidylinositol and phosphatidylethanolamine (Rittenhouse-Simmons, 1981; Prescott and Majerus, 1981; McKean *et al.*, 1981). AA is metabolized to prostaglandin G_2 (PGG_2) by cyclo-oxygenase, an enzyme that is irreversibly inhibited through acetylation by aspirin. PGG_2 is further metabolized via PGH_2 to thromboxane A_2 (TXA_2) which is released and reacts with receptors on the platelet surface leading to platelet activation through the G protein-PLC-pathway (Armstrong *et al.*, 1983; FitzGerald, 1991). PGH_2 can also be converted to the inhibitory PGD_2, to the inactive or minimally inhibitory $PGF_2α$ and PGE_2. Some of the AA is converted to 12-H(P)ETE (hydroxy(peroxy)eicosatetraenoic acid) and 15-H(P)ETE by 12- or 15-lipoxygenase (Hamberg and Samuelson, 1974; Wong *et al.*, 1985). The cyclic endoperoxides also leave the platelet and enter the endothelial cells where they are converted to the platelet inhibitory and vasodilating PGI_2 (Schafer *et al.*, 1984). 12-HETE from platelets are metabolized by unstimulated netrophils to 12,20-di-HETE (Marcus *et al.*, 1984; 1988), which the platelets are unable to synthesize themselves. Platelets has also been found to secrete PLA_2 upon activation (Mueller *et al.*, 1993) which may be important in the regulation of thrombosis.

9.4.9 Protein tyrosine kinases

We have previously described the serine-threonine kinases, the CCAMK, MLCK and PKC, for which MLC pleckstrin are the major substrates, respectively. Platelet activation is also accompanied by an increase in tyrosine phosphorylation of many proteins (Clark and Brugge, 1994), the role of which has not yet been fully elucidated. Inhibitors of protein tyrosine kinases (PTK) have been shown to block platelet secretion and aggregation (Dhar and Shukla, 1993) which suggests that PTKs and tyrosine-phosphorylation play a primary role in platelet activation. Except for the PDGF-α receptor (Vassbotn *et al.*, 1994), other platelet transmembrane receptor tyrosine kinases have not been detected, but they possess several non-receptor tyrosine kinases (NRTK, as described

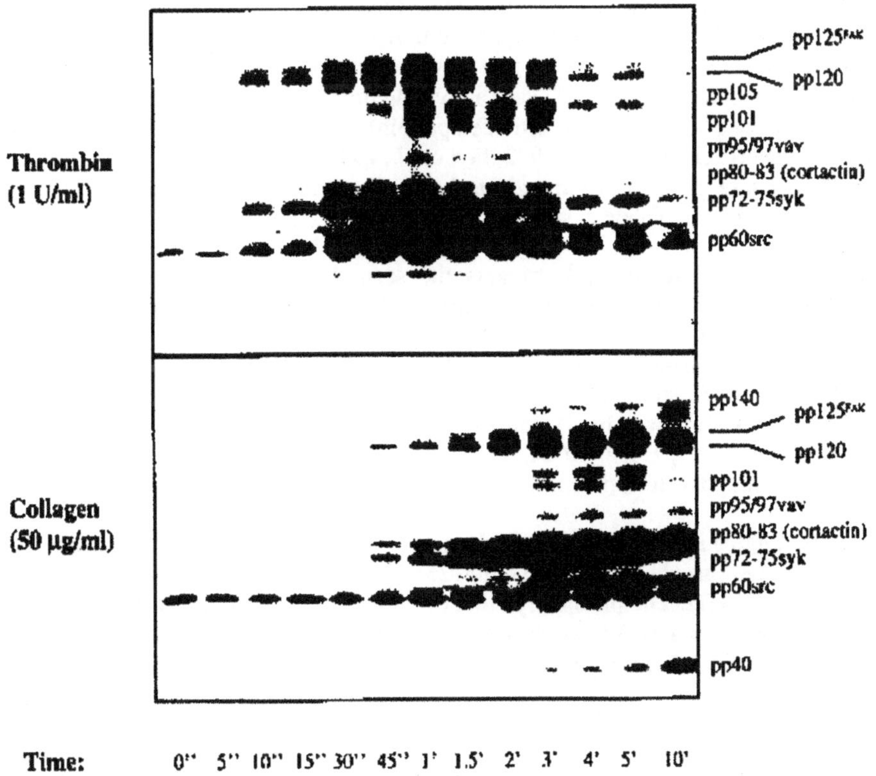

Figure 9.4.9.1. Time courses of tyrosine phosphorylation in intact platelets stimulated with thrombin (1 U/ml) and collagen (50 µg/ml) in the absence of inhibitors of autocrine stimulation. Thrombin induce a fast and transient (maximal at 1 min) tyrosine phosphorylation (TF), while collagen is a slow inducer of TF (maximal around 3-5 min). Collagen induces specifically TF of proteins of 40, 140 kDa, where the 40 kDa protein after 10 min of incubation still have not reached its maximal levels. Thrombin and collagen induces most of the same proteins.

in 9.4.10). Five members of the Src family of tyrosine kinases (Src, Fyn, Yes, Lyn and Hck) are present (Golden *et al.*, 1986; Horak *et al.*, 1990; Huang *et al.*, 1991) which all have a SH-1-containing catalytic domain (SH-1). Furthermore, they have a SH2-domain which interacts with phosphotyrosine residues intra- and intermolecularly and a SH3-domain which interacts with proline-rich aminoacid sequences in other cellular proteins (Koch *et al.*, 1991). The most abundant NRTK is Src, which comprises 0.4 % of the total platelet protein (Golden *et al.*, 1986). Other abundant protein tyrosine kinases found in platelets are Syk and FAK (Clark and Brugge, 1994).

Tyrosine phosphorylation of proteins have different time courses depending on the agonist and occurs in several "waves" which suggests regulation with several protein tyrosine kinases and/or phosphatases (Figure 9.4.9.1). The tyrosine phosphorylation can further be divided into four phases: resting platelets, early integrin-independent, integrin-engagement-dependent and aggregation-dependent. The tyrosine phosphorylated proteins in the resting platelets include five members of the Src-family, all with molecular weights between 54 and 62 kDa. In addition, a protein of 120 kDa is also phosphorylated. Agonist stimulation of platelets triggers the first wave of tyrosine phosphorylation, independent on $\alpha_{IIb}\beta_3$, and include GAP (the GTPase activating protein for the small molecular weight G protein p21ras), MAP (mitogen activated protein), pp72syk (cytosolic PTK) and a protein of 110 kDa (Cichowski *et al.*, 1992; Fox *et al.*, 1993a; Papkoff *et al.*, 1994; Clark *et al.*, 1994a). These proteins are thought to be involved in the events that occur rapidly in agonist-stimulated platelets, such as shape change, conformational change of $\alpha_{IIb}\beta_3$.

Fibrinogen binding to $\alpha_{IIb}\beta_3$ triggers the second wave of agonist-induced tyrosine phosphorylation of pp72syk and proteins of 140 and 50-68 kDa. These proteins are not phosphorylated in platelets from in patients with Glanzmann thrombasthenia (deficient in $\alpha_{IIb}\beta_3$) or in normal platelets treated with the tetrapeptide RGDS which inhibits fibrinogen binding. Specific stimulation of $\alpha_{IIb}\beta_3$ with LIBS6 (induces conformational change in $\alpha_{IIb}\beta_3$ and lets fibrinogen bind) induces phosphorylation of the mentioned proteins (Clark and Brugge, 1994; Huang *et al.*, 1993). Phosphorylation of these proteins is probably necessary for adhesive and/or in cytoskeletal events that follows integrin engagement.

Platelet aggregation induced by fibrinogen cross-linking of $\alpha_{IIb}\beta_3$ on adjacent platelets stimulates tyrosine phosphorylation of pp125FAK and two proteins with molecular weight of 95 and 97 kDa. The phosphorylation of these proteins is inhibited by cytochalasin D treatment and dependent on cytoskeletal rearrangements, PKC and a rise in $[Ca^{2+}]_i$ (Lipfert *et al.*, 1992; Shattil *et al.*, 1994).

Caution should however be taken in determining the proteins that are specifically tyrosine phosphorylated in response to various agonists. This can be illustrated with thrombin, where studies have shown that removal of the autocrine stimulation in platelets during activation strongly reduces the tyrosine phosphorylation stimulated by thrombin (Figure 9.4.9.2).

Not so much is known about the significance of protein tyrosine phosphatases (PTP). Inhibition of PTPs (by vanadate) leads to increase in tyrosine phosphorylation, secretion and aggregation (Inazu *et al.*, 1990; Pumiglia *et al.*, 1992). At least two PTPs appear to be involved in platelet aggregation. PTP1B is proteolytically cleaved and activated by calpain and released into the cytosol during aggregation. SH-PTP1 associates with the cytoskeleton during aggregation (Frangioni *et al.*, 1993; Li *et al.*, 1994).

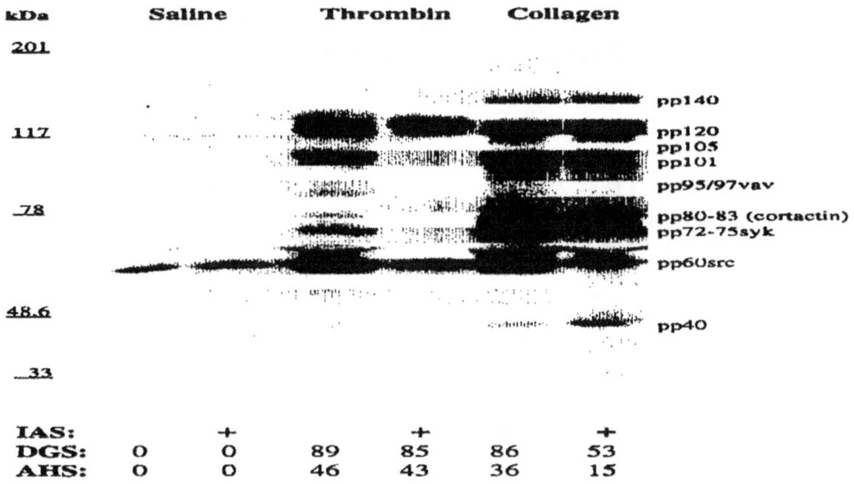

Figure 9.4.9.2. Protein tyrosine phosphorylation and secretion in platelets stimulated with thrombin and collagen in the absence and presence of inhibitors of autocrine stimulation (IAS). Lane 1 and 2 shows that proteins with Mw 60 and 120 kDa are phosphorylated in resting platelets. Lane 3 and 4 show platelets that are stimulated with 1 U/ml thrombin. In the absence of IAS, thrombin induces additional phosphorylation of numerous proteins with Mw 68, 72-75, 80-83, 95, 101, 105, 130 and 140 (weak). IAS (lane 4) inhibits this phosphorylation dramatically, except for the IAS-induced phosphorylation of a protein at 125-130 kDa. Collagen (100 μg/ml) phosphorylates most of the same proteins as for thrombin, except for the 40 and 140 kDa proteins. Interestingly, IAS did not inhibit the collagen-induced tyrosine phosphorylation to the same degree as for thrombin, which strongly suggests that tyrosine phosphorylation is not part of the signal transduction pathway initiated by thrombin *per se*, but represents an integral part of signal transduction initiated by collagen. Secretion of ADP+ATP from the dense granula and β-N-acetylglucosaminidase from the lysosomes (both final platelet responses) in thrombin-stimulated platelets are not affected by IAS, while the secretion by collagen is inhibited (Ryningen and Holmsen, 1998). Thrombin and collagen induce a similar profile of tyrosine phosphorylation, suggesting that structurally distinct receptors must converge on common pathways leading to tyrosine phosphorylation (discussed further in 9.4.10).

9.4.10 The cytoplasmic protein kinase cascade

In platelets, signaling is initiated primarily through members of the heterotrimeric G protein-coupled family of receptors (seven transmembrane domains) and through adhesion receptors, and the signaling involves activation of both Ser/Thr kinases and tyrosine kinases. Neither G protein-coupled receptors nor adhesion receptors have intrinsic tyrosine kinase activity. However, NRTKs (with SH2-, SH3-, and proline-rich domains) are activated and initiate tyrosine phosphorylation reactions that in turn lead to the recruitment of signaling molecules to certain locations in the cell. These tyrosine kinases may phosphorylate submembranous proteins including receptors for cytoplasmic domains or components of the submembranous cytoskeleton of adhesion receptor-cytoskeleton complexes that can then recruit and activate appropriate signaling molecules. Figure 9.4.10 summarizes what we know to-day about the signaling molecules involved in the two pathways, how they may converge to activate Ras, Rac and Rho proteins, and how the pathways are involved in platelet functions. The scheme is mainly based on information from Clark and Brugge (1995), Fox (1996) and Brass *et al.* (1997).

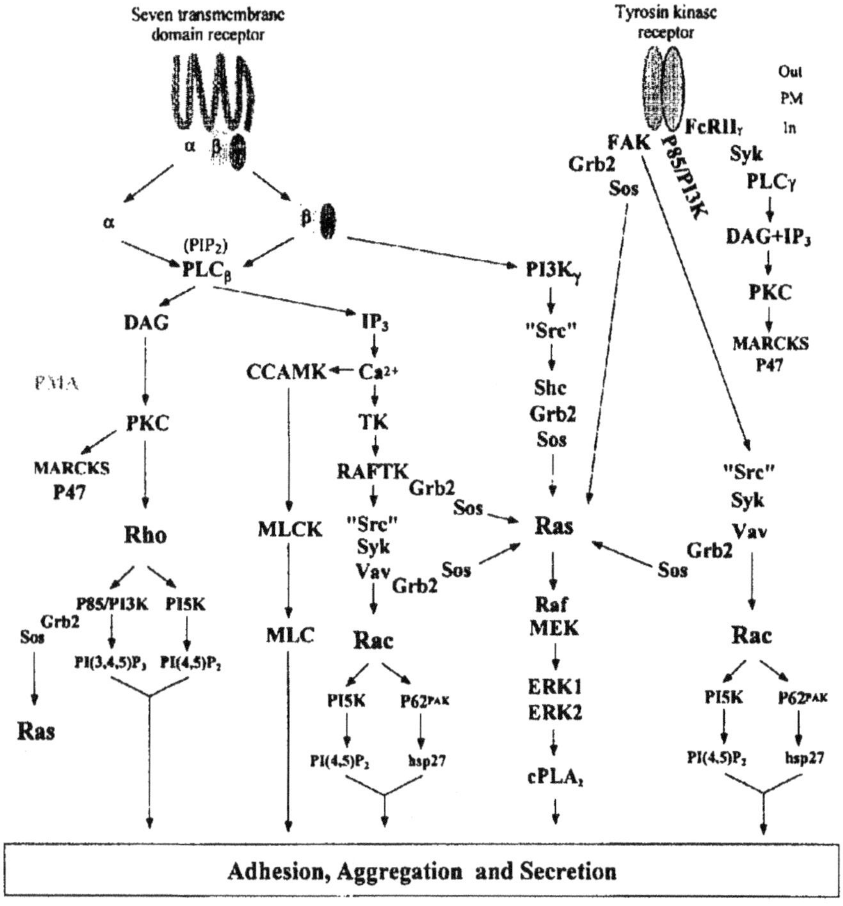

Figure 9.4.10. The cytoplasmic protein kinase cascades. Se text for explanations.

Thrombin will be used as an example of an activator of a seven transmembrane domain receptor. Upon activation of the receptor, the G_i-protein splits into G_α and $G_{\beta\gamma}$, which both can activate PLC_β (Exton, 1994; Brass *et al.*, 1997). G leads to phosphoinositide hydrolysis and formation of DAG and IP_3 which activates PKC and empty Ca^{2+}-stores in the endoplasmatic reticulum, respectively. The rise in $[Ca^{2+}]_i$ causes tyrosine phosphorylation of the related adhesion focal tyrosine kinase (RAFTK) by an unidentified tyrosine kinase (TK) (Raja *et al.*, 1997). RAFTK can also be tyrosine phosphorylated by an unknown tyrosine kinase (TK) and activated when PKC is stimulated by phorbol esters such as PMA (indicated in gray in Figure 9.4.10), but Ca^{2+}-induced activation seems to be the most important. It should also be mentioned that the rise in $[Ca^{2+}]_i$ activates CCAMK, which phosphorylates MLCK. MLCK then phosphorylates MLC which is involved in the cytoskeletal rearrangements in stimulated platelets. RAFTK is a cytosolic tyrosine kinase which lacks SH2 and SH3 domains, but has two proline-rich domains which may interact with SH3 domains in other proteins and at least two tyrosines that become phosphorylated, Y882 (a Grb2 binding site) and Y402 (a binding site for members of the Src family) (Lev *et al.*, 1995; Dikic *et al.*, 1996; Brass *et al.*, 1997). This makes RAFTK to a signaling-complex, where binding of Grb2 (an SH2/SH3 adapter protein) to RAFTK causes complexation of Sos which can activate Ras, and binding of Src to RAFTK activates Syk, a 72 kDa tyrosine kinase in platelets. Syk has shown to associate with the 95/97 kDa Vav which become phosphorylated during platelet activation (Deckert *et al.*, 1996; Cichowski *et al.*, 1996; Miyakawa *et al.*, 1997). Vav is constituitively associated with Grb2 and can therefore activate Ras via Sos (Ye and Baltimore, 1994). Vav contains several structural motifs, including one SH2- and two SH3 domains (Bustelo *et al.*, 1992; Margolis *et al.*, 1992), a PH domain (Mussachio *et al.*, 1993), a leucine-rich and cysteine-rich domain (Coppola *et al.*, 1991) and a central Dbl homology (DH) domain (Boguski and McCormick, 1993).) DH domains are responsible for the exchange of GDP for GTP in Rho proteins (Hart *et al.*, 1991; 1994). Thus, Vav might activate Rho-family proteins. Phosphorylated Vav do exchange protein for Rac-1, a member of the Rho-family (Crespo *et al.*, 1996; 1997). The exact steps from Vav - Rac are still not clear in platelets. Downstream targets for Rac are PI(4,5)Kinase and $P62^{PAK}$ (Hartwig *et al.*, 1995; Teo *et al.*, 1995). Activated PI4P 5-Kinase generates $PI(4,5)P_2$ which induces uncapping of actin filament ends (Hartwig *et al.*, 1995). Phosphorylated $p62^{PAK}$ activates hsp27, which become associated with the cytoskeleton (Zhu *et al.*, 1994). Both Rac and Rho are thought to be involved in cytoskeletal rearrangements in different locations during platelet activation.

There are two different forms of phosphoinositide 3-kinase (PI3K). The P85/PI3K is a heterodimer containing a 85 kDa regulatory subunit and a 110 kDa catalytic subunit. The other PI3K contains only the 110 kDa subunit, and is known as $PI3K_\gamma$ (Stoyanov *et al.*, 1995). $PI3K_\gamma$ is activated by the $\beta\gamma$ subunits of heterotrimeric GTP binding proteins (Thomason *et al.*, 1994), while P85/PI3K is activated by a mechanism involving Rho (Zhang *et al.*, 1993; 1995; 1996). Ras can be activated in a $G_{\beta\gamma}$-mediated process involving $PI3K_\gamma$ (Luttrell *et al.*, 1996; Hawes *et al.*, 1996; Lopez-Ilasaca *et al.*, 1997), Shc phosphorylation, Shc/Grb2 association and Sos activation (van Biesen *et al.*, 1995). Inhibitors of Src family of tyrosine kinase activity abrogate G_i-coupled and $G_{\beta\gamma}$-mediated Shc phosphorylation and MAP kinase activation in COS-7 cells (van Corven *et al.*, 1993; Hordijk *et al.*, 1994; Hawes *et al.*, 1995; Touhara *et al.*, 1995) suggesting that a Src family tyrosine kinase is involved upstream of Src. The downstream effectors of Ras are well characterized. Activated Ras recruits the Ser/Thr kinase Raf to the membrane where it is activated by unknown mechanisms (Moodie *et al.*, 1993; Stokoe *et al.*, 1994). Raf then phosphorylates MAP kinase kinases (MEKs) which in turn phosphorylate MAP kinases (ERKs) on serine, threonine and tyrosine residues (Pelech and Sanghera, 1992; Nishida and Gotoh, 1993; Cobb and Goldsmith, 1995). Both tyrosine and threonine residues have to be phosphorylated on ERKs for activation (Anderson *et al.*, 1990). Finally, there is evidence that ERKs activates phospholipase A_2 ($cPLA_2$) in platelets (Kramer *et al.*, 1995).

The DAG produced in the PLC_β reaction activates PKC together with Ca^{2+} and PS.

Activated PKC's main substrate is pleckstrin (P47) as described in 9.4.6 and the PKC-specific substrate MARCKS (major alanine-rich C-kinase). The RhoA-specific ADP-ribosyltransferase, C3 toxin, inhibits platelet aggregation (Morii *et al.*, 1992; Nemoto *et al.*, 1992; Zhang *et al.*, 1993, 1995), which suggests that Rho is necessary for platelet activation. RhoA activates P85/PI3K (Zhang *et al.*, 1993, 1995, 1996) and PI(4,5)K (Chong *et al.*, 1994; Tolias *et al.*, 1995). P85/PI3K, PI(4,5)K and Rho associates with the membrane skeleton in activated platelets. Grb2 is directly associated with the P85 subunit of P85/PI3K (Wang *et al.*, 1995) and can potentially serve as a pathway for Ras activation.

In contrast to the signaling initiated via G proteins in platelets, agonist-induced integrin activation is in general proposed to initiate formation of phosphotyrosine signaling complexes. The non-receptor, focal adhesion kinase FAK has been shown to interact with the cytoplasmic domain of β_1, β_2 and β_3 integrins (Schaller *et al.*, 1995) and becomes autophosphorylated following integrin engagement. FAK interacts with Grb2 (Schlaepfer *et al.*, 1994), and it is suggested that Grb2 in turn binds Sos and activates Ras. Src is bound via a SH2 domain at Y397 to FAK (Clark and Brugge, 1995) and is further postulated to activate Rac via Syk and Vav (Brass *et al.*, 1997). Vav, as mentioned earlier, activates Ras via Grb2-Sos. P85/PI3K is also found to bind to FAK, though the binding site has not been established. Collagen (type I) stimulates platelets through phosphorylation of PLC_γ (Blake *et al.*, 1994), and Syk kinase is reported to be upstream of PLC_γ (Keely and Parise, 1996) at the level of GPIa-IIa occupation. Stimulation of platelets with collagen induces tyrosine phosphorylation of the Fc receptor γ-chain (FcR γ-chain) which contains an immunoreceptor tyrosine-based activation motif (ITAM) (Gibbins *et al.*, 1996). This promotes association between the FcR γ-chain and Syk. PLC_γ liberates DAG and IP_3 which PKC phosphorylate P47 and MARCKS (Aderem, 1992).

It should be clear to the reader that the different receptors on the platelet do not function alone, but synergize with other receptor pathways to enhance or dampen signals elicited by each receptor.

9.5 PLATELET AGONISTS AND THEIR SIGNALING SYSTEMS

Platelets can be activated by a variety of distinct substances including coagulation factors (e.g. thrombin), hormones (e.g. adrenaline, vasopressin), low molecular weight substances (e.g. serotonin, ADP), lipid derivatives (e.g. PAF, thromboxane A_2), insoluble high molecular weight proteins with repetitive structures (collagen and immune complexes) and to surfaces normally not in contact with platelets, e.g. nonphysiological agonists like artificial surfaces. All these agonists exert their effects through interaction with specific receptors on the platelet plasma membrane. The receptors for thrombin, TXA_2, adenosine, PAF, vWF and fibrinogen are well described elsewhere in this book and will only be briefly described in this chapter. The receptors of human platelets can be subdivided as shown in Table 9.5.1.

9.5.1 Platelet receptors

Thrombin is the most potent physiologic platelet agonist, and induce all the platelet responses mentioned in section 9.3 and suppresses cAMP synthesis. Although thrombin is an enzyme, its ability to activate platelets depends on its concentration (Detwiler and Feinman, 1973). Harmon and Jamieson (1986) suggested that there are both high affinity

Table 9.5.1 Platelet receptors have been divided into receptors that stimulate or inhibit platelet activation through heterotrimeric G proteins, receptors that stimulate platelets through glycoproteins or integrins and other types of receptors. Their agonists are mentioned as well as their main receptor-coupled effector systems. The agonists are listed as strong, intermediate or weak platelet agonists according to their ability of inducing all or just some of the platelet responses mentioned in section 9.3. Adrenaline is not a true agonist and is denoted with "—" in this regard. Because of the complexness of the complement receptors, these are handled alone beneath. Abbreviations: Thrombospondin (TS), Fibrinogen (FG), Fibronectin (FN), Vitronectin (VN), Phospholipase C (PLC), Adenylyl cyclase (AC), Guanylyl cyclase (GC), Protein kinase C (PKC), Protein tyrosine kinase (PTK), stimulation of effectorsystem (+), inhibition of effectorsystem (-).

	Receptors	Agonist	Receptor-coupled effector system	Platelet activating potential of agonist
G protein linked receptors (activating platelet responses)	Thrombin-receptor	Thrombin	PLC (+), AC (-)	Strong
	TP-receptor	TXA_2, PGH_2	PLC (+), AC (-)	Strong/Intermediate
	Serotonin (5-HT) -receptor	Serotonin	PLC (+)	Weak
	PAF-receptor	PAF	PLC (+)	Weak
	α_2-receptor	Adrenaline	AC (-), PKC (+)	—
	V_1A-receptor	Vasopressin	PLC (+)	Weak
G protein linked receptors (inhibiting platelet activation)	PGI_2-receptor	PGI_2, PGE_1	AC (+)	Strong
	PGD_2-receptor	PGD_2	AC (+)	Intermediate
	Adenosine-receptor	Adenosine	AC (+)	Intermediate
	cGC	NO	GC (+)	Weak
Glycoproteins and Integrins	GPIa/IIa ($\alpha_2\beta_1$)	Collagen	PTK	Strong
	GPIV	Collagen, TS	PTK	Intermediate
	GPVI	Collagen	PTK	Intermediate
	GPIb/IX	vWF	PTK	Unknown
	GPIIb/IIIa ($\alpha_{IIb}\beta_3$)	FG, FN, VN, vWF	PTK	Unknown
	($\alpha_v\beta_3$)	VN, TS	PTK	Unknown
	GPIc/IIa ($\alpha_5\beta_1$)	FN	PTK	Unknown
	GPIc'/IIa ($\alpha_6\beta_1$)	Laminin	PTK	Unknown
Other receptors	Purine receptor ($P2T_{AC}$)	ADP	AC (-)	Intermediate
	Purine receptor ($P2T_{PLC}$)	ADP	PLC (+)	Intermediate
	Purine receptor (P2X)	ADP	Ca^{2+}-channel	Intermediate
	α_2-PDGF-receptor	PDGF	PTK	Weak
	Complement-receptors	(see beneath)		Unknown

and moderate affinity sites for thrombin binding, and that the high affinity binding sites are associated with membrane GPIb (De Marco *et al.*, 1991; Brass *et al.*, 1993). Although the platelet thrombin-receptor has not yet been detected, thrombin-receptors are cloned (Vu *et al.*, 1991; Rasmussen *et al.*, 1991), but the platelet receptor resembles the one found in a Dami cell line (Vu *et al.*, 1991; Brass *et al.*, 1993). Thrombin is proposed to cleave at a site on the N-terminus of the receptor (Vu *et al.*, 1991) and form a tethered ligand capable of activating the receptor. Thrombin and thrombin-receptor activation is reviewed by Grand *et al.* (1996).

Serotonin (5-HT) binds to the platelet 5-HT_2 receptor coupled to PLC via a G protein. The human serotonin receptor has been cloned and have the usual seven transmembrane domain form (Saltzman *et al.*, 1991). Serotonin is rapidly taken up by passive diffusion and by a high affinity energy-dependent carrier, and is released upon platelet activation by secretion from the dense granules (Pletscher, 1987).

ADP was the first specific platelet agonist that was identified (Gaarder *et al.*, 1961). It is secreted from the dense granula when the platelets are activated, or it is released from the red cells or damaged tissue (Gaarder *et al.*, 1961; Holmsen and Weiss, 1979). The ADP receptors activated by extracellular nucleotides have been classified as P2-receptors and are divided into two subclasses: P2X intrinsic ion channels and P2Y receptors coupled to heterotrimeric G proteins (Fredholm *et al.*, 1994) Three different receptors for ADP is found on human platelets: the $P2T_{AC}$-receptor which is coupled to inhibition of adenylyl cyclase, the $P2T_{PLC}$-receptor which is coupled to mobilization of intracellular Ca^{2+}-stores through inositol phosphate production and the P2X-receptor coupled to rapid Ca^{2+}-influx (MacKenzie *et al.*, 1996; Daniel *et al.*, 1998; Jin *et al.*, 1998).

TXA_2 and PGH_2 are products of cyclooxygenase-catalyzed metabolism of arachidonic acid and bind to TP-receptors on platelets (Hung *et al.*, 1983; Buch *et al.*, 1985; Saussy *et al.*, 1986; Hanasaki and Arita, 1991). Different species exist and studies with platelets have postulated two separate receptor coupling pathways, one which is responsible for shape change and the other for aggregation (Takahara *et al.*, 1990; Hanasaki and Arita, 1991).

Platelet-activating factor (PAF) is formed from the membrane phospholipid 1-O-alkyl-2-acyl-*sn*-glycero-3-phosphocholine by the consecutive actions of PLA_2 and acyltransferase (Chao and Olson, 1993) and binds to its receptor on human platelets (Kloprogge and Akkerman, 1985; Kloprogge *et al.*, 1986). Rat and mouse platelets lack this receptor (Iñarrea, 1984; Siess 1989). It has been shown that 87 % of the PAF produced is contained in the microvesicles that are shed from the platelets during the activation process (Iwamoto *et al.*, 1996).

Vasopressin bind to human platelets via V_1 vasopressin receptors and mediates activation of the polyphosphoinositol metabolism via PLC, increase in $[Ca^{2+}]_i$ and aggregation (Pollock and MacIntyre, 1986; Inaba *et al.*, 1988).

Adrenaline is only capable of producing platelet aggregation in the presence of other platelet agonists (Steen *et al.*, 1993; Lanza *et al.*, 1988), and is therefore not a true agonist. Platelets have stimulatory α_2 adrenoreceptors and inhibitory β adrenoreceptors, but the α_2 adrenoreceptors predominates. The α_2 adrenoreceptor is coupled via G proteins to adenylyl cyclase, while the α_1 adrenoreceptor (present in about 30 %) has been proposed to be coupled to PLC (Grant and Scrutton, 1979). It has been demonstrated that the α_{2A}-adrenergic receptors activate PKC via a G_i protein, independent of the PLC pathway and DAG formation (Nieuwland *et al.*, 1994).

PGE_1 and PGI_2 PGE_1 acts via the PGI_2 receptor (Miller and Gorman, 1979) which has two binding sites, one with high affinity and one with low affinity (MacIntyre and Armstrong, 1987). The receptor is coupled via G_s protein to adenylyl cyclase, thereby increasing the $[cAMP]_i$ which inhibits platelet activation.

PGD_2 binds to its discrete platelet (DP) receptor (Siegl *et al.*, 1979; Cooper and Ahern, 1979; Bonne *et al.*, 1981) and is coupled to adenylyl cyclase via a G_s protein, thereby increasing the $[cAMP]_i$ which inhibits platelet activation.

Adenosine binds to its adenosine receptor of the A_2 subtype that stimulates adenylyl cyclase via a G_s protein (Hütteman *et al.*, 1984).

Nitric oxide (NO) diffuses readily across cell membranes. Within the cell it binds to the heme moiety of the soluble guanylyl cyclase (cGC) thereby increasing the $[cGMP]_i$ which inhibits platelet activation due to effects of protein kinase G (PKG) (Welch *et al.*, 1995; Jensen and Holmsen, 1995).

PDGF is a dimeric disulfide-bonded dimeric protein secreted from the α-granules (Witte *et al.*, 1978; Kaplan *et al.*, 1979) upon platelet activation, and exists as three different isoforms, PDGF-AA, -AB and -BB, of which the PDGF-AB isoform is the most abundant (Hammacher *et al.*, 1988). Other growth factors are also found in the α-granules: EGF, TGF-β and PD-ECGF (Miyazono and Takaku, 1989). There are α- and β-types of PDGF-receptors, of which human platelets only have the PDGF α-receptor (Vassbotn *et al.*, 1994). PDGF-binding induced tyrosine phosphorylation of the receptor, cytoplasmic proteins binds and start a signal cascade.

The role of the integrin type of receptors in mediating signals from particulate agonists and surfaces, natural or artificial

Adhesive signaling in platelets has been extensively reviewed by Nurden & Nurden (1993) and Shattil *et* al, (1994). vWF is found in the platelet α-granules, plasma and subendothelium. Upon binding to subendothelial structures (type VI collagen), vWF changes its conformation and binds to GPIb on the platelet surface. GPIb consists of two disulfide-linked subunits, GPIbα and GPIbβ, which are non-covalently bound to GPIX (Blockmans *et al.*, 1995) and GPV (López and Dong, 1997). The binding sites for vWF and thrombin (López and Dong, 1997) are located on the α-chain. The interaction between vWF and GPIb results in platelet activation which turns on the GPIIb/IIIa so that vWF will bind to this receptor via its RGD-sequence. GPIIb/IIIa ($\alpha_{IIb}\beta_3$) is the receptor for fibrinogen, fibronectin, vitronectin and vWF, while the other isoform of GPIIb/IIIa ($\alpha_V\beta_3$) binds vitronectin and thrombospondin. Fibronectin also binds to GPIc/IIa ($\alpha_5\beta_1$), whereas thrombospondin also binds to GPIV. Collagen binds to three receptors on the platelets: GPIa/IIa ($\alpha_2\beta_1$), GPVI (P62) and GPIV (CD36) (Moroi and Jung, 1997). The quaternary structure of collagen, with the triple helix fibres, is crucial for activation of the platelets. Platelet adhesion to collagen is enhanced by Mg^{2+} (Dickeson *et al.*, 1997). Stimulation of platelets by collagen leads to activation of a tyrosine kinase cascade resulting in secretion and aggregation.

Platelets adhere to and become activated by nearly all artificial surfaces surfaces (McManna, 1986; Steen, 1987; Mason, 1974), and this is a problem in the development of new particulate medicaments or artificial organs. Plasma proteins adsorb to the surface before the platelets adhere, i.e. the platelets adhere to the "coat" of plasma proteins. The physiochemical properties of the surface determine the composition of the surface/liquid interface of plasmaproteins and therefore its platelet activating potential. (McManama *et al.*, 1986). Artificial surfaces therefore induce adhesion signaling in platelets via the receptors described above.

The platelet complement receptors

Early in evolution platelets and leukocytes was combined within one cell type. Later platelets and leukocytes were formed from these cells and are now primarily involved in hemostasis and inflammation, respectively. However, platelets are also involved in immunological reactions. They interact with immunoglobulins or complement directly through receptors located at their surface. Platelets contain and release adhesive proteins, activate the complement system, interact with bacteria and viruses, enhance vascular permeability, take up, store or metabolize vasoactive substances, produce eicosanoids and supply tissue proteases and glycosidases (Packham *et al.*, 1968). Platelets interact with leukocytes and endothelium (see later) (Jungi *et al.*, 1986; Ware and Heistad, 1993), and provide growth factors to the endothelium. Activated platelets as a part of the coagulation system initiate the growth of the blood clot by exposing phospholipids on the outer leaflet of the plasma membrane, which primes the enzyme complex that generates thrombin from prothrombin (Tracy and Mann, 1986).

Platelets contain different receptors for binding immune complexes with foreign antigens: of the three different FcRs observed so far, platelets contain only FcRII (Cdw32) (Rosenfeld *et al.*, 1987). Human platelets express C1qR and bind C1q, a glycoprotein which circulates in plasma as one of three subunits of the first component of the complement system (C1) (Peerschke and Ghebrehiwet, 1986, 1988; Peerschke *et al.*, 1994). The function of C1q in the C1 complex is to interact with polymeric structures, e.g. immune complexes, bacteria and viruses, and activate the classical complement pathway. Also receptors for complement factors C2 (Nunez *et al.*, 1987) and C4 (Vik and Fearon, 1987) are found on the platelets.

All members of the membrane attack complex (C5b-9) are found to be released after activation of the platelets (Houle *et al.*, 1989). Assembly of the C5b-9 complex on platelets initiate transient and reversible depolarization of the plasma membrane potential, a rise in the cytosolic [Ca^{2+}], thromboxane synthesis is induced, protein kinases are activated (Wiedmer and Sims, 1985; Wiedmer *et al.*, 1987; Hänsch *et al.*, 1986; Hattori *et al.*, 1989). They undergo shape change, release the contents of both α- and dense granula and shed microparticles (Wiedmer and Sims, 1985; Wiedmer *et al.*, 1987; Hattori *et al.*, 1989; Sims and Wiedmer, 1986; Sims *et al.*, 1988). On the surface of the platelets, although to a higher degree on the microparticles, there is *de novo* expression of membrane receptor sites for coagulation factor Va, initiating assembly of the prothrombinase enzyme complex and accelerating plasma clotting (Sims *et al.*, 1988; Wiedmer *et al.*, 1986a, 1986b) and C5b-9. This might be a mechanism for the platelets to avoid cytolysis after formation of the complement pore and to reverse the Na^+ conductance mediated by the membrane-inserted complex (Wiedmer and Sims, 1985; Sims and Wiedmer, 1986). Platelet activation by C5b-9 is directly linked to influx of Ca^{2+} and Ca^{2+} -dependent activation of protein kinases, bypassing receptor-coupled activation of PLC (Wiedmer *et al.*, 1987). This activation occurs without expression of the active fibrinogen-receptor (GPIIb/IIIa) and without aggregation (Ando *et al.*, 1988).

The membrane attack complex is controlled by several factors also found on platelets, namely the decay-accelerating factor (DAF, CD55) which promotes the decay of the C3 convertases of the classical and alternative complement pathways (Nicholson-Weller *et al.*, 1985a, 1985b; Kinoshita *et al.*, 1985), a membrane cofactor protein (MCP, CD46) which acts as a cofactor of the factor I-induced cleavage of C3b (Seya *et al.*, 1986), a membrane inhibitor of reactive lysis (MIRL, CD59) (Sims *et al.*, 1989) and C8bp which protects the platelet from complement-mediated lysis and regulates C5b-9-mediated activation of cells (Blaas *et al.*, 1988; Hänsch, 1988). Interaction of C8 and C9 with histidine-rich glycoprotein (HRGP) was described to inhibit complement activation (Chang *et al.*, 1992) and could represent another mechanism of the inhibition of C5b-9-induced platelet activation.

Stored complement factors that are released upon platelet activation (Table 9.3) are: C5-C9 (Houle *et al.*, 1989), the regulatory protein for the formation of C5b-9, Vitronectin (Suzuki *et al.*, 1987), Clusterin, a complement lysis inhibitor, present in the α-granules (Tschopp *et al.*, 1993), also present in the α-granules is C1-INH (Schmaier *et al.*, 1985), Factor D (Kenney and Davis, 1981) and Factor H (Devine and Rosse, 1987).

9.6 INHIBITION OF PLATELET ACTIVATION

Activation of platelets is counteracted by biochemical processes which prevent agonist-induced responses. These are mainly mediated by the signaling molecules cAMP and cGMP, and all substances that induce an increase in one or both of these are defined as platelet antagonists; some of these are listed in table 9.5.1.

9.6.1 The adenylyl cyclase pathway

In many cell systems, cAMP is a stimulatory intracellular messenger. In platelets, however, an elevation of cAMP (cyclic adenosine-3',5'-monophosphate) has an inhibitory effect (Mills and Smith, 1971). cAMP is formed from ATP by the agonist-stimulated membrane-bound enzyme adenylyl cyclase and broken down by various cAMP phosphodiesterases to keep the cAMP level at a steady state concentration. Both types of enzymes are activated and inhibited by agonists that bind to specific membrane receptors or by agents that act directly on the enzymes. Formation of cAMP is stimulated through surface receptors by PGI_2 (prostaglandin), PGD_2, PGE_1 and adenosine (Holmsen, 1976; Mills, 1982). PGI_2 is produced in the endothelium and is the most potent stimulator known for cAMP production. It is 10-30 times more potent than PGE_1 and PGD_2, and it has been suggested that PGI_2 might be a circulating hormone which regulates the responsivity of circulating platelets (Vane and Moncada, 1980). Most platelet agonists counteract formation of cAMP (Zavoico and Feinstein, 1984; Brass *et al.*, 1986; Salzman, 1977; Miller *et al.*, 1977; Mills and Smith, 1972; Vanderwei *et al.*, 1983) and this is thought be with stimulatory (Gs) and inhibitory (Gi) G proteins. See figure 9.6.1.

Figure 9.6.1 Formation and destruction of cAMP in platelets. Explanations are given in the text.

Forskolin stimulates the catalytic subunit of adenylyl cyclase directly, bypassing the G proteins, which makes forskolin a much used substance in experiments with elevated platelet cAMP-levels. A number of agents has been demonstrated to inhibit the cAMP phosphodiesterases in platelets, e.g. papaverine, dipyridamole and the methylxantines

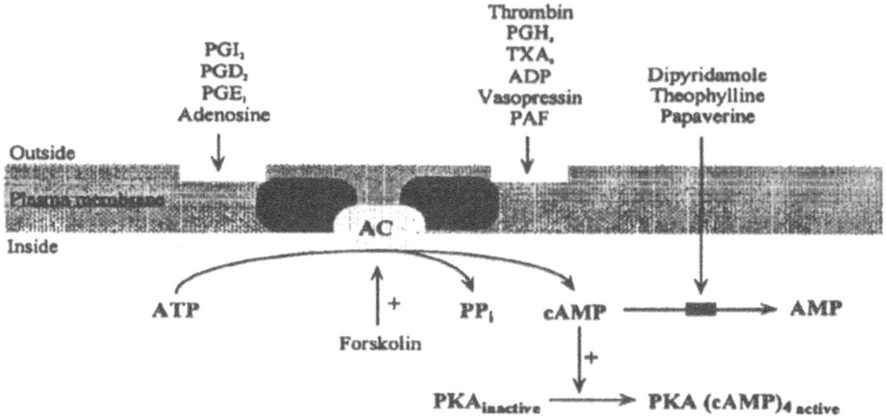

theophylline, caffeine and aminophyllin (Mills and Smith, 1971; Markwardt and Hofman, 1970; Smith and Mills, 1970). Dipyridamole has been used clinically as an anti-platelet drug. Platelets contain three different types of cyclic nucleotide phosphodiesterases (PDEs) with different K_m for cAMP and cGMP: a cGMP-specific PDE, a cGMP-stimulated PDE and a cGMP-inhibited PDE (Holmsen, 1987; Seth and Colman, 1995). Most of the PDE activity is found in the soluble/cytosolic fraction (Grant, 1992). cGMP-inhibited PDE is activated by phosphorylation by PKA as the cAMP-concentration rises (Grant *et al.*, 1992). Platelets contain both a cytosolic and membrane bound cAMP-dependent protein kinase (PKA) (Salama and Haslam, 1984).

cAMP exerts multiple inhibitory actions on platelet activation mechanisms. Increased levels of intracellular cAMP ($[cAMP]_i$) are associated with a reduced ability of thrombin to bind to its receptor and inhibition of several thrombin-induced responses (Lerea *et al.*, 1987), downregulation of the exposed vWF and fibrinogen receptors (Fujimoto *et al.*, 1982; Graber & Hawiger, 1982), antagonization of the activation of PKC (Takai *et al*, 1982; de Chaffoy de Courcelles *et al.*, 1987) and inhibition of the thrombin-induced formation of phosphatidic acid (PA) (Imai *et al.*, 1983; Knight and Scrutton, 1984; Lapetina, 1986) and of inositol 1,4,5-*tris*phosphate (IP_3) (Nakashima *et al.*, 1986). cAMP, via activation of protein kinase A (PKA), inhibits activation of PLC by phosphorylating a 240 kDa membrane protein which uncouples the G protein from PLC (Yada *et al.*, 1989). An elevation of cAMP in unstimulated, $^{32}P_i$-prelabeled platelets also increases the level of [^{32}P]phosphatidylinositol 4-phosphate (PIP) (de Chaffoy de Courcelles *et al.*, 1986; Suga *et al.*, 1986; Suzuki *et al.*, 1994). After a given time of platelet activation the

[^{32}P]PA/[^{32}P]PIP (or [^{32}P]PIP$_2$) ratio has been shown to be constant for a specific concentration of thrombin (Steen *et al.*, 1989). PGE$_1$, which stimulates adenylate cyclase and increases [cAMP]$_i$, markedly changed these ratios indicating that cAMP/PKA interferes with PPI metabolism upstream of PLC (Tysnes *et al.*, 1990). Increased [cAMP]$_i$ lowers cytoplasmic free [Ca^{2+}] (Yamanishi *et al.*, 1983; Cavallini *et al*, 1996) both by stimulating Ca^{2+} extrusion from the cells (Johansson *et al.*, 1992) and by stimulating reuptake of cytosolic Ca^{2+} in the dense tubular system (Tao *et al.*, 1992). Phosphorylation of the type 1 IP$_3$ receptor in platelets can either stimulate or inhibit IP -mediated Ca^{2+}-release rates, depending upon the phosphorylation state of the receptor (Quinton and Dean., 1992, Quinton *et al.*, 1996). Enouff *et al.* (1987) postulate that PKA stimulates uptake of Ca^{2+} from the cytosol into platelet microsomes through phosphorylation of a 24 kDa Ca^{2+}-ATPase. It should be mentioned finally that an increase in [cAMP]$_i$ induces phosphorylation of several proteins which functions are still not known (Haslam *et al.*, 1980; Siess and Lapetina, 1990; Butt and Walter, 1996).

9.6.2 The guanylyl cyclase pathway

In parallel to the cAMP-system, cyclic GMP (cyclic guanosine-3',5'-monophosphate) is produced and degraded by guanosine cyclase and cGMP-phosphodiesterase, respectively. Apart from PGI$_2$, endothelial cells also synthesize and release endothelium-derived releasing factor (EDRF) which is identical to nitric oxide (NO) (Gryglewsli *et al.*, 1988; Palmer *et al.*, 1987). NO is not as potent as PGI$_2$ as a platelet inhibitor. NO diffuses through the plasma membrane and activates cytosolic guanylyl cyclase (cGC) which results in an increase of cGMP (Gryglewsli *et al.*, 1988). cGMP activates a cGMP dependent protein kinase (PKG) (Waldmann *et al.*, 1986) and is accompanied by phosphorylation of a 46/50 (VASP) and 130 kDa proteins (Nakashima *et al.*, 1986; Waldmann *et al.*, 1986, 1987; Reinhard *et al.*, 1992; Butt *et al.*, 1994). cGMP and cAMP elevating agents act synergistic on platelet inhibition (Gryglewski, 1993), possibly because of an inhibition of cGMP-inhibited PDE due to increased [cGMP]$_i$ which results in accumulation of cAMP (Bowen and Haslam, 1991; Nolte *et al.*, 1994).

Inhibition of platelet activation by cGMP differs somewhat from that of cAMP, but many effects are similar (Radomski *et al.*, 1987), probably because many of the effects observed indirectly are effects of cAMP, due to inhibition by cGMP of cGMP-inhibited PDE (Maurice and Haslam, 1991).

9.7 AUTOCRINE STIMULATION AND INHIBITION

Platelet activation results in secretion of a number of substances as described in section 9.3.4 and *de novo* production of active compounds which all are able to act as platelet agonists and promote full response to the original extracellular signal. Two of the substances secreted from the dense granules, ADP and serotonin, are themselves platelet agonists and when secreted they will bind to their respective receptors and lead to further stimulation via mechanisms described in section 9.5. There is no effective ADP-receptor antagonists (maybe because of the plurality of these receptors), but a combination of *creatinphosphate (CP)/creatinphosphate-kinase (CPK)* will catabolize ADP to AMP, where AMP itself is a platelet antagonist. The serotonin-receptor could be antagonized with *cyproheptadine*. The α-granules contain the adhesion molecules fibrinogen, fibronectin, thrombospondin and vWF. GPIIb-IIIa, the fibrinogen-receptor, is not accessible for fibrinogen on resting platelets, but changes its conformation upon platelet activation and binds fibronectin, vWF and thrombospondin in addition to fibrinogen. Common for all these ligands are the Arg-Gly-Asp-Ser (RGDS)-sequence that are recognized by the receptor. It should be mentioned that platelet aggregation is dependent on extracellular Ca^{2+}, fibrinogen, active platelet collision, expression of the active GPIIb-IIIa-complex and active energy metabolism. The tetrapeptide, *RGDS*, is therefore a useful antagonist to all these platelet agonists that binds to GPIIb-IIIa and inhibits aggregation of the platelets. Aggregation is dependent on binding of fibrinogen to the agonist-exposed GPIIb-IIIa receptor and bound fibrinogen may serve to link platelets together in the aggregates. The primary agonist will also activate PLA_2 (through signalpathways indicated in Figure 9.4.10) which liberates arachidonate (AA) from phospholipids. Free AA is rapidly oxygenated by cyclooxygenase and thromboxane synthase to the platelet agonists PGG_2, PGH_2 and TXA_2 which bind to their receptors and cause further activation of the platelet (FitzGerald, 1991). An inhibitor of the TXA_2 receptor on human platelets is *SQ 29,548*. Platelet activating factor (PAF), a phospholipid with the structure 1-O-alkyl-2-acetyl-glycerophosphocholine is produced upon platelet activation (Hanahan *et al.*, 1980). It is shown that 87 % of the PAF produced is contained in the microvesicles that are shed from the platelets during the activation process (Iwamoto *et al.*, 1996). *BN 52021* is a PAF receptor antagonist. Platelet derived growth factor (PDGF) is also liberated from the α-granules upon activation. PDGF, in contrast to the other substances described, act as an platelet autocrine inhibitor with mechanisms still not clear. So far, no specific and effective PDGF-receptor inhibitors have been developed. Figure 9.7.1 shows some of the platelet autocrine effectors and their effects on platelet activation.

All the inhibitors of autocrine stimulation (IAS) mentioned act from the outside of the platelet to block binding of agonist to its receptor binding-site. They are not enzyme blockers like inhibitors of cyclooxygenase or lipoxygenase as aspirin, indomethacin or BW 755C, nor elevators of intracellular cAMP, since cAMP can change protein phosphorylation and Ca^{2+}-mobilization which may interfere with the effects of the primary agonist of interest. This should be kept in mind when one are designing experiments where the specific cellular effects of *one* agonist is of interest. Figure 9.4.9.2 shows the profound effect of inhibitors of autocrine stimulation (CP/CPK, SQ 29,548, RGDS, BN 52021 and cyproheptadine) on total tyrosine-phosphorylation (TP) in thrombin-stimulated platelets. While thrombin causes massive TP in the absence of IAS, practically all (97 %) of the TP is vanished when IAS are present. However the degree of secretion is the same with or without IAS. Thus thrombin *itself* induce little TP in the presence of IAS, which means that some of the other substances liberated as an effect of the primary agonist, are responsible for most of the TP seen. By removing one by one of the inhibitors of autocrine stimulation, fibrinogen was shown to induce most of the TP (Ryningen and Holmsen, 1998).

9.8 CROSSTALK BETWEEN DIFFERENT SIGNALING SYSTEMS

The stimulus-response mechanisms in platelets are complex and involve the variety of substances that stimulate different effector-systems in the platelets, the autocrine stimulation in response to platelet activation by a primary agonist as well as by "cross-talk" between signal transduction pathways. Cross-talk can happen at several stages of the signal transduction pathway. The most studied are the competition between stimulatory and inhibitory pathways, and examples are the action of thrombin via G_i and PGI_2 via G_s

in controlling IP_3/DAG and cAMP, respectively (section 9.6.1). In the case of adrenaline, platelets have α_2 adrenoreceptors coupled to AC and α_1 adrenoreceptors coupled to PLC (section 9.5). We have also described the cross-talk between the seven transmembrane domain receptors coupled to G proteins and the adhesion type of receptors (9.4.10) which have numerous points of convergence. The adhesion-receptors co-operate with the G protein mediated signaling pathways to stimulate platelets, while the PDGF-receptors inhibit the platelet activation from the G protein mediated signaling pathway via several tyrosine-phosphorylations and adapter proteins. The future will probably show us even more cross-roads in intracellular platelet signaling.

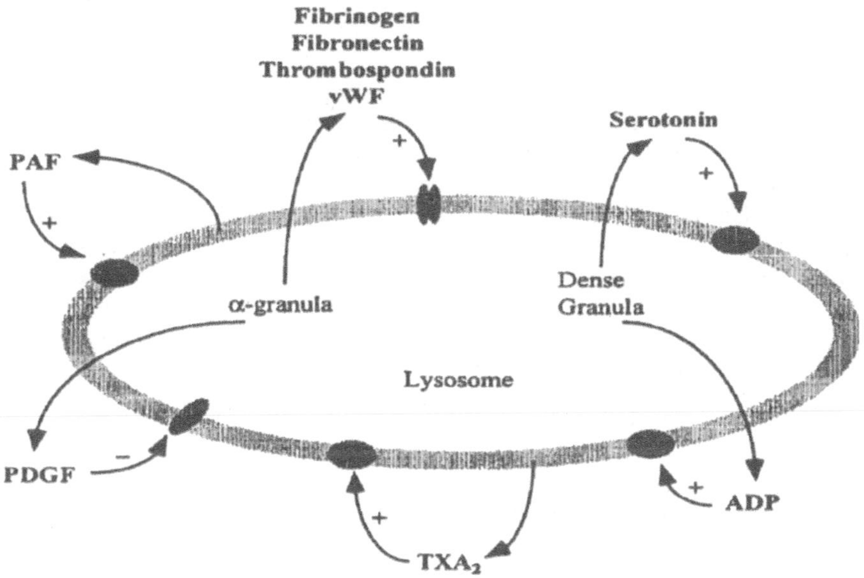

Figure 9.7. Platelet autocrine stimulation and inhibition. See text for explanations.

9.9 COMMUNICATION BETWEEN PLATELETS AND OTHER BLOOD CELLS

Platelets communicates not only internally by different signaling pathways, but also with other blood cells in the circulation and the endothelium, to interact in hemostasis and inflammation.

9.9.1. Adhesion receptors

Hemostasis and inflammation involves adhesion of cells in the flowing blood to the vessel wall. Three main families of receptors are involved.

Firstly, the Immunoglobulin Superfamily (IgSF) which is the most abundant family of cell surface molecules, and includes receptors for growth factors, receptors for the Fc region of Ig (where the FcRII-type is specific for platelets) and adhesion molecules. An important adhesion receptor is CD31/PECAM-1 (Platelet Endothelial Cell Adhesion Molecule-1) which is expressed predominantly on endothelial cells and platelets Wu and Lian, 1997).

Secondly, there is an integrin family of receptors which are heterodimers consisting of noncovalently associated α and β subunits. The largest number of integrins are members of the β1 or VLA (Very Late Antigen) subfamily, expressed on a variety of cell types (Albelda and Buck, 1990). Platelet membrane integrin molecules include the collagen receptor (the GPIa-IIa complex (VLA-2) ($\alpha_2\beta_1$)), the fibronectin receptor (the GPIc-IIa complex (VLA-5) ($\alpha_5\beta_1$)), the laminin receptor (the GPIc'-IIa complex (VLA-6) ($\alpha_6\beta_1$)) and von Willebrand factor (the GPIb-IX complex). The β_2 subfamily of integrins consists of leucocyte receptors, whereas the β_3 subfamily of integrins are expressed primarily on endothelial cells and platelets, e.g. the fibrinogen-receptor GPIIb/IIIa ($\alpha_{IIb}\beta_3$) and the vitronectin receptor ($\alpha_V\beta_3$).

Thirdly, there are the selectins, a family of Ca^{2+}-dependent carbohydrate-binding proteins which function to control leukocyte interactions with vascular endothelium (Tedder *et al.*, 1995). There are three closely related members of the selectin family, namely L (leukocyte)-, P (platelet)-, and E (endothelial)-selectin. P-selectin is expressed in the Weibel-Palade bodies of endothelial cells and in the α-granules of platelets. It is rapidly mobilized to the surface of endothelial cells and platelets in response to a variety of inflammatory agents such as thrombin, histamine, complement factors, free radicals and cytokines (Tedder *et al.*, 1995). P-selectin mediates the adhesion of neutrophils and monocytes to activated platelets and endothelial cells. Platelets can be bound to the endothelium indirectly by adhering to leukocytes bound to the endothelium.

9.9.2 Release of platelet-derived products

Without adhering to the endothelium, platelets can affect the endothelial function by the release of vasoactive substances from their storage granules (Houston *et al.*, 1986). These are ADP and serotonin (5-HT) that mediate vasodilation. Platelets also produce TXA_2 which stimulates formation of prostaglandin I_2 (prostacyclin, PGI_2) by enzymes localized in the endothelium. Endothelial cells may have receptors for TXA_2 that causes receptor-mediated increase in Ca^{2+} and finally release of vasodilator substances. It is reported that platelets express CD154 (CD40 ligand) upon activation. This transmembrane protein is structurally related to the cytokine TNF-α, and induces endothelial cells to secrete chemokines and to express adhesion molecules. This generates signals for recruitment of leukocytes at the site of injury and thereby initiate the inflammatory response of the vessel wall (Henn *et al.*, 1998).

9.9.3 Release of endothelial-derived products

Endothelial cells respond to a variety of stimuli that are delivered by activated platelets. The endothelium reacts by expressing receptors for cytokines and adhesive proteins and by releasing substances that prevent platelet aggregation. Both PGI_2 and nitric oxide exert their effects locally and cause vasodilation and inhibit platelet aggregation. PGI_2 inhibits platelet activation by increasing intracellular [cAMP] while NO increases intracellular [cGMP] and are capable to act synergistically to inhibit platelet function (Jensen and Holmsen, 1995; Bowen and Haslam, 1991). Endothelial cells synthesize tissue plasminogen activator (t-PA), which catalyzes the formation of plasmin, a protease that can have both aggregatory and inhibitory effects on platelets (Coller, 1990), but activated platelets also release plasminogen-activator inhibitor type-1 (PAI-1) which neutralizes t-PA. The endothelial surface has an ADPase that converts ADP released from erythrocytes and activated platelets to platelet-inactive AMP, inosine and to adenosine which inhibits platelet activation through cAMP (Marcus *et al.*, 1991).

9.9.4 Cooperation between platelets and leukocytes

Platelets have been shown to cooperate with leukocytes in production of chemotactic factors which the cells are unable to synthesize themselves. 12-HETE from platelets are metabolized by unstimulated neutrophils to 12,20-di-HETE (Marcus *et al.*, 1984, 1988). Leukocytes produces increased amounts of leukotrienes, because 12-HPETE produced in activated platelets stimulates the activity of leukocyte 5-lipoxygenase (Maclouf *et al.*, 1982; Romano and Serhan, 1992). Neutrophils convert arachidonic acid released from stimulated platelets to produce 5-HETE and leukotriene B_4 (LTB_4) (Marcus *et al.*, 1982). Platelet activating factor (PAF) are shown to stimulate production of LTB_4 from these cells (Lin *et al.*, 1982). Platelets can also produce the vasoconstrictor leukotriene C4 (LTC4) from leukotriene A4 (LTA4) synthesized by leukocytes (Maclouf and Murphy, 1988). Activated platelets and neutrophils release PAF (Lynch and Henson, 1986) which is a potent mediator of inflammation and asthma.

9.10 SUMMARY

Platelets play a central role in hemostasis and also function in inflammation, initiation of coagulation and activation of the complement system. This review deals with the biochemical mechanisms underlying those platelet responses that can be studied *in vitro* and are believed to be operative *in vivo*: adhesion, shape change, aggregation, secretion, neoepitope exposure, microvesicle shedding, development of procoagulant activity and clot retraction. Particular emphasis is made concerning the participation of heterotrimeric and monomeric G proteins in the early and late steps in activation and inhibition of platelet responses. The numerous soluble and particulate agonists that induce these responses as well as their corresponding receptors have been systematized and linked to specific effector systems. The roles of phospholipase C and protein kinase C isoenzymes, inositolphosphates, Ca^{2+} and the phosphorylation of pleckstrin and myosin light chain are discussed in detail, and the central roles of the polyphosphoinositide cycle and activation of phospholipase A_2 are highlighted. Particular emphasis has been made regarding the roles and mechanisms of serine/threonine and tyrosine protein phosphorylation cascades in platelet activation and of the cAMP- and cGMP-linked pathways in inhibition of platelet activation. Finally, the important roles of autocrine stimulation in platelet activation, crosstalk between signaling systems and communication between platelets and other blood cells are reviewed. The biochemical mechanisms of platelet activation are extremely complex, but although we are on the bottom of the ladder leading to a complete understanding of this fascinating machinery, their pieces are indeed starting to fall into their right places and hopefully will form a solid platform for future research.

REFERENCES

ABDEL-LATIF AA. "Metabolism of phosphoinositides." In: *Handbook of Neurochemistry.* Ed: A Lajtha. Plenum Publ Corp. (New York) pp 91-131, 1983.

ABRAMS C, SHATTIL SJ. Immunological detection of activated platelets in clinical disorders. *Thromb Haemost* **65**: 467-473, 1991.

ABRAMS CS, WU H, ZHAO W, BELMONTE E, WHITE D, BRASS L. Pleckstrin inhibits phosphoinositide hydrolysis initiated by G-protein coupled and growth factor receptors-a role for pleckstrin PH domains. *J Biol Chem* **270**: 14485-14492, 1995a.

ABRAMS CS, ZHAO W, BELMONTE E, BRASS LF. Protein kinase C regulates pleckstrin by phosphorylation of sites adjacent to the N-terminal pleckstrin homology domain. *J Biol Chem* **270**: 23317-23321, 1995b.

ABRAMS CS, ZHANG W, DOWNES CP, TANG X-W, ZHAO W, RITTENHOUSE S. Phospho-pleckstrin inhibits Gβγ-activatible platelet phosphatidylinositol (4,5) bisphosphate-3-kinase. *J Biol Chem* **271**: 25192-25197, 1996.

ADEREM A. Signal transduction and the actin cytoskeleton: the roles of MARCKS and profilin. *Trends Biochem Sci* **17**: 438-443, 1992.

AIKEN ML, GINSBERG MH, PLOW EF. Mechanisms for expression of thrombospondin on the platelet cell surface. *Semin Thromb Haemost* **13**: 307-316, 1987.

AKKERMAN JWN, HOLMSEN H, LOUGHNANE M. Simultaneous measurement of aggregation, secretion, oxygen uptake, proton production, and intracellular metabolites in the same platelet suspension. *Anal Biochem* **97**: 387-393, 1979.

AKKERMAN JWN, HOLMSEN H. Interrelationships among platelet responses: studies on the burst in proton liberation, lactate production and oxygen uptake during platelet aggregation and Ca^{2+} secretion. *Blood* **57**: 956-966, 1981.

AKKERMAN JWN, VERHOEVEN AJM. "Energy metabolism and function." In: *Platelet responses and metabolism, Vol III*. Ed: H Holmsen. CRC Press (Boca Raton, FL), pp 69-99, 1987.

ALBELDA SM, BUCK CA. Integrins and other cell adhesion molecules. *FASEB J* **4**: 2868-2880, 1990.

AMATRUDA TTI, STEELE DA, SLEPAK VZ, SIMON MI. $G_{\alpha 16}$ a G protein α subunit specially expressed in hematopoietic cells. *Proc Natl Acad Sci USA* **88**: 5587-5593, 1991.

ANDERSON NG, MALLER JL, TONKS NK, STURGILL TW. Requirements for integration of signals from two distinct phosphorylation pathways for activation of MAP kinase. *Nature* **343**: 651-653, 1990.

ANDO B, WIEDMER T, HAMILTON KK, SIMS PJ. Complement proteins C5b-9 initiate secretion of platelet storage granules without increased binding of fibrinogen or von Willebrand Factor to newly expressed cell surface GPIIb-IIIa. *J Biol Chem* **263**: 11907-11914, 1988.

ARMSTRONG RA, JONES RL, WILSON NH. Ligand binding to thromboxane receptors on human platelets: correlation with biological activity. *Br J Pharmacol* **79**: 953-964, 1983.

AUETHAVEKIAT V, ABRAMS CS, MAJERUS PW. Phosphorylation of platelet pleckstrin activates inositol polyphosphate 5-phosphatase I. *J Biol Chem* **272**: 1786-1790, 1997.

BALDASSARE JJ, HENDERSON PA, BURNS D, LOOMIS C, FISHER GJ. Translocation of protein kinase C isoenzymes in thrombin-stimulated human platelets. Correlation with 1,2-diacylglycerol levels. *J Biol Chem* **267**, 15585-15590, 1992.

BANNO Y, YADA Y, NOZAWA Y. Purification and characterization of membrane-bound phospholipase C specific for phosphoinositides from human platelets. *J Biol Chem* **263**: 11459-11465, 1988.

BANNO Y, YU A, NAKASHIMA T, HOMMA Y, TAKENAVA T, NOZAWA Y. Purification and characterization of a cytosolic phosphoinositide-phospholipase C (γ2-type) from human platelets. *Biochem Biophys Res Commun* **167**: 396-401, 1990.

BANNO Y, NAKASHIMA T, KUMADA T, EBISAWA K, NONOMURA Y, NOZAWA Y. Effects of gelsolin on human platelet cytosolic phosphoinositide-phospholipase C isozymes. *J Biol Chem* **267**: 6488-6494, 1992.

BANNO Y, NAKASHIMA T, HASHIYA T, NOZAWA Y. Endogenous cleavage of phospholipase C-β3 by agonist-induced activation of calpain in human platelets. *J Biol Chem* **270**: 4318-4324, 1995.

BAUMGARTNER HR, TSCOPP TB, MEYER D. Shear rate dependent inhibition of platelet adhesion and aggregation on collagenous surfaces by antibodies to Factor VIII/von Willebrand factor. *Br J Haematol* **44**: 127-139, 1980.

BENCKE-MARTI K, LAPETINA EG. Epinephrine suppresses rap 1B. GAP-activated GTPase activity in human platelets. *Proc Natl Acad Sci USA* **89**: 2784-2788, 1992.

BERMAN CL, YEO EL, WENCEL-DRAKE JD, FURIE BC, GINSBERG MH, FURIE B. A platelet alphagranule membrane protein that is associated with the plasma membrane after activation: characterization and subcellular localization of platelet activation-dependent granule -external membrane protein. *J Clin Invest* **78**: 130-137, 1986.

BERRIDGDE MJ. Calcium oscillations. *J Biol Chem* **265**: 9583-9586, 1990.

BERRIGDE MJ. Inositol trisphosphate and calcium signalling. *Nature* **361**, 315-317, 1993.

BEVERS EM, COMFURIUS P, ZWAAL RFA. Changes in membrane phospholipid distribution during platelet activation. *Biochim Biophys Acta* **736**: 57-66, 1983.

BEVILACQUA M, BUTCHER E, FURIE B, FURIE B, GALLATIN M, GIMBRONE M, HARLAN J, KISHIMOTO K, LASKY L, MCEVER R. Selectins: a family of adhesion receptors. *Cell* **67**: 233, 1991.

BEVILACQUA MP, NELSON RM. Selectin. *J Clin Invest* **91**: 379-387, 1993.

BHULLAR RP, CHARDIN P, HASLAM RJ. Identification of multiple *ral* gene products in human platelets that account for some but not all of the platelet G_n proteins. *FEBS Lett* **260**: 48-52, 1990.

BISCHOFF FP, PONSTINGL H. Mitotic regulator protein RCC1 is complexed with a nuclear ras-related polypeptide. *Proc Natl Acad Sci USA* **88**: 10831-10834, 1991.

BLAAS P, BERGER B, WEBER S, PETER HH, HÄNSCH GM. Paroxysmal nocturnal hemoglobinuria. Enhanced stimulation of platelets by the terminal complement components is related to the lack of C8bp in the membrane. *J Immunol* **140**: 3045-3051, 1988.

BLAKE RA, SCHIEVEN GL, WATSON SP. Collagen stimulates tyrosine phosphorylation of phospholipase C-gamma 2 but not phospholipase C-gamma 1 in human platelets. *FEBS Lett* **353**: 212-216, 1994.

BLOCKMANS D, DECKMYN H, VERMYLEN J. Platelet activation. *Blood Rev* **9**: 143-156, 1995.

BOGUSKI MS, MCCORMICK F. Proteins regulating Ras and its relatives. *Nature* **366**: 643-654, 1993.

BOKOCH GM, DER CJ. Emerging concepts in the ras superfamily of GTP-binding proteins. *FASEB J* **7**: 750-759, 1993.

BONNE C, MARTIN B, WATADA M, REGNAULT F. The antagonism of prostaglandins I_2, E_1 and D_2 by prostaglandin E_2 in human platelets. *Thromb Res* **21**: 13-22, 1981.

BOOTMAN MD, BERRIDGDE MJ. The elemental principles of calcium signalling. *Cell* **83**: 675-678, 1995.

BORSCH-HAUBOLD AG, KRAMER RM, WATSON SP. Cytosolic phospholipase A_2 is phosphorylated in collagen- and thrombin-stimulated human platelets independent of protein kinase C and mitogen-activated protein kinase. *J Biol Chem* **270**: 25885-25892, 1995.

BOWEN R, HASLAM RJ. Effects of nitrovasodilators on platelet cyclic nucleotide levels in rabbit blood: role for cyclic AMP in synergistic inhibition of platelet function by SIN-1 and prostaglandin E1. *J Cardiovasc Pharmacol* **17**: 424-433, 1991.

BRASS LF, LAPOSATA M, BANGA HS, RITTENHOUSE SE. Regulation of the phosphoinositide hydrolysis pathway in thrombin-stimulated platelets by pertussis toxin-sensitive guanine nucleotide-binding protein: Evaluation of its contribution to platelet activation and comparisons with the adenylate cyclase inhibitory protein, G1. *J Biol Chem* **261**: 16838-16847, 1986.

BRASS LF, HOXIE JA, MANNING DR. Signalling through G proteins and G protein-coupled receptors during platelet activation. *Thromb Haemost* **70**: 217-223, 1993.

BRASS LF. "Molecular basis for platelet activation." In: *Hematology: basic principles and practice.* 2nd ed. Ed: R. Hoffman. Churchill Livingstone (New York). pp 1536-1552, 1995.

BRASS LF, MANNING DR, CISHOWSKI K, ABRAMS CS. Signalling through G proteins in platelets: to the integrins and beyond. *Thromb Haemost* **78**: 581-589, 1997.

BURCH RM, MAIS DE, SAUSSY DL JR, HALUSHKA PV. Solubilization of a thromboxane A_2/prostaglandin H_2 antagonist binding site from human platelets. *Proc Natl Acad Sci USA* **82**: 7434-7438, 1985.

BUSTELO XR, LEDBETTER JA, BARBACID M. Product of vav proto-oncogene defines a new class of tyrosine protein kinase substrates. *Nature* **356**: 68-71, 1992.

Butt E, Walter U. *Platelets: a practical approach*. Eds: Watson SP & Authi KS. IRL Press (Oxford, UK), 1996.

Butt E, Abel K, Krieger M, Palm D, Hoppe V, Hoppe J, Walter U. cAMP- and cGMP-dependent protein kinase phosphorylation sites of the focal adhesion vasodilator-stimulated phosphoprotein (VASP) in vitro and in intact human platelets. *J Biol Chem* **269**: 14509-14517, 1994.

Capitanio AM, Niewiarowski S, Rucinski B, Tuszynski GP, Cierniewski CS, Hershock D, Kornecki E. Interaction of platelet factor 4 with human platelets. *Biochim Biophys Acta* **839**: 161-173, 1985.

Carlson KE, Brass LF, Manning DR. Thrombin and phorbol esters cause the selective phosphorylation of a G protein other than G_i in human platelets. *J Biol Chem* **264**: 13298-13305, 1989.

Carty DJ, Padrell E, Codina J, Birnbaumer L, Hildenbrandt JD, Iyengar RJ. Distinct guanine nucleotide binding and release properties of the three G_i proteins. *J Biol Chem* **265**: 6268-6273, 1990.

Casey PJ. Lipid modifications of G proteins. *Curr Opinion in Cell Biol* **6**: 219-225, 1994.

Casey PJ. Protein lipidation in cell signalling. *Science* **268**: 221-225, 1995.

Castle AG, Crawford N. Platelet microtubule subunit proteins. *Thromb Haemostas* **42**: 1630-1633, 1979.

Cavallini L, Coassin M, Borean A, Alexandre A. *J Biol Chem* **271**: 5541-5551, 1996.

Chang N-S, Leu RW, Rummage JA, Anderson JK, Mole JE. Regulation of complement functional efficiency by histidine-rich glycoprotein. *Blood* **79**: 2973-2980, 1992.

Chao W, Olson MS. Platelet-activating factor: receptors and signal transduction. *Biochem J* **292**: 617-629, 1993.

Chardin P, Boquet P, Madaule P, Popoff MR, Rubin EJ, Gill DM. The mammalian G protein rhoC is ADP ribosylated by *Clostridium botulinum* exoenzyme C3 and affects actin microfilaments in Vero cells. *EMBO J* **8**: 1087-1092, 1988.

Chen JW, Murphy TL, Willingham MC, Pastan I, August JT. Identification of two lysosomal membrane glycoproteins. *J Cell Biol* **101**: 85-95, 1985.

Chong LD, Traynor-Kaplan A, Bokoch GM, Schwartz MA. The small GTP-binding protein Rho regulates a phosphatidylinositol 4-phosphate 5-kinase in mammalian cells. *Cell* **79**: 507-513, 1994.

Cichowski K, McCormick F, Brugge JS. $p21^{ras}$GAP association with Fyn, Lyn and Yes in thrombin-activated platelets. *J Biol Chem* **267**: 5025-5028, 1992.

Cichowski K, Brugge JS, Brass LF. Thrombin receptor activation and integrin engagement stimulate tyrosine phosphorylation of the proto-oncogene product, $p95^{vav}$, in platelets. *J Biol Chem* **271**: 7544-7550, 1996.

Clapham DE, Neer EJ. New roles for G-protein $\beta\gamma$-dimers in transmembrane signalling. *Nature* **365**: 403-406, 1993.

Clark EA, Shattil SJ, Ginsberg MH, Bolen J, Brugge JS. Regulation of the protein tyrosine kinase $pp72^{syk}$ by platelet agonists and the integrin $\alpha_{IIb}\beta_3$. *J Biol Chem* **269**: 28859-28864, 1994a.

Clark EA, Brugge JS. Regulation of protein tyrosine kinases in platelets. *Trends Biochem Sci* **19**: 464-469, 1994b.

Clark EA, Brugge JS. Integrins and signaltransduction pathways: the road taken. *Science* **268**: 233-239, 1995.

Cobb MH, Goldsmith EJ. How MAP kinases are regulated. *J Biol Chem* **270**: 14843-14846, 1995.

Cockcroft S, Thomas GMH: Inositol-lipid-specific phospholipase C isoenzymes and their differential regulation by receptors. *Biochem J* **288**: 1-9, 1992.

Cohen GB, Ren R, Baltimore D. Modular Binding Domains in Signal Transduction Proteins. *Cell* **80**: 237-248, 1995.

Cohen I, Gerrard JM, White JG. Ultrastructure of clots during isometric contraction. *J Cell Biol* **93**: 775-787, 1982.

Coller BS. Platelets and thrombolytic therapy. *N Engl J Med* **322**: 33-42, 1990.

Conti MA, Adelstein RS. Phosphorylation by cyclic adenosine 3',5'-monophosphate-dependent protein kinase regulates myosin light chain kinase. *Fed Proc* **39**: 1569-1573 1980.

COOPER B, AHERN D. Characterization of the platelet PGD_2 receptor. Loss of PGD_2 receptors in platelets with myeloproliferative disorders. *J Clin Invest* **64**: 586-590, 1979.

COPPOLA J, BRYANT S, KODA T, CONWAY D, BARBACID M. Mechanism of action of the vav protooncogene, *Cell Growth Diff* **2**: 95-105, 1991.

COX AD, GOODALL AH. "Activation-specific neo-antigen on platelet detected by monoclonal antibodies." In: *Current Studies in Haematology and Blood Transfusion, Vol 58*. Eds: Albertini A, Lenfant CL, Mannucci PM, Sixma JJ. (Karger, Basel), pp 194-199, 1991.

CRESPO P, XU N, SIMONDS WF, GUTKIND JS. Ras-dependent activation of MAP kinase pathway mediated by G-protein βγ subunits. *Nature* **369**: 418-420, 1994.

CRESPO P, BUSTELO XR, AARONSON DS. Rac-1 dependent stimulation of the JNK/SAPK signaling pathway by Vav. *Oncogene* **13**: 455-460, 1996.

CRESPO P, SCHUEBEL KE, OSTROM AA, GUTKIND JS, BUSTELO XR. Phosphotyrosine-dependent activation of Rac-1 GDP/GTP exchange by the vav proto-oncogene product. *Nature* **385**: 169-172, 1997.

CROSS MJ. Effect of fibrinogen on the aggregation of platelets by adenosine diphosphate. *Thrombs Diath Haemorrh* **12**: 521, 1964.

DACHARY-PRIGENT J, TOTI F, SATTA N, PASQUET JM, UZAN A, FREYSSINET JM. Physiopathological significance of catalytic phospholipids in the generation of thrombin. *Semin Thromb Hemostas* **22**: 157-164, 1996.

DANIEL JL, MOLISH IR, HOLMSEN H. Myosin phosphorylation in intact platelets. *J Biol Chem* **256**: 7510-7514, 1981.

DANIEL JL, MOLISH IR, RIGMAIDEN M, STEWARD G. Evidence for a role of myosin phosphorylation in the initiation of the platelet shape change response. *J Biol Chem* **259**: 9826-9831, 1984.

DANIEL JL, DANGELMAIER C, JIN J, ASHBY B, SMITH JB, KUNAPULI SP. Molecular basis for ADP-induced platelet activation. I. Evidence for three distinct ADP receptors on human platelets. *J Biol Chem* **273**: 2024-2029, 1998.

DE-CATERINA M, STRAZZULLO P, IACONE R, POMPEO F, VARRIALE V, GRIMALDI E, SCOPACASA F. Determination of the kinetics of Na^+/H^+ exchange in platelets using the Coulter S-plus cell counter. *Eur J Clin Biochem* **32**: 57-60, 1994.

DE CHAFFOY DE COURCELLES D, ROEVENS P, BELLE H. Agents that elevate platelet cAMP stimulate the formation of phosphatidylinositol 4-phosphate in intact human platelets. *FEBS Lett* **195**: 115-118, 1986.

DE CHAFFOY DE COURCELLES D, ROEVENS P, BELLE H. Prostaglandin E_1 and forskolin antagonize C-kinase activation in the human platelet. *Biochem J* **244**: 93-99, 1987.

DECKERT M, TARTARE-DECKERT S, COUTURE C, MUSTELIN T, ALTMAN A. Functional and physical interactions of Syk family kinases with the Vav proto-oncogene product. *Immunity* **5**: 591-604, 1996.

DE MARCO L, MAZZUCATO M, MASOTTI A, FENTON JW, RUGGERI ZM. Function of glycoprotein Ibα in platelet activation induced by α-thrombin. *J Biol Chem* **266**: 23776-23783, 1991.

DERENLEAU DA, LUTHY R, LUSCHER EF. Stochastic response of human platelets to stimulation of shape changes and secretion. *Proc Natl Acad Sci USA* **83**: 2076-2080, 1986.

DERENLEAU DA. Blood platelet shape change ABCs. *Trends Biochem Sci* **12**: 439-442, 1987.

DETWILER TC, FEINMAN RD. Kinetics of the thrombin-induced release of calcium (II) by platelets. *Biochemistry* **12**: 282-289, 1973.

DEVINE DV, ROSSE WF. Regulation of the activity of platelet-bound C3 convertase of the alternative pathway of complement by factor H. *Proc Natl Acad Sci USA* **84**: 5873-5877, 1987.

DEVIVO M, IYENGAR R. G protein pathways: signal processing by effectors. *Mol Cell Endocrinol* **100**: 65-70, 1994.

DHAR A, SHUKLA SD. Tyrosine kinases in platelet signalling. *Br J Haematol* **84**: 1-7, 1993.

DICKESON SK, WALSH JJ, SANTORO SA. Contributions of the I and EF hand domains to the divalent cation-dependent collagen binding activity of the alpha2beta1 integrin. *J Biol Chem* **272**: 7661-7668, 1997.

DIKIC I, TOKIWA G, LEV S, COURTNEIDGE SA, SCHLESSINGER J. A role for Pyk2 and Src in linking G-protein-coupled receptors with MAP kinase activation. *Nature* **383**: 547-550, 1996.

DOWNES CP, CARTER AN. Phosphoinositide 3-kinase - A new effector in signal transduction. *Cell Signal* **3**: 501-513, 1991.

DOWNWARD J. The ras superfamily of small GTP-binding proteins. *Trends Biochem Sci* **15**: 469-472, 1990.

DOWNWARD J. "Role of phosphoinostide-3-OH kinase in ras signaling." In: *Signal Transduction in Health and Disease, Advances in Second Messenger and Phosphoprotein Research, Vol 31*. Eds: J Corbin and S Francis. Lippincott-Raven Publishers (Philadelphia), 1997.

ENOUFF J, GIRAUD F, BREDOUX R, BORDEAU N, SARKADI B, LEVY-TOLEDANO S. Stimulation of the 23-Kd protein cAMP dependent phosphorylation by inositol 1,4,5 trisphosphate in human platelet membrane vesicles. *Biochem Biophys Commun* **145**: 139-145, 1987.

EXTON JH. Phosphoinositide phospholipases and G proteins in hormone action. *Annu Rev Physiol* **56**: 349-369, 1994.

FEBBRAIO M, SILVERSTEIN RL. Identification and characterization of LAMP-1 as an activation-dependent platelet surface glycoprotein. *J Biol Chem* **265**: 18531-18537, 1990.

FISHER TH, GATLING MN, LACAL JC, WHITE GC. Rap 1B, a cAMP-dependent protein kinase substrate, associates with the platelet cytoskeleton. *J Biol Chem* **265**: 19405-19408, 1990.

FITZGERALD GA. Mechanisms of platelet activation: thromboxane A_2 as an amplifying signal for other agonists. *Am J Cardiol* **68**: 11B-15B, 1991.

FOX JEB, BOYLES JK, REYNOLDS CC, PHILLIPS DR. Actin filament content and organization in unstimulated platelets. *J Cell Biol* **98**: 1985-1991, 1984.

FOX JEB. Linkage of a membrane skeleton to integral membrane glycoproteins in human platelets. Identification of one of the glycoproteins as glycoprotein Ib. *J Clin Invest* **75**: 1673-1683, 1985.

FOX JEB, LIPFERT L, CLARK EA, REYNOLDS CC, AUSTIN CD, BRUGGE JS. On the role of the platelet membrane skeleton in mediating signal transduction. *J Biol Chem* **268**: 25973-25984, 1993a.

FOX JEB. The platelet cytoskeleton. *Thromb Haemost* **70**: 884-893, 1993b

FOX JEB. Platelet activation: New aspects. *Haemostasis* **26**: 102-131, 1996.

FRANGIONI JV, ODA A, SMITH M, SALZMAN EW, NEEL BG. Calpain-catalyzed cleavage and subcellular relocation of protein phosphotyrosine phosphatase 1B (PTP-1B) in human platelets. *EMBO J* **12**: 4843-4856, 1993.

FREDHOLMI BB, ABBRACCHIO MP, BURNSTOCK G, DALY JW, HARDEN TK, JACOBSON KA, LEFF P, WILLIAMS M. Nomenclature and classification of purinoceptors. *Pharmacol Rev* **46**: 143-156, 1994.

FRØLICH KW, AARBAKKE GM, HOLMSEN H. Chlorpromazine increases the turnover of metabolically active phosphoinositides and elevates the steady-state level of phosphatidylinositol-4-phosphate in human platelets. *Biochem Pharmacol 44*: 2013-2020, 1992.

FUJIMOTO T, OHARA S, HAWIGER J. Thrombin-induced exposure and prostacyclin inhibition of the receptor for factor VIII/von Willebrand factor on human platelets. *J Clin Invest* **69**: 1212-1222, 1982.

FUKAMI M, HOLMSEN H. Diacylglycerol elevations in control platelets are unaccompanied by pleckstrin phosphorylation. Implications for the role of diacylglycerol in platelet activation. *Eur J Biochem* **228**: 579-586, 1995.

FUKUDA M, VIITALA J, MATTESON J, CARLSSON SR. Cloning of cDNAs encoding human lysosomal membrane glycoproteins, h-Lamp-1 and h-Lamp-2. Comparison of their deduced amino acid sequences. *J Biol Chem* **263**: 18920-18928, 1988.

GAARDER A, JONSEN J, LALAND S, HELLEM AJ, OWREN P. Adenosine diphosphate in red cells as a factor in the adhesiveness of human blood platelets. *Nature* **192**: 531-532, 1961.

GAGNON WA, MANNING DR, CATANI L, GERWIRTZ A, PONCZ M, BRASS LF: Identification of G_{zz} as a pertussis toxin-insensitive G protein in human platelets and megakaryocytes. *Blood* **78**: 1247-1253, 1991.

GARTNER TK, OGILVIE ML. Peptides and monoclonal antibodies which bind to platelet glycoproteins IIb and/or IIIa inhibit clot retraction. *Thromb Res* **49**: 43-53, 1988.

GEMMEL CH, SEFTON MV, YEO EL. Platelet-derived microparticle formation involves glycoprotein IIb-IIIa. *J Biol Chem* **268**: 14586-14589, 1993.

GEORGE PY, HELKAMP JR GM. Purification and characterization of a phosphatidylinositol transfer protein from human platelets. *Biochim Biophys Acta* **836**: 176-184, 1985.

GERRARD JM, WHITE JG, PETERSON DA. The platelet dense tubular system: its relationship to prostaglandin synthesis and calcium flux. *Thromb Haemost.* **40**: 224-231, 1978.

GERRARD JM, LINT D, SIMS PJ, WIEDMER T, FUGATE RD, MCMILLAN E, ROBERTSON C, ISRAELS SJ. Identification of a platelet dense granule membrane protein that is deficient in a patient with the Hermansky-Pudlak syndrome. *Blood* **77**: 101-112, 1991.

GIBBINS J, ASSELIN J, FARNDALE R, BARNES M, LAW CL, WATSON SP. Tyrosine phosphorylation of the Fc receptor gamma-chain in collagen-stimulated platelets. *J Biol Chem* **271**: 18095-18099, 1996.

GILBERT GE, SIMS PJ, WIEDMER T, FURIE B, FURIE BC, SHATTIL SJ. Platelet-derived microparticles express high affinity receptors for factor VIII. *J Biol Chem* **266**: 17261-17268, 1991.

GILMAN AG. G proteins: transducers and receptor generated signals. *Annu Rev Biochem* **56**: 615-623, 1987.

GOLDEN A, NEMETH SP, BRUGGE JS. Blood platelets express high levels of the $pp60^{c-src}$-specific tyrosine kinase activity. *Proc Natl Acad Sci USA* **83**: 852-856, 1986.

GOLDSCHMIDT-CLERMONT PJ, MACHESKY LM, BALDASSARE JJ, POLLARD TP. The actin-binding protein binds PIP_2 and inhibits its hydrolysis by phospholipase C. *Science* **247**: 1575-1578, 1990.

GOODY RS. Signal transduction: How G proteins turn off. *Nature* **372**: 220-221, 1994.

GRABAREK J, RAYCHOWDHURY M, RAVID K. Identification and functional characterization of protein kinase C isoenzymes in platelets and HEL cells. *J Biol Chem* **267**: 10011-10020, 1992.

GRABER SE, HAWIGER J. Evidence that changes in platelet cyclic AMP levels regulate the fibrinogen receptor on human platelets. *J Biol Chem* **257**: 14606-14609, 1982.

GRAND RJA, TURNELL AS, GRABHAM PW. Cellular consequences of thrombin-receptor activation. *Biochem J* **313**: 353-368, 1996.

GRANT JA, SCRUTTON MC. Novel alpha2-adrenoreceptors primarily responsible for inducing platelet aggregation. *Nature* **227**: 659-661, 1979.

GRANT PG, DECAMP DL, BAILEY JM, COLMAN RF, COLMAN RW. Low-K_m cyclic AMP phosphodiesterase from human platelets. *Adv second Messenger Phosphoprotein Res* **25**: 73-85, 1992.

GRONDIN P, PLANTAVID M, SULTAN C, BRETON M, MAUCO G, CHAP H. Interaction of $pp60^{c-src}$, phospholipase C, Inositol-lipid, and diacylglycerol kinases with the cytoskeletons of thrombin-stimulated platelets. *J Biol Chem* **266**: 15705-15709, 1991.

GRYGLEWSKI RJ, BOTTING RM, VANE JR. Mediators produced by the endothelial cell. *Hypertension* **12**: 530-548, 1988.

GRYGLEWSKI RJ. Interactions between nitric oxide and prostacyclin. *Seminars Thromb Heam* **19**: 158-166, 1993.

HALENDA SP, VOLPI M, ZAVOICO GB, SHA'AFI RI, FEINSTEIN MB. Effects of thrombin, phorbol myristate acetate and prostaglandin D_2 on 40-41 kDa protein that is ADP ribosylated by pertussis toxin in human platelets. *FEBS Lett* **204**: 341-346, 1986.

HAMBERG M, SAMUELSSON B. Prostaglandin endoperoxides. Novel transformations of arachidonic acid in human platelets. *Proc Natl Acad Sci USA* **71**: 3400-3408, 1974.

HAMMACHER A, HELLMAN U, JOHNSSON A, ÖSTMAN A, GUNNARSSON K, WESTERMARK B, WASTESON Å, HELDIN C-H. A major part of platelet-derived growth factor purified from human platelets is a heterodimer of one A and one B chain. *J Biol Chem* **263**: 16493-16498, 1988.

HANAHAN DJ, DEMOPOULOS CA, LIEHR J, PINKARD RN. Identification of platelet activating factor isolated from rabbit basophils as glyceryl ether phosphorylcholine. *J Biol Chem* **255**: 5514-5516, 1980.

HANASAKI K, ARITA H. Recent aspects of TXA_2 action on platelets and blood vessels. *Platelets* **2**: 69-76, 1991.

HÄNSCH GM, GEMSA D, RESCH K. Induction of prostanoid synthesis in human platelets by the late complement components C5b-9 and the channel forming antibiotic nystatin: inhibition of the recyclation of liberated arachidonic acid. *J Immunol* **135**: 1320-1324, 1986.

HÄNSCH GM. The homologous species restriction of the complement attack: structure and function of the C8 binding protein. *Curr Top Microbiol Immunol* **140**: 109-118, 1988.

HARLAN JE, HAJDUK PJ, YOON HS, FESIK SW. Pleckstrin homology domains bind to phosphatidylinositol-4,5-bisphosphate. *Nature* **371**: 168-170, 1994.

HARMON JT, JAMIESON GA. Activation of platelets by alpha-thrombin is a receptor-mediated event. *J Biol Chem* **261**, 15928-15933, 1986.

HARRIS HE, WEEDS A. Platelet actin: subcellular distribution and association with profilin. *FEBS Lett* **90**: 84-88, 1978.

HARTWIG JH, BOKOCH GM, CARPENTER CL, JANMEY PA, TAYLOR LA, TOKER A, STOSSEL TP. Thrombin receptor ligation and activated Rac uncap actin filament barbed ends through phosphoinositide synthesis in permeabilized human platelets. *Cell* **82**: 643-653, 1995.

HARTWIG JH, BARKALOW K. Polyphosphoinositide synthesis and platelet shape change. *Curr Opinion Hematology* **4**: 351-356, 1997.

HART MJ, EVA A, EVANS T, AARONSON SA, CERIONE RA. Catalysis of guanine nucleotide exchange on the CDC42Hs protein by the dbl oncogene product. *Nature* **354**: 311-314, 1991.

HART MJ, EVA A, ZANGRILLI D, AARONSON SA, EVANS T, CERIONE RA, ZHENG Y. Cellular transformation and guanine nucleotide exchange activity are catalyzed by a common domain on the dbl oncogene product. *J Biol Chem* **269**: 62-5, 1994.

HASLAM RJ, LYNHAM JA, FOX JEB. Effects of collagen, ionophore A23187 and prostaglandin E_1 on the phosphorylation of specific proteins in blood platelets. *Biochem J* **178**: 397-406, 1979.

HASLAM RJ, SALAMA SE, FOX JEB, LYNHAM JA, DAVIDSON MML. Platelets: Cellular response mechanisms and their biological significance, J. Wiley & Sons (NY), 1980.

HASLAM RJ, KOLDE BJ, HEMMINGS BA. Pleckstrin domain homology. *Nature* **363**: 309-310, 1993.

HATTORI R, HAMILTON KK, MCEVER RP, SIMS PJ. Complement proteins C5-9 induce secretion of high molecular weight multimers of endothelial von Willebrand Factor and translocation of granule membrane protein GMP-140 to the cell surface. *J Biol Chem* **264**: 9053-9060, 1989.

HAWES BE, VAN BIESEN T, KOCH WJ, LUTTRELL LM, LEFKOWITZ RJ. Distinct pathways of G_i and G_q-mediated mitogen-activated protein kinase activation. *J Biol Chem* **270**: 17148-17153, 1995.

HAWES BE, LUTTRELL LM, VAN BIESEN T, LEFKOWITZ RJ. Phosphatidylinositol 3-kinase is an early intermediate in the Gβγ-mediated mitogen-activated protein kinase signalling pathway. *J Biol Chem* **271**: 12133-12136, 1996.

HEEMSKERK JWM, SAGE SO. Calcium signalling in platelets and other cells. *Platelets* **5**: 295-316, 1994.

HENN V, SLUPSKY JR, GRÄFE M, ANAGNOSTOPOULOS I, FÖRSTER R, MÜLLER-BERGHAUS G, KROCZEK RA. CD40 ligand on activated platelets triggers an inflammatory reaction of endothelial cells. *Nature* **391**: 591-594, 1998.

HOFFMAN M, MONROE DM, ROBERTS HR. Coagulation factor IXa binding to activated platelets and platelet-derived microparticles: a flow cytometric study. *Thromb Haemost* **68**: 74-78, 1992.

HOLME R, SIXMA JJ, MURER EH, HOVIG T. Demonstration of platelet fibrinogen secretion via the surface connecting system. *Thromb Res* **3**: 347-356, 1973.

HOLME R, SIXMA JJ, MURER EH, HOVIG T. Errata: Demonstration of platelet fibrinogen secretion via the surface connecting system. *Thromb Res* **4**: 377-382, 1974.

HOLMSEN H, DAY HJ. The selectivity of the thrombin-induced platelet release reaction: subcellular localization of released and retained constituents. *J Clin Med* **75**: 840-855, 1970.

HOLMSEN H, SETKOWSKY CA, DAY HJ. Effects of antimycin and 2-deoxyglucose on adenine nucleotides in human platelets. Role of metabolic adenosine trisphosphate in primary aggregation, secondary aggregation and shape change in platelets. *Biochem J* **144**: 385-396, 1974.

HOLMSEN H. Classification and possible mechanisms of action of drugs that inhibit platelet aggregation. *Ser Haematol* **8**: 50-80, 1976.

HOLMSEN H, WEISS HJ. Secretable storage pools in platelets. *Annu Rev Med* **30**: 119-134, 1979.

HOLMSEN H, KAPLAN KL, DANGELMAIER CA. Differential energy requirements for platelet responses: a simultaneous study of aggregation, three secretory processes, arachidonate liberation, phosphatidylinositol breakdown and phosphatidate production. *Biochem J* **208**: 9-18, 1982.

HOLMSEN H. "Cyclic AMP-dependent protein kinases and protein kinase C." In: *Platelet responses and Metabolism. Response-Metabolism Relationships, Vol III.* Ed: H Holmsen. CRC Press (Boca Raton, FL), pp 51-68, 1987.

HOLMSEN H, TYSNES OB, VERHOEVEN AJM. "Polyphosphoinositide metabolism in resting and stimulated platelets." In: *The platelet amine storage granule.* Eds. KM Meyers and CD Barnes. CRC Press (Boca Raton), pp 97-116, 1992.

HORAK ID, CORCORAN ML, THOMPSON PA, WAHL LM, BOLEN JB. Expression of $p60^{fyn}$ in human platelets. *Oncogene* **5**: 597-602, 1990.

HORDIJK PL, VERLAAN I, VAN CORVEN EJ, MOOLENAAR WH. Protein tyrosine phosphorylation induced by lysophosphatidic acid in rat-1 fibroblasts. Evidence that phosphorylation of MAP kinase is mediated by the G_i-$p21^{ras}$ pathway. *J Biol Chem* **269**: 645-651, 1994.

HORNE WC, NORMAN NE, SCHWARTZ DB, SIMONS ER. Changes in cytoplasmic pH and membrane potential in thrombin-stimulated human platelets. *Eur J Biochem* **120**: 295-302, 1981.

HOULE JJ, LEDDY JP, ROSENFELD SI. Secretion of the terminal complement proteins, C5-C9, by human platelets. *Clin Immunol Immunopathol* **50**: 385-393, 1989.

HOUSTON DS, SHEPHERD JT, VANHOUETTE PM. Aggregating human platelets cause direct contraction and endothelium-dependent relaxation of isolated canine coronary arteries: role of serotonin, thromboxane A_2 and adenine nucleotides. *J Clin Invest* **78**: 539-544, 1986.

HSU-LIN S, BERMAN CL, FURIE BC, AUGUST D, FURIE B. A platelet membrane protein expressed during platelet activation and secretion. Studies using a mono-clonal antibody specific for thrombin-activated platelets. *J Biol Chem* **259**: 9121-9126, 1984.

HUANG M-M, BOLEN JB, BARNWELL JW, SHATTIL SH, BRUGGE JS. Membrane glycoprotein IV (CD36) is physically associated with the Fyn, Lyn and Yes protein tyrosine kinases in human platelets. *Proc Natl Acad Sci USA* **88**: 7844-7848, 1991.

HUANG M-M, LIPFERT L, CUNNINGHAM M, BRUGGE JS, GINSBERG MH, SHATTIL SH. Adhesive ligand binding to integrin alpha IIb beta 3 stimulates tyrosine phosphorylation of novel protein substrates before phosphorylation of pp125FAK. *J Cell Biol* 122: 473-483, 1993.

HUNG SC, GHALI NI, VENTON DL, LE BRETON GC. Specific binding of the thromboxane A_2 antagonist 13-azaprostanoic acid to human platelet membranes. *Biochem Biophys Acta* **728**: 171-178, 1983.

HÜTTEMAN E, UKENA D, LENSCHOW V, SCHWABE U. Adenosine receptors in human platelets. *Naunyn Schmieedebergs Arch Pharmacol* **325**: 226-233, 1984.

IMAI A, HATTORI H, TAKAHASHI M, NOZAWA Y. Evidence that cyclic AMP may regulate Ca^{2+}-mobilization and phospholipases in thrombin-stimulated human platelets. *Biochem Biophys Res Commun* **112**: 693-700; 1983.

INABA K, UMEDA Y, YAMANA Y, URAKAMI M, NADA M. Characterization of human platelet vasopressin-induced platelet aggregation and vasopressin binding to platelets. *Clin Endocrinol* **29**: 377-386, 1988.

INAGAKI M, KAWAMOTO S, HIDAKA H. Serotonin secretion from human platelets may be modified by Ca^{2+}-activated, phospholipid-dependent myosin phosphorylation. *J Biol Chem* **259**: 14321-14323, 1984.

IÑARREA P, GOMEZ-CAMBRONERO J, NIETO M, SANCHEZ CRESPO M. Characteristics of the binding of platelet-activating factor to platelets of different animal species. *Eur J Pharmacol* **105**: 309-315, 1984.

INAZU T, TANIGUCHI T, YANAGI S, YAMAMURA H. Protein-tyrosine phosphorylation and aggregation of intact human platelets by vanadate with H_2O_2. *Biochem Biophys Res Commun* **170**: 259-263, 1990.

INGLESEJ, KOCH WJ, TOUHARA K, LEFKOWITZ RJ. G beta gamma interactions with PH domains and Ras-MAPK signaling pathways. *Trends Biochem Sci* 20: 151-156, 1995.

INGLEY E, HEMMINGS BA. Pleckstrin homology (PH) domains in signal transduction. *J Cell Biochem* **56**: 436-443, 1994.

ISENBERG WM, BAINTON DF. "Megacarocyte and platelet structure." In: *Hematology: basic principles and practice*, 2nd ed. Ed: R Hoffman, Churchill Livingstone (New York), pp 1516, 1995.

ISRAELS SJ, GERRARD JM, JACQUES YV, MCNICOL A, CHAM B, NISHIBORI M, BAINTON DF. Platelet dense granule membranes contain both granulophysin and P-selectin (GMP-140). *Blood* **80**: 143-152, 1992.

IWAMOTO S-I, KAWASAKI T, KAMBAYASHI J-I, ARIYOSHI H, MONDEN M. Platelet microparticles: A carrier of platelet-activating factor? *Biochem Biophys Res Commun* **218**: 940-944, 1996.

JANMEY PA, STOSSEL TP. Modulation of gelsolin function by phosphatidylinositol 4,5-bisphosphate. *Nature* **325**: 362-364, 1987.

JANMEY PA, MATSUDAIRA PT. Functional comparison of villin and gelsolin. *J Biol Chem* **263**: 16738-16743, 1988.

JENSEN BO, HOLMSEN H. Nitric oxide (NO)-platelet interactions: inhibition is independent of the prostanoid and ADP pathways. *Platelets* **6**: 83-90, 1995.

JIN J, DANIEL JL, KUNAPULI SP. Molecular basis for ADP-induced platelet activation. II. The P2Y1 receptor mediates ADP-induced intracellular calcium mobilization and shape change in platelets. *J Biol Chem* **273**, 2030-2034, 1998.

JOHANSSON JS, NIED LE, HAYNES DH. Cyclic AMP stimulates Ca(2+)-ATPase-mediated Ca^{2+} extrusion from human platelets. *Biochim Biophys Acta* **1105**: 19-28, 1992.

JONES TLZ, SIMONDS WF, MERENDINO JJ, BRANN MR, SPIEGEL AM. Myristoylation of an inhibitory GTP-binding protein a subunit is essential for its membrane attachment. *Proc Natl Acad Sci* **87**: 568-572, 1990.

JUNGI TW, SPÜCHER MO, NYDEGGER UE, BARANDUN S. Platelet-leukocyte interaction: selective binding of thrombin-stimulated platelets to human monocytes, polymorphonuclear leukocytes and related cell lines. *Blood* **647**: 629-636, 1986.

JY W, HORSTMAN L, WANG F, ARCE M, MAO WW, AHN YS. Platelet factor 3 in plasma fractions: Its relation to microparticle size and thrombosis. *Thromb Res* **80**: 471-482, 1995.

KAHN RA, DER CJ, BOKOCH GM. The ras superfamily of GTP-binding proteins: Guidelines on nomenclature. *FASEB J* **6**: 2512-1513, 1992.

KAIBUCHI K, TAKAI Y, SAWAMURA M. Synergistic functions of protein phosphorylation and calcium mobilization in platelet activation. *J Biol Chem* **258**: 6701-6709, 1983.

KANE WH, DAVIE EW. Blood coagulation factor V and VIII: Structural and functional similarities and their relationship to hemorrhagic and thrombotic disorders. *Blood* **71**: 539-555, 1988.

KANOH H, BANNO Y, HIRATA M, NOZAWA Y. Partial purification and characterization of phosphatidylinositol kinases from human platelets. *Biochim Biophys Acta* **1046**: 120-126, 1990.

KAPLAN DR, CHAO FC, STILES CD, ANTONIADES HN, SCHER CD. Platelet alpha granules contain a growth factor for fibroblasts. *Blood* **53**: 1043-1052, 1979.

KAPLAN KL, BROEKMAN MJ, CHERNOFF A, LESZNIK GR, DRILLINGS M. Platelet alpha-granule proteins - studies on release and subcellular localization. *Blood* **53**: 604-618, 1979.

KARNIGUIAN A, ZAHRAOUI A, TAVITIAN A. Identification of small GTP-binding rab proteins in human platelets: thrombin-induced phosphorylation of rab3B, rab6 and rab8 proteins. *Proc Natl Acad Sci USA* **90**: 7647-7651, 1993.

KAZIRO Y, ITOH H, KOZASA T, NAKAFUKU M, SATOH T. Structure and function of signal-transducing GTP-binding proteins. *Annu Rev Biochem* **60**: 349-400, 1991.

KEELY P-J, PARISE L-V. The alpha-2-beta-1 integrin is a necessary co-receptor for collagen-induced activation of Syk and the subsequent phosphorylation of phospholipase C-gamma-2 in platelets. *J Biol Chem* **271**: 26668-26676, 1996.

KENNEY DM, DAVIS AE. Association of alternative complement pathway components with human blood platelets: secretion and localization of factor D and B1H globulin. *Clin Immunol Immunopathol* **21**: 351-363, 1981.

KIM D, LEWIS DL, GRAZIADEI L. G-protein βγ-subunits activate the cardiac muscarinic K^+ channel via phospholipase A_2. *Nature* **337**: 557-559, 1989.

KINOSHITA T, MEDOF ME, SILBER R, NUSSENZWEIG V. Distribution of decay-accelerating factor in the peripheral blood of normal individuals and patients with paroxysmal nocturnal hemoglobinuria. *J Exp Med* **162**: 75-92, 1985.

KITCHENS CS, NEWCOMB TF. Factor XIII. *Medicine (Baltimore)* **58**: 413, 1979.

KLOPROGGE E, AKKERMAN JWN. Binding kinetics of PAF-acether (1-O-alkyl-s-acetyl-*sn*-glycerol-3-phosphocholine) to intact platelets. *Biochem J* **223**: 901-909, 1985.

KLOPROGGE E, MOMMERSTEG M, AKKERMAN JWN. Kinetics of platelet-activating factor 1-O-alkyl-2-acetyl-*sn*-glycerol-3-phosphocholine-induced fibrinogen binding to human platelets. *J Biol Chem* **261**: 11071-11076, 1986.

KNIGHT DE, SCRUTTON MC. Cyclic nucleotides control a system which regulates Ca^{2+} sensitivity of platelet secretion. *Nature* **309**: 66-68, 1984.

KOCH CA, ANDERSON D, MORAN MF, ELLIS C, PAWSON T. SH2 and SH3 domains. Elements that control interactions of cytoplasmic signalling proteins. *Science* **252**: 668-674, 1991.

KOERNER TA, CUNNINGHAM MT, ZHANG DS. The role of membrane lipid in the platelet storage lesion. *Blood Cells* **18**: 481-97; discussion 498-500, 1992.

KOMETANI M, SATO T, FUJI T. Platelet cytoskeletal components involved in shape change and secretion. *Thromb Res* **41**: 801-810, 1986.

KRAMER RM, ROBERTS EF, HYSLOP PA, UTTERBACK BG, HUI KY, JAKUBOWSKI JA. Differential activation of cytosolic phospholipase A_2 ($cPLA_2$) by thrombin and thrombin receptor agonist peptide in human platelets. *J Biol Chem* **270**: 14816-14823, 1995.

KUCERA GL, RITTENHOUSE SE. Human platelets form 3-phosphorylated phosphoinositides in response to_α-thrombin, U 46619 or GTPγS. *J Biol Chem* **265**: 5345-5348, 1990.

LANZA F, BERETZ A, STIERLE A, HANAU D, KUBINA M, CAZENAVE JP. Epinephrine potentates human platelet activation but is not an aggregating agent. *Am J Physiol* **255**: 1276-1288, 1988.

LAPETINA EG, REEP B, GANONG BR, BELL RM. Exogenous sn-1,2-diacylglycerols containing saturated fatty acids function as bioregulators of protein kinase C in human platelets. *J Biol Chem* **260**: 1358-1365, 1985.

LAPETINA EG, REEP B, CHANG KJ. Treatment of human platelets with trypsin, thrombin or collagen inhibits the pertussis toxin-induced ADP-ribosylation of a 41-kDa protein. *Proc Natl Acad Sci USA* **83**: 5880-5883, 1986a.

LAPETINA EG. Incorporation of synthetic 1,2-diacylglycerol into platelet phosphatidylinositol is increased by cyclic AMP. *FEBS Lett* **195**: 111-114, 1986b.

LAPETINA EG, LACAL JC, REEP BR, MOLINA Y, VEDIA L. A *ras*-related protein is phosphorylated and translocated by agonists that increase cAMP levels in human platelets. *Proc Natl Acad Sci USA* **86**: 3131-3134, 1989.

LASSING I, LINDERG U. Specific interaction between phosphatidylinositol 4,5-bisphosphate and profilactin. *Nature* **314**: 472-474, 1985.

LASSING I, LINDERG U. Evidence that the phosphatidylinositol cycle is linked to cell motility. *Exp Cell Res* **174**: 1-15, 1988.

LASSING I, LINDERG U. Polyphosphoinositide synthesis in platelets stimulated with low concentrations of thrombin is enhanced before the activation of phospholipase C. *FEBS Lett* **262**: 231-233, 1990.

LEBOWITZ EA, COOKE R. Contractile properties of actomyosin from human blood platelets. *J Biol Chem* **253**: 5443-.5447, 1978.

LEREA KM, GLOMSET JA, KREBS EG. Agents that elevate cAMP levels in platelets decrease thrombin binding. *J Biol Chem* **262**: 282-288; 1987.

LEV S, MORENO H, MARTINEZ R, CANOLL P, PELES E, MUSACCHIO JM, PLOWMAN GD, RUDY B, SCHLESSINGER J. Protein tyrosine kinase PYK2 involved in Ca^{2+}-induced regulation of ion channel and MAP kinase functions. *Nature* **376**: 739-744, 1995.

LI RY, GAITS F, RAGAB A, RAGAB-THOMAS JMF, CHAP H. Translocation of an SH2-containing protein tyrosine phosphatase (SH-PTP1) to the cytoskeleton of thrombin-activated platelets. *FEBS Lett* **343**: 89-93, 1994.

LIN AH, MORTON DR, GORMAN RR. Acetyl glyceryl ether phosphorylcholine stimulates leukotriene B_4 synthesis in human polymorphonuclear leukocytes. *J Clin Invest* **70**: 1058-1065, 1982.

LINDER ME, MIDDELTON P, HEPLER JR, TAUSSIG R, GILMAN AG, MUMBY SM. Lipid modification of G proteins: α subunits are palmitoylated. *Proc Natl Acad Sci USA* **90**: 3675-3679, 1993.

LIPFERT L, HAIMOVICH B, SCHALLER MD, COBB BS, PARSONS JT, BRUGGE JS. Integrin-dependent phosphorylation and activation of the protein tyrosine kinase $pp125^{FAK}$ in platelets. *J Cell Biol* **119**: 905-912, 1992.

LOGOTHETIS DE, KURACHI Y, GALPER J, NEER EJ, CLAPHAM DE. The $\beta\gamma$ subunits of GTP-binding proteins activate the muscarinic K^+ channel in heart. *Nature* **325**: 321-326, 1987.

LOPEZ-ILASACA M, CRESPO P, PELLICI PG, GUTKIND JS, WETZKER R. Linkage of G protein-coupled receptors to the MAPK signaling pathway through PI 3-kinase gamma. *Science* **275**: 394-397, 1997.

LÓPEZ JA, DONG JF. Structure and function of the glycoprotein Ib-IX-V complex. *Curr Opinion Hematol* **4**: 323-329, 1997.

LOWY DR, WILLUMSEN BM. Function and regulation of ras. *Annu rev Biochem* **62**: 851-891, 1993.

LUTTRELL LM, HAWES BE, VAN BIESEN T, LUTTRELL DK, LANSING TJ, LEFKOWITZ RJ. Role of c-Src tyrosine kinase in G protein-coupled receptor- and $G_{\beta\gamma}$ subunit-mediated activation of mitogen-activated protein kinases. *J Biol Chem* **271**: 19443-19450, 1996.

LYNCH JM, HENSON PM. The intracellular retention of newly synthesized platelet-activating factor. *J Immunol* **137**: 2653-2661, 1986.

MACINTYRE DE, ARMSTRONG RA. "Agonists and receptors: prostaglandins and thromboxanes." In: *Platelet Responses and Metabolism.* Receptors and Metabolism, Vol II. Ed. H Holmsen, CRC Press (Boca Raton FL), pp 93-103, 1987

MACKENZIE AB, MAHAUT-SMITH MP, SAGE SO. Activation of receptor-operated cation channels via P_{2X1} not P_{2T} purinoceptors in human platelets. *J Biol Chem* **271**: 2879-2881, 1996.

MACLOUF J, FRUTEAU DE LACLOS B, BORGEAT P. Stimulation of leukotriene biosynthesis in human blood leukocytes by platelet-derived 12-hydroperoxyeicosatetraenoic acid. *Proc Natl Acad Sci USA* **79**: 6042-6046, 1982.

MACLOUF JA, MURPHY RC. Transcellular metabolism of neutrophil-derived leukotriene A_4 by human platelets: a potential source for leukotriene C_4. *J Biol Chem* **263**: 174-181, 1988.

MADSHUS IH. Regulation of intracellular pH in eukaryotic cells. *Biochem J* **250**: 1-8, 1988.

MARCUS AJ, BROEKMAN MJ, SAFIER LB, ULLMAN HL, ISLAM N, SHERHAN CN, RUTHERFORD LE, KORCHAK HM, WEISSMANN G. Formation of leukotrienes and other hydroxy acids during platelet-neutrophil interactions in vitro. *Biochem Biophys Res Commun* **109**: 130-137, 1982.

MARCUS AJ, SAFIER LB, ULLMAN HL. 12S-20-dihydroxyeicosatetraenoic acid: a new eicosanoid synthesized by neutrophils from 12S-hydroxyeicosatetraenoic acid produced by thrombin- or collagen-stimulated platelets. *Proc Natl Acad Sci USA* **81**: 903-907, 1984.

MARCUS AJ, SAFIER LB, ULLMAN HL. Platelet neutrophil interaction. *J Biol Chem* **263**: 2223-2229, 1988.

MARCUS AJ, SAFIER LB, HAJJAR KA. Inhibition of platelet function by an aspirin-insensitive endothelial cell ADPase: thromboregulation by endothelial cells. *J Clin Invest* **88**: 1690-1696, 1991.

MARGOLIS B, HU P, KATZAV S, LI W, OLIVER JM, ULLRICH A, WEISS A, SCHLESSINGER J. Tyrosine phosphorylation of vav proto-oncogene product containing SH2 domain and transcription factor motifs. *Nature* **356**: 71-74, 1992.

MARKWARDT F, HOFFMAN A. Effects of papaverine on cyclic AMP phosphodiesterase in human platelets. *Biochem Pharmacol* **19**: 2519, 2520, 1970.

MASON RG, ZUCKER WH, SHIMODA BA, CHUANG HY, KINGDON HS, CLARK HG. Study of the reactions of blood with artificial surfaces. Use of the thrombogenerator. *Lab Invest* **31**: 143-155, 1974.

MATSUMOTO T, BEACH D. Premature initiation of mitosis in yeast lacking RCC1 or an interacting GTPase. *Cell* **66**, 347-360, 1991.

MAUCO G, DANGELMAIER CA, SMITH JB. Inositol lipids, phosphatidate and diacylglycerol share stearoylarachidonoylglycerol as a common backbone in thrombin-stimulated human platelets. *Biochem J* **244**: 933-940, 1984.

MAUCO G, DAJEANS P, CHAP H, DOUSTE-BLAZY L. Subcellular localization of inositol lipids in blood platelets as deduced from the use of labeled precursors. *Biochem J* **244**: 757-761, 1987.

MAURICE DH, HASLAM RJ. Molecular basis of the synergistic inhibition of platelet function by nitrovasodilators and activators of adenylate cyclase: inhibition of cyclic AMP breakdown by cyclic GMP. *Mol Pharmacol* **37**: 671-681, 1991.

MARKWARDT F, HOFFMAN A. Effects of papaverine on cyclic AMP phosphodiesterase in human platelets. *Biochem Pharmacol* **19**: 2519-2520, 1970.

MAYERBJ, REN R, CLARK KL, BALTIMORE DL. A putative modular domain present in diverse signalling proteins. *Cell* **73**: 629-630, 1993.

MCEVER RP, MARTIN MN. A monoclonal antibody to a membrane glycoprotein binds only to activated platelets. *J Biol Chem* **259**: 9799-9804, 1984.

MCKEAN ML, SMITH JB, SILVER MJ. Formation of lysophosphatidylcholine by human platelets in response to thrombin. Support for the phospholipase A_2 pathway for the liberation of arachidonic acid. *J Biol Chem* **156**: 1522-1524, 1981.

MCLEAN JR, MAXWELL RE, HERTLER D. Fibrinogen and adenosine diphosphate-induced aggregation of platelets. *Nature* **202**: 605-606, 1964.

MCMANAMA GP, JOHNSON PC, SALZMAN EW. "In vitro platelet responses: adhesion to artificial surfaces." In: *Platelet Responses and Metabolism, Vol I.* Ed: H Holmsen. CRC Press (Boca Raton, FL), pp 63-79, 1986.

MENCHE D, ISRAEL A, KARPATKIN K. Platelets and microtubules. Effect of colchicine and D_2O on platelet aggregation and release induced by ionophore A23187. *J Clin Invest* **66**: 284-291, 1980.

METZELAAR MJ, HEIJNEN HF, SIXMA JJ, NIEUWENHUIS HK. Identification of a 33-Kd protein associated with the alpha-granule membrane (GMP-33) that is expressed on the surface of activated platelets. *Blood* **79**: 372-379, 1992.

MILLER OV, JOHNSON RA, GORMAN RR. Inhibition of PGE_1-stimulated cyclic AMP accumulation by thromboxane A_2. *Prostaglandins* **13**: 599-609, 1977.

MILLER OV, GORMAN RR. Evidence for distinct prostaglandin I_2 and D_2 receptors in human platelets. *J Pharm Exp Ther* **210**: 134-140, 1979.

MILLIGAN G, MULLANEY I, MCCALLUM JF. Distribution and relative levels of expression of the phosphoinositidase-C-linked G-proteins $G_{q\alpha}$ and $G_{11\alpha}$: Absence of $G_{11\alpha}$ in human platelets and haemopoietically derived cell lines. *Biochem Biophys Acta Mol Cell Res* **1179**: 208-212, 1993.

MILLIGAN G, PARENTI M, MAGEE AI. The dynamic role of palmitoylation in signal transduction. *Trends Biochem Sci* **20**: 181-186, 1995.

MILLS DCB, ROBB IA, ROBERTS CCK. The release of nucleotides, 5-hydroxytryptamine and enzymes from human blood platelets during aggregation. *J Physiol* **195**: 715-729, 1968.

MILLS DCB, SMITH JB. The influence on platelet aggregation of drugs that effect the accumulation of adenosine 3',5'-cyclic monophosphate in platelets. *Biochem J* **121**: 185-196, 1971.

MILLS DCB, SMITH JB. The control of platelet responsiveness by agents that influence cyclic AMP levels. *Ann NY Acad Sci* **201**: 391-399, 1972.

MILLS DCB. Changes in the adenylate energy charge in human blood platelets induced by adenosine diphosphate. *Nature New Biol* **243**: 220, 1973.

MILLS DCB. "The mechanism of action of antiplatelet drugs." In: *Hemostasis and thrombosis.* Eds: RW Colman, J Hirsh, VJ Marder, EW Salzman. JB Lippincott (Philadelphia) pp 1058-1067, 1982.

MIYAKAWA Y, ODA A, DRUKER BJ, OZAKI K, HANDA M, OHASHI H, IKEDA Y. Thrombopoietin and thrombin induce tyrosine phosphorylation of Vav in human blood platelets. *Blood* **89**: 2789-2798, 1997.

MIYAZONO K, TAKAKU F. Platelet-derived growth factors. *Blood Rev* **3**: 269-276, 1989.

MONACO ME, GERSHENGORN MC. Subcellular organization of receptor-mediated phosphoinositide turnover. *Endoocrine Rev* **13**: 707-718, 1992.

MOODIE SA, WILLUMSEN BM, WEBER MJ, WOLFMAN A. Complexes of Ras-GTP with Raf-1 and mitogen-activated protein kinase. *Science* **260**: 1658-1661, 1993.

MOORE MS, BLOBEL G. The GTP binding protein ran/TC4 is required for protein import into the nucleus. *Nature* **365**: 661-663, 1993.

MORII N, TERU-UCHI T, TOMINAGA T, KUMAGI N, KOZAKI S, USHIKUBI F, NARUMIYA S. A Rho gene product in human blood platelets. II. Effects of the ADP-ribosylation by botulinum C3 ADP-ribosyltransferase on platelet aggregation. *J Biol Chem* **267**: 20921-20926, 1992.

MOROI M, JUNG SM. Platelet receptors for collagen. *Thromb Haemost* **78**: 439-444, 1997.

MUELLER HW, PRITZKER CR, KUBIK A, DEYKIN D. Characterization of phospholipase A_2 secretion from human platelets. *Thromb Res* **72**: 519-530, 1993.

MUSACCHIO A, GIBSON T, RICE P, THOMPSON J, SARASTE M. The PH domain: a common piece in the structural patchwork of signalling proteins. *Trends Biochem Sci* **18**: 343-348, 1993.

NAHAS N, PLANTAVID M, MAUCO G, CHAP H. Association of phosphatidylinositol kinase and phosphatidylinositol 4-phosphate kinase activities with the human platelets. *FEBS Lett* **246**: 30-34, 1989.

NAKA M, NISHIKAWA M, ADELSTEIN RS, HIDAKA H. Phorbol ester-induced activation of human platelets is associated with protein kinase C phosphorylation of myosin light chains. *Nature* **306**: 490-492, 1983.

NAKASHIMA S, TOHMATSU T, HATTORI H, OKANO Y, NOZAWA Y. Inhibitory action of cyclic GMP on secretion, polyphosphoinositide hydrolysis and calcium mobilization in thrombin-stimulated human platelets. *Biochem Biophys Res Commun* **135**: 1099-1104, 1986.

NEER EJ. Heterotrimeric G proteins: organizers of transmembrane signals. *Cell* **80**: 249-257, 1995.

NEMOTO Y, NAMBA T, TERU-UCHI T, USHIKUBI F, MORII N, NARUMIYA S. A *rho* gene product in human blood platelets. I. Identification of the platelet substrate for botulinum C3 ADP-ribosyltransferase as *rhoA* protein. *J Biol Chem* **267**: 20916-20920, 1992.

NICHOLSON-WELLER A, SPICER DB, AUSTEN KF. Deficiency of the complement regulatory protein, "decay-accelerating factor," on membranes of granulocytes, monocytes, and platelets in paroxysmal nocturnal hemoglobinuria. *N Engl J Med* 312: 1091-1097, 1985a.

NICHOLSON-WELLER A, MARCH JP, ROSEN CE, SPICER DB, AUSTEN KF. Surface membrane expression by human blood leukocytes and platelets of decay-accelerating factor, a regulatory protein of the complement system. *Blood* **65**: 1237-1244, 1985b.

NIEUWENHUIS HK, VAN OOSTERHOUT JJG, ROZEMULLER E, VAN IWAARDEN F, SIXMA JJ. Studies with a monoclonal antibody against activated platelets: Evidence that a secreted 53,000-molecular weight lysosome-like granule protein is exposed on the surface of activated platelets in the circulation. *Blood* **70**: 838-845, 1987.

NIEUWLAND R, WIJBURG OLC, VAN WILLIGEN G, AKKERMAN JWN. α_{2A}-adrenergic receptors activate protein kinase C in human platelets via a pertussis toxin-sensitive G-protein. *FEBS Lett* 339: 79-83, 1994.

NISHIDA E, GOTOH Y. The MAP kinase cascade is essential for diverse signal transduction pathways. *Trends Biol Sci* **18**: 128-131, 1993.

NOLTE C, EIGENTHALER M, HORSTRUP K, HÖNIG-LIEDL P, WALTER U. Synergistic phosphorylation of the focal adhesion-associated vasodilator-stimulated phosphoprotein in intact human platelets in response to cGMP- and cAMP-elevating platelet inhibitors. *Biochem Pharmacol* **48**: 1569-1575, 1994.

NOVICK PJ, BRENNWALD P. Friends and family: the role of the rab GTPases in vesicular traffic. *Cell* **75**, 597-602, 1993.

NUNEZ D, CHARRIAUT-MARLANGUE C, BAREL M, BENVENISTE J, FRADE R. Activation of human platelets through gp140, the C3d/EBV receptor (CR2). *Eur J Immunol* **17**: 515-520, 1987.

NURDEN AT, NURDEN P. A review of the role of platelet membrane glycoproteins in the platelet-vessel wall interaction. *Baillière's Clinical Haematology* **6**: 653-690, 1993.

OHMORI T, KIKUCHI A, YAMAMOTO K, KIM K, TAKAI Y. Small molecular weight GTP binding proteins in human platelet membranes. Purification and characterization of a novel GTP-binding protein with a molecular weight of 22.000. *J Biol Chem* **264**: 1877-1881, 1989.

OHMSTEDE C-A, FARELL FX, REEP BR, CLEMETSON KJ, LAPETINA EG. Rap2B: a ras-related GTP-binding protein from platelets. *Proc Natl Acad Sci USA* **87**: 6527-6531, 1990.

O'ROURKE F, MATTHEWS E, FEINSTEIN MB. Purification and characterization of the human type 1 Ins(1,4,5)P_3 receptor from platelets and comparison with receptor subtypes in other normal and transformed blood cells. *Biochem J* **312**: 499-503, 1995.

PACKHAM MA, NISHIZAWA E, MUSTARD JF. Response of platelets to tissue injury. *Biochem Pharmacol* **17**: 171-184, 1968.

PAINTER RG, PRODOUZ KN, GAARDE W. Isolation of a subpopulation of glycoprotein IIbIIIa from platelet membranes that is bound to membrane actin. *J Cell Biol* **100**: 652-657, 1985.

PALMER RMJ, FERRIGE AG, MOCADA S. Nitric oxide release accounts for the biological activity of endothelium-derived relaxing factor. *Nature* **327**: 524-526, 1987.

PAPKOFF J, CHEN R-H, BLENIS J, FORSMAN J. p42 mitogen-activated protein kinase and p90 ribosomal S6 kinase are selectively phosphorylated and activated during thrombin-induced platelet activation and aggregation. *Mol Cell Biol* **14**: 463-472, 1994.

PEERSCHKE EIB, GHEBREHIWET B. Human blood platelets possess specific binding sites for C1q. *J Immunol* **138**: 1537-1541, 1986.

PEERSCHKE EIB, GHEBREHIEWET B. Identification and partial characterization of human platelet C1q binding sites. *J Immunol* **141**: 3505-3511, 1988.

PEERSCHKE EIB, REID KBM, GHEBREHIWET B. Identification of a novel 33-kDa C1q-binding site on human blood platelets. *J Immunol* **152**: 5896-5901, 1994.

PELECH SL, SANGHERA JS. MAP kinases: Charting the regulatory pathways. *Science* **257**: 1355-1356, 1992.

PLETSCHER A. The 5-hydroxytryptamine system of blood platelets: physiology and pathophysiology. *Int J Cardiol* **14**: 177-188, 1987.

POLAKIS PG, WEBER RF, NEVINS B, DIDSBURY JR, EVANS T, SNYDERMAN RJ. Identification of the *ral* and *rac1* gene products, low molecular mass GTP-binding proteins from human platelets. *J Biol Chem* **264**: 16383-16389, 1989.

POLLOCK WK, MACINTYRE DE. Desensitization and antagonism of vasopressin-induced phosphoinositide metabolism and elevation of cytosolic free calcium. *Biochem J* **234**: 67-73, 1986.

PRESCOTT SM, MAJERUS PW. The fatty acid composition of phosphatidylinositol from thrombin-stimulated human platelets. *J Biol Chem* **256**: 579-582, 1981.

PRIBLUDA V, ROTMAN A. Dynamics of membrane-cytoskeleton interactions in activated blood platelets. *Biochem J* **21**: 2825-2832, 1982.

PRONIN AN, GAUTAM N. Interaction between G-protein β and γ subunit types is selective. *Proc Natl Acad Sci USA* **89**: 6220-6224, 1992.

PUMIGLIA KM, LAU L-F, HUANG C-K, BURROUGHS S, FEINSTEIN MB. Activation of signal transduction in platelets by the tyrosine phosphatase inhibitor pervanadate (vanadyl hydroperoxide). *Biochem J* **286**: 441-449, 1992.

QUINTON TM, DEAN WL. Cyclic AMP-dependent phosphorylation of the inositol-1.4.5-trisphosphate receptor inhibits Ca^{2+} release from platelet membranes. *Biochem Biophys Commun* **184**: 893-899; 1992.

QUINTON TM, BROWN KD, DEAN WL. Inositol 1.4.5-trisphosphate-mediated Ca^{2+} release from platelet internal membranes is regulated by differential phosphorylation. *Biochemistry* **35**: 6865-6871; 1996.

RADOMSKI MW, PALMER RMJ, MONCADA S. The role of nitric oxide and cGMP in platelet adhesion to vascular endothelium. *Biochem Biophys Res Commun* **148**: 1482-1489, 1987.

RAJA S, AVRAHAM S, AVRAHAM H. Tyrosine phosphorylation of the novel protein-tyrosine kinase RAFTK during an early phase of platelet activation by an integrin glycoprotein IIb-IIIa-independent mechanism. *J Biol Chem* **272**: 10941-10947, 1997.

RASMUSSEN H, ISALES CM, CALLE R, THROCKMORTON D, ANDERSON M, GASALLA-HERRAIZ J, MCCARTHY R. Diacylglycerol production, Ca^{2+} influx, and protein kinase C activation in sustained cellular responses. *Endocrine Rev* **16**, 649-681, 1995.

RASMUSSEN UB, VOURET-CRAVIARI V, JALLAT S, SCHLESINGER Y, PAGÈS G, PAVIRANI A, LECOCQ J-P, POUSSÉGUR J, VAN OBBERGHEN-SCHILLING E. cDNA cloning and expression of a hamster α-thrombin receptor coupled to Ca^{2+} mobilization. *FEBS Lett* **288**: 123-128, 1991.

RAY K, KUNSCH C, BONNER LM, ROBISHAW JD. Isolation of cDNA clones encoding eight different human G protein γ subunits, including three novel forms designated the γ4, γ10 and γ11 subunits. *J Biol Chem* **270**: 21765-21771, 1995.

REINHARD M, HALBRÜGGE M, SCHEER U, WIEGAND C, JOCKUSCH BM, WALTER U. The 46/50 kDa phosphoprotein VASP purified from human platelets is a novel protein associated with actin filaments and focal contacts. *EMBO J* **11**: 2063-2070, 1992.

RIDLEY A, HALL A. The small GTP-binding protein rho regulates the assembly of focal adhesion and actin stress fibres in response to growth factors. *Cell* **70**: 389-399, 1992a.

RIDLEY AJ, PATERSON HF, JOHNSTON CL, DIEKMAN D, HALL A. The small GTP binding protein rac regulates growth-factor-induced membrane ruffling. *Cell* **70**: 401-410, 1992b.

RINK TJ, SAGE SO. Calcium signalling in human platelets. *Annu Rev Physiol* **52**: 431-349, 1990.

RITTENHOUSE SE, HORNE WC. Ionomycin can elevate intraplatelet Ca^{2+} and activate phospholipase A_2 without activating phospholipase C. *Biochem Biophys Res Commun* **123**: 393-397, 1984.

RITTENHOUSE-SIMMONS SE. Production of diglyceride from phosphatidylinositol in activated platelets. *J Clin Invest* **63**: 580-587, 1979.

RITTENHOUSE-SIMMONS S. Differential activation of platelet phospholipases by thrombin and ionophore A23187. *J Biol Chem* **256**: 4153-4155, 1981.

ROMANO M, SERHAN CN. Lipoxin generation by permeabilized human platelet. *Biochemistry* **31**: 8269-8277, 1992.

ROSENBERG S, STRACHER A, LUCAS RC. Isolation and characterization of actin and actin-binding protein from human platelets. *J Cell Biol* **91**: 201-211, 1981.

ROSENFELD SI, RYAN DH, LOONEY RJ, ANDERSON CL, ABRAHAM GN, LEDDY JP. Human Fcγ receptors: stable inter-donor variation in quantitative expression on platelets correlates with functional responses. *J Immunology* **138**: 2869-2873, 1987.

ROSS EM. Protein modification: Palmitoylation in G-protein signalling pathways. *Curr Biol* **5**: 107-109, 1995.

RUAN C, TOBELEM G, MCMICHAEL AJ, DROUET L, LEGRAND Y, DEGOS L, KIEFFER N, LEE H, CAEN JP. Monoclonal antibody to human glycoprotein I: II Effects on human platelet function. *Br J Haematol* **49**: 511-519, 1981.

RUGGERI ZM, ZIMMERMAN TS. von Willebrand factor and von Willebrand disease. *Blood* **70**: 895-904, 1987.

RYNINGEN A, HOLMSEN H. Thrombin *per se* does not induce tyrosine phosphorylation in human platelets as judged by Western blotting, while collagen does: the significance of synergistic, autocrine stimulation. *Biochem Biophys Res Commun*. In Press, 1998.

SALAMA SE, HASLAM RJ. Characterization of the protein kinase activities of human platelet supernatant and particulate fractions. *Biochem J* **218**: 285-294, 1984.

SALTZMAN AG, MORSE B, WHITMAN MM, IVANSCHENKO Y, JAJE M, FELDER S. Cloning of the human serotonin 5-HT2 and 5-HT1C receptor subtypes. *Biochem Biophys Res Commun* **181**: 1469-1478, 1991.

SALZMAN EW. Interrelation of prostaglandin endoperoxide PGG_2 and cyclic 3',5' adenosine monophosphate in human blood platelets. *Biochim Biophys Acta* **499**: 48-60, 1977.

SAUSSY DL JR, MAIS DE, BURCH RM, HALUSHKA PV. Identification of a putative thromboxane A_2/prostaglandin H_2 receptor in human platelet membranes, *J Biol Chem* **261**: 3025-3029, 1986.

SCHAFER AI, CRAWFORD DD, GIMBRONE MA. Unidirectional transfer of prostaglandin endoperoxides between platelets and endothelial cells. *J Clin Invest* **73**: 1105-1112, 1984.

SCHALLER MD, OTEY CA, HILDEBRAND JD, PARSONS JT. Focal adhesion kinase and paxillin bind to peptides mimicking β integrin cytoplasmic domains. *J Cell Biol* **130**: 1181-1187, 1995.

SCHLAEPFER DD, HANKS SK, HUNTER T, VAN DER GEER P. Integrin-mediated signal transduction linked to Ras pathway by GRB2 binding to focal adhesion kinase. *Nature* **372**: 786-791, 1994.

Schmaier AH, Smith PM, Colman RW. Platelet C1 inhibitor. A secreted alpha-granule protein. *J Clin Invest* **75**: 242-250, 1985.

Serafini T, Orci L, Amherdt M, Brunner M, Kahn R, Rothman JE. ADP-ribosylation factor is a subunit of the coat of golgi-derived COP-coated vesicles: a novel role for a GTP binding protein. *Cell* **67**: 239-253, 1991.

Seth SB, Colman RW. Regulatory and catalytic domains of platelet cAMP phosphodiesterases: target for drug design. *Semin Hematol* **32**: 110-119, 1995.

Seya T, Turner J, Atkinson JP. Purification and characterization of a membrane protein (gp45-70) which is a cofactor for cleavage of C3b and C4b. *J Exp Med* **163**: 837-855, 1986.

Shattil SJ, Hoxie JA, Cunningham M, Brass LF. Changes in the platelet membrane glycoprotein IIb/IIIa complex during platelet activation. *J Biol Chem* **260**: 11107-11114, 1985.

Shattil SJ, Brass LF. Induction of the fibrinogen receptor on human platelets by intracellular mediators. *J Biol Chem* **262**: 992-999, 1987.

Shattil SJ, Haimovich B, Cunningham M, Lipfert L, Parsons JT, Ginsberg MH, Brugge JS. Tyrosine phosphorylation of $pp125^{FAK}$ in platelets requires coordinated signalling through integrin and agonist receptors. *J Biol Chem* **269**: 14738-14745, 1994a.

Shattil SJ, Ginsberg MH, Brugge JS. Adhesive signaling in platelets. *Curr Opin Cell Biol* **6**: 695-704, 1994b.

Shaw G. Identification of novel pleckstrin homology (PH) domains provides a hypothesis for PH domain function. *Biochem Biophys Res Commun* **195**: 1145-1151, 1993.

Shuman MA, Greenberg CS. "Platelet regulation of thrombus formation." In: *Biochemistry of Platelets*. Eds. DR Phillips and MA Shuman. Academic (Orlando), pp 319, 1986.

Siegl AM, Smith JB, Silver MJ. Specific binding sites for prostaglandin D_2 on human platelets. Biochem *Biophys Res Commun* **90**: 291-296, 1979.

Siess W. Molecular mechanisms of platelet activation. *Physiol Rev* **69**: 58-178, 1989.

Siess W, Lapetina EG. Functional relationship between cyclic AMP-dependent protein phosphorylation and platelet inhibition. *Biochem J* **271**: 815-819; 1990.

Siffert W, Fox G, Munckenhoff K, Scheid P. Thrombin stimulates Na^+/H^+ exchange across the human platelet plasma membrane. *FEBS Lett* **172**: 272-274, 1984.

Siffert W, Siffert G, Scheid P. Activation of Na^+/H^+ exchange in human platelets stimulated by thrombin and a phorbol ester. *Biochem J* **241**: 301-303, 1987.

Siffert W. Regulation of platelet function by sodium-hydrogen exchange. *Cardiovasc Res* **29**: 160-166, 1995.

Simon MI, Strathmann MP, Gautam N. Diversity of G proteins in signal transduction. *Science* **252**: 802-08, 1991.

Sims PJ, Wiedmer T. Repolarization of the membrane potential of blood platelets after complement damage: evidence for a Ca^{2+}-dependent exocytotic elimination of C5b-9 pores. *Blood* **68**: 556-561, 1986.

Sims PJ, Faioni EM, Wiedmer T, Shattil SJ. Complement proteins C5b-9 cause release of membrane vesicles from the platelet surface that are enriched in the membrane receptor for coagulation factor Va and express prothrombinase activity. *J Biol Chem* **263**: 18205-18212, 1988.

Sims PJ, Rollins SA, Wiedmer T. Regulatory control of complement on blood platelets. Modulation of platelet procoagulant responses by a membrane inhibitor of the C5b-9 complex. *J Biol Chem* **264**: 19228-19235, 1989.

Smith JB, Mills DCB. Inhibition of adenosine 3',5' cyclic monophosphate phosphodiesterase. *Biochem J* **120**: 20, 1970.

Smrcka AV, Sternweis PC. Regulation of purified subtypes of phosphatidylinositol-specific phospholipase Cβ by G protein α and βγ subunits. *J Biol Chem* **268**: 9667-9674, 1993.

Steen VM, Holmsen H. Current aspects on human platelet activation and responses. *Eur J Haematol* **38**: 383-399, 1987.

STEEN VM, TYSNES OB, HOLMSEN H. Evidence for tight metabolic control of the receptor-activated polyphosphoinositide cycle in human platelets. *Biochem J* **263**: 621-624, 1989.

STEEN VM, AARBAKKE G, HOLMSEN H. The platelet-stimulating effect of adrenaline through alpha 2-adrenergic receptors requires simultaneous activation by a true stimulatory platelet agonist. Evidence that adrenaline *per se* does not induce human platelet activation in vitro. *Thromb Haemost* **70**: 506-513, 1993.

STENBERG PE, SHUMAN MA, LEVINE SP, BAINTON DF. Redistribution of alpha-granules and their contents in thrombin-stimulated platelets. *J Cell Biol* **98**: 748-60, 1984

STENBERG PE, MCEVER RP, SHUMAN MA, JACQUES YV, BAINTON DF. A platelet alpha-granule membrane protein (GMP140) is expressed on the plasma membrane after activation. *J Cell Biol* **101**: 880-886, 1985.

STOKOE D, MACDONALD SG, CADWALLADER K, SYMONS M, HANCOCK JF. Activation of Raf as a result of recruitment to the plasma membrane. *Science* **264**: 1463-1467, 1994.

STOYANOV B, VOLINIA S, HANCK T, RUBIO I, LOUBTCHENKOV M, MALEK D, STOYANOVA S, VANHAESEBROECK B, DHAND R, NURNBERG B, GIERSHIK P, SEEDORF K, HSUAN JJ, WATERFIELD MD. Cloning and characterization of a G protein-activated human phosphoinositide-3 kinase. *Science* **269**: 690-693, 1995.

SUGA K, KAMBAYASHI J, KAWASAKI T, SAKON M, MORI T. Phosphorylation of phosphoinositides in human platelets. *Thromb Res* **44**: 155-163, 1986.

SUZUKI S, ARGRAVES WS, ARAI H, LANGUINO LF, PIERSCHBACHER MD, RUOSLATHI E. Aminoacid sequence of the vitronectin receptor alpha subunit and comparative expression of adhesion receptor mRNAs. *J Biol Chem* **262**: 14050-14058, 1987.

SUZUKI T, BANNO Y, NOZAWA Y. Partial purification and characterization of two forms of phosphatidylinositol 4-phosphate 5-kinase from human platelet membrane. *Thromb Res* **64**: 45-56, 1991.

SUZUKI T, NAKASHIMA S, NOZAWA Y. Inhibition of phosphatidylinositol 4-phosphate 5-kinase by cyclic AMP in human platelets. *Platelets* **5**: 258-265, 1994.

TAKAHARA K, MURRAY R, FITZGERALD GA, FITZGERALD DJ. The response to thromboxane A_2 analogues in human platelets. *J Biol Chem* **265**: 6836-6844, 1990.

TAKAI Y, SASAKI T, TANAKA K, NAKANISHI H. Rho as a regulator of the cytoskeleton. *Trends In Biochem Sci* **20**: 227-230, 1995.

TAO J, JOHANSSON JS, HAYNES DH. Stimulation of dense tubular Ca^{2+} uptake in human platelets by cAMP. *Biochim Biophys Acta* **1105**: 29-39, 1992.

TANS G, ROSING J, THOMASSEN MC, HEEB MJ, ZWAAL RF, GRIFFIN JH. Comparison of anticoagulant and procoagulant activities of stimulated platelets and platelet-derived microparticles. *Blood* **77**, 2641-2648, 1991.

TEDDER TF, STEEBER DA, CHEN A, ENGEL P. The selectins: vascular adhesion molecules. *FASEB J* **9**: 866-873, 1995.

TEO M, MANSER E, LIM L. Identification and molecular cloning of a $p21^{cdc42/rac1}$-activated serin/threonine kinase that is rapidly activated by thrombin in platelets. *J Biol Chem* **270**: 26690-26697, 1995.

THOMASON PA, JAMES SR, CASEY PJ, DOWNES CP. A G-protein βγ-subunit-responsive phosphoinositide 3-kinase activity in human platelet cytosol. *J Biol Chem* **269**: 16525-16528, 1994.

TOLIAS KF, CANTLEY LC, CARPENTER CL. Rho family GTPases bind to phosphoinositide kinases. *J Biol Chem* **270**: 17656-17659, 1995.

TORTI M, SINGAGLIA F, RAMASCHI G, BALDUINI C. Platelet glycoprotein IIb-IIIa is associated with 21-kDa GTP-binding protein. *Biochim Biophys Acta* **1070**: 20-26, 1991.

TORTI M, LAPETINA EG. Role of rap1B and p21ras GTPase activating protein in the regulation of phospholipase C-g1 in human platelets. *Proc Natl Acad Sci USA* **89**:7796-77800, 1992.

TORTI M, LAPETINA EG. Structure and function of rap proteins in human platelets. *Thromb Haemost* **71**: 533-543, 1994.

TOUHARA K, INGLESE J, PITCHER JA, SHAW G, LEFKOWITZ RJ. Binding of G protein βγ-subunits to pleckstrin homology domains. *J Biol Chem* **269**: 10217-10220, 1994.

TOUHARA K, HAWES BE, VAN BIESEN T, LEFKOWITZ RJ. G protein βγ subunits stimulate phosphorylation of the Shc adapter protein. *Proc Natl Acad Sci USA* **92**: 9284-9287, 1995.

TRACY PB, MANN KG. "A model for assembly of coagulation factor complexes on cell surfaces: Prothrombin activation on platelets." In *Biochemistry of Platelets*. Eds: DR Phillips and MA Shuman, Academic (Orlando), pp 296, 1986.

TSCHOPP J, JENNE DE, HERTIG S, PREISSNER KT, MORGENSTERN H, SAPINO A-E, FRENCH L. Human megakaryocytes express clusterin and package it without apolipoprotein A-1 into α-granules. *Blood* **82**: 118-125, 1993.

TSUKUDA M, ASAOKA Y, SEKIGUCHI K, KIKKAWA U, NISHIZUKA Y. Properties of protein kinase C subspecies in human platelets. *Biochem Biophys Res Commun* **155**: 1387-1395, 1988.

TURITTO VT, WEISS HJ, BAUMBARTNER HS. Decreased platelet adhesion on vessel segments in von Willebrand's disease: A defect in initial platelet attachment. *J Lab Clin Med* **102**: 551-564, 1983.

TURITTO VT, WEISS HJ, ZIMMERMAN TS, SUSSMAN II. Factor VIII/von Willebrand factor in subendothelium mediates platelet adhesion. *Blood* **65**: 823-831, 1985.

TYERS M, RACHUBINSKI RA, STEWART MI, VARRICHIO AM, SHORR RG, HASLAM RJ, HARLEY CB. Molecular cloning and expression of the major protein kinase C substrate of platelets. *Nature* **333**: 470-473, 1988.

TYSNES OB, VERHOEVEN AJM, AARBAKKE GM, HOLMSEN H. Phosphoinositide metabolism in resting and thrombin-stimulated human platelets. *FEBS Lett* **218**: 68-72, 1987.

TYSNES OB, VERHOEVEN AJM, HOLMSEN H. Rates of production and consumption of phosphatidic acid upon thrombin stimulation of human platelets. *Eur J Biochem* **174**: 75-79, 1988.

TYSNES OB, STEEN VM, FRØLICH KW, HOLMSEN H. Evidence that chlorpromazine and prostaglandin E_1 but not neomycin interfere with the inositol phospholipid metabolism intact human platelets. *FEBS Lett* **264**: 33-36, 1990.

UEDA N, INIGUEZ-LLUHI JA, LEE E, SMRCKA AV, ROBISHAW JD, GILMAN AG. G protein βγ subunits. *J Biol Chem* **269**: 4388-4395, 1994.

VAN BIESEN T, HAWES BE, LUTTRELL DK, KRUEGER KM, TOUHARA K, PORFIRI E, SAKAUE M, LUTTRELL LM, LEFKOWITZ RJ. Receptor-tyrosine-kinase-and Gβγ-mediated MAP kinase activation by a common signalling pathway. *Nature* **376**: 781-784, 1995.

VAN CORVEN EJ, HORDIJK PL, MEDEMA RH, BOS JL, MOOLENAAR WH. Pertussis toxin-sensitive activation of $p21^{ras}$ by G protein-coupled receptor agonists in fibroblasts. *Proc Natl Acad Sci USA* **90**: 1257-1261, 1993.

VANDERWEI M, LUM DS, HASLAM RJ. Vasopressin inhibits the adenylate cyclase activity of human platelet particulate fraction through V_1-receptors. *FEBS Lett* **164**: 340-344, 1983.

VANE JR, MONCADA S. Prostacyclin. *Ciba Found Symp* **71**: 79-97, 1980.

VAN WILLIGEN G, AKKERMAN JWN. Protein kinase C and cyclic AMP regulate reversible exposure of binding sites for fibrinogen on the glycoprotein IIb/IIIa complex of human platelets. *Biochem J* **273**: 115-121, 1991.

VAN WILLIGEN G, DONATH J, LAPETINA EG, AKKERMAN JWN. Identification of α-subunits of trimeric GTP-binding proteins in human platelets by RT-PCR. *Biochem Biophys Res Commun* **214**: 254-262, 1995.

VASSBOTN FS, HAVNEN OK, HELDIN C-H, HOLMSEN H. Negative feedback regulation of human platelets via autocrine activation of the platelet-derived growth factor α-receptor. *J Biol Chem* **269**: 13874-13879, 1994.

VERHOEVEN AJM, MOMMERSTEG ME, AKKERMAN JWN. Quantification of energy consumption in platelets during thrombin-induced aggregation and secretion: tight coupling between platelet responses and the increment in energy consumption. *Biochem J* **221**: 777-787, 1984.

VERHOEVEN AJM, TYSNES OB, AARBAKKE GM, COOK CA, HOLMSEN H. Turnover of the phosphomonoester groups of polyphosphoinositol lipids in unstimulated human platelets. *Eur J Biochem* **166**: 3-9, 1987.

VICKERS JD, MUSTARD JF. The phosphoinositides exist in multiple pools in rabbit platelets. *Biochem J* **238**: 411-417, 1986.

VIK DP, FEARON DT. Cellular distribution of complement receptor type 4 (CR4): expression on human platelets. *J Immunol* **138**: 254-258, 1987.

VOYNO-YASENETSKAYA T, CONKLIN BR, GILBERT RL, HOOLEY R, BOURNE HR, BARBER DL. $G\alpha_{13}$ stimulates Na-H exchange. *J Biol Chem* **269**: 4721-4724, 1994.

VUT-KH, HUNG DT, WHEATON VI, COUGHLIN SR. Molecular cloning of a functional thrombin receptor reveals a novel proteolytic mechanism of receptor activation. *Cell* **64**: 1057-1068, 1991.

WAGNER DD, MARDER VJ. Biosynthesis of von Willebrand protein by human endothelial cells: Processing steps and their intracellular localization. *J Cell Biol* **99**: 2123-2130, 1984.

WALDMANN R, BAUER S, GÖBEL C, HOFMANN F, JACOBS KH, WALTER U. Demonstration of cGMP-dependent protein kinase and cGMP-dependent phosphorylation in cell-free extracts of platelets. *Eur J Biochem* **158**: 203-210, 1986.

WALDMANN R, NIEBERDING M, WALTER U. Vasodilator-stimulated protein phosphorylation in platelet is mediated by cAMP- and cGMP-dependent protein kinases. *Eur J Biochem* **167**: 441-448, 1987.

WANG F, NAIK UP, EHRLICH YH, FREYBERG Z, OSADA S, OHNO S, KUROKI T, SUZUKI K, KORNECKI E. A new protein kinase C, nPKCη', and nPKCθ are expressed in human platelets: Involvement of nPKCη' and nPKCθ in signal transduction stimulated by PAF. *Biochem Biophys Res Commun* **191**: 240-246, 1993.

WANG J, AUGER KR, JARVIS L, SHI Y, ROBERTS TM. Direct association of Grb2 with the p85 subunit of phosphatidylinositol 3-kinase. *J Biol Chem* **270**: 12774-12780, 1995.

WARE JA, HEISTAD DD. Platelet-endothelium interactions. *N Engl J Med* **328**: 628-635, 1993.

WEDEGAERTNER PB, CHU DH, WILSON PT, LEVIS MJ, BOURNE HR. Palmitoylation is required for signalling functions and membrane attachment of $G_{q\alpha}$ and $G_{s\alpha}$. *J Biol Chem* **268**: 25001-25008, 1993.

WEDEGAERTNER PB, BOURNE HR. Activation and depalmitoylation of $G_{s\alpha}$. *Cell* **70**: 1063-1070, 1995.

WEISS HJ, TSCHOPP TB, BAUMGARTNER HR, SUSSMAN II, JOHNSON MM, EGAN JJ. Decreased adhesion of giant (Bernard-Soulier) platelets to subendothelium. *Am J Med* **57**: 920-925, 1974.

WELCH GN, UPCHURCH GR, LOSCALZO J. Nitric oxide as a vascular modulator. *Blood Rev* **9**: 262-269, 1995.

WHITE JG. A search for the platelet secretory pathway using electron-dense tracers. *Am J Pathol* **58**: 31-49, 1970.

WHITE JG. Electron microscopic studies of platelet secretion. *Prog Hemost Thromb* **2**: 49, 1974.

WHITE JG, KRUMWIEDE M. Further studies of the secretory pathway in thrombin-stimulated human platelets. *Blood* **69**: 1196-1203, 1987a.

WHITE JG. Views at the platelet cytoskeleton at rest and at work. *Ann NY Acad Sci* **509**: 156-176, 1987b.

WHITE JG. "Platelet Ultrastructure." In. *Haemostasis and Thrombosis, 2nd. ed.* Eds: AL Bloom & DP Thomas. Churhill Livingstone (UK) pp 20-46, 1987c.

WIEDMER T, SIMS PJ. Effect of complement proteins C5b-9 on blood platelets. Evidence for reversible depolarization of membrane potential. *J Biol Chem* **260**: 8014-8019, 1985.

WIEDMER T, ESMON CT, SIMS PJ. Complement proteins C5b-9 stimulate procoagulant activity through platelet prothrombinase. *Blood* **68**: 875-880, 1986a.

WIEDMER T, ESMON CT, SIMS PJ. On the mechanism by which complement proteins C5b-9 increase platelet prothrombinase activity. *J Biol Chem* **261**: 14587-14592, 1986b.

WIEDMER T, ANDO B, SIMS PJ. Complement C5b-9-stimulated platelet secretion is associated with a Ca^{2+}-initiated activation of cellular protein kinases. *J Biol Chem* **262**: 13674-13681, 1987.

WITTE LD, KAPLAN KL, NOSSEL HL, LAGES BA, WEISS HJ, GOODMAN DS. Studies of the release from human platelets of the growth factor for cultured human smooth muscle cells. *Circ Res* **42**: 402-409, 1978.

WONG PY, WESTLUND P, HAMBERG M. 15-lipoxygenase in human platelets. *J Biol Chem* **260**: 9162-9171, 1985.

WRIGHT JH. The origin and nature of blood platelets. *Boston Med surg J* **154**: 643, 1906.

WRIGHT JH. The histogenesis of the blood platelets. *J Morphol* **21**: 203, 1910.

WU XW, LIAN EC..Binding properties and inhibition of platelet aggregation by a monoclonal antibody to CD31 (PECAM-1). *Arterioscler Thromb Vasc Biol* **17**: 3154-3158, 1997.

YADA Y, NAGAO S, OKANO Y, NOZAWA Y. Inhibition by cyclic AMP of guanine nucleotide-induced activation of phosphoinositide-specific phospholipase C in human platelets. *FEBS-Lett* **242**: 368-372, 1989.

YAMANISHI J, KAWAHARA Y, FUKUZAKI H. Effect of cyclic AMP on cytoplasmic free calcium in human platelets stimulated by thrombin: direct measurement with quin2. *Thromb Res* **32**: 183-188, 1983.

YANO Y, KAMBAYASHI J, SHIBA E, SAKON M, OIKI E, FUKUDA K, KAWASAKI T, MORI T. The role of protein phosphorylation and cytoskeletal reorganization in microparticle formation from the platelet plasma membrane. *Biochem J* **299**: 303-308, 1994.

YE ZS, BALTIMORE D. Binding of Vav to Grb2 through dimerization of Src homology 3 domains. *Proc Natl Acad Sci USA* **91**: 12629-12633, 1994.

YOON HS, HAJDUK PJ, PETROS AM, OLEJNICZAK ET, MEADOWS RP, FESIK SW. Solution structure of a pleckstrin-homology domain. *Nature* **369**: 672-675, 1994.

YOSHIDA K, DUBYAK K, NACHIMAS VT. Rapid effects on phorbol ester on platelet shape change, cytoskeleton and calcium transient. *FEBS Lett* **206**: 273-278, 1986.

ZAVOICO GB, FEINSTEIN MB. Cytoplasmic Ca^{2+} in platelets is controlled by cyclic AMP: Antagonism between stimulators and inhibitors of adenylate cyclase. *Biochem Biophys Res Commun* **120**: 579-585, 1984.

ZHANG J, KING WG, DILLON S, HALL A, FEIG L, RITTENHOUSE SE. Activation of platelet phosphatidylinositide 3-kinase requires the small GTP-binding protein Rho. *J Biol Chem* **268**: 22251-22254, 1993.

ZHANG J, BENOVIC JL, SUGAI M, WETZKER R, GOUT I, RITTENHOUSE SE. Sequestration of a G-protein βγ subunit or ADP-ribosylation of rho can inhibit thrombin-induced activation of platelet phosphoinositide 3-kinases. *J Biol Chem* **270**: 6589-6594, 1995.

ZHANG J, SHATTIL SJ, CUNNINGHAM MC, RITTENHOUSE SE. Phosphoinositide 3-kinase γ and p85/phosphoinositide 3-kinase in platelets. *J Biol Chem* **271**: 6265-5272, 1996.

ZIGMOND SH. Signal transduction and actin organization. *Curr Opin Cell Biol* **8**: 66-73, 1996.

ZHU Y, O'NEILL S, SAKLATVALA J, TASSI L, MENDELSOHN ME. Phosphorylated HSP27 associates with the activation-dependent cytoskeleton in human platelets. *Blood* **84**: 3715-3723, 1994.

10

GTP BINDING PROTEINS IN PLATELETS

Bruce R. Lester
Knowledge Frontiers, LLC and
Biomedical Engineering Center
University of Minnesota
Minneapolis, MN
55455, USA

10.1 INTRODUCTION

10.1.1 Signal Transduction

Cells are constantly engulfed in an environment of electrical and chemical stimuli. Specific receptors enable cells to differentiate among the stimuli and relay a specific message that leads to a defined cellular response. The process that translates the stimulus-receptor interaction into an intracellular chemical message is termed signal transduction. Signal transduction pathways are involved in the regulation of cell function in four broad physiological functions: 1) Reproduction, 2) Growth and Development,. 3) Maintenance of Internal Environment (e.g.

Bruce Lester, Ph. D., has more than 20 years of research experience in biological and biomedical research working in both industry and academic environments. In conjunction with several other faculty members at the University of Minnesota, he founded Knowledge Frontiers, which specializes in technology assessment and scientific due diligence. Over the years, Dr. Lester has been involved in studying the role of G-proteins in the adhesion, motility and invasion of metastatic cancer cells.

fluid volume, blood pressure, electrolyte balance, etc.) and 4) Regulation of Energy Availability (i.e. an organism must be able to convert the calories contained in food into energy).
There are four primary signaling mechanisms, each employing a unique strategy to breach the barrier posed by the lipid bilayer of the cell plasma membrane. These four established signaling mechanisms do not account for all the signals conveyed across cell membranes, however, they do transduce many of the most important signals used in nature.

10.1.2. Intracellular Receptors that Regulate Gene Expression

Ligands for this class of receptors are sufficiently lipid-soluble to cross the plasma membrane and act on intracellular receptors. Receptors for the corticosteroids, mineralocorticoids, sex steroids, vitamin D thyroid hormones and retinoids belong to this family of receptors. This class of ligand-receptors regulate transcription of genes in the cell nucleus by binding to specific DNA sequences (enhancers) near the gene to be regulated.

10.1.3 Ligand-gated Ion Channels

Members of this family are transmembrane receptors which are gated (opened or closed) by extracellular hormones, neurotransmitter, etc. and include channels for Na^+, K^+ and Ca^{2+}. These receptors transmit their signal across the plasma membrane by changing the transmembrane conductance of the relevant ion and thereby alter the electrical potential across the membrane.

10.1.4 G-protein coupled receptors

Ligands that activate the intracellular generation of small second messenger molecules such as cAMP, Inositol trisphosphate (IP3) and ion channels bind to receptors that contain seven transmembrane domains. Transmembrane signaling by this class of receptors consists of three components: 1) a seven membrane spanning serpentine receptor, 2) a heterotrimeric guanine nucleotide regulatory protein (G-protein) and 3) an effector enzyme or molecule which generates a secong messenger.

10.1.5 Ligand-regulated transmembrane enzymes

Members of this receptor family have bifunctional receptors that span the plasma membrane and possess ligand binding domains on the extracellular portion of the receptor and an effector domain exhibiting an enzyme activity on the intracellular face. The signaling pathway begins with a ligand induced allosteric activation of the enzyme located on the cytoplasmic domain of the receptor. The most prevalent enzyme activity found associated with this class of receptors is protein tyrosine kinase. In addition to protein tyrosine kinase, two other enzyme activities have also been found to be associated with receptors of this family including protein tyrosine phosphatases and guanylyl cyclase.

10.2 G-proteins and Signal Transduction

10.2.1. GTP-binding superfamily

Guanine Nucleotide Binding Proteins (GTP-binding proteins) are a superfamily of proteins that regulate diverse cellular functions. Members of this superfamily sort transmembrane signals, direct the fidelity of protein synthesis, guide vesicular transport through the cytoplasm, catalyze the polymerization of tubulin and control growth and differentiation of cells. GTP-binding proteins include a group of small (i.e., 21-28 kDa) proteins that regulate intracellular functions, e.g. $p21^{ras}$ and EF-Tu, as well as a group of larger proteins, i.e. 39-52 kDa, that regulate transmembrane signal transduction.

The signal transduction G-proteins or heterotrimeric GTP binding proteins, are interposed between cell surface receptors and intracellular effectors. The signal transduction G-proteins are a family of GTP-binding proteins in which the GTP-GDP switch serves to propagate and amplify regulatory signals from activated cellular membrane receptors to effector channels and enzymes.

G-protein linked receptors, sometimes referred to as "seven membrane spanning serpentine receptors" include, as their name suggests, seven membrane-spanning sequences. Sites of receptor/G-protein interaction are within cytoplasmic and possibly membrane spanning sequences. Recent evidence indicates that the carboxyl terminal portion of the third intracellular loop and the carboxyl terminal portion of the receptor are the likely sites of interaction.(Birnbaumer, Birnbaumer, 1995).

Although none of the G-protein subunits contains regions that might obviously associate with a lipid bilayer, heterotrimeric G-proteins are associated with the cytoplasmic surface of the plasma membrane as is Ras p21 and some of the other low molecular weight GTP-binding proteins. This is apparently due to the fact that the γ-subunits of the heterotrimeric G-proteins are prenylated, as is ras, and at least some of the G-protein α-subunits, e.g. G_i subfamily.

10.2.2. G-protein Cycle

Hormone or ligand binding to receptors activates G-proteins by facilitating the exchange of bound GDP for GTP on the α-subunit, thus initiating dissociation of the α-subunit from the $\beta\gamma$-dimer. Both the free α-GTP and $\beta\gamma$ subunits interact with, and regulate the activity of, specific signal effector enzymes or ion channels. Both an intrinsic GTPase and extrinsic factors terminate the interaction of the α-subunit with its effector by hydrolyzing GTP to GDP (Brass *et al.*, 1997). The GDP-liganded α chain reassociates with the $\beta\gamma$-dimer and the heterotrimeric G-protein is ready for another cycle of receptor activation. The interaction of additional proteins that regulate the rate of GDP dissociation and GTP hydrolysis add another level of control to G-protein activity.

Some of the G-proteins are substrates for ADP-ribosylation by bacterial toxins. Specifically, toxins from *Vibrio cholera*, cholera toxin (CT) or *Bordetella pertussis*, pertussis toxin (PT), can covalently modify the G-proteins by addition of an ADP-ribose group to the α-subunit. Both toxins transfer ADP-ribose from NAD to the G-proteins. G-protein α-subunits contain specific acceptor sites for ADP-ribosylation by CT (G_s) or by PT (G_i and G_o) or by both toxins (G_t) while some of the G-proteins are substrates for neither (G_q, G_z and G_{12}). Modification of the α-subunits of G_i, G_o and G_t by PT requires intact heterotrimers while CT modification of G_s and G_t utilizes the dissociated α-subunit as substrate and requires a soluble or membrane bound protein for its activity, ADP-ribosylation factor (ARF). Cholera toxin causes activation of G_s and G_t while PT attenuates the ability of G_i, G_o and G_t to respond to agonist-occupied receptors or photolyzed rhodopsin.

10.2.3. Heterotrimeric G-proteins in Platelets

G-proteins possess a heterotrimeric structure. G-proteins contain α-, β- and γ-subunits, each distinct gene products. At this time cDNAs that encode 23 distinct G-protein α-subunits (the product of 18 genes) have been cloned. These G-proteins can be divided into four major subfamilies according to amino acid sequence relationships, which include G_s, G_i, G_q and G_{12} (Birnbaumer, Birnbaumer, 1995). Evidence is accumulating which indicates that a similar structural heterogeneity of the β- (five isoforms) and γ-subunits (ten isoforms) exists, and that this heterogeneity contributes to the specificity of these interactions (Coleman, Sprang, 1996; Fields, Casey, 1997). More than 300 possible combinations of G-protein heterotrimers consisting of αβγ-subunit structure can be generated. Given its interposition between receptor and effector, much of the fidelity of signal transmission must reside in the subunit structure of the G-proteins.

Nine heterotrimeric G-proteins have been identified in platelets. One ($G_{s\alpha}$) from the G_s family, four ($G_{i\alpha1}$, $G_{i\alpha2}$, $G_{i\alpha3}$ and $G_{z\alpha}$) from the G_i family, two ($G_{q\alpha}$ and $G_{11\alpha}$) from the G_q family and two ($G_{12\alpha}$ and $G_{13\alpha}$) from the G_{12} family (Akkerman, Van Willigen, 1996; Brass *et al.*, 1997).

10.2.3.1. α_s molecules

There are four forms of α_s cDNA differing in the presence, or absence, of a 42 or 45 nucleotide-fragment at position 242. These cDNA's arise from two types of alternative splicing reactions. Translation of the four forms indicates that the two short forms of α_s consist of 380 and 381 amino acids and that the two long forms of α_s consist of 394 and 395 amino acids. Arg-202 is the site of ADP-ribosylation by CT of the 395 amino acid α_s. Elevation of intracellular cAMP is one of the most potent ways to inhibit platelet function. Agonists which cause an increase in cAMP are thought to bind to receptors which activate $G_{\alpha s}$ subunits. They include receptors for prostaglandins, adenosine and β_2-adrenergic receptors. The order of potency for inhibiting platelets is PGI_2>PGD_2>PGE_1>adenosine>β adrenergic agonists (Haslam *et al.*, 1978). Eight adenylyl cyclase genes have been identified and, functionally, they can all be stimulated by GTP liganded α_s and by the diterpene, forskolin. However, they differ in their responses to calcium and calmodulin. Types-I and III are stimulated by Ca^{++}/CaM while types-V and VI are inhibited by Ca^{++}. Type-I is also inhibited by βγ dimers while types-II and IV are stimulated by βγ dimers, however, only if they have been simultaneously stimulated by α_s. As can be seen, some of the adenylyl cyclase enzymes are effectors for βγ dimers (Birnbaumer, Birnbaumer, 1995). There is little known about the various forms of adenylyl cyclase present in platelets except that type I is absent (Berg, Hornbeck, 1993) and types III and VI have been detected in a megakaryoblastic cell line (HEL) using PCR (Hellevuo *et al.*, 1993). Finally, the α_s subunit also regulates the intracellular pH of platelets by stimulating a Na^+/H^+ exchanger pump (Van Willigen *et al.*, 1995).

10.2.3.2. α_i molecules

Three highly homologous cDNAs, from independent genes, have been cloned. They encode α-subunits that are substrates for PT and predict amino acid sequences which represent 40 to 41 kDa proteins. This coincides with observations from sequencing of purified protein or from immunoblots with antibodies raised against consensus sequences predicted from cDNAs. These α-subunits have been called α_{i1}, α_{i2} and α_{i3} on the basis of the order of cloning. They are 354, 355 and 354 amino acids long, respectively and are between 87 and 93% identical. Platelets contain all three of the α_i subunits in the rank order $G_{i\alpha 2}$>>$G_{i\alpha 3}$>$G_{i\alpha 1}$ (Williams *et al.*, 1990). G-proteins which activate (G_s) and inhibit (G_i) adenylyl cyclase have become the classic example of cross-communication in signal transduction mechanisms. Agonists that decrease or lower cAMP formation in platelets, such thrombin and epinephrine, are thought to act on receptors linked to G_i.

10.2.3.3. α_z molecules

This α-subunit was cloned from brain libraries and has been placed in the G_i class of G-proteins. It has 355 amino acids and is 40 to 67% homologous in amino acid composition with those of other known α subunits, being most similar to the α_i family. It lacks an ADP-ribosylation site for PT and it is not known whether it possesses a CT modification site. It differs from the above G-protein α-subunits in that it has amino acid substitutions within a number of regions that are strongly conserved among the other G-proteins (the GTP binding domains) including a substitution of three amino acids within the identity box of the α-subunits. The consequence of this amino acid substitution results in a slower rate of intrinsic GTP hydrolysis (Casey *et al.*, 1990). C-kinase activation in platelets results in the phosphorylation of α_z at ser-27 (Lounsbury *et al.*, 1993). This causes an inability to interact with βγ subunits (Wong *et al.*, 1992) the biological consequences of which are not well understood since the functions of α_z itself are not well understood. G_z has been shown to inhibit cAMP accumulation in other cell types (Wong *et al.*, 1992) but not platelets (Brass *et al.*, 1997). Platelet receptors known to activate α_z include those for thrombin and α_2-adrenergic agonists.

10.2.3.4. α_q molecules

Compelling evidence for the involvement of PT and CT-insensitive G-proteins in receptor-mediated activation of phospholipase-Cβ (PLC-β) has been obtained. Recent reports have identified members of the G_q family as regulators of this pathway. Reconstitution of purified G_q or G_{11} with purified PLC-β results in specific and marked stimulation of the enzyme. G_{14} exhibits the same action. In addition to PLC-β, platelets are also known to have PLC-γ, an enzyme regulated by receptor tyrosine kinases (RTK) (see below).

Phospholipase C (both β and γ) catalyzes the hydrolysis of Phosphatidyl inositol bisphosphate (PIP_2) leading to the generation of diacylglycerol (DAG) and inositol trisphosphate (IP_3). In platelets, thrombin and thromboxane-A_2 (TxA_2) stimulate PLC-β. PLC-β has three isotypes. It is known that PLC-β1 and PLC-β3 are activated by G-protein

α_q subunits while PLC-β2 is primarily activated by $G_{\beta\gamma}$ subunits (Brass *et al.*, 1997). This G-protein dichotomy may explain the differential effects of pertussis toxin on the thrombin (sensitive) and TxA_2 (insensitive) stimulation of PLC-β in platelets (Brass *et al.*, 1988; Brass *et al.*, 1987; Brass *et al.*, 1986). Thrombin receptors use $G_{\beta\gamma}$ subunits (from G_i) to stimulate PLC-β2 while TxA_2 receptors use α_q or α_{11} subunits to stimulate PLC-β1 (Shenker *et al.*, 1991).

10.2.3.5. βγ subunit

It is important to recognize that the β- and γ-subunits exist as a complex of tightly associated proteins in the physiological setting. Only by the use of strong detergents used in extraction and purification, such as 1% SDS or 8M urea is this complex broken down. Therefore from a functional point of view, G-proteins should be thought of as a "heterodimer" consisting of an α-subunit and a βγ-subunit. Traditionally, the α subunit has been viewed as the business end of the heterotrimeric G-protein because it binds and hydrolyses GTP and interacts with effectors. In contrast, the βγ-complex has been viewed variously as a regulatory component which stabilizes the GDP-bound form of α, as a 'presenter' of α to receptors and as a membrane anchor for the heterotrimer (Birnbaumer, Birnbaumer, 1995). However, a growing body of evidence supports the idea that free βγ can itself interact functionally with effector proteins like adenylyl cyclase (see above) and as an activator of PLC-β1 through the thrombin receptor.

10.3. Low Molecular Weight GTP-binding Proteins

The second group of GTP-binding proteins present in platelets are the low molecular weight G-proteins. The discovery that ras genes code for proteins that bind GTP and GDP suggested, by analogy, that these genes coded for a novel class of G-proteins involved in the control of growth and cell proliferation. Ras, which migrates as a 21 kDa protein (p21), also serves as a prototype for what has been termed the superfamily of Ras-related proteins, possessing more than 50 mammalian members (Herrmann, Nassar, 1996). These G-proteins have molecular weights of 20-30 kDa and share a structural homology with Ras.

The Ras-related or LMWG proteins function as critical regulators of a diverse spectrum of intracellular processes, including cellular proliferation and differentiation (Ras and Rap), intracellular vesicular trafficking (Rab and Arf), oxidase generation (Rac and Rap) and cytoskeletal control (Rho and Rac). Based on sequence homology the Ras superfamily can be cataloged into the Ras, Rac/Rho, Ran/TC4, Rab, Arf, Rad and the newly elucidated Rag subfamilies (Herrmann, Nassar, 1996). Ras can be positively controlled by GDS's (guanine nucleotide dissociation stimulators) or GES (guanine nucleotide exchange factors), and negatively controlled by GAPs (GTPase activating proteins) as well as GDIs (guanine nucleotide dissociation inhibitors). Although the well-characterized heterotrimeric G-proteins have been suggested as models for how ras-related proteins might work, genetic and biochemical analyses suggest that the small G-proteins may have many novel features.

The fact that mutant Ras genes have been associated with a wide variety of human tumors suggests that they are responsible for the uncontrolled growth of tumor cells. Studies involving in vitro cell culture and transgenic animals have demonstrated the ability of mutated Ras proteins to perturb both cellular growth and differentiation.

Finally, Ras proteins are also involved in the control of normal growth and are activated in response to diverse extracellular stimuli that influence cell growth and differentiation. Such factors include those that stimulate the growth of fibroblasts, e.g. EGF and PDGF, or hematopoietic cells e.g. IL-2, IL-3 and GM-CSF. The cell surface receptors for these extracellular stimuli are typically receptor tyrosine kinases (RTKs) such as the EGF receptor or receptors that have associated nonreceptor tyrosine kinases (Herrmann, Nassar, 1996; Khosravi-Far, Der, 1994).

The transformation of rodent fibroblasts by a variety of tyrosine kinase oncogenes also causes chronic elevation of Ras-GTP levels. It thus seems obvious that Ras proteins are essential components of tyrosine kinase mediated mitogenic signaling pathways. Ras proteins are activated in response to a wide variety of extracellular stimuli that activate receptor and receptor-associated tyrosine kinases (RTKs).

10.3.1 Genes coding for ras

The first ras genes were discovered as the transforming genes of murine sarcoma viruses Harvey (H or Ha) and Kirsten (K or Ki). These genes encode homologous proteins with a molecular weight of 21 kDa, which differ from normal cellular ras homologs by having amino acid substitutions at positions 12 and 59 or 61. Mutated forms of the Harvey and Kirsten ras genes as well as that of a third homologous gene, N-ras, were also isolated as transforming genes by DNA transfection experiments using DNA isolated from various human tumors. Substitution of the normal Gly12 or Gln61 by a wide range of amino acids reduces the intrinsic GTPase activity of Ras and renders its GTPase resistant to acceleration by GAP. The resulting Ras protein is characteristic of the oncogenically transformed phenotype. The oncogenic or activated forms of ras p21 can transform mammalian cells when expressed at normal levels. In contrast, high levels of expression of normal ras p21 are necessary for the transformation of mammalian cells by normal or cellular ras.

10.3.2. Primary Structure of Ras Proteins

The structural properties of Ras family indicate that they are regulatory G-proteins (Milburn *et al.*, 1990). For example, the three-dimensional structure of Ras has many features in common with elongation factor, EF-Tu and both structures have been used to identify homologous sites in the larger heterotrimeric G-proteins. Subsequent X-ray crystolographic data on the transducin α-subunit has confirmed the homologous three dimentional structures of these G-proteins (Bourne, 1993).

The Ras proteins encoded by all three human ras genes are homologous in their first 164 amino acids but their last 25 residues are divergent, except for a cystein (residue 186) four amino acids from the C-terminus. Four GTP binding domains can clearly be defined in the structure of the Ras proteins.

In addition, three functional domains are apparent. The first one includes the first 86 amino acids and is almost completely conserved among all mammalian ras gene products. In this domain the activating point mutations have been found to occur in natural tumors.

In addition, the ras genes of lower eukaryotes show a greater than 80% homology in this domain. The next 78 amino acids define a second domain that shows slightly less conservation (70 to 80% homology between any pair of vertebrate gene products. The rest of the molecule, except for the last four amino acids, constitutes the hypervariable region, with sequences specific for each ras gene. The hypervariable region probably contributes to function and/or target specificity.

Several site directed mutation studies have indicated that residues 26 to 45, which are conserved in all Ras proteins, are likely to specify an interacting surface for (an) effector molecule(s), since neither guanine nucleotide nor membrane association was affected by these mutations while transforming activity was abolished.

The carboxyl terminal four amino acids constitute a CAAX sequence (where A is any aliphatic amino acid and X is any amino acid) that is highly conserved among all ras and ras-related genes. This sequence is present not only in ras and related proteins but also in the carboxyl terminus of other proteins, including the α and γ-subunits of several heterotrimeric G-proteins, nuclear lamins and unprocessed yeast alpha mating factor. This motif has been shown to be responsible for carboxyl terminal modification via isoprenylation for attachment of ras (and other proteins) to the inner leaflet of the plasma membrane (Birnbaumer, Birnbaumer, 1995; Khosravi-Far, Der, 1994).

10.3.3 Biochemical Activities of Ras

The basic biochemical sequence of events constituting signal transduction by Rasproteins has been worked out (Herrmann, Nassar, 1996). Ras is activated by a complex consisting of a guanine nucleotide exchange factor (GEF) (mSOS, cdc25 etc) and an adaptor protein containing one SH2 domain and two SH3 domains (sem, Grb2 etc). The GEF binds to the SH3 domains on the adaptor protein. Mitogenic or hormonal signals acting through receptor tyrosine kinases (RTK) recruit the adapter protein/(GEF) complex to the plasma membrane. Binding of this complex via the SH2 domain on the adaptor to the autophosphorylated RTK localizes the GEF or GDS (mSOS) so that it can activate Ras. The adaptor/GEF complex catalyzes the dissociation of GDP from Ras which immediately binds GTP thus activating the Ras protein. The intrinsic or spontaneous rate of GDP dissociation is very low, $\sim 10^{-5}$ moles/sec dissociating per mole of complex. The Ras that does dissociate from nucleotide will very rapidly rebind GTP, since the intracellular concentration of GTP is much greater than that of GDP. This means that in the normal Ras cycle, dissociation of bound GDP leads to Ras activation.

While GTP-bound, the activated Ras interacts with target proteins like Raf, MEK and MAP kinase. Another set of proteins stimulate the intrinsic GTP hydrolytic activity of Ras and result in the termination of its activity. These proteins include GAP and neurofibromin. They recognize and interact with the effector region of p21 Ras, i.e. Raf binding sites, and may have target functions themselves.

An initial examination of this pathway suggests that it is a well-defined, sequence of events, however, recent evidence suggests a greater complexity. For example, heterotrimeric G_i-mediated MAP kinase activation involves βγ subunits activation of Ras (Inglese *et al.*, 1995; Koch *et al.*, 1994). In addition, Ras activates phosphatidylinositol-3-kinase and can interfere with the Rac/Rho pathways leading to cytoskeletal changes.

10.3.4. Ras Involvment in Adhesion, Motility and theCytoskeleton

Recent evidence suggests that members of the Ras superfamily of proteins are involved in the control of cell adhesion, motility, formation of the actin cytoskeleton and activation of intracellular signaling cascades (Fox, 1996; Parsons, 1996). In addition to the MAPK cascade which is stimulated by Ras, a second down stream component of Ras signaling, which is composed of members of the Ras-related Rho family of proteins, has been described. Rho family proteins (Rac and Rho) are involved in signal transduction pathways that control the organization of the actin cytoskeleton. In fact, their activation by oncogenic Ras may contribute to the aberrant morphology and invasive properties of malignant cells (Khosravi-Far, Der, 1994).

The Rho proteins constitute one of the major branches of the Ras superfamily and are represented by seven distinct proteins (Rho A, B, C and G, Rac1 and 2 , TC10 and CDC42Hs). Like Ras, they bind GTP and catalyze its hydrolysis to GDP, being active in the GTP liganded state. As in the case with Ras, this hydrolysis is also stimulated by GAP proteins such as RhoGAP, Bcr, N-chimaerin and p190, a protein that associates with RasGAP and has GAP activity towards Rho and Rac. Importantly, these GAPs are negative regulators of the Rho family proteins.

In addition, recent data has shown that a serum growth factor (or factors) acting through Rho and Rac can regulate the assembly of focal adhesions and actin stress fibers. The evidence is now quite strong that Rho/Rac regulates actin microfilament organization and assembly. Microinjection of a constitutively activated form of Rho resultes in drastic cytoskeletal reorganization and formation of focal adhesions, while addition of constitutively activated Rac1 to cells stimulated drastic increases in membrane ruffling accompanied by increases in actin accumulation in these ruffles.

10.3.5. Ras Proteins in Platelets

Platelets contain most, if not all, of the Ras superfamily of proteins found in other cells (Brass *et al.*, 1997; Fox, 1996). These include Ras and several members of the Rap proteins, e.g. Rap1A (Klinz *et al.*, 1992), Rap1B (Hoshijima *et al.*, 1988; Lapetina *et al.*, 1989; White *et al.*, 1990) and Rap2B (Berger *et al.*, 1994). In addition, members of the Rho family including RhoA, Rac1, Rac2 and CDC42Hs are present (Nada *et al.*, 1991; Nemoto *et al.*, 1992; Polakis *et al.*, 1989). Platelets also contain the adaptor proteins, Grb, Shc and the GAP protiens as well as the exchange protein SOS1 (Fox, 1996).

The presence of Ras proteins in platelets was noted when Morii et al., (Morii *et al.*, 1992) found that the Botulinum toxin, C3, inhibited thrombin-stimulated platelets. It is well known that most of the bacterial toxins that have ADP-ribosyltransferase activity act on GTP-binding proteins, including heterotrimeric G-proteins (cholera and pertussis), protein elongation factors, (ricin) and the Ras family proteins like Rho (Botulinum). Since platelets do not appear to express any receptor tyrosine kinase (RTK) on their cell surface it appears that the low molecular weight GTP binding proteins in platelets may play a major role in phenomena such as adhesion, motility and cytoskeletal rearrangement (Fox, 1996). The fact that platelets do contain several cytoplasmic tyrosine kinases that could be involved in the formation signaling complexes supports this contention. These tyrosine kinases include several members of the Src family (Src, Fyn, Yes, Lyn and Hck) (Golden *et al.*, 1986; Horak *et al.*, 1990; Huang *et al.*, 1991), Csk (Bergman *et al.*, 1992; Nada *et al.*, 1991; Okada *et al.*, 1991), Syk (Ohta *et al.*, 1992; Taniguchi *et al.*, 1991) and focal adhesion kinase (FAK) (Parsons *et al.*, 1994). Although the subject of Ras family participation in platelet function is beyond this scope of this chapter there are several excellent recent reviews of the subject (Brass *et al.*, 1997; Fox, 1996; Nobes, Hall, 1995; Parsons, 1996; Weiss *et al.*, 1997) including Chapter 9 of the present volume.

10.4 Summary

Platelet activation begins with the adhesion of platelets to one of several cellular substituents including: von Willebrand factor, collagen, fibrinogen and fibronectin, and culminates in the events of platelet aggregation, secretion and clot formation. Platelets can also be activated by soluble agonists such as thrombin, cathepsin G, epinephrine, thromboxane A_2 and ADP. Each of these agonists acts on its heterotrimeric G-proteins coupled receptor to induce a unique intracellular signaling pathway (Brass *et al.*, 1993). A second group of GTP binding proteins much smaller in size than the heterotrimeric G-proteins and ranging in size from 21 to 28 kDa plays a crucial role in platelet function. These proteins compose the Ras superfamily of GTP binding proteins and many if not all of its members are present in platelets. In cells other than platelets the Ras proteins have been implicated in protein transport, cell activation, cell growth and malignant transformation. Ras has also been shown to transduce extracellular signals impinging on the RTK to downstream effectors such Raf and the MAP kinase cascade of events which leads ultimately to the control of key nuclear transcription factors (Pulverer *et al.*, 1991; Sistonen *et al.*, 1989). In addition, Ras family members have been shown to act as key regulators of a cascade of kinases that control cytoskeletal organization of a cell, and consequently, may participate in the control of cell adhesion, cell motility and cell-cell interactions. It appears that the tyrosine kinases in platelets may serve to phosphorylate submenbranous proteins including other kinases (FAK) or integrins that contribute to adhesion, and which may then recruit and activate additional appropriate signaling molecules (Fox, 1996; Nozawa *et al.*, 1993). Future studies will be required to determine if the intracellular signals initiated by activated kinases and phosphorylated integrins serve to sustain and complete the transformation of the circulating platelet into a component of the hemostatic plug initiated by platelet receptor/G-protein activation.

References

AKKERMAN JW, VAN WILLIGEN G (1996) Platelet activation via trimeric GTP-binding proteins. *Haemostasis*, **26 Suppl 4**, 199-209.

BERG W, HORNBECK P (1993) A micro-radioimmunoassay for the measurement of intracellular cAMP. *BioTechniques*, **15**, 56-59.

BERGER G, QUARCK R, TENZA D, LEVY-TOLEDANO S, DE GUNZBURG J, CRAMER EM (1994) Ultrastructural localization of the small GTP-binding protein Rap1 in human platelets and megakaryocytes. *British Journal of Haematology*, **88**, 372-382.

BERGMAN M, MUSTELIN T, OETKEN C, PARTANEN J, FLINT NA, AMREIN KE, et al (1992) The human p50csk tyrosine kinase phosphorylates p56lck at Tyr-505 and down regulates its catalytic activity. *EMBO Journal*, **11**, 2919-2924.

BIRNBAUMER L, BIRNBAUMER M (1995) Signal transduction by G proteins: 1994 Edition. *Journal of Receptor and Signal Transduction Research*, **15**, 213-252.

BOURNE HR (1993) A turn-on and a surprise. *Nature*, **366**, 628-629.

BRASS L, WOOLKALIS M, MANNING D (1988) Interactions in Platelets between G proteins and the agonists that stimulate phospholipase C and inhibit adenylyl cyclase. *J.of.Biol.Chem.*, **263** , 5348-5355.

BRASS LF, HOXIE JA, KIEBER-EMMONS T, MANNING DR, PONCZ M, WOOLKALIS M (1993) Agonist receptors and G Proteins as Mediators of Platelet Activation. In: *Mechanisms of Platelet Activation and Control.* Edited by SA Kalwant, SP Watson, VV Kakkar. New York: Plenum Press. 17-37.

BRASS LF, LAPOSATA M, BANGA HS, RITTENHOUSE SE (1986) Regulation of the phosphoinositide hydrolysis pathway in thrombin-stimulated platelets by a pertussis toxin-sensitive guanine nucleotide-binding protein. Evaluation of its contribution to platelet activation and comparisons with the adenylate cyclase inhibitory protein, Gi. *Journal of Biological Chemistry*, **261**, 16838-16847.

BRASS LF, MANNING DR, CICHOWSKI K, ABRAMS CS (1997) Signaling through G proteins in platelets: to the integrins and beyond. *Thrombosis & Haemostasis*, **78**, 581-589.

BRASS LF, SHALLER CC, BELMONTE EJ (1987) Inositol 1,4,5-triphosphate-induced granule secretion in platelets. Evidence that the activation of phospholipase C mediated by platelet thromboxane receptors involves a guanine nucleotide binding protein-dependent mechanism distinct from that of thrombin. *Journal of Clinical Investigation*, **79**, 1269-1275.

CASEY PJ, FONG HK, SIMON MI, GILMAN AG (1990) Gz, a guanine nucleotide-binding protein with unique biochemical properties. *Journal of Biological Chemistry*, **265**, 2383-2390.

COLEMAN DE, SPRANG SR (1996) How G proteins work: a continuing story. *Trends in Biochemical Sciences*, **21**, 41-44.

FIELDS TA, CASEY PJ (1997) Signalling functions and biochemical properties of pertussis toxin-resistant G-proteins. *Biochemistry Journal*, **321**, 561-571.

FOX JE (1996) Platelet activation: new aspects. *Haemostasis*, **26 Suppl 4**, 102-131.

GOLDEN A, NEMETH S, BRUGGE J (1986) Blood platelets express high levels of the pp60c-src-specific tyrosine kinase activity. *Proceedings of the National Academy of Science*, **83**, 852-856.

HASLAM RJ, DAVIDSON MM, DAVIES T, LYNHAM JA, MCCLENAGHAN MD (1978) Regulation of blood platelet function by cyclic nucleotides. *Advances in Cyclic Nucleotide Research*, **9**, 533-552.

HELLEVUO K, YOSHIMURA M, KAO M, HOFFMAN PL, COOPER DM, TABAKOFF B (1993) A novel adenylyl cyclase sequence cloned from the human erythroleukemia cell line. *Biochemical & Biophysical Research Communications*, **192**, 311-318.

HERRMANN C, NASSAR N (1996) Ras and its effectors. *Progress in Biophysics & Molecular Biology*, **66**, 1-41.

HORAK ID, CORCORAN ML, THOMPSON PA, WAHL LM, BOLEN JB (1990) Expression of p60fyn in human platelets. *Oncogene*, **5**, 597-602.

HOSHIJIMA M, KIKUCHI A, KAWATA M, OHMORI T, HASHIMOTO E, YAMAMURA H, et al (1988) Phosphorylation by cyclic AMP-dependent protein kinase of a human platelet Mr 22,000 GTP-binding protein (smg p21) having the same putative effector domain as the ras gene products. *Biochemical & Biophysical Research Communications*, **157**, 851-860.

HUANG MM, BOLEN JB, BARNWELL JW, SHATTIL SJ, BRUGGE JS (1991) Membrane glycoprotein IV (CD36) is physically associated with the Fyn, Lyn, and Yes protein-tyrosine kinases in human platelets. *Proceedings of the National Academy of Sciences of the United States of America*, **88**, 7844-7848.

INGLESE J, KOCH WJ, TOUHARA K, LEFKOWITZ RJ (1995) G beta gamma interactions with PH domains and Ras-MAPK signaling pathways. *Trends in Biochemical Sciences*, **20**, 151-156.

KHOSRAVI-FAR R, DER CJ (1994) The Ras signal transduction pathway. *Cancer and Metastasis Reviews*, **13**, 67-89.

KLINZ FJ, SEIFERT R, SCHWANER I, GAUSEPOHL H, FRANK R, SCHULTZ G (1992) Generation of specific antibodies against the rap1A, rap1B and rap2 small GTP-binding proteins. Analysis of rap and ras proteins in membranes from mammalian cells. *European Journal of Biochemistry*, **207**, 207-213.

KOCH WJ, HAWES BE, ALLEN LF, LEFKOWITZ RJ (1994) Direct evidence that Gi-coupled receptor stimulation of mitogen-activated protein kinase is mediated by G beta gamma activation of p21ras. *Proceedings of the National Academy of Sciences of the United States of America*, **91**, 12706-12710.

LAPETINA EG, LACAL JC, REEP BR, MOLINA (1989) A ras-related protein is phosphorylated and translocated by agonists that increase cAMP levels in human platelets. *Proceedings of the National Academy of Sciences of the United States of America*, **86**, 3131-3134.

LOUNSBURY K, SCHLEGEL B, PONCZ M, BRASS L, MANNING D (1993) Analysis of GzAlpha by site-directed mutagenesis. Sites and specificity of protein kinase C-dependent phosphorylation. *J.Biol.Chem.*, **268**, 3494-3498.

MILBURN M, TONG L, DEVOS A, BRUNGER A, YAMAIZUMI Z, NISHIMURA S, et al (1990) Molecular switch for signal transduction: structural differences between active and inactive forms of protooncogenic ras proteins. *Science*, **247**, 939-945.

MORII N, TERU-UCHI T, TOMINAGA T, KUMAGAI N, KOZAKI S, USHIKUBI F, et al (1992) A rho gene product in human blood platelets. II. Effects of the ADP-ribosylation by botulinum C3 ADP-ribosyltransferase on platelet aggregation. *Journal of Biological Chemistry*, **267**, 20921-20926.

NADA S, OKADA M, MACAULEY A, COOPER JA, NAKAGAWA H (1991) Cloning of a complementary DNA for a protein-tyrosine kinase that specifically phosphorylates a negative regulatory site of p60c-src. *Nature*, **351**, 69-72.

NEMOTO Y, NAMBA T, TERU-UCHI T, USHIKUBI F, MORII N, NARUMIYA S (1992) A rho gene product in human blood platelets. I. Identification of the platelet substrate for botulinum C3 ADP-ribosyltransferase as rhoA protein. *Journal of Biological Chemistry*, **267**, 20916-20920.

NOBES CD, HALL A (1995) Rho, rac and cdc42 GTPases: regulators of actin structures, cell adhesion and motility. *Biochemical Society Transactions*, **23**, 456-459.

NOZAWA Y, BANNO Y, NAGATA K (1993) Regulation of Phosphoinositide-specific Phospholipase C Activity in Human Platelets. In: *Mechanisms of Platelet Activation and Control*. Edited by SA Kalwant, SP Watson, VV Kakkar. New York: Plenum Press. 37-47.

OHTA S, TANIGUCHI T, ASAHI M, KATO Y, NAKAGAWARA G, YAMAMURA H (1992) Protein-tyrosine kinase p72syk is activated by wheat germ agglutinin in platelets. *Biochemical & Biophysical Research Communications*, **185**, 1128-1132.

OKADA M, NADA S, YAMANASHI Y, YAMAMOTO T, NAKAGAWA H (1991) CSK: a protein-tyrosine kinase involved in regulation of src family kinases. *Journal of Biological Chemistry*, **266**, 24249-24252.

PARSONS JT (1996) Integrin-mediated signalling: regulation by protein tyrosine kinases and small GTP-binding proteins. *Current Opinion in Cell Biology*, **8**, 146-152.

PARSONS JT, SCHALLER MD, HILDEBRAND J, LEU TH, RICHARDSON A, OTEY C (1994) Focal adhesion kinase: structure and signalling. *Journal of Cell Science - Supplement*, **18**, 109-113.

POLAKIS PG, WEBER RF, NEVINS B, DIDSBURY JR, EVANS T, SNYDERMAN R (1989) Identification of the ral and rac1 gene products, low molecular mass GTP-binding proteins from human platelets. *Journal of Biological Chemistry*, **264**, 16383-16389.

PULVERER BJ, KYRIAKIS JM, AVRUCH J, NIKOLAKAKI E, WOODGETT JR (1991) Phosphorylation of c-jun mediated by MAP kinases. *Nature*, **353**, 670-674.

SHENKER A, GOLDSMITH P, UNSON C, SPIEGEL A (1991) The G protein coupled to the thromboxane A2 receptor in human platelets is a member of the novel G_q family. *J.Biol.Chem.*, **266**, 9309-9313.

SISTONEN L, HOLTTA E, MAKELA TP, KESKI-OJA J, ALITALO K (1989) The cellular response to induction of the p21 c-Ha-ras oncoprotein includes stimulation of jun gene expression. *EMBO Journal*, **8**, 815-822.

TANIGUCHI T, KOBAYASHI T, KONDO J, TAKAHASHI K, NAKAMURA H, SUZUKI J, et al (1991) Molecular cloning of a porcine gene syk that encodes a 72-kDa protein-tyrosine kinase showing high susceptibility to proteolysis. *Journal of Biological Chemistry*, **266**, 15790-15796.

VAN WILLIGEN G, NIEUWLAND R, GIESBERTS AN, GORTER G, SIFFERT W, AKKERMAN J (1995) Inhibition of Na^+/H^+ exchange in human platelet plasma membranes by the trimeric G-proteins G_s and G_B. *Thromb.Haemost.*, **73**, 994

WEISS FU, DAUB H, ULLRICH A (1997) Novel mechanisms of RTK signal generation. *Current Opinion in Genetics & Development*, **7**, 80-86.

WHITE TE, LACAL JC, REEP B, FISCHER TH, LAPETINA EG, WHITE GC (1990) Thrombolamban, the 22-kDa platelet substrate of cyclic AMP-dependent protein kinase, is immunologically homologous with the Ras family of GTP-binding proteins. *Proceedings of the National Academy of Sciences of the United States of America*, **87**, 758-762.

WILLIAMS A, WOOLKALIS M, PONCZ M, MANNING D, GEWIRTZ A, BRASS L (1990) Identification of the pertussis toxin-sensitive G proteins in platelets, megakaryocytes, and human erythroleukemia cells. *Blood*, **76**, 721-730.

WONG Y, CONKLIN B, BOURNE G (1992) G_z-mediated hormonal inhibition of cyclic-AMP accumulation. *Science*, **255**, 339-342.

11

PLATELET CYCLIC NUCLEOTIDE PHOSPHODIESTERASES

Robert W. Colman, M.D.
Sol Sherry Thrombosis Research Center
Temple University School of Medicine
Philadelphia, PA
19140, USA

11.1 INTRODUCTION

Dr. Robert W. Colman has been contributing to our knowledge of platelet pathophysiology and biochemistry for 25 years. At Harvard Medical School, he investigated the interaction of platelet collagenase and collagen and described platelet hyperactivity in patients with type IIa hyperlipoproteinemia. At the University of Pennsylvania he investigated the interaction of cholesterol and platelets and described platelet activation by anti-platelet antibodies. At Temple University School of Medicine, he has concentrated his efforts on identifying an ADP receptor on platelets as well as the elucidation of the biochemistry and molecular biology of platelet cyclic nucleotide phosphodiester-ases. Dr. Colman is the lead editor of the major textbook in the field: Hemostasis and Thrombosis: Basic Principles and Clinical Practice, the 4th edition of which is now being prepared for publication.

Currently, the two antiplatelet agents with proven efficacy are aspirin, which inhibits cyclooxygenase -dependent synthesis of thromboxane A_2 (TXA_2), and ticlopidine, which blocks the ability of ADP to inhibit stimulated adenyl cyclase. Both of these drugs have proven prophylactic uses in reducing the risk of thrombo -occlusive and thromboembolic complications for all major arterial beds in individuals with a previous history of such episodes. Controlled trials show that both aspirin and ticlopidine are indicated in the secondary prevention of stroke, myocardial infarction and peripheral vascular occlusion. However, there are limitations to their efficacy. No net changes in vascular events are seen with primary prevention. Moreover, antiplatelet drugs do not alter thrombocytopenia or impairment of hemostasis seen in cardiopulmonary bypass, and neither drug appears to prevent reocclusion of coronary arteries after either thrombo-lytic therapy or angioplasty. The failure of aspirin and ticlopidine in these situations is largely due to their inability to inhibit thrombin-induced platelet activation. Although, at low concentrations of thrombin, platelet aggregation is dependent on ADP and TXA_2, high concentrations of thrombin aggregate and activate platelets by pathways independent of ADP and TXA_2. In contrast, elevation of cAMP or cGMP block all activating pathways in platelets including increase in intracellular Ca^{2+}, shape change, aggregation, secretion, and the effects of phospholipases A_2 and C. Cyclic nucleotide phosphodiesterases hydrolyze cAMP and/or cGMP, lowering their intracellular concentration. Inhibitors could function as potent anti-platelet agonists.

11.2 REGULATION OF PLATELET ACTIVATION BY cAMP AND cGMP

All agonist-induced platelet responses, including shape change, aggregation, adhesion and secretion are inhibited by increased intracellular cAMP levels (1). Synthesis of cAMP in the platelet is stimulated by the binding of mediators, such as prostacyclin and adenosine, to cell surface receptors (Rs) coupled to GTP binding proteins (Fig. 1).

G proteins mediate the interaction of agonist-occupied seven transmembrane spanning cell surface receptors to regulate intracellular membrane-bound enzymes or ion channel activity. Gs forms a link between purinergic or prostaglandin receptors and adenylate cyclase (AC), leading to stimulation of the latter. On the other hand, activation of platelets by thrombin, ADP or epinephrine diminishes the elevated intracellular cAMP levels via a Gi-coupled receptor (Ri) (2). cAMP levels are also regulated by the degradation of cAMP via the cyclic-nucleotide phosphodiesterases, a group of enzymes that catalyze the hydrolysis of 3',5'-cyclic nucleotides to inactive 5'-nucleotides by cleaving a phosphodiester bond. The levels of cAMP are tightly controlled and are ultimately dependent on its rate of synthesis by AC and its rate of hydrolysis by cAMP-PDEs. In vitro intracellular cAMP levels can be increased by stimulating AC or by inhibiting cAMP-PDEs (3-6). Increased intracellular levels of cGMP also serve as an inhibitor of platelet activation by various agonists. Platelets possess both membrane bound and soluble guanylate cyclase activity, and have the ability to synthesize cGMP in response to certain stimuli, such as nitric oxide. The cGMP level is further regulated by the cGMP-specific PDE present in platelets (cGB-PDE).

11.3 CLASSIFICATION OF CYCLIC NUCLEOTIDE PDES

The existence of distinct families of mammalian cyclic nucleotide phosphodiesterases, which are expressed in a cell-specific manner, has encouraged the further development of drugs that selectively inhibit specific cAMP-PDEs (Table 1, next page). The mammalian PDEs are currently classified into seven distinct families based on a combination of amino acid sequence homology and a variety of biochemical properties including substrate specificity, response to selective inhibitors, mode of regulation, and kinetic properties (7). The sequencing of the cDNA from at least one member of each PDE family suggests that the seven families are coded for by related but distinct genes. They share a conserved region of about 250 amino acids at the C-terminal end that contains the catalytic domain; however, the extent of homology is only 28-40% (8). In contrast, the N-terminal region of each class is distinct and in certain cases appears to contain the regulatory domain(s).

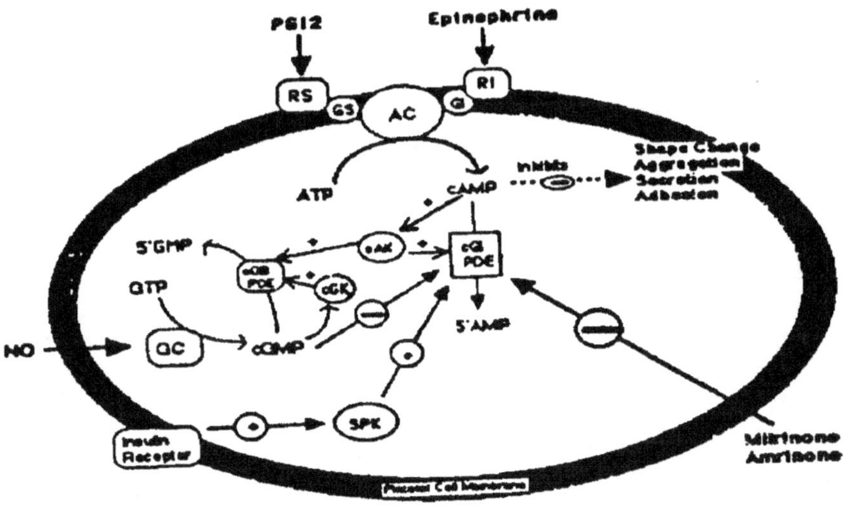

Fig. 1. Regulation of cyclic AMP metabolism in the platelet. Agents, such as PGI_2, elevate cAMP levels by binding to Rs and activating adenylate cyclase through Gs, a GTP-binding protein. AC increases cAMP synthesis from ATP. Intracellular cAMP levels can be down-regulated by agents, such as epinephrine, which binds to Ri resulting in inhibition of adenylate cyclase through Gi. Alternatively, cAMP can be degraded to 5'AMP by the action of cGMP inhibited phosphodiesterase (cGI-PDE). cGI-PDE activity is stimulated by cAMP-dependent protein kinase (cAK). It can be inhibited by cGMP which may be increased, for instance, by nitric oxide stimulation of soluble guanylate cyclase. cGMP, in turn, is regulated by the cGMP binding, cGMP-specific PDE (cGB-PDE). The cGB-PDE can be phosphorylated by both cAK or cGMP kinase. Drugs, such as milrinone and cilostamide, will inhibit the cGI-PDE specifically, resulting in an increase in intracellular levels of cAMP. Increased cAMP will inhibit shape change, aggregation, secretion, and adhesion. Reproduced by permission from Sheth SB and Colman RW, Platelets 6: 61-70, 1995.

Each family is further divided into subclasses which, in the conserved domain, demonstrate up to 70-90% identity to each other within a family and are encoded by distinct but homologous genes. In addition, many of these subclass proteins have several members which are products of alternative mRNA splicing.

11.3.1 CYCLIC NUCLEOTIDE PDES IN PLATELETS

Hidaka and Asano reported in 1976 the isolation of three distinct cyclic nucleotide PDEs by DEAE-cellulose chromatography from the soluble fraction of platelet extracts (9). These three forms were affected differently by various PDE inhibitors (10,11) and also demonstrated distinct biochemical properties. Form I had a high Km for cAMP and a low Km for cGMP, and is now known as the cGMP-binding cGMP-specific PDE (cGB-PDE or PDE 5). The other two forms are specific for cAMP and are known as the cGMP-stimulated cAMP-PDE (cGS-PDE or PDE 2), and the low K_m cGMP-inhibited cAMP-PDE (cGI-PDE or PDE 3). The cGS-PDE hydrolyzes both cGMP and cAMP at

Table 1: Mammalian Phosphodiesterase Families and Their Class-Specific Inhibitors

Short Name	Family	# of Gene Products	Family-Specific Inhibitors	cAMP (mM)	cGMP (mM)
PDE 1	Ca++-Calmodulin-dependent (CM-PDE)	3	Vinpocetin, phenothia-zines	50	35
PDE 2	cGMP-stimulated (cGS-PDE)	1	Erythro-9-(2-hydroxyl-3-nonyl)-adenine (EHNA)	50	35
PDE 3	Low Km cGMP-inhibited (cGI-PDE)	2	Milrinone, amrinone, cilostamide	0.5	0.02
PDE 4	Low Km cAMp-specific (cAS-PDE)	4	Rolipram	3.0	---
PDE 5	cGMP-binding, cGMP-specific (cGB-PDE)	2	Zaprinast	40	0.2
PDE 6	Rod and cone	3	--	--	30
PDE 7	Low Km cAMP-specific	1	Rolipram-insensitive	1.0	--

approximately equal rates and is allosterically stimulated by cGMP (12) as we have shown. The cGI-PDE, the most abundant of the three in platelets, has a low K_m for cAMP and is competitively inhibited by cGMP. PDE inhibitors inhibit platelet aggregation in response to all common agonists (4,6,13). The mode of action of these agents is the inhibition of platelet PDE activity, which leads to an increase in intracellular cAMP (14). The cAMP-PDE enzymes in platelet extracts (15) offer novel targets for the design of anti-platelet agents.

11.3.2 CLINICAL SIGNIFICANCE

Recent work has identified potent and specific inhibitors such as the bipyridines (milrinone and amrinone) (16), the flavonoids (17), and the imidazoquinolone derivatives (anagrelide) (4,13,18,19). These drugs do not increase the basal level of intracellular cAMP in the resting platelet, but they potentiate the ability of PGI2 or other AC agonists to increase cAMP (13,20). The drug concentrations required for inhibition of the cGI-PDE and of platelet activation are similar. Combination of agents which inhibit thromboxane synthase and/or serve as a thromboxane receptor antagonists with a cGI-PDE inhibitor (21,22) result in an additive effect in vitro. Several studies have demonstrated potential clinical usefulness of cGI-PDE inhibitors. The specific cGI-PDE inhibitor, cilostazol, was found to prevent graft occlusion in a dose-dependent manner in an experimental model of thrombosis in venous prosthesis in rabbits (23). Cilostazol was also found to be as effective as ticlopidine, an inhibitor of ADP-induced inhibition of AC, in maintaining graft patency in patients who had undergone an aortoiliac bypass graft (23). In a dog model, cilostazol effectively prevents coronary artery reocclusion after thrombolysis with recombinant tissue-type plasminogen activator (24). These studies suggest that cGI-PDE inhibitors may be effective antiplatelet agents in vivo.

11.4 PLATELET cGI-PDE (PDE3A)

11.4.1 Isolation and characterization

We have purified the cGI-PDE from outdated blood bank platelets and reported it to have a molecular weight of 61 kDa (25). The 61 kDa enzyme hydrolyzes cAMP with a Km value of 0.18 (M and cGMP, a competitive inhibitor, inhibits this reaction with a Ki of 0.02 mM. Despite a low Km for both cAMP and cGMP, the cGI-PDE has a kcat that is 10-fold lower for cGMP than for cAMP, allowing this enzyme to more effectively hydrolyze cAMP. Subsequently, it was determined that the 61 kDa enzyme is actually a proteolytic cleavage product of a 110 kDa enzyme (26). We have performed N-terminal amino acid analysis of the 61 kDa fragment which yielded the sequence APETMMFLDKPILAPEPLVMDN, indicating that a chymotryptic-like cleavage of a Phe-Ala bond was responsible for the truncated protein. Comparison of this sequence with the heart enzyme suggests that the 61 kDa fragment is derived from the C-terminal end of the intact enzyme and includes the domain conserved among the various PDE classes. Since this fragment hydrolyzes cAMP (25), it must contain the active site.

11.4.2 Metal cation requirements of cGI-PDE

We have shown that cGI-PDE activity is dependent on the presence of divalent metal cation (27). The inactivation of cGI-PDE by the metal chelators EDTA and 1,10-phenanthroline was time- and concentration-dependent, supporting the actual removal of critical metal cation(s). The activation of enzyme activity of the inactive aopenzyme is specific and limited to three divalent metal cations. Mn^{2+} is most effective since it requires the lowest concentration (0.1 mM) followed by Ca^{2+} (1 mM) and Mg^{2+} which requires 10 mM for maximum reactivation. At all concentrations, Zn^{2+} is inhibitory and zinc is the tightest bound metal. This conclusion is based on the fact that exposing the cGI-PDE first to Zn^{2+} significantly inhibits the activation effect of Mn^{2+}, Co^{2+}, or Mg^{2+} even when these cations are present in several thousand-fold excess. For example, only 30-40% activation was observed with 10 mM manganese-, magnesium-, or 5 mM cobalt chloride when the enzyme was first exposed to 1 mM Zn^{2+}. The Zn (and Cd) binding to cGI-PDE competes with the activation effect of Mn, Mg, and Co. The data implicates the presence of metal binding site(s) on cGI-PDE which are inhibitory when occupied by Zn or Cd, but stiimulatory with Mn, Co, and Mg. The presence of Mn^{2+}, Mg^{2+}, or Co^{2+} at their most effective concentrations partially prevented the inhibitory effect of Zn^{2+}. In the presence of increasing concentration of activating divalent metal cation, the effective Zn^{2+} required to inhibit 50% of the enzyme activity (IC50) increases. For example, when the cGI-PDE was first preincubated with Mn^{2+}, the Zn^{2+} IC50 is increased several-fold and in a manner dependent on the Mn2+ concentration. The competition data might mean that these cations influence enzyme activity by binding to the protein at the same motif(s).

11.4.3 Characterization of catalytic region of cGI-PDE

Differences in the binding requirements of a series of cGMP- and cAMP-analogs suggests that different PDE classes can distinguish between the two nucleotides. Exploration of these structure-activity relationships of the individual PDE classes has led to the design of second generation inhibitors which are class specific, unlike the relatively nonspecific xanthine derivatives. The structure of the second generation inhibitors, such as milrinone, enoximone and imazodan (28) is generally made up of three subunits: a heterocyclic portion, a phenyl, and an imidazole. Erhardt proposed a topographical model of the enzyme's active site which is complementary and contains three functional moieties (28). The model predicts an excellent fit not only for cAMP but also for the second generation inhibitors which are not nucleotide analogs. The model also predicts that cGMP will bind at a site distinct from cAMP, a prediction supported by our most recent study.

Affinity labeling exploits an enzyme's specificity for a natural ligand or substrate by utilizing a structurally similar reagent to the ligand, but which, in addition, contains a functional group which reacts covalently with many amino acid side chains in order to identify the amino acid residues which contribute to the binding site. Three cAMP analogues, 2-[(4-bromo-2,3,-dioxobutyl)thio] adenosine 3',5'-cyclic mono- phosphate (2-BDB-TcAMP), 2-[(3-bromo-2-oxopropyl)thio] adenosine 3',5'-cyclic monophosphate (2-BOP-TcAMP), and 8-[(4-bromo-2,3,-dioxobutyl)thio] adenosine 3',5'-cyclic monophosphate (8-BDB-TcAMP) were synthesized in collaboration with Dr. Roberta F. Colman, University of Delaware-Newark, to take advantage of this principle (29). We found that the 2-BDB-TcAMP and 2-BOB-TcAMP were competitive inhibitors of cAMP hydrolysis by cGI-PDE; however, they do not irreversibly inactivate the enzyme. The 8-substituted compound is positioned to interact with nucleophiles in the active-site pocket since incubation of purified cGI-PDE with this inhibitor resulted in a rapid, time-dependent and irreversible inactivation of the enzyme. The rate of inactivation of the enzyme by 8-BDB-TcAMP was decreased by the addition of cAMP, cGMP, or AMP to the reaction mixture. This inactivation is not prevented by NAD^+, however, which is not a ligand for this enzyme. The active site protection by substrate, competitive inhibitors, and product indicates a reaction at the active site of the enzyme. Studies at the molecular level have further localized the active site. By expressing a 54 kDa fragment containing the conserved region from the C-terminal end of the cardiac cGI-PDE as a fusion protein in E. coli, Meacci et al. (30) were able to demonstrate catalytic activity characteristic of the cGI-PDE. This finding confirmed our observation of active 61, 53 and 51 kDa fragments of cGI-PDE. This activity was inhibited by both OPC 3911, a cGI-PDE specific inhibitor and cGMP, but not rolipram, an inhibitor of the low Km cAMP-specific PDE (PDE4). This protein also reacted with the anti-platelet cGI-PDE antibody. The 44 amino acids specific to the cGI-PDE class and ~100 amino acids NH2-terminal to the conserved catalytic domain are required for the expression of cGI-PDE activity (31).

Cyclic GMP-inhibited phosphodiesterase (cGI-PDE), the predominant enzyme hydrolyzing cAMP, has a pH rate profile plot which yields pKa values of 6.5 and 9.0, consistent with histidine and cysteine (32). Diethyl pyrocarbonate (DEP) inactivates cGI-PDE in a time- and concentration-dependent manner, and was rapidly reversed by hydroxylamine. Two histidine residues per mol of the enzyme were responsible for the loss of catalytic activity as deduced from the correlation of the difference spectrum at 240 nm of the DEP-modified cGI-PDE with the enzymatic activity. N-ethylmaleimide (NEM) and 5,5'-dithiobis-(2-nitrobenzoicacid)(DTNB) inactivate cGI-PDE in a time-and concentration-dependent manner, suggesting the selective modification of a cysteine residue. A M P protects the enzyme against DEP, N E M, and DTNB, suggesting the presence of histidine and cysteine residues at the active site of cGI-PDE. [14C]DEP incorporation in the presence of AMP or cGMP indicate protection of two histidine residues, by each nucleotide. These residues are different since the combination of AMP and cGMP protects four histidine residues. [3H]NEM incorporation shows that 1 mol of cysteine per mol of cGI-PDE was protected by AMP, but not by cGMP. cGI-PDE possesses two essential histidine residues for activity, two additional histidines for cGMP inhibition, and one cysteine residue at the active site.

11.5 CGI-PDE REGULATORY DOMAIN

Stimulation of cAMP-PDE activity has been reported in human platelets treated with the prostaglandins, PGE_1 and PGI_2 (3). The increase in activity appears to parallel the rise in intracellular cAMP. The plant diterpene, forskolin, which binds to the catalytic subunit of adenylate cyclase and stimulates cAMP formation, has also been shown to increase cAMP-PDE activity (33). Both Macphee et al. (34,35) and our laboratory (33) have demonstrated that phosphorylation of the platelet enzyme is accompanied by an increase in PDE activity. We showed that forskolin elevated cAMP levels in human platelets and caused phosphorylation of the cGI-PDE concomitant with a 10-fold rise in activity (33). Incubation of intact platelets with iloprost, a stable PGI2 analog which also increases cAMP levels, resulted in a 2-fold increase in cGI-PDE activity in platelet lysates. The increased activity was inhibited when a protein kinase inhibitor, H-8, was added prior to treatment with forskolin, suggesting that the stimulation of PDE activity involved phosphorylation. Incubation of intact platelets with both PGE1 and PGI2 or platelet lysate with the catalytic subunit of cAMP-dependent protein kinase results in an increase in activity and phosphorylation of the cGI-PDE (35). Protein kinase inhibitor (PKI), a specific inhibitor of cAMP-dependent protein kinase, abolished the observed increase in enzyme activity and phosphorylation suggesting that the regulatory subunit of the cGI-PDE is phosphorylated by cAMP-dependent protein kinase.

11.5.1 Isolation and regulation of the cGMP-inhibited cyclic-AMP phosphodiesterase from human erythroleukemia (HEL) cells)

The predominant PDE in platelets is the cGI-PDE, a protein of 110 kilodaltons (kDa). Isolation of significant quantities of cGI-PDE from outdated platelets invariably resulted in purification of a 61 kDa form due to proteolysis during platelet storage. Therefore, we have isolated cGI-PDE from HEL cells (36). Both cAS-PDE and the cGI-PDE were found in HEL cells. PDE activity was present in the supernatant of the centrifuged extract of the thawed and lysed HEL cells. The cGI-PDE was separated from the cAS-PDE by adsorption on blue dextran Sepharose and eluted with cGMP. The enzyme was identified by Western blot using a monospecific polyclonal antibody. The 110 kDa form of the enzyme was the predominant species. The enzyme appears to hydrolyze cAMP with a Km of 0.5 (M and is inhibited by cGMP with a Ki of 0.06 (M, identical to the 61 kDa form of the platelet cGI-PDE. Incubation of cell supernatant with cAMP-dependent protein kinase resulted in an immediate increase in activity. HEL cells grown for 24 h in the presence of 50 (M forskolin, an adenylate cyclase activator, demonstrate an increase in cGI-PDE of 274% of control ($p = 0.03$). Cells incubated with forskolin and cycloheximide or actinomycin D demonstrated no increase, suggesting that cAMP stimulates cGI-PDE synthesis by transcriptional regulation. Changes in cAMP levels correlate with changes in PDE activity. The results indicate that cAMP affects both the short- and long-term regulation of cGI-PDE. The latter effect may play a role in the developing hematopoietic cell and the cardiovascular system to regulate cAMP levels.

11.5.2 Cloning of human cGI-PDE from HEL cells and platelets, expression in yeast and localization of catalytic domain by deletion mutagenesis

To obtain structural information on platelet PDE3, we cloned the enzyme cDNA from a human erythroleukemia cell (HEL) library since the cell line expresses many platelet proteins (37). This sequence of this clone is identical to the full-length human myocardial PDE3 cDNA, spanning from nucleotides 456 to 4606 (38). Partial cDNA sequences obtained by RT-PCR of platelet RNA are also identical to both the HEL and myocardial sequences, indicating that the HEL, myocardial, and platelet PDE3s are the same. Northern blot analysis of HEL cell RNA detected two mRNAs of 7.5 and 4.4 kb. Two deletion mutants, encoding amino acids 665 to 1141 and amino acids 79 to 1141, respectively, were expressed in a PDE-deficient yeast. They displayed PDE activities of 172 and 844 pmol/mg/min, respectively, comparable to purified platelet PDE3. Two other mutants, encoding amino acids 686 to 1141 and 700 to 1141, had no detectable PDE activity all mutant proteins were expressed as determined by Western blot analysis. These findings localize the PDE3 catalytic domain to within amino acid residues 679 to 1141. Two distinct but related cGI-PDEs have been cloned (31). One of these enzymes, PDE3B, hybridized strongly with RNA isolated from adipocyte, whereas the second enzyme, PDE3A, was found to be highly homologous to the corresponding region of the human cardiac cGI-PDE. The cGI-PDE family appears to have a conserved region of 44 amino acids at the C-terminal end within the conserved catalytic domain, but not present in other PDE families (39). This region demonstrates ~90% sequence identity in similar tissues of different species (e.g. rat and human cardiac tissue) but only ~70% identity between tissues in the same species (e.g. rat adipose and cardiac tissue) suggesting a substantial evolutionary conservation of certain cGI-PDE isoforms.

11.5.3 Structural characteristics of cAMP and cGMP binding sites in proteins

Sites for cAMP

The primary receptors for cAMP are the regulatory (R) subunits of cAMP-dependent protein kinases (cAPK). Upon binding of cAMP to R subunits, they dissociate from the catalytic (C) subunits. Thus, while the holoenzyme R2C2 is inactive, on addition of cAMP, a R2 (cAMP)4 dimer forms, releasing 2 active C subunits. The R subunit contains a N-terminal dimerization domain, a central autoinhibitory site resembling a substrate and an inhibitor which binds via arginines 94 and 95 (40), the active site of the catalytic subunit and a C-terminal region consisting of two tandem homologous cAMP binding domains A and B (41). For site A, if R209 is mutated to K, cAMP binding is lost. An analogous mutation of R233K results in decreased binding of site B. The initial binding of cAMP to B triggers a conformational change, allowing binding to A which dissociates R2C2 (42). Site A prefers N6 substrate analogs and site B, C2 and C8 analogs. Although the full R subunits have not been crystallized, the D1-91 deletion mutant has been crystallized and is structurally defined at 2.9Å resolution. Each of the tandem cAMP binding domains provides a network of hydrogen bonds that buries the cyclic phosphate and the ribose between two B strands that are linked by a short (helix. The H-bonds are contributed by a series of critical aspartic and glutamic residues (43). These structures provide the best model for understanding a cAMP binding site and could be considered as a template for molecular modeling of PDEs.

Sites for cGMP

In the PDEs, cGMP may be a substrate (PDE1 and PDE5), a competitive inhibitor (PDE3) or an allosteric activator (PDE2). Moreover, unlike cAMP, where cAPK is the predominant receptor, there are numerous binding sites for cGMP. For example, photoaffinity labeling of platelets with [^{32}P] cGMP resulted in radiolabeled PDE3, cGMP-protein kinase (cGPK) and a novel 65 kDa cGMP binding protein (44). cGPK has a similar domain orgainization as cAPK with a dimerization, autoinhibitory and 2 cGMP binding sites (45) followed by a catalytic domain. Unfortunately, no crystal structure currently exists for cGPK. There are two major types of cGPK: type I located in the cytosol, and type II in the membrane, which bind 2 and 1 moles of cGMP per monomer of enzyme, respectively (45). A conserved threonine in cGPK which replaces an alanine in cAPK, is critical for forming a H-bond with the 2-NH2 group of cGMP (46). The phosphorothioate derivative has been a useful nonhydrolyzable probe. The Rp analogs (with the sulfur equatorial to the sugar ring) are antagonists, while Sp analogs (with the sulfur axial to the sugar ring) are agonists (47). The structure of the cGMP binding sites has been deduced by modeling based on the crystalline bacterial protein, cAP. The general features of the binding sites include 3 (-helices and an 8-stranded anti-parallel (-barrel which, with the (C helix, forms a pocket for cGMP to bind. Six invariant amino acids include 3 glycines and an alanine as well as an arginine interacting with the phosphate group and a glutamine which forms a H-bond with 2'-OH. These combine with the aforementioned threonine, which confers selectivity for cGMP and cAMP.

The cGMP sites in PDEs are less well understood. From homology considerations, the re are two regions that uniquely identify cGMP sites (48). The first is the long general sequence L(X)2PIXN(X)6GVA(X)3N(X) ,G, and the second is the more specific sequence L(C,S)(F,L,M)PI(K,V)NXX(E,Q)(E,D)(I,V)(IV)GUAX-(A,F,L)(I,V,Y)N(K,R)(I,K)XG.

11.6 PLATELET cGMP-STIMULATED PDE (PDE2)

Isolation and characterization

The cGS-PDE has been purified in our laboratory to near homogeneity from human platelets by a combination of DEAE-cellulose and cyclic GMP affinity chromatography (49). This enzyme exists as a homodimer, has a native molecular weight of 210 kDa, and hydrolyzes cAMP and cGMP wi h similar maximal rates. The enzyme shows positive homotropic cooperativity with a half saturation value ($S_{0.5}$) 0f 50±12 (μM for cAMP and 35±15 (μM for cGMP, and Hill coefficients of 1.2 to 1.5 for both nucleotides. The activity of cGS-PDE is stimulated 3- to 10-fold by low levels (< 10 (M) of cGMP. Whereas either Mg^{2+} or Mn^{2+} is required for optimal activity of the cGS-PDE, the activity is not increased by calcium or calmodulin. Recently, a specific inhibitor, erythro-9-(2-hydroxyl -3-nonyl) adenine (EHNA), of the cGS-PDE has been described (50). The IC50 for cGMP to stimulate cGS-PDE is 1 (M (51). This enzyme bears some similarity to the rod and cone cGMP-binding PDE. Most of the studies concerning cGS-PDE have focused on the bovine heart enzyme. Whereas the platelet enzyme has not been cloned, the molecular weight and kinetic parameters of this and the bovine heart cGS-PDE are comparable, suggesting that the enzymes are identical or very similar.

Characterization of catalytic region of cGS-PDE

Like the cGI-PDE, the catalytic domain of the cGS-PDE occurs in the COOH-terminal region in the conserved area (52), Stroop and Beavo identified the catalytic site of the cGS-PDE by affinity photolabeling the bovine heart enzyme, using a high-specific-activity [32P]cGMP (53). The labeled cGS-PDE was cleaved, yielding two fragments, a 60- and a 36 kDa fragment, both labeled (Fig. 2, next page). The addition of cAMP before photolabeling prevented the labeling of the 36 kDa fragment, implicating this fragment as the catalytic site. NH2-terminal amino acid sequencing placed this fragment in the COOH-terminal one-third of the cGS-PDE. In addition, they were able to show that the enzyme activity co-isolated with the photolabeled 36 kDa fragment, and that the 60 kDa fragment constituted an independent functional domain of the enzyme. Whereas the catalytic activity of the 36 kDa fragment was little changed from the native molecule, the allosteric cooperativity displayed by low levels of cGMP was ablated by proteolysis, suggesting that the N-terminal or portions of the C-terminal was required for this action.

Characterization of the allosteric region of cGS-PDE

A second domain of the cGS-PDE seems conserved and homologous to that of two other cGMP binding PDEs, the rod and cone enzymes, but not those of other PDE classes (54,55). These three PDEs share the common ability to bind cGMP at a specific non-catalytic site, a conserved 340 residue cGMP-binding region, which is unique to, and located in the middle of, all three proteins (56). This site does not participate in catalytic activity but may regulate it. Evidence suggests that cGMP is an allosteric activator of the cGS-PDE (57). Photolabeling experiments of these enzymes (55) in the presence of cAMP, followed by digestion with cyanogen bromide (CNBr), yields a 28 kDa fragment that is directly NH2-terminal to the 36 kDa catalytic fragment (Fig. 2). The sequence of this fragment is within the defined conserved cGMP binding region. cGMP binding studies using the native bovine heart cGS-PDE and chymotryptic fragment suggest that the enzyme contains two non-catalytic cGMP-binding sites per homodimer. cGMP binds to the allosteric site with a dissociation constant (Kd) of ~2 (M.

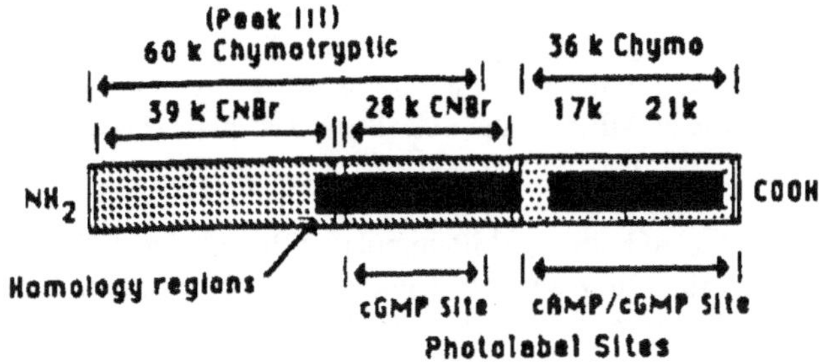

Fig. 2. Schematic linear sequence diagram of fragment alignment illustrating the major fragments produced by either limited chymotryptic proteolysis or digestion with cyanogen bromide. The two homologous cGMP-biding sites are indicated by the black rectangles. Reprinted from (52).

Regulatory domain of cGS-PDE

The 60 kDa NH2-terminal two-thirds of the enzyme also contains a dimerization site. There appears to be a dynamic "bridge region" between the conserved cGMP binding area and the conserved catalytic area of the cGS-PDE (53). Interestingly, the native enzyme is re latively resistant to cleavage by chymotrypsin in the absence of cGMP, with proteolysis resulting only in fragments larger than approximately 90 kDa. However, in the presence of a low concentration of cGMP (1 μM), treatment with chymotrypsin results in several discrete fragments at 60, 36, 21 and 1 kDa. Treatment of the enzyme with dicyclohexylcarbodiimide (DCC) in the presence of, but not in the absence of, cGMP causes an irreversible activation of the enzyme. The proteolytic pattern given by the DCC-activated enzyme is similar to that of the cleaved cGMP-activated enzyme. These data suggest that the "bridge region" is a driving force in the activation of the cGS-PDE by cGMP, and implies that allosteric conformational signals are conveyed directly via the peptides in this area.

Regulation of cyclic AMP activity in the platelets

To date, the various tissues studied seem to contain multiple classes of PDEs. It is possible that the presence of multiple PDEs differing in their modes of regulation and kinetic characteristics is necessary to choreograph cell responses to a variety of signals in a single cell type. Alternatively, the efficiency of PDE regulation of cAMP activity may lie in its subcellular localization and ability to compartmentalize cAMP activity. Whereas the cGI-PDE occurs in both the cytosolic and particulate fractions of cells, in platelets, the majority of cAMP-PDE activity is found in the cytosolic fraction (58,59). The particulate pool of platelet PDE may be involved in inhibition of platelet aggregation, whereas the larger cytosolic pool is associated with inhibition of cytosolic Ca2+ rise and 5HT release (58). The endogenous nitrovasodilator, nitric oxide (NO), has been shown to inhibit platelet activation (60,61). One mechanism of this inhibition may involve the stimulation of guanylate cyclase and the subsequent increase in cGMP levels. Indeed, nitroprusside, a nitrovasodilator that stimulates the formation of cGMP by soluble guanylate cyclase, causes a dose-dependent increase in cGMP and cAMP (62). The latter nucleotide is elevated by a mechanism in which cGMP inhibits the cGI-PDE, causing an elevation of AMP levels. This effect is augmented by the activator, PGE1. These findings suggest that one nucleotide is involved in the regulation of the other. In addition, the cGMP level is further regulated by the cGMP-binding, cGMP-specific PDE (PDE5) found in platelets (see below).

The presence in platelets of both a cGMP-stimulated PDE and a cGMP-inhibited PDE is more difficult to explain. Both seem to have a primarily cytosolic localization. It is possible that their response to cGMP may be regulated differentially by cGMP concentration. The cGI-PDE is inhibited at concentrations more than 10-fold lower than those at which the cGS-PDE is stimulated, suggesting that the cGI-PDE is initially inhibited, allowing the cAMP concentration to rise. Then , as the cGMP level increases, the cGS-PDE may be stimulated, and the cAMP level is finally allowed to fall back to baseline. Thus, cGI-PDE may regulate basal levels of cAMP and cGS-PDE may regulate the elevated levels of cAMP mediated by stimulation of AC. Although inhibition of the major cAMP-PDE in platelets, the cGI-PDE, has clearly been shown to inhibit platelet activation in vitro and has prevented platelet activation in vivo, the relative importance of the cGS-PDE remains to be determined. The lack of specific inhibitors of cGS-PDE had prevented the delineation of the exact role of that enzyme in the platelet, but the discovery of EHNA should obviate this problem.

11.7 PLATELET cGMP-BINDING, cGMP-SPECIFIC PHOSPHODI-ESTERASE (cGB-PDE, PDE5)

Isolation and characterization of cGB-PDE (PDE5)

A cGB-PDE has been purified from both rat (Mr = 93 kDa) and bovine (Mr = 95 kDa) platelets (63,64). In the platelet, the cGB-PDE appears to be primarily cytosolic (63,64). This enzyme is dimeric and is composed of two homologous monomers. The bovine cGB-PDE displays Michaelis-Menten kinetics and hydrolyzes cGMP with a Km of 0.22 μM and cAMP with a K_m of 40 μM. The K_{cat}/K_m is more than twice as high for cGMP as for cAMP. Thus, the Kcat/Km favor cGMP over 400-fold. Optimal hydrolysis occurs at pH of 8 in a Mg^{2+}-dependent reaction. The binding of cGMP is specific and reversible, with optimal binding occurring at an alkaline pH (8.5). The relative potencies of PDE inhibitors is as follows: zaprinast (IC_{50} = 0.1 μM) > dipyridamole (IC_{50} = 0.4 μM) > isobutyl-methylxanthine (IBMX) (IC_{50} = 6 μM) > cilostamide (IC_{50} = 200 μM).

Regulatory domain of cGB-PDE

Characteristic of the cGB-PDEs is the ability of isobutyl-methylxanthine (IBMX) to promote cGMP binding while inhibiting the enzyme due to enhanced binding at a non-catalytic site. IBMX acts as a competitive inhibitor of cGMP hydrolysis. In fact, a low affinity binding site (K_d = 1.5 x 106 M) in the presence of IBMX has been described for the platelet cGB-PDE (64). Analogs specific for the catalytic site also stimulated cGMP labeling of the cGB-PDE, suggesting a communication between a cGMP hydrolytic site and a cGMP-binding site (65). Binding of cGMP produces an electronegative shift on HPLC chromatography in accordance with a conformational change in the enzyme.

Both cAMP (cAK) and cGMP-dependent kinases (cGK) serine-phosphorylate the cGB-PDE at the same or adjacent sites in a cGMP-dependent reaction. cGMP regulates cGB-PDE phosphorylation through both kinase and substrate effects. However, cGK-catalyzed phosphorylation occurs at a rate 10-fold greater than with cAK (66). The stoichiometry of the reaction approaches 1 mol of phosphate incorporated/mol of cGB-PDE dimer (66,67). The conformational change associated with the binding of cGMP to a noncatalytic bnding site may be required to expose a phosphorylation site. The cGK phosphorylation site is located at the N terminal of the protein (65), outside of the binding and dimerization domains (68). Phosphorylation of the bovine platelet cGB-PDE did not result in a change in the specific activity of the enzyme (64). However, phosphorylation of guinea pig lung cGB-PDE by the catalytic subunit of PKA resulted in a 10-fold increase in the Vmax for cGMP hydrolysis (67). cGMP enhanced the enzyme 's ability to act as a substrate for protein kinase A. A decrease in the sensitivity of the enzyme to inhibition by Zaprinast was also noted with phosphorylation.

Regulation of cGMP in the platelet

The stimulation of guanylate cyclase also provides an important pathway for cell-cell interaction. Endothelial cells interact with platelets by releasing nitric oxide which, in turn, stimulates soluble guanylate cyclase within the platelet, resulting in increased levels of cGMP within the cell (Fig. 1) (69,70). The cGMP-binding, cGMP-specific PDE (cGB-PDE) also participates in this regulation by degrading cGMP.

The indirect regulation of the PDEs via phosphorylation suggests that the nucleotides are involved in cross-communication. Elevation of the cAMP levels may trigger the activation of cAK and result in the phosphorylation of the cGI-PDE which, in turn, increases its rate of cAMP hydrolysis, thereby acting in a negative feedback mode. At the same time, the cGB-PDE is phosphorylated and also activated, thereby increasing hydrolysis of cGMP, allowing the cGI-PDE, the major PDE in platelets, to hydrolyze cAMP. These actions will result in the returning of the cAMP levels back to baseline. On the other hand, an increase in cGMP concentration will result in elevated cAMP levels by inhibiting the cGI-PDE. However, cGMP may also serve in a feedback loop by activation of the cGMP-dependent kinase and phosphorylation of the cGB-PDE.

REFERENCES

1. Salzman EW, Weisenberger H: Role of cyclic AMP in platelet function. Adv Cyclic Nucleotide Res 1972; 1: 231-247.

2. Simonds WF, Goldsmith PK, Codina J, Unson CG, Spiegel AM: Gi2 mediates (2-adrenergic inhibition of adenylyl cyclase in platelet membranes: In situ identification with Ga C-terminal antibodies. Proc Natl Acad Sci USA 1989; 86: 7809-7813.

3. Alvarez R, Taylor A, Fazzari J, Jacobs JR: Regulation of cyclic AMP metabolism in human platelets. Sequential activation of adenylate cyclase and cyclic AMP phosphodiesterase by prostaglandins. Molec Pharmacol 1981; 20: 302-309.

4. Seiler S, Gillespie E, Arnold AJ, Brassard CL, Meanwell NA, Fleming JS: Imidazoquinoline derivatives: Potent inhibitors of platelet cAMP phosphodiesterase which elevate cAMP levels and activate protein kinase in platelets. Thromb Res 1991; 62: 31-42.

5. Siegl AM, Daly JW, Smith JB: Inhibition of aggregation and stimulation of cyclic AMP generation in intact human platelets by the diterpene forskolin. Molec Pharmacol 1982; 21: 680-687.

6. Tani T, Sakurai K, Kimura Y, Ishikawa T, Hidaka H: Pharmacological manipulation of tissue cyclic AMP by inhibitors. Effects of phosphodiesterase inhibitors on the functions of platelets and vascular endothelial cells. Adv Second Messenger Phosphoprotein Res 1992; 25: 215-227.

7. Beavo JA, Conti M, Haeship RJ: Multiple cyclic nucleotide phosphodiesterases. Molec Pharmacol 1994; 46: 399-405.

8. Charbonneau H: Structure-function relationships among cyclic nucleotide phosphodiesterases. In: Cyclic Nucleotide Phosphodiesterases: Structure, Regulation and Drug Action (Beavo J, Houslay MD, eds.). New York: John Wiley and Sons, 1990: pp. 267-296.

9. Hidaka H, Asano T: Human blood platelet 3': 5'-cyclic nucleotide phosphodiesterase. Isolation of low-Km and high-Km phosphodiesterase. Biochim Biophys Acta 1976; 429: 485-497.

10. Asano T, Ochiai Y, Hidaka H: Selective inhibition of separated forms of human platelet cyclic nucleotide phosphodiesterase by platelet aggregation inhibitors. Molec Pharmacol 1977; 13: 400-406.

11. Hidaka H, Endo T: Selective inhibitors of three forms of cyclic nucleotide phosphodiesterase: Basic and potential clinical applications. Adv Cyclic Nucl Protein Phosphorylation Res 1984; 16: 245-259.

12. Grant PG, Mannarino AF, Colman RW: Purification and characterization of a cyclic GMP-stimulated cyclic nucleotide phosphodiesterase from the cytosol of human platelets. Thromb Res 1990; 59: 105-119.

13. Muggli R, Tschopp TB, Mittelholzer E, Baumgartner HR 7-Bromo-1,5-dihydro-3,6-dimethyl-imidazo[2,1-b]quinazolin-2(3H)- one (Ro 15-2041), a potent antithrombotic agent that selectively inhibits platelet cyclic AMP-phosphodiesterase. J Pharmacol Exp Ther 1985; 235: 212-219.

14. Mills DC, Smith JB: The influence on platelet aggregation of drugs that affect the accumulation of adenosine 3':5'-cyclic monophosphate in platelets. Biochem J 1971; 121: 185-196.

15. MacPhee CH, Harrison SH, Beavo JA. Immunological identification of the major platelet low-Km cAMP phosphodiesterase: Probable target for anti-thrombotic agents. Proc Natl Acad Sci USA 1986; 83: 6660-6663.

16. Honerjager P: Pharmacology of bipyridine phosphodiesterase III inhibitors. Am Heart J 11991; 21: 1939-1944.

17. Beretz A, Briancon Scheid F, Stierle A, Corre G, Anton R, Cazanave JP: Inhibition of human platelet cyclic AMP phosphodiesterase and of platelet aggregation by a hemisynthetic flavonoid, amentoflavone hexaacetate. Biochem Pharmacol 1986; 35: 257-262.

18. Meanwell NA, Roth HR, Smith EC, Wedding JJ, Fleming JS, Gillespie E: 1,3-Dihydro-2H-imidazo [4,5-b]quinolin-2-one- inhibitors of blood platelet cAMP phosphodiesterase and induced aggregation. J Med Chem 1991; 34: 2906-2916.

19. Gillespie E: Anagrelide: A potent and selective inhibitor of platelet cyclic AMP phosphodiesterase enzyme activity. Biochem Pharmacol 1988; 37: 2866-2868.

20. Mikashima H, Nakao T, Goto K, Ochi H, Yasuda H, Tsumagari T. Y-590 (a new pyridazinone derivative), a potent anti-thrombotic agent-II. Inhibition of platelet phosphodiesterase. Thromb Res 1984; 35: 589-594.

21. Martinez GR, Walker KA, Hirschfeld DR, Bruno JJ, Yang DS, Maloney PJ: 3,4-Dihydroquinolin -2(1H)-ones as combined inhibitors of thromboxane A2 synthase and cAMP phospho- diesterase. J Med Chem 1992; 35: 620-628.

22. Hoet B, Arnout J, Deckmyn H, Vermylen J. Synergistic antiplatelet effect of ridogrel, a combined thromboxane receptor antagonist and thromboxane synthase inhibitor and UDCG-212, a cAMP-phosphodiesterase inhibitor. Thromb and Haemos 1993; 70: 822-825.

23. Kambayashi J, Watase M, Kawasaki T, Shiba E, Sakon M, Mori T, Isaka Y, Kimura K, Kamada T: Phosphodiesterase inhibitors as anti-platelet agents in vascular surgery. Adv Second Messenger Phospho-protein Res 1992; 25: 383-393.

24. Saitoh S, Saito T, Otake A, Owada T, Mitsugi M, Hashimoto H, Maruyama Y: Cilostazol, a novel cyclic AMP phosphodiesterase inhibitor, prevents reocclusion after coronary arterial thrombolysis with recombinant tissue-type plasminogen activator. Arterioscler Thromb 1993; 13: 563-570.

25. Grant PG, Colman RW: Purification and characterization of a human platelet cyclic nucleotide phosphodiesterase. Biochemistry 1984; 23: 1801-1807.

26. Grant PG, Colman RW: Purification of cAMP phosphodiesterase from platelets. Methods Enzymol 159: 772-777, 1988.

27. Omburo GA, Brickus T, Ghazaleh FA, Colman RW. Divalent metal cation requirement and possible classification of cGMP-inhibited phosphodiesterase as a metallohydrolase. Arch Biochem Biophys 232: 1-5, 1995.

28. Erhardt PW. Second-generation phosphodiesterase inhibitors: Structure-activity relationships and receptor models. In: Cyclic Nucleotide Phosphodiesterases: Structure, Regulation and Drug Action Beavo J, Houslay MD, eds.). New York: John Wiley and Sons, 1990: pp. 317-332.

29. Grant PG, DeCamp DL, Bailey JM, Colman RW, Colman RF: Three new potential cAMP affinity labels. Inactivation of human platelet low Km cAMP phosphodiesterase by 8-[(4-bromo-2,3-dioxobutyl)-thio]adenosine 3',5'-cyclic monophosphate. Biochemistry 1990; 29: 887-894.

30. Meacci E, Taira M, Moos M, Jr., Smith CJ, Movsesian MA, Degerman E, Belfrage P, Manganiello V: Molecular cloning and expression of human myocardial cGMP-inhibited cAMP phosphodiesterase. Proc Natl Acad Sci USA 1992; 89: 3721-3725.

31. Taira M, Hockman SC, Calvo JC, Taira M, Belfrage P, Manganiello VC. Molecular cloning of the rat adipocyte hormone-sensitive cyclic GMP-inhibited cyclic nucleotide phosphodiesterase. J Biol Chem 1993; 268: 18573-18579.

32. Ghazaleh FA, Omburo GA, Colman RW. Evidence for the presence of essential histidine and cysteine residues in platelet cGMP-inhibited phosphodiesterase. Biochem J 317: 495-501, 1996.

33. Grant PG, Mannarino AF, Colman RW: cAMP-mediated phosphorylation of the low-Km cAMP phosphodiesterase markedly stimulates its catalytic activity. Proc Natl Acad Sci USA 1988; 85: 9071-9075.

34. Macphee CH, Reifsnyder DH, Moore TA, Beavo JA: Intact cell and cell-free phosphorylation and concomitant activation of a low Km, cAMP phosphodiesterase found in human platelets. J Cyclic Nucleotide Protein Phosphor Res 1986; 11: 487-496.

35. Macphee CH, Reifsnyder DH, Moore TA, Lerea KM, Beavo JA: Phosphorylation results in activation of a cAMP phosphodiesterase in human platelets. J Biol Chem 1988; 263: 10353-10358.

36. Sheth SB, Biradavolu R, Brennan KJ, Colman RW: Isolation and regulation of the cGMP-inhibited cAMP phosphodiesterase in human erythroleukemia cells. Thromb Haemost 77: 155-162, 1997.

37. Tabilio A, Rosa JP, Testa U, Kieffer N, Nurden AT, Del Canizo MC, Breton-Gorius J, Vainchenker W: Expression of platelet membrane glycoproteins and alpha-granule proteins by a human erythroleukemia cell line (HEL). EMBO J 1984; 3: 453-459.

38. Cheung PP, Xu H, McLaughin MM, Ghazaleh FA, Livi GP, Colman RW. Human platelet cGI-PDE: Expression in yeast and localization of the catalytic domain by deletion mutagenesis. Blood 88: 1321-1329, 1996.

39. Butt E, Beltman J, Becker D, Jensen G, Jastorff, Beavo JA: Characterization of cyclic nucleotide phosphodiesterases: topology of the catalytic domains. FASEB J 1994; 8: A86.

40. Buechler YJ, Herberg FW, Taylor SS. Regulation-defective mutants of type I cAMP-dependent protein kinase. J Biol Chem 1993; 268:16495-16503.

41. Su Y, Dostmann WR, Herberg FW, et al. Regulatory subunit of protein kinase A: Structure of deletion mutant with cAMP binding domains. Science 1995; 269:807-813.

42. Herberg FW, Taylor SS, Dostmann WR. Active site mutations define the pathway for the cooperative activation of cAMP-dependent protein kinase. Biochemistry 1996; 35:2934-2942.

43. Gibson RM, Buechler YJ, Taylor SS. Interaction of the regulatory and catalytic subunits of cAMP-dependent protein kinase. J Biol Chem 1997; 272:16343-16350.

44. Tang KM, Sherwood JL, Haslam RJ. Photoaffinity labeling of cyclic GMP-binding proteins in human platelets. Biochemistry 1993; 294:329-333.

45. Francis SH, Corbin JD. Progress in understanding the mechanism and function of cyclic GMP-dependent protein kinase. Adv Pharmacol 1994; 26:115-170.

46. Reed RB, Sandberg M, Jahnsen T, Lohmann SM, Francis SH, Corbin JD. Fast and slow cyclic nucleotide-dissociation sites in cAMP-dependent protein kinase are transposed in type Ib cGMP-dependent protein kinase. J Biol Chem 1996; 271:17570-17575.

47. Butt E, Pohler D, Genesier HG, Huggins JP, Bucher B. Inhibition of cyclic GMP-dependent protein kinase-mediated effects by (T-)-8-bromo-PET-cylic GMPS. Br J Pharmacol 1995; 116:3110-3116.

48. Sonnenburg WK, Beavo JA. Cyclic GMP and regulation of cyclic nucleotide hydrolysis. Adv Pharmacol 1994; 26:87-114.

49. Grant PG, Mannarino AF, Colman RW. Purification and characterization of a cyclic GMP-stimulated cyclic nucleotide phosphodiesterase from the cytosol of human platelets. Thromb Res 59: 105-199, 1990.

50. Podzuweit T, Neunstiel P, Muller A. Isozyme selective inhibitor of cGMP-stimulated cyclic nucleotide phosphodiesterases by erythro-9-(2-hydroxy-3-nonyl)adenosine. Cellular Signalling 7: 733-738, 1995.

51. Martins TJ, Mumby MC, Beavo JA. Purification and characterization of a cyclic GMP-stimulated cyclic nucleotide phosphodiesterase from bovine tissues. J Biol Chem 257: 1973-1979, 1982.

52. Stroop SD, Beavo JA. Sequence homology and structure: Function studies of the bovine cyclic GMP-stimulated and retinal phosphodiesterases. Adv Second Messenger Phosphoprotein Res 25: 55-71, 1992.

53. Stroop SD, Beavo JA. Structure and function studies of the cGMP-stimulated phosphodiesterase. J Biol Chem 266: 23802-23809, 1991

54. Charbonneau H, Prusti RK, LeTrong H, Sonnenburg WK, Mullaney PJ, Walsh KA, Beavo JA. Identification of a non-catalytic cGMP-binding domain conserved in both the cGMP-stimulated and photoreceptor cyclic nucleotide phosphodiesterase. Proc Natl Acad Sci 87: 288-292, 1990.

55. Stroop SD, Charborreau H, Beavo JA. Direct photolabeling cGMP-stimulated cyclic nucleotide phosphodiesterase. J Biol Chem 264: 13718-13725, 1989.

56. Trong HL, Beier N, Sonnenburg WK, Stroop SD, Walsh KA, Beavo JA, Charborreau H. Amino acid sequence of the cyclic GMP stimulated cyclic nucleotide phosphodiesterase from bovine heart. Biochemistry 29: 10280-10288, 1990.

57. Wada H, Manganiello VC, Osborne Sr. JC. Analysis of the kinetics of cyclic AMP hydrolysis by the cyclic GMP-stimulated cyclic nucleotide phosphodiesterase. J Biol Chem 262: 13938-13945, 1987.

58. Ashida S, Sakuma K. Demonstration of functional compartments of cyclic AMP in rat platelets by the use of phosphodiesterase inhibitors. Adv Second Messenger Phosphoprotein Res 25: 229-239, 1992.

59. Grant PG, Colman RW. Purification of cAMP phosphodiesterase from platelets. Methods Enzymol 159: 772-777, 1988.

60. Radomski MV, Moncada S. Regulation of vascular homeostasis by nitric oxie. Thromb Haemost 70: 36-41, 1993.

61. Radomski MV, Palmer RM, Moncada S. The anti-aggregating properties of vascular endothelium: Interactions between prostacyclin and nitric oxide. Br J Pharmacol 92: 639-646, 1987.

62. Maurice DH, Haslam R. Molecular basis of the synergistic inhibition of platelet function by nitrovasodilators and activators of adenylate cyclase: Inhibition of cyclic AMP breakdown by cyclic GMP. Mol Pharmacol 37: 671-681, 1990.

63. Coquil JF, Franks DJ, Wells JN, Dupuis M, Hamet P. Characteristics of a new binding protein distinct from the kinase for guanosine 3":5'-monophosphate in rat platelets. Biochim Biophys Acta 1980; 631:148-165.

64. Robichon A. A new cGMP phosphodiesterase isolated from bovine platelets is substrate for cAMP- and cGMP-dependent protein kinases: Evidence for a key role in the process of platelet activation. J Cell Biochem 1991; 47:147-157.

65. Thomas MK, Francis SH, Corbin JD. Characterization of a purified bovine lung cGMP phosphodiesterase. J Biol Chem 1990; 265:14964-14970.

66. Thomas MK, Francis SH, Corbin JD. Substrate- and kinase-directed regulation of phosphorylation of a cGMP-binding phosphodiesterase by cGMP. J Biol Chem 1990; 265:14971-14978.

67. Burns F, Rodger IW, Pyne NJ. The catalytic subunit of protein kinase A triggers activation of the type V cyclic GMP-specific phosphodiesterase from guinea-pig lung. Biochem J 1992; 283:487-491.

68. McAllister-Lucas LM, Sonnenburg WK, Kadlecek A. The structure of a bovine lung cGMP-binding, cGMP-specific phosphodiesterase deduced from a cDNA clone. J Biol Chem 1993; 268:22863-22873.

69. Vallance P, Benjamin N, Collier J. The effect of endothelium-derived nitric oxide on ex vivo whole blood platelet aggregation in man. Eur J Clin Pharmacol 1992; 42:37-41.

70. Sneddon JM, Vane JR. Endothelium-derived relaxing factor reduces platelet adhesion to bovine endothelial cells. Proc Natl Acad Sci U S A 1988; 85:2800-2804.

12

POLYENOIC FATTY ACIDS AND PLATELET FUNCTION

Mahadev Murthy, Ph. D.
Division of Endocrinology,
Metabolism & Nutrition
Department of Medicine
Hennepin County Medical Center
Minneapolis, MN
55404, USA

12.1 INTRODUCTION

Over the last 2-3 decades, platelets have been extensively investigated because of their close association with the incidence of coronary artery disease (CAD), which is still the number one killer in the United States and in other industrialized countries. Although platelets are non-nucleated (lack of nucleus), their functional instruments continue to impact not only the CAD but also many other diseases. Thus, platelet research has clearly led the way in the continuing development of new strategies for the prevention of arterial thrombosis. In this regard, considerable progress has been made in the development of new antiplatelet agents. These newer agents are based on interrupting specific sites in the sequence of platelet activation. Inhibitors of specific platelet agonist-receptor interactions include antithrombins,

Over the years, Dr. Murthy has served on the Faculty of the University of Guelph, ONT, and Manitoba, MB, Canada. His research interests include antioxidants, lipid and fatty acid metabolism, eicosanoids and calcium, and their impact on health and disease. More recently, he led the development of a Canadian Center for Nutraceuticals while at the University of Manitoba. He continues to be interested in lipid biochemistry and nutrition. Dr. Murthy is currently working in the Division of Endocrinology, Metabolism and Nutrition, Department of Medicine, Hennepin County Medical Center (HCMC), Minneapolis, Minnesota, USA.

thromboxane A_2 (TxA_2) receptor antagonists, and adenosine diphosphate (ADP) receptor blockers, including ticlopidine and clopidogrel. Inhibitors of arachidonic acid (AA; 20:4, ω-6) metabolism and TxA_2 include ω-3 fatty acids, aspirin and other nonsteroidal antiinflammatory drugs that inhibit cyclooxygenase and thromboxane synthase. The clinical efficacy of these agents may be limited by their actions, as they are restricted to single, specific platelet receptors or metabolic pathways. However, newer drugs that block ligand binding to the platelet glycoprotein IIb/IIIa complex (i.e. tirofiban) will likely be more effective (1). The overview presented by Dr. Rao (Chapter 1) in this book highlights some of these advances that represent the hallmark of platelet physiology, biochemistry and pharmacology.

12.2 Platelet function and its relevance to thrombosis

Circulating platelets neither adhere to the vessel wall nor aggregate unless they encounter a zone of injury. Upon encountering such a zone of injury, they become almost instantly activated, which leads to their adhesion and aggregation, both reactions of fundamental importance for hemostasis and thrombosis (2, 3). Adhesion of blood platelets therefore constitute one of the first reactions after endothelial injury. Arterial thrombi are composed of predominantly platelets (Fig. 1). The mechanism by which a platelet thrombus is formed involves at least two adhesive macromolecules: von Willebrand factor (vWF) and fibrinogen, and their receptors on platelets (membrane glycoproteins). vWF is thought to provide a molecular anchor between the subendothelium and platelets. Fibrinogen, on the other hand, is considered to provide interplatelet linkages after activation of platelets with ADP and rearrangement of glycoproteins IIb/IIIa (GPIIb/IIIa), which constitute the receptor for fibrinogen. A deficiency of these glycoproteins results in impaired formation of a platelet thrombus on the subendothelium. This aspect has been discussed in depth by other contributors in this book.

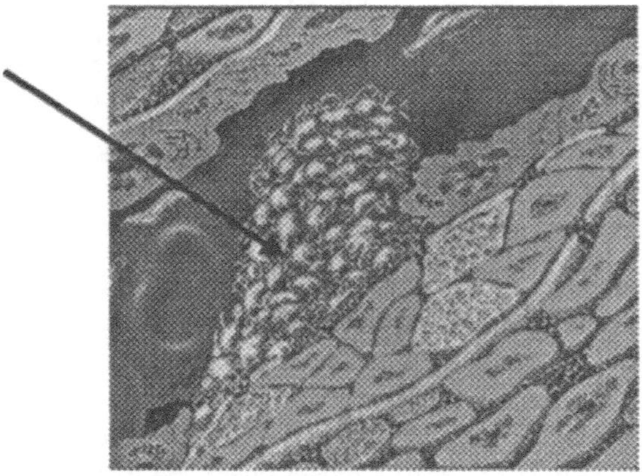

Figure 1: Massive platelet thrombus formation clearly showing narrowing of the artery and adherence of platelets to the endothelium.

12.3 Polyunsaturated fatty acids (PUFAs)

The fatty acid, specifically AA (20:4, ω-6), has a key functional role in platelet-mediated hemostasis and thrombosis because of its rapid conversion to TxA_2 upon platelet activation. Over the last few years, we have seen a tremendous interest on AA metabolism within the scientific community. Certainly the interest on PUFAs, has

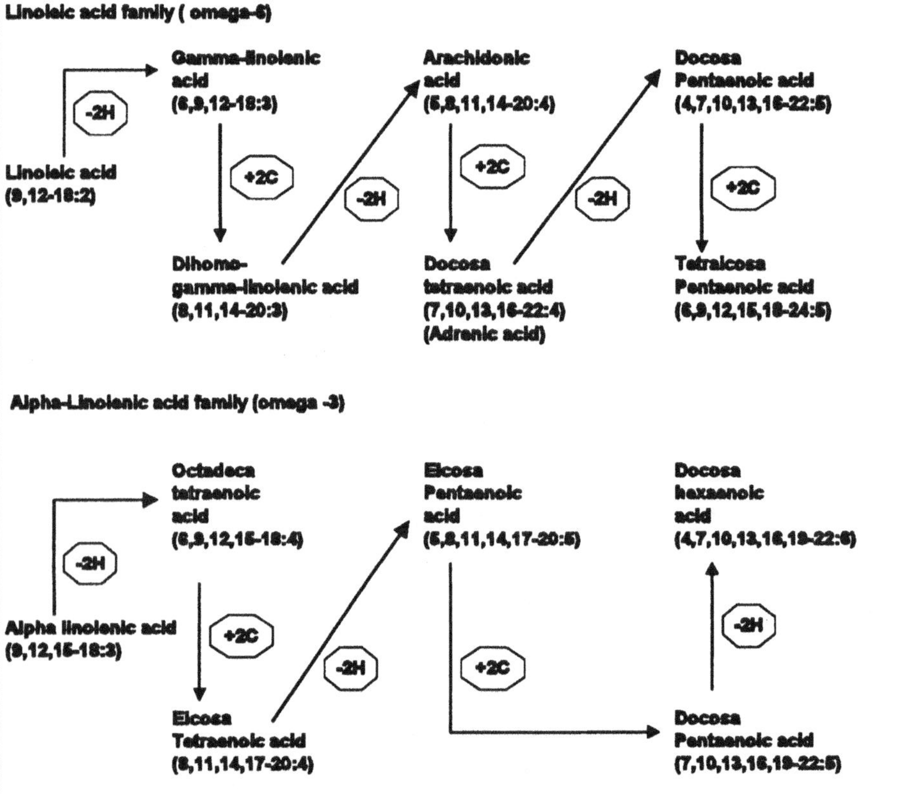

Figure 2. Interconversion pathways for polyenoic fatty acids

grown beyond their simple requirement to maintain growth and proper nutrition. It is beyond the scope of this chapter to discuss in detail every aspect of PUFAs metabolism. Cited references are therefore restricted to key reviews and original publications that highlight major contributions to the metabolism of PUFAs in relation to platelet function and thrombosis. As illustrated in Figure 2, the PUFAs taken in the diet are desaturated and elongated in the liver to a longer chain and more unsaturated PUFAs of highly physiological and pathophysiological relevance. These fatty acids are actively taken up and metabolized by platelets (4-10).

Dietary linoleic acid (18:2, ω-6) is the indirect precursor for both the eicosanoid 1 and

and eicosanoid 2 series. Alpha-linolenic acid (18:3, ω-3) is the indirect precursor of the eicosanoid 3 series. Eicosanoids are oxygenated metabolites of AA (20:4, ω-6) and other PUFAs of 20 carbons in length ("eicosa" indicating "20"). However, the metabolites derived from higher length PUFAs are also included in the eicosanoid

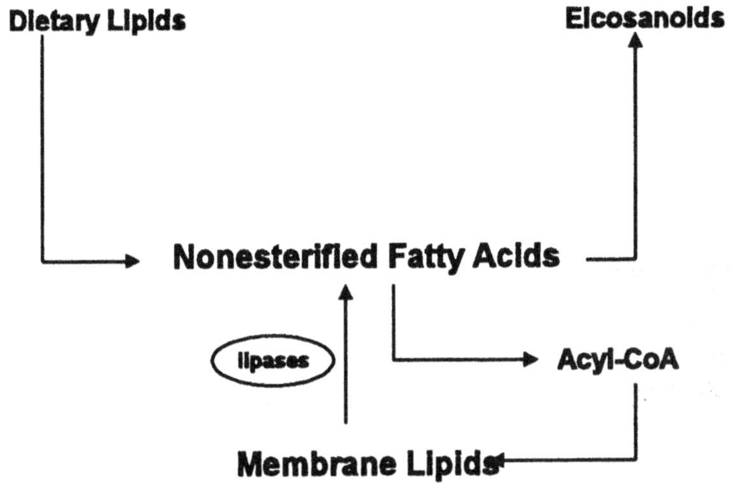

Figure 3. The fatty acid turnover pathway, showing the release, uptake and conversion of fatty acids

terminology. The eicosanoids derived from AA (20:4, ω-6) represent the most abundantly occurring group, with wide array of potent pathophysiological consequences in mammalian cells/tissues. It is recognized that the indirect precursor fatty acids taken with the diet not only contribute to the assembly of biological membranes with appropriate fatty acids but also to the increased flux of eicosanoids that impact both local and global environments (Fig. 3). In general, AA (20:4, ω-6) is one of the most essential and predominant PUFAs present in its esterified form in membrane phospholipids. This fatty acid seldomly accumulates as free (non-esterified) in cells, tissues, interstitial fluids and body fluids and its concentrations remain very low under physiological conditions. But this fatty acid is released from membrane phospholipids in zones of injuries via the activation lipases and once it is free, it is either rapidly oxidized to eicosanoids or it is reincorporated into phospholipids via the highly selective acyl-CoA-mediated pathway (Lands pathway) or the exchange pathway (summarized in Figure 3).

12.4 Platelet membranes and their lipid composition

Platelet phospholipid and fatty acid composition was first reported in some detail by two groups (11, 12). Subsequently, our laboratory fully characterized individual fatty acid specific molecular species in human platelet phospholipids (13), in addition to documenting the molar concentration of individual phospholipids (Table 1). This work was of significant interest in view of several reports regarding the origin of the selectively released AA (20:4, ω-6) from platelet membrane phospholipids, for

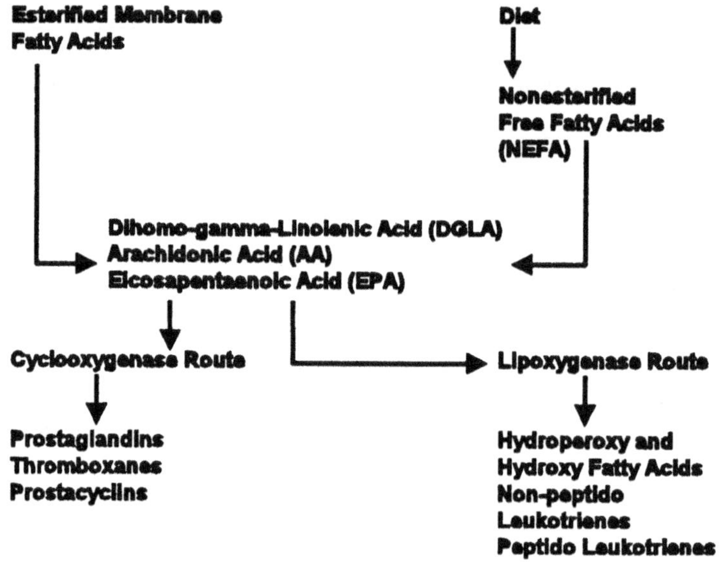

Figure 4: The pathways showing the conversion of different eicosanoid precursors

eicosanoid synthesis. This work also established through stereospecific analysis that AA (20:4, ω-6) is exclusively esterified to the sn-2 position of the individual platelet membrane phospholipids. Other cell and tissue phospholipids also contain AA (20:4, ω-6) exclusively in the sn-2 position. Similarly, PUFAs of both the ω-6 and ω-3 family selectively occupy the sn-2 position in membrane phospholipids (14). In platelets, phosphatidylcholine (PC), phosphatidylethanolamine (PE), phosphatidylserine (PS) and phosphatidylinositol (PI), contain 31, 35, 17 and 17 % 1-acy 2-arachidonoyl species, respectively (13). Although platelets contain only 11 of PS and 6 % PI, together they contain disproportionate 34 % of 1-acyl-2-arachidonoyl species. In other words, platelet PS and PI are preferentially enriched with 1-acy 2-arachidonoyl species as compared to PE and PC. This selective enrichment of arachidonoyl species in PS and PI, appears to be important for many membrane associated physiological functions, including immune function. Platelets also contain plasmalogens, which serve as a storage depot for highly unsaturated fatty acids, including eicosanoid precursors. We are now beginning to realize the importance of such highly selective and differential fatty acid assembly among platelet phospholipids that impact many functional instruments, including signal transduction, metabolic regulation and gene expression.

Table 1. Phospholipid Composition of Normal Human Platelets[b]

Phospholipid	% of Total[a]	
	nmol/mg Protein	mol % of Total
Phosphatidylcholine (PC)	70.8±1.2	39.9±0.8
Phosphatidylethanolamine (PE)	41.7±3.1	23.3±1.1
Phosphatidylserine (PS)	19.6±0.4	11.1±0.5
Phosphatidylinositol (PI)	10.9±0.8	6.1±0.4
Sphingomyelin (SPH)	34.9±1.5	19.6±0.4

[a]Values are given as means±S.E., for four subjects. Total phospholipid was 178 nmol/mg protein or 42 nmol/10^8 platelets. [b] Data with permission from V. G. Mahadevappa and B. J. Holub, Biochim. Biophys. Acta 713, 73-79 (1982).

12.5 Arachidonic acid and platelet eicosanoids

Platelet AA metabolism was first reported by Smith and Willis (1971, 15). Subsequent studies demonstrated both the AA conversion and formation of endoperoxides and thromboxnes from exogenous AA in platelets, leading to the conclusion that TxA_2 is an active aggregating compound (16-19). It is only the non-esterified AA that can be a substrate for oxygenase (20). The subsequent discovery of prostacyclin (PGI_2) by Moncada and co-workers (1976) led to the new concept of considering a balance between proaggregatory compounds such as platelet thromboxanes and antiaggregatory compounds such as prostacyclins from a vascular tissue (21-24). Platelets also contain 12-lipoxygenase that oxygenates AA at C12, leading to the formation of 12-hydroperoxyeicosatetraenoic acid (12-HPETE) and then,12-hydroxyeicosatetraenoic acid (12-HETE) (18). Since its discovery, platelet TxA_2 is recognized as an important intracellular mediator in activated platelets. More specifically, TxA_2 appears more important for initial collagen-induced platelet responses, including shape change, aggregation, secretion, phosphoinositide turnover, Ca^{2+} mobilization from intracellular stores and protein phosphorylations (24-28; Chapter 9 in this book). In addition, TxA_2 amplifies its own production in a novel manner through its own receptor present on platelet membranes. Platelet adhesion to collagen appears to be independent of TxA_2. Nevertheless, TxA_2 is an important mediator of pathophysiological functions, ranging from hemostasis to smooth muscle contraction by virtue of its potent vasoconstricting and platelet aggregating properties. The pathways of AA metabolism and their significance in pathophysiology have been extensively discussed by others and can be found in a number of reviews (4, 5, 9, 10, 18, 23, 24, 29). These pathways are summarized in Figure 5.

The enzymatic pathways that are responsible for the initial mobilization of AA (20:4, ω-6) as well as other PUFAs for eicosanoid synthesis, are being discussed in detail in chapter 13 of this book. Once the AA (20:4, ω-6) is released from platelet membrane phospholipids, it is rapidly metabolized via the cyclooxygenase and lipoxygenase pathways (Figure 5). Therefore, the initial release of AA/other precursors serves as the rate limiting step in the biosynthesis of eicosanoids.

Cyclooxygenase contains oxygenase and peroxidase activities within a single polypeptide chain. The oxygenase activity catalyses the addition of two O_2 molecules to AA, giving rise to a cyclic endoperoxide, PGG_2. This unstable metabolite is then

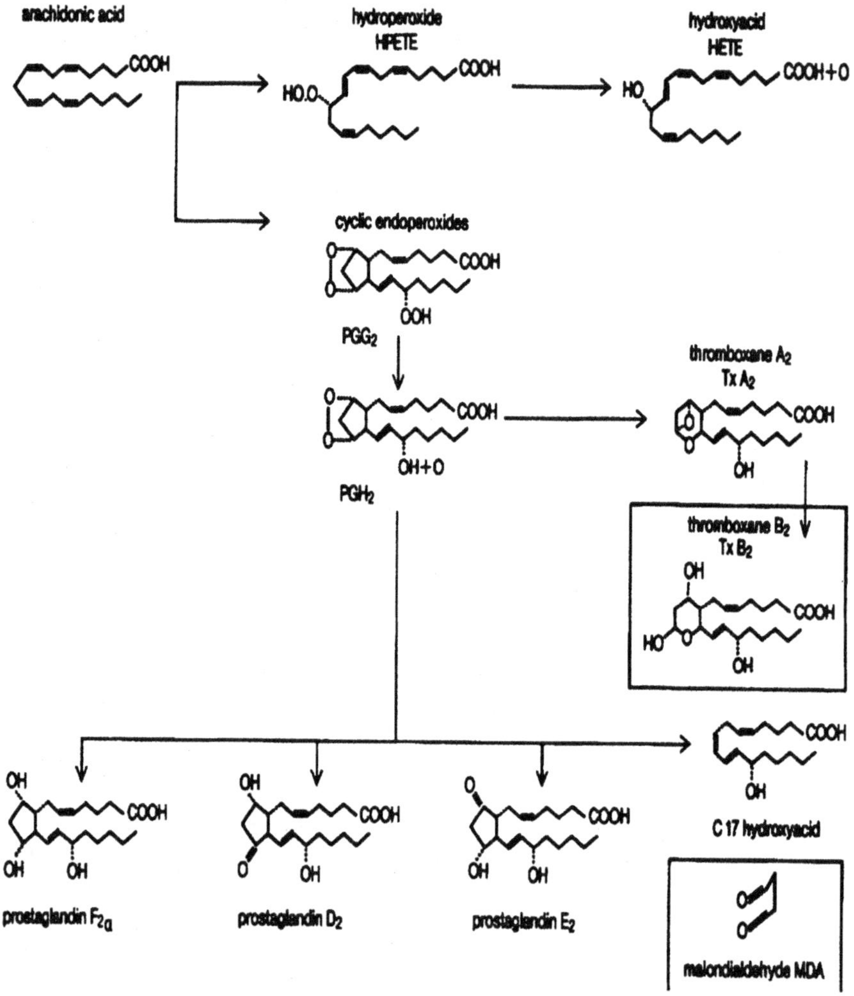

Figure 5. The pathway showing the formation different eicosanoids from arachidonic acid (20:4, ω-6) from human platelets. When eicosapentaenoic acid (EPA: 20:5, ω-3) is substituted in the membrane, TxA_3 is formed in addition to TxA_2. TxA_3 is biologically less potent compared to TxA_2.

converted to PGH_2 by the peroxidase activity. The platelet cyclooxygenase pathway also contributes to the synthesis of prostacyclin (PGI_2) in the vessel wall (30). Cyclooxygenase is a membrane-bound glycoprotein (M_r 72,000), with a single heme group that mediates both oxygenase and peroxidase activity (31). Thromboxane B_2

(TxB_2) is the biologically inactive and stable end product of thromboxane synthesis in platelets. It is formed from TxA_2, a labile ($t_{1/2}$ about 30 sec, 37°C), but biologically very potent aggregation and secretion initiating agent and vasoconstrictor. TxA_2 and malondialdehyde are formed in equal amounts and the reaction is catalysed by thromboxane synthase. Platelets also produce other prostaglandins such PGD_2, PGE_2 and $PGF_{2\alpha}$ (Figure 4). It is beyond the scope of this chapter to go into the various aspects of eicosanoid metabolism as they can be found elsewhere in this book and in other relevant articles and reviews.

12.5.1 Other omega-6 polyunsaturated fatty acids and platelets

As shown in Figure 2, γ-linolenic acid (GLA; 18:3, ω-6) is an intermediate between dietary linoleic acid (18:2, ω-6) and AA (20:4, ω-6). There has been considerable interest on γ-linolenic acid (GLA; 18:3, ω-6) and its metabolic 20-carbon product, dihomo-γ-linolenic acid (DGLA; 20:3, ω-6), which is a precursor of the 1-series eicosanoids. γ-linolenic acid (GLA; 18:3, ω-6) is present in evening primrose oil (EPO) and borage oil that contain 8-9 % and 18-19 % GLA, respectively. These two oils have been extensively used as a source of GLA for studying its metabolism in animal and human models. The interpretation of GLA on platelet fatty acid metabolism is quite complex for a variety of valid scientific reasons. Many dietary studies have shown that the degree of incorporation of GLA and/or DGLA into platelet membrane phospholipids is extremely small, with little change in the AA (20:4, ω-6) level, which is a precursor of the 2-series eicosanoids. Secondly, it is argued that an increased intake of GLA may in fact increase the levels of AA, thereby providing more substrate for the 2-series eicosanoids. Furthermore, the interpretation of its interaction with other fatty acid pathways such as those of ω-3 fatty acids without clear cut data becomes much more difficult even though it may produce small amounts of PGE_1 (32-34). However, both platelet aggregation and thromboxane formation have been shown to be significantly affected by γ-linolenic acid (GLA; 18:3, ω-6) (35).

With respect to GLA effects on platelet fatty acid metabolism and function, it has been shown that mold oil containing γ-linolenic acid inhibits platelet thrombus formation in mesenteric microvessels of the rat (36). In addition, dietary intake of GLA appears to improve vascular resistance and lower blood pressure (37-39). Although there is inconsistent data to conclude that GLA either reduces platelet aggregation/secretion or platelet thrombus formation, its ability to improve vascular resistance and blood pressure, may provide significant cardiovascular protection. This has led to the argument that we must find an appropriate ratio of ω-3/ω-6 fatty acids in order to reduce the risk of CAD. In order to accomplish an appropriate ratio, we would need to reduce the total intake of dietary ω-6 fatty acids. This issue raises a critical question as to whether we should decrease the amount of dietary linoleic acid (18:2, ω-6) while increasing the amount of GLA (18:3, ω-6), which would increase the ratio of GLA (18:3, ω-6) and linoleic acid (18:2, ω-6). Although this idea merits further consideration, it may be easier to incorporate this concept into the food chain through genetic engineering that could lead to the development of new vegetable oils, with higher levels of GLA and lower levels of linoleic acid (18:2, ω-6).

12.6 Omega-3 fatty acids

As shown in Figure 2, α-linolenic acid (18:3, ω-3) is a precursor of long chain fatty acids, including eicospentaenoic acid (EPA) and docosahexaenoic acid (DHA). Early literature on this fatty acid can be found in the review by Tinoco (1982) (40). A developing fetus uses dietary α-linolenic acid to synthesize DHA, an essential component of the Central Nervous System (CNS). There is now significant evidence in the literature showing that a deficit of DHA in the CNS, may result in altered cognitive and visual functions. Until 1982, its status as a dietary essential nutrient in humans was uncertain. However, the reported case of apparent α-linolenic acid (18:3, ω-3) deficiency in a child on total parenteral nutrition (TPN) set the stage for recognizing this fatty acid as an essential nutrient for pregnant women, infants and specific groups, who depend on TPN (41).

> "Death from cardiovascular disease is rare among Eskimos. Haemostasis was investigated in twenty-one Greenland Eskimos and twenty-one age and sex matched Danish controls. Platelet lipid analysis demonstrated that a high consumption of ω-3 polyunsaturated fatty acids (such as cis 5, 8, 11, 14, 17-eicosapentaenoic acid [20:5]) by Eskimos increased the proportion of ω-3 polyunsaturated fatty acids in the platelets. The Eskimos had a significantly longer bleeding time due to a reduction in platelet aggregation. It is suggested that C20:5 in the platelets is converted by the vascular-wall tissue to an anti-aggregatory prostacyclin. Partial dietary substitution of arachidonic acid by eicosapentaenoic acid may reduce the incidence of thrombotic disorders, including myocardial infarction".

In addition to its essentiality, long chain ω-3 fatty acids (EPA & DHA) have attracted a great deal of attention spurred by the classical paper on "haemostatic function and platelet polyunsaturated fatty acids in Eskimos. 1979", published in 1979 by Dyerberg and Bang (The Lancet, ii, 433-435). This paper has changed the face of fatty acid biochemistry, moving the discussion much closer to the use of fatty acids for not only reducing the risk-associated factors as part of primary prevention but also for therapeutic intervention. The level of scientific interest and the number of papers that have appeared over the last few years clearly illustrates the importance of this classical paper.

Since this publication, numerous original research papers and reviews have been written on this subject, specifically in relation to platelet function (Reviews: 42-51). Similarly, many reviews on topics such as blood lipids and lipoproteins, cardiovascular disease, diabetes, cytokine products and microcirculation (Reviews: 52-59). Because of the enormity of the number of papers in the area, a sincere attempt has been made to include key citations that summarize new findings in regards to the effects of ω-3 fatty acids on platelet-mediated hemostatic and thrombotic functions.

12.6.1 Omega-3 fatty acids and bleeding times

In their classical paper, Dyerberg and Bang (1979) documented that Eskimos have significantly longer bleeding times (8.1 vs. 4.8, as shown in Table 2) (60). Their platelets also contained high levels of ω-3 fatty acids, with EPA (20:5, ω-3) almost equal to that of AA (20:4, ω-6) and the total long chain ω-3 fatty acids slightly over twice that of AA (20:4, ω-6) (Table 2).

The data on Eskimos living in rural southwestern Alaska who depend on fish and marine mammals as major sources of subsistence food, supports a similar trend in plasma ω-3 fatty acids and bleeding times. The ratios of EPA (20:5, ω-3) to AA (20:4, ω-6) in adult coastal- and river-village participants were 14 and 9 times those of non-Native adults, respectively (61). Human studies conducted in populations of Eskimos, Japanese, and Caucasians since 1970 have been reviewed elsewhere (43). It is generally observed that intake of long chain ω-3 fatty acids do affect bleeding times in human subjects.

Table 2. Differences in Platelet Polyenoic Fatty Acids (ω-6 and ω-3) and Bleeding Times Between Greenland Eskimos and Danes[b]

Fatty acids	% of Total[a]	
	Greenland Eskimos	Danes
Linoleic acid (18:2, ω-6)	3.9 ± 0.34	8.2 ± 0.33
Arachidonic acid (20:4, ω-6)	8.5 ± 0.21	22.1 ± 0.47
Alpha-linolenic acid (18:3, ω-3)	NR	NR
Eicosapentaenoic acid (20:5,ω-3)	8.0 ± 0.36	0.5 ± 0.20
Docosapentaenoic acid (22:5, ω-3)	3.3 ± 0.16	1.0 ± 0.14
Docosahexenoic acid (22:6, ω-3)	5.8 ± 0.36	1.5 ± 0.09
Ratio of ω-3/ω-6	1.38	0.10
Ratio of EPA/AA	0.94	0.02
Ratio of EPA & DHA/AA	1.6	
Ratio of DPA & DHA/AA	1.07	
Ratio of EPA, DPA & DHA/AA	2.01	
Bleeding times (BT)	8.05 ± 2.56 (SD)	4.76 ± 1.39 (SD)

[a]Values are given as means±SEM. NR, Not Reported. AA, Arachidonic acid; EPA, Eicosapentaenoic acid; DPA, Docosapentaenoic acid; DHA, Docoshexaenoic acid. [b]Data are taken from J. Dyerberg and H. O. Bang, Lancet, II, 433-435 (1979).

In a double-blind controlled and randomized study with subjects taking 12 g of flax oil containing 55 % α-linolenic acid (18:3, ω-3), a precursor of EPA (20:5, ω-3) and DHA (22:6, ω-3) and identical wheat germ oil capsules for 3 months, we found no significant prolongation of bleeding times (mean BT were 4.5 and 4.8 recorded in the placebo and flax oil groups, respectively, and were not significantly different from those recorded prior to the study in the same subjects). However, EPA and DHA levels in platelet membranes in the flax oil group were detectable but low (62). Compared to long chain ω-3 fatty acids, the supplementation of α-linolenic acid (18:3, ω-3) even at very high doses, appears to have very little effect on bleeding times. It is known that the relative conversion of α-linolenic acid (18:3, ω-3) to long chain fatty acids is quite small in humans. In addition, it is consistently observed that intake of α-linolenic acid (18:3, ω-3) exerts very minimal effects on bleeding times as well as on the biosynthesis of eicosanoids. These findings lead to an obvious conclusion that this fatty acid is least effective with respect to its therapeutic functions.

In a more recent double-blind controlled crossover trial with healthy subjects receiving 8 g of ω-3 fatty acids or identical-looking olive oil capsules for 21 days, the overall percent change in bleeding times after ω-3 fatty acid supplementation was significantly prolonged compared with olive oil supplementation. In the same study, post-aspirin bleeding times after ω-3 fatty acid supplementation were not prolonged significantly when compared with those after olive oil administration (63). In other words, giving a single dose of aspirin to patients eating fish may not further worsen bleeding times. If we were to assume that the ω-3 PUFAs work predominantly through cyclooxygenase, we can also reach a reasonable conclusion that dietary ω-3 PUFAs cannot be as effective as aspirin in modifying cyclooxygenase-dependent platelet activation as the later inactivates the enzyme in an irreversible manner (48).

Table 3. Relative Distribution of AA and EPA among individual phospholipids of platelets from Subjects receiving MaxEPA capsules[b]

Phospholipid	% of Total[a]	
	AA	EPA
Phosphatidylcholine (PC)	28.6	43.2
Phosphatidylethanolamin (PE)	45.0	52.1
Phosphatidylserine (PS)	11.6	3.3
Phosphatidylinositol (PI)	14.9	1.4

[a]Absolute amount in phospholipid (PC + PE + PS + PI): AA, 268 nmol/2 x 10 platelets and EPA, 51 nmol/2x10[9] platelets. [b]Data from V. G. Mahadevappa and B. J. Holub, J. Lipid Res. 28, 1275-1280 (1987).

These findings on bleeding times enforce the concept that it is the relative concentrations of platelet membrane long chain ω-3 fatty acids (EPA & DHA) that serve as key determinants for decreasing platelet adhesiveness and prolonging bleeding times.

It is worth noting that not even a single human feeding study has produced the levels of long chain ω-3 fatty acids that are comparable to those of Greenland Eskimo platelets. It is very likely that the intake of these fatty acids over many generations, might be responsible for retaining such high levels of EPA and DHA in Greenland Eskimos. In one of our own studies (64), subjects who received 20 MaxEPA capsules (3.6 g of EPA and 2.4 g of DHA) per day for 22 days incorporated significant amounts of EPA and DHA into platelet membranes but not to the levels seen in Greenland Eskimos (Table 3).

Based on our calculation (Table 3), EPA represented only 1/5 of AA and the ratio of EPA/AA is not even near to one (see legend in Table 3). This study also demonstrated that over 95 % of EPA and DHA was incorporated into phosphatidylcholine (PC) and phosphatidylethanolamine (PE). This seems to be preferred route for these fatty acids for their in vivo incorporation into membrane phospholipids. This has been documented in numerous similar studies. However, there is general agreement that the intake of 3-3.5 g of long chain ω-3 fatty acids containing EPA and/or DHA, per day may be safe and effective. Relative to Greenland Eskimo platelets, the levels of EPA and DHA in platelet membranes are significantly low even at this relatively large dose. As a result, it is unlikely that we would see bleeding times anywhere close to those observed in Greenland Eskimos. However, we might see a small shift in bleeding times with the above dose. Long term human studies are still lacking to determine if we can reduce the dosage over a period of time and still retain these fatty acids in cells/tissues at the same level and what happens to the benficial effects when discontinued. In a recent study, the long lasting impairment in platelet aggregation was accompanied by the retention of ω-3 fatty acids in platelet phospholipids (65). However, the long chain ω-3 fatty acids must be retained in membrane lipids in order to have an effect via the reduction of AA-derived eicosanoids. A short term intake fish oil may be insufficient to facilitate the retainment of these fatty acids in membrane phospholipids as they are removed readily in the absence of dietary intake.

12.7 Omega-3 fatty acids and platelet function

The obeserved low incidence of CAD in Greenland Eskimos, has been attributed to reduced platelet aggregation (Reviews: 42-51, 60, 66, 67). The metabolites formed from EPA (20:5, ω-3) (in vivo as well as exogenous) and their effects, have been documented in various tissues, including platelets, umbilical blood vessels and calf aorta (68-79). Platelets with incorporated EPA, produce TxA_3, which is weakly proaggregatory (71). In addition, PGI_3 formation from EPA has been doumented (71, 74, 77). On the other hand, Corey et al. (73) have demonstrated that DHA is a strong inhibitor of prostaglandin synthesis but not leukotriene biosynthesis. In summary, EPA metabolites are less potent biologically compared to AA metabolites and they may also exert significant inhibitory effects on AA metabolism in a competitive manner, causing a shift towards a less thrombotic state. However, it is less clear as to the precise in vivo kinetics of functionally opposed TxA_2 and PGI_2, and TxA_3 and PGI_3, in human subjects receiving long chain ω-3 fatty acids despite a large number of studies. Fish oil is also a potent inhibitor of platelet adhesiveness (80). It is therefore reasonable to conclude that the ultimate benefits of fish oil fatty acids result from mutiple and cumulative effects in an in vivo situation, which makes the interpretation much more difficult compared to other antiplatelet drugs, including aspirin. Numerous studies have also evaluated the in vitro effects of different fatty acids on platelet responses. The in vitro effects fatty acids are often difficult to interpret and relate to their pathophysiological functions due to a number of factors, including types of fatty acids and use of unphysiological concentrations.

12.8 Docoshexaenoic acid and platelets

Rao and his colleagues (1983) demonstrated that exogenous DHA could block both platelet aggregation and the second wave platelet response to the action of agonists such as epinephrine, adenosine diphosphate and thrombin (81). They also showed that the inhibition resulted in the presence of DHA (150 μM-450 μM) could be overcome by increasing the concentrations of the agonists. DHA also inhibited the conversion of labeled AA to thromboxane in intact and washed platelet suspensions. They speculated that the polyenoic acids, if released in sufficient quantities in the vicinity of cyclooxygenase, could effectively compete for the heme site and inhibit the conversion of arachidonic acid. Subsequently, Fischer et al. (1984) studied the uptake, release and metabolism of DHA in human platelets and neutrophils using labeled exogenous DHA (82). They demonstrated the formation of both 14- and 11-hydroxydocosahexaenoic acid (HDHE) in platelets and only, 7-HDHE in neutrophils from DHA. In comparison with that of AA and EPA, very little DHA was released even though it was actively taken up by phospholipids. Using platelets and neutrophils enriched by dietary means, they found very little release of DHA from phospholipids upon stimulation with agonists. Corset et al. (1992) also studied the turnover of EPA and DHA in human platelets in vitro (83). We have observed similar findings in both platelets (64) and neutrophils (Conquer and Murthy, Unpublished, 1998) that DHA is not released from membrane phospholipids. We find no detectable levels of free DHA in activated platelets/neutrophils that are enriched with EPA/DHA by dietary means and it suggests a poor release of DHA from membrane phospholipids for further metabolism. More recently, Shikano et. al. (1994) has provided evidence that DHA-diacylGPE is a poor substrate for $cPLA_2$, which is known to be the key enzyme in the mobilization of eicosanoid precursors (116, Chapter 13). This is an important finding because dietary DHA preferentially incorporates into PE (both diacyl and alkenylacyl GPE). However, there is some recent evidence that alkenylacyl GPE may serve as a major source for eicosanoid precursors (Chapter 13). Nevertheless, we find no significant PLA_2-mediated release of DHA from activated platelets (unpublished results). It is therefore still not clear as to the precise regulation of DHA metabolism in activated platelets. We are also not clear on the in vivo concentrations of HDHEs formed from DHA and their pathophysiological impact on platelet function.

In our laboratory, we have compared the in vitro effects of DHA on platelet responses in the presence of physiological concentrations of albumin with several other fatty acids. The hypothesis was that if this fatty acid were to circulate in plasma bound to the albumin fraction at reasonable concentrations in people receiving fish oil fatty acids, it might still influence platelet function and thus, the thrombotic tendency. We have summarized our results in Table 4.

Compared to other fatty acids at similar concentrations (100 μM), DHA was found be the most potent inhibitor of platelet aggregation. Its inhibitory effect on TxA_2 formation was less pronounced with no effect on 12-lipoxygenase activation (Wilkinson T, and Mahadev Murthy, 1998; Unpublished). Recently, Conquer and Holub (1998) have reported an increased circulation of nonesterified DHA (NE-DHA) (up to 13 μM) in subjects receiving different dosages of DHA. Enrichment of the plasma-free fatty acid fraction with ω-3 lipids may likely offer a therapeutic regimen to suppress the synthesis of the potent proaggregatory and vasoconstrictory agent TxA_2 and thus, arterial thrombosis (84, 85). A direct effect of DHA may affect endoperoxide receptor and/or post-receptoral events as well (86-88).

Table 4. Fatty acid effects (in vitro) on collagen-induced platelet aggregation and eicosanoid synthesis[a]

Fatty Acids	Aggregation	TxB_2 Formation	12-HETE Formation
CONTROL	100	100	100
18:0	No change	No change	No Change
18:1, ω-9	No change	No change	No change
18:2, ω-6	Slight inhibition (2 %)	No change	Elevated
18:3, ω-6	Slight inhibition (6 %)	Slight inhibition (6 %)	No change
20:3, ω-6	Slight inhibition (2 %)	No change	Slight elevation
18:3, ω-3	Moderate inhibition (9 %)	No change	Slight elevation
20:3, ω-3	Moderate inhibition (9 %)	No change	Slight elevation
20:5, ω-3	Significant Inhibition (31 %)	Slight inhibition	Slight elevation
22:6, ω-3	Significant Inhibition (87 %)	Significant inhibition (19 %)	Slight inhibition

[a]Data with permission from a thesis on fatty acids and platelet function by Timothy Wilkinson, University Guelph, Guelph, Ontario, Canada (Wilkinson T and Mahadev Murthy M, 1998; unpublished data). Fatty acid effects on platelet aggregation and eicosanoid synthesis, were evaluated in the presence of physiological concentration of albumin.

In addition, it has been shown recently that sub-micromolar levels of the lipoxygenase products of ω-3 fatty acids can antagonize both the contractile effects of thromboxane (U46619) and its platelet aggregating effect (89). Furthermore, OH-22:6, ω-3 inhibits thromboxane-induced decreases in cerebral blood flow of the rat. The 22-carbon ω-3 hydroxy fatty acids are the most potent biological antagonists of thromboxane in comparison to the ω-6 hydroxy fatty acids and their parent fatty acids. Also, submicromolar levels of the docosapentaenoic (22:5) and docosahexaenoic (22:6, ω-3) hydroperoxy and hydroxy derivatives have been shown to inhibit AA-induced platelet aggregation but not that of ADP or collagen. DHA hydroperoxy and hydroxy derivatives exhibited the greatest potency (90). Interestingly, all the DHA and EPA epoxides appear to inhibit platelet aggregation at concentrations below those that affected thromboxane synthesis, in contrast to AA epoxides (91).

12.9 PUFAs and their newly discovered roles

A recent study suggests that molecular species of PC with DHA (22:6, ω-3) at the sn-2 position inhibits TxA_2/PGH_2 receptors, in a highly selective manner. It has been suggested that the docosahexaenoyl PC, located on the outer surface of the plasma membrane, selectively inhibits TxA_2/PGH_2 receptors. This selective modulation does not appear to be caused by a generalized perturbation of the membrane due to DHA (22:6, ω-3 incorporation (92). These inhibitory effects of DHA are unique because they are not mediated by its metabolites nor by free DHA. In other words, DHA through its incorporation into membrane phospholipids could significantly alter TxA_2-

mediated platelet function.

Table 5: A summary of the biological effects of different polyunsaturated fatty acids

Fatty acid	Biological Effects
Dietary linoleic acid (18:2, ω-6)	A precursor of AA, an essential component of cell membranes, is the substrate for the 2-series eicosanoids, which are the most biologically potent eicosanoids.
γ-Linolenic acid (18:3, ω-6)	A precursor of AA. It is also a precursor of DGLA, which is the substrate for the 1-series eicosanoids.
Arachidonic acid (20:4, ω-6)	Small amounts of this are consumed in the diet in the form of meat. It is a precursor for the 2-series eicosanoids.
Linolenic acid (18:3, ω-3)	A precursor of EPA and DHA. DHA is an essential component of neural membranes. EPA is the substrate for the series-3 eicosanoids, which are relatively less active compared to those from AA.
	It is known that the relative conversion of dietary α-linolenic acid (18:3, ω-3) to EPA and DHA, is quite small in humans, but it is readily oxidized.
	It is considered essential for pregnant women, infants and special population groups, who depend on TPN. A developing fetus utilizes this fatty acid during the last trimester of pregnancy for the synthesis of DHA, which is then incorporated into the fetal brain phospholipids. Therapeutic effects such as antithrombotic and antitriglyceridemic are minimal compared to EPA and DHA.
Eicosapentaenoic acid (20:5, ω-3)	The precursor for the series-3 eicosanoids. It replaces AA in membrane phospholipids. EPA is selectively incorporated into the eicosanoid precursor pool of phospholipids. Like AA, it is released from membrane phospholipids. It competes with AA for oxygenases.
	Antithrombotic, Antinflammatory and Antitriglyceridemic
Docosahexaenoic acid (22:6, ω-3)	It is retroconverted to EPA. DHA also replaces AA in membrane phospholipids. It is poorly released from membrane phospholipids (poorly accessible to PLA_2). Exogenous DHA is converted to hydroxyfatty acids. It inhibits cyclooxygenase but not lipoxygenase. Exogenous DHA is a potent inhibitor of platelet aggregation.
	Antithrombotic, Antinflammatory and Antitriglyceridemic DHA has unique effects that differ from those of EPA.

Lipid peroxides are known to modulate cellular functions. We have shown that fatty acid hydroperoxides exhibit a novel dual regulatory role in platelet AA metabolism (93). While cyclooxygenase/lipoxygenase enzymes are inhibited by low concentrations of hydroperoxides, the phospholipases would still be active allowing a continuous removal of esterified hydroxy/hydroperoxy fatty acids from membrane phospholipids. The work of va Kuijk et al. (1987) has provided evidence for preferential removal of the peroxidised fatty acids by phospholipase A_2 (94). The results from a recent study also indicate that physiologically relevant concentrations of HPETEs potentiate platelet aggregation, which appears to be mediated via a stimulation of cyclooxygenase activity (95). Although the role of hydroperoxides derived from ω-3 fatty acids appears less clear, it is conceivable that fatty acid hydroperoxides play a significant role in platelet fatty metabolism and function. Fatty acids of marine origin may alter the redox status in platelets. They may for instance increase the platelet glutathione peroxidase (GPx) activity, an effect that can be prevented by antioxidants. Among the ω-3 fatty acids, DHA (22:6, ω-3) is the most potent activator of GPx activity (96). As the enhanced activity GPx induced by DHA is abolished in the presence of cycloheximide at a concentration known to inhibit platelet protein synthesis and platelets are devoid of nucleus, DHA (22:6, ω-3) might act at a translational level.

On the other hand, the increased GPx activity (possibly via protein synthesis) might be associated with an oxidative stress induced by DHA (22:6, ω-3) and/or AA (20:4, ω-6) released from a platelet endogenous pool in the course of the DHA (22:6, ω-3) enrichment (97). Similar to the findings in our laboratory, DHA (22:6, ω-3) esterified to triglycerides is rapidly redistributed within blood lipoproteins. The DHA (22:6, ω-6) bound and circulated with the albumin fraction not only inhibit platelet aggregation but also influences its uptake into phospholipid species by target tissues (98). DHA therefore seems to impact platelet fatty acid metabolism through unique and novel mechanisms.

The findings that more than 20 platelet proteins, including glycoprotein IX beta chain of glycoprotein Ib, components of the von Willebrand factor receptor on the platelet surface, P-selectin, and alpha subunits of Gz, Gq, and Gi, can be posttranslationally acylated with fatty acids such as palmitic and myristic acids via thioester linkages, clearly provide a new functional role for fatty acids. In particular, thioesterification of platelet proteins with PUFAs of both the ω-6 and ω-3 series may have significant functional consequences for reversible binding of proteins to membranes (99). Direct modification of proteins by fatty acid can occur as cotranslational N-myristoylation of an N-terminal glycine residue or as posttranslational thioesterification of cysteine residue(s). Platelets provide an excellent model for studying the post-translational type modification in the absence of active protein synthesis and in the absence of protein synthesis-related protein modifications with lipids. Using this model system, it has been shown recently that thioesterification of proteins with fatty acid is less specific for palmitate than it was thought earlier and that other saturated, mono- and even polyunsaturated long chain fatty acids can also participate. The chain length and the extent of unsaturation of the protein-linked fatty acid moiety can, very likely, modulate hydrophobic protein-membrane lipid and protein-protein interactions. CD9, HLA class

I glycoprotein, glycoproteins Ib, IX and IV, P-selectin and alpha subunits of G proteins have been demonstrated unequivocally as S-fatty acid acylated platelet proteins (100).

A recent demonstration that dietary ω-3 fatty acids down regulate gene expression of both PDGF-A (-66%), and PDGF-B (-70%), represents a novel mechanism for the antifibrotic and antiatherosclerotic action of ω-3 fatty acids (101). Furthermore, a high glucose and osmotic pressure-induced increase in PDGF (platelet-derived growth factor) production is involved in the development and progression of diabetic macroangiopathy. Since EPA (20:5, ω-3) inhibits the PDGF production induced by high glucose concentration, it may have an unique role as an anti-arteriosclerotic agent (102). These findings suggest a possible role for PUFAs within the PDGF-dependent signal transduction pathway, which may be the direct result of decreased receptor dimerization and/or kinase activity (103). It has been suggested that the inactivation of Gzalpha by unsaturated fatty acids results from an interaction of an acidic lipid micelle with the nucleotide-free form of the protein. Although the physiologic significance of this finding is unclear, similar effects of unsaturated fatty acids on other proteins that are key to cell signaling, can be anticipated. In addition, the ability of AA (20:4, ω-6) to inactivate this adenylyl cyclase-inhibitory G protein provides a molecular mechanism for previous findings that treatment of platelets with AA results in elevated cAMP levels (104). DHA enrichment has also been reported to attenuate enzymatic reactions for PAF (platelet activating factor) synthesis, mainly the initial reaction catalyzed by AA-specific phospholipase, and thereby reduces PAF synthesis in Eol-1 (105).

Concluding comments

The extensive scientific data that has been gathered over the last three decades is contrary to the belief that platelets may have a very limited role in human health and disease. The discovery of new antiplatelet drugs based on polyenoic fatty acids acids, has clearly impacted the treatment of CAD (106). We have now added long chain ω-3 fatty acids to the list of antiplatelet agents. In a recent study, it has been shown that a significant positive correlation exists between the content of ω-3 PUFAs, particularly DHA in platelets and heart rate variability (107). In addition, there is some beleif that ω-3 PUFAs may increase mean platelet volume and thus, affect platelet reactivity. A similar observation has been made in women with HRT (hormonal replacement therapy) with poor correlation between the changes in mean platelet volume and membrane fatty acids (108). In addition, estrogen replacement therapy seems to affect the metabolism of PUFAs in platelets (109). Although the mechanism of protection conferred by estrogen is partly attributable to changes in serum lipoproteins, the changes in platelet membrane fatty acid composition may be a primary effect of hormone replacement therapy, especially estrogen (109).

Conflicting results have been reported about the absorption of EPA and DHA either as an ethyl ester (EE) or in a triglyceride (TG) formula. Based on a randomized double-blind study on the effects of EE and TG on plasma fatty acids, platelet function and haemostasis, it has been concluded that TG and EE fish oils are well incorporated into plasma lipids and exert similar influence on platelet function in men (110). Interestingly, the results from a recent study on blood clotting parameters and in vitro platelet aggregation, suggest that adding 6 g/d of dietary DHA for 90 d to a typical Western diet containing less than 50 mg/d of DHA produces no observable physiological changes in blood coagulation, platelet function, or thrombotic tendencies in healthy, adult males (111). However, there is increasng evidence that circulating DHA may offer significant antithrombotic benefits (112).

While there are many reports of studies that fed AA to animals, there are very few reports of AA feeding to humans under controlled conditions. In a series of studies, several indices, including the fatty acid composition plasma, red blood cells, platelets, and adipose tissue, in vitro platelet aggregation, bleeding times, clotting factors; immune response as measured by delayed hypersensitivity skin tests, cellular proliferation of peripheral blood mononuclear cells in response to various mitogens and antigens, natural killer cell activity, and response to measles/mumps/rubella and influenza vaccines, the metabolic conversion of deuterated linoleic acid to AA and the metabolic fate of deuterated AA in the subjects on and off the high-AA diet, and the production of eicosanoids as measured by excretion of 11-DTXB_2 and PGI_2-M in urine, have been studied in healthy adult male volunteers. They lived in the metabolic research unit (MRU) of the Western Human Nutrition Research (WHNRC) for the entire duration of the study (113). The results from this study on blood clotting parameters and in vitro platelet aggregation suggest that adding 1.5 g/d of dietary AA for 50 d to a typical Western diet containing about 200 mg of AA produced no observable physiological changes in blood coagulation and thrombotic tendencies in healthy, adult males compared to the unsupplemented diet. This has led to the conclusion that moderate intakes of foods high in AA have few effects on blood coagulation, platelet function, or platelet fatty acid composition (114). The assumption that consumption of AA produces obeservable physiological changes is misleading simply because the intake of dietary AA just increases the turnover with very minimal changes in the fatty acid composition of plasma, cells and tissues. A continuous flux of potent oxidized metabolites produced from dietary AA over a period of time is considered responsible for chronic tissue damage and disease. It is conceivable that a long term intake of dietary AA as part of the Western diet may produce results that would support AA-induced pathophysiology over a period of time. Therefore, short term studies with dietary AA, will not support the premise that they do not affect pathophysiological functions. Nevertheless, an increase in the ratio of ω-3/ω-6 in the diet needs to be recognized. The ratio that ensures moderate suppression in platelet aggregation without affecting bleeding times may be useful to determine the ratio of these fatty acid groups. However, the suggested ratio for rats (at least 0.2 and no more than 0.5) may not work for humans as there are significant differences between rat and human platelet responses (115).

Acknowledgements

The support received over the years from several Canadian agencies, including the Heart and Stroke Foundations of Canada, Natural Sciences and Engineering Research

Council of Canada (NSERC), Canola Council of Canada, is sincerely acknowledged. My sincere appreciation must also go my wife and children, for their understanding and support during the preparation of this manuscript.

REFERENCES

1. Schafer AI. (1996). Antiplatelet therapy. Am. J. Med. 101(2):199-209.

2. Hawiger J, Kloczewiak M, and Timmons S. (1985). Platelet receptor mechanisms for adhesive macromolecules. In: Interaction of Platelets With The Vessel Wall (Oats J, Hawiger J, and Ross R: eds), American Physiological Society, Bethesda, MD).

3. Sixma JS and Wester J. (1977). Semin. Hematol. 14, 265-269.

4. Needleman P, Turk J, Jakschik BA, Morrison AR, Lefkowith JB. (1986). Arachidonic acid metabolism. Annu. Rev. Biochem. 55: 69-102.

5. Willis, AL. (1987). The eicosanoids: An introduction and an overview. In: Handbook of Eicosanoids: Prostaglandins and Related Lipids (Willis, AL; Ed.), Vol. 1, CRC Press, Boca Raton, Fl.

6. Mead, JF, and Willis, AJ. (1987). The essential fatty acids: Their derivation and role. In: Handbook of Eicosanoids: Prostaglandins and Related Lipids (Willis, AL; Ed.), Vol. 1, CRC Press, Boca Raton, Fl.

7. Brenner, RR. (1987). Biosynthesis and the interconversion the essential fatty acids. In: Handbook of Eicosanoids: Prostaglandins and Related Lipids (Willis, AL; Ed.), Vol. 1, CRC Press, Boca Raton, Fl.

8. Holmsen, H. (1987). Uptake and release of ecicosanoid precursors from phospholipids and other pools: Platelets. In: Handbook of Eicosanoids: Prostaglandins and Related Lipids (Willis, AL; Ed.), Vol. 1, CRC Press, Boca Raton, Fl.

9. Roth GI. (1986). Platelet arachidonic acid metabolism and platelet-activating factor. In: Biochemistry of Platelets (Phillips DR, and Shuman MA; Eds.), PP. 69-100, Academic Press Inc. (London) Ltd., London.

10. Samuelsson, B. (1983). Leudotrienes: mediators of immediate hypersensitivity reactions and inflammation. Science. **221**, 568-575.

11. Marcus AJ, Ullman HL, and Safier LB. (1969). Lipid composition of subcellular particles of human blood platelets. J. Lipid Res. 10, 108-114.

12. Cohen P and Derksen A. (1969). Comparison of phospholipid and fatty acid composition of erythrocytes and platelets. Br. J. Haematol. 17, 359-371.

13. Mahadevappa VG, and Holub BJ. (1982). The molecular species composition of individual diacyl phospholpids in human platelets. Biochim. Biophys. Acta 713, 73-79 (1982).

14. Mahadevappa VG and Holub BJ. (1987). Quantitative loss of individual eicosapentaenoyl-relative to arachidonyl-containing phospholipids in thrombin-stimulated human platelets. J. Lipid Res. 28, 1275-1280.

15. Smith JB, and Willis AL. (1971). Aspirin selectively inhibits prostaglandin production in human platelets. Nature (Lond.), New Biol. 231, 235-237.

16. Silver MJ, Smith JB, Ingerman CM, and Kocsis JJ. (1973). Arachidonc acid-induced human platelet aggregation and prostaglandin fromation. Prostaglandins 4, 863-875.

17. Vergaftig BB, and Zirinis P. (1973). Platelet aggregation induced by arachidonic acid is accompanied by release of potential inflammatory mediators distinct from PGE_2 and $PGF_{2\alpha}$. Nature (Lond.), New Biol. 244, 114-116.

18. Hamberg M, and Samulesson B. (1974). Prostaglandin endoperoxides: Novel transformations of arachidonic acid in human platelets. Proc. Natl. Acad. Sci. U.S.A. 71, 3400-3404.

19. Hamberg M, Svensson J, and Samuelsson B. (1975). Thromboxanes: A new group of biologically active compounds derived from prostaglndin endoperoxides. Proc. Natl. Acad. Sci. U.S.A. 72, 2994-2998.

20. Lands WEM, and Samuelsson B. (1968). Phospholipid precursors of prostaglndins. Biochim. Biophys. Acta 164, 426-429.

21. Moncada S, Gryglewski R, Bunting S, and Vane JR. (1976). The enzyme isolated from arteries transforms prostaglandin endoperoxides to an unstable substance that inhibits platelet aggregation. Nature (Lond.) 263, 663-665.

22. Moncada S, and Vane JR. (1979). Arachidonic acid metabolites and the interactions between platelets and blood-vessel walls. New Engl. J. Med. 300, 1142-1147.

23. Bunting S, Moncada S, and Vane JR. (1983). The prostacyclin-thromboxane A_2 balance: Pathophysiological and therapeutic implications. Br. Med. Bull. 39, 271-276.

24. Majerus PW. (1983). Arachidonic acid metabolism in vascular disorders. J. Clin. Invest. 72, 1521-1525.

25. MacIntyre, DE. (1981). Platelet prostaglandin receptors. In: Platelets in Biology and Pathology (Gordon JL, ed.), vol. 2, pp. 211--247, Elsevier/North-Holland, Amsterdam, The Netherlands.

26. Lewis GP, Watts, IS. (1982). Prostaglandin endoperoxides, thromboxanes A_2 and adenosine diphosphate in collagen-induced aggregation in man and the cat. Br. J. Pharmacol. 75: 623-663.

27. Lapetina EG. (1986). Inositide-Dependent and Independent Mechanisms in Platelet Activation. In: Receptor Biochemistry and Methodology (Putney JW, Ed.). Vol 7. New York: Alan R Liss , 271-286.

28. Siess W. Molecular mechanisms of platelet activation. Physiol Rev 1989; 69: 58-178.

29. Samuelsson B, Goldyne M, Granstrom E, Hamberg M, Hammarstrom S, and Malmsten C. (1978). Prostaglandins and thromboxanes. Ann. Rev. Biochem. 47, 997-1029.

30. Needleman P, Wyche A, and Raz A. (1979). Platelet and blood vessel arachidonate metabolism and interactions. J. Clin. Invest. 63, 345-349.

31. Roth, GJ, Machuga ET, and Strittmatter P. (1981). The hem-binding properties of prostaglndin synthetase from sheep vascular gland. J. Biol. Chem. 256, 10018-10022.

32. Horrobin, DF. (1990). Cardiovascular and inflammatory diseases: interactions between n-3 and n-6 essential fatty acids (EFAs). Omega-3 News 5, 1.

33. Horrobin DF. (1980). A new concept of life-style related cardiovascular disease: the importance of interactions between cholesterol, essential fatty acids, prostaglandin E_1 and thromboxane A_2. Med. Hypotheses 6, 785.

34. Takahashi R, Nassar BA, Huang YS, Begin ME, and Horrobin DF. (1987). Effect of different ratios of dietary n-6 and n-3 fatty acids on fatty acid composition, prostaglandin formation, and platelet aggregation in the rat. Thromb. Res. 47, 135.

35. Guivernau M, Meza N, Barja P, and Roman O. (1994). Clinical and experimental study on the long-term effect of dietary gamma-linolenic acid on plasma lipids, platelet aggregation, thromboxane formation, and prostacyclin production. Prost. Leukotrienes & Essential Fatty Acids. 51, 311-316.

36. Nakahara T, Yokochi T, Kamisaka Y, Yamaoka M, Suzuki O, Sato M, Okazaki S, and Ohshima N. (1990). Inhibitory effects of mold oil including gamma-linolenate on platelet thrombus formation in mesenteric microvessels in the rat. Thromb. Res. 57, 371.

37. Mills DE, Summers MR, and Ward RP. (1985). Gamma-linolenic acid attenuates cardiovascular responses to stress in borderline hypertensive rats. Lipids 20, 573.

38. Mills DE, Mah M, Ward RP, Morris BL, and Floras JS. (1990). Alteration of baroreflex control of forearm vascular resistance by dietary fatty acids. Am. J. Physiol. 259, R1164.

39. Mills DE, Prakachin KM, Harvey KA, and Ward RP. (1989). Dietary fatty acid supplementation alters stress reactivity and performance in man. J. Hum. Hyperten. 3, 111.

40. Tinoco J. (1982). Dietary requirements and functions of α-linolenic acid in animals. Prog. Lipid Res. 21, 1-45.

41. Holman RT, Johnson SB, and Hatch TF. (1982). A case of human linolenic acid deficiency involving neurological abnormalities. Am. J. Clin. Nutr. 65, 617-623.

42. Goodnight SH, Jr., Harris WS, Conner WE, and Illingworth DR. (1982). Polyunsaturated fatty acids, hyperlipidemia, and thrombosis. Arteriosclerosis 2, 87-113.

43. Kantha SS. (1987). Dietary effects of fish oils on human health: a review of recent studies. [Yale J. Biol. Med. 60, :37-44, 1987.

44. Goodnight SH, Jr. (1988). Effects of dietary fish oil and omega-3 fatty acids on platelets and blood vessels. Sem. Thromb. Hemost. 14:285-289.

45. Mueller BA, and Talbert RL. (1988). Biological mechanisms and cardiovascular effects of omega-3 fatty acids. Clinical Pharmacy. 7, 795-807.

46. Malle E, and Kostner GM. (1993). Effects of fish oils on lipid variables and platelet function indices. Prost. Leukotrienes & Essential Fatty Acids. 49, 645-663, 1993.

47. Mutanen M, and Freese R. (1996). Polyunsaturated fatty acids and platelet aggregation. Current Opinion Lipidology. 7, 14-19.

48. Heemskerk JW, Vossen RC, and van Dam-Mieras MC. (1996). Polyunsaturated fatty acids and function of platelets and endothelial cells. Current Opinion Lipidology. 7, 24-29.

49. Dupont J, Holub BJ, Knapp HR, and Meydani M. (1996). Fatty acid-related functions. Am. J. Clin. Nutri. 63, 991S-993S.

50. Ponte E, Cafagna D, and Balbi M. (1997). Cardiovascular disease and omega-3 fatty acids. Minerva Medica. 88:343-353.

51. Mutanen M. (1997). Cis-unsaturated fatty acids and platelet function. Prost. Leukotrienes Essential Fatty Acids. 57, 403-410.

52. Holman RT. (1998). The slow discovery of the importance of omega 3 essential fatty acids in human health. J. Nutri. 128(2 Suppl), 427S-433S.

53. Harris WS. (1996). n-3 fatty acids and lipoproteins: comparison of results from human and animal studies. Lipids. 31, 243-252.

54. Calder PC. n-3 polyunsaturated fatty acids and cytokine production in health and disease. Annals Nutrition Metabolism 41, 203-234.

54. Simopoulos AP. (1997). Omega-6/omega-3 fatty acid ratio and trans fatty acids in non insulindependent diabetes mellitus. Annals New York Acad. Sciences 827, 327-338.

55. Rustan AC, Nenseter MS, and Drevon CA. (1997). Omega-3 and omega-6 fatty acids in the insulin resistance syndrome. Lipid and lipoprotein metabolism and atherosclerosis. Annals New York Acad. Sciences 827, 310-326.

56. Harris WS. (1997). n-3 fatty acids and serum lipoproteins: animal studies. Am. J. Clinical Nutri. 65(5 Suppl), 1611S-1616S.

57. Bruckner G. (1997). Microcirculation, vitamin E and omega 3 fatty acids: an overview. Adv. Expl. Med. Biol. 415, 195-208..

58. Simopoulos AP. (1997). Omega-3 fatty acids in the prevention-management of cardiovascular disease. Can. J. Physiol. Pharmacol. 75, 234-239, 1997.

59. Prince MJ., and Deeg MA. (1997). Do n-3 fatty acids improve glucose tolerance and lipemia in diabetics?. Current Opinion In Lipidology 8, 7-11.

60. Dyerberg J, and Bang HO. (1979). Haemostatic function and platelet polyunsaturated fatty acids in Eskimos. The Lancet ii, 433-435; Nutrition. 11, 475; discussion 474, 476, 1995.

61. Parkinson AJ, Cruz AL, Heyward WL, Bulkow LR, Hall D, Barstaed L, and Connor WE. (1994). Elevated concentrations of plasma omega-3 polyunsaturated fatty acids among Alaskan Eskimos. Am. J. Clin. Nutri. 59, 384-388, 1994.

62. Murray E, Murray RP, and Murthy M. (1996). Clinical and biochmical effects of flax oil in patients with psoriasis. Presented at the Canadian Society for Dermatology.

63. Mueller BA, Talbert RL, Tegeler CH, and Prihoda TJ. (1991). The bleeding time effects of a single dose of aspirin in subjects receiving omega-3 fatty acid dietary supplementation. J. Clin. Pharmacol. 31, 185-190.

64. Mahadevappa VG, and Holub BJ. (1987). Quantitative loss of individual eicosapentaenoyl- relative to arachidonoyl-containing phospholipids in thrombin-stimulated human platelets. J. Lipid Res. 28, 1275-1280.

65. Tremoli E, Maderna P, Marangoni F, Colli S, Eligini S, Catalano I, Angeli MT, Pazzucconi F. Gianfranceschi G, Davi G. et al. (1995). Prolonged inhibition of platelet aggregation after n-3 fatty acid ethyl ester ingestion by healthy volunteers. Am. J. Clin. Nutri. 61, 607-613.

66. Dyerberg J, Bang HO, and Hjorne N. (1975). Fatty acid composition of the plasma lipids in Greenland Eskimos. Am. J. Clin. Nutr. 28, 958-966.

67. Seiss W, Scherer B, Bohlig B, Roth P, Kurzman I, and Weber PC. (1980). Platelet-membrane fatty acids, platelet aggregation, and thromboxane formation during Mackrel diet. Lancet I, 441-444.

68. Whitaker MO, Wyche A, Fitzpatrick F, Sprecher H, and Needleman P. (1979). Triene prostaglandins: prostaglandin D_3 and eicosapentaenoic acid as potent antithrombotic substances. Proc. Natl. Acad. Sci. USA 76, 5919-5923.

69. Culp BR, Titus BG, and Lands WEM. (1979). Inhibition of protaglandin biosynthesis by eicosapentaenoic acid. Prostaglandins Med. 3, 269-278.

70. Dyerberg J, and Bang HO. (1980). All cis-5,8,11,14,17--eicosapentaenoic acid triene prostaglandins: potential antithrombotic agents. Scand. J. Clin. Lab. Invest. 40,589-593. 69.

71. Dyerberg J, Jorgensen KA, and Arnfred T. (1981). Human umbilical blood vessel converts all cis-5,8,11,17-eicosapentaenoic acid to prostaglandin I_3. Prostaglandins 22, 857-862.

72. Fischer S, and Weber, PC. (1983). Thromboxane A3(TXA3) is formed in human platelets after dietary eicosapentaenoic acid (C20:5w3). Biochem. Biophys. Res. Commun. 116, 1091-1099.

73. Corey EJ, Shih C, and Cashman JR. (1983). Docosahexaenoic acid is a strong inhibitor of prostaglandin but not leukotriene biosynthesis. Proc. Natl. Acad. Sci. USA 80, 3581-3584.

74. Fischer S, and Weber PC. (1984). Prostaglandin I_3 is formed in vivo in man after dietary eicosapentaenoic acid. Nature (Lond.) 303, 165-168.

75. Morita I, Takahashi R, Saito Y, and Murota S. (1983). Stimulation of eicosapentaenoic acid metabolism in washed human platelets by 12- hydroperoxyeicosatetraenoic acid. J. Biol. Chem. 258, 10197-10199.

76. Takenaga M, Hirai A, Terano T, Tamura Y, Kitagawa H and Yoshida S. (1986). Comparison of the in vitro effect of eicosapentaenoic acid (EPA) -derived lipoxygenase metabolites on human platelet function with those of arachidonic acid. Throm. Res. 37, 373-384.

77. Powell WS, Gravelle F. (1984) Metabolism of icosapentaenoic acid and docosahexaenoic acid by fetal calf aorta and their effects on prostacyclin formation. Proc. 2nd Int. Symp. 94-100.

78. Terano T, Salmon JA, Higgs GA, and Moncada S. (1986) Eicosapentaenoic acid as a modulator of inflammation: Effect on prostaglandin and leukotriene synthesis. Biochem. Pharm. 35, 779-785.

79. Juan, H, Peskar BA and Simmet T. (1987) Effect of exogenous 5, 8, 11, 14, 17-eicosapentaenoic acid on cardiac anaphylaxis. Br. J. Pharmac. 90, 315-325.

80. Li, X., Steiner, M. (1990) Fish oil: a potent inhibitor of platelet adhesiveness. Blood. 76, 938-945.

81. Rao GH, Radha E, and White JG. (1983). Effect of docosahexaenoic acid (DHA) on arachidonic acid metabolism and platelet function. Biochem. Biophys. Res. Commun. 117, 549-555.

82. Fischer S, von Schacky C, Siess W, Strasser T, and Weber PC. (1984). Uptake, release and metabolism of docosahexaenoic acid (DHA, c22:6 omega 3) in human platelets and neutrophils. Biochem. Biophys. Res. Commun. 120, 907-918.

83. Croset M, Bayon Y, and Lagarde M. (1992). Incorporation and turnover of eicosapentaenoic and docosahexaenoic acids in human blood platelets in vitro. Biochem. J. 281, 309-316.

84. Conquer JA, and Holub BJ. (1998). Effect of supplementation with different doses of DHA on the levels of circulating DHA as non-esterified fatty acid in subjects of Asian Indian background. J. Lipid Res. 39, 286-292.

85. Kramer HJ, Stevens J, Grimminger F, and Seeger W. (1996). Fish oil fatty acids and human platelets: dose-dependent decrease in dienoic and increase in trienoic thromboxane generation. Biochem. Pharmacol. 52, 1211-1217.

86. Swann PG, Parent CA, Croset M, Fonlupt P, Lagarde M, Venton DL, and Le Breton GC. (1990) Enrichment of platelet phospholipids with eicosapentaenoic acid and docosahexaenoic acid inhibits thromboxane A2/prostaglandin H2 receptor binding and function. J. Biol. Chem. 265, 21692-21697.

87. Parent CA, Lagarde M, Venton DL and Le Breton G. (1992). Selective modulation of the human platelet thromboxane A2/prostaglandin H2 receptor by eicosapentaenoic and docosahexaenoic acids in intact platelets and solubilized platelet membranes. J. Biol. Chem. 267, 6541-6446.

88. Scheurlen M, Kirchner M, and Clemens MR., and Jaschonek K. (1993). Fish oil preparations rich in docosahexaenoic acid modify platelet responsiveness to prostaglandin-endoperoxide/thromboxane A2 receptor agonists. Biochem. Pharmacol. 46, 245-249.

89. Karanian JW, and Salem N Jr. (1995). Hydroxylated 22-carbon fatty acids in platelet and vascular smooth muscle function: interference with TXA2/PGH2 receptors. Agents & Actions (Supplements) 45, 39-45.

90. Karanian JW, Kim HY, and Salem N Jr. (1996). The structure-activity relationship of lipoxygenase products of long-chain polyunsaturated fatty acids: effects on human platelet aggregation. Lipids. 31 (Suppl), S305-308.

91. VanRollins M. (1995). Epoxygenase metabolites of docosahexaenoic and eicosapentaenoic acids inhibit platelet aggregation at concentrations below those affecting thromboxane synthesis. J. Pharmacol. Exptl. Therapeut. 274, 798-804.

92. Bayon Y, Croset M, Daveloose D, Guerbette F, Chirouze V, Viret J, Kader JC, and Lagarde M. (1995). Effect of specific phospholipid molecular species incorporated in human platelet membranes on thromboxane A2/prostaglandin H2 receptors. J. Lipid Res. 36, 47-56.

93. Vedelago HR, and Mahadevappa VG. (1988a). Differential effects of 15-HPETE on arachidonic acid metabolism in collagen-stimulated human platelets. Biochem. Biophys. Res. Commun. 150, 177-184.

94. Van Kuijk FJCM, Sevanian A, Handelman GJ, and Dratz RA. (1987). A new role for phopholipase A_2: protection of membrane from lipid peroxide damage. Trends Biochem. Sci. 12, 31-36.

95. Calzada C, Vericel E, and Lagarde M. (1997). Low concentrations of lipid hydroperoxides prime human platelet aggregation specifically via cyclo-oxygenase activation. Biochem. J. 325 (Pt 2), 495-500.

96. Lagarde M, Lemaitre D, Calzada C, and Vericel E. (1997). Involvement of lipid peroxidation in platelet signalling. Prost. Leukotrienes & Essential Fatty Acids. 57(4-5):489-491.

97. Lemaitre D, Vericel E, Polette A, and Lagarde M. (1997). Effects of fatty acids on human platelet glutathione peroxidase: possible role of oxidative stress [published erratum appears in Biochem Pharmacol 1997 Jun 15;53(12):1945]. Biochem. Pharmacol. 53(4):479-86.

98. Brossard N, Croset M, Lecerf J, Pachiaudi C, Normand S, Chirouze V, Macovschi O, Riou JP, Tayot JL, and Lagarde M. (1996). Metabolic fate of an oral tracer dose of [13C]docosahexaenoic acid triglycerides in the rat. Am. J. Physiol. 270(4 Pt 2), R846-854.

99. Laposata M., and Muszbek L. (1996). Thioesterification of platelet proteins with saturated and poly-unsaturated fatty acids. Lipids. 31 Suppl, S217-S221.

100. Muszbek L, Racz E, and Laposata M. (1997). Posttranslational modification of proteins with fatty acids in platelets. Prost. Leukotrienes & Essential Fatty Acids. 57(4-5), 359-366.

101. Kaminski WE, Jendraschak E, Kiefl R, and von Schacky C. (1993). Dietary omega-3 fatty acids lower levels of platelet-derived growth factor mRNA in human mononuclear cells. Blood. 81, 1871-1879.

102.Mizutani M, Okuda Y, Suzuki S, Sawada T, Soma M, and Yamashita K. (1995). High glucose increases platelet-derived growth factor production in cultured human vascular endothelial cells and preventive effects of eicosapentaenoic acids. Life Sciences. 57, PL31-35.

103.Tomaska L., and Resnick RJ. (1993). Suppression of platelet-derived growth factor receptor tyrosine kinase activity by unsaturated fatty acids. J. Biol. Chem. 268, :5317-5322.

104.Glick J, Santoyo G, and Casey PJ. (1996). Arachidonate and related unsaturated fatty acids selectively inactivate the guanine nucleotide-binding regulatory protein, Gz. J. Biol. Chem. 271, 2949-2854.

105.Shikano M, Masuzawa Y, and Yazawa K. (1993). Effect of docosahexaenoic acid on the generation of platelet-activating factor by eosinophilic leukemia cells, Eol-1. J. Immunol. 150(8 Pt 1), 3525-3533.

106. Farstad M. (1998). The role of blood platelets in coronary atherosclerosis and thrombosis. Scand. J. Clin. Lab. Invest. 58, 1-10, 1998.

107. Christensen JH, Korup E, Aaroe J, Toft E, Moller J, Rasmussen K, Dyerberg J, and Schmidt EB. (1997). Fish consumption, n-3 fatty acids in cell membranes, and heart rate variability in survivors of myocardial infarction with left ventricular dysfunction. Am. J. Cardiol. 79, 1670-1673.

108. Ranganath LR, Christofides J, and Semple MJ. (1996). Increased mean platelet volume after oestrogen replacement therapy. Annals Clin. Biochem. 33 (Pt 6), 555-560.

109. Ranganath LR, Christofides JA, Wright JW, and Marks V. (1996). Effects of hormone replacement therapy on platelet membrane fatty acid composition. J. Endocrinol. 148, 207-212.

110. Hansen JB, Olsen JO, Wilsgard L, Lyngmo V, and Svensson B. (1993). Comparative effects of prolonged intake of highly purified fish oils as ethyl ester or triglyceride on lipids, haemostasis and platelet function in normolipaemic men. Eur. J. Clin. Nutri. 47, 497-507.

111. Nelson GJ, Schmidt PS, Bartolini GL, Kelley DS, and Kyle D. (1997). The effect of dietary docosahexaenoic acid on platelet function, platelet fatty acid composition, and blood coagulation in humans. Lipids. 32, 1129-1136.

112. Conquer JA, and Holub BJ. (1996). Supplementation with an algae source of docosahexaenoic acid increases (n-3) fatty acid status and alters selected risk factors for heart disease in vegetarian subjects. J. Nutrition 126, 3032-3039.

113. Nelson GJ, Kelley DS, Emken EA, Phinney SD, Kyle D, and Ferretti A. (1997). A human dietary arachidonic acid supplementation study conducted in a metabolic research unit: rationale and design. Lipids. 32, 415-420.

114. Nelson GJ, Schmidt PC, Bartolini G, Kelley DS, and Kyle D. (1997). The effect of dietary arachidonic acid on platelet function, platelet fatty acid composition, and blood coagulation in humans. Lipids. 32, 421-425.

115.Yamada N, Takita T, Wada M, Kannke Y, and Innami S. (1996). Effects of dietary n-3/n-6 and polyunsaturated fatty acid/saturated fatty acid ratios on platelet aggregation and lipid metabolism in rats. J. Nutri. Sci. Vitaminol. 42, 423-434.

116.Shikano M, Masuzawa Y, Yazawa K, Takayama K, Kudo I, Inoue K. (1994). Complete discrimination of docosahexaenoate from arachidonate by 85 kDa cytosolic phospholipase A2 during the hydrolysis of diacyl- and alkenylacylglycerophosphoethanolamine. Biochim. Biophys. Acta. 1212, 211-216.

13

PHOSPHOLIPASE A_2 IN PLATELETS

Mahadev Murthy, Ph.D.
Division of Endocrinology,
Metabolism & Nutrition
Department of Medicine
Hennepin County Medical Center
Minneapolis, MN
55404, USA

13.1 INTRODUCTION

Platelet activation begins with the binding of an agonist to the cell surface, leading to aggregation, secretion and clot formation. The thromboxane A_2 (TxA_2) formed from arachidonic acid (AA, ω-6) serves as an important intracellular mediator in platelet function (1-5). More specifically, TxA_2 appears more critical for platelet responses such as shape change, aggregation, secretion, phosphoinositide turnover, Ca^{2+} mobilization from intracellular stores and protein phosphorylations induced by collagen (1-9). However, platelet adhesion to collagen appears to be independent of TxA_2 (10-12). Thromboxane$_2$A also plays a major role in pathophysiological functions, ranging from hemostasis to smooth muscle contraction by virtue of its potent vasoconstricting and platelet aggregating properties (13, 4, 6). Thromboxane A_2 (TxA_2) is formed via prostaglandin (PG) endoperoxide intermediates (PGG_2/PGH_2) by a specific

> Dr. Murthy's other area of research interest involves studies to understand the underlying mechanisms of eicosanoid precursors release by different enzymatic pathways and their impact on cellular function. He believes that PLA_2 inhibitors may one day be new effective therapies. Dr. Murthy also has interests on the effects of dietary fatty acids and their role in health and disease.

thromboxane synthase. This has been discussed in greater detail in chapter 13 and elsewhere, in this book.

The rate-limiting step in the biosynthesis of eicosanoids is the availability of free precursor, unesterified AA (20:4, ω-6), for both cyclooxygenase (COX) and lipoxygenase enzymes (13). The initial mobilization cellular AA (20:4, ω-6) is an essential step in the synthesis of eicosanoids (14, 15). Cellular AA is known to be exclusively associated with membrane phospholipids (16-18). It is also tightly regulated through enzymes of the Lands cycle. The enzymes such as phospholipase A_2, arachidonoyl-CoA synthetase and lysophosphatidyl acyltransferases appear to be simultaneously active in order to maintain a steady turnover of AA (20:4, ω-6) (19). Platelets contain arachidonoyl CoA synthetase (20).

13.2 Pathways of arachidonic acid release in platelets

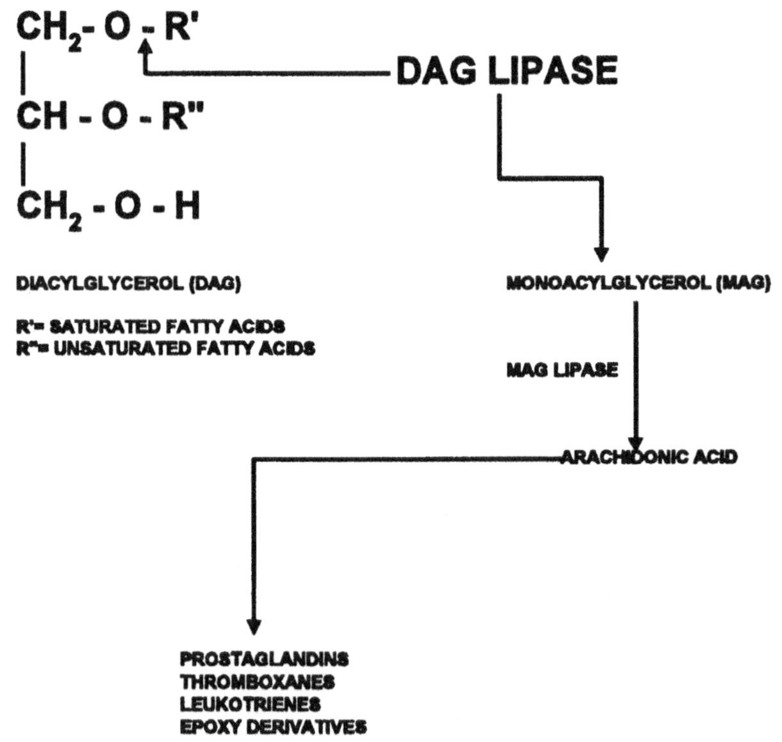

Figure 1: The diacylglycerol pathway

Thromboxane A_2 synthesis in platelets is triggered by the release of AA from platelet membrane phospholipids (14). This initial release appears to occur through activation of two major pathways: a) the first pathway involves a sequential hydrolysis of 1, 2-diacylglycerol (DAG) formed during platelet receptor-coupled activation of phospholipase C/D (PLC/PLD), by DAG- and monoacylglycerol (MAG) lipases, respectively (21, 22). We have shown that hydrolysis of DAG by this pathway results equimolar amounts of saturated fatty acids (stearic & palmitic acids) from the sn-1 position and unsaturated fatty acids [oleic acid (18:1, ω-9), linoleic acid (18:2, ω-6) and

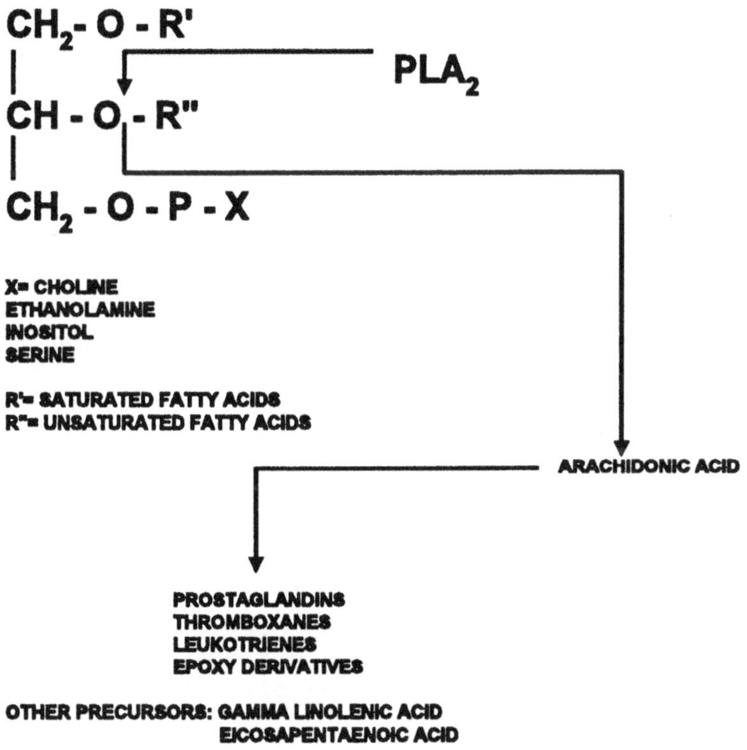

Figure 2: The PLA_2 Pathway

AA (ω-6)] from the sn-2 position of platelet phospholipids (22). The major source of DAG appears to be predominantly derived from phosphoinositides, which are rich in 1-stearoyl 2-arachidonoyl species (18, 22). Although 1-stearoyl 2-arachidonoyl DAG species may also come from a PLD-mediated hydrolysis of phospholipids, it is less clear as to the extent of contribution from this pathway for the release of AA. Furthermore, the DAG lipase pathway appears to be an unlikely route for the release of other eicosanoid precursors (DGLA, dihomo-γ-linolenic acid; 20:3, ω-6); EPA, eicospentaenoic acid (20:5, ω-3).

Alternatively, the release of platelet phospholipid AA is mediated by the activation of phospholipase A_2 (PLA_2) that is coupled to PLC, intracellular Ca^{2+} rise and protein kinase C (PKC) and tyrosine kinase-mediated protein phosphorylations (23, 24) and/or by the activation of PLA_2 that is not coupled to these events in platelets (25-29). Furthermore, the hydrolysis of phosphatidic acid (PA) by a PA-specific PLA_2 may also contribute to eicosanoid synthesis (30, 31). It is evident that the differential sensitivity of platelets to multiple agonists and their signals, have contributed to the complexity in understanding the regulation of PLA_2.

13.3 Phospholipid breakdown measurements in stimulated platelets

Over the last 20 years, a variety of agonists and methodologies have been employed to assess a number of parameters, including specific platelet phospholipid and fatty acid changes, sources of phospholipid for AA release, selectivity of the enzymes for AA and other fatty acids (32-46). Despite numerous studies, the amount and source of AA mass mobilized from membrane phospholipids, has remained a subject of intense controversy. We and others have proposed the existence of multiple fatty acid-specific phospholipid pools that are susceptible to hydrolyzing enzymes such as DAG lipase and PLA_2. For example, phosphatidylcholine (PC) and phosphatidylinositol (PI) have been considered as major sources for AA release in platelets. However, the results from a recent study suggest that plasmenylethanolamines which are the major storage depot of AA in resting platelets, as the major source of AA mobilized after thrombin stimulation of human platelets (46). In platelets, plasmenylethanolamines serve as a AA storage depot. The ability of these phospholipid species to conserve AA is a very unique mechanism, which allows platelets to maintain their at least suboptimal levels of AA for subsistence of membrane integrity. Phosphatidylethanolamine (PE) along with its plasmologen species contain 45-50 % of the total platelet AA, followed by phosphatidylcholine (PC), phosphatidylserine (PS) and phosphatidylinositol (PI). Plasmenylethanolamine species also store other long chain fatty acid eicosanoid precursors, including eicosapentaenoic acid (EPA) and docoshexaenoic acid (DHA).

13.4 Phospholipase A_2 in platelets

A number of laboratories have been successful in obtaining semipurified/purified PLA_2 preparations from a number of sources, including platelets (47-70). These studies have been helpful to understand the molecular size as well as the nature of active sites of these enzymes. Several reports have demonstrated the presence of two substrate specific and distinct enzymes, including secretory PLA_2 ($sPLA_2$) and cytosolic PLA_2 ($cPLA_2$) in platelets. $cPLA_2$ appears to have a molecular mass of 85 kDa that is structurally distinct from that of the mammalian Type II, a 14 kDa non-pancreatic PLA_2, which exists in an extracellular form in many tissues, including platelets, inflammatory joint fluid, spleen and placenta (47-70). The human non-pancreatic PLA_2 enzyme purified is also present in platelets and is enriched in rheumatoid synovial fluid.

The enzyme is calcium-dependent, has a pH optimum of 8-10, and shows a striking preference for substrate presented in the form of Escherichia coli membranes. Using oligonucleotide probes based on amino-terminal sequence data, the corresponding human gene from a genomic DNA library, has been cloned and expressed in animal cells. The protein is secreted from cells in an active form. The deduced amino acid sequence of the human protein consists of 124 amino acids, contains structural features common to all known PLA_2s, and has a half-cysteine pattern that is characteristic of the snake venom group II enzyme (60).

The 85-90-kDa $cPLA_2$ is considered to be a novel enzyme, distinct from the 14-kDa one previously purified from human platelets and it is suggested to participate mainly in AA metabolism of platelets (65, 66). Several studies have shown that Asp-49 is the residue that controls calcium binding and thus, plays a critical role in the calcium-mediated activation of, low-M(r) group I-III PLA_2s. However, more recent evidence shows that Asp-49 is not an absolute prerequisite for the enzymic activity of PLA_2s, and that proteins with amino acid(s) other than Asp at position 49 can exhibit significant phospholipase activity (69). The purification, complete amino acid sequence and characterization of ecarpholin S, a PLA_2 from Echis carinatus sochureki (saw-scaled viper) venom, has been recently described (69). This single-chain, 122-amino-acid, basic (pI 7.9) protein is a group II PLA_2. Although Asp-49 was replaced by Ser and Tyr-28 by Phe (both of these positions are being involved in the Ca^{2+}- binding site of PLA_2s), the lipolysis of soybean phosphatidylcholine and egg yolk in the presence of 10 mM $CaCl_2$ was 1.5 times and 2.9 times greater respectively with ecarpholin S than with recombinant human group II PLA2. Computer modelling of the Ca^{2+}-binding site of Ser-49 PLA_2 compared with the Asp-49 and Lys-49 forms, for which crystallographic data exist, shows that the Ca^{2+}-binding site is sterically blocked by Lys-49 but not by Ser-49; in the latter, the Ser hydroxy group may replace the Asp carboxylate in stabilization of Ca2+ binding. Sequence comparisons of ecarpholin S and other low-M(r) PLA_2 predicts the presence of a Ser-49 group in the protein family of low-M(r) PLA_2s that is distinct from the Asp-49 and Lys-49 groups (69)

A kinetic analysis of recombinant hs-PLA_2 has shown that this enzyme strongly prefers phosphatidic acid (PA) as a substrate over other phospholipids found in the mammalian plasma membrane including PS, PC and PE (70). The order of preference is PA >> PE approximately PS > PC. It appears that Lys-69 is at least partially involved in the PA specificity of hs-PLA_2 and Glu-56 in the distinction between PE and PC. These studies suggest that hs-PLA_2 can rapidly hydrolyse PA molecules exposed to the outer layer of cell-derived microvesicles and thereby produce LPA (70).

In a recent study, a serine-phospholipid-selective phospholipase A has been purified and cloned its cDNA from rat platelets, which secrete two types of phospholipases upon stimulation (71). The purified enzyme from extracellular medium of activated rat platelets, yielded a 55-kDa protein band on SDS-polyacrylamide gel electrophoresis. The presence of active serine residues was confirmed by labeling the 55-kDa protein with [^{3}H]Diisopropyl fluorophosphate, an inhibitor of the enzyme. Based on cDNA for the enzyme cloned from a rat megakaryocyte cDNA library, the 456-amino acid sequence contains a putative short N-terminal signal sequence and a GXSXG sequence, which is a motif of an active serine residue of serine esterase. Amino acid sequence homology analysis revealed that the enzyme shares about 30% homology with mammalian lipases (lipoprotein lipase, hepatic lipase, and pancreatic lipase). Regions surrounding the putative active serine, histidine, and aspartic acid, which may form a "lipase triad," were highly conserved among these enzymes. The recombinant protein, which was expressed in Sf9 insect cells using the baculovirus system, hydrolysed a fatty acyl residue at the sn-1 position of lysophosphatidylserine and phosphatidylserine, but did not appreciably hydrolyse phosphatidylcholine, phosphatidylethanolamine, phosphatidylinositol, phosphatidic acid, and triglyceride. The enzyme, named phosphatidylserine-phospholipase A_1, is the first phospholipase that exclusively hydrolyses the sn-1 position and has a strict head group specificity for the substrate (71).

Human group II secretory PLA_2 ($sPLA_2$) is an enzyme found in the α-granules of platelets and at inflammatory sites. A total chemically synthesised PLA_2 exhibits kcat and K_m values identical to those of recombinant $sPLA_2$, with 1, 2-bis (heptanoylthio)-PC as substrate. Synthetic $sPLA_2$, like that of the recombinant $sPLA_2$, inhibits thrombin generation from prothrombinase complex (factors Xa, V, II, Ca^{2+}, and phospholipids). This chemically synthesised $sPLA_2$ yields a fully active native-like enzyme and may offer a tool to provide s PLA_2 analogs for structure-activity studies of anticoagulant, lipolytic, or inflammatory activities (72).

Although PLA_2 ($sPLA_2$) protein has been described in human platelets, its presence in human megakaryocytes had not been established until recently. Using human erythroleukaemia (HEL) cell line that expresses megakaryoblastic features and an anti-rh $sPLA_2$ monoclonal antibody (mAb BA11) and dot-blot detection, it has been recently shown that HEL cells and platelets release $sPLA_2$. Double labelling experiments using mAb BA11 and an anti-GPIIb antiserum revealed the presence of $sPLA_2$ in human bone-marrow megakaryocytes. The use of reverse transcription-polymerase chain reaction conjugated with hybridization analysis further demonstrated the presence of mRNA encoding for $sPLA_2$ in platelets and HEL cells. Expression of sPLA2 in platelets and megakaryocytes at both transcriptional and post-translational levels strongly argues in favour of a megakaryocytic origin of platelet $sPLA_2$ (73).

13.5 Calcium and phospholipase A_2

Increases of intracellular free Ca^{2+} concentration that are initiated by changes in plasma membrane potential or receptor-stimulated polyphosphoinositide (PIs) hydrolysis are astonishingly complex. The key to understanding these complex Ca^{2+} signals lies in understanding the interactions between different cellular pools from which Ca^{2+} rapidly enters the cytosol and their transporting systems (74-76). These cellular events are summarized in Figure 3.

Both inositide-dependent and -independent mechanisms are present in platelets although their involvement may vary with the type of stimulus (77). It is now believed that both cytosolic- and other cellular pools of Ca^{2+}, in conjunction with the flow of external Ca^{2+} are indeed essential for the activation of PLA_2 in platelets (76). In platelets, prostaglandin synthesis is known to be associated with the dense tubular system (DTS) (78). Based on this observation, it had also been suggested that AA releasing enzymes are either localized in the DTS or become associated with the DTS in order provide substrate for eicosanoid synthesis. We now have a better understanding of how PLA_2 might function in activated platelets. Although the signalling mechanisms that regulate PLA_2 still remain complex, it is clear from in vitro studies that $cPLA_2$ is a highly Ca^{2+}-dependent enzyme and requires Ca^{2+} in millimolar concentrations for its activation (55, 56, 61, 67, 68, 79). This Ca^{2+}-dependent activation

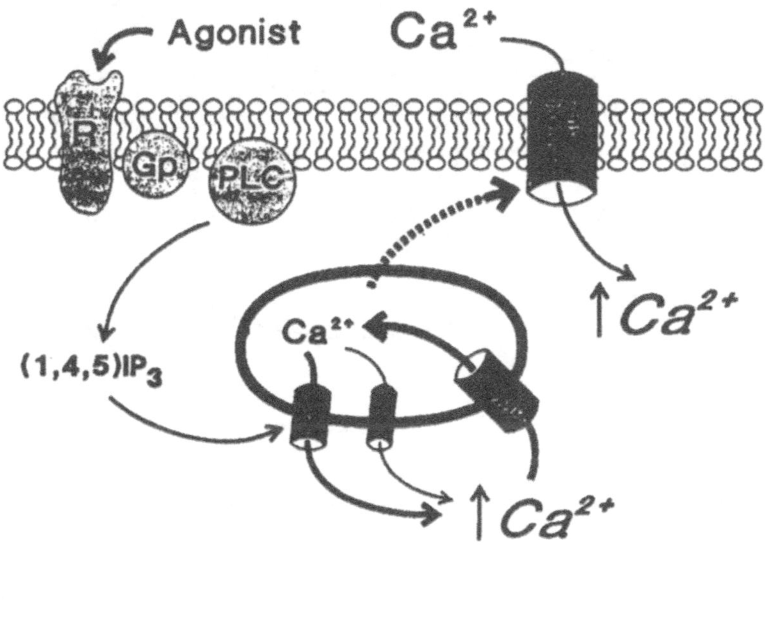

Figure 3: Calcium and inositol 1,4,5-trisphophate (IP_3) relationship.

involves the movement of HMM (high molecular mass)-$cPLA_2$ from the cytosol to membrane fractions. This has lead to the suggestion that agonist-induced Ca^{2+} elevation in the cytosol may constitute an important step in the translocation of PLA_2 to the plasma membrane (80). In these glioma cells, PLA_2 activation is coupled to the influx of external Ca^{2+} rather than to the elevation of cytosolic Ca^{2+} following its release from internal stores (80). The enzyme purified from other sources appear to involve very similar mechanisms (81, 82). The translocation of this HMM-PLA_2 under intracellularly relevant concentrations of Ca^{2+} and its specificity to AA suggests a critical role in the cellular mobilization of eicosanoid precursors and the formation of lipid mediators following stimulation of cells with agonists. Even though there is sufficient evidence for the existence of two structurally distinct isoforms of PLA_2 and their differential responsiveness to multiple agonists and cytosolic Ca^{2+}, the precise underlying biochemical mechanisms that may be involved in the activation of these enzymes in agonist-stimulated cells still remain to be more clearly defined.

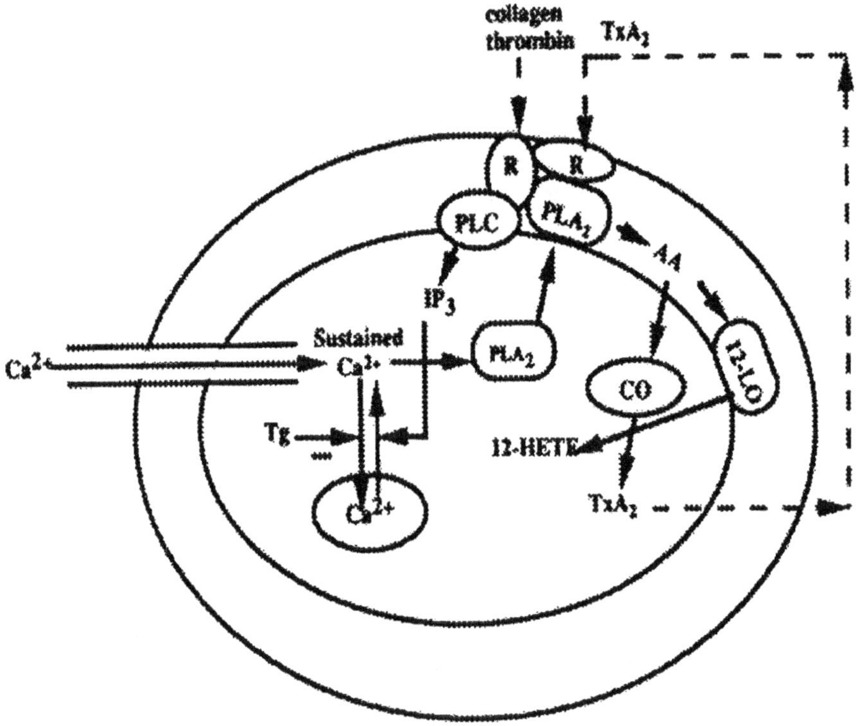

Figure 4: Postulated mechanism for PLA$_2$ regulation in stimulated human platelets.

Thapsigargin, a plant derived sesquiterpene lactone, causes a sustained increase in cytosolic Ca^{2+} without the formation of Inositol 1, 4, 5-trisphosphate (IP_3) in different cell-types (83). This non-phorbol ester tumor promoter is known to inhibit cytosolic Ca^{2+}-ATPases followed by depletion of cytosolic Ca^{2+} stores (83-88). Furthermore, the thapsigargin-induced depletion of internal calcium stores is coupled to the opening of Ca^{2+} channels, which results in the influx of external Ca^{2+}. Thapsigargin also stimulates eicosanoid synthesis, histamine and serotonin secretion in various cell-types including platelets, neutrophils and mast cells (83-88). We have recently reported a novel mechanism for PLA$_2$ activation, which is highly dependent upon the formation of TxA$_2$ in activated human platelets (summarized in Figure 4). This TxA$_2$-dependent PLA$_2$ requires and influx of extracellular Ca^{2+} coupled to its release from internal stores. Our results have led to the conclusion that the transient and possibly, IP_3-dependent cytosolic Ca^{2+} elevation without the influx of extracellular calcium may not be sufficient enough to activate this distinct enzyme (89, Figure 4). In other words, cPLA$_2$ requires a sustained elevation of cytosolic calcium for its activation and this requires external calcium. In addition, we have reported a similar mechanism for LTB$_4$-dependent PLA$_2$ regulation in human neutrophils (90).

13.5.1 Protein kinase C and phospholipase A_2

Protein kinase C is a ubiquitous enzyme involved in the transduction of biological signals. Having recognized the importance of phospholipids and their breakdown products in cell signalling, there has been an intense effort to define the role of intracellular messenger signals such as sustained elevation of DAG and protein kinase C activation, both of which are coupled to phosphatidylinositol hydrolysis (91, 92). It is now clear that the PLD-mediated release of DAG from PC hydrolysis also contributes to the regulation of PKC and cellular responses. In a similar way, PLA_2 has also been shown to be activated by most of the signals that affect phosphoinositide hydrolysis (93). In addition, the reaction products of PC hydrolysis by PLA_2, cis-unsaturated fatty acids and lysoPC, also enhance PKC and PLA_2 activation by feed back mechanisms (94, 95).

Phorbol 12-myristate (PMA), a tumor promoter, is a potent activator of PKC and it mimics the effects of 1, 2-DAG, a product PLC/PLD and also, an intracellular activator of PKC. Interestingly, it has been shown that both PMA and DAG also prime PLA_2 in human platelets stimulated with A23187, a calcium ionophore. In contrast, these agents, which prime PLA_2 in A23187-stimulated platelets, have no effect or inhibit the metabolism of phosphoinositides in platelets stimulated with agonists such as thrombin, PAF, and vasopressin (96-102). This has led to the suggestion that the decreased turnover of phosphoinositides observed in response to these agonists in the presence of PMA, which leads to the inhibition of inositol 1,4,5-trisphosphate (IP_3)-Ca^{2+}-mediated mechanisms, may decrease PLA_2 activation in stimulated platelets. We have recently provided evidence that PLA_2 activation can occur independent of thromboxane A_2-sensitive and IP_3-coupled intracellular Ca^{2+} rise and aggregation (PLC- and PKC-mediated) and PKC activation in both platelets and neutrophils (103-105).

A role for tyrosine kinases (mitogen-activated protein kinase; MAP kinases) in the regulation of $cPLA_2$ in activated platelets, has also been suggested (106, 107). Activated MAP kinases may not be involved in aggregation and secretory responses but may play a role in the late AA release via the activation of $cPLA_2$ in human platelets. Human platelets pretreated with indomethacin or other cyclooxygenase inhibitors release AA, suggesting a TxA_2- and aggregation-independent mechanism for PLA_2 activation (108, 109). The evidence now seems to suggest that this release of AA occurs predominantly through the activity of $cPLA_2$, an 85-kDa protein (110). Interestingly, $cPLA_2$ activation appears independent of PKC in thrombin- and collagen-stimulated platelets. The evidence points to the existence of both PKC- and MAPK-independent phosphorylation of $cPLA_2$ (85kDA) in platelets stimulated by the physiological activators such as collagen and thrombin (110, 111). PKC is necessary for MAPK stimulation in platelets. This enzyme is regulated by an increase in intracellular Ca^{2+}, a necessary condition for AA liberation. Phosphorylation of $cPLA_2$ enhances the Ca^{2+}-induced AA release in platelets stimulated by the ionophore A23187 and phorbol ester (phorbol 12,13-dibutyrate; PDBu). In thrombin-stimulated platelets, however, phosphorylation appears not to be necessary for AA release. However, collagen, thrombin, and PDBu induce phosphorylation of platelet $cPLA_2$ as well as activation of mitogen-activated protein kinase (MAPK; p42mapk and p44mapk) (112). Accordingly, p38 kinase activation by thrombin suggests a role for this MAP kinase in thrombin- and collagen-mediated signaling events during platelet activation (113, 114). $cPLA_2$ is a physiological target of p38 kinase and it is involved in the early phosphorylation of $cPLA_2$ in stimulated platelets (113). However, phosphorylation of $cPLA_2$ by proline-directed kinases is not required for its receptor-mediated release of AA by $cPLA_2$. In a recent report, Ser-505 and Ser-727 have identified as common phosphorylation sites on 85-kDa $cPLA_2$ in human platelets and HeLa cells (115). It has been further shown that this kinase is identical with or closely related to p38 (the mammalian homolog of HOG1 from yeast), a recently discovered protein kinase, typically activated by inflammatory cytokines and environmental stress. Their results also demonstrate that activation of this kinase by thrombin is transient (with maximal stimulation at 1 min), and accompanied by tyrosine phosphorylation, and precedes the activation of the ERK kinases.

13.6 Hydroperoxides and phospholipase A_2

Mammalian cells produce hydroperoxides from unsaturated fatty acids via the action of oxygenases. At lower concentrations, they serve to activate cyclooygenase (116). However, increases in cellular hydroperoxides inactivate oxygenases (117). It has been suggested that hydroperoxides may also be involved in the regulation of PLA_2 in activated cells (118, 119). We and others have shown that hydroperoxides may be associated with collagen-induced platelet responses, including aggregation and PLA_2 activation (119, 120). There is increasing evidence that the hydroperoxide-induced modification of membrane phospholipids may affect both $sPLA_2$ (secretory PLA_2) and $cPLA_2$ enzymes in platelets although the precise mode of action still remains unclear (121, 122). Evidence also suggests that both hydroperoxides and hydroxyl radicals generated within cells and tissues, may differentially affect AA metabolizing enzymes, including PLA_2 and other oxygenases (117-125). It has been shown that H_2O_2 may also inhibit arachidonoyl-CoA synthetase, which facilitates the selective uptake of AA in intact cells. Intracellular messengers such as Ca^{2+} and DAG do not seem to directly affect arachidonoyl-CoA synthetase and lysophospholipid acyltransferase enzymes (126).

13.7 Phosphatidic acid and platelets

In platelets, phosphatidic acid (PA) and lysophosphatidic acid (LPA) are formed during platelet activation as a result of the combined activities of PLC and DAG kinase. It has been suggested that PA formed may be hydrolyzed further by a PA-specific PLA_2 in platelets to release AA for eicosanoids (Figure 2; 127-130). There is some indication that $sPLA_2$ may generate LPA as observed in cell-derived microvesicles (131). A loss of phospholipid asymmetry may provide PA as a preferential substrate for $sPLA_2$. Phosphatidic acid seems to affect specific membrane receptors 3T3 cells (132). Lysophosphatidic acid (LPA) is known to have growth factor-like activities and it causes platelet aggregation, in activating phosphoinositide 3-kinase (133). PA derived from membrane phospholipids potentiates PLA_2 activation and thus, contributes to the amplification of platelet activation (134). In vivo, LPA derived from platelets may be an important mediator of the transport barrier formed by the vascular endothelium

(135). Both PA and LPA may also affect Ca^{2+} homeostasis in stimulated platelets and thus, many Ca^{2+}- mediated cellular signalling events.

13.8 PAF and phospholipase A_2

PAF-acether, first discovered in 1971-72, is now recognized as a 1-O-alkyl-2-acetyl-sn-glyceryl-3-phosphorylcholine released from various cells and

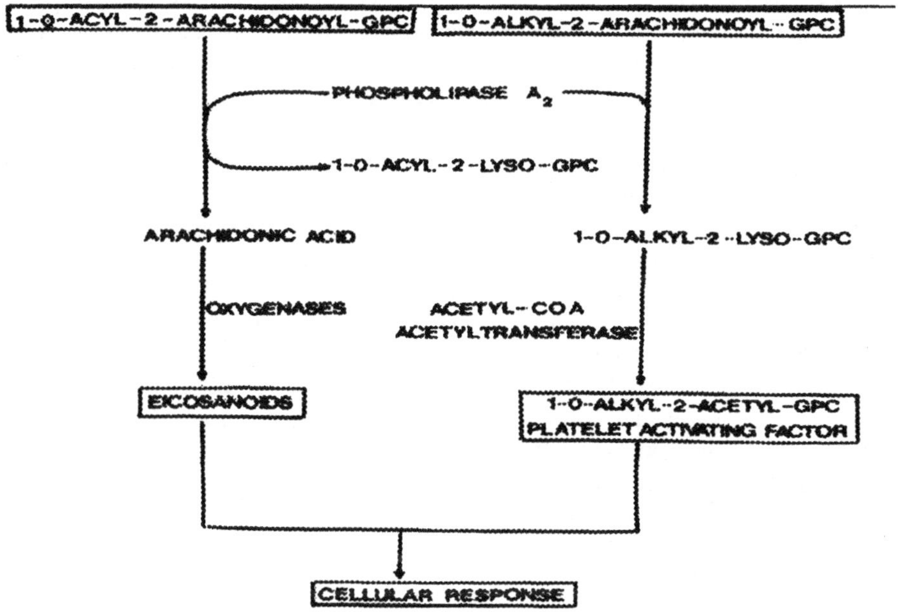

Figure 5. A common PLA_2 pathway releasing both Platelet-activating factor (PAF) and AA, in activated platelets.

organs, including macrophages, neutrophils, platelets, the kidney and heart. PAF-acether is formed by murine macrophages through a two-step process implicating successively a PLA_2 enzyme and an acetyltransferase. The ether linkage at position 1, the short acyl chain at position 2 and the natural optical configuration, are critical for the biological activity of PAF-acether. Besides platelet aggregation, PAF-acether induces a platelet-dependent, aspirin-independent bronchoconstriction in the guinea-pig and the monkey. It exhibits also a potent antihypertensive action in the rat, and triggers platelet-independent hemodynamic changes in perfused guinea-pig heart (136, 137). In stimulated platelets, alkylarachidonoyl-GPC serves as the substrate and rapidly hydrolyzed by PLA_2 and may serve as a precursor of both PAF and eicosanoids (138) PAF biosynthesis and its biological functions have been discussed in greater detail elsewhere in this book.

13.9 LDL and platelet function

There is considerable interest on the interaction between low density lipoproteins (LDL) and platelet surface because the oxidized LDL (oxLDL) bounded to specific receptors on the platelet surface are able to activate platelets (139). It appears that the inhibition of PLC-, PKC- and Ca^{2+}-mediated phosphoinositide pathways are responsible for the inhibition of LDL-induced platelet aggregation (139). However, there is also evidence that oxidized LDL may activate platelets through a PLA_2/cyclooxygenase -dependent pathway (140). In addition, glycoprotein IIb/IIIa has been proposed as the platelet receptor for high density lipoproteins (HDL3). The second messenger response induced by HDL3 appears to be mediated through via the glycoprotein IIb/IIIa, PLD and PA signalling pathway (141). Cholesterol enrichment of human platelets may selectively promote G-protein/PLA_2 interaction as well as the Ca^{2+}-release mechanism, without significantly altering G-protein/PLC interaction (142). Von Willebrand factor (vWF) is a large glycoprotein which plays a central role in thrombus formation and blood clotting. The type IIB variant of vWF is characterized by an abnormally high affinity for the platelet receptor GPIb. Platelet exposure to IIB vWF causes AA release, TxA_2 and IP_3 formation. LJIB1, a monoclonal antibody against GPIb, completely suppressed all platelet responses, whereas LJCP8, an antibody against the receptor GPIIb-IIIa (alpha IIb beta 3 integrin), or the tetrapeptide RGDS, caused a complete inhibition of the aggregation but a partial inhibition of the activation-linked parameters. This has led to the conclusion that type IIB vWF-binding to GPIb induces PLA_2 activation, AA release and GPIIb-IIIa dependent cellular Ca^{2+} influx. These events may lead to platelet secretion and aggregation (143).

Concluding remarks

PLA_2 plays a crucial role in platelet function. In addition to the other factors being discussed above, this enzyme may also be regulated by a G-protein-mediated mechanism (144). Because of the complex nature involving multiple signals in the activation of this enzyme(s), specific targeting by drugs has been more difficult compared to other targets such as COX. The specific targeting of c PLA_2 by drugs may provide a new class of drugs for the treatment of coronary heart disease (CHD). In any event, the importance PLA_2 in platelet-mediated hemostasis and arterial thrombosis, can never be diminished because of its central role in the metabolism of AA. In addition, its highly selective effects on arachidonoyl species of certain phospholipid pools may indeed be essential for protecting platelets from losing their other phospholipid species, which might be critical to maintain cellular membrane structure and integrity. Platelets do conserve AA in the form of plasmalogens and these appear to be less susceptible to PLA_2 attack that is coupled to PKC/PLD- and PKC-mediated cellular signals. Furthermore, the poor preference exhibited by PLA_2 to phospholipid species containing long chain fatty acids such as DGLA (a precursor of the 1-series eicsanoids) and docosahexaenoic acid (DHA) during platelet activation clearly illustrates the complexity of PLA_2 regulation. More recent studies clearly point to the new direction of recognizing the importance of this enzyme and its multifunctional roles in a number of diseases, including atherosclerosis, thrombosis, inflammation and cancer.

Acknowledgements

I sincerely acknowledge the support received from a number of Canadian funding agencies in the form of research grants, during my tenure at the University of Guelph and University of Manitoba, Canada. My sincere appreciations must also go my wife and children, for their understanding and support during the preparation of this manuscript. Many of the my earlier papers (until 1990) have been published under the previous name, V. G. Mahadevappa.

REFERENCES

1. Siess W. Molecular mechanisms of platelet activation. (1989). Physiol. Rev. 69, 58-178.

2. Ashby B, Daniel JL, and Smith JB. (1990). Mechanisms of platelet activation and inhibition. Hematology-Oncology Clinics of North America. 4, 1-26.

3. Manning DR, and Brass LF. (1991). The role of GTP-binding proteins in platelet activation. Thromb. Haemost. 66, 393-399.

4. FitzGerald GA. (1991). Mechanisms of platelet activation: thromboxane A2 as an amplifying signal for other agonists. Am. J. Cardiol. 68, 11B-15B.

5. Chen YQ, Liu B, Tang DG, and Honn KV. (1992). Fatty acid modulation of tumor-cell-platelet-vessel wall interaction. Cancer Metastasis Rev. 11, 389-409.

6. Reilly M, and Fitzgerald GA. (1993). Cellular activation by thromboxane A2 and other eicosanoids. Eur. Heart J. 14 Suppl K:88-93.

7. Rittenhouse SE, and Allen CL. (1982). Synergistic activation by collagen and 15-hydroxy-9a, 11a-peroxidoprosta-5, 13-dienoic acid (PGH_2) of phosphatidylinositol metabolism and arachidonic acid release in human platelets. J. Clin. Invest. 70, 1216-1224.

8. Lewis GP, and Watts IS. (1982). Prostaglandin endoperoxides, thromboxanes A_2 and adenosine diphosphate in collagen-induced aggregation in man and the cat. Br. J. Pharmacol . 75, 623-663.

9. Sano K, Takai Y, Yamanishi J, and Nishizuka Y. (1983). A role of calcium-activated phospholipid-dependent protein kinase in human platelet activation. Comparison of thrombin and collagen. J. Biol. Chem. 258, 2010-2013.

10. Zucker M, and Nachmias VT. (1985). Platelet activation. Arteriosclerosis. 5, 2-18.

11. Lapetina EG, Reep B, Read NG, and Moncada S. (1986). Adhesion of human platelets to collagen in the presence of prostacyclin, indomethacin and BW755C. Thromb. Res. 44: 325-335.

12. Rao GHR, Escolar G, and White JG. (1986). Epineprine reverses the inhibitory influence of aspirin in platelet-vessel wall interactions. Thromb. Res. 44, 65-74.

13. Needleman P, Turk J, Jakschik BA, Morrison AR, and Lefkowith JB. (1986). Arachidonic acid metabolism. Annu. Rev. Biochem. 55: 69-102.

13. Lands WEM, and Samuelsson B. (1968). Phospholpid precursors of prostaglandins. Biochim. Biophys. Acta 164, 426-429.

14. Irvine RF. (1982). How is the level of free arachidonic acid controlled in mammalian cell ? Biocehm J. 204, 3-16.

15. Roth GJ. (1986). Platelet arachidonic acid metabolism and platelet-activating factor. In: Biochemistry of Platelets (Phillips DR, and Shuman MA, eds.), Academic Press Inc. (London) Ltd.

16. Marcus AJ, Ullman HL, and Safier LB. (1969). Lipid composition of subcellular particles of human blood platelets. J. Lipid Res. 10, 108-114.

17. Cohen P and Derksen A. (1969). Comparison of phospholipid and fatty acid composition of erythrocytes and platelets. Br. J. Haematol. 17, 359-371.

18. Mahadevappa VG, and Holub BJ. (1982). The molecular species composition of individual diacyl phospholpids in human platelets. Biochim. Biophys. Acta 713, 73-79 (1982).

19. Preuss I. And Patscheke H. (1992). Regulation of the concentration of free arachidonic acid in homogenates of human platelets. Agents Actions (Supplements) 37, 34-40.

20. Wilson DB, Prescott SM, and Majerus PW. (1982). Discovery of arachidonyl coenzyme A synthetase in human platelets. J. Biol. Chem. 257, 3510-3515.

21. Bell RL, Kennerly DA, Stanford N, and Majerus PW. (1979). Diglyceride lipase: A pathway for arachidoncate release from human platelets. Proc. Natl. Acad. Sci. U. S. A. 76, 3238-3244.

22. Mahadevappa VG, and Holub BJ. (1986). Diglyceride pathway is a minor source of released arachidonic acid in thrombin-stimulated platelets. Biochem. Biophys. Res. Commun. 134, 1327-1333.

23. Berridge MJ. (1987). Inositol triphosphate and diacylglycerol: Two interacting second messengers. Annu. Rev. Biochem. 56, 159-193.

24. Mayer RJ, Marshall LA. (1993). New insights on mammalian phospholipase A_2 (s); comparison of arachidonoyl-selective and -nonslective enzymes. FASEB J. 7, 339-347.

25. Pollock WK, Rink TJ, Irvine RF. (1986). Liberation of [^{3}H]arachidonic acid and changes in cytosolic free calcium in fura-2-loaded human platelets stimulated by ionomycin and collagen. Biochem. J. 28, 1275-1280.

26. Mahadevappa VG, and Holub BJ. (1987). Quantitative loss of individual eicosapentaenoyl-relative to arachidonyl-containing phospholipids in thrombin-stimulated human platelets. J. Lipid Res. 28, 1275-1280.

27. Takamura H, Narita H, Park HJ, Tanaka K, Matsuura T, and Kito M. (1987). Differential hydrolysis of phospholipid molecular species during activation of human platelets with thrombin and collagen. J. Biol. Chem. 262, 2262-2269.

28. Vedelago HR, and Mahadevappa VG. (1988). Mobilization of arachidonic acid in collagen-stimulated human platelets. Biochem. J. 256, 981-987.

29. Piche LA, and Mahadevappa VG. (1989). Aggregation and/or oxygenated products of arachidonic acid are not required for collagen-induced deacylation of phosphatidylcholine in human platelets. Biochem. J. 263, 143-148.

30. Lapetina EG, Billah MM, and Cuatrecasas, P. (1981). The phosphatidylinositol cycle and the regulation of arachidonic acid production. Nature (Lond) 292, 367-369.

31. McCrea JM, Robinson P, and Gerrard JM. (1985). Mepacrine (quinacrine) inhibition of thrombn-induced plletlet responses can be overcome by lysophosphatidic acid. Biochim. Biophys. Acta 842: 189-194.

32. Derksen A, and Cohen P. (1975). Patterns of fatty acid release from endogenous substrates by human platelet homogenates and membranes. J. Biol. Chem. 250, 9342-9347.

33. Schoene NW, and Iacono JM. (1976). The influence of phospholipase A2 on prostaglandin production in platelets. Adv. Prostaglandin Thromboxane Res. 2, 763-766.

34. Pickett WC, Jesse RL, and Cohen P. (1976). Initiation of phospholipase A2 activity in human platelets by the calcium ion ionophore A23187. Biochim. Biophys. Acta. 486, 209-213.

35. Blackwell GJ, Duncombe WG, Flower RJ, Parsons MF, and Vane JR . (1977). The distribution and metabolism of arachidonic acid in rabbit platelets during aggregation and its modification by drugs. Br. J. Pharmacol. 59, 353-366.

36. Bills TK, Smith JB, and Silver MJ. (1977). Selective release of archidonic acid from thephospholipids of human platelets in response to thrombin. J. Clin. Inves. 60, 1-6.

37. Rittenhouse-Simmons S, and Deykin D. (1977). The mobilization of arachidonic acid in platelets exposed to thrombin or ionophore A23187. Effects of adenosine triphosphate deprivation. J. Clin. Invest. 60, 495-498.

38. Vincent JE, and Zijlstra FJ. (1978). Formation by phospholipase A2 of prostaglandins and endoperoxides in platelets of normal and essential fatty acid-deficient rats. Adv. Prostaglandin Thromboxane Res. 3, 143-146.

39. Billah MM, and Lapetina EG. (1982). Formation of lysophosphatidylinositol in platelets stimulated with thrombin or ionophore A23187. J. Biol. Chem. 257, 5196-5200.

40. Mahadevappa VG, and Holub BJ. (1983). Degradation of different molecular species of phosphatidyl-inositol in thrombin stimulated human platelets: Evidence for preferential deradation of 1-acyl 2-arachidonoyl species. J. Biol. Chem. 258, 5337-5339.

41. Mahadevappa VG, and Holub BJ. (1984). Relative degradation of different molecular species of phosphatidylcholine in thrombin-stimulated human platelets. J. Biol. Chem. 259, 9369-9373.

42. Holmsen H, Daniel JL, Dangelmaier CA, Molish I, Rigmaiden M, and Smith JB. (1984). Differential effects of trifluoperazine on arachidonate liberation, secretion and myosin phosphorylation in intact platelets. Thromb. Res. 36, 419-428.

43. Broekman MJ. (1986). Stimulated platelets release equivalent amounts of arachidonate from phosphatidylcholine, phosphatidylethanolamine, and inositides. J. Lipid Res. 27, 884-891.

44. Wang CT, Tsai WJ, Chang SM, Shiao YJ, and Yang CC. (1987). Different susceptibilities of platelet phospholipids to various phospholipases and modifications induced by thrombin. Possible evidence of rearrangement of lipid domains. Biochimi. Biophys. Acta. 899, 205-212.

45. Colard O, Breton M, and Bereziat G. (1987). Hydrolysis of endogenous phospholipids by rat platelet phospholipase A2: ether or acyl bond and polar head group selectivity. Biochimica et Biophysica Acta. 921, 333-340.

46. Han X, Gubitosi-Klug RA, Collins BJ, and Gross RW. (1996). Alterations in individual molecular species of human platelet phospholipids during thrombin stimulation: electrospray ionization mass spectrometry-facilitated identification of the boundary conditions for the magnitude and selectivity of thrombin-induced platelet phospholipid hydrolysis. Biochemistry 35, 5822-5832.

47. Ballou LR, DeWitt LM, and Cheung WY. (1986). Substrate-specific forms of human platelet phospholipase A2. J. Biol. Chem. 261, 3107-3111.

48. Hayakawa M, Horigome K, Kudo I, Tomita M, Nojima S, and Inoue K. Amino acid composition and NH2-terminal amino acid sequence of rat platelet secretory phospholipase A2. J. Biochem. 101, 1311-1314..

49. Horigome K, Hayakawa M, Inoue K, and Nojima S. (1987). Selective release of phospholipase A2 and lysophosphatidylserine-specific lysophospholipase from rat platelets. J. Biochemistry 101, 53-61.

50. Horigome K, Hayakawa M, Inoue K, and Nojima S. (1987). Purification and characterization of phospholipase A2 released from rat platelets. J. Biochemistry 101, 625-631.

51. Pierik AJ, Nijssen JG, Aarsman AJ, and Van den Bosch H. (1988). Calcium-independent phospholipase A2 in rat tissue cytosols. Biochim. Biophys. Acta. 962, 345-353.

52. Hayakawa M, Kudo I, Tomita M, and Inoue K. (1988). Purification and characterization of membrane-bound phospholipase A2 from rat platelets. J. Biochemistry. 103, 263-266.

53. Kim DK, Kudo I, and Inoue K. (1988). Detection in human platelets of phospholipase A2 activity which preferentially hydrolyzes an arachidonoyl residue. J. Biochemistry. 104, 492-494.

54. Lai CY, and Wada K. (1988). Phospholipase A_2 from human synovial fluid: purification and structural homology to the placental enzyme. Biochem Biophys Res Commun 157: 488-493.

55. Leslie CC, Voelker DR, Channon JY, Wall MM, and Zelarney PT. (1988). Properties and purification of an arachidnoyl-hydrolyzing phospholipase A_2 from a macrophage line, RAW 2647. Biochim Biophys Acta 963, 476-492.

56. Seilhamer JJ, Pruzanski W, and Vada P. (Other authors). (1989). Cloning and recombinant expression of phospholipase A_2 present in rheumatoid arthritic synovial fluid. J Biol Chem 264; 5335-5338.

57. Mizushima H, Kudo I, Horigome K, Murakami M, Hayakawa M, Kim DK, Kondo E, Tomita M, and Inoue K. (1989). Purification of rabbit platelet secretory phospholipase A2 and its characteristics. J. Biochemistry. 105, 520-525.

58. Murakami M, Kudo I, and Inoue K. (1989). In vivo release and clearance of rat platelet phospholipase A2. Biochim. Biophys. Acta. 1005, 270-276.

59. Aarsman AJ, Leunissen-Bijvelt J, Van den Koedijk CD, Neys FW, Verkleij AJ, and Van den Bosch H. (1989). Phospholipase A2 activity in platelets. Immuno-purification and localization of the enzyme in rat platelets. J. Lipid Mediators 1, 49-61.

60. Kramer RM, Hession C, Johansen B, Hayes G, McGray P, Chow EP, Tizard R, and Pepinsky RB. (1989). Structure and properties of a human non-pancreatic phospholipase A2. J. Biol. Chem. 264, 5768-5775.

61. Channon JY, and Leslie CC. (1990). A calcium-dpendent mechanism for associating a soluble arachidonoyl-hydrolyzing phospholipase A_2 with membrane in the macrophage cell line RAW 2647. J. Biol. Chem. 265, 5409-5413.

62. Kim DK, Kudo I, Fujimori Y, Mizushima H, Masuda M, Kikuchi R, Ikizawa K, and Inoue K. (1990). Detection and subcellular localization of rabbit platelet phospholipase A2 which preferentially hhydrolyzes an arachidonoyl residue. J. Biochemistry. 108, 903-906.

63. Schalkwijk CG, Marki F, and Van den Bosch H. (1990). Studies on the acyl-chain selectivity of cellular phospholipases A2. Biochim. Biophys. Acta. 1044, 139-146.

64. Murakami M, Kudo I, Natori Y, and Inoue K. (1990). Immunochemical detection of 'platelet type' phospholipase A2 in the rat. Biochim. Biophys. Acta. 1043, 34-42.

65. Takayama K, Kudo I, Kim DK, Nagata K, Nozawa Y, and Inoue K. (1991). Purification and characterization of human platelet phospholipase A2 which preferentially hydrolyzes an arachidonoyl residue. FEBS Lett. 282, 326-330.

66. Kim DK, Kudo I, and Inoue K. (1991). Purification and characterization of rabbit platelet cytosolic phospholipase A2. Biochim. Biophys. Acta. 1083, 80-88.

67. Clark JD, Lin L, Kriz R (Other Authors ?). (1991). A novel arachidonic acid-selective cytosolic PLA_2 contains a Ca^{2+}-dependent translocation domain with homology to PKC and GAP. Cell 65, 1043-1051.

68. Sharp JD, White DL, Chiou XG (Other Authors ?). (1991). Molecular cloning and expression of human Ca^{2+}-sensitive cytosolic phospholipase A_2. J. Biol. Chem. 266, 14850-14853.

69. Polgar J, Magnenat EM, Peitsch MC, Wells TN, and Clemetson KJ. (1996). Asp-49 is not an absolute prerequisite for the enzymic activity of low-M(r) phospholipases A2: purification, characterization and computer modelling of an enzymically active Ser-49 phospholipase A2, ecarpholin S, from the venom of Echis carinatus sochureki (saw-scaled viper). Biochem. J. 319 (Pt 3), 961-968. .

70. Snitko Y, Yoon ET, and Cho W. (1997). High specificity of human secretory class II phospholipase A2 for phosphatidic acid. Biochem. J. 321 (Pt 3), 737-741.

71. Sato T, Aoki J, Nagai Y, Dohmae N, Takio K, Doi T, Arai H, and Inoue K. (1997). Serine phospholipid-specific phospholipase A that is secreted from activated platelets. A new member of the lipase family. J. Biol. Chem. 272, 2192-2198.

72. Hackeng TM, Mounier CM, Bon C, Dawson PE, Griffin JH, and Kent SB. (1997). Total chemical synthesis of enzymatically active human type II secretory phospholipase A2. Proc. Nat. Acad. Sci. U. S. A. 94, 7845-7850.

73. Emadi S, Mirshahi M, Elalamy I, Nicolas C, Vargaftig BB, and Hatmi M. (1998). Cellular source of human platelet secretory phospholipase A2. Br. J. Haematol. 100, 365-373.

74. Taylor CW, and Marshall ICB. (1992). Calcium and inositol 1,4,5-trisphosphate receptors: a complex relationship. TIBS 17, 403-407.

75. Gerrard JM, Butler AM, Graff G, Stoddard SF, and White JG. (1978). Prostaglandin endoperoxides promote calcium release from platelet membrane fraction in vitro. Prostaglandins Med. 1, 373-385.

76. Gerrard JM, Peterson DA, and White JG. (1981). Calcium mobilization. In: Platelets in Biology and Pathology (Gordon JL, ed.), Vol. 2, pp. 407-436, Elsevier North-Holland, Amsterdam.

77. Lapetina EG. (1986). Inositide-dependent and -independent mechanisms in platelet activation. In: Receptor Biochemistry and Methodology (Putney JW, Jr., ed.), New York, A. R. Liss, vol. 7, pp. 271-286).

78. Gerrard JM, White JG, Rao GHR, and Townsend D. (1976). Localization of platelet prostaglandin production in the platelet dense tubular system. Am. J. Pathol. 83, 283-298.

79. Billah MM, Lapetina EG, and Cuatrecasas P. (1980). Phospholipase A_2 and phopsholipase C activities in platelets. Differential substrate specificity, Ca^{2+} requirement, pH dependence, and cellular localization. J. Biol. Chem. 255, 10227-10231.

80. Brooks RC, McCarthy KD, Lapetina EG, and Morrel P. (1989). Receptor-stimulated phospholipase A_2 activation is coupled to influx of external calcium and not to mobilization of intracellular calcium in C62 glioma cells. J. Biol. Chem. 264, 20147-20153.

81. Ulevitch RJ, Watanabe Y, Sano M, Lister MD, Deems RA, and Dennis EA. (1988). Solubalization, purification, and characterization of a membrane-bound phospholipase A_2 from the P388D1 macrophage-like cell line. J. Biol. Chem. 263, 3079-3085.

82. Ono T, Tojo H, Kuramitsu S, Kagamiyama H, and Okamoto M. (1988). Purification and characterization of a membrane-associated phospholipase A_2 from rat spleen. Its comparison with a cytosolic phospholipase A_2 S-1. J. Biol. Chem. 263, 5732-5738.

83. Jackson TR, Patterson SI, Thastrup O, and Hanley MR. (1988). A novel tumor promoter, thapsigargin, transiently increases cytoplasmic free Ca^{2+} without generation of inositol phosphates in NG115-401L neuronal cells. Biochem. J. 253, 81-86.

84. Thastrup O, Foder B, and Scharff O. (1987). The calcium mobilizing tumor promoting agent, thapsigargin, elevates the platelet cytoplasmic free calcium concentration to a higher steady state level. A possible mechanism of action for the tumor promotion. Biochem. Biophys. Res. Commun. 142, 654-660.

85. Brune B, Ullrich V. (1991). Calcium mobilization in human platelets by receptor antogonists and calcium-ATPase inhibitors. FEBS Lett. 284: 1-4.

86. Brune B, and Ullrich V. (1991). Differential calcium pools in human platelets and their role in thromboxane A_2 formation. J. Biol. Chem. 266, 19232-19237.

87. Malcolm KC, and Fitzpatrick FA. (1992). Indirect actions thapsigargin in human platelets: activation of eicosanoid biosynthesis and cellular signalling pathways. J. Pharm. Exp. Therap. 260: 1244-1249.

88. Ohuchi K, Sugawara T, Watanabe M (Other Authors ?). (1988). Analysis of the stimulative effect of thapsigargin, a non-TPA-type tumor promoter, on arachidonic acid metabolism in rat peritonial macrophages. Br. J. Pharmacol. 94: 917-923.

89. Murthy M, Rao GH, Robinson P, and Reddy R. (1995). Influx of extracellular calcium and agonist-coupling appear essential for the activation of thromboxane A_2-dependent phospholipase A in human platelets. Prostaglandin Leuk. Essen. Fatty Acids. 53, 31-39.

90. Reddy S, Bose R, Rao GH, and Murthy M. (1995). Phopsholipase A_2 activation in human neutrophils requires influx of extracellular Ca^{2+} and leukotriene B_4. Am. J. Physiol. 268 (Cell Physiol. 37): C138-C146.

91. Nishizuka Y. (1988). Nature (Lond.) 334, 661-665.

92. Asaoka Y, Nakamura S, Yoshida K, and Nishizuka Y. (1992). Protein kinase C, calcium and phospholipid degradation. TIBS 17, 414-417.

93. Axelrod H, Burch RM, and Jelsema CL. (1988). Trends Neurosci. 11, 117-123.

94. Yoshida K, Asaoka Y, and Nishizuka Y. (1992). Proc. Natl. Acad. Sci. U.S.A. 89, 6443-6446.

95. Ballou LR, and Cheung WY. (1985). Inhibition of human platelet phospholipase A2 activity by unsaturated fatty acids. Proc. Natl. Acad. Sci. U.S.A. 82, 371-375.

96. Watson SP, and Lapetina EG. (1985). !,2-Diacylglycerol and phorbol ester inhibit agonist-induced formation of inositol phosphates in human platelets.: Possible implications for negative feedback regulation of inositol phospholipid hydrolysis. Proc. Natl. Acad. Sci. U.S.A. 82, 2623-2626.

97. Halenda SP, Zavoico GB, and Feinstein MB. (1985). Phorbol esters and oleoyl acetoyl glycerol enhance relase of arachidonic acid in platelets stimulated with Ca^{2+} ionophore A23187. J. Biol. Chem. 260, 12484-12491.

98. Mobely A, and Tai HH. (1985). Synergistic stimulation of thromboxane biosynthesis by calcium ionophore and phorbol ester or thrombin in human platelets. Biochem. Biophys. Res. Commun. 130, 717-723.

99. Zavoico GB, Halenda SP, Sha'afi RI, and Feinstein MB. (1985). Phorbol myristate acetate inhibits thrombin-stimulated Ca^{2+} mobilization and phosphatidylinositol 4, 5-bisphosphate hydrolysis in human platelets. Proc. Natl. Acad. Sci. U.S.A. 82, 3859-3862.

100. Krammer RM, Chocani GC, and Deykin D. (1987). Stimulation of Ca^{2+}-activated human platelet phospholipase A_2 by diacylglycerol. Biochem. J. 248, 779-783.

101. Krishnamurti S, Joseph S, and Kakkar VV. (1987). Effect of phorbol 12-myristate 13-acetate (PMA). on agonsit-induced arachidonate release and 5-hydroxytryptamine secreation in human platelets: Dependence of effects on agonist type and time of incubation with PMA. Biochim. Biophys. Acta 927, 429-436.

102. Halenda SP, Banga HS, Zavoico GB, Lau LF, and Feinstein MB. (1989). Synergistic release of arachidonic acid from platelets by activators of protein kinase C and Ca^{2+} ionophores. Evidence for the role of protein phosphorylation in the activation of phospholipase A_2 and independence from `the Na^+/H^+ exchanger. Biochemistry 28, 7356-7363.

103. Reddy S, Rao GH, and Murthy M. (1994). Differential effects of phorbol 12-myristate and diacylglycerols on thromboxane A_2 activation in collage-stimulated human platelets. Biochem. Med. Metabolic Biol. 51, 118-128.

104. Murthy M, Rao GH, and Reddy and S. (1994). !, 3-Dooctanoylglycerol (1, 3-DiC_8) is as effective as 1, 2-dioctanoylglycerol (1, 2-DiC_8) in priming phospholipase $_2$A in human platelets. Biochem. Med. Metabolic Biol. 52, 89-96.

105. Conquer J, and Mahadevappa VG. (1991). Evidence for the possible involvement of protein kinase C in the activation of non-specific phospholipase A_2. J. Lipid Med. 3, 113-123.

106. Hargreaves PG, Licking EF, Sargeant P, Sage SO, Barnes MJ, and Farndale RW. (1994). The tyrosine kinase inhibitors, genistein and methyl 2,5-dihydroxycinnamate, inhibit the release of (3H)arachidonate from human platelets stimulated by thrombin or collagen. Thromb. Haemost. 72, 634-642.

107. Nakashima S, Chatani Y, Nakamura M, Miyoshi N, Kohno M, and Nozawa Y. (1994). Tyrosine phosphorylation and activation of mitogen-activated protein kinases by thrombin in human platelets: possible involvement in late arachidonic acid release. Biochem. Biophys. Res. Commun. 198, 497-503.

108. Vedelago HR, and Mahadevappa VG. (1988). Mobilization of arachidonic acid in collagen-stimulated human platelets. Biochem. J. 256, 981-987.

109. Piche LA, and Mahadevappa VG. (1989). Aggregation and/or oxygenated products of arachidonic acid are not required for collagen-induced deacylation of phosphatidylcholine in human platelets. Biochem. J. 263, 143-148.

110. Borsch-Haubold AG, Kramer RM, and Watson SP. (1995). Cytosolic phospholipase A2 is phosphorylated in collagen- and thrombin-stimulated human platelets independent of protein kinase C and mitogen-activated protein kinase. J. Biol. Chem. 270, 25885-25892.

111. Kramer RM, Roberts EF, Hyslop PA, Utterback BG, Hui KY, and Jakubowski JA. (1995). Differential activation of cytosolic phospholipase A2 (cPLA2) by thrombin and thrombin receptor agonist peptide in human platelets. Evidence for activation of cPLA2 independent of the mitogen-activated protein kinases ERK1/2. J. Biol. Chem. 270, 14816-4823.

112 Kramer RM, Roberts EF, Strifler BA, and Johnstone EM. (1995). Thrombin induces activation of p38 MAP kinase in human platelets. J. Biol. Chem. 270, 27395-27398.

113. Kramer RM, Roberts EF, Um SL, Borsch-Haubold AG, Watson SP, Fisher MJ, and Jakubowski JA. (1996). p38 mitogen-activated protein kinase phosphorylates cytosolic phospholipase A2 (cPLA2) in thrombin-stimulated platelets. Evidence that proline-directed phosphorylation is not required for mobilization of arachidonic acid by cPLA2. J. Biol. Chem. 271, 27723-27729.

114. Borsch-Haubold AG, Kramer RM, and Watson SP. (1997). Phosphorylation and activation of cytosolic phospholipase A2 by 38-kDa mitogen-activated protein kinase in collagen-stimulated human platelets. Eur. J. Biochem. 245, 751-759.

115. Borsch-Haubold AG, Bartoli F, Asselin J, Dudler T, Kramer RM. Apitz-Castro R, Watson SP, Gelb MH. (1998). Identification of the phosphorylation sites of cytosolic phospholipase A2 in agonist-stimulated human platelets and HeLa cells. J. Biol. Chem. 273, 4449-4458.

116. Hemler ME, and Lands WEM. (1980). J. Biol. Chm. 255, 6253-6261.

117. Pace-Asciak CR, and Smith WL. (1983). In: The Enzymes XVI, 3rd Edition (Boyer PD., ed.), pp. 543-603, Academic Press, Toronto.

118. Van Kuijk FJGM, Sevanian A, Handelman GJ, and Dratz EA. (1987). Trends Biochem. Sci. 12, 31-34.

119. Vadelago HR, and Mahadevappa VG. (1988). Differential effects of 15-HPETE on arachidonic acid metabolism in collagen-stimulated human platelets. Biochem. Biophys. Res. Commun. 150, 177-184.

120. Pignatelli P, Pulcinelli FM, Lenti L, Gazzaniga PP, and Violi F. (1998). Hydrogen peroxide is involved in collagen-induced platelet activation. Blood. 91, 484-490.

121. Akiba S, Nagatomo R, Hayama M, and Sato T. (1997). Lipid peroxide overcomes the inability of platelet secretory phospholipase A2 to hydrolyze membrane phospholipids in rabbit platelets. J. Biochemistry. 122, 859-864, 1997.

122. Akiba S, Nagatomo R, Hayama M, and Sato T. (1997). Stimulation of cytosolic phospholipase A2-catalyzed arachidonic acid liberation by low dose tert-butyl hydroperoxide without an influence on the enzyme activity in rabbit platelets. Biochem. Mol. Biol. International. 43, 799-806.

123. Muller M, and Sorrell TC. (1997). Oxidative stress and the mobilisation of arachidonic acid in stimulated human platelets: role of hydroxyl radical. Prostaglandins 54, 493-509.

124. Leo R, Pratico D, Iuliano L, Pulcinelli FM, Ghiselli A, Pignatelli P, Colavita AR, FitzGerald GA, and Violi F. (1997). Platelet activation by superoxide anion and hydroxyl radicals intrinsically generated by platelets that had undergone anoxia and then reoxygenated. Circulation. 95, 885-891; Comment in: Circulation 18, 95:787-789

125. van Gorp RM, van Dam-Mieras MC, Hornstra G, and Heemskerk JW. (1997). Effect of membrane-permeable sulfhydryl reagents and depletion of glutathione on calcium mobilisation in human platelets. Biochem. Pharmacol. 53, 1533-1542.

126. Hornberger W, and Patscheke H. (1990). Primary stimuli of icosanoid release inhibit arachidonoyl-CoA synthetase and lysophospholipid acyltransferase. Mechanism of action of hydrogen peroxide and methyl mercury in platelets. Eur. J. Biochem. 187, 175-181.

127. Lapetina EG, and Billah MM. (1981). The importance of the production of phosphatidic acid for the release of arachidonic acid in stimulated platelets. Agents & Actions. 11, 536-537.

128. Lapetina EG, Billah MM, and Cuatrecasas P. (1981) Lysophosphatidic acid potentiates the thrombin-induced production of arachidonate metabolites in platelets. J. Biol. Chem. 256, 11984-11987.

129. Lapetina EG, Billah MM, and Cuatrecasas P. (1981). The initial action of thrombin on platelets. Conversion of phosphatidylinositol to phosphatidic acid preceding the production of arachidonic acid. J. Biol. Chem. 256, 5037-5040.

130. Billah MM, Lapetina EG, and Cuatrecasas P. (1981). Phospholipase A2 activity specific for phosphatidic acid. A possible mechanism for the production of arachidonic acid in platelets. J. Biol. Chem. 256, 5399-5403.

131. Fourcade O, Simon MF, Viode C, Rugani N, Leballe F, Ragab A, Fournie B, Sarda L, and Chap H. (1995). Secretory phospholipase A2 generates the novel lipid mediator lysophosphatidic acid in membrane microvesicles shed from activated cells. Cell. 80, 919-827.

132. Murayama T, and Ui M. (1987). Phosphatidic acid may stimulate membrane receptors mediating adenylate cyclase inhibition and phospholipid breakdown in 3T3 fibroblasts. J. Biol. Chem. 262, 5522-5529.

133. Zhang J, and Rittenhouse SE. (1995). Lysophosphatidic acid activates phosphoinositide 3-kinase and phospholipase C in human platelets: inhibitory effects of Wortmannin on phosphoinositide 3-kinase and aggregation. Biochem. Biophys. Res. Commun. 211, 484-490.

134. Hashizume T, Taniguchi M, Sato T, and Fujii T. (1994). Arachidonic acid liberation induced by phosphatidic acid endogenously generated from membrane phospholipids in rabbit platelets. Biochim. Biophys. Acta. 1221, 179-184.

135. Alexander JS, Patton WF, Christman BW, Cuiper LL, and Haselton FR. (1998). Platelet-derived lysophosphatidic acid decreases endothelial permeability in vitro. Am. J. Physiol. 274, H115-22.

136. Benveniste J. Platelet-activating factor (PAF-acether). (1981). Agents & Actions. 11, 542-544.

137. Benveniste J, Chignard M, Le Couedic JP, and Vargaftig BB. (1982). Biosynthesis of platelet-activating factor (PAF-ACETHER). II. Involvement of phospholipase A2 in the formation of PAF-ACETHER and lyso-PAF-ACETHER from rabbit platelets. Thromb. Res. 25, 375-385.

138. Kramer RM, Jakubowski JA, and Deykin D. (1988). Hydrolysis of 1-alkyl-2-arachidonoyl-sn-glycero-3-phosphocholine, a common precursor of platelet-activating factor and eicosanoids, by human platelet phospholipase A2. Biochim. Biophys. Acta 959, 269-279.

139. Vlasova II, Azizova OA, and Lopukhin YuM. (1997). Inhibitor analysis of LDL-induced platelet aggregation. Biochemistry. 62, 307-311.

140. Weidtmann A, Scheithe R, Hrboticky N, Pietsch A, Lorenz R, and Siess W. (1995). Mildly oxidized LDL induces platelet aggregation through activation of phospholipase A2. Arteriosclerosis, Thromb. Vascular Biol. 15, 1131-138.

141. Nofer JR, Walter M, Kehrel B, Seedorf U, and Assmann G. (1995). HDL3 activates phospholipase D in normal but not in glycoprotein IIb/IIIa-deficient platelets. Biochem. Biophys. Res. Commun. 207, 148-154.

142. Sorisky A, Kucera GL, and Rittenhouse SE. (1990). Stimulated cholesterol-enriched platelets display increased cytosolic Ca2+ and phospholipase A activity independent of changes in inositol trisphosphates and agonist/receptor binding. Biochem. J. 265, 747-754.

143. Francesconi M, Casonato A, Pontara E, Dalla Via L, Girolami A, and Deana R. (1995). Type IIB von Willebrand factor induces phospholipase A2 activation and cytosolic Ca2+ increase in platelets. Biochem. Biophys. Res. Commun. 214, 102-109.

144. Iorio P, Gresele P, Stasi M, Nucciarelli F, Vezza R, Nenci GG, and Goracci G. (1996). FEBS Lett. 38, 244-248.

14

PLATELET BIORHEOLOGY: ADHESIVE INTERACTIONS IN FLOW

M. M. Frojmovic
Departments of Physiology and Medicine
McGill University
Montreal, Quebec
Canada

14.1 INTRODUCTION: GENERAL OVERVIEW FOR FLOW STUDIES OF PLATELET AGGREGATION

> Dr. Mony Frojmovic is Professor of Physiology and Associate member of Dept. of Medicine at McGill University, Canada. He is currently the chairman of the Biorheology subcommittee of the International Society of Hemostasis and Thrombosis. He founded and chaired the Gordon Conference on the Biorheology of Cell Adhesion in 1993. He has studied the physiology of the human platelet in hemostasis and thrombosis for the past 31 years, with innovative contributions in platelet size and shape changes accompanying platelet activation, and in elucidating the role of adhesive ligands and receptors in mediating flow-dependent platelet aggregation.

Platelets will normally adhere to 'damaged' vessel walls and to other 'activated' platelets in flowing blood in the presence of predominantly red blood cells present in greater than 99% by volume of all blood particles, and in the presence of white cells potentially 'reactive' with the platelets. There are many excellent reviews dealing with the flow properties of whole blood, including the macroscopic and microscopic behaviour of cells (1). There are additionally many reports on studies of platelet adhesion from flowing whole blood onto vessel segments (2,3) and onto solid surfaces in flow chambers coated with extracellular matrix or purified adhesive proteins (4-8). Even though platelet deposition onto surfaces has been directly observed by epifluorescence microscopy (7,8), these studies have been limited to a description of cell deposit growth and have not been analyzed to-date in terms of efficiencies of capture of platelets, which requires a precise knowledge of collision frequencies of platelets onto the surface. Such analyses have not been done since they are too complicated by the dependence for platelet collisions with the surface on the presence of a high concentration of red blood cells needed for transport of platelets in flow onto the surfaces (2). We will therefore concentrate on the aspects of platelet biorheology associated with platelet aggregation in suspensions of pure platelets as a function of variable flow conditions. We will consider the adhesive behaviour of platelets , and model particles with select platelet receptors and ligands, in suspensions with flow conditions varying from 'stasis': diffusion-controlled platelet motion; to 'stirred' or 'shaken' (gyrotation) suspensions; and finally to suspensions evaluated in laminar flow regimes varying from 'physiologic' through to 'pathologic' at extremes of shear rates. We will outline the use of flow cytometry for measuring the quality and quantity of adhesive receptors and ligands on platelets and model cells and particles which can then be related to the molecular determinants of platelet aggregation measured in parallel studies with flow devices. We will concentrate on our own reports on the flow-dependent differential roles of the platelet receptors, GPIb and GPIIb-IIIa, in platelet aggregation, mediated by adenosine diphosphate (ADP) and soluble fibrinogen (Fg), by ristocetin and soluble von Willebrand factor (vWF), and by ADP and thrombin in the absence of exogenous ligands. Finally, we will also point to some new directions in bio-rheological research on platelet aggregation, including the existence and great need for rheologic bedside devices for diagnosing and monitoring cardiovascular diseases and their management with the fast-growing array of anti-platelet drugs.

14.2 General physiology of platelet activation and aggregation in flow

A variety of physiological activators such as adenosine diphosphate (ADP), thrombin or collagen cause human blood platelets, circulating as smooth-surfaced discoid cell fragments, to rapidly change their shape (9-11). Adhesion of these activated platelets to each other requires fibrinogen (Fg), calcium, and activated GPIIb-IIIa receptors (12-14) to form aggregates which are critical to the normal hemostatic process (Fig. 1). There is additionally a role for von Willebrand factor (vWF) and its receptors (GPIIb-IIIa and GPIb, shown in Fig. 1) in appropriate conditions of flow and cell activation which will be discussed in later sections, where it will be shown that platelet can undergo aggregation either i) directly via shear-dependent GPIb/vWF binding at "pathological" high flow conditions, associated with GPIIb-IIIa activation, or ii) via agonist activation of platelet (e.g. ADP or thrombin) and surface-secreted vWF occurring at physiological flow conditions. At the low shear rates associated with stirred suspensions of platelet largely used in aggregation studies ($G<100\ s^{-1}$), soluble Fg, present in blood or plasma at 7 μM, predominates over soluble vWF in driving aggregation via binding to activated GPIIb-IIIa, essentially because there is >1000-fold more Fg than soluble vWF; in fact, in the absence of Fg, vWF readily drives stir-associated aggregation of ADP-activated platelet (vWF replaces Fg in Fig. 1).

We and others have shown that platelet aggregation in flow occurs in three steps: 1) platelet activation and formation of pseudopods initiated within hundreds of milliseconds

(10,11,17); 2) microaggregate (doublets to less than 20-cell aggregates) formation after about 1 sec of being activated (11,18), initially via pseudopod-mediated contacts (19); and 3) build-up of microaggregates to form large aggregates depending on activator and flow conditions (11,18,20,21). Thus, weak activation of normal platelet yields large visible aggregates (TA) at very low shear (<30 s^{-1}) which dissociate to microaggregates (PA) at more normal shear (~100 s^{-1}, as seen in stirred suspensions in the aggregometer), while strong activation of such platelet yields both PA and TA in stirred suspensions (21). Platelet activation and aggregation have also been found to vary with size- and density-dependent subpopulations, with reported heterogeneities in structure and biochemistry (9,20,22-24).

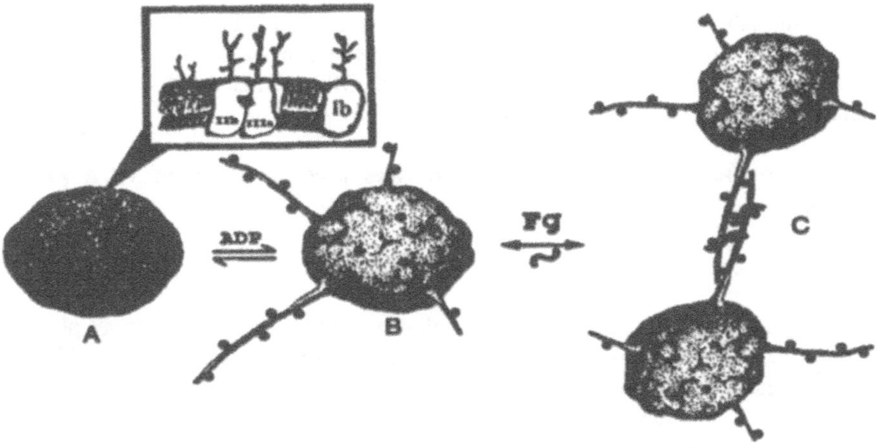

Fig. 1. Schematic for platelet activation, expression of GpIIb-IIIa receptors and aggregation. The resting platelet is a smooth discoyte (A) containing ~50,000 copies of GPIIb-IIIa and 25,000 copies of GPIb in the surface membrane (inset), the latter normally associated with GPIX and V. Activation causes shape change (B) and "transformation" of the "resting" GPIIb-IIIa complexes into activated receptors (•), which, at physiological shear rates, will bind fibrinogen (Fg,~) and allow platelet doublet and multiplet formation (C) to occur by 1 sec after ADP addition to normal citrated platelet-rich plasma. The arrangement of receptors and selection of platelet contacts is only hypothetical (Modified from ref. 64).

14.3 Range of shear rates in normal and pathological settings

The rate and extent of the aggregation in flowing suspensions in vitro and in the circulation in vivo are governed by the local flow conditions. Hydrodynamic properties of flowing suspensions are described by the shear rate (G), related to the volume flow and size of the blood vessel, and the shear stress which is simply the product of G and the viscosity of the suspension, where the viscosity is typically 4 centipoise for whole blood and closer to 1 for suspensions in plasma or buffer (25). Time average values of wall shear stresses in the normal blood circulation found in post-capillary venules range from approximately 1 dyn/cm^2 to 20 dyn/cm^2 (26), corresponding to about 25-500 s^{-1} shear rates. Values of 500-5000 s^{-1} have been reported for small arteries/ arterioles (27), and mean values around 1700 s^{-1} (25). However, in "pathological" vessels, wall shear rates readily exceed 10,000 s^{-1}, generated at the top of plaques occluding the lumen of diseased coronary arteries by 50->80% (28). At the other end of the spectrum are the near-stasis conditions found in separated flow in diseased vessels at bifurcations, known as "disturbed" flow, which is neither laminar nor turbulent (25).

Shear rates, and associated shear stress, can have multiple effects on aggregation by i) regulating the collision frequency between the platelets in the suspension; and (ii) the efficiency of platelet adhesion which generally decreases with increasing shear rate. It should be noted that the shear rate determines the duration of contact between two colliding platelets, given by 2.6/G (29), with a critical time required for adhesive receptor/ligand organization and cross-bridging allowing 'capture' of two platelets before they are forced to separate by the acting shear stresses. At 'pathological' shear rates, typically >6000 / sec, shear stresses can also cause i) the release of ADP from dense granules which has been shown to be associated with high shear stress-mediated platelet activation and subsequent aggregation (30,31); and (ii) the activation of the GPIb receptor and/or von Willebrand factor (vWF), where vWF is present in plasma, but can also be released from α-granules, and can bind to either GPIb or activated GPIIb-IIIa, also associated with platelet aggregation (6,32-35).

14.4 Flow regimes and corresponding devices used to study in vitro platelet aggregation

14.4.1 General overview

Platelet aggregation in suspension requires collisions of platelets and 'reactive' platelets. We will describe studies of activated platelet aggregation in suspensions which are unstirred, where diffusion-driven collisions occur, representing the most 'efficient' capture system once two platelets meet because they have the longest contact times for optimal cross-bridging; however, we shall see the anomalous appearance of a major role for pseudopods in this setting. We will next briefly describe the 'flow' system most commonly-used because of its convenience, namely stirred platelet suspensions (1000 rpm) and light transmission or direct particle counting (18), where the shear regime is complex and ill-defined, though on average representing the equivalent effects of low 'venal' shear rates (< 100/sec); i.e. comparable rates of platelet aggregation were observed for suspensions stirred at 1000 rpm or sheared at <100/sec (Frojmovic unpublished observations 1995). Finally, devices with well-defined laminar flow regimes will be described.

14.4.2 Types of flow devices for studying aggregation of platelet suspensions

The kinetics of aggregation in flow has largely been studied under stir in aggregometers, where the hydrodynamic conditions cannot be specified (36-38). Moreover, the conventional light transmission method of quantifying aggregates was not sufficiently sensitive to detect micro-aggregates composed mainly of doublets and triplets (11), although the interpretation of turbidimetric measurements has been improved based on light scattering theory (39,40). Electronic particle counting (41,42) is a more straight forward and accurate technique to determine aggregation; equally simple and accurate is the use of the flow cytometer for particle counting (43).

Platelet aggregation has also been studied under well-defined hydrodynamic conditions, for instance, in capillary tubes (44) where local or average shear rate are known, or in a cone-and-plate viscometer (31,42,45) or couette apparatus (39) where shear rates are uniform. Although these well-characterized shear fields enable the determination of frequency of collision between platelet, and consequently, the efficiency of aggregation, the measurement of the effect of shear on efficiency of aggregation has been limited, until very recently, only to relatively low shear rates (<100 s^{-1}) (39,44). Determination of capture efficiency has essentially been based on the Smoluchowski model of two-body collision kinetics, assuming that hydrodynamic drag forces on the colliding platelet are balanced by the attractive colloidal forces (e.g. van der Waals forces). This model, although restricted to the early stage of aggregation, has been accepted to describe a wide variety of systems, including latex spheres (46,47), and hydrosols (48), under conditons of both Brownian diffusion and flow (theory presented below).

14.2 Ligands and receptors involved in platelet aggregation.

14.2.1 Overview

The platelet glycoprotein IIb-IIIa is an integrin present as ~50,000 copies uniformly distributed on the surface of the resting platelet (Fig. 1), with another 30,000 present in internal membrane pools (49). There are also about 25,000 copies of the GPIb-IX-V membrane complex in the resting platelet, containing the heterodimer GPIb alpha (143 kDa) and a disulfide-linked beta-chain (22 kDa) (Fig. 1). In contrast to resting GPIIb-IIIa complexes, GPIb is largely cytoskeletally-attached, and may effectively decrease in number of functional receptors with thrombin activation due to surface alterations or "internalization" (50). Activation of platelet, by agonists such as ADP, adrenaline, thrombin, or phorbol myristate acetate (PMA), is now widely believed to cause a change in the microenvironment and/or in the conformation of the GPIIb-IIIa complex. such that it can serve as a receptor for fibrinogen, vWF and other select adhesive proteins (14,49,51-53). Fibrinogen (and other adhesive proteins) binding requires continued occupancy of the platelet receptor for physiological activators such as ADP (13,14,54,55). The number of Fg molecules binding to competent receptors expressed per platelet on the surface of the platelet can vary from about 20,000 generated by "weak" activators such as ADP, to ~80,000 for "strong" activators like thrombin or PMA (49). The source of receptors may include internal pools of GPIIb-IIIa recruited from the surface-connecting canalicular membrane system for ADP activation (56) and/or from α-granule membranes for PMA activation (57,58). There appears to be one class of these receptors with equilibrium dissociation constants $K_d = 0.1\text{-}0.5\ \mu M$, in the range of concentrations supporting both micro (PA) and turbidimetrically-measured macro-aggregation (TA) (14,54,55).

14.2.2 'Physiologic' platelet activators

Three distinct activators have been chosen for most of our studies with soluble Fg binding to platelet: i) ADP, representing a receptor-specific agonist widely evaluated in platelet physiology causing minimal granule secretion, ii) PMA acting as an "intracellular signal mimetic" which sustains activation of protein kinase C, causing maximal expression of activated GPIIb-IIIa, maximal platelet secretion, and maximal platelet aggregation (51,52,59), and iii) we have also used thrombin for platelets in buffer free of added plasma proteins:- at low thrombin concentrations (≤ 0.05 U/ml thrombin), where thrombin binds to its high affinity site on the GPIbα receptor, and minimally to the thrombin seven-transmembrane receptor (PAR-1), there is nevertheless an increased Ca^{2+} response (60), with secretion of granule proteins (61), and aggregation (62), and with little downregulation of GPIb receptors (50).

14.2.3 Flow cytometry for measuring receptors/ligands number and quality

There is a general consensus that all activators causing Fg-mediated platelet aggregation will yield the activated GPIIb-IIIa receptor (GPIIb-IIIa*) on the activated platelet surface which will become occupied with Fg, with both the receptor and the bound Fg undergoing further conformational changes, and lead to cross-bridging for capture of platelet into doublets and multiplets, as shown schematically in Fig. 2 for hypothetical stoichiometries and idealized cross-bridging. The dynamics of expression of the transformed GPIIb-IIIa receptor for Fg on such shape-changed platelet (Figs.1 and 2) must be assessed between 0-10 sec of ADP activation. Such rapid and time-dependent measurements can uniquely be made with flow cytometry, which can readily detect subpopulation behaviour

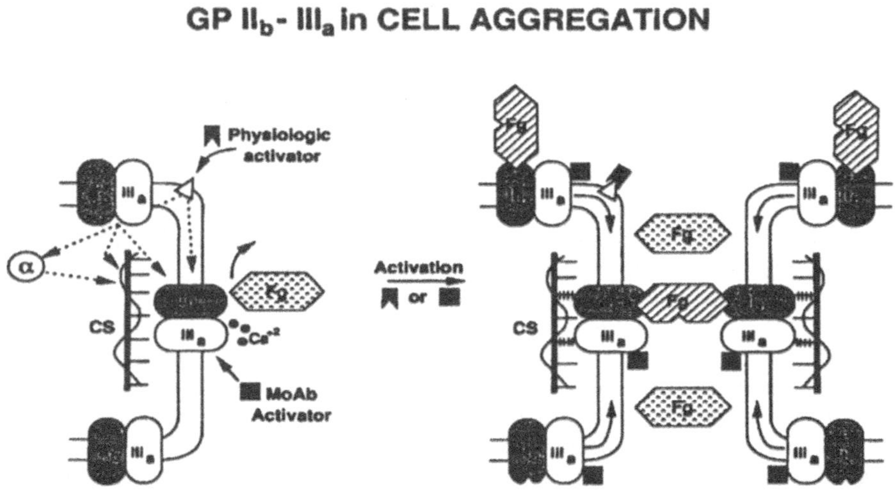

Fig. 2. Schematic for the transformation of resting GPIIb-IIIa into a calcium-dependent receptor for fibrinogen, occupancy by Fg and its subsequent conformational changes and hypothetical cross-bridging by Fg of two activated platelets. Activation can occur via physiological agonists binding to their receptors (ADP or thrombin), via intracellular signalling mediated by phorbol myristate acetate (PMA) or with direct activation via binding of a monoclonal antibody (mAb) to GPIIb-IIIa, with calcium required for the functional receptor. Cytoskeletal proteins (CS) are believed to bind to activated receptors, while α - granules can provide further GPIIb-IIIa and ligands like Fg and vWF (from ref. 64).

(extensively reviewed in ref.63). Flow cytometry is routinely used to measure the maximal numbers of platelet surface molecules under equilibrium binding conditions: 30-60 min of incubation with activators and reporting probes. In order to assess the rapid expression and time-dependent reorganization of surface molecules following platelet activation, we developed the theoretical and experimental base for determining such changes in non-equilibrium conditions. We could thus determine the numbers and "accessibility" of surface molecules on platelet in real time (64,65), using highly-specific fluorescein isothiocyanate (FITC)-labeled monoclonal antibodies, as well as directly labeled ligands like Fg (66,67) and von Willebrand factor (vWF) (68). We showed that measures of initial on-rates (v) or maximal extents of binding (Fl_{max}) of reporting probes or ligands as a function of pre-activation times (τ) of platelet suspensions can yield information on (i) the onset time, (ii) both the rate and extent of receptors expressed and (iii) the organization and steric accessibility of these receptors (64,65).

14.2.3.1 Parallel studies of platelet surface receptors/ligands and platelet aggregation

When we compared parallel measurements of the kinetics of platelet recruitment into microaggregates (for stirred platelet suspensions in poorly defined flow conditions) with the expression and Fg occupancy of GPIIbIIIa* as a function of ADP concentration, we discovered an unexpected new model for platelet activation and recruitment. This model was based once again on the selective recruitment of platelet subpopulations and an all or none, quantal response of any single platelet in expressing all of its receptors (69) and becoming recruitable for aggregation (70), but at a critical ADP concentration dependent on its own subpopulation characteristics. It would thus appear that a "protective" mechanism exists to limit the numbers of platelet that can be activated at any given ADP concentration, with a "full and maximal" expression of receptors and aggregability of the

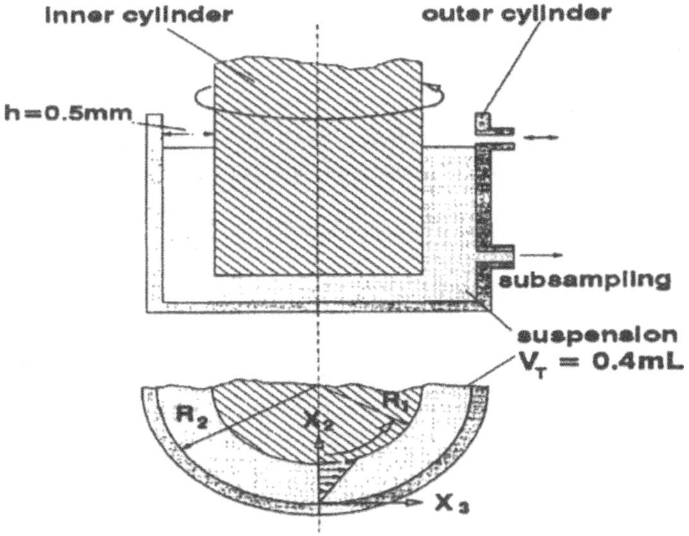

Fig. 3. Micro-couette. A micro-couette composed of two concentric plexi-glass cylinders of respective diameters of 10 mm (o.d.) and 11 mm (i.d.), with a gap h = 0.5 mm. The inner cylinder, driven by a high precision step motor, rotates at a desired angular velocity, with respect to the stationary outer cylinder, to yield a simple shear flow in the space between the confining cylinders, with a mean shear rate ranging from 1 - 1000 s^{-1}. Suspensions of 400 l are added from the top of the couette after lifting the inner cylinder, whereas the subsamples (40 l) are collected from the outlet on the side, after inserting the inner cylinder, and discarding a dead volume (20 l) before each subsampling. Since the outlet is above the bottom surface of the inner cylinder, the volume at the bottom of the couette is not sampled (from ref. 67).

"responders". We have also used FITC-labeled human and bovine Fg to demonstrate a delayed expression of receptors and Fg binding to ADP-activated platelet from bleeding Simmental cattle, although the final expression of numbers and accessibility of receptors were normal (71,72).

14.3 Quantitation of Aggregation: Theoretical and experimental approaches.

14.3.1 Determination of the kinetics of platelet aggregation

Platelet activated either by ADP or GRGDSP were subjected to shear in a micro-couette at a desired shear rate (100-1000 s^{-1}) (Fig. 3). At time t, the suspensions were subsampled and analyzed by measuring the change in platelet particle number per unit volume with the Elzone resistive particle counter or by flow cytometry, as previously described (43). The fraction of particles recruited into aggregates, PA, was calculated using the equation 1: PA= $(1 - (N_t / N_o)$, where N_t and N_0 are platelet number concentrations determined by counting particles in the subsamples taken at t = t and t = 0, respectively.

14.3.2 Determination of capture efficiency

Capture efficiency, α, is defined as the fraction of shear-induced collisions between two platelets that result in doublets, i.e., the ratio of the rate of the measured initial aggregate formation for dilute suspensions (dN_t/dt) to the frequency of two-body collisions in the suspension.

14.3.2.1. Diffusion-driven aggregation

The rate of platelet aggregation for spherical particles of equal size in suspension and colliding only via Brownian motion is related to the capture efficiency, α (B), as

$$dN_t/dt = (1.33\ kT)/ u\ .\ N_t 2.\ \alpha\ (B) \qquad (2)$$

where dN_t/dt is the change in particle number (per unit volume) with time; k,T, u are the Boltzman constant, temperature, and viscosity of the medium, respectively; and N_t as in eq. 1. This simple theory has been confirmed for spherical particles in the size range of platelets (0.01-2 μm diameter) aggregating to form multiplets up to six particles per aggregate (46-48). Experimentally obtained α (B) values were generally < 0.5-0.8 i.e the observed rate could correspond to up to about 50-80% of all diffusion-driven collisions resulting in capture of latex spheres with similar size as human platelets (46).

14.3.2.2. Shear-driven aggregation

The Smoluchowski collision frequency of a suspension, CF, is given by equation 2: CF = 16/3 ($G N_t^2 a^3$), where G is the shear rate and a is the equivalent spherical radius of platelet. Therefore the capture efficiency can be expressed as Eq. 3:

$$\alpha = \frac{\frac{dN_t}{dt}}{\frac{16}{3} G a^3 N_t^2}$$

Combining Eq. 1 and the integrated form of Eq. 3 gives the initial rate of aggregation as

$$\frac{dPA}{dt}\bigg|_{t\to 0} = \frac{16}{3}\alpha G a^3 N_o \tag{4}$$

The initial rate of aggregation can be obtained experimentally by finding the best fit of the PA values at time t to the equation

$$PA = PA_{max}(1 - e^{-t/B}) \tag{5}$$

where the initial rate of aggregation is PA_{max}/B. α values can be calculated from Eq. 4.

Plots of PA versus time for varying N_o values have been shown to yield initial aggregation rates in accordance with Eq. 4.

14.4 Platelet aggregation in non-stirred platelet suspensions: role of pseudopods

Chang and Robertson (39) first reported that rabbit platelets in non-stirred suspensions of platelet-rich plasma (PRP), activated with ADP, underwent aggregation due to Bownian motion with an α (B) of about 0.4, comparable to slowly aggregating latex spheres. We confirmed these results for rabbit platelets, with early time α (B) vcalues found to be as high as 0.9, but surprisingly found that human and dog platelets, which have comparable sizes, showed apparent α (B) values of >>1 and 2 respectively (73). In contrast, continuously stirred suspensions of ADP-activated PRP showed identical kinetics of platelet aggregation (same apparent capture efficiencies) for all three species. The differences in α (B) values for the three species' platelets aggregating in Brownian diffusion was largely attributable to species'differences in pseudopod length (74). We indeed found that >90% of all platelet contacts in the doublets and triplets formed were via at least one pseudopod, as was indeed largely the case for early aggregation for the stirred platelets. Theoretical calculations assessing the effects of pseudopod length and number on the collision frequency enhanced by an increased radius of a collision sphere supported the experimental observations that about 3-4 pseudopods per human or canine platelet yielded optimal α (B) values, with the average pseudopod length of about 3,2 and 1 μm respectively for human,canine and rabbit platelets, paralelling the α (B) differences. The persistent diferences may be related to biochemical differences in dynamics and surface density of adhesive sites (GPIIbIIIa* and Fg) present on the pseudopods from the different species. The similarity in aggregation kinetics for the three species' platelets in stir points to a less significant role for pseudopod length and number for rapidly colliding platelets, but this issue still needs to be resolved. Particular attention must be paid to the biochemistry of adhesion and the more complex challenge related to the observed enrichment of GPIIbIIIa* and bound Fg on pseudopods of activated human platelets (75).

14.5 Model cell aggregation in near-stasis versus stirred suspensions

We have reported that Chinese hamster ovary (CHO) cells transfected with GPIIb-IIIa (A5) and activated with a mAb (A5*) would only aggregate under the near-static conditions of gyrorotation (76). We have estimated shear rates in this poorly defined 'flow' system to be near-stasis or <$10s^{-1}$. When these A5* cells were with Ca /Fg stirred at 1,000 rpm, no aggregation whatsoever occured. However, the large aggregates formed at near static conditions, were completely stable if subsequently stirred at 1,000 rpm (unpublished observation), suggesting that in shear there is insufficient organization of ligands and receptors, and too low a dynamic avidity, to allow the cross-bridging necessary in the tens of msecs of contact time before colliding CHO cells separate. We confirmed the absence of and aggregation with these A5* cells at shear rates in our microcouette apparatus at 50 $sec.^{-1}$ (unpublished observations, 1997). We also reported (77) that aggregation of ~12% of all particles occured in gyrotation (almost static conditions) between resting A5 cells and CHO cells containing constitutively active GPllb-llla plus Fg, with only 2% co-aggregation with blockage of free receptors. Thus, the shear conditions clearly determine whether or not any aggregation will occur between potentially competent receptors and ligands, as will be further discussed in sect. 8.1 with model particle studies.

14.6 Dynamics of soluble Fg binding (receptor occupancy) and platelet aggregation as function of shear rate (66,67)

14.6.1 Fibrinogen receptor occupancy and capture efficiencies

The effects of shear rate on protein adsorption and platelet aggregation were evaluated in well-defined shear field in a micro-couette for shear rate G = 10-1000 s^{-1}. The rate of protein binding was independent of G, shown for adsorption of albumin to latex beads and the large IgM, PAC1, to activated platelets. The initial aggregation rate for ADP-activated platelet in citrated platelet-rich plasma followed second order kinetics at the initial platelet concentrations between 20,000 and 60,000/µl. α values, which dropped nearly five-fold for a 10-fold increase in G, were approximately proportional to G^{-1}; this is a general observation for fixed colloidal particles, since the contact time between two colliding particles decreases inversely with the shear rate (2.62/G) (25). Varying ADP concentrations did not change the capture efficiency (α) of maximally activated platelet subpopulations, suggesting that aggregation between unactivated and activated platelet is negligible. At a given shear rate, tested at 300 s^{-1}, α for AFP remained maximal (~0.24) over 25 to 75% Fg occupancy, otherwise decreasing rapidly, with a half maximum occurring at around 2% occupancy, suggesting that very few bound Fg were required to cause significant aggregation (Fig. 4). Similar α values were obtained for fixed activated platelets (AFP) as for ADP-activated platelets in plasma (PRP), with the shear-dependent drop shown for PRP (Table 1).

The apparent rapid decay in α at high Fg occupancy (80-90%) suggested that free GPIIb-IIIa receptors on one platelet actually cross-bridge Fg bound to its receptor on a second platelet. This cross-bridging model, though speculatively proposed by many researchers, has never been directly proven, and was confirmed by the following experiment (see also sect. 6.2). Adding 1 mM GRGDSP to PRP (1:10) immediately before ADP, or to AFP suspensions before 500 nM Fg, resulted in a five-fold decrease in PA measured after 30 sec of shear at 300 s^{-1}. These low PA values were comparable to those of the resting PRP (1:10) and Fg-free AFP suspensions, but this procedure blocked specific Fg binding in both cases. However, adding GRGDSP at the end of ADP or Fg incubation equally

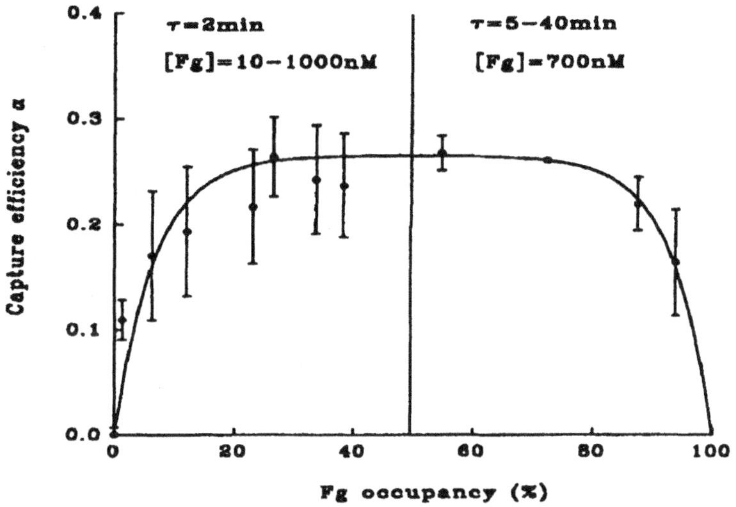

Fig. 4. Capture efficiency of platelets with varying fibrinogen occupancy at a fixed shear rate (300 s^{-1}). Fg occupancy was defined as the ratio of the mean fluorescent intensity (Fl) of the bound FITC-Fg of the suspensions to that of maximum Fl. Low Fg occupancy (<50%) as achieved by incubating AFP with Fg of various concentrations (10-1000 nM) for 2 minutes, whereas high occupancy (>50%) was obtained by varying incubation time (τ) at a fixed Fg concentration (700 nM). The line is the best of fit to an exponential equation to the experimental values (mean of two experiments, each having duplicates) for α >50% and its mirror image for α <50% (mean of six experiments, each having duplicates) (from ref. 67).

minimized PA; however, there was no significant reversal in the Fg occupancy due to inhibitor addition: ~50% occupancy for either PRP (1:10) or AFP under the condition used. Similar inhibition effects were achieved by replacing GRGDSP with 500 nM non-peptide analog Ro 43-5054 for PRP (1:10), and with 100 nM Ro 43-5054 for AFP suspensions. Thus, allowing about 50% of receptors to continue binding Fg while blocking all the remaining free activated/resting receptors with GRGDSP or its more specific Ro analogue, led to a complete block of platelet aggregation. This result rules out a primary mechanism for two bound Fg to bridge two platelets via side-side interactions. Indeed, the GPRP peptide which blocks such interactions leading to fibrin polymerization (78), did not affect the above Fg-mediated aggregation (Xia and Frojmovic, unpublished observation, 1995).

14.6.2 The Domain(s) of Platelet-Bound Fibrinogen Needed for Platelet Aggregation (79)

Three different sites on each half of the Fg molecule can interact with platelet membrane GPIIb-IIIa: an RGDS site at Aα 572-575, an RGDF site at Aα 95-98, and a dodecapeptide HHLGGAKQAGDV (H12) on the carboxyl terminus of the γ chain (γ400-411) (80-84). The RGD sites are shared by other GPIIb-IIIa ligands, such as von Willebrand factor (vWF) (85), fibronectin (86-88) and vitronectin (89), while the H12 sequence is unique to Fg. It has been demonstrated (90-96) that for the initial binding of Fg from solution to activated GPIIb-IIIa, only the normal carboxyl terminus of the γ chain is needed, while the RGD sequences on the Aα chain do not seem to play a role. However, the above studies did not distinguish between the binding of Fg from solution to activated platelet GPIIb-IIIa, and the subsequent cross-bridging of platelets by the platelet GPIIb-IIIa-bound Fgs. Platelet-bound Fg can be seen as surface immobilized Fg, and surface-immobilized Fgs behave differently from soluble Fgs (97,98). Thus it is possible that structures not required for binding might be required for aggregation, while those required for binding might not be required for cross-bridging (aggregation). In addition, new sites such as RIBS (receptor-induced binding site) are exposed upon binding of Fg from solution to its GPIIb-IIIa receptors (78), and an antibody to the RIBS-I epitope (γ373-385), which is not involved in the initial binding of soluble Fg to activated GPIIb-IIIa, was reported to inhibit aggregation without inhibiting Fg binding (78). This further raised doubts about the role of the three putative GPIIb-IIIa binding domains on Fg (H12, RGDS, and RGDF) in cross-bridging.

TABLE 1. Capture Efficiencies (α_c) for Fg and vWF-Platelet Aggregation

		α_c Values at		
P +Activator	Ligand[4]	$100s^{-1}$	$300\ s^{-1}$	$1000s^{-1}$
PRP + ADP[1]	Fg	0.32	0.18	0.05
AFP[1a]	Fg_s	-	0.25	-
WP + ADP[2]	vWF_{sb}	-	0.09±0.01 (4)[5]	0.07±0.01 (5)
WP + thrombin[3]	vWF_{sb}	-	0.16±0.02 (7)[6]	0.08±0.01 (6)
WP + ristocetin	vWF_s	0.45±0.01 (9)	0.43±0.01 (8)	0.38±0.02 (11)

[1]100μM (67); [2]5μM (43); [3]0.05U/ml, for 10 mins. prior to shear study (68); 1[a]AFP= "RGD"- activated platelets + 500 nM Fg (67)
[4]Subscripts s and sb refer to soluble and platelet surface-bound Fg or vWF, respectively
[5]Mean± SEM, with # experiments shown in brackets
[6]Unpublished data

We therefore examined the role of the putative GPIIb-IIIa binding domains (H12, RGDS, RGDF) on platelet-bound Fg in platelet aggregation. The experiments were done under controlled Fg occupancies and a homogeneous, well-defined shear rate ($300\ s^{-1}$), as previously reported (99), in contrast to most previous studies where Fg concentrations were used instead of Fg occupancies, while shear rates in stirred platelet suspensions were generally low, heterogeneous, and poorly defined. Activated platelets were first incubated with Fg to let the Fg bind, then with monoclonal antibodies (mAb) to block the putative adhesive domains, and the platelet suspension was then sheared or stirred to induce aggregation. Fg occupancies were always maintained at around 16% to ensure optimal sensitivity of platelet aggregation to changes in Fg occupancy, given that maximal and unchanging aggregation occurs between about 25-75% occupancy (67).

The mAb 4A5, which recognizes the last four amino acid residues (AGDV) in a dodecapeptide (H12) on the carboxyl terminus of the Fg γ chain, markedly inhibited platelet aggregation. Z69/8, a mAb whose epitope is also on the dodecapeptide but does not recognize the AGDV residues, did not have any inhibitory effect on aggregation. It seems paradoxical that the large IgM molecule, Z69/8, binding to a structure so close to the AGDV sequence, neither inhibits platelet aggregation nor the binding of 4A5 to platelet-bound Fg, both of which involve the AGDV residues. However, X-ray diffraction of a peptide from the carboxyl terminus of Fg γ chain including the H12 showed that the AGDV residues on the H12 are not in the same plane with the other residues, with a wide bend between Gln 407 and Ala 408 (100). This would allow the binding of Z69/8 (to residues on the H12 other than the AGDV) without inhibiting platelet aggregation, as well as the observed simultaneous binding of 4A5.

The anti-RGDS and anti-RGDF mAbs did not at all affect macro- or micro-aggregation, whether tested singly or together. These mAbs have previously been reported (i) to require the RGD sequence in the target antigen for binding; (ii) to block endothelial cell ($\alpha_v\beta_3$-mediated) and platelet (GPIIb-IIIa-mediated) adhesion to their corresponding RGDS or RGDF containing peptides immobilized on a plastic surface; and (iii) in the case of mAb 134B29 (anti-RGDS), it blocked endothelial cell binding to surface-immobilized Fg, but as in our aggregation studies, it did not inhibit platelet adhesion to surface immobilized Fg (101).

The above results were also confirmed for platelet-bound Fg modeled as Fg adsorbed to polystyrene latex beads, where coaggregation of Fg-coated beads and activated platelets was readily quantitated by flow cytometry. A recombinant Fg mutant lacking the AGDV γ terminal domain was adsorbed to latex beads and shown not to support coaggregation, in contrast to positive results for the recombinant type Fg (102). These results demonstrate that, similar to the situation for the initial binding of soluble Fg, only the γ chain carboxyl terminus with the AGDV residues are needed for platelet or surface-bound Fg to support aggregation, while the RGD sites on the Aα chain do not seem to be required.

14.7 Dynamics of von Willebrand Factor-mediated Platelet aggregation

14.7.1 Background

Von Willebrand factor (vWF) can readily cross-bridge platelets to other platelets or to vessel walls via repeating subunits containing binding sites to extracellular matrix (ECM) proteins and to platelet glycoprotein receptors GPIb and GPIIb-IIIa. As the molecular weight varies from about 1-20 million Da, with each subunit 240,000 Da, there are up to 80 binding sites for each receptor per molecule of vWF; moreover, the up to 1.3 μm long molecule can cover entire pseudopods or wrap the entire minor circumference of a single platelet (103). Metabolically active platelets adhere transiently to immobilized vWF, at shear rates up to $G = 6000\ s^{-1}$ via GPIb, with activated GPIIb-IIIa required for stable attachment (6), or through only the GPIb receptor if vWF is immobilized on a collagen-coated surface (104). Soluble vWF can mediate platelet aggregation in the absence of chemical agonists only at high shear rates of G≥6000 s^{-1} via both GPIb and GPIIb-IIIa (32,105,106). However, below these "pathological" shear rates, an external agonist is required. Ristocetin, or botrocetin mediate the binding of soluble vWF to GPIb on formalin-fixed platelets, while platelets activated with agonists such as ADP, thrombin or collagen, have been shown to bind soluble vWF via activated GPIIb-IIIa receptors, with vWF-mediated agglutination/aggregation in stir (85,107-110). Thus, heparinized whole blood sheared at 100-500 s^{-1} showed aggregation within 60 sec, but this was entirely associated with released ADP (likely from red cells), and reversed at 1000-2000 s^{-1} shear rates; the active receptors and ligands in this aggregation were not identified (111,112). The role of secretion of granule-stored proteins such as vWF, in mediating platelet aggregation as a function of shear rates, had not been evaluated in the above studies.

At the low shear rates associated with stirred suspensions of platelets largely used in aggregation studies (G<100 s^{-1}), soluble Fg predominates over soluble vWF in driving aggregation via binding to activated GPIIb-IIIa (Fig.1), essentially because there is >1000-fold more Fg than soluble vWF. At very high shear rates, Fg is no longer efficient at mediating aggregation, while soluble vWF can bind directly to GPIb (shown in Fig.1 inset) and cross-bridge platelets via subsequent platelet activation and participation of activated GPIIb-IIIa receptors. (For a good review see ref. 6). The binding of vWF to the GPIb-IX-V membrane complex, (GPIb in Fig.1), is governed by the GPIb alpha chain via a hinge region in GPIb speculated to expose a binding site directly induced by high shear rates (113), but the molecular mechanism is still unclear (6). The role of surface-secreted Fg and vWF in shear-dependent aggregation is discussed below.

14.7.2 Surface-secreted vWF mediates aggregation of ADP-activated platelets at physiological (moderate) shear stress: facilitation by GPIb but control by GPIIb-IIIa

In the absence of exogenous fibrinogen, ADP-induced activation of multiple-washed and centrifuged human platelets suspended in Tyrodes-albumin supported aggregation at moderate shear rates in Poiseuille flow in a glass tube (mean values, 335 and 1335 s^{-1}) corresponding to shear stresses of 0.4 and 1.6 Nm^{-2} (114). Although 54% of the resting, and >75% of the ADP-activated, washed cells expressed maximal fibrinogen-occupied GPIIb-IIIa receptors, the aggregation appeared to be independent of the bound fibrinogen, though blocked by the hexapeptide GRGDSP which did not alter the prebound fibrinogen. The critical shear stress for the onset of aggregation with 0.7 μM ADP was about 20-fold lower than that observed by Ikeda et al. (32). If vWF plays a major role in the ADP-induced aggregation at the much lower shear stress, one would expect a significantly increased secretion and surface expression of the adhesive protein on the ADP-activated washed platelets. We therefore prepared and tested single-washed, relatively unactivated platelets, before and after addition of agonist, for α-granule release and secretion and surface expression of vWF, and compared the results with those of platelets in cPRP, in which freshly-prepared cells are known to be unactivated (67). In addition, we studied the effect on ADP-induced aggregation at moderate shear stress, of monoclonal antibodies (mAbs) LJ-152B/6 (152B/6) and NMC4, respectively blocking the GPIIb-IIIa and GPIb binding domains on vWF (32,114), of mAbs 6D1 and 10E5, respectively blocking the vWF binding domains on GPIb (115) and GPIIb-IIIa (116), and of aurin tricarboxylic acid (ATA) reported to block the GPIb binding domain on vWF (117), using both a variable shear (Poiseuille flow) and a uniform shear field (couette flow) at moderate shear stresses, from 0.3-1.1 Nm^{-2} (G= 300-1000 s^{-1}) (43). This above approach led to the hypothesis that vWF, secreted during the isolation and activation of the single centrifuged and washed platelets, and surface-bound on the activated cells, is required for ADP-induced aggregation at moderate shear stress in the absence of exogenously added Fg, with a supporting (facilitating) role for GPIb and an obligatory role for GPIIb-IIIa.

The initial collision capture efficiencies, α_G, in the experiments with washed platelets activated by 5 μM ADP were 0.09 and 0.07 at G = 300 and 1000 s^{-1}, respectively (Table 1). In a previous study of *soluble fibrinogen-driven* ADP-induced aggregation in couette flow, we found α_G = 0.18 and 0.05 at G = 300 and 1000 s^{-1}, respectively (67). Thus, α_G for *surface-secreted vWF-mediated* aggregation at G = 300 s^{-1} is already 50% of the maximal value for fibrinogen-driven aggregation, and at G = 1000 s^{-1} is even 40% more efficient. These results are consistent with an increasingly important role for vWF in mediating aggregation at higher shear rates (32,105,106,118).

At shear rates <100 s^{-1}, in stirred aggregometer vials or in tube flow (43), the addition of Fg is an absolute requirement for ADP-induced aggregation in washed platelet suspensions. Thus, the surface expressed vWF resulting from washing and centrifugation of the platelets, appears to be unable to mediate aggregation at these low shear rates. However, above a critical threshold shear rate ~300 s^{-1}, the platelets will aggregate. At this moderate shear rate, it seems that the interaction between vWF and GPIb can largely be bypassed, since inhibition of the initial rates of aggregation by mAbs 6D1 and NMC4 is only minor. With increasing shear rate, however, the role of GPIb becomes increasingly important, as evidenced by the increased inhibition of the initial rate of aggregation, from only 22% at G = 300 s^{-1} to 52% at G = 1000 s^{-1}.

We therefore propose a model for the shear dependent involvement of vWF in the aggregation of platelets in the absence of exogenously added Fg, based on differences in the strength and lifetime of the bonds between the respective binding domains of vWF for the GPIb and GPIIb-IIIa receptors. First, it should be pointed out that much of the vWF secreted from the α-granules is likely associated with, and anchored to, the GPIIb-IIIa receptors. Thus, it has been reported that freshly secreted vWF from thrombin-activated platelets cannot be prevented from binding to the cell surface using mAb 6D1 directed against the vWF binding site on GPIb, and is only ~72% inhibited from binding to GPIIb-IIIa by mAb 10E5 (119). Thus, it appears that vWF is partly secreted in association with GPIIb-IIIa, and mAb 10E5 is unable to inhibit this subpopulation of vWF from being released and surface bound to its receptors. Upon activation with ADP, it is proposed that the first reaction between the surface expressed and bound vWF on one platelet with a GPIb receptor on an adjacent platelet results in the formation of rapidly reversible bonds having fast on and off rates, but which are relatively resistant to rupture by externally applied shear stress. The second reaction between an altered domain on vWF on one platelet and a GPIIb-IIIa receptor on an adjacent platelet, results in the formation of a strong bond resistant to rupture at high shear stress, but with relatively slow on and off rates. Thus, with increasing shear rate, the first reaction becomes increasingly important in order to facilitate bond formation in the second reaction, and thus provide efficient capture of platelets, as seen at G = 1000 s^{-1}. There is, therefore, an obligatory role for GPIIb-IIIa at all shear rates, and a facilitative role for GPIb especially at high shear rates.

There are, in fact, two good analogues for the above two-step "roll and capture" model, perhaps appropriately called "kiss and embrace" in more anthropomorphic terms: (i) the rapid reversible bond formation of L-selectin on neutrophils to counter receptors on endothelial cells enabling the neutrophils to roll along the vessel wall. In turn, this facilitates capture of the leukocyte via the firm bond formed between the activated integrin Mac-1 on the neutrophil, and its counter receptor on activated endothelial cells, required at modest shear rates (<300 s^{-1}) (120); and (ii) the translocation of resting platelets via rapidly reversible bond formation between GPIb on resting platelets and vWF immobilized on a plastic surface. Again, this process facilitates the subsequent firm arrest of the platelets through activated GPIIb-IIIa (6).

14.7.3 Ristocetin and thrombin-induced platelet aggregation at physiological shear rates: Differential roles for GPIb and GPIIb-IIIa receptor

We have extended the above studies to "activation" with low ristocetin and thrombin concentrations (68). Washed platelet suspensions with added soluble vWF and ristocetin (0.3-0.75 mg/ml), or activated with low concentrations of thrombin (0.01-0.5 U/ml) but no added ligand, were sheared in a coaxial cylinder device at uniform shear, G=1000 s^{-1}. Capture efficiency (α_G) of aggregating platelets was correlated with vWF platelet binding assessed by flow cytometry. The vWF-GPIb interaction was exclusively able to support ristocetin-mediated shear aggregation of metabolically active platelets, with very few vWF monomer equivalents (representing ≤ 10 molecules of 20 million Da) bound per platelet required to yield high capture efficiencies $\alpha_G = 0.38$ compared to 0.05 for ADP and Fg-mediated studies (Table 1), suggesting rapid and stable bond formations between vWF and GPIb. However, surface-bound secreted vWF, generated in a threshold response to about 0.02-0.05 U/ml thrombin addition to WP in parallel with a threshold response for aggregation, can mediate platelet aggregation with $\alpha_G = 0.08$, surprisingly comparable to that previously reported for WP and ADP activation (Table 1). Blocking the GPIIb-IIIa receptor decreased α_G by 95 ± 3% (n = 3), while a monoclonal antibody to the vWF site on GPIb caused a 49 ± 7% (n = 8) decrease in α_G.The partial role for GPIb reflects a facilitative function for increasing contact time between flowing platelets, and allowing engagement of the GPIIb-IIIa receptor to yield stable attachment.

14.8 Some new directions

14.8.1 Studies with model particles: latex spheres and liposomes

Despite extensive research into the mechanisms of platelet crossbridging, it is still not clear as to what are all the 'cofactors' which provide normal physiological conditions for the interaction between Fg and GPIIb-IIIa sufficient to yield platelet aggregation in flow conditions. This question has been addressed through the use of systems less complex than the native platelet, such as latex beads coupled to purified GPIIb-IIIa (unpublished, 1997) or coated with platelet membranes (121), which exhibit Fg-dependent aggregation in stirred suspensions, but these models do not adequately mimic the flexibility of the platelet membrane. Chinese hamster ovary cells expressing recombinant GPIIb-IIIa (76), have been shown to aggregate under the near-static conditions of gyrorotation, but the receptor had to be externally activated with an antibody to bind Fg, and likely had different characteristics from the platelet protein (such as altered glycosylation). Moreover, no aggregation occured for stirred suspensions of these cells (76).

Another approach to studies of the Fg-GPIIb-IIIa interaction has been made by incorporating the purified receptor into the bilayer of artificial lipid vesicles known as liposomes (89,122-124), which are commonly used as model membrane systems. Previous work has shown that liposomes containing GPIIb-IIIa are able to bind fibrinogen and other ligands, but only one study has examined at the ability of such liposomes to be crosslinked by Fg, with inconclusive results; at best, extremely shear-sensitive flocculation was observed under non-stir conditions (diffusion-driven) (124). We recently (unpublished, 1998) prepared more 'physiologic' phospholipid/cholesterol vesicles of a size and composition similar to that of platelets and inserted purified GPIIb-IIIa into the lipid membrane. We found that these GPIIb-IIIa-liposomes are able to bind Fg in a manner similar to platelets, but cannot specifically aggregate in the presence of Fg, though the receptor-bound Fg does express the $\gamma^{408-411}$ terminus required. A probable role for receptor "anchoring" is expected, predicted to be absent in the liposomes, but present in activated platelets and in latex sphere model systems.

A recent report using co-transfection of GPllb-llla and β3-endonexin in CHO cells showed that coaggregation of these particles plus Fg, with resting A5 cells, yielded only 7% aggregation for suspensions stirred for 20 mins., versus 3.5% with blockage of free receptors (125). It is of note that the β3-endonexin which activated the GPllb-llla to yield "normal" Fg binding but very poor aggregation of these CHO cells, neither "strongly nor consistently concentrated in focal adhesions". We therefore suggest that "anchoring" (eg. via cytoskeletal attachments (125)) of GPllb-llla is necessary for efficient cross-bridging via receptor-bound Fg and resulting aggregation in flow conditions. Such "anchoring" may occur on latex beads coupled to GPllb-llla (121) or attached to platelet membranes (unpublished, 1997) (which have a much more rigid membrane than that of liposomes or CHO cells), leading to Fg-specific aggregation with both stir and shear (121). Another explanation for failure of aggregation related to lack of "anchoring" of Fg receptors may lie in the possible extraction or 'fracture' of GPIIb-IIIa-Fg from the lipid bilayer during initiation of liposome cross-bridging in doublet formation -- as occurs for " unanchored" red blood cell receptors (126,127). The relation of receptor clustering and anchoring (to cytoskeleton) to the shear-dependant ability of particles to undergo Fg and Gpllb-llla-mediated cross-bridging into stable aggregates, remains to be determined. The use of model cells and particles such as liposomes and latex microspheres should facilitate this venture.

14.8.2 Thrombin-induced secretion in flow (128)

We have reported on initial lag times for shear-associated aggregation of washed platelets evaluated in tubular flow for both ADP and thrombin-induced activation (128). Aggregation with ADP activation, without any added exogenous ligands, did not occur at 42 sec^{-1} shear rates (as it does not occur in stirred suspensions), but did with thrombin (0.35 nM) with a lag time of about 30 s, predicted to reflect the slow release and re-uptake of secreted ligands like Fg and vWF. At the higher shear rates of 335 and 1335 sec^{-1}, delayed aggregation was observed for both ADP and thrombin (0.17-0.26) nM activation, far below shear rates (>6000 sec^{-1}) at which aggregation occurs in the absence of exogenously added activators, for which vWF also plays a predominant role in cross-bridging the platelets (105,106). The shorter lag time of about 4 s observed for thrombin above about 300 sec^{-1} shear rates, is predicted to reflect the time needed for secretion and surface-expression of alpha-granule derived ligands such as vWF or TSP. Markedly longer lag times of 34 and 17 s were observed for the thrombin receptor tethered ligand mimetic (SFLLRN) at 2 and 5 μM concentrations, requiring further studies on relative roles of distinct thrombin agonist types and concentrations on the delay and aggregate build-up at 'physiological' shear rates.

14.8.3 Flow devices and monitoring anti-thrombotic drugs

There are currently a growing number of anti-platelet drugs being used and evaluated in the management of cardiovascular diseases, centered on the platelet GPIIb/IIIa blockade (129-131). There is widespread recognition of the urgent need for ex vivo/in vito tests of platelet function in flow conditions which will allow the monitoring of the 'functional' efficacy of drug administration to patients, particularly in selecting appropriate drug concentrations (131). There are a growing number of commercial devices on the market or in progress potentially serving as bedside flow devices for monitoring the effects of these drugs on patients centered on platelet deposition from whole blood onto surfaces or through apertures in defined flow regimes. A few examples include analyzers utilizing deposition of platelets from whole blood at high shear stresses (> 6000 sec^{-1}) (132,133), as well as deposition onto plastic surfaces which spontaneously adsorb Fg and vWF conducted at normal arterial shear rates (134).

14.9 Summary

We have shown that vWF secreted and surface expressed on ADP- or thrombin-activated, washed platelets can mediate aggregation at moderate, physiological shear rates (300-1000 s^{-1}), with capture efficiencies for aggregation surprisingly comparable to that seen for purely Fg-mediated aggregation in the absence of platelet secretion, summarized in Table 1.These studies are in marked contrast to the much higher shear rates (>6000 s^{-1}) required for aggregation of resting platelets with soluble plasma vWF (32). With increasing shear rate, the vWF adhesive domain on GPIb apparently becomes increasingly essential for aggregation. The activated GPIIb-IIIa domain on secreted and surface-bound vWF is also involved in aggregation. Thus, GPIb–vWF interactions facilitate and are increasingly important at moderate and high shear rates, to permit effective capture of platelets. We postulate that these most likely occur via GPIIb-IIIa- vWF cross-bridging between cells. Surprisingly, particular "activation" of vWF/GPIb, as occurs with low concentrations of ristocetin (Table 1), can yield efficient capture of platelets at physiological shear, exclusively via GPIb receptor interactions with multivalent vWF. The potential contributions of other surface-expressed proteins, especially P-selectin and thrombospondin (135) to aggregation at moderate shear rate, remain to be explored. It should be clear from these observations that flow conditions will be critically important in designing and selecting anti-thrombotic drugs directed against the appropriate adhesive domains on receptors and ligands, with bedside flow devices increasingly available for this challenge.

14.9.1 Concluding Remarks

The cell and molecular biology of platelet function in flow (molecular biorheology) has evolved rapidly with in vitro models centered on adhesion to planar surfaces and on studies in suspension undergoing well-defined shear in micro-flow devices. Optimal molecular modelling of efficiencies of platelet adhesion and aggregation require a precise knowledge of the "rolling" or "collision" history of the platelets, most amenable to-date with studies of platelets and particles in suspension. It will be of great interest to reconstitute whole blood stepwise and to evaluate pulsatile flow effects (26). It must be noted that the topic on platelet biorheology is much broader than that presented in this chapter whose focus was on platelet activation and aggregation in suspensions. Thus, some areas of omission include flow-dependent interactions of other blood cells with platelets, such as neutrophils adhering to platelets in suspension (136) or adhering to platelets immobilized on planar surfaces (137,138). The flow-dependent role of other adhesive ligands in platelet aggregation , such as P-selectin and thrombospondin (135), will likely add colorful features to the platelet adhesive horizon.

Acknowledgements

The effective contributions of my students to the above research, particularly Q. Liu and A. Kasirer-Friede, and my post-doctoral fellow, Dr. Z. Xia, are particularly appreciated. The active collaborations of Drs. H.L. Goldsmith, T. van de Ven, S. Lord, and G. Matsueda are gratefully acknowledged. The expert technical assistance of Ms. E. A. Brown, Mr. T. Wong and Ms. F.A. McIntosh is gratefully acknowledged. The following colleagues generously provided: monoclonal antibody NMC4 from Drs. Zaverio Ruggeri and Akira Yoshioka, and DLJ-152B/6, and 2.2.9 from Dr. Ruggeri; the anti-RIBS-I from Dr. E. F. Plow; mAb S12 and aurin tricarboxylic acid, respectively, from Drs. R. McEver and M. Weinstein; mAb 9F9 at a nominal charge from Dr. A. Budzynski; and antibodies 6D1 and 10E5 from Dr. B.S. Coller. The research and fellowship support from the Medical Research Council of Canada and the Heart and Stroke Foundation of Quebec is gratefully acknowledged.

REFERENCES

1. Goldsmith HL. Flow-induced interactions in the circulation. Chapter 1. In: Advances in the flow and rheology of non- newtonian fluids. D.A. Siginer, D. De Klee & R.P. Chlabra (eds), Elsevier Science, 1998 ; Amsterdam, in press.

2. Slack SM, Cui Y, Turitto VT. The effects of flow on blood coagulation and thrombosis. Thrombos Haemost 1993 ; **70**: 129-134.

3. Weiss HJ. Flow-related platelet deposition on subendothelium. Thrombos Haemost 1995; **74**: 117-122.

4. André P, Arbeille B, Drouet V, Hainaud P, diSollier CB, Caen JP, Drouet LO. Optimal antagonism of GpIIb/IIIa favors platelet adhesion by inhibiting thrombus growth. Arterioscler Thromb Vasc Biol 1996; **16**: 56-63.

5. Sixma JJ, van Zanten GH, Saelman EUM,Verkleij M, Lankhof H, Nieuwenhuis HK, de Groot PG. Platelet adhesion to collagen. Thrombos Haemost 1995; **74**: 454-459.

6. Ruggeri, ZM. Mechanisms initiating platelet thrombus formation. *Thromb. Haemost.* 1997; **78**: 611-616.

7. Jen CJ, Lin JS. Direct observation of platelet adhesion onto fibrinogen-coated and fibrin- coated surfaces. Am J Physiol 1991; **261**: H1457-H1463.

8. Zaidi TN, McIntire LV, Farrell DH, Thiagarajan P. Adhesion of platelets to surface-bound fibrinogen under flow. Blood 1996; **88**: 2967-2972.

9. Frojmovic MM, Milton JG. Human platelet size, shape and related functions in health and disease. *Physiological Reviews* 1982; **62**: 185-261.

10. Gear ARL. Rapid platelet morphological changes visualized by scanning-electron microscopy: kinetics derived from a quenched-flow approach. *Br J Haematol* 1984; **56**: 387-398.

11. Pedvis LG, Wong T, Frojmovic MM. Differential inhibition of the platelet activation sequence: shape change, micro- and macro-aggregation, by a stable prostacyclin analogue (Iloprost). *Thromb Haemost* 1988; **59**: 323-328.

12. Frojmovic MM, Milton JG, Duchastel A. Microscopic measurements of platelet aggregation reveal a low ADP-dependent process distinct from turbidometrically measured aggregation. *J Lab Clin Med* 1983; **101**:964-976.

13. Mustard JF, Packham MA, Kinlough-Rathbone RL, Perry DW, Regoeczi E. Fibrinogen and ADP-induced platelet aggregation. *Blood* 1978; **52**: 453-466.

14. Peerschke EIB. The platelet fibrinogen receptor. *Semin Haematol* 1985; **22**: 241-259.

15. Timmons S, Hawiger J. Von Willebrand factor can substitute for plasma fibrinogen in ADP-induced platelet aggregation. *Trans Assoc Am Phys* 1986; **99**: 226-235.

16. De Marco L, Girolami A, Zimmermann TS, Ruggeri, Z. Von Willebrand factor interaction with the glycoprotein IIb/IIIa complex; its role in platelet function as demonstrated in patients with congenital afibrinogenemia. *J Clin Invest* 1986; **77**:1272-1277.

17. Derenleau DA, Dubler D, Rothen C, Luscher EF. Transient kinetics of the rapid shape change of unstirred human blood platelets stimulated with ADP. *PNAS* 1982; **79**: 7297-7301.

18. Frojmovic MM, Milton JG, Gear AL. Platelet aggregation measured in vitro by microscopic and electronic particle counting. *Methods Enzymol* 1989; **169**: 134-147.

19. Frojmovic MM, Longmire K, van de Ven TGM. Long-range interactions in mammalian platelet aggregation - the role of platelet pseudopod number and length. *Biophys J* 1990; **58**: 309-318.

20. Carty DJ, Gear ARL. Fractionation of platelets according to size: functional and biochemical characteristics. *Am J Haematol* 1986; **21**: 1-6.

21. Yung W, Frojmovic MM. Platelet aggregation in laminar flow. I Adenosine diphosphate concentration, time and shear rate dependence. *Thrombos Res* 1982; **28**:361-378.

22. Penington DG, Streatfield K, Roxburgh AE. Megakaryocytes and the heterogeneity of circulating platelets. *Br J Haematol* 1976; **34**: 639-653.

23. Thompson CB. Selective consumption of large platelets during massive bleeding. *Br Med J* 1985; **291**: 95-96.

24. Wong TW, Pedvis LG, Frojmovic MM. Platelet size affects both micro- and macro-aggregation: contributions of platelet number, volume fraction and cell surface. *Thrombos Haemost* 1989; **62**: 733-741.

25. Goldsmith HL, Turitto VT. Rheological aspects of thrombosis and haemostasis: Basic principles and applications. *Thrombos Haemost* 1986; **55**: 415-435.

26. Milnor, WR. Hemodynamics. Williams and Wilkins, Baltimore, USA. 1982; 53 pp.

27. Tangelder GJ, Slaaf DW, Arts T, Reneman RS. Wall shear rate in arterioles in vivo: Least estimates from platelet velocity profiles. *Am J Physiol* 1988; **254**: H1059-1064.

28. Back, CH, Radbill, JR and Crawford, DW Analysis of pulsatile viscous flow through diseased coronary arteries of man. *J Biomech* 1977; **10**: 339-353.

29. Goldsmith HL and Mason SG. The microrheology of dispersions. In Rheology, Theory, and Applications. FR Eirich, Ed., Academic Press, New York. 1967; **4**: 85-250.

30. Anderson, GH, JD Hellums, JL Moake, CP Alfrey. Platelet response to shear stress: Changes in serotonin uptake, serotonin release, and ASP induced aggregation. *Thromb Res* 1978; **13**: 1039-1047.

31. Belval TK, JD Hellums, RT Solis. The kinetics of platelet aggregation induced by fluid-shearing stress. *Microvasc Res* 1984; **28**: 279-288.

32. Ikeda Y, Handa M, Kawano K, Kamata T, Murata M, Araki Y, AnboH, Kawai Y, Watanabe K, Itagaki I, Kiyotaka S, Ruggeri Z. The role of von Willebrand factor and fibrinogen in platelet aggregation under varying shear stress. *J Clin Invest* 1991; **87**: 1234-1240.

33. Joist, JH, Bauman EJ, Speer M, Sutera SP. Role of platelet surface interaction in fluid shear-induced platelet aggregation. *Thromb Haemost* 1985; **54**: 109 (abstract).

34. Peterson, DM, Stathopoulos NA, Giotgio TD, Hellums JD, Moake JL. Shear-induced platelet aggregation requires von Willebrand factor and platelet membrane glycoproteins Ib and IIb-IIIa *Blood* 1987; **69**: 625-628.

35. Iijima K, Murata M, Nakamura K, Kitaguchi T, Handa M, Watanabe K, Fujimura Y, Yoshioka A, and Ikeda Y. High shear stress attenuates agonist-induced, glycoprotein llb/llla-mediated platelet aggregation when von Willebrand factor binding to glyco-lb/lX is blocked. BBRC 1997; **233**:796-800.

36. Lumley P, Humphrey PPA. A method for quantitating platelet aggregation and analyzing drug-receptor interactions on platelets in whole blood in vitro. *J Pharmacol Methods* 1981; **6**: 153-166.

37. Peerschke E, Zucker MB. Fibrinogen receptor exposure and aggregation of human blood platelets produced by ADP and chilling. *Blood* 1981; **57**: 663-670.

38. Ruf A, Frojmovic MM, Patcheke H. Platelet Aggregation. In Handbook of Experimental Pharmacology: Platelet and their factors. Von Bruckhausen F, Walter U, eds. Berlin, Springer-Verlag 1997; pp. 83-98.

39. Chang HM, Robertson CR. Platelet aggregation by laminar shear and Brownian motion. *Ann Biomed Eng* 1976; 4: 151-183.

40. Hantgan RR. A study of the kinetics and mechanism of ADP-triggered platelet aggregation. *Biochim Biophys Acta* 1985; **846**: 64-75.

41. Frojmovic MM, Milton JG, Gear AL. Platelet aggregation measured in vitro by microscopic and electronic particle counting. *Methods Enzymol* 1989; **169**: 134-149.

42. Huang PY, Hellums JD. Aggregation and disaggregation kinetics of human blood platelet: Part II. Shear-induced platelet aggregation. *Biophys J* 1993; **65**: 344-353.

43. Frojmovic MM, Kasirer-Friede A, Goldsmith HL, Brown EA. Surface-secreted von Willebrand factor mediates aggregation of ADP-activated platelets at moderate shear stress: facilitated by GPIb but controlled by GPIIb-IIIa. *Thromb Haemost* 1997; **77**: 568-576.

44. Bell D N, Goldsmith HL. Platelet aggregation in Poiseuille flow: II. Effect of shear rate. *Microvasc Res* 1984; **27**: 316-330.

45. Klose HJ, Rieger H, Schmid-Schönbein H. A rheological method for the quantification of platelet aggregation (PA) in vitro and its kinetics under defined flow conditions. *Thromb Res* 1975; **7**: 261-272.

46. Higuchi WI, Okada T, Stelter GA. Kinetics of rapid aggregation in suspensions (comparison of experiments with the Smoluchowski theory). *J Pharm Sci* 1963; **52**: 49-54.

47. Varennes S, van de Ven TGM. Effects of neutral polymers on the kinetics of particle deposition. *Physicochem Hydrodyn* 1988; **10**: 229-238.

48. Swift DL, Friedlander SK. The coagulation of hydrosols by Brownian motion and laminar shear flow. *J Colloid Sci* 1964; **19**: 621-647.

49. Plow EF, Ginsberg M. Cellular adhesion: GPIIb-IIIa as a prototypic adhesion receptor. *Progress Hemost Thrombos* 1989; **9**: 117-156.

50. Michelson AD, Benoit SE, Kroll MH, Li JM, Rohrer MJ, Kestin AS, Barnard MR. The activation-induced decrease in the platelet surface expression of the glycoprotein Ib-IX complex is reversible. *Blood* 1994; **83**: 3562-3573.

51. Shattil SJ, Brass LF. Induction of the fibrinogen receptor on human platelets by intracellular mediators. *J Biol Chem* 1987; **262**: 992-1000.

52. Shattil SJ, Cunningham M, Hoxie JA. Detection of activated platelets in whole blood using activation-dependent monoclonal antibodies and flow cytometry. *Blood* 1987; **70**: 307-315.

53. Shattil SJ, Budzynski A, Scrutton MC. Epinephrine induces platelet fibrinogen receptor expression, fibrinogen binding, and aggregation in whole blood in the absence of other excitatory agonists. *Blood* 1989; **73**: 150-158.

54. Marguerie GA, Edgington TS, Plow EF. Interaction of fibrinogen with its platelet receptor as part of a multistep reaction in ADP-induced platelet aggregation. *J Biol Chem* 1980; **255**: 154-161.

55. Niewiarowski S, Kornecki E, Budzynski AZ, Morinelli TA, Tuszynski GP. Fibrinogen interaction with platelet receptors. *Ann N Y Acad Sci* 1983; **408**: 536-555.

56. Marguerie GA, Thomas-Maison N, Ginsberg MH, Plow EF. The platelet-fibrinogen interaction. Evidence for proximity of the A alpha chain of fibrinogen to platelet membrane glycoproteins IIb/IIIa. *Eur J Biochem* 1984; **139**: 5-11.

57. Isenberg WM, McEver RP, Phillips DR, Shuman MA, Bainton DF. The platelet fibrinogen receptor: An immunogold-surface replica study of agonist-induced ligand binding and receptor clustering. *J Cell Biol* 1987; **104**: 1655-1663.

58. Woods VL Jr, Woolff LE, Keller DM. Resting platelets contain a substantial centrally located pool of glycoprotein IIb-IIIa complex which may be accessible to some but not other extracellular proteins. *J Biol Chem* 1986; **261**: 15242-15251.

59. Kroll MH, Schafer AI. Biochemical mechanisms of platelet activation. *Blood* 1989; **74**: 1181-1195.

60. Greco NJ, Jones GD, Tandon NN, Kornhauser R, Jackson B, Jamieson GA. Differentiation of the two forms of GPIb functioning as receptors for alpha-thrombin and von Willebrand factor: Ca^{2+} responses of protease-treated human platelets activated with alpha-thrombin and the tethered ligand peptide. *Biochem* 1996; **35**: 915-921.

61. Yamamoto N, Greco NJ, Barnard MR, Tanoue K, Yamazaki H, Jamieson GA, Michelson AD. Glycoprotein Ib (GPIb)-independent pathways of thrombin-induced platelet activation. *Blood* 1991; **8**: 1740-1748.

62. De Marco L, Mazzucato M, Masotti A, Ruggeri ZM. Localization and characterization of an α-thrombin-binding site on platelet glycoprotein Ibα. *J Biol Chem* 1994; **269**: 6478-6484.

63. Frojmovic MM. Flow cytometric analysis of platelet activation and fibrinogen binding. Platelets 1996; 7: 9-21.

64. Frojmovic MM, Wong T, van de Ven T. Dynamic measurements of the platelet membrane glycoprotein Iib-IIIa receptor for fibrinogen by flow cytometry. *Biophys J* 1991; **59**: 815-827.

65. Frojmovic Mony M, Wong Truman. Dynamic measurements of the platelet membrane glycoprotein II_b-III_a receptor for fibrinogen by flow cytometry: II. Platelet-size dependent subpopulations. *Biophys J* 1991; **59**: 828-837.

66. Xia Z, Wong T, Liu Q, Kasirer-Friede A, Brown E, Frojmovic MM. Optimally-functional fluorescein isothiocyanate (FITC)-labelled fibrinogen for quantitative studies of binding to activated platelets and platelet aggregation. *Br J Haematol* 1996; **93:** 204-214.

67. Xia Z, Frojmovic MM. Aggregation efficiency of activated normal or fixed platelets in a simple shear field: Effect of shear and fibrinogen occupancy. *Biophys J* 1994; **66**: 2190-2201.

68. Kasirer-Friede A, Frojmovic MM. The role of GPIb and GPIIb-IIIa in vWf-mediated platelet aggregation at intermediate shear conditions. *Thromb Haemost* 1997; **Suppl**: 341 (abstract) [Note: manuscript in review]

69. Frojmovic MM, Mooney RF, Wong T. Dynamics of platelet glycoprotein IIb-IIIa receptor expression and fibrinogen binding. I. Quantal activation of platelet subpopulations varies with adenosine diphosphate concentration. *Biophys J* 1994; **67**: 2060-2068.

70. Frojmovic MM, Mooney RF, Wong T. Dynamics of platelet glycoprotein IIb-IIIa receptor expression and fibrinogen binding. II. Quantal activation parallels platelet capture in stir-associated microaggregation. *Biophys J* 1994; **67**: 2069-2075.

71. Searcy GP, Frojmovic MM, McNicol A, Robertson C, Wong T, Gerrard JM. Platelets from bleeding simmental cattle mobilize calcium, phosphorylate myosin light chain and bind normal numbers of fibrinogen molecules but have abnormal cytoskeletal assembly and aggregation in response to ADP. *Thrombos Haemost* 1994; **71**: 240-246.

72. Frojmovic MM, Wong T, Searcy GP. Platelets from bleeding Simmental cattle have a long delay in both ADP-activated expression of Gp IIb-IIIa receptors and fibrinogen-dependent platelet aggregation. *Thrombos Haemost* 1996; **76**: 1047-1052.

73. Longmire KL and Frojmovic MM. Long range interactions in mammalian platelet aggregation. I. Evidence from kinetic studies in Brownian diffusion. Biophys J 1990; **58**: 299-307.

74. Frojmovic MM, Longmire KL, and van de Ven TGM. Long range interactions in mammalian platelet aggregation. II. The role of platelet pseudopod number and length. Biophys J 1990; **58**: 309-318.

75. Hensler ME, Frojmovic MM, Taylor RG, Hantgan RR, and Lewis JC. Platelet morphological changes and fibrinogen receptor localization: initial responses in ADP-activated human platelets. Am J Pathol 1992; **141**: 707-719.

76. Frojmovic MM., O'Toole TE, Plow EF, Loftus JC, and Ginsberg MH. Transfection of the α_{IIb}-β_3 integrin (platelet GPIIb-IIIa) confers fibrinogen and activation dependent aggregation on heterologous cells. Blood 1991; **78**: 369-376.

77. Gawaz MP, Loftus JC, Bajt ML, Frojmovic MM, Plow EF, & Ginsberg MH. Ligand bridging mediates integrin α_{IIb}-β_3 (platelet GPIIb-IIIa) dependent homotypic and heterotypic cell-cell interactions. J Clin Invest 1991; **88**:1128-1134.

78. Zamarron C, Ginsberg MH, Plow EF. A receptor-induced binding site in fibrinogen elicited by its interaction with platelet membrane glycoprotein IIb-IIIa. *J Biol Chem* 991; **266**: 16193-16199.

79. Liu Q, Matsueda G, Brown E, Frojmovic MM. The AGDV residues on the gamma chain carboxyl terminus of platelet-bound fibrinogen is needed for platelet aggregation. *BBA* 1997; **1343**: 316-326.

80. Hawiger J, Timmons S, Kloczewiak M, Strong DD, Doolittle RF. Gamma and α chainof human fibrinogen possess sites reactive with human platelet receptors. *PNAS USA* 1982; **79**: 2068-2071.

81. Kloczewiak M, Timmons S, Hawiger J. Localization of a site interacting with human platelet receptor on the carboxyl-terminal segment of human fibrinogen. *Biochem Biophys Res Communic* 1982; **107**: 181-187.

82. Kloczewiak M, Timmons S, Lukas TJ, Hawiger J. Platelet receptor recognition site on human fibrinogen. Synthesis and structure function relationship of peptides corresponding to the carboxyl-terminal segment of the γ chain. *Biochemistry* 1984; **23**: 1767-1774.

83. Kloczewiak M, Timmons S, Bednarek MA, Sakon M, Hawiger, J. Platelet receptor recognition domain on the γ chain of human fibrinogen and its synthetic peptide analogues. *Biochemistry* 1989; **28**: 2915-2919.

84. Hawiger J, Kloczewiak M, Bednarek MA, Timmons S. Platelet receptor recognition domains on the α chain of human fibrinogen: Structure-function analysis. *Biochemistry* 1989; **28**: 2909-2914.

85. Ruggeri ZM, De Marco LD, Gatti L, Bader R, Montgomery RR. Platelets have more than one binding site for von Willebrand Factor. *J Clin Invest* 1983; **72**: 1-12.

86. Pierschbacher MD, Ruoslahti E. Cell attachment activity of fibrinectin can be duplicatedby small synthetic fragments of molecule. *Nature (London)* 1984; **309**: 30-33.

87. Haverstick DM, Cowan JF, Yamada KM, Santoro SA. Inhibition of platelet adhesion to fibronectin, fibrinogen, and von Willebrand Factor substrates by a synthetic tetrapeptide derived from the cell binding domain of fibrinectin. *Blood* 1985; **66**: 946-952.

88. Plow EF, McEver RP, Coller BS, Woods VL Jr, Marguerie GA, Ginsberg, MH. Related binding mechanisms for Fibrinogen, fibrinectin, von Willebrand Factor, and thrombospondin on thrombin-stimulated human platelets. *Blood* 1985; **66**: 724-727.

89. Pytela RP, Pierschbacher MD, Ginsberg MH, Plow EF, Ruoslahti E. Platelet membrane glycoprotein Iib/IIIa: Member of a family of Arg-Gly-Asp-specific adhesion receptors *Science* 1986; **231**: 1559-1562.

90. Peerschke EIB, Galanakis DK. Binding of fibrinogen to ADP-treated platelets: Comparison of plasma fibrinogen fractions and early plasmic fibrinogen derivatives. *J Lab Clin Med* 1983; **101**: 453-460.

91. Peerschke EIB, Francis CW, Marder VJ. Fibrinogen binding to human blood platelets: effect of γ chain carboxyterminal structure and length. *Blood* 1986; **67**: 385-390.

92. Amrani DL, Newman PJ, Meh D, Mosesson MW. The role of fibrinogen Aα chains in ADP-induced platelet aggregation in the presence of fibrinogen molecules containing γ' chains. *Blood* 1988; **72**: 919-924.

93. Lawrence SO, Wright TW, Francis CW, Fay PJ, Haidaris PJ. Purification and functional characterization of homodimeric γB-γB fibrinogen from rat plasma. *Blood* 1993; **82**: 2406-2413.

94. Farrell DH, Thiagarajan P, Chung DW, Davie EW. Role of fibrinogen α and γ chain sites in platelet aggregation. *PNAS USA* 1992; **89**: 10729-10732.

95. Farrell DH, Thiagarajan P. Binding of recombinant fibrinogen mutants to platelets. *J Biol Chem* 1994; **269**: 226-231.

96. Rooney MM, Parise LV, Lord ST. Dissecting clot retraction and platelet aggregation. *J Biol Chem* 1996; **271**: 8553-8555.

97. Gartner TK, Amrani DL, Derrick JM. Characterization of adhesion of non-exogenously stimulated and resting platelets in normal plasma to fibrinogen and its fragments. *Blood Coagulation Fibrinolysis* 1994; **5**: 747-754.

98. Gartner TK, Amrani DL, Derric JM, Kirschbaum NE, Matsueda GR, Taylor DB. Characterization of adhesion of "resting" and stimulated platelets to fibrinogen and its fragment. *Thromb Res* 1993; **71**: 47-60.

99. Xia Z, Frojmovic MM. Aggregation efficiency of activate normal and fixed platelets in a simple shear field: Effect of shear and fibrinogen occupancy. *Biophysical J.* 1994; **66**: 2190-2201.

100. Donahue JP, Patel H, Anderson WF, Hawiger J. Three dimensional structure of the platelet integrin recognition segment of the fibrinogen γ chain obtained by carrier protein-driven crystallization. *PNAS USA* 1996; **91**: 12178-12182.

101. Cheresh DA, Berliner SA, Vicente V, Ruggeri ZM. Recognition of distinct adhesive sites on fibrinogen by related integrins on platelet and endothelial cells. *Cell* 1989; **58**: 945-953.

102. Liu Q, Rooney MM, Kasirer-Friede A, Brown E, Lord ST, and Frojmovic MM. Role of the g Chain Ala-Gly-Asp-Val and Aa chain Arg-Gly-Asp-Ser Sites of Fibrinogen in Coaggregation of Platelets and Fibrinogen-Coated Beads. BBA 1998; (Protein structure and molecular enzymology), in press.

103. Ruggeri ZM, Ware J. The structure and function of von Willebrand factor. *Thrombos Haemost* 1992; 67: 594-599.

104. Lankhof H, Wu YP, Vink T, Schiphorst ME, Zerwes HG, deGroot P, Sixma JJ. Role of the glycoprotein Ib-binding A1 repeat and the RGD sequence in platelet adhesion to human recombinant von Willebrand factor. *Blood* 1995; **86**: 1035-1042.

105. Ikeda Y, Handa M, Kamata T, Kawano K, Kawai Y, Watanabe K, Kawakami K, Sakai K, Fukuyama M, Itagaki I. Transmembrane calcium influx associated with von Willebrand factor binding to GPIb in the initiation of shear-induced platelet aggregation. *Thromb Haemost* 1993; **69**: 496-502.

106. Moake JL, Turner NA, Stathopoulos NA, Nolasco LH, Hellums JD. Shear-induced platelet aggregation can be mediated by vWF released from platelets, as well as by exogenous large or unusually large vWF multimers, requires adenosine diphosphate and is resistant to aspirin. *Blood* 1988; **71**: 1366-74.

107. Fujimoto T, Hawiger J. Adenosine diphosphate induces binding of von Willebrand factor to human platelets. *Nature* 1982; **297**: 154-156.

108. Gralnick HR, Williams SB, Coller B. Fibrinogen competes with von Willebrand factor for binding to the Glycoprotein IIb /IIIa complex when platelets are stimulated with thrombin. Blood 1984; **64**: 797-800.

109. Scott JP, Montgomery RR, Retzinger GS. Dimeric ristocetin flocculates proteins, binds to platelets, and mediates von-Willebrand factor-dependent agglutination of platelets. *J Biol Chem* 1991; **266**: 8149-8155.

110. Goto S, Salomon DR, Ikeda Y, Ruggeri ZM. Characterization of the unique mechanism mediating the shear-dependent binding of soluble von Willebrand factor to platelets. *J Biol Chem* 1995 **270**: 23352-23361.

111. Dewitz TS, Martin RR, Solis RT, Hellums JD, McIntire LV. Microaggregate formation in whole blood exposed to shear stress. *Microvasc Res* 1978; **16**: 263-271.

112. Jen CJ, McIntire LV. Characteristics of shear-induced aggregation in whole blood. *J Lab Clin Med* 1984; **103**: 115-124.

113. Roth GJ. Developing relationships: Arterial platelet adhesion, glycoprotein Ib and leucine-rich glycoproteins. *Blood* 1991; **77**: 5-19.

114. Berliner S, Niiya K, Roberts JR, Houghton RA, Ruggeri Z. Generation and characterization of peptide-specific antibodies that inhibit von Willebrand factor binding to glycoprotein IIb-IIIa without interacting with other adhesive molecules. Selectivity is conferred by Pro1743 and other amino acid residues adjacent to the sequence Arg1744-Gly1745-Asp1746. *J Biol Chem* 1988; **263**: 7500-7505.

115. Coller BS, Peerschke EI, Scudder LE, Sullivan CA. Studies with a murine monoclonal antibody that abolishes ristocetin-induced binding of von Willebrand factor to platelets: additional evidence in support of GPIb as a platelet receptor for von Willebrand factor. *Blood* 1983; **61**: 99-110.

116. Coller BS, Peerschke EI, Scudder LE, Sullivan CA. A murine monoclonal antibody that completely blocks the binding of fibrinogen to platelets produces a thromasthenic-like state in normal platelets and binds to glycoproteins IIb and/or IIIa. *J Clin Invest* 1983; **72**: 325-338.

117. Girma J-P, Fressinaud E, Christophe O, Rouault C, Obert B, Takahashi Y, Meyer D. Aurin tricarboxylic acid inhibits platelet adhesion to collagen by binding to the 509-695 disulphide loop of von Willebrand factor and competing with glycoprotein Ib. *Thromb Haemost* 1992; **68**: 707-13.

118. Moake JL, Turner NA, Stathopoulos NA, Nolasco LH, Hellums JD. Involvement of large plasma von Willebrand factor (vWF) multimers and unusually large vWF forms derived from endothelial cells in shear stress-induced platelet aggregation. *J Clin Invest* 1986; **78**: 1456-1461.

119. Parker RI, Gralnick HR. Identification of platelet glycoprotein IIb/IIIa as the major binding site for released platelet-von Willebrand factor. *Blood* 1986; **68**: 732-6.

120. Lawrence MB, Springer TA. Leukocytes roll on a selectin at physiological flow rates: Distinction from and prerequisite for adhesion through integrins. *Cell* 1991; **65**: 859-73.

121. Gawaz M, Otto I, Reininger AJ, Heinzmann U, and Neumann FJ. Agglutination of isolated platelet membranes. Arterioscler Thromb Vasc Biol 1996; **16**: 621-627.

122. Baldassare JJ, Kahn RA, Knipp MA, and Newman PJ. Reconstitution of platelet proteins into phospholipid vesicles: functional proteoliposomes. J Clin Invest 1985; **75**: 35-39.

123. Parise LV and Phillips DR. Reconstitution of the purified platelet fibrinogen receptor: fibrinogen binding properties of the glycoprotein llb-llla complex. J biol chem 1985; **260**: 10698-10707.

124. Rybak ME. Glycoproteins llb and llla and platelet thrombospondin in a liposome model of platelet aggregation. Thromb Haemost 1986; **55**: 240-245.

125. Kashiwagi H, Schwartz MA, Eigenthaler M, Davis KA, Ginsberg MH, Sanford JS. Affinity modulation of platelet integrin $\alpha_{IIb}\beta_3$ by β_3-endonexin, a selective binding partner of the β_3 integrin cytoplasmic tail. J Cell Biol 1997; **137**:1433-1443.

126. Xia Z, Goldsmith HL, and van de Ven TG. Flow-induced detachment of red blood cells adhering to surfaces by specific antigen-antibody bonds. Biophys J 1994; **66**: 1222-1230.

127. Xia Z, and van de Ven TGM. Adhesion kinetics of phosphatidylcholine liposomes by evanescent wave light scattering. Langmuir 1992; **8**: 2938-2946.

128. Goldsmith HL, McIntosh FA, and Frojmovic MM. The kinetics of thrombin- and SFLLRN-induced aggregation of human platelets in flow through tubes. Biorheology 1998; **35**: in press.

129. Tcheng JE. Platelet glycoprotein ll_b/lll_a integrin blockade: recent clinical trials in international cardiology. Thrombos Haemost 1997; **78**:205-209.

130. Coller BS. Gpllb/llla antagonists: pathophysiologic and therapeutic insights from studies of c7E3 fab. Thrombos Haemost 1997; **78**: 730-735.

131. Théroux P (editor). Platelet Function and monitoring with long term inhibition of the Gpllb-llla receptor with oral antagonists. Am Heart J 1998; **Suppl. 135**: in press.

132. Fressinaud E, Veyradier A, Truchaud F, Martin I, Boyer-Neumann C, Trossaert M, Meyer D. Screening for von Willebrand disease with a new analyzer utilizing high shear stress. A study of 60 cases. Blood 1998; in press.

133. Muga KM, Melton LG, and Gabriel DA. A flow dynamic technique used to asses global hemostasis. Blood Coag and Fibrinolysis 1995; **6**: 73-78.

134. Varon D, Beyer R, Brener B, et al.. Cone and Plate (let) Analyzer (CPA). Am Heart J (Suppl.) 1998; **135**: in press.

135. Agbanyo FR, Sixma JJ, De Groot PG, Languino LR, Plow EF. Thrombospondin-platelet interactions: role of divalent cations, wall shear rate, and membrane Glycoproteins.J Clin Invest 1993; **92**: 288-296.

136. Ruf A, patscheke H. Platelet induced neutrophil activation: platelet-expressed fibrinogen induces the oxidative burst in neutrophils by an interaction with CD11c/CD18. Brit J Haematol 1995; **90**: 791-797.

137. Lalor P, Nash GB. Adhesion of flowing leukocytes to immobilized platelets. Br J Haematol 1995; **89**: 725-732.

138. Weber C, Springer TA. Neutrophil accumulation on activated, surface-adherent platelets in flow is mediated by interaction of MAC-1 with fibrinogen bound to aIIbb3 and stimulated by platelet-activating factor. J Clin Invest 1997; **100**: 2085-2093.

15

PLATELET VESSEL WALL INTERACTIONS

Ginés Escolar and
Maribel Díaz-Ricart
Servicio de Hemoterapia y Hemostasia
Hospital Clínic, IDIBAPS
Universidad de Barcelona
Villarroel 170
Barcelona 08036

15.1 INTRODUCTION

Platelets are contractile blood elements with a specialized task in the arrest of bleeding. The fundamental mechanisms of this specific function involve: adhesion, aggregation, internal contraction and secretion. Furthermore, the phospholipids present in the platelet membranes play a key role in the development of coagulation mechanisms.

Over the years, Dr. Gines Escolar, M. D., has been actively pursuing research in relation to platelet-vessel wall interactions. He has published extensively in this area. His work has made significant contributions to the understanding of platelet-vessel wall interactions.

Studies with blood circulating through vascular segments mounted in specially designed chambers [1] have dramatically contributed to the understanding of the specific mechanism involved in the interaction of platelets with subendothelial components (figure 1). Experiments can be performed with blood from normal individuals or from patients with congenital defects of platelets or plasma proteins. Blood can also be manipulated in the laboratory to remove a specific adhesive protein, and the flow conditions modified to reproduce specific situations present at the vascular tree. The information obtained from such studies, combined with clinical experience have helped to delimit the precise roles of the different platelet, plasma and vascular proteins in the regulation of platelet interactions with vessel subendothelium.

Thanks to the investigations using perfusion systems, it is well established that binding of VWF to subendothelium and to platelet GPIb is of critical importance for platelet attachment onto damaged vascular surfaces [2-4]. Interaction of VWF bound to the sub-endothelium with platelet GPIIb-IIIa mediates further platelet spreading onto the exposed vascular surface [4,5]. Interactions of platelet GPIIb-IIIa with plasma fibrinogen play a major role in platelet-platelet interactions necessary for platelet aggregate formation [4]. The correct sequence of interactions among platelet, plasma and vascular proteins is crucial for maintenance of correct hemostasis.

Figure 1. Light micrograph illustrating different stages of platelets interacting with subendothelium. The morphometric criteria described by Baumgartner (1973) would classify platelets or group of platelets as: contact (C); adhesion (A) and thrombi (T). A percentage of the vessel surface covered by each category and a total surface coverage can be easily calculated by using a graduated eye-piece in the microscope or by applying computerized image analysis to the histological sections.

Nevertheless, the interest in platelet function goes beyond their physiological role in the arrest of bleeding. It is widely accepted that thrombotic events appear as a result of dysregulation of mechanisms involved in primary hemostasis. Cardiovascular disease accounts for about 35% of mortality in developed countries. Coronary artery disease and its complications are the main contributors to this pathological condition, and platelets are known to play a key role at different stages of its development [6].

15.2. INTERACTION OF PLATELETS WITH VASCULAR SUBENDOTHELIUM

Adhesion of platelets onto a damaged vascular surface is the primary event in the formation of the hemostatic plug. For perfect hemostatic performance, this initial adhesion step should be followed by further deposition of platelets onto the initial adherent layer until an effective microthrombi is formed at the site of damage. Formation of this hemostatic plug is the result of a well recognized series of interactions among platelet glycoproteins, plasma adhesive proteins and components of the subendothelium.

15.2.1 Platelet Receptors

Glycoprotein Ib

Primary attachment of platelets onto damaged vascular subendothelium is a process which is essentially dependent on exposure of collagen and VWF, already present in the vascular wall, and interaction of platelet receptors specific for these proteins. The interaction of vascular VWF with platelet membrane GPIb results in the primary attachment of platelets to subendothelium [7]. GPIb is the receptor for VWF and forms a complex with GPIX and

GPV (GPI-IX-V) on platelet membranes [8]. The receptor for VWF is located in the α subunit of GPIb (GPIbα). GPIX is necessary for a correct expression of the GPIb molecule on the platelet surface and may act by stabilizing covalent bonds between α and β subunits of GPIb [9]. GPV has been proposed as a receptor for thrombin and would also function as a stabilizing molecule for the GPIb-IX structure [10].

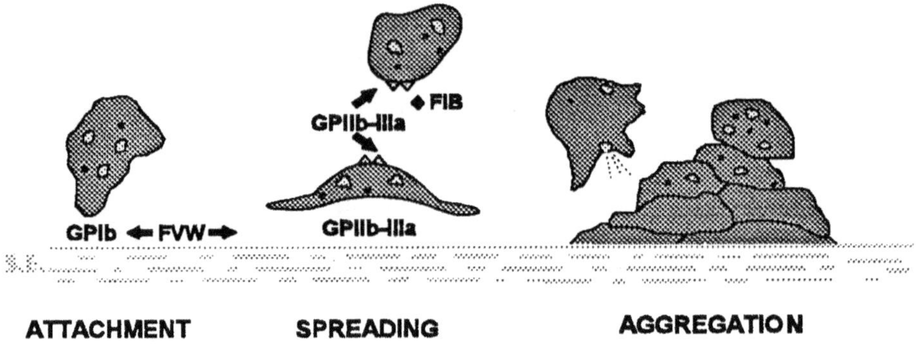

Figure 2. Membrane and adhesive proteins with a major role in the regulation of platelet adhesive and cohesive functions. Glycoprotein Ib (GPIb) through its interaction with von Willebarnd factor (VWF) deposited on the subendothelium is responsible for primary platelet attachment. Interaction of GPIIb-IIIa with VWF facilitates platelet spreading. Fibrinogen interacting with GPIIb-IIIa promotes platelet aggregate formation at low/intermediate shear rate. VWF would also participate in platelet aggregate formation under high shear rate conditions.

The first evidence that GPIb plays a role as a receptor for VWF was obtained from studies on Bernard-Soulier syndrome patients, whose platelets lack GPIb [11]. GPIb does not recognize plasma VWF but does avidly interact with VWF bound to the subendothelium [2]. Binding of VWF to platelet GPIb, under shear stress conditions, triggers activation signalling pathways in platelets [12]. There are about 20,000 copies of GPIb per platelet. The severe deficiency of this glycoprotein is manifested in perfusion studies by a dramatic impairment of platelet adhesion (figure 3).

GPIa-IIa

Platelets possess several collagen receptors (see Chapter 4 on Collagen Receptors). Perhaps the most widely characterized is glycoprotein Ia (GPIa). Although the inherited defect of this glycoprotein is highly infrequent, it is associated with severe hemorrhagic diathesis. Only one case report has been fully described [13], which suggests that GPIa is always present in platelets. GPIb, through its interaction with VWF, is therefore still to be considered the main receptor responsible for platelet adhesion under flow conditions [14].

Besides GPIa, which is the most important, there are other collagen receptors: glycoproteins IV and VI (GPIV and GPVI, respectively). Perfusion studies under flow conditions, conducted in patients with inherited deficiencies show that these glycoproteins could also participate by favoring platelet adhesion processes [15,16].

GPIIb-IIIa

Once the activation signal is initiated through GPIb-VWF interactions, the binding site for other adhesive proteins will be expressed in GPIIb-IIIa. GPIIb-IIIa plays an essential role in platelet spreading and aggregate formation. This receptor interacts with the amino acid sequence arg-gly-asp (RGD) which is present in several adhesive proteins including VWF, fibrinogen and fibronectin.

Studies with monoclonal antibodies have facilitated the recognition of resting and activated conformations in GPIIb-IIIa [17]. The binding site for RGD remains inaccessible in resting GPIIb-IIIa and becomes accessible in activated GPIIb-IIIa. Interaction of GPIIb-IIIa with the RGD sequence in VWF facilitates platelet spreading onto subendothelium [5,18]. Subsequent interactions between GPIIb-IIIa and fibrinogen are involved in platelet aggregation and formation of an effective hemostatic plug. Regulation of the expression of resting versus activated stages on GPIIb-IIIa may play a role in the limitation of thrombus growth. An immunocytochemical study at ultrastructural level has demonstrated a differential rate in the expression of resting or activated GPIIb-IIIa receptors in luminal versus abluminal surfaces in platelets present in aggregates [14].

There are about 45,000 copies of the GPIIb-IIIa per platelet although a recent study has revealed that the actual figure could be doubled if appropriate monoclonal antibodies are used [19]. Severe deficiency of this receptor (Glanzmann's thrombasthenia) produces a characteristic pattern of platelet interaction with subendothelium. Platelets from these patients attach on the subendothelium through GPIb-VWF, but fail to spread or form aggregates (figure 3).

15.2.2 Adhesive Proteins

Von Willebrand factor is the adhesive protein with an essential role in promoting interactions between platelets and vascular subendothelium. VWF is synthesized by endothelial cells and megakaryocytes. Endothelial cells release their synthesized VWF to the plasma and the vessel subendothelium, while the VWF produced by megakaryocytes is stored in the platelet alpha granules [20]. VWF is thus physiologically distributed in plasma, subendothelium and platelets. Adequate hemostasis requires balanced interactions among the VWF located in these three compartments.

Disruption of the endothelial cell layer results in the exposure of VWF already bound to subendothelium but also in the exposure of type I and III collagen fibers and other subendothelial components involved in initial stages of platelet adhesion. Plasma VWF also has the ability to bind to the exposed subendothelial elements. Platelets can either interact with the exposed VWF or with the collagenous surface (through the collagen receptors). If the activation of platelets is intense, the contents of platelet alpha granules are released. This VWF is attached to the surface of activated platelets. A conformational change in GPIIb-IIIa allows the interaction of platelets with VWF at the level of the open canalicular system membranes, which merge with granules during platelet activation [21].

Plasma von Willebrand Factor

Platelets from patients with quanti or qualitative defects of plasma VWF show a delayed interaction with the subendothelium (figure 3) which often results in mucosal hemorrhagic episodes. Perfusion studies [7] have confirmed a marked decrease in platelet adhesion in von Willebrand's patients (figure 3). This decrease could be corrected by adding VWF to the blood of these patients. Incubation of blood with antibodies against human VWF resulted in a similar adhesion defect [22]. VWF-mediated adhesion is dependent on shear rate [23], the effect of VWF already being observed at shear rates above 700 sec^{-1} and is maximum at 2500 sec^{-1}. The presence of free Ca^{++} is essential for VWF mediated platelet adhesion [24].

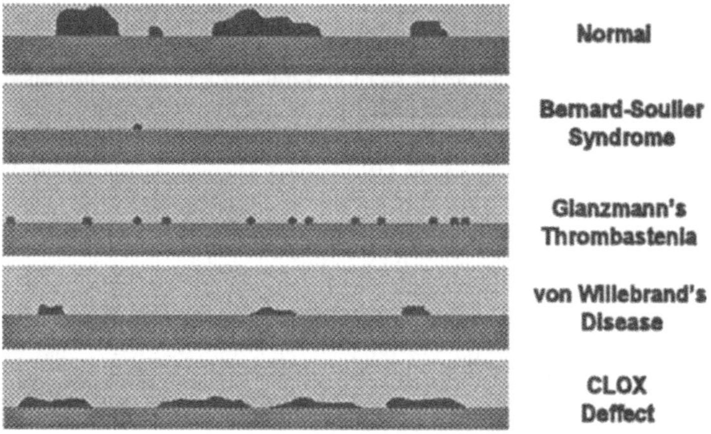

Figure 3. Schematic diagram of different patterns of platelet-subendothelium interactions in congenital disorders of platelet function. The diagrams illustrate profiles of the interactons in perfusions performed at moderately high shear rates (800-1,000 sec^{-1}). Platelets from atients with the Bernard-Soulier syndrome lack glycoprotein Ib (GPIb) and are unable to attach on the subendothelium. GPIIb-IIIa is lacking or defective in Glanzmann's thrombasthenia. Platelets from the latter patientsdo not spread or form aggregates. Plasma levels of VWF are reduced or absent in von Willebrand's disease (VWD) resulting in delayed platelet interaction with the subendothelium. In the severest forms of VWD the deficiency in intraplatelet pool of VWF has an additional negative impact on hemostasis. The formation of aggregates is reduced in storage pool defects and congenital or acquired cycloxygenase deficiencies.

The precise understanding of the role of VWF in platelet adhesion is relatively recent, i.e. dating from the late 70's. At that time, it was known that VWF could bind to collagen [25], but the significance of this binding was unknown. Sakariassen et al [2] conducted studies that confirmed that the binding of VWF to the components of subendothelium preceded platelet adhesion. It has been shown that VWF has specific domains for purified types I and III collagens, which are located differentially from the binding site for platelet GPIb [26].

The capacity of plasma VWF to bind to collagen depends on its multimeric composition. Changes in the multimeric composition of VWF also affect the adhesive capacity on platelets. The type 2a variant of von Willebrand disease, in which high and medium molecular weight multimers are missing, present a platelet adhesion defect [27]. It is known, however, that proteolytic fragments of VWF containing only low molecular weight multimers support a certain degree of adhesion onto collagen [28]. This fact reveals that VWF has alternative binding sites for subendothelial elements other than collagens [29].

Plasma VWF does not interact with resting platelets unless ristocetin is added [30]. VWF bound to subendothelium would undergo a conformational change in the presence of Ca^{++} that would make it reactive to circulating platelets [24,31]. This hypothesis has been demonstrated in perfusion studies using monoclonal antibodies, specifically designed against circulating VWF (with no conformational change) or subendothelial VWF [22].

Subendothelial von Willebrand Factor

Endothelial cells release the synthesized VWF towards the luminal portion as well as towards the abluminal (subendothelial) portion. Several studies suggest that subendothelial VWF can support platelet adhesion regardless of plasma VWF.

VWF in the subendothelium is richer in high molecular weight multimers than the plasma VWF, and is already present in an active conformation for platelets. It has been estimated that subendothelial VWF could guarantee about 40% of total platelet adhesion [32]; the plasma VWF, however, would still be necessary to support optimal platelet adhesion [3]. As may readily be deduced, in type 1 von Willebrand disease, the platelet adhesion defect and the severity of the hemorrhagic episodes would be affected by decreased plasma levels, but also by the decrease in subendothelial VWF.

Levels of VWF in the subendotelium are variable and depend on the origin of endothelial cells and their growth conditions. It is likely that the activation status of endothelial cells is what determines the amount of VWF secreted towards the subendothelium. Studies on cultured endothelial cells have shown that activation induced by phorbol-myristate or thrombin modify the reactivity of the extracellular matrix towards platelets [33]. Although in these *in vitro* studies, endothelial cell activation is followed by the release of VWF to the supernatant, it may be speculated that, *in vivo*, chronic endothelial activation could result in an increase in subendothelial VWF which, together with rapid de-endothelization could result in thrombotic episodes [34].

Intraplatelet von Willebrand Factor

In addition to endothelial cells, megakaryocytes also synthesize VWF. This VWF is stored in the alpha granules of platelets and it is released during platelet activation. The concentration of VWF in platelets may be up to 50 fold its concentration in plasma.

Gralnick et al [35] demonstrated a lack of correlation between plasma VWF levels and bleeding times, in a group of type 1 patients. However, the same authors found a virtually complete inverse correlation between intraplatelet VWF levels and bleeding times in the same group of patients. There is clinical evidence showing that in patients with a severe deficiency of VWF (type 3 VWD), correction of bleeding time and control of hemorrhagic episodes is incomplete regardless of transfusional normalization of plasma VWF levels [36,37].

Studies conducted on pigs with type 3 von Willebrand disease showed that the bleeding time was shortened when they were engrafted with bone marrow from normal animals [38]. *In vitro* studies have also shown that the internal pool of VWF contributes to platelet adhesion on collagen-rich surfaces [39,40]. Platelets possess collagen receptors that would facilitate the initial interaction of platelet with the exposed surface. As a result of this interaction, the vWF released from normal platelets could facilitate the spreading and contact with other platelets [39].

Two clinical studies [41,42] have confirmed the importance of the correction of the platelet pool of VWF in the control of bleeding episodes in type 3 von Willebrand patients. The clinical evidence published by Castillo et al [42] suggests that full control of hemorrhagic episodes in type 3 patients can be attained with a 20% correction of plasma VWF levels provided there has been previous correction of the platelet VWF pool (5%) by transfusing normal platelets.

Fibrinogen and Fibronectin

As mentioned earlier, plasma fibrinogen plays a major role in mediating platelet-platelet interactions during aggregate formation. The alpha chain of fibrinogen contains the RGD sequence which interacts with the binding site on activated GPIIb-IIIa. The role of fibrinogen facilitating platelet aggregate formation is more evident at intermediate shear rates. At shear rates above 1,000 sec^{-1}, VWF would actually be a more important ligand than fibrinogen in mediating platelet aggregate formation [43,44]. Studies which will be commented on in section 3 of this chapter discuss the role of fibrinogen supporting platelet adhesion.

Fibronectin is present in plasma, subendothelium and platelet alpha granules. Despite containing the RGD sequence, the role of fibronectin in adhesive and cohesive properties of platelets seems less important. It has been suggested that plasma fibronectin would not be required for platelets to adhere on vascular extracellular matrices [45]. However, studies with plasma depleted of fibronectin seem to indicate that this adhesive protein could play a role in mediating adhesion of platelet and thrombus formation on collagen rich surfaces [46]. As will be commented on later, the more recent studies suggest that vascular fibronectin would mediate platelet adhesion under certain experimental conditions [47].

15.2.3 Regulatory Mechanisms

Red Blood Cells and Flow Rate

Platelet vessel interactions are highly dependent upon physical factors. The flow rate and the presence of red blood cells are the two factors with a major influence. Platelet adhesion to the vessel wall increases with the shear rate [23], this shear rate being proportional to the velocity gradient between two fluid layers. The shear rate is directly dependent on flow rate and inversely correlated with the diameter of the vessel [48]. Low shear rates (50-300 sec^{-1}) are found at the level of large veins (low flow) but also in large arteries (large diameter). Intermediate shear rates (300-1,200 sec^{-1}) are found in large to medium sized arteries. High shear rates (1,000-2,000 sec^{-1}) can be observed in the physiologic microcirculation. Very high shear rates (>2,500 sec^{-1}) are uncommon in normal circulation (except in microcapillaries), but can be observed under pathological conditions (stenosed arteries).

Platelet transport to the vessel wall is also regulated by red blood cells [49]. Under current laminar flow conditions such as those taking place in circulating blood, red blood cells displace particles of smaller size towards the boundary layer (subendothelium). In an experimental study, comparing the deposition of platelets in whole blood vs platelet rich plasma, Turitto and Baumgartner [49] demonstrated that the presence of red blood cells increased 50 fold the overall platelet deposition. Clinical studies have also confirmed a beneficial effect of moderate corrections in hematocrit in patients with compromised hemostasis [50,51].

Thrombin Generation

Experimental studies in perfusion devices can also be performed with non-anticoagulated blood [52] or with blood anticoagulated with low molecular weight heparin [53]. In contrast with the studies with citrated blood, the latter experimental systems allow the generation of thrombin and fibrin deposition on the damaged vessel due the concurrence of several mechanisms: exposure of tissue factor in the subendothelium [54], presence of plasma coagulation factors [55] and the expression of a procoagulant surface on activated platelets.

The deposition of fibrin on the subendothelium is inversely correlated with the shear rate [56]. Although less fibrin would be locally deposited at higher flow rates [52], thrombin would still be generated, facilitating platelet deposition and growth of platelet aggregates at both low or high shear rates [57-59].

Platelet Secretion

Patients with storage pool deficiency who are selectively deficient in dense granules produce smaller thrombi in perfusion studies [4], though some patients may also present an adhesion defect probably related to reduced ADP release [60].

Pharmacological inhibition of arachidonic acid metabolism by aspirin impairs platelet-platelet interactions but does not affect platelet adhesion (figure 3). The antiplatelet effects of aspirin are evident in blood anticoagulated with citrate [61,62] and less obvious in studies with native (non anticoagulated) blood [63].

Endothelial Cells

There is evidence raised from perfusion studies with citrated blood suggesting that local generation of PGI2 could inhibit platelet adhesion [64]. This anti-adhesive effect might have also been magnified by the use of anticoagulated blood and is not observed when platelets have been previously exposed to aspirin [65].

Nitric oxide (NO) has been shown to decrease platelet adhesion [66]. To counteract the inhibitory action of hemoglobin on NO the previous study had to be performed in a specially devised chamber in which a positive NO flow could be maintained during the experiments. Our personal experience with NO-donors (i.e. nitroprusside) is that such agents have little effect on platelet deposition when studies are conducted at arterial P_{O2} and blood is perfused through a thrombogenic surface (personal observations). It has been suggested that NO and PGI2 could potentiate their antiplatelet effects *in vivo* by acting coordinately on cGMP and cAMP, though this action has not been fully demonstrated.

15.3. INTERACTION OF PLATELETS WITH EXTRACELLULAR MATRICES, ISOLATED COMPONENTS OF THE VESSEL WALL OR PURIFIED PLASMA PROTEINS

The development of parallel-plate perfusion chambers [67,68] made possible the study of platelet interaction with the extracellular matrix (ECM) generated by cells in culture or with isolated subendothelial components under defined experimental conditions. The use of the ECM produced by human umbilical vein endothelial cells (HUVEC) in culture as adhesive substrate has facilitated the understanding of the mechanisms involved in primary hemostasis [68]. HUVECs are immature and not subjected to flow conditions during their culture, two factors which may influence the reactivity of their ECM towards platelets [69]. Interestingly, the properties and reactivity of the underlying ECM can be modified by exposure of HUVECs to different stimuli, an experimental approach which has favored the investigation of basic mechanisms of thrombosis [33].

Parallel-plate perfusion system have been also useful to evaluate the relative roles of purified plasma adhesive proteins or isolated components of the vessel wall. Receptors for all these proteins are present in the membrane of platelets. Potentially, platelets can adhere to all of them, although this ability does not depend exclusively on the recognition of the specific proteins by its receptor but also on rheological factors. In fact, there may be more than one receptor for each of the adhesive proteins. Recognition by one or another receptor may also be affected by the presence of divalent cations and by the degree of platelet activation. The use of the parallel-plate perfusion chamber together with manipulation of the perfusates and variations in experimental conditions allow the detailed study of basic pathophysiological mechanisms involved in platelet-vessel interactions.

In this section, we will summarize the present knowledge of major adhesive proteins present in plasma and the subendothelium, which may play a role in mediating platelet deposition under flow conditions.

15.3.1 Collagen

Collagen is the most important platelet-reactive protein present in basement membranes, including the subendothelium. There are at least nine well characterized collagen types known to be present in the subendothelium. Differences in structure, relative proportions and their differential thrombogenicity have been extensively reviewed elsewhere [70]. Studies of platelet adhesion to purified collagens indicate that collagen types I, II and III are strongly thrombogenic, while types VI, VII and VIII show weaker reactivity towards platelets and type V does not support platelet adhesion under flow conditions.

Numerous putative receptors have been suggested to mediate platelet adhesion to collagen although few, if any, fully meet the criteria proposed to validate one as the specific collagen receptor [71]. However, clinical and experimental evidence strongly supports roles in platelet adhesion for integrin $a_2\beta_1$ (GPIa-IIa) [72], and the non-integrin receptors GPIV [73] and GPVI [74].

The role of GPIa-IIa as a platelet receptor is clear since GPIa-deficient platelets fail to adhere to collagen and do not aggregate in response to collagen [13]. Severe deficiency in this glycoprotein results in hemorrhagic diathesis. An extensive review of the relative role of this receptor can be examined in a previous chapter.

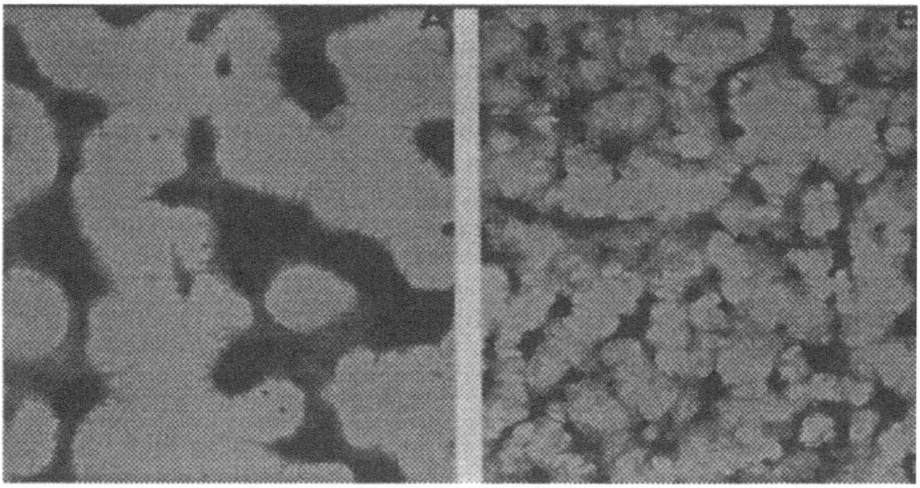

Figure 4. Light micrographs show platelet adhesion on collagen type III (A) and von Willebrand factor (B). Perfusions were performed using the parallel-plate perfusion chamber were coverslips coated with the proteins were inserted and perfused at 800 s^{-1} for 5 minutes. Patterns of platelet adhesion differ significantly depending on the protein perfused.

CD36 (GPIV) is a member of a newly recognized family of receptor molecules that may be involved in adhesive processes and protein trafficking. The role of CD36 mediating platelet adhesion on collagen has been discussed since GPIV-deficient individuals (with a high incidence in the Japanese blood donor population) do not present clinically detectable hemostatic disorders. Several studies have shown that CD36 accelerates the interaction of platelets with collagen both under static and flow conditions these effects being more apparent in the absence of divalent cations [15,73]. Further studies indicate that CD36 plays a role in the early stages of platelet adhesion to physiologically relevant subendothelial surfaces [75]. However, studies performed with low-molecular-weight heparin as anticoagulant have failed to recognize a role for CD36 in platelet-collagen interaction [76]. Under the latter conditions, platelet activation due to tissue factor-dependent thrombin generation and physiological levels of divalent cations may have undervalued the role of CD36 in platelet-collagen interactions [75].

Platelet GPVI has also been identified as a platelet receptors for collagen, since GPVI-deficient platelets exhibit abnormal aggregation when activated with collagen and patients with this deficiency present a mild bleeding tendency [74]. GPVI was initially reported to be involved in platelet adhesion to a collagen-coated surface under static conditions [77]. Studies under flow conditions using a parallel-plate chamber and a newly developed perfusion method, the fluorescent adhesion assay, have demonstrated that platelets lacking GPVI present deficient deposition on collagen types I and III [16]. Studies performed using annular perfusion chambers show that GPVI-deficient platelets exhibit significantly less adhesion and aggregation onto collagen-rich subendothelial structures exposed to citrated blood (figure 5). GPVI therefore seems to play a major role in the aggregation step occurring in platelet adhesion to collagen [78].

The interaction of platelets with collagen is complex and there is some controversy on the sequence of events taking place during this interaction. The mechanisms involved in this process have been divided into two categories [71]: primary adhesion and secondary reinforcing adhesion. Primary adhesion would imply two consecutive steps: an initial divalent cation-independent step followed by a Mg^{2+}-dependent phase. The secondary reinforcing adhesion would involve bridging proteins such as von Willebrand factor and fibronectin, which bind both to collagen and to receptors on the platelet surface.

Following a similar approach, the group led by G.A. Jamieson proposed the hypothesis that platelet interaction with collagen proceeds as an adhesion cascade in two (or more) stages [79]. An initial recognition and adhesion in which CD36 would be directly involved. This primary adhesion process would lead to changes in the platelet membrane conformation with the expression of other receptors, possibly GPVI or integrins such as GPIa-IIa which would then mediate platelet anchorage to the collagen substrate.

Figure 5. Micrographs showing cross-sections of denuded vascular segments perfused with blood from a control donor (A) and from a patient whose platelets are deficient in GPVI (B). Perfusions were performed using the annular chamber at a shear rate of $800s^{-1}$ for 10 minutes. Aggregate formation is almost absent when blood samples from a GPVI-deficient patient are perfused.

Recently, Moroi et al [16,80] using a newly developed fluorescent adhesion assay have also described a biphasic interaction process on collagen, in which GPIb-von Willebrand interaction would play a primary role. In a first step, called temporary arrest, platelets would interact with collagen, transiently stopping on the surface. The work by Moroi et al proposes that this temporary arrest seems to be mediated by the interaction of platelet GPIb/IX and collagen-von Willebrand factor complexes on the collagen surface. In the second phase, mainly dependent on GPIa-IIa and GPVI, platelets adhere more rapidly and permanently on the surface leading to final aggregate formation.

15.3.2 Von Willebrand Factor

Adhesion to isolated VWF under flow conditions requires the presence of two receptors, GPIb and GPIIb-IIIa. It has been recently reported that immobilized von Willebrand factor (VWF) initially supports a slow and continuous movement of platelets on the surface under flow conditions. Under high shear stress conditions, this capability results in an efficient capture of platelets, leading to irreversible attachment [80,81]. The latest hypotheses regarding the role of VWF threfore propose that VWF acts first, by slowing platelet motion, and then by facilitating their adhesion to collagen and other adhesive proteins present in subendothelium.

15.3.3 Fibronectin

Fibronectin (FN) is present in plasma, the subendothelium of the vessel wall, and in the a-granules of platelet. Vessel-wall fibronectin differs structurally from plasma FN, but there is no evidence of differences between the two forms in supporting platelet adhesion. In fact, evidences suggest an important role for vessel wall fibronectin [45]. Platelets can also adhere to a fibronectin-coated surface at low shear rates through mechanisms requiring integrin receptors [47]. Platelets possess at least two receptors for fibronectin: glycoprotein Ic-IIa, corresponding to the very late antigen 5 (VLA-5), and GPIIb-IIIa, which are both members of the superfamily of integrins [82]. Like many of the adhesive proteins, fibronectin present the peptide sequence arginine-glycine-aspartic acid (RGD), which is located in the cell-binding domain of the molecule. This sequence is essential for the recognition of adhesive proteins by integrin receptors [83]. In fact, platelet adhesion to fibronectin seems to be RGD-mediated. Although interaction with fibronectin through GPIIb-IIIa requires previous platelet activation, interaction with VLA-5 does not. Several studies performed under both static and flow conditions show that fibronectin supports platelet adhesion, spreading and aggregation [46]. In addition, fibronectin supports platelet adhesion in flow to other surfaces such as non-fibrillar collagens type I and III [84], the extracellular matrix of cultured endothelial cells [85] and the subendothelium of the vessel wall[86].

15.3.4 Fibrinogen and Fibulin-1

Fibrinogen is a plasma protein that plays a major role in hemostatic plug formation. It exists as a homodimer of a, ß and t chains. The RGD sequences at the a chain and a dodecapeptide sequence at the carboxy terminus of the t chain are essential for fibrinogen function. The receptor for fibrinogen on the platelet membrane is GPIIb-IIIa. Platelets adhere to surface-bound fibrinogen [87]. Adhesion on immobilized fibrinogen is inhibited by plasma fibrinogen [88]. It has been reported by several groups that the adhesion of platelets to fibrinogen occurs at moderate shear rates and decreases markedly under high flow. The mechanism proposed for platelet adhesion on fibrinogen is that fibrinogen supports irreversible attachment of platelets that is less efficient when shear stress increases [81].

Fibulin-1 is a glycoprotein that has been shown to be a component of the extracellular matrix that surrounds vascular smooth muscle. Fibulin-1 can bind to a number of matrix constituents including fibronectin, laminin, nidogen and fibrinogen. In fact, studies performed under flow conditions on fibulin-1 coated surfaces showed that this protein promotes adhesion of platelets via a bridge of fibrinogen [89]. This mechanism was proposed as a putative one to mediate fibrinogen immobilization at sites of vascular injury.

15.3.5 Laminin

Laminin is a ubiquitous basement membrane component which exhibits a number of biological activities including the promotion of cell attachment, migration and differentiation. It is a large glycoprotein composed of three chains (A, B1 and B2) arranged in a cross-shaped structure [90]. At least two laminin binding proteins have been identified in platelet: the glycoprotein Ic-IIa complex (VLA-6) which is a member of the integrin family of cell adhesion receptors [91], and the non-integrin receptor of 67kD, also described in other mammalian cell types [92,93]. Both the VLA-6 and the 67kD receptors present in platelets support platelet adhesion to laminin. Studies under flow conditions, performed by blocking either laminin present in the extracellular matrix or the 67kD receptor in platelets, resulted in a slight though significant decrease in platelet attachment to the ECM [94]. Moreover, platelets adhere to surface-immobilized purified laminin, under flow conditions, through the VLA-6 receptor. This adhesion seems to be directly related to the presence of divalent cations and inversely related to the shear rate [95].

15.3.6 Thrombospondin

Thrombospondin (TSP) is a large glycoprotein (450kD) present in the extracellular matrix of the vessel wall and in platelet a-granules [96]. It has been described as a multifunctional protein showing similarities with other ECM constituents that mediate platelet adhesion. TSP forms molecular complexes with many adhesive proteins, including fibrinogen, fibronectin, collagen, laminin and von Willebrand factor. Both adhesive and antiadhesive properties have been attributed to this protein [97,98]. Studies performed under static and flow conditions indicate that TSP supports platelet adhesion, without spreading, in a calcium-dependent manner. Many receptors have been proposed for TSP, including GPIV (CD36), although specific antibodies have not been able to inhibit platelet deposition on this protein when immobilized in a surface [99]. TSP seems to modulate the adhesive role of other proteins under specific microenvironmental conditions.

15.3.7 Vitronectin

A role for vitronectin in mediating platelet adhesion and thrombus formation is less clear. Collins et al [100] reported that vitronectin adsorbed on various polymers increased platelet deposition in a canine ex vivo model. However, antibodies either to vitronectin or to the specific platelet vitronectin receptor (a_vb_3) did not decrease platelet adhesion or thrombus formation in citrated blood at any shear rate.

15.4. CONCLUDING REMARKS

Interaction of platelets with the vessel wall is regulated by a complex series of interactions among membrane glycoproteins, subendothelial components and plasma adhesive proteins. Platelets possess several receptors for collagen (GPIa-IIa, GPIV and GPVI). Severe congenital deficiencies of such collagen receptors are highly infrequent. These receptors would therefore guarantee a certain level of platelet-subendothelium interactions.

Plasma VWF modifies its conformation once bound to the subendothelium. VWF bound to the subendothelium enhances primary platelet adhesion through its binding to platelet GPIb. This primary adhesion of platelets through GPIb or collagen receptors leads to series of signal transduction processes that result in platelet activation, which implies changes in the conformation of membrane GPIIb-IIIa. Interaction of activated GPIIb-IIIa with VWF plays a role on platelet spreading. Subsequent platelet-platelet interactions are supported by the interaction of GPIIb-IIIa with plasma fibrinogen. VWF could also promote platelet-platelet interactions through its interaction with GPIIb-IIIa, though this mechanism would be more evident at higher shear rates.

Receptors for adhesive proteins present in plasma and subendothelium are present in the membrane of platelets. Potentially, platelets can adhere to all of them, although this ability depends not only on the recognition of the specific proteins but also on rheological factors. The understanding of specific mechanisms involved in the regulation of primary hemostasis is already contributing to the development of new antithrombotic strategies.

ACKNOWLEDGEMENTS

IDIBAPS (Institut D'Investigacions Biomèdiques August Pi i Sunyer). This work has been partially supported by FIS 98/321 from Fondo de Investigaciones Sanitarias de la Seguridad Social, DGICYT PM95/103 from Dirección General de Investiaciones Científicas y Técnicas and SGR 97-133 from CIRIT.

REFERENCES

1. Baumgartner HR: The role of blood flow in platelet adhesion, fibrin deposition and formation of mural thrombi. Microvasc Res 5:167-179, 1973.

2. Sakariassen KS, Bolhuis PA, Sixma JJ: Human platelet adhesion to artery subendothelium is mediated by factor VIII-von Willebrand factor bound to the subendothelium. Nature 279:636-638, 1979.

3. Stel HV, Sakariassen KS, de Groot PG, van Mourik JA, Sixma JJ: Von Willebrand factor in the vessel wall mediates platelet adherence. Blood 65:85-90, 1985.

4. Weiss HJ, Turitto VT, Baumgartner HR: Platelet adhesion and thrombus formation on subendothelium in platelets deficient in glycoproteins IIb-IIIa, Ib, and storage granules. Blood 67:322-330, 1986.

5. Weiss HJ, Turitto VT, Baumgartner HR: Further evidence that glycoprotein-IIb-IIIa mediates platelet spreading on subendothelium. Thromb Haemost 65:202-205, 1991.

6. Fuster V, Badimon L, Badimon JJ, Chesebro JH: The pathogenesis of coronary artery disease and the acute coronary syndromes. N Engl J Med 326:242-250, 1992.

7. Tschopp TB, Weiss HJ, Baumgartner HR: Decreased adhesion of platelets to subendothelium in von Willebrand's disease. J Lab Clin Med 83:296-300, 1974.

8. Li CQ, Dong JF, Lanza F, Sanan DA, Saetung G, Lopez JA: Expression of platelet glycoprotein (GP) V in heterologous cells and evidence for its association with GP Ib alpha in forming a GP Ib-IX-V complex on the cell surface. J Biol Chem 270:16302-16307, 1995.

9. Lopez JA, Leung B, Reynolds CC, Li CQ, Fox JEB: Efficient plasma membrane expression of a functional platelet glycoprotein-Ib-IX complex requires the presence of its 3 subunits. J Biol Chem 267:12851-12859, 1992.

10. Clemetson KJ, Clemetson JM: Platelet GPIb-V-IX complex: Structure, function, physiology, and pathology. Seminars in Thromb Hemostasis 21:130-136, 1995.

11. Nurden AT: Platelet membrane glycoproteins and their clinical aspects. Thromb Haemostasis 6-29, 1987.

12. Kroll MH, Harris TS, Moake JL, Handin RI, Schafer AI: von Willebrand Factor binding to platelet GpIb initiates signals for platelet activation. J Clin Invest 88:1568-1573, 1991.

13. Nieuwenhuis HK, Sakariassen KS, Houdijk WP, Nievelstein PF, Sixma JJ: Deficiency of platelet membrane glycoprotein Ia associated with a decreased platelet adhesion to subendothelium: a defect in platelet spreading. Blood 68:692-695, 1986.

14. Heynen HFG, Molero ML, Degroot PG, Nieuwenhuis HK, Sixma JJ: Absence of ligands bound to glycoprotein IIb-IIIa on the exposed surface of a thrombus may limit thrombus growth in flowing blood. J Clin Invest 94:1098-1112, 1994.

15. Diaz-Ricart M, Tandon NN, Carretero M, Ordinas A, Bastida E, Jamieson GA: Platelet lacking functional CD36 (Glycoprotein IV)show reduced adhesion to collagen in flowing whole blood. Blood 2:491-496, 1993.

16. Moroi M, Jung SM, Shinmyozu K, Tomiyama Y, Ordinas A, Diaz-Ricart M: Analysis of platelet adhesion to a collagen-coated surface under flow conditions: The involvement of glycoprotein VI in the platelet adhesion. Blood 88:2081-2092, 1996.

17. Ginsberg MH, Frelinger AL, Lam SCT, Forsyth J, Mcmillan R, Plow EF, Shattil SJ: Analysis of platelet aggregation disorders based on flow cytometric analysis of membrane glycoprotein IIb-IIIa with conformation-specific monoclonal antibodies. Blood 76:2017-2023, 1990.

18. Turitto VT, Weiss HJ, Baumgartner HR: Decreased platelet adhesion on vessel segments in von Willebrand's disease: a defect in initial platelet attachment. J Lab Clin Med 102:551-564, 1983.

19. Wagner CL, Mascelli MA, Neblock DS, Weisman HF, Coller BS, Jordan RE: Analysis of GPIIb/IIIa receptor number by quantification of 7E3 binding to human platelets. Blood 88:907-914, 1996.

20. Wagner DD: Cell Biology of von Willebrand factor. Annu Rev Cell Biol 6:217-246, 1990.

21. Adelman B, Carlson P, Powers P: von Willebrand factor is present on the surface of platelets stimulated in plasma by ADP. Blood 5:1362-1366, 1987.

22. Stel HV, Sakariassen KS, Scholte BJ, Veerman EC, van der Kwast TH, de Groot PG, Sixma JJ, van Mourik JA: Characterization of 25 monoclonal antibodies to factor VIII-von Willebrand factor: relationship between ristocetin-induced platelet aggregation and platelet adherence to subendothelium. Blood 63:1408-1415, 1984.

23. Weiss HJ, Turitto VT, Baumgartner HR: Effect of shear rate on platelet interaction with subendothelium in citrated and native blood. I. Shear dependent decrease of adhesion in von Willebrand's disease and the Bernard-Soulier syndrome. J Lab Clin Med 92:750-764, 1978.

24. Sakariassen KS, Ottenhof-Rovers M, Sixma JJ: Factor VIII-von Willebrand factor requires calcium for facilitation of platelet adherence. Blood 63:996-1003, 1984.

25. Legrand YJ, Rodriguez-Zeballos A, Kartali G, Fauvel F, Caen JP: Adsorption of factor VIII antigen-activity complex by collagen. Thromb Res 13:909-911, 1978

26. Sixma JJ, Sakariassen KS, Stel HV, Houdijk WP, In der Maur DW, Hamer RJ, de Groot PG, van Mourik JA: Functional domains on von Willebrand factor. Recognition of discrete tryptic fragments by monoclonal antibodies that inhibit interaction of von Willebrand factor with platelets and with collagen. J Clin Invest 74:736-744, 1984.

27. Sakariassen KS, Cattaneo M, v.d.Berg A, Ruggeri ZM, Mannucci PM, Sixma JJ: DDAVP enhances platelet adherence and platelet aggregate growth on human artery subendothelium. Blood 64:229-236, 1984.

28. Sixma JJ, Sakariassen KS, Beeser-Visser NH, Ottenhof-Rovers M, Bolhuis PA: Adhesion of platelets to human artery subendothelium: effect of factor VIII-von Willebrand factor of various multimeric composition. Blood 63:128-139, 1984.

29. de Groot PG, Ottenhof Rovers M, van Mourik JA, Sixma JJ: Evidence that the primary binding site of von Willebrand factor that mediates platelet adhesion on subendothelium is not collagen. J Clin Invest 82:65-73, 1988.

30. Zucker MB, Kim S-J, McPherson J, Grant RA: Binding of factor VIII to platelets in the presence of ristocetin. Br J Haematol 35:535-549, 1977.

31. Sakariassen KS, Bolhuis PA, Sixma JJ: Platelet adherence to subendothelium of human arteries in pulsatile and steady flow. Thromb Res 19:547-559, 1980.

32. Sixma JJ, de Groot PG: von Willebrand Factor and the blood vessel wall. Mayo Clin Proc 66:628-633, 1991.

33. de Groot PG, Reinders JH, Sixma JJ: Perturbation of human endothelial cells by thrombin or PMA changes the reactivity of their extracellular matrix towards platelets. J Cell Biol 104:697-704, 1987.

34. Fuster V, Griggs TR: Porcine von Willebrand disease: implications for the pathophysiology of atherosclerosis and thrombosis. Prog Hemost Thromb 8:159-183, 1986.

35. Gralnick HR, Rick ME, Mckeown LP, Williams SB, Parker RI, Maisonneuve P, Jeanneau C, Sultan Y: Platelet von Willebrand factor: An important determinant of the bleeding time in type I von Willebrand's disease. Blood 68:58-61, 1986.

36. Mannucci PM, Moia M, Rebulla P, Altieri D, Monteagudo J, Castillo R: Correction of the bleeding time in treated patients with severe von Willebrand disease is not solely dependent on the normal multimeric structure of plasma von Willebrand factor. Am J Hematol 25:55-65, 1987.

37. Cattaneo M, Moia M, DellaValle P, Castellana P, Mannucci PM: DDAVP shortens the prolonged bleeding times of patients with severe von Willebrand disease treated with cryoprecipitates. Evidence for a mechanism of action independent of released von willebrand factor. Blood 74:1972-1975, 1989.

38. Bowie EJW, Solberg LA, Fass DN, Johnson CM, Knutson GJ, Stewart ML, Zoecklein LJ: Transplantation of normal bone marrow into a pig with severe von Willebrand's disease. J Clin Invest 78:26-30, 1986.

39. Fressinaud E, Baruch D, Rothschild C, Baumgartner HR, Meyer D: Platelet von Willebrand factor: evidence for its involvement in platelet adhesion to collagen. Blood 70:1214-1217, 1987.

40. d'Alessio P, Zwaginga JJ, de Boer HC, Federici AB, Rodeghiero F, Castaman G, Mariani G, Mannucci PM, de Groot PG, Sixma JJ: Platelet adhesion to collagen in subtypes of type I von Willebrand's disease is dependent on platelet von Willebrand factor. Thromb Haemost 64:227-231, 1990.

41. Castillo R, Monteagudo J, Escolar G, Ordinas A, Magallon M, Martin Villar J: Hemostatic effect of normal platelet transfusion in severe von Willebrand disease patients. Blood 77:1901-1905, 1991.

42. Castillo R, Escolar G, Monteagudo J, Aznar-Salatti J, Reverter JC, Ordinas A: Hemostasis in patients with severe von Willebrand disease improves after normal platelet transfusion and mormalizes with further correction of the plasma defect. Transfusion 37:785-790, 1997.

43. Fressinaud E, Baruch D, Girma JP, Sakariassen KS, Baumgartner HR, Meyer D: von Willebrand factor-mediated platelet adhesion to collagen involves platelet membrane glycoprotein IIb-IIIa as well as glycoprotein Ib. J Lab Clin Med 112:58-67, 1988.

44. Ruggeri ZM: Mechanisms initiating platelet thrombus formation. Thromb Haemostasis 78:611-616, 1997.

45. Houdijk WP, de Groot PG, Nievelstein PF, Sakariassen KS, Sixma JJ: Subendothelial proteins and platelet adhesion. von Willebrand factor and fibronectin, not thrombospondin, are involved in platelet adhesion to extracellular matrix of human vascular endothelial cells. Arteriosclerosis 6:24-33, 1986.

46. Bastida E, Escolar G, Ordinas A, Sixma JJ: Fibronectin is required for platelet adhesion and for thrombus formation on subendothelium and collagen surfaces. Blood 70:1437-1442, 1987.

47. Beumer S, Ijsseldijk MJW, Degroot PG, Sixma JJ: Platelet adhesion to fibronectin in flow: Dependence on surface concentration and shear rate, role of platelet membrane glycoproteins GP IIb/IIIa and VLA-5, and inhibition by heparin. Blood 84:3724-3733, 1994.

48. Turitto VT, Weiss HJ, Baumgartner HR, Badimon L, Fuster V: Cells and aggregates at surfaces. Ann N Y Acad Sci 516:453-467, 1987.

49. Turitto VT, Baumgartner HR: Platelet interaction with subendothelium in a perfusion system: Physical role of red blood cell. Microvasc Res 9:335-344, 1975.

50. Livio M, Gotti E, Marchesi D, Remuzzi G, Mecca G, de Gaetano G: Uraemic bleeding: Role of anemia and beneficial effect of red cell transfusion. Lancet 2:1013-1015, 1982.

51. Escolar G, Garrido M, Mazzara R, Castillo R, Ordinas A: Experimental basis for the use of red cell transfusion in the management of anemic-thrombocytopenic patients. Transfusion 28:406-411, 1988.

52. Baumgartner HR, Turitto V, Weiss HJ: Effect of shear rate on platelet interaction with subendothelium in citrated and native blood. II. Relationships among platelet adhesion, thrombus dimensions, and fibrin formation. J Lab Clin Med 95:208-221, 1980.

53. Zwaginga JJ, Sixma JJ, de Groot PG: Activation of endothelial cells induces platelet thrombus formation on their matrix: Studies of new in vitro thrombosis model with low molecular weight heparin as anticoagulant. Arteriosclerosis 10:49-61, 1990.

54. Weiss H, Turitto V, Baumgartner H, Nemerson Y, Hoffmann T: Evidence for the presence of tissue factor activity on subendothelium. Blood 73:968-975, 1989.

55. Weiss HJ, Turitto VT, Vicic WJ, Baumgartner HR: Fibrin formation, fibrinopeptide A release, and platelet thrombus dimensions on subendothelium exposed to flowing native blood: greater in factor XII and XI than in factor VIII and IX deficiency. Blood 63:1004-1014, 1984.

56. Weiss HJ, Turitto VT, Baumgartner HR: Role of shear rate and platelets in promoting fibrin formation on rabbit subendothelium Studies utilizing patients with quantitative and qualitative platelet defects. J Clin Invest 78:1072-1082, 1986.

57. Inauen W, Baumgartner HR, Bombeli T, Haeberli A, Straub PW: Dose- and shear rate-dependent effects of heparin on thrombogenesis induced by rabbit aorta subendothelium exposed to flowing human blood. Arteriosclerosis 10:607-615, 1990.

58. Orvim U, Roald HE, Stephens RW, Roos N, Sakariassen KS: Tissue factor-induced coagulation triggers platelet thrombus formation as efficiently as fibrillar collagen at arterial blood flow conditions. Arterioscler Thromb 14:1976-1983, 1994.

59. Salatti JA, Fenton J, Anton P, Sakariassen KS: alpha-Thrombin bound to extracellular endothelial matrix induces pronounced fibrin deposition and platelet thrombus growth in flowing non-anticoagulated human blood. Blood Coagul Fibrin 5:561-566, 1994.

60. Lages B, Weiss HJ: Enhanced increases in cytosolic Ca2+ in ADP-stimulated platelets from patients with delta-storage pool deficiency - a possible indicator of interactions between granule-bound ADP and the membrane ADP receptor. Thromb Haemost 77:376-382, 1997.

61. Weiss HJ, Tschopp TB, Baumgartner HR: Impaired interaction (adhesion-aggregation) of platelets with subendothelium in storage-pool disease and after aspirin ingestion: A comparison with von Willebrand's disease. N Engl J Med 293:619-623, 1975.

62. Escolar G, Bastida E, Ordinas A, Castillo R: Interaction of platelets with subendothelium in humans treated with aspirin and dipyridamole alone or in combination. Thromb Res 40:419-424, 1985.

63. Weiss HJ, Turitto VT, Vicic WJ, Baumgartner HR: Effect of aspirin and dipyridamole on the interaction of human platelets with sub-endothelium: studies using citrated and native blood. Thromb Haemost 45:136-141, 1981.

64. Tschopp TB, Baumgartner HR: Platelet adhesion and mural platelet thrombus formation on aortic subendothelium of rats, rabbits, and guinea pigs correlate negatively with the vascular PGI2 production. J Lab Clin Med 98:402-411, 1981.

65. Alemany M, Hernandez MR, Bozzo J, Bono A, Escolar G, Ordinas A, Bastida E: A deleterious effect of aspirin on the antithrombotic properties of endothelial cells is not observed with platelets previously exposed to aspirin: Studies in a flow system. Platelets 7:29-34, 1996.

66. De Graaf JC, Banga JD, Moncada S, Palmer RM, de Groot PG, Sixma JJ. Nitric oxide functions as an inhibitor of platelet adhesion under flow conditions. Circulation 85:2284-2290, 1992.

67. Muggli R, Baumgartner HR, Tschopp TB, Keller H: Automated microdensitometry and protein assays as a measure for platelet adhesion and aggregation on collagen-coated slides under controlled flow conditions. J Lab Clin Med 95:195-207, 1980.

68. Sakariassen KS, Aarts PA, de Groot PG, Houdijk WP, Sixma JJ: A perfusion chamber developed to investigate platelet interaction in flowing blood with human vessel wall cells, their extracellular matrix, and purified components. J Lab Clin Med 102:522-535, 1983.

69. Escolar G, Garrido M, Aznar-Salatti J, Ordinas A, Bastida E: Comparison between human umbilical artery and rabbit abdominal aorta as substrata for platelet adhesion and platelet thrombus formation under flow conditions. Blood Vessels 28:520-531, 1991.

70. Saelman EUM, Nieuwenhuis HK, Hese KM, de Groot PG, Heijnen FG, Sage EH, Williams S, Mckeown L, Gralnick HR, Sixma JJ: Platelet adhesion to collagen types I through VIII under conditions of stasis and flow is mediated by GPIa/IIa(a2B1-Integrin). Blood 83:1244-1250, 1994.

71. Santoro SA: Molecular basis of platelet adhesion to collagen. In Jamieson GA: (ed): "Platelet membrane receptors: molecular biology, immunology biochemistry and pathology". New York: Alan R.Liss, 291-314, 1988.

72. Nieuwenhuis HK, Akkerman JW, Houdijk WP, Sixma JJ: Human blood platelets showing no response to collagen fail to express surface glycoprotein Ia. Nature 318:470-472, 1985.

73. Tandon NN, Kralisz U, Jamieson GA: Identification of GPIV (CD36) as a primary receptor for platelet-collagen adhesion. J Biol Chem 264:7576-7583, 1989.

74. Sugiyama T, Okuma M, Ushikubi F, Sensaki s, Kanaji K, Uchino H: A novel platelet aggregating factor found in a patient with defective collagen-induced platelet aggregation and autoimmune thrombocytopenia. Blood 69:1712-1720, 1987.

75. Diaz-Ricart M, Tandon NN, Gomez-Ortiz G, Carretero M, Escolar G, Ordinas A, Jamieson GA: Antibodies to CD36 (GPIV) inhibit platelet adhesion to subendothelial surfaces under flow conditions. Arterios Thromb Vasc Biol 16:883-888, 1996.

76. Saelman EUM, Kehrel B, Hese KM, de Groot PG, Sixma JJ, Nieuwenhuis HK: Platelet adhesion to collagen and endothelial cell matrix under flow conditions is not dependent on platelet glycoprotein IV. Blood 83:3240-3244, 1994.

77. Moroi M, Jung SM, Okuma M, Shinmyozu K: A patient with platelets deficient in glycoprotein VI that lack both collagen-induced aggregation and adhesion. J Clin Invest 84:1440-1445, 1989.

78. Moroi M, Jung SM: Platelet receptors for collagen. Thromb Haemostasis 78:439-444, 1997.

79. Greenwalt DE, Lipsky RH, Ockenhouse CF, Ikeda H, Tandon NN, Jamieson GA: Membrane glycprotein CD36: a review of its roles in adherence, signal transduction, and transfusion medicine. Blood 80:1105-1115, 1992.

80. Moroi M, Jung SM, Nomura S, Sekiguchi S, Ordinas A, Diaz-Ricart M: Analysis of the involvement of the von Willebrand factor glycoprotein Ib interaction in platelet adhesion to a collagen-coated surface under flow conditions. Blood 90:4413-4424, 1997.

81. Savage B, Saldivar E, Ruggeri ZM: Initiation of platelet adhesion by arrest onto fibrinogen or translocation on von Willebrand factor. Cell 84:289-297, 1996.

82. Hynes RO: Integrins: Versatility, modulation, and signalling in cell adhesion. Cell 69:11-25, 1992.

83. Pytela R, Pierschbacher MD, Ginsberg MH, Plow EF, Ruoslahti E: Platelet membrane glycoprotein Iib/IIIa: Member of a family of Arg-Gly-Asp-specific adhesion receptors. Science 231:1559-1562, 1986.

84. Houdijk WP, Sakariassen KS, Nievelstein PF, Sixma JJ: Role of factor VIII-von Willebrand factor and fibronectin in the interaction of platelets in flowing blood with monomeric and fibrillar human collagen types I and III. J Clin Invest 75:531-540, 1985.

85. Houdijk WP, de Groot PG, Nievelstein PF, Sakariassen KS, Sixma JJ: Subendothelial proteins and platelet adhesion. von Willebrand factor and fibronectin, not thrombospondin, are involved in platelet adhesion to extracellular matrix of human vascular endothelial cells. Arteriosclerosis 6:24-33, 1986.

86. Houdijk WP, Sixma JJ: Fibronectin in artery subendothelium is important for platelet adhesion. Blood 65:598-604, 1985.

87. Zaidi TN, Mcintire LV, Farrell DH, Thiagarajan P: Adhesion of platelets to surface-bound fibrinogen under flow. Blood 88:2967-2972, 1996.

88. Endenburg SC, Lindeboomblokzijl L, Zwaginga JJ, Sixma JJ, deGroot PD: Plasma fibrinogen inhibits platelet adhesion in flowing blood to immobilized fibrinogen. Arterios Thromb Vasc Biol 16:633-638, 1996.

89. Godyna S, Diaz-Ricart M, Argraves WS: Fibulin-1 mediates platelet adhesion via a bridge of fibrinogen. Blood 88:2569-2577, 1996.

90. Iwamoto Y, Robey FA, Graf J, Sasaki M, Kelinman HK, Yamada YM, G.R. YIGRS, a synthetic laminin pentapeptide, inhibits experimental metastasis formation. Science 238:1132-1134, 1987.

91. Sonnenberg A, Modderman PW, Hogervorst F: Laminin receptor on platelets is the integrin VLA-6. Nature 336:487-489, 1988.

92. Rao CN, Barsky SH, Terranova VP, Liotta LA: Isolation of a tumor cell laminin receptor. Biochem Biophys Res Commun 111:804-808, 1983.

93. Kleinman HK, Ogle RC, Cannon FB, Little CD, Sweeny TM, Luckenbill-Edds L: Laminin receptors for neurite formation. Proc Natl Acad Sci U S A 85:1282-1286, 1988.

94. Ordinas A, Diaz-Ricart M, Bastida E, Escolar G, Castillo R, Tandon NN, Jamieson GA: The role of subendothelial laminin and platelet laminin receptors in haemostasis. Nouvelle Revue Francaise D Hematologie 34:61-65, 1992.

95. Hindriks G, Ijsseldijk MJW, Sonnenberg A, Sixma JJ, de Groot PG: Platelet adhesion to laminin - role of Ca^{2+} and Mg^{2+} ions, shear rate, and platelet membrane glycoproteins. Blood 79:928-935, 1992.

96. Baenziger N, Brodie G, Majerus P: Isolation and properties of a thrombin-sensitive protein of human platelets. J Biol Chem 247:2723-2731, 1972.

97. Tuszynski GP, Rothman VL, Murphy A, Siegler K, Smith L, Smith S, Karczewski J, Knudsen KA: Thrombospondin promotes cell-substratum adhesion. Science 236:1570-1573, 1987.

98. Lahav J: Thrombospondin inhibits adhesion of platelets to glass and protein-covered substrata. Blood 71:1096-1099, 1988.

99. Agbanyo FR, Sixma JJ, de Groot PG, Languino LR, Plow EF: Thrombospondin platelet interactions - role of divalent cations, wall shear rate, and platelet membrane glycoproteins. J Clin Invest 92:288-296, 1993.

100. Collins WED, Mosher F, Tomasini Br, Cooper SL: A preliminary comparison of the thrombogenic activity of vitronectin and other RGD-containing proteins when bound to surfaces. Ann N Y Acad Sci 516:291-292, 1988.

16

PLATELET-BIOMATERIAL INTERACTIONS

Thomas Chandy[2] and Gundu. H. R. Rao[1,2]
Laboratory Medicine and Pathology[1] and
Biomedical Engineering Institute[2]
University of Minnesota
Box: 609, 420 Delaware ST. SE.
Minneapolis, MN 55455
USA

16.1 INTRODUCTION

There has been a long standing interest in thrombus formation on foreign surfaces and this interest has been enhanced by the growth of biomaterials and the increased use of such materials in clinical applications involving the contact of blood with artificial surfaces. It is generally believed that a clear understanding of the blood response to artificial surfaces would promote a better utilization of existing materials and the development of new materials with improved properties.

Dr. Thomas Chandy is a research associate in the Division of Chemical Engineering & Material Sciences, Biomedical Engineering Institute and Interventional Cardiology Laboratories at the University of Minnesota. He has over two decades research experience at Sri Chitra Tirunal Institute for Medical Sciences & Technology , Trivandrum, India, in the area of biomaterial surface engineering and blood biomaterial interactions. More recently, Dr. Chandy and Dr. Rao have focused their research on platelet biomaterial interactions and development of assist devices for cardiovascular applications. They continue to be active in this newly evolving area of research.

In many blood-contacting biomaterial applications, such as catheters, artificial heart valves, vascular grafts, hemodialysers and oxygenators, the adhesion of platelets to material surfaces leads to thrombus formation and embolization, which in turn can cause the failure of the implant. Upon exposure of biomaterial to blood, a film of plasma proteins is immediately deposited on the surface, making the foreign material attractive for either platelet adhesion and subsequent platelet aggregation and/or promotion of endothelial cell growth. Platelet adhesion to surfaces occurs, in various steps including initial attachment, spreading, release of granule contents, and platelet aggregation. Fibrinogen and other adhesive proteins, such as fibronectin, von Willebrand Factor (vWF) and Ig G mediates platelet attachment and spreading through their integrin or non-integrin receptors on the platelet surface. In clinical application, foreign surfaces represent only one of the factors influencing the blood response, with the response also influenced by the presence of antithrombotic agents, the blood condition and the nature of the application. A brief overview on the current concept of platelet-biomaterial interactions are presented in this chapter.

16.2 Contribution of platelets to thrombus formation

The adhesion of circulating platelets at sites of injury on the vessel wall where subendothelium, in particular collagen, is exposed to the flowing blood, is known to be an important step in a sequence of events leading to the formation of a thrombus and perhaps also its subsequent incorporation into an atherosclerotic plaque(1,2). Circulating platelets rapidly adhere to the subendothelial matrix, become activated spread, release the contents of various storage organelles and become cohesive, leading to formation of an aggregate or plug of platelets that prevents blood loss from the damaged vessel. This initial event in hemostasis and thrombosis is the contact adhesion of blood platelets to vessel wall subendothelial matrix at high shear flow (3). Whether thrombus formation is akin to hemostasis where tissue damage initiates platelet deposition, or other factors present in the plasma provide the initial stimulus is unknown.

Hemostatic plugs and arterial thrombi are largely composed of platelet aggregates adherent to the injury site, with fibrin in and around the platelets acting to stabilize the plug or thrombus (4). Platelet that interact with the subendothelium change from their normal disc shape to a more round form, extend pseudopods, and become adherent to the surface (4,5). Alternatively, they may release their granule contents into the surrounding medium, which include ADP, ATP, calcium, serotonin, platelet factors, B-thromboglobulin, fibrinogen, fibronectin etc.(6,7).

There are several ways in which platelets can contribute towards the coagulation process. First, the platelet surface protects active clotting factors from inactivating by their natural inhibitors (8). Second, platelets contribute through aggregation, during which platelet phospholipids (platelet factor 3) are made available and stimulate activation of factor X and the conversion of prothrombin to thrombin (9). Finally, during the release reaction fibrinogen, platelet factor 4 and a variety of proaggregatory compounds are made available. The amount of released fibrinogen is small in relation to plasma fibrinogen level (10). However, the exact role of this newly released fibrinogen in platelet/biomaterial interaction and the subsequent promotion of thrombus growth is not clear.

16.3 Platelet adhesion on biomaterials

Activation of platelets by contact with artificial surfaces is a key event in the thromboembolic complications of prosthetic devices in contact with the blood (11,12), but the exact mechanism of these events is not fully understood. It is known that a film of plasma protein adsorbs on artificial materials exposed to blood and that this event proceeds interaction of the surface with blood cells (12,13). The composition of the protein film and the configuration of its molecular constituents must relay to the blood elements information describing the nature of the underlying surface.

Platelet adhesion to surfaces occurs in various steps including initial attachment, spreading, release of granule contents, and platelet aggregation (14). These changes are collectively termed surface-induced platelet activation. During the process of platelet activation on surfaces, platelets continuously change shape while the platelet ventral membrane makes transient surface contacts. Platelet adhesion and shape change are mediated by cytoskeletal reorganization, as observed by high voltager electron microscopy (11,15). It is also believed that platelets release their granule contents into the surrounding medium , then providing proaggregatory compounds for the promotion of platelet aggregate formation (15).

Platelets are seen to adhere rapidly to various surfaces upon contact with blood, either as a monolayer or as aggregates and this process is influenced by a combination of factors including surface smoothness, surface charge, wettability, surface tension and the flow conditions (16). Baier et al (17) studied the effect of surface energetics of substrates towards platelet attachment, their spreading and correlated towards bloodcompatibility. He suggested that minimal platelet spread areas were found on substrata with critical surface tension between 20 and 30 dynes/cm and increased spreading and morphological changes occured on substrata of both higher and lower critical surface tension. So the adhesion and spreading of platelets to artificial substrates have upmost importance to predict the subsequent process of thrombus formation at the interface.

16.4 Role of plasma proteins on platelet adhesion

Immediately following implantation the prosthetic surface is exposed to flowing blood, and plasma proteins are adsorbed on the surface of the artificial matrix (6). The adsorption of three specific plasma proteins, fibrinogen, Ig G and albumin on artificial surfaces has been well documented (13). The proteins are rapidly deposited on artificial surfaces and are partially replaced over time by high-molecular weight kininogen, and hageman factor (18). Ziats et al (19) have shown that fibronectin, hemoglobin and von Willebrand factor are also deposited on artificial surfaces exposed to flowing blood. The platelet-surface interactions that can be described as consisting of three consecutive events (20). 1). An initial platelet-to-surface adhesion occurs between platelet surface receptors and adsorbed matrix proteins. The information provided by the interactive domains of cell matrix components dictate degree of activation. For instance , laminin induces only focal adhesion, whereas, fibronectin promotes spreading and collagen supports aggregation and secretion. 2). This adhesion results in platelet activation, with subsequent degranulation and release of platelet activating factors. 3). Thrombus formation results from the deposition of additional blood elements and recruitment of platelets.

16.4.1 Role of fibrinogen

Fibrinogen appears to be an important component of the adsorbed protein film and may be an essential player in the early events of platelet activation (4,21), adsorbing prefentially on many surfaces in a concentration relatively higher than in the bulk plasma and participating in the platelet response to the prosthetic surface (21,22). Fibrinogen is recognized as an indispensible element in platelet aggregation. An early feature of activation of platelets by soluble agonists such as ADP or thrombin is exposure of specific membrane glycoproteins (GPIIb-IIIa), to which fibrinogen molecule bind with high affinity (23). It should be noted that fibrinogen dissolved in plasma does not induce platelet aggregation but rather acts as a cofactor in the process, once platelet GP11b/111a receptors are activated. Fibrinogen normally circulates in peaceful coexistence with unactivated platelets without any obvious interaction (24).

It is claimed that platelets have membrane receptors of low affinity for fibrinogen (23,24), which would permit reversible interactions at physiologic fibrinogen concentrations, but the prevailing view is that it is only in the "activated" platelet (ie, the platelet stimulated by agonists such as thrombin or ADP) that fibrinogen forms a detactable complex with its glycoprotein membrane receptors, which are "masked" or otherwise unavailable in intact nonactivated circulating platelets(25). Surface bound fibrinogen is involved in platelet adhesion to surfaces by bridging the gap between platelet and surface in a manner analogous to its action in platelet-platelet aggregates. It has been suggested that fibrinogen molecules adsorbed on artificial surfaces undergo conformational changes and interact with unactivated platelets to promote their adhesion on the surface. They may even be responsible for the progression from platelet adhesion to aggregation and secretion (6,26).

Vroman et al (28) extensively studied the initial phase of blood-surface interactions. They have postulated that on hydrophilic surfaces (eg. Glass), Fg is deposited within seconds along with traces of high molecular weight kininogen (HMWK) and factor XII. Then more HMWK and factor XII arrive and displace Fg. Platelets adhere most, where fibrinogen remains. Thus, Fg plays a key role in platelet-surface attachment and subsequent thrombus formation.

16.4.2 Role of other cell adhesive proteins

The platelet receptor-endothelial matrix protein interactions have been well characterized. Under high shear conditions, this attachment is mediated by the binding of von Willebrand Factor (vWF) to the glycoprotein receptor GP1b-1X (27). Secondary attachment between platelets results in thrombus formation, which is caused by Fg or vWF-mediated crosslinking of GP11b-111a receptors on adjacent platelets(28). Several other matrix protein ligands such as laminin, fibronectin, collagen and their associated platelet receptors have been shown to contribute to platelet adhesion and aggregation to the subendothelium.

While the receptors involved in platelet adhesion to the subendothelium have been studied extensively, the receptors involved in platelet adhesion to prosthetic materials are less well identified. The biomaterials adsorb a layer of plasma proteins on their surface upon exposure to blood. The content of this protein layer varies with time and induces Fg, Ig G, Fn, vWF and vibronectin (18,19) etc.. For which there are specific surface receptors on platelet surface (28).

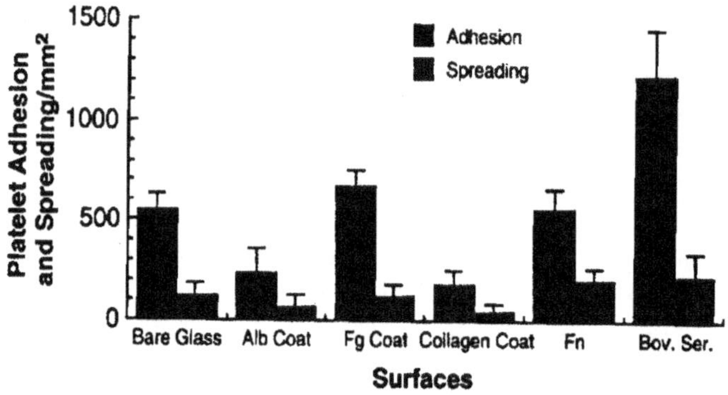

Fig: 1. Platelet adhesion and spreading to protein coated glass, for 15 minutes exposure to washed platelets t $1000S^{-1}$ shear.

We have monitored platelet attachment and spreading on glass and various protein coated glass, under shear ($1000S^{-1}$) with washed platelets, platelet rich plasma (PRP) and whole blood using fluorescence optimas imaging system and software (30). Platelet adhesion and spreading to protein coated glass, for 15 minutes exposure to whole blood, at $1000S^{-1}$ shear is represented in figure 1. Platelet adhesion and spreading has been higher with Fg, Fn and vWF coated glass and least on albuminated surfaces. Platelet spreading is quantitated through platelet size, where greater than 3 u M (diameter) are considered as spread cells. The binding of Fg and other platelet adhesive proteins to a surface may induce conformational changes in the proteins , which could result in higher affinity binding of these proteins with platelet adhesion receptors. Thus the observed changes in platelet adhesion and spreading to albuminated glass and other adhesive protein coated glass may be due to the status and availability of platelets and other protein molecules in solution at the interface.

The photomicrographs of platelet adhesion on protein coated glass, after exposing with washed cells and whole blood, under $1000S^{-1}$ shear, for 15 minutes are shown in figures 2 and 3 respectively.

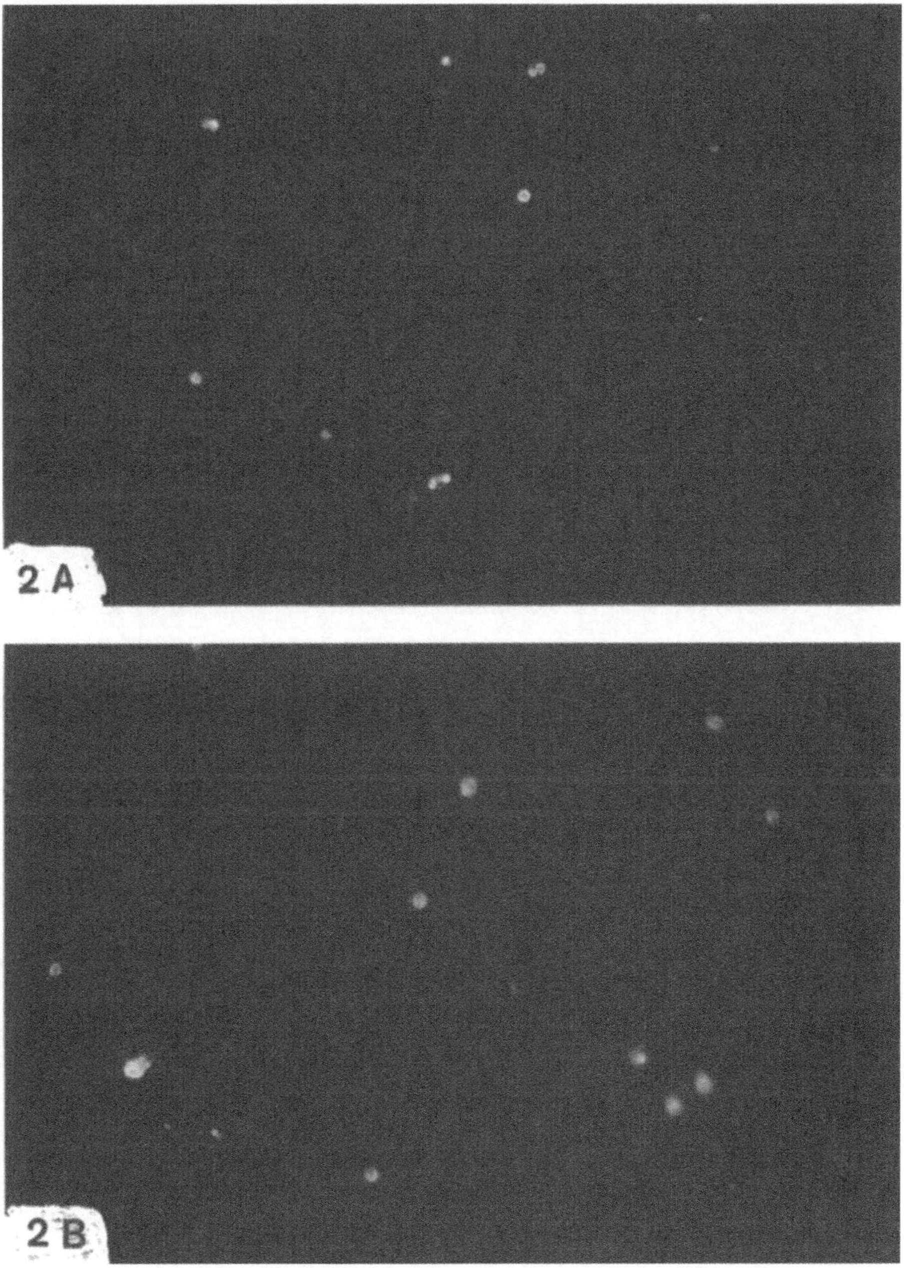

Fig:2. Photomicrograph of the platelet adhesion to Glass (A) and Fg coated glass (B) for 15 minutes from washed cells under $1000S^{-1}$ shear.

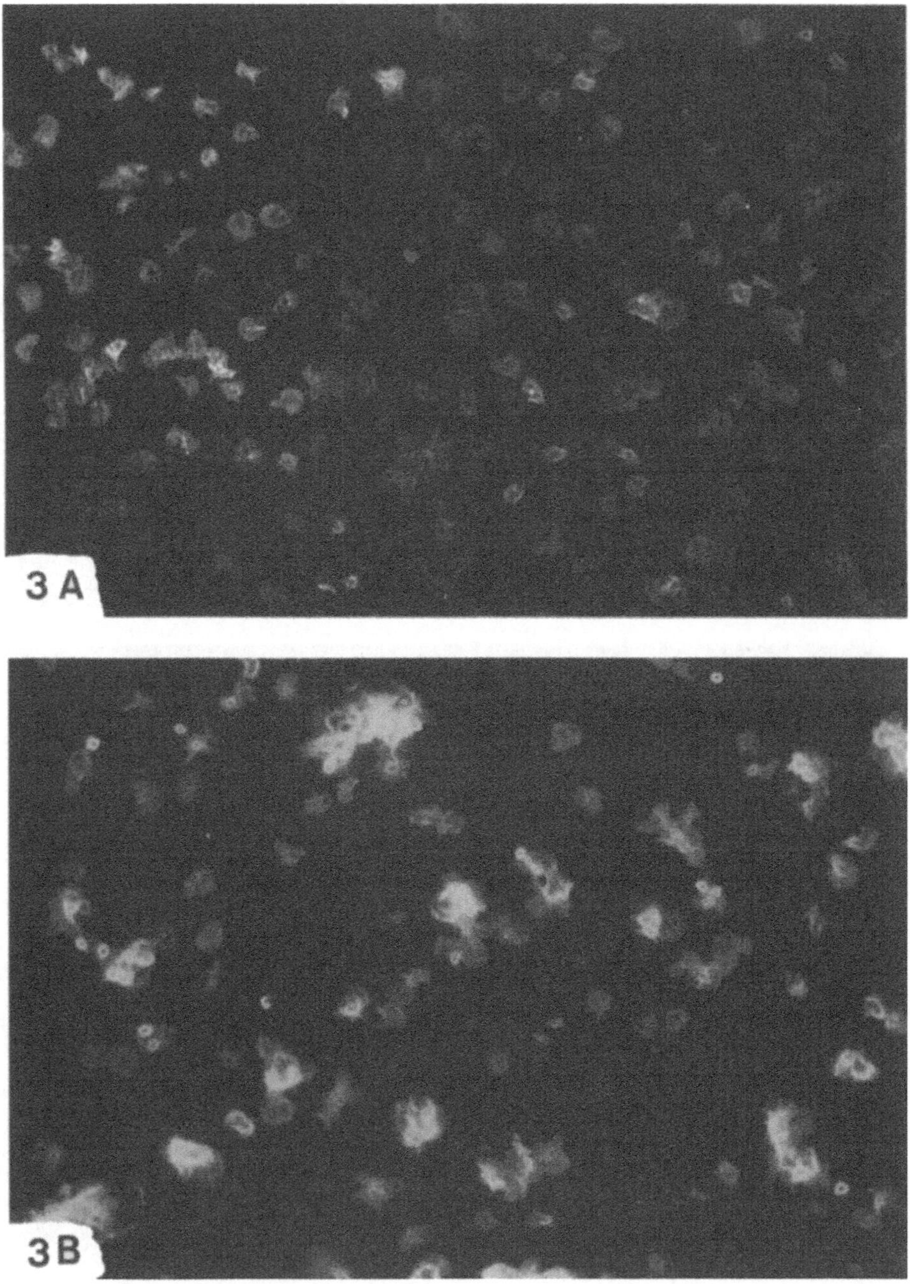

Fig:3. Photomicrograph of the platelet adhesion to glass (A) and Fg coated glass (B) for 15 minutes from whole blood under $1000S^{-1}$ shear.

Goodman etal (31) have suggested that GP11b/111a may be involved in the spreading process of platelet-fibrinogen interaction. Because the adhesive plasma proteins in general possess RGD binding domain which interacts with GP11b-111a (Fg, Fn and perhaps von Willebrand factor), and induce actin polymerization and receptor-cytoskeleton tethering, as occurs when Fg binds GP11b-111a (32). Based upon these observations, it is reasonable to speculate how integrin receptors could influence subsequent platelet spreading on synthetic surfaces. The cells spread differently to material surfaces, which had been modified with denatured and conformationally altered plasma proteins. Because platelet surfaces are largely glycoprotein-rich, one possibility is that adsorption-induced alterations in the secondary structure of these proteins somehow leads to cellular response. The plasma proteins like Fg and vWF are unable to bind their respective receptors when in solution, yet once adsorbed to surfaces the proteins become "active" and are capable of receptor-ligand coupling (31,33).

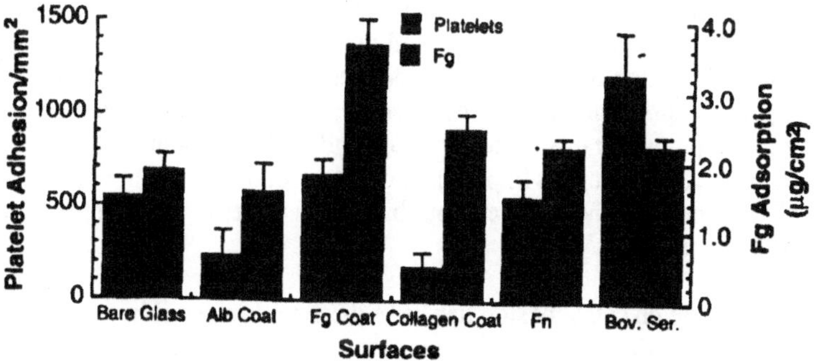

Fig:4. Fibrinogen adsorption from plasma vs platelet adhesion from blood to protein coated glass for 15 minutes exposure at 1000S-1 shear.

Fibrinogen adsorption from plasma and platelet attachment from blood to various protein coated glass substrates are measured as a function of exposure time, under shear ($1000S^{-1}$). It appears from figure 4 that Fg-surface binding is maximum on fibrinogen precoated glass, followed by other platelet adhesive proteins and least on albuminated surfaces. However, it is likely that loosely adsorbed albumin molecules are displaced by plasma to enhance its surface concentration. These results suggest that there is no direct correlation with Fg surface attachment and platelet adhesion (Figs 2-4). Nit all bound Fg may be in a state in which platelet receptors can interact. Platelet GP11b/111a can only recognize the bound fibrinogen that has undergone appropriate conformational changes and has the RGD site available for interaction with the receptor (34). Studies of Lindon et al (35) have shown that platelet adhesion is related to the amount of antibody-recognizable fibrinogen domains rather than total amount of Fg on biomaterials. Thus, it is reasonable to speculate that the surface properties affects the conformation of adsorbed Fg and other platelet-adhesive protein molecules in a variety of ways and thereby influences subsequent platelet adhesion and spreading to a different degree (36).

Receptors expressed on the platelet surface have been shown to mediate platelet binding to immobilized proteins. Fg binds to the platelet-receptor glycoprotein (GP)11b-111a (25,28), Ig G binds to the platelet Fc r R11 receptor (37), vWF binds to receptors GP11b-111a and Gp 1b-1X (27), and Fn binds to GP 1c-11a (32). Of the plasma proteins that are deposited on artificial surfaces, albumin and hemoglobin do not appear to bind to specific platelet GP receptors (38). The present observations (Figs 1-4) have concluded that platelet adhesion and spreading appears to be higher on surfaces coated with cell adhesive proteins such as vWF, Fn, collagen and Fg. On albumin coated glass, platelet adhesion is less, when exposed with washed cells (contains no plasma proteins), but the spreading and attachment appear to increase on these substrates on exposure to PRP or whole blood. These results indicate that multiple adhesion receptors can mediate platelet adhesion and spreading to matrix proteins immobilized on biomaterial surfaces.

16.5 Effect of shear on platelet-surface interaction

The in vitro study of the hemocompatibility of biomaterials requires the consideration of many parameters, static or dynamic contact, flow rate, wall shear rate, form of biomaterial to be tested, pathway to consider (platelet adhesion, platelet activation, complement activation, contact phase activation etc..) and duration of contact(39). It has previously been demonstrated that hemodynamic circumstances play a significant role in determining localization, growth and fragmentation of thrombi and platelet adhesion in vivo, and that flow rate controls platelet transport to a surface and their adhesion (40). This evidence is supported by observed differences in platelet activity predominance in venous and arterial flow (41). Clearly, defining the blood compatibility of a material is a compromise between a number of these factors.

The effect of flow rate on cell adhesion has been discussed in detail by several researchers in terms of shear rate (42,43). Generally increased flow rate diminished platelet adhesion until some critical shear rate is reached at which erythrocyte hemolysis occurs. The release of hemoglobin then appears to promote increased cell adhesion (44). The previous reports of the effects of fluid mechanical trauma, low to moderate levels of shear stress (100 to 300 dyn/cm^2) has produced altered polymorphonuclear neutrophil leukocyte morphology and function (44). Thus the shear related changes include an increased adhesion, aggregation and cell lysis, moderate decrease in chemiluminescence accompanying phagocytosis, and substantial loss of lysosomal enzymes. These changes also cause an increased platelet-platelet sticking and the formation of microemboli. Studies of Rhodes et al (39) have indicated that platelet adhesion and activation ancreased with increasing flow rate to various polymeric surfaces. A greater cell adhesion at high flow rate is as expected, since platelet transport to a surface is greater at high flow rates, a greater number of collisions results in a greater degree of adhesion. Thus the hemodynamic parameters have great influence on platelet-surface interaction and subsequent thrombus formation.

16.6 Role of erythrocytes and white cells on platelet-biomaterial interactions

Red cell-surface interactions may play a major role in the dynamics of protein and platelet behaviour at the interface. The effect of red cells on platelet sticking has been noted widely, causing an augmentation of the rate of adhesion, probably by a combination of physical and biochemical mechanisms (40). Hellem et al (45) have observed that red cell ghosts restore retension of platelets in column almost as effectively as whole red cells, which may be due to the presence of ADP. The role of blood cells and their interaction process with an artificial surface to modulate protein-surface binding and subsequent platelet attachment have been indicated (46). The adsorption of albumin to the polymer surface has negative influence on adhesion of all blood cells, while the Fg-surface attachment enhances adhesion of blood cells (46). Brash et al (47) have indicated that the addition of red cells to buffered solutions of plasma proteins caused a decrease in the quantity of protein adsorbed to a polyethylene surface.

Vroman et al (18) have proposed that platelets adhere, where Fg remains adsorbed, and in the absence of erythrocytes, the ring of platelets deposited is smaller. So it is possible that the blod cells can cause an augmented fibrinogen adsorption to the surface, which may be cell membrane related effect, probably via the receptors. On many surfaces of dialysers, where dog blood will deposit platelets, human blood will deposit granulocytes, probably as a result of complement activation (48). It is also found that granulocytes adhere preferentially where r- globulins have been predeposited onto a hydrophobic substrate. These studies suggest that while platelets are in a race with fibrinogen/high molecular weight kininogen interactions, granulocytes are in a race with r- globulin/plasma protein interactions in a similar way. So the identification of various components deposited from formed elements of blood, to material surface and the knowledge of subsequent interactions at this modified interface towards the formation of thrombi needs to be better understood.

16.7 Platelet activation and morphological changes

Platelet adhesion to surfaces occurs in various steps including initial attachment, spreading and release of granule contents and platelet aggregation (3). During the process of such activation sequences on surfaces, platelets continuously change shape while the platelet ventral membrane makes transient surface contacts. Platelet adhesion and shape change are mediated by cytoskeletal reorganization as observed by high voltage electron microscopy (11,15). The reorganization of the platelet cytoskeleton, upon surface activation, is critical for pseudopod extension and retraction, centripetal movement of the platelet fibrinogen receptors, and the centralization and secretion of granules (15).

Scanning electron microscopy (SEM) has been used to visualize specific regions of intimate contact between the ventral surface of adherent cells and an underlying surface. Polycarbonate and silicone wafer surfaces were coated with sulphonated polyelectrolytes (SPE) to improve their blood compatibility. Figure 5 indicates the scanning electron micrographs of the adhered platelets (from PRP, 1 hour) on polycarbonate and SPE modified polycarbonate.

The platelet adhesion and spreading was higher with bare polymer (Fig: 5A), which had substantially reduced due to SPE (Fig:5B) treatments. The adhesion and spreading of platelets (from PRP, 2 hours) on silated silicone wafer was similarly inhibited by pretreatment with sulphonated polyelectrolytes (Fig:6). Regardless of treatment with SPE, all platelets adherent to polycarbonate and silicone wafer were pseudopodial (Fig: 5 & 6).

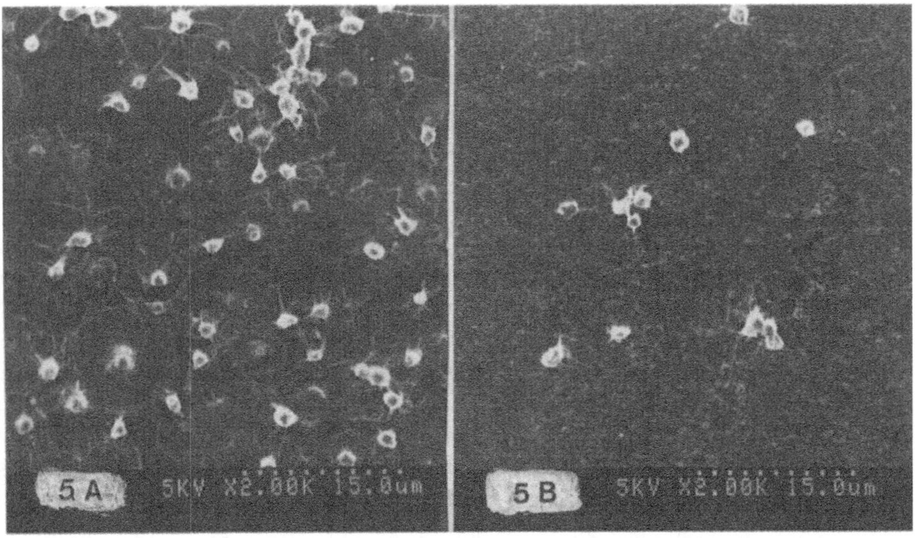

Fig:5. Scanning electron micrographs of adhered platelets to (A) Polycarbonate and (B) Polyelectrolyte coated polycarbonate, on exposure to PRP for 1 hour.

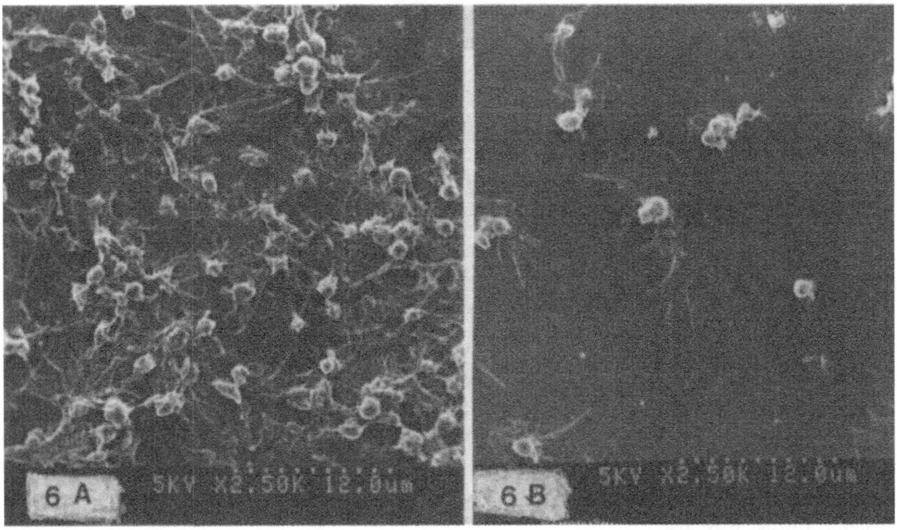

Fig:6. Scanning electron micrographs of adhered platelets to (A) Silicone wafer and (B) Polyelectrolyte coated silicone wafer, on exposure to PRP for 2 hours.

Polyelectrolytes, having sulfamate and carboxylic groups arranged in a steric manner as that of heparin, have been synthesized. These synthetic polyelectrolytes are employed on biomaterials for improving the antithrombotic properties (49). The polyelectrolyte has shown antiplatelet and anticoagulant activity. The protein-surface binding in presence of polyelectrolyte has been investigated, from a pretein mixture, using polyacrylamide gel electrophoresis (50). The synthetic polyelectrolyte has enhanced the albumin and has reduced the fibrinogen and r- globulin bands tremendously. The observation that SPE reduces platelet-surface attachment may partly be due to the reduced availability of platelet receptors for Fg on the modified material.

Polyethylene glycol (PEG) is a water soluble polymer that exhibits properties such as protein resistance, low toxicity, and immunogenicity (51). Further, it has been observed that the PEGs can abrogate the immunogenicity of proteins and are capable of preserving their biological properties. The grafting of hydrophilic polyethylene glycol at the blood-material interface can increase surface hydrophilicity, reduce protein adsorption, platelet adhesion and develop a nonthrombogenic interface (52). We have grafted PEG on bovine pericardium via glutaraldehyde linkages for improving their blood compatibility (52, 53). Scanning electron micrographs of the adhered platelets on glutaraldehyde treated pericardium (GABP) and PEG modified pericardium (PEG-GABP), on exposure to washed platelets for 30 minutes is represented in figure 7.

The platelet adhesion and spreading was substantially reduced due to PEG grafting on GABP. The adhered platelets on GABP had extented pseudopods (Fig.7A), while no morphological changes were observed to platelets adhered on PEG substrate (Fig.7B). Thus , in the absence of exogenous adhesive proteins, or if the adhesive protein affinity to substrates are removed via surface modifications, platelet adhesion to nonbiological substrates can be reduced.

16.7.1 Effect of antiplatelets on platelet-surface attachments

In clinical applications, foreign surfaces represent only one of the factors influencing the blood response. With the response also influenced by the presence of antithrombotic agents, the blood condition and the nature of the application. It is important in the clinical arena to remember that blood-material interaction depends not only on the material being used but also on constituents of the blood which may alter in disease states, ie the blood may be more hypercoagulable and platelets may be activated (54). Many of the platelet reactions which have been previously described may be altered by appropriate antithrombotic therapy, in particular by heparin, warfarin and fibrinolytic agents and it is a combination of these factors that determines the long-term viability of the procedure which is being undertaken.

Earlier studies have indicated that certain vitamins, antibiotics, antihypertensive drugs, antiplatelet agents, etc.. can modulate the protein-surface binding and subsequent cellular attachment towards an artificial substrate (50,55). It seems vitamin B_6, C, E combinations of aspirin-vit.C, certain antibiotics etc.. have inhibited the fibrinogen-surface concentration or enhanced the albumin surface-binding to variable degrees (50,55). This may be one of the parameters for a reduced platelet-surface binding in presence of these agents. So it is possible that the antiplatelets arriving as mediators at the interface may alter the sequence of proteins deposited or displaced by plasma ; and favour albumin binding in several cases,

Platelets arrive in the later stages to find Fg to adhere. In summary, the clinical use of artificial surfaces in contact with blood leads to a response influenced by the material/device, antithrombotic agents, blood condition and application.

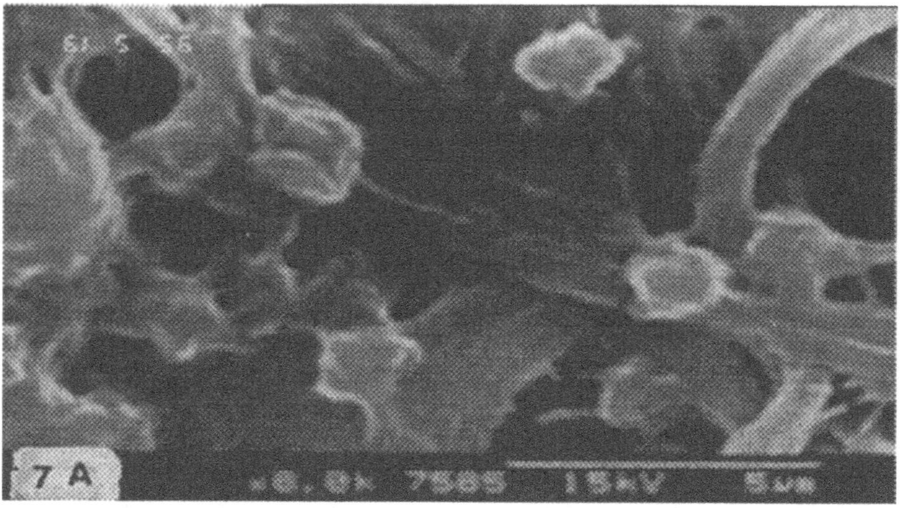

Fig:7. Scanning electron micrographs of adhered platelets to (A) Glutaraldehyde fixed bovine pericardial tissue and (B) Polyethylene glycol grafted tissue, on exposure to washed platelets for 15 minutes.

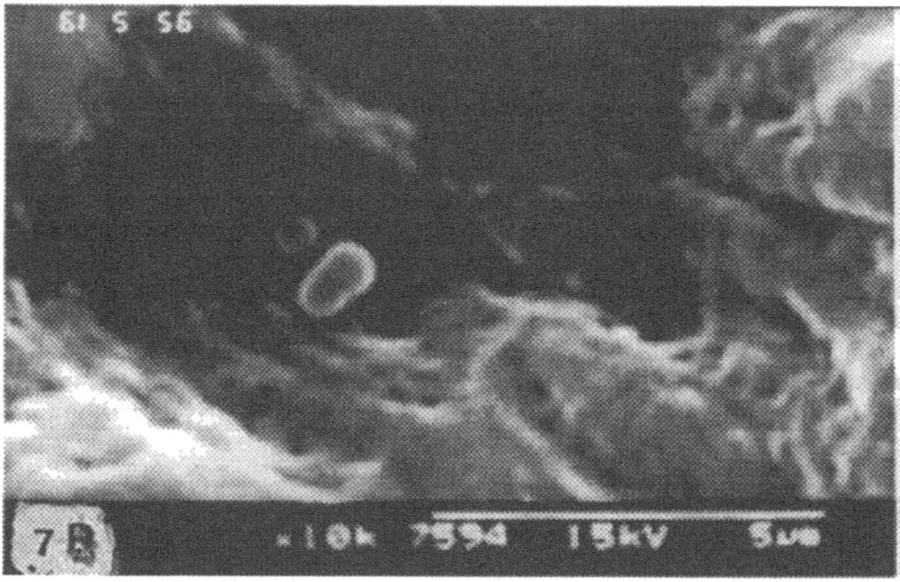

16.9 Concluding Remarks

The increasing use of biomaterials in blood-contacting applications underlines the importance of an enhanced understanding of the interactions of blood with foreign surfaces. Platelet adhesion, shape change and release their constituents upon exposure to biomaterial surfaces in flowing blood is the initial and critical reaction for in vivo thrombus formation. The activation-independent and dependent integrin receptors-glycoproteins (GP1c-11a) and GP 11b-111a - are involved in platelet adhesion and thrombus growth on biomaterial interfaces through interactions with fibrinogen, fibronectin, vWF and other adhesive proteins. More detailed studies are needed to understand the chemical changes in polymers due to the use of new antithrombotic agents to suppress platelet-surface interactions and also coagulation for long-term acceptability of new clinical devices.

Acknowledgements

The authors thank Dr. Yuanyuan Zhang for some of the SEM photomicrographs and Dr. Mahadev Murthy for technical assistance in the preparation of this manuscript. Authors also thank Departments of Laboratory Medicine and Pathology and Biomedical Engineering Institute.

References

1. Mustard. J. F and Packham. M. A, The role of blood and platelets in atherosclerosis and the complications of atherosclerosis, Thromb. Diathes. Haemorrh., 33, 444-456 (1975).

2. Karino. T and Goldsmith .H.L, Adhesion of human platelets to collagen on the walls distal to a tubular expansion, Microvas. Research, 17, 238-262 (1979).

3. Baumgartner. H.R, Muggli. R, Tschopp. T.B, and Turitto. V.T, Platelet adhesion, release and aggregation in flowing blood: Effects of surface properties and platelet function, Thromb. Haemost., 35, 124-138 (1976).

4. Packham. M.A and Mustard. J.F, in "Blood platelet function and Medicinal Chemistry", Lasso. A, (ed), New York, Elsevier Biomedical, 1984, pp. 61-128.

5. Chesterman. C.N and Berndt. M.C, Platelet and vessel wall interaction and the genesis of atherosclerosis, Clin. Haematol, 15, 323-353 (1986).

6. Salzman. E.W, Lindon. J, Brier. D and Merrill. E.W, Surface -induced platelet adhesion, aggregation and release, Ann. N.Y. Acad. Sci., 283,114-124 (1977).

7. Holmsen. H, Day. H. J, and Stormorken. H, The blood platelet release reaction, Scand. J. Haematol, Supp., 8, 220-225 (1969).

8. Jackson. C. M and Nemerson. Y, Blood coagulation, Ann. Rev. Biochem., 49, 765-811 (1980).

9. Chesney. C. M, Pifer. D.D, Byers. L.W, and Muirhead. E. E, Effect of platelet-activating factor (PAF) on human platelets, Blood, 59, 582-585 (1982).

10. White. A. M, and Heptinstall. S, Contribution of platelets to thrombus formation, Br. Med. Bull., 34, 123-128 (1978).

11. Park. K, and Park. H, Application of video-enhanced interference reflection microscopy to the study of platelet-surface interactions, Scann. Microscopy. Supplement, 3, 137-146 (1989).

12. Goodan. S.L, Cooper. S.L, and Albrecht. R.M, The effect of substrate-adsorbed albumin on platelet spreading, J. Biomater. Sci. Polym. Ed., 2, 147-159 (1990).

13. Feuerstein. I.A and Sheppard. J. I, States in adherent platelet morphology and the processing of adsorbed protein on biomaterials, Biomaterials, 14, 137-146 (1993).

14. Friedman. L.I, Liem. H, Grabowski. E.F, Leonard. E.F and McCord. C.W, Inconsequentiality of surface properties for initial platelet adhesion, Trans. Am. Soc. Artif. Intern. Organs, 16, 63-73 (1970).

15. Lewis. J.C, White. M.S, Prater. T, Porter. K.R and Steele. R.J, Cytoskeletal changes during adhesion and release: a comparison of human and nonhuman primate platelets., Scan. Electron. Microsc., 1, 199-209 (1986).

16. Mohammad. S.F, Hardison. M.D, Glenn. C.H, Morton. B. D, Bolan. J.C and Mason. R.G, Adhesion of blood platelets to glass and polymer surfaces, Haemostasis, 3, 257-270 (1974).

17. Baier. R.E, Depalma. V.A, Goupil. D. W and Cohen. E, Human platelet spreading on substrata of known surface chemistry, J. Biomed, Mater, Res., 19, 1157-1167 (1985).

18. Vroman. L, Adams. A.I, Fischer. G. C, Munoz. P.C and Stanford. M, Proteins, plasma and blood in narrow spaces of clot promoting surfaces, Adv. Chem. Ser., 199, 266-276 (1982).

19. Ziats. N.P, Pankowsky. D.A, Tierney. B.P, Ratnoff. O.D and Anderson. J. M, Absorption of Hageman factor (factor XII) and other human plasma proteins to biomedical polymers, J. Lab. Clin. Med., 116, 687-693 (1990).

20. Difazio. L.T, Stratoulias. C, Greco. R.S and Haimovich. B, Multiple platelet surface receptors mediate platelet adhesion to surafce scoated with plasma proteins, J. of Surgical Res., 57, 133-137 (1994).

21. Brash. J.L and Suniyal. S, Dependence of albumin-fibrinogen simple and competitive adsorption on surface properties of biomaterials, J. Polym. Sci., 66, 377-389 (1979).

22. Chandy. T and Sharma. C.P, Changes in protein adsorption on polycarbonate-due to L-ascorbic acid, Biomaterials, 6, 416-420 (1985).

23. Bailly. A.L, Laurent. A, Lu. H, Elalami. I, Jacob. P, Mundler. O, Merland. J.J, Lautier. A, Soria. J and Soria. C, Fibrinogen binding and platelet retention: relationship with the thrombogenicity of catheters, J. Biomed. Mater. Res., 30, 101-108 (1996).

24. Peerschke. E.I, Zucker. M.B, Grant. R.A, Egan. J.J and Johnson. M.M, Correlation between fibrinogen binding to human platelets and platelet aggregability, Blood, 55, 841-847 (1980).

25. Marguerie. G. A, Plow. E. F, and Edgungton. T.S, Human platelets possess an inducible and saturable receptor specific for fibrinogen, J. Biol. Chem., 254, 5357-5363 (1979).

26. Tomikawa. M, Iwamoto,.M, Olsson. P, Soderman. S and Blomback. B, On the platelet-fibrinogen interaction, Thromb. Res., 19, 869-876 (1980).

27. Fressinaud. E, Baruch. D, Girma. J.P, Sakariassen. K.S, Baumgartner. H. R and Meyer. D, Von Willebrand factor-mediated platelet adhesion to collagen involves platelet membrane 11b/111a as wel l as glycoprotein 1b., J. Lab. Clin. Med., 112, 58-66 (1988).

28. Andrews. R.K and Fox. J.E.B, Platelet receptors in hemostasis, Curr. Opin. Cell. Biol., 2, 894-903 (1991).

29. Vroman. L, Adams. A. L, Fischer. G.C and Munoz. P.C, Interaction of high molecular weight kininogen, factor XII and fibrinogen in plasma at interfaces, Blood, 55, 156-168(1980).

30. Chandy. T, Mooradian. D. L and Rao. G.H.R, Platelet adhesion and sprewading on protein coated surfaces: Variations in behavior in washed cells, PRP and whole blood, J. Biomat. Applications (In Press, 1998).

31. Goodman. S. L, Cooper. S.L and Albrecht. R.M, Integrin receptors and platelet adhesion to synthetic surfaces, J. Biomed. Mater. Res., 27, 683-695 (1993).

32. Philips. D.R, Jennings, L.K and Edwards. H.H, Identification of membrane proteins mediating the interactions of human platelets, J. Cell. Biol, 86, 77-86 (1980).

33. Roth. G.J, Developing relationships: Arterial platelet adhesion, glycoprotein 1b and leucine-rich glycoproteins, Blood, 77, 5-19 (1991).

34. Kloczewiak. M, Timmons. S, Bednarek. M.A, Sakon. M and Hawiger. J, Platelet receptor recognition domain on the -chain of human fibrinogen and its synthetic peptide analogues, Biochemistry, 28, 2915-2919 (1989).

35. Lindon. J.N, McManama. G, Kushner. L, Merrill. E.W and Salzman. E.W, Does the conformation of adsorbed fibrinogen dictate platelet interactions, Blood, 68, 355-362 (1986).

36. Park. K, Mosher. D.E and Cooper. S.L, Acute surface-induced thrombosis in the canine ex vivo model: Importance of protein composition of the initial monolayer and platlet activation, J. Biomed. Mater. Res., 20, 589-612 (1986).

37. Rosenfeld. S.I, Leddy. L. J.P, Abraham. G.N and Anderson. C. L, Human platelet Fc receptor for immunoglobulin G. J. Clin. Inves., 76, 2317-2326 (1985).

38. Ryu. G, Han. D, Kim. Y and Min. B, Albumin immobilized polyurethane and its blood compatibility, ASAIO. J, 38, M644-M650 (1992).

39. Rhodes. N.P, Kumary. T.V and Williams. D.F, Influence of wall shear rate on parameters of blood compatibility of intravascular catheters, Biomaterials, 17, 1995-2002 (1996).

40. Goldsmith. H.L, The flow of model particles and blood cells and its relation to thrombogenesis, Prog. Haemost., Thromb., 1, 97-127 (1972).

41. Harker. L.A and Slichter. S. J, Arterial and venous thromboembolism: Kinetic characterization and evaluation of therapy, Thromb. Diath. Haemorrh. 31, 188- 203 (1974).

42. Absolom. D.R, Zingg.w, VanOss. C.J and Newmann. A.W, Protein and platelet interactions with polymer surfaces, Biomater. Med. Devices. Artif. Organs, Sharma. C.P (ed), 12, 235-266 (1984-85).

43. Salzman. E.W, Role of platelets in blood-surface interactions, Fed. Proc., 30, 1503-1509 (1971).

44. Stewart. G.J, Lynch. P.R, reichle. F.A, Ritchie. W.G.M, Smith.A and Schaub. R.G, The adhesion of leukocytes, erythrocytes and noncellular material to the luminal surface of natural and artificial blood vessels, in vivo, Ann. N.Y. Acad. Sci., 283, 179-207 (1977).

45. Hellem. A.J, The adhesiveness of human blood platelets in vitro, Scand, J. Clin. Lab. Invest., 12, 51-60 (1960).

46. Sharma. C.P and Chandy. T, Influence of L-ascorbic acid, blood cells and components on protein adsorption/desorption on polycarbonate, Haemostasis, 17, 70-78 (1987).

47. Brash. J.L and Uniyal. S, Adsorption of albumin and fibrinogen to polyethylene in presence of red cells, Trans. Am. Soc. Artif. Intern. Organs, 22, 253-259 (1976).

48. Flaherty. J.T.O, Craddock. P.R, and Jacob. H.S, Effect of intravascular complement activation on granulocyte adhesiveness and distribution, Blood, 51, 731-739 (1978).

49. Sharma. C.P , Chandy. T and Ashalatha. P.V, Surface modification of polycarbonate with synthetic polyelectrolyte-anticoagulant activity, Biomater. Med. Devices. Artif. Organs, Sharma. C.P (ed), 12 215-233 (1984-85).

50. Chandy.T and Sharma. C.P, Protein-polymer interaction- Changes with plasma components, vitamins and antiplatelet drugs at the interface, Polym. Plast. Tech. Engg., 26, 143-227 (1987).

51. Harris. J.M, Polyethylene glycol: Biotechnical and Biomedical applications, Plenum Press, New York, 1992.

52. Vasudev. S.C, Chandy.T and Sharma. C. P, Influence of polyethylene glycol graftings on the in vitro degradation and calcification of bovine pericardium, J of Biomat. Applications, 11, 430-452 (1997).

53. Vasudev. S.C, and Chandy. T, Effect of alternative crosslinking techniques on the enzymatic degradation of bovine pericardia and their calcification, J. Biomed. Mater. Res., 35, 357-369 (1997).

54. Courtney. J.M and Forbes. C.D, Thrombosis on foreign surfaces, Brit. Medical Bulletin, 4, 966-981 (1994).

55. Chandy. T and Sharma. C. P, Changes in albumin-platelet interaction with an artificial surface due to antibiotics, pyridoxal phosphate and lymphocytes, Artif. Organs, 12, 143-151 (1988).

17

COMPARATIVE PHYSIOLOGY OF PLATELETS FROM DIFFERENT SPECIES

Douglas J. Weiss, DVM, Ph. D.
Department of Pathobiology
School of Veterinary Sciences
University of Minnesota
St. Paul. MN
55108, USA

17.1 INTRODUCTION

The mammalian blood platelet contributes to hemostasis as a result of its ability to adhere to exposed subendothelial collagen and to respond to stimuli, including, thromobin and adenosine diphosphate, with formation of platelet aggregates. In general, platelets from companion and food-producing animals, and rodents have similar morphology but vary considerably in size (Clemmons et al. 1983). Platelet volume varies from large (15.1 fl) in cats, to intermediate (7.6 to 8.3 fl) in pig and dog, to small (3.2 to 5.4 fl) in cattle, horse, sheep, rat, guinea pig, and mouse (Jain 1993). With few exceptions, platelets of various species contain a plasma membrane, circumfrential microtubular system, mitochondria, alpha granules, dense bodies, glycogen granules, and an open canalicular system. However, ruminant platelets lack an organized microtubular system

Dr. Douglas J. Weiss is a Professor at the College of Veterinary Medicine at the University of Minnesota. Over the years, he has been extensively involved in both teaching and research at the College. He has received a number of awards for his teaching. His research interests include platelet activation mechanisms in thrombotic diseases, mechanisms of acute lung injury and mucosal inflammation. He has published extensively on his research.

and an open canalicular system. Species differences in the number and relative abundance of membrane glycoproteins occur. Unlike human platelets, that contain 7 major glycoproteins, canine, porcine, bovine, and rabbit platelets contain only 3 or 4 major glycoproteins and feline platelets contain only 2. Feline platelets lack the glycoprotein I band found in all other species (Nurden et al. 1977). This variation in glycoprotein number and relative abundance may, in part, explain the species-dependent variation in adhesion, and in aggregation responses to various agonists (Sinakos et al. 1977).

Species difference occur in aggregation and secretion responses to various agonist. However, all species appear to possess similar activation and metabolic pathways and differences are due to the relative importance of each pathway in the activation process. In general, ruminants have relatively unreactive platelets whereas, carnivores and horses have more reactive platelets. Adenosine diphosphate (ADP), thrombin, and collagen are universal agonist but response to epinephrine, arachidonic acid, serotonin, 5-hydroxytryptamine, and platelet-activating factor (PAF) are species dependent. Considerable variation exists in the inportance of thromboxane in the activation of platelets. Although the arachidonic acid pathway is present in platelet of all species and results in thromboxane production, species-dependent differences in activity of the pathway exists. The pathway is most prominent in humans, but is poorly developed in bovine, mink, and pig platelets. Activated dog platelets produce high concentrations of thromboxane while ruminant platelets produce low concentrations (McKeller et al. 1990). There are also species differences in the requirement for extracellular calcium in thrombin-induced dene granule secretion. Extracellular calcium is not required for secretion of human and bovine platelets, whereas it is required for secretion of pig, rabbit, and rat platelets (Meyers 1986).

This chapter will review and contrast platelet physiology of various animal species. Additionally, where it contributes to understanding physiologic responses of platelets, platelet function defects will be discussed. Specific species discussed include horse, cow, sheep, goat, dog, cat, pig, rabbit, rat, mouse, Guinea pig, and bird.

17.2 HORSE

17.2.1 Platelet structure

The ultrastructure of horse platelets is similar to that of human platelets (White et al. 1976). Features include a flattened discoid shape, a circumfrential band of microtubules, an open canalicular system, alpha granules, mitochondria, dense bodies, and glycogen particles. Compared to human platelets, horse dense bodies are smaller and alpha granules are larger and structurally more complex. The serotonin content of horse platelets is similar to that of human platelets. Horse platelet membrane proteins have been evaluated by polyacrylamide gel electrophoresis and glycoprotein IIbIIIa has been characterized and monoclonal antibodies have been produced (Lipscomb et al. 1995; Pintado et al. 1995).

17.2.2 Adhesion

Adhesion of horse platelets to subendothelial collagen has been investigated using the Baumgartner perfusion method (Weiss et al. 1990). Both the total surface area of subendothelium covered by platelets and the surface area covered by platelet thrombi were greater than that for human platelets. These data indicate that horse platelets are highly reactive with subendothelium. Despite this, bleeding times for horses are 3 to 4 times longer than those of other species (Weiss et al. 1997a).

17.2.3 Aggregation

Horse and pony platelets readily aggregate in response to thrombin, collagen, and ADP (White et al. 1976, Weiss et al. 1990, Boudreaux et al. 1988; Meyers et al. 1979). As in humans, thrombin produces a rapid irreversible aggregation response. Collagen produces a delay phase followed by shape change and a slow irreversible aggregation phase. Low concentration of ADP cause reversible aggregation, whereas, high concentrations produces irreversible aggregation responses. Newborn foals are less responsive to ADP and collagen compared to adult horses but responses increase over the first week of life. Horse platelet fail to aggregate in response to epinephrine and epinephrine does not enhanse the response to other agonists. Weak reversible aggregation responses have been reported for serotonin and arachadonic acid (Meyers 1979), however, another paper reported irreversible aggregation with serotonin (White et al. 1976). Platelet-activating factor is a potent aggregating agent for horse platelets. Horse platelets aggregate at much lower PAF concentrations (3.58×10^{-16}M) than other species including sheep, rabbit, rat, guinea pig, and dog (Shinozaki et al. 1992; Suquet and Leid 1983). The arachidonic acid pathway does not appear to be involved in PAF-induced platelet aggregation (Suquet and Leid 1983).

Endotoxin induces horse platelet aggregation, however, the effect of endotoxin is indirect requiring the presence of neutrophils (Jarvis et al. 1994). The leukocyte-dependent initiator of endotoxin-induced platelet aggregation is PAF (Jarvis et al. 1996).

Equine platelet aggregation responses appears to be less dependent on thromboxane generation. Blocking the cyclooxygenase pathway with aspirin slightly decreases the aggregation response to collagen but has no effect on aggregation in response to other agonists (Heath et al. 1994; Weiss et al. 1997). Further, the amount of thromboxane released by activated equine platelets is much less than that released by other species and arachidonic acid is a weak reversible agonist for equine platelets (McKeller et al. 1990). Activation of horse platelets results in early activation of phospholipase C (Lapetina et al 1979; Lapetina 1982). Phospholipase C induces formation of diacylglycerol, however, diacylglycerol is largely converted to phosphatadic acid.

17.2.4 Secretion

Horse platelets contain a large nonmetabolic pool of serotonin and adenine nucleotides within dense bodies (White et al. 1976). Calcium ionophore (A23187) and high doses of thrombin (0.4 units/ml), but not collagen, ADP, or arachadonic acid, induced secretion of ATP and/or ^{14}C-serotonin (Weiss et al. 1990, Boudreaux et al. 1988).

17.2.5 Pathophysiologic considerations

Equine platelets appear to be highly reactive in vivo, however, the bleeding time for horses is 3 to 4 times longer than that of other species. The horse suffers from several thrombotic diseases including thromboembolic colic and iliac thromboembolism. Horses also readily develop thrombosis associated indwelling intravenous catheters. Microvascular thrombosis has been associated with endotoxemia and acute laminitis (Weiss et al. 1994; Weiss et al. 1995; Duncan et al. 1985). A platelet function defect has been proposed as a cause of pulmonary hemorrhage associated with intense exercise (i.e. exercise-induced pulmonary hemorrhage) but its existence remains controversial (White et al. 1987; Weiss et al. 1990; Johnstone et al. 1991).

Intravascular platelet activation with formation of platelet-neutrophil aggregates has been documented in the prodromal stages of acute laminitis and during intense exercise (Weiss et al. 1997; Weiss et al. 1998, Weiss et al. in press). Activated platelets appear to be involved in the pathogenesis of laminitis because administration of a peptide that competitively blocks fibrinogen binding to glycoprotein IIbIIIa prevents onset of experimental laminitis (Weiss et al. in press). Formation of platelet-neutrophil aggregates during intense exercise may result from shear-induced platelet activation (unpublished observation).

A platelet function disorder, termed thrombasthenia, has been described in a Thoroughbred foal (Miura et al. 1987). The disorder is characterized by bleeding tendencies, absence of clot retraction, and lack of aggregation in response to ADP, collagen, and thrombin.

17.3 RUMINANTS

17.3.1 Platelet structure

Large ruminant platelets, including cattle, goats, and sheep, are similar in structure and function. Bovine platelets have been studied most extensively. Resting bovine platelets are discoid in shape and vary between 1 and 5 μm in diameter. Bovine platelets contain a few large alpha granules and dense bodies and have a dense tubular system, but lacks an open canalicular system (Zucker-Franklin et al. 1985; White 1987). Lack of an open canalicular system probably is responsible for altered function of bovine platelets. Surface-activated bovine platelets produce pseudopods but do not spread as do platelets of other species (Grouse et al. 1990). Further, glycoprotein IIbIIIa molecules do not concentrate in the center of surface-activated bovine platelet as they do in human platelets.

17.3.2 Adhesion

Bovine platelets adhere to subendothelial collagen less readily than do platelets of other species. The combination of reduced adhesiveness and inability to spread suggests that ruminant platelets are inherently less functional than are platelets of other species. Some investigators have hypothesized that ruminant platelets have evolved so that they can function as inflammatory cells, by releasing granule contents, without acting as a potent thrombogenic agent (Cheryk et al. 1997).

17.3.3 Aggregation

In general, ruminant platelets are less responsive to agonists compared to platelets of other species (Bondy and Gentry 1989). Bovine platelets are more sensitive to stimulation by PAF than by ADP or thrombin. ADP-induced aggregation does not induce dense granule secretion. Epinephrine, arachadonic acid, and serotonin do not aggregate bovine platelets but do enhance the aggregation in response of other agonists (Meyers et al. 1980). Bovine platelets produce minimal amounts of malondialdehyde and thromboxane B_2 in response to thrombin. Additionally, incubation of bovine platelets with aspirin or thromboxane A_2 receptor antagonists do not inhibit platelet aggregation in response to ADP but incubation with calcium channel blockers do inhibit ADP-induced aggregation. Therefore, bovine platelet activation is more dependent on calcium mobilization than on arachidonic acid metabolism. Activated bovine and ovine platelets form relatively small amount of arachadonic acid (Meyers 1986).

17.3.4 Pathophysiologic considerations

Despite the fact that large ruminant platelets are relatively less functional than those of other species, platelet-related bleeding is a rare occurrence in ruminants. Alternatively, thrombotic disorders are also infrequently documented in cattle.

A familial platelet function defect has been documented in Simmental cattle (Searcy et al. 1990). Affected cattle have spontaneous epistaxis, and hematomas and hemorrhage following injury. Platelets from affected cattle do not aggregate in response to ADP, collagen, or calcium ionophore. The defect may result from delayed expression of functional fibrinogen binding sites on activated platelets (Frojmovic et al. 1996).

Chediak-Higashi syndrome has been reported to occur as an autosomal recessive trait in Hereford, Brangus, and Japanese black cattle. Giant lysosomal granules occur in both platelets and megakaryocytes (Menard et al 1990). Affected cattle have hemorrhagic tendencies and contain significantly reduced quantities of ADP, serotonin, calcium, and magnesium. Platelets are essentially devoid of dense granules.

17.4 DOG

17.4.1 Platelet structure

The ultrastructural features of dog platelets are similar to those of human platelets (Jain 1993).

17.4.2 Aggregation

Platelet aggregation responses have been studied using platelet-rich plasma and whole blood (Thomas 1996). Aggregation in response to ADP was less in whole blood than in platelet-rich plasma, but maximum aggregation responses to collagen and PAF were greater in whole blood. Aggregation responses of dog platelets to ADP and collagen are similar to those of human platelets, however, the lag phase for collagen-induced aggregation is 2 to 3 times longer for dog platelets (Feingold et al. 1986). Dog platelets are less responsive to arachidonic acid and epinephrine than are human platelets and do not respond to ristocetin. Platelets from some dogs aggregate irreversibly in response to arachidonic acid while platelets from other dogs do not aggregate or aggregate reversibly (Meyers 1986). The greyhound breed has a high incidence reversible aggregation in response to arachidonic acid while English terriers and Scottish terriers have a high incidence of irreversible aggregate (Meyers 1986). Platelets from dogs that aggregate reversibly in response to arachidonic acid aggregate irreversibly after exposure to epinephrine. High concentrations of arachidonic acid inhibit canine platelet activation.

17.4.3 Secretion

Dog platelets readily secrete granule contents in response to agonists including ADP, collagen, thrombin, and PAF (Boudreaux et al. 1994; Thomas 1996).

17.4.4 Pathophysiologic considerations

Several congenital defects in platelet function have been described in dogs. Three of the disorders resemble Glanzmann's thrombasthenia. Glanzmann's thrombasthenia is an intrinsic defect of human platelets which results from marked reduction or altered function of glycoprotein IIbIIIa. Platelets aggregate poorly in response to all agonists because aggregation is dependent on binding of fibrinogen to glycoprotein IibIIIa.

Thrombasthenic thrombopathia in Otterhounds, is similar to Type I Glanzmann's thrombasthenia. The disorder results from a marked reduction in the b_3 subunit of glycoprotein IIbIIIa (Bordeaux 1996). Unlike Glanzmann's thrombasthenia, 30 to 80% of platelets consist of large bizarre forms resembling platelets seen in Bernard-Soulier syndrome. However, these platelets have increased numbers of glycoprotein Ib rather than decreased numbers as is seen in Bernard-Soulier syndrome.

In Basset hound hereditary thrombopathy, platelets change shape but do not aggregate in response to ADP, PAF, collagen, thromboxane, and low concentrations of thrombin (McConnell et al. 1995). Platelet from affected dogs also have defective contact activation in plasma (Estry et al. 1995). A defect in glycoprotein IIbIIIa was inferred by reduced capacity of platelets from affected dogs to agglutinate fibrinogen-coated beads after stimulation with high doses of thrombin or calcium ionophore (Catalfamo et al. 1986). However, a more recent report describes normal fibrinogen binding and receptor distribution on adherent, spread platelets (Estry et al 1995).

Type I Glanzmann's thrombasthenia has been described in a female Great pyrenees (Boudreaux et al. 1996a). Platelets have minimal aggregation in response to ADP, collagen, thrombin, and PAF, and clot retraction is severely impaired. Surface expression of glycoprotein IIbIIIa is markedly reduced. Therefore, this disorder closely resembles human type I Glanzmann's thrombasthenia.

Spitz thrombopathia has been described in two female Spitz dogs. Platelets fail to aggregate in response to ADP, collagen, or PAF but do respond to thrombin after a lag phase (Boudreaux et al. 1994).

A inherited storage pool disease has been described in American cocker spaniels (Callan et al. 1995). Platelet aggregation responses to ADP and collagen were abnormal in 3 of 5 dogs studied. The ADP content of platelets is decreased but ATP content is normal. The number and morphology of dense granules are similar to those of unaffected dogs.

Several types of von Willebrand's disease have been described in more than 30 dog breeds (Johnson et al. 1988). Counterparts of all 3 major types of the human disease have been recognized in dogs. In Doberman pinschers and Airedales, the disease resembles human type I von Willebrand's disease. The prevalence of von Willebrand's disease in Doberman pinschers in North America has been estimated to be 58 to 65% (Dodds et al. 1981). von Willebrand's factor can be detected by autoradiographic methods in all affected Dobermans but concentrations may be as low as 4%. Spontaneous bleeding episodes are relatively rare in affected dogs. The prevalence of von Willebrand's disease in Airedales is similar to that in Doberman pinschers, however, no reports of abnormal bleeding have been reported in the breed (Johnson et al. 1988). von Willebrand's disease in German Shorthair Pointers resembles human type II von Willebrand's disease. The disease in Scottish terriers and Chesapeake Bay retrievers resembles human type III von Willebrand's disease (Meyers et al. 1972). Scottish terriers that are homozygous for the trait have severe bleeding episodes and very prolonged bleeding times. von Willebrand's factor cannot in detected in blood even with sensitive autoradiographic techniques. Factor VIII coagulant activity is 15 to 50% of normal.

Acquired platelet function defects have been associated with several canine diseases. Decreased platelet aggregation responses have been reported for dog with liver disease, pancreatitis, myeloproliferative disease, and ehrlichiosis (Harrus et al. 1996; Willis et al. 1989).

17.5 CAT

17.5.1 Platelet structure

Cats have large platelets with a mean platelet volume of 15.1 fl. These large platelets overlap in size with erythrocytes making automated platelet counts subject to error (Tschopp 1970). Additionally, platelet aggregates are frequently encountered in blood samples collected for routine hematologic tests.

17.5.2 Aggregation

Cat platelets tend to aggregate spontaneously in citrate-anticoagulated platelet-rich plasma (Tschopp 1970). Spontaneous aggregation was hypothesized to result from release of serotonin from platelets (Tschopp 1970).

Cat platelets readily aggregate in response to ADP, collagen, thrombin, serotonin, and arachadonic acid, and frequently exhibit a biphasic response (Meyers 1986). A biphasic response to ADP is consistently seen. Cat platelets generate relatively large amounts of thromboxane B_2 when exposed to arachidonic acid and consistently aggregate irreversibly (Meyers et al. 1981). Arachidonic acid-induced aggregation appears to be independent of dense granule secretion.

Serotonin is a strong agonist for cat platelets and induces a biphasic aggregation response (Tschopp 1970). Response to serotonin is coupled to formation of thromboxane B_2 and dense granule secretion. The serotonin receptor on cat platelets binds more serotonin than do platelets other species (Leysen et al. 1983). Epinephrine potentiates aggregation responses to other agonists.

17.5.3 Pathophysiologic considerations

Chediak-Higashi syndrome has been reported in smoke gray Persian cats (Meyers 1981). Bleeding times are markedly prolonged in affected cats but whole blood clotting time is normal. Affected cats have reduced amounts of stored ADP, ATP, serotonin, magnesium, and calcium (Meyers 1981). Serotonin and ADP are virtually absent from platelets of affected cats. Platelets aggregate in response to ADP, collagen, and serotonin but the amount of agonist required is considerably higher than for normal cat platelets.

Acquired alterations on platelet function have been detected in cats. Taurine deficient cats have decreased ADP-induced platelet aggregation and release reaction (Welles et al. 1993). Cats with cardiomyopathy and those infected with feline infectious peritonitis have increased platelet aggregation responses (Helenski et al. 1987; Boudreaux et al 1990).

17.6 PIG

17.6.1 Platelet structure

Pig platelets have most of the structural elements seen in platelets of other species including circumfrential bands of microtubules, an open canalicular system, glycogen particles, and mitochondria. Dense bodies are rare. Dense bodies contain significant amounts of histamine.

17.6.2 Adhesion

More than 70% of pig platelets are retained after one passage through a glass bead column (Fass et al. 1976). Normal bleeding time is 2.1 ± 0.7 minutes (Fass et al. 1976).

17.6.3 Aggregation

Aggregation of pig platelets in response to ADP is reversible, whereas, collagen-induced aggregation is irreversible (Meyers 1986). Pig platelets fail to aggregate in response to thrombin and arachidonic acid. Pig platelets release relatively small amounts of malondialdehyde and thromboxane B_2 in response to thrombin but release large amounts of magnesium (Meyers et al. 1980). Epinephrine does not potentate aggregation in response to any agonists.

17.6.4 Secretion

Pig platelets secrete contents of dense granules in response to collagen but not ADP, or epinephrine (Addonizio et al. 1978).

17.6.5 Pathophysiologic considerations

Porcine von Willebrand's disease has been studied extensively (Owens et al. 1974). The disease is characterized by severe bleeding tendencies, markedly prolonged bleeding time, decreased factor VIII coagulant activity and von Willebrand's factor, decreased platelet adhesiveness to glass beads, and failure to aggregate in response to ristocetin. The von Willebrand antigen is virtually undetectable in pig with the severe form of the disease (Fass et al. 1976). Pigs that are heterozygous for von Willebrand's trait have approximately 60% reduction in von Willebrand antigen.

A platelet storage pool deficiency has been reported in a colony of pigs with von Willebrand's disease and in heterozygous carriers (Daniels et al. 1986). Platelet aggregation in response to collagen is reduced in most affected pigs. Dense bodies are markedly reduced in number and total platelet content of ADP, ATP, and serotonin is decreased.

17.7 RABBIT

17.7.1 Platelet structure

Rabbit platelets are small ranging from 1 and 3 μm in diameter. The ultrastructure of rabbit platelets resembles that of humans and other species.

17.7.1 Aggregation

Rabbit platelets aggregate in response to ADP, collagen, thrombin, arachadonic acid, and PAF (Meyers 1986). Rabbit platelets aggregate reversibly in response to ADP and do not secrete dense granule contents. ADP-induced aggregation appears to be independent of arachidonic acid metabolism. Rabbit platelets generate large amounts of thromboxane B_2 when exposed to arachidonic acid and aggregate irreversibly. Platelet-activating factor is a potent agonist for rabbit platelets, whereas, serotonin is a weak agonist. Epinephrine potentiates aggregation in response to other agonists.

17.7.2 Secretion

Dense granules contain a large nonmetabolic pool of serotonin and histamine and large amounts of serotonin, ATP, and magnesium are released from activatied platelets. Endotoxin induces clumping and secretion of rabbit platelets (Morrison et al. 1978)

17.8 RAT AND MOUSE

17.8.1 Platelet structure

Rat and mouse platelets contain circumfrential microtubules, dense bodies, alpha granules, glycogen particles, and an open canalicular system. Ultrastructural features of mouse and rat platelets are similar to that of human platelets.

17.8.2 Adhesion

Platelet retention in glass bead columns for Wistar rat platelets is similar to that for human platelets.

17.8.3 Aggregation

Rat and mouse platelets aggregate more vigorously when heparin is used as the anticoagulant than when citrate is used (Sinakos and Caen 1967; Rosenblum et al. 1983). Collagen and arachidonic acid induce aggregation in heparin-anticoagulated platelet-rich plasma but not in citrated-anticoagulated platelet-rich plasma. ADP induced greater aggregation in heparinized platelet-rich plasma (Dwyer and Meyers 1986).

Platelet aggregation responses vary between stains of rats. In general, rat platelets aggregate well with ADP, moderately well with arachidonic acid, and hardly at all with collagen, thrombin, and serotonin. Additionally, serotonin does not potentiate responses to other agonists. Aggregation responses to arachidonic acid is irreversible probably because rat platelets generate relatively large amounts of thromboxane B_2 in response to arachidonic acid. Unlike human, pig, and rabbit platelets, adenosine monophosphate does not inhibit rat platelet aggregation and ATP induces rat platelet aggregation (Meyers 1986). Rat and mouse platelets do not respond to PAF.

Mouse platelets aggregate in response to arachidonate, thrombin, collagen, and ADP and undergo release reaction in response to arachidonate, collagen, and thrombin (Rosenblum et al. 1983).

17.8.4 Pathophysiologic considerations

A hereditary platelet function defect occurs in fawn-hooded rats (Tschopp and Zucker 1972). The defect is characterized by prolonged bleeding time, reduced platelet retention in glass bead columns, normal aggregation in response to ADP, failure to aggregate or release ^{14}C-serotonin in response to collagen, failure to release ADP and ATP in response to thrombin, and reduced concentration of ADP, ATP, and serotonin in dense granules. Therefore, the disorder resembles human storage pool deficiency.

Chediak-Higashi-like syndrome has been described in beige mice (Holland 1976). Larged granules occur in many cell types including pituitary, kidney, spleen, scent glands, iris, pancreas, liver, adrenal cortex, and gastric and duodenal mucosa. Beige mice have a mild bleeding disorder characterized by prolonged bleeding time that is corrected by administration of serotonin. Beige mouse platelets lack dense granules. Platelet concentrations of ADP, ATP, and serotonin are significantly reduced and uptake and binding of exogenous serotonin by beige mouse platelets is also impaired.

17.9 GUINEA PIG

17.9.1 Platelet structure

Guinea pig platelets contain circumfrential microtubules, dense bodies, alpha granules, glycogen particles and an open canalicular system. Ultrastructural features of guinea pig platelets are similar to that of human platelets.

17.9.2 Adhesion

Platelet glass retention indices for Guinea pig platelets is similar to that for human platelets.

17.9.3 Aggregation

Guinea pig platelets aggregate readily in response to ADP and consistently have a biphasic response. ATP induces rather than inhibits platelet aggregation. The response probably results from conversion of ATP to ADP (Meyers 1986). Platelet aggregation responses to collagen and thrombin are poor and platelets do not respond to serotonin. Arachidonic acid consistently produces irreversible aggregation. Epinephrine does not potentate aggregation in response to other agonists

17.9.4 AVIAN

Platelet structure

Bird and reptile thrombocytes are larger than mammalian platelets and are nucleated (Maxwell 1974). Thrombocytes are derived from mononuclear precursors in the bone marrow. Avian thrombocytes are spindle-shaped and contain large oval nuclei with condensed nuclear chromatin. The cytoplasm contain many small vacuoles that probably represent an open canalicular system. Few organelles are apparent in avian thrombocyte. A few glycogen granules, mitochondria, and dense bodies have been observed. The Golgi apparatus is well developed. Microtubules are present only at the poles of the cells. Avian thrombocytes are involved in hemostasis but may also function in clearing the blood of foreign material in that they actively ingest colloidal carbon and kill bacteria (Grecchi et al. 1980).

Aggregation

Avian thrombocytes aggregate poorly in response to ADP, collagen, arachidonic acid, and thrombin.

CONCLUSIONS

With few exceptions, platelets of all species have similar ultrastructural features and contain similar metabolic pathways. Therefore, species-dependent differences in platelet function tend to be quantitative rather than quantitative. Platelets from larger ruminants, including cattle, sheep, and goats, do not contain an open canalicular system and appear to be less functional than platelets from other species. Platelet function may be less important to these species because they are not frequently exposed to trauma in the wild. Alternatively, carnivores are regularly exposed to trauma in the process of hunting and defending themselves. These species tend to have highly functional platelets and also can enhance platelet function at times of excitement or fear as a result of epinephrine release.

When compared to humans, none of the animal species are predisposed to thrombotic disorders or arterial vascular pathology. Recent studies of in vivo platelet activation and platelet-dependent thrombosis provide evidence that the relationships between intrinsically highly active platelets and thrombotic diseases is not coincidental.

References

Addonizio VP, Edmunds LH, Coleman RW. The function of monkey (M. mulatta) platelets compared to platelets of pig, sheep, and man. J Lab Clin Med 1978;91:989-997

Bondy GS, Gentry PA. Characteristics of the normal bovine platelet aggregation response. Comp Biochem Physiol 1989;92C:67-72

Boudreaux MK, Wagner-Mann C, Purohit R, Hankes G, Spano J, Pablo L, Lee S, Conti J. Platelet function testing in the pony. Lab Animal Sci 1988;38:448-451

Bordeaux MK, Weiss RC, Toivio-Kinnucan, Cox N, Spano JS. Enhanced platelet reactivity in cats experimentally infected with feline infectious peritonitis virus. Vet Pathol 1990;27:269-273

Boudreaux MK, Crager C, Dillon AR, Stanz K, Toivio-Kinnucan M. Identification of an intrinsic platelet function defects in Spitz dogs. J Vet Intern Med 1994;8:93-98

Boudreaux MK. Platelets and coagulation: an update. Vet Clin North Am 1996a;26:1065-1083

Boudreaux MK, Kvam K, Dillon AR, Bourne C, Scott M, Schwartz KA, Toivio-Kinnucan. Type I Glanzman's thrombasthenia in a Great Pyrinees dog. Vet Pathol 1996;33:503-511

Callan MB, Bennett JS, Phillips DK, Haskins ME, Hayden JE, Anderson JG, Giger U. Inherited platelet delta-storage pool disease causing severe bleeding: An animal model for a specific ADP deficiency. Thrombosis and Haemostasis 1995;74:949-953

Catalfamo JL, Raymond SL, White JG, Dodds J. Defective platelet-fibrinogen interaction in hereditary canine thrombopathy. Blood 1986;67:1568-1577

Cheryk LA, Gentry PA, Tablin F. Morphological characteristics of bovine platelets activated with platelet-activating factor or thrombin. Comparat Haematol Inter 1997;7:88-95

Clemmons RM, Bliss EL, Dorsey-Lee MR, Deachord CL, Meyers KM. Platelet function, size, and yeild in whole blood and in platelet-rich plasma prepared uising differing centrifugation force and time in domestic and food-producing animals. Thrombo Haemostas 1983;50:838-843

Daniels TD, Fass DN, White JG, Bowie EJW. Platelet storage pool deficiency in pigs. Blood 1986;67:1043-1047

Dodds WJ, Moynihan AC, Fischer TM. The frequency of inherited blood and eye diseases as determined by genetic screening programs. J Am Anim Hosp Assoc 1981;17:697-704

Duncan SG, Meyers KM, Reed SM, Grant B. Alterations in coagulation and hemograms of horses given endotoxin for 24 hours via hepatic portal infusion. Am J Vet Res 1985;46:1287-1292

Dwyer SD, Meyers KM. Anesthetics and anticoagulants used in the preparation of rat platelet-rich plasma alter rat platelet aggregation. Thrombosis Res 1986;42:139-151

Estry DW, Mattson JC, Oesterle JR, Mahoney GJ, Patterson WR. Bassett hound hereditary thrombopathy - an inherited disorder with defective platelet aggregation despite normal fibrinogen binding and receptor mobility. Compar Haematol Inter 1995;5:227-236

Fass DN, Brockway WJ, Owen CA, Bowie EJW. Factor VIII (Willebrand) antigen and ristocetin-Willebrand factor in pigs with von Willebrand's disease. Thrombosis Res 1976;8:319-327

Feingold HM, Pivacek LE, Melaragno AJ, Valeri R. Coagulation assays and platelet aggregation patterns in human, baboon, and canine blood. Am J Vet Res 1986;47:2197-2199

Frojmovic MM, Wong T, Searcy GP. Platelets from bleeding Simmental cattle have a long delay in both ADP-activated expression of GPIIbIIIa receptors and fibrinogen-dependent platelet aggregation. Thrombosis and Haemostasis 1996;76:1047-1052

Grouse LH, Rao GH, Weiss DJ, Perman V, White JG. Surface-activated bovine platelets do not spread, they unfold. Am J Pathol 1990;136:399-408

Harrus S, Waner T, Eldor A, Zwang E, Bark H. Platelet dysfunction associated with experimental acute canine ehrlichiosis. Vet Rec 1996;139:290-293

Heath MF, Evans RJ, Poole AW, Hayes LJ, McEvoy RJ, Littler RM. The effects of aspirin and paracxetamol on the aggregation of equine blood platelets. J Vet Pharmacol Therap 1994;17:374-378

Helenski CA, Ross JN. Platelet aggregation in feline cardiomyopathy. J Vet Intern Med 1987;1:24-28

Holland JM. Serotonin deficiency and prolonged bleeding time in beige mice (39137). Proc Soc Exp Biol Med 1976;151:32-39

Jain, Nemi. *Essentials of Veterinary Hematology.* Philadelphia: Lea & Febiger, 1993.

Jarvis GE, Evans RJ. Endotoxin-induced platelet aggregation in heparinized equine whole blood in vitro. Res Vet Sci 1994;57:317-324

Jarvis GE, Evans RJ. Platelet-activating factor and not thromboxane A (2) is an important mediator of endotoxin-induced platelet aggregation in equine heparinized whole blood in vitro. Blood Coag & Fibrinolysis 1996;7:194-198

Johnstone IB, Viel L, Crane S, Whiting T. Hemostatic studies in racing standardbred horses with exercise-induced pulmonary hemorrhage. Hemostatic parameters at rest and after moderate exercise. Can J Vet Res 1991;55:101-106

Johnson GS, Turrentine MA, Kraus KH. Canine von Willebrand's disease: a heterogeneous group of bleeding disorders. Vet Clin North Amer 1988;18:195-229

Lapetina EG, Cuatrecasas P. Stimulation of phosphatadic acid production in platelets produces the formation of arachidonate and parallels the release of serotonin. Biochem Biophys Acta 1979;573:394-401

Lapetina EG. Regulation of arachidonic acid production: role of phospholipase C and A2. Trends Pharmacol 1982;115-119

Leysen JE, Gommeren W, DeClerck F. Demonstration of S2-receptor binding sites on cat blood platelets using [3H] Kanserin. Euro J Pharmacol 1983;88:125-134

Maxwell MH. An ultrastructural comparison of the mononuclear leukocytes and thrombocytes in six species of domestic bird. J Anat 1974;117:69-80.

McKeller QA, Nolan AM, Galbraith EA. Serum thromboxane generation by platelets in several domestic species. Brit Vet J 1990;146:398-404

Menard M, Meyers KM, Prieur DJ. Primary and secondary lysosomes in megakaryocytes and platelets of cattle with the Chediak-Higashi syndrome. Throm Haemostasis 1990;64:156-160

Meyers KM, Pierce KR, Gowing GM. Hemorrhagic diathesis resembling pseudohemophilia in a dog. 1972;161:1028-1029

Meyers KM, Katz JB, Clemmons RM, Smith JB, Holmsen H. An evaluation of the arachadonate pathway of platelets from companion and food-producing animals, mink, and man. Thrombosis Res 1980;20:13-24

Meyers KM, Seachord CL, Holmsen H, Prieur DJ. Evaluation of the platelet storage pool deficiency in the feline counterpart of Chediak-Higashi syndrome. Am J Hematol 1981;11:241-248

Meyers KM. "Species Differences" in Platelet Responses and Metabolism, H Holmsen, ed. Boca Raton Fl: CRC Press pp 209-227, 1986.

Meyers KM, Lindner C, Grant B. Characterization of the equine platelet aggregation response. Am J Vet Res 1979;40:260-264

Meyers, Kenneth and Wardrop, K. Jane. "Platelets and Coagulation." In *Comparative Transfusion Medicine,* Susan M. Cotter, ed. New York: Academic Press, 1991.

McConnell MF, Thomas JS, Dipinto MN, Bell TG. Circumvention of the Basset hound hereditary thrombopathy by platelet activation with phorbol myristate acetate. Platelets 1995;6:131-138

Miura N, Senba H, Ogawa H, Sasaki N, Oishi H, Fumihito F, Takeuchi A, Usui K. A case of equine thrombasthenia. Jpn J Vet Sci 1987;49:155-158

Morrison DC, Kline LF, Oades ZG. Mechanisms of lipopolysaccharide-initiated rabbit platelet responses. Infect Immun 1978;20:744-751

Nurden AT, Butcher PD, Hawkey CM. Comparative studies on the glycoprotein composition of mammalian platelets. Comp Biochem Physiol 1977; 89:407-413

Owens CA, Bowie EJW, Zollman PE, Fass DN, Gordon H. Carriers of porcine von Willebrand's disease. Am J Vet Res 1974;35:245-248

Rosenblum WI, Nelson GH, Cockrell CS, Ellis EF. Some properties of mouse platelets. Thrombosis Res 1983;30:347-355

Searcy GP, Sheridan D, Dobson KA. Preliminary studies of a platelet function disorder in simmental cattle. Can J Vet Res 1990;54:394-396

Sinakos Z, Caen JP. Platelet aggregation in mammalians (human, rat, rabbit, guinea-pig, horse, dog) a comparative study. Thrombosis and Haemostasis 1967;17: 99-111

Shinozaki K, Kawasaki T, Kambayashi J, Sakon M, Shiba E, Ou M, Mori T. Species differences in platelet aggregation induced by platelet-activating factor (PAF). Meth Find Exp Clin Pharmacol 1992;14:663-665

Suequet CM, Leid RW. Aggregation of equine platelets by PAF (platelet-activating factor). Inflammation 1983;7:197-203

Thomas JS. Comparison of platelet aggregation and adenosine triphosphate secretion in whole blood and platelet-rich plasma from normal dogs. Compar Haematol Inter 1996;6:70-76

Tschopp TB. Aggregation of cat platelets in vitro. Thromb Diath Haemorrh 1970;23:601-620

Tschoff TB, Zucker MB. Hereditary defects in platelet function in rats. Blood 1972;40:217-225

Weiss DJ, Smith CM, Rao GH, White JG. Platelet function in the racing Thoroughbred: implications for exercise-induced pulmonary hemorrhage. Vet Clin Path 1990;19:35-39

Weiss DJ, Geor RJ, Johnston G, Trent A. Microvascular thrombosis associated with onset of acute laminitis in ponies. Am J Vet Res 1994;55:606-612

Weiss DJ, Trent AM, Johnston G. Prothrombotic events in the prodromal stages of acute laminitis in horses. Am J Vet Res 1995;56:986-991

Weiss DJ, Evanson OA, Wells RE. Evaluation of arginine-glycine-aspartate-containing peptides as inhibitors of equine platelets function. Am J Vet Res 1997a;58:457-460

Weiss DJ, Evanson OA, McClenahan D, Fagliari JJ. Evaluation of platelet activation and platelet-neutrophil aggregates in ponies with alimentary laminitis. Am J Vet Res 1997b;58:1376-1380

Weiss DJ, Evanson OA, Fagliari JJ, Valberg S. Evaluation of platelet activation and platelet-neutrophil aggregates in Thoroughbreds undergoing near-maximal treadmill exercise. Am J Vet Res (in press)

Weiss DJ, Evanson OA, McClenahan D, Fagliari JJ, Dunnwiddie CT, Wells RE. A competitive inhibitor of platelet aggregation prevents experimental equine laminitis. Am J Vet Res (in press)

Welles EG, Boudreaux MK, Tyler JW. Platelet antithrombin, and fibrinolytic activities in taurine-deficient and taurine-replete cats. Am J Vet Res 1993;54:1235-1242

White JG, Matlack KL, Mundschenk D, Rao GHR. "Platelet studies in normal and a bleeder horse." In *First international symposium on equine hematology*, H. Kitchen and JD Krehbiel ed. Golden Co, 1976.

White JG. The secretory pathway of bovine platelets. Blood 1987;69:878-885

Willis SE, Jackson ML, Meric SM, Rousseaux CG. Whole blood platelet aggregation in dogs with liver disease. Am J Vet Res 1989;50:1893-1896

Zucker-Franklin D Benson KA, Meyers KM. Absence of a surface-connecting canalicular system in bovine platelets. Blood 1985;65:241-244

18

THE MOLECULAR PATHOLOGY OF GLANZMANN'S THROMBASTHENIA

Deborah French, M. D.
Department of Medicine
Mount Sinai Hospital & Medical School
One Gustave L. Levy Place
New York, NY
10029, USA

18.1 INTRODUCTION

The glycoprotein (GP) IIb-IIIa receptor (integrin $a_{IIb}b_3$) is expressed at high concentrations on the platelet surface, representing approximately 15% of the total surface protein. Glanzmann thrombasthenia is a well-characterized inherited disorder of platelet GPIIb-IIIa receptors and the hallmark of this disease is severely reduced or absent platelet aggregation in response to multiple physiologic agonists. Underlying this disorder are mutations in the GPIIb or GPIIIa genes resulting in qualitative or quantitative abnormalities of the platelet membrane GPIIb (a_{IIb}: CD41) and/or GPIIIa (b_3: CD61) subunits. The identification and characterization of naturally occuring DNA mutations causing Glanzmann thrombasthenia has provided a wealth of information on the biosynthesis and structure-function relationships of the platelet GPIIb/IIIa receptor. In addition, the molecular characterization of patients with this disorder has enabled DNA-based carrier detection and prenatal diagnoses to be

The laboratory of Dr. Deborah French is focused on platelet integrin biology. She has published extensively on the molecular characterizations of mutations in Glanzmann's thrombasthenia. Dr. French has recently published a database describing the clinical, biochemical, and molecular defects in patients with Glanzmann thrombasthenia and a review characterizing the molecular mutations in this disease. In addition to her work on integrin receptors, Dr. French has expertise in the areas of immuno-globulin genes, antibody production, and matrix metalloproteinases.

performed. More than 200 individuals with Glanzmann thrombasthenia have been described in the literature and to date, 54 molecular defects have been identified in 45 patients. The identified mutations include deletions, insertions, inversions, point mutations, and splicing mutations that produce abnormalities in the GPIIb or GPIIIa subunits resulting in undetectable to normal levels of protein expression. The clinical, biochemical, and molecular information of 41 of these patients and one carrier of disease have been summarized in a published database[1] and the molecular characterization of 50 mutations identified in 43 patients has been described in a review of Glanzmann thrombasthenia[2]. This chapter describes the current total of 54 mutations that have been identified and characterized at the molecular level. The 30 GPIIb and 24 GPIIIa mutations are described in separate tables and the mutations are listed according to patient designation and exon number. Within the text, the mutational defects are described in groups according to the biochemical characterization of each molecular abnormality[2]. For additional reading, excellent reviews have been published describing the historical and clinical spectrum of Glanzmann thrombasthenia[3,4], the structure and function of the platelet GPIIb/IIIa complex[5-7], and the molecular identification of patients with Glanzmann thrombasthenia[8]

18.2 Glanzmann Thrombasthenia

The disease was originally described in 1918 by a Swiss pediatrician named Glanzmann who identified a group of patients with hemorrhagic symptoms that had normal platelet counts, but abnormal clot retraction[9]. He identified this group of disorders as "thrombasthenie" or "weak platelets". Significant findings that advanced our understanding of Glanzmann thrombasthenia came from the observations that platelets from these patients failed to spread onto a surface[10] and failed to aggregate[11,12]. The association of this disease with platelet GPIIb/IIIa receptors became apparent with the discovery that platelets from a number of patients with Glanzmann thrombasthenia were deficient in GPIIb and GPIIIa subunits[13,14]. Subsequent studies analyzing the biosynthesis of the receptor showed that both subunits were required to associate and form a surface-expressed complex that functions as a receptor for fibrinogen and other adhesive ligands[15-17]. With the identification of more patients having this disorder, a list of common diagnostic criteria have evolved which include: 1) failure of platelets to aggregate in response to common physiologic agonists such as ADP, epinephrine, collagen, and thrombin; 2) prolonged bleeding times; 3) failure of surface platelet spreading; 4) nearly undetectable to near normal levels of platelet fibrinogen; 5) absent to normal clot retraction; 6) normal platelet count, morphology and size; and 7) normal ristocetin-induced platelet agglutination.

18.2.1 Clinical features of the disease

Glanzmann thrombasthenia is a life-long disease that is characterized by mucocutaneous bleeding. This disease is often diagnosed soon after birth in which purpura can be observed at sites of excessive pressure or trauma. Epistaxis, gingival bleeding, and petechaie are common features with epistaxis being a common bleeding problem in childhood. Menorrhagia is a serious problem in female patients and normal bleeding that occurs during pregnancy, tooth extractions, severe trauma, and surgical procedures can be excessive and require transfusion. The severity of bleeding in this disease is not predictable, even among groups of patients with similar mutational defects. The occurrence of intracranial and gastrointestinal bleeding are rare events, but have been reported in some patients[3].

18.2.2 Frequency and distribution of disease

Glanzmann thrombasthenia is a rare disorder with a worldwide distribution. Although this disease does not occur with a high frequency, there is a high incidence among certain geographically restricted populations where consanguinous marriages are common. For this reason, a high carrier frequency for this disease exists among Iraqi Jews in Israel, an Arab population in Israel and Jordon, a gypsy population in France, and native groups in South India[18-22]. Carriers or obligate heterozygotes of this disease have 50-60% the normal number of GPIIb/IIIa receptors and these individuals do not have abnormalities of platelet function or clinically significant bleeding[3,20,23]. The occurrence of abnormal GPIIb and GPIIIa genes in the human gene pool is low, but to date, 17 out of the 45 patients with mutations identified at the molecular level are compound heterozygotes. Only within the past several years has this number of identified compound heterozygotes increased from two[24-27] to seventeen[1,28,29] raising the interesting question of whether these mutations arose spontaneously in the population or emerged from ancestral gene pools. Since this disease is an autosomal recessive disorder, all of the identified homozygous mutations result from consanguineous or intermarriages and a putative founder effect has been described for two of the mutant genes identified in the Iraqi-Jewish population in Israel[30].

18.2.3 Classification of the disease

The classification of this disease into subtypes has been difficult because the number of identified molecular biologic defects has been as numerous as the number of patients. However, with an increasing number of mutations characterized at the biochemical and molecular levels, groups of mutational defects are beginning to emerge. The first classification of Glanzmann thrombasthenia was proposed in 1972 by Caen and was based on platelet fibrinogen content and the ability of platelets to retract a fibrin clot[31]. The platelets of type I patients had undetectable platelet fibrinogen levels and absent clot retraction and the platelets of type II patients had detectable platelet fibrinogen levels and evidence of clot retraction. The association of this disease with platelet GPIIb/IIIa receptors became evident when Nurden and Caen[13] and Phillips et al.[14,32] discovered a deficiency of GPIIb and GPIIIa subunits in the platelets from patients with Glanzmann thrombasthenia. Observations were made that patients with type I disease had undetectable levels of GPIIb/IIIa and patients with type II disease had moderate levels of expression[33-35]. In addition, a new class of patients with a "variant" form of the disease was identified in which the patients' platelets expressed dysfunctional GPIIb/IIIa receptors that were present on the platelet surface in normal number or in the range of asymptomatic heterozygous individuals[36-38]. A classification scheme was formed in which patients were further characterized by the surface expression of GPIIb/IIIa receptors on their platelets[3]. However, with the identification and characterization of an increasing number of patients with Glanzmann thrombasthenia, the defects in fibrinogen concentration, clot retraction, and bleeding symptoms have been found to vary dramatically. These criteria no longer fit into the three traditional categories of the original classification scheme and more elaborate classifications have been devised such as one based on expression and trafficking of the GPIIb and GPIIIa subunits through the Golgi to the cell surface[39]. In this chapter, the mutations will be categorized according to the biochemistry of the resulting defect at the levels of RNA accumulation, subunit biosynthesis, and surface expression of the receptor complex[2]. In spite of the numerous differences of the mutational defects, the common thread amongst all patients with Glanzmann thrombasthenia is that their platelets do not aggregate in response to agonists such as ADP, thrombin, and epinephrine. The lack of platelet aggregation is caused by defective GPIIb/IIIa receptors that are required for mediating the binding and cross-linking to the adhesion ligands fibrinogen and von Willebrand factor.

18.3 Genetics and expression of the platelet GPIIb/IIIa receptor

The GPIIb and GPIIIa subunits are transcribed separately[40,41] from genes that have been mapped to chromosome 17q21[42,43] at reported distances of 250 kb[42] and 4 cM[44]. The gene encoding GPIIb spans ~17 kb and is comprised of 30 exons[45], whereas the gene encoding GPIIIa spans ~46 kb and is comprised of 15 exons[46,47]. The sequencing of cDNA and genomic DNA encoding GPIIb[45,48] and GPIIIa[46,47,49] identified them as separate subunits that are members of the integrin family of adhesion receptors[50,51].

The GPIIb (a_{IIb}) subunit is within the group of integrin a-chains that are synthesized as propeptides and processed into heavy and light chains in the Golgi complex[52]. ProGPIIb has a molecular weight of approximately 145,000 and the mature heavy and light chains have molecular weights of approximately 120,000 and 25,000, respectively. The GPIIb subunit contains 18 cysteine residues in which the nearest-neighbor-cysteines have been arranged into eight short-range intradisulfide bonds and one interdisulfide bond linking the heavy and light chains of the mature GPIIb subunit[53]. GPIIb, like most integrin a-chain subunits, contains four calcium-binding domains that have similar sequences[48,54] and characteristic calcium binding b-turns similar to proteins such as tropomyosin and calmodulin[55]. The occupation of these calcium-binding sites in GPIIb has been shown to result in optimal fibrinogen binding[56]. Recently, a structural model of the amino-terminal ligand-binding region of integrin a-chain subunits has been generated by computer modeling in which a b-propeller domain folding pattern has been predicted[57]. The calcium-binding motifs in this predicted model are on the lower face of the b-propeller while putative ligand-binding sites are located on the upper face. The importance of a number of regions in the normal functioning of the GPIIb subunit have been identified by mutations from patients with Glanzmann thrombasthenia and those mutations that are located within the b-propeller will be discussed in the context of this structural model in the section describing the mutations.

The GPIIIa (b_3)subunit is characteristic of other integrin β-chains and has a molecular weight of approximately 95,000 by SDS-PAGE under non-reduced conditions and increases to approximately 110,000 under reduced conditions. GPIIIa contains five cysteine-rich regions and like other integrin b-chain subunits, contains 56 highly conserved cysteine residues[58,59] that form 26 short-range intradisulfide bonds and two long-range disulfide bonds between C5-C435 and C406-C655[60]. The role of a number of these cysteine residues in the biosynthesis and function of the GPIIb/IIIa receptor has been studied by characterization of naturally occurring mutations and by mutagenesis. Interestingly, the removal of C655[61] which is involved in one of the long-range disulfide bonds and addition of C636[62] resulting in a free sulfhydryl group have no apparent effects on synthesis, surface expression, or function of GPIIb/IIIa. On the other hand, a C374Y mutation has a major effect on the biosynthesis of the GPIIb/IIIa receptor[63]. These data indicate that some but not all of the highly conserved cysteine residues in GPIIIa are critical for the function of the receptor complex[61]. The importance of this observation in the normal functioning of the GPIIIa subunit will be discussed in the section describing the identified mutations in patients with Glanzmann thrombasthenia.

GPIIIa, like most of the other integrin b–chain subunits, interacts with divalent cations that are bound to a highly conserved site[64]. This cation binding motif in the integrin b-subunits is also highly conserved in the inserted (I) domain of six integrin a-subunits[65]. High resolution crystal structures of isolated I-domains have established this motif as part of a unique metal coordination site designated the metal ion-dependent adhesion site (MIDAS)[66]. A structural model of the MIDAS domain in GPIIIa has been generated by computer modeling showing the association of critical amino acid residues that are predicted to play a direct role in ligand binding function[67]. These data are supported by the identification of naturally occurring mutations in patients with Glanzmann thrombasthenia and will be discussed in the context of this model in the section describing the mutations.

The GPIIb subunit is normally tissue-specific and restricted to cells of the megakaryocyte lineage[68,69]. This restricted expression is due to enhancer and repressor binding sites within the promoter that are specific for nuclear factors present in megakaryocyte and erythroid cell lineages[70-72]. The GPIIIa subunit is more widely expressed and present not only on megakaryocytes, but on endothelial cells, osteoclasts, smooth muscle cells, and many cultured cells. The expression of GPIIIa occurs in association with GPIIb, but also another a-chain subunit, a_v, forming the $a_v b_3$ vitronectin receptor[73]. The GPIIb/IIIa receptor is abundantly expressed on the platelet surface (~80,000 molecules per platelet)[74], whereas $a_v b_3$ receptors are expressed in very low numbers (50-100 molecules per platelet)[75]. The $a_v b_3$ vitronectin receptor can bind fibrinogen and von Willebrand factor as well as vitronectin and these receptors may contribute to the development of coagulant activity after platelet activation[76].

18.3.1 Biochemical identification of defects in the GPIIb and GPIIIa subunits

Since GPIIb and GPIIIa are encoded by separate genes, a mutation in either gene can result in a defective subunit and cause disease. The most common methods for determining platelet GPIIb/IIIa levels are flow cytometry[77] and immunoblot analysis[78] . By using immunoblot analysis of whole platelet lysates, even trace amounts of one integrin subunit in patients' platelets with a severe deficiency of GPIIb/IIIa can be instructive as to the nature of the molecular defect[79]. A reliable indicator of GPIIb and GPIIIa defects is by surface expression of $a_v b_3$ receptors on either the patients' platelets[75] or on the patients' B-lymphocytes that have been EBV-transformed[80]. Detection of $a_v b_3$ receptors on the patients' platelets is performed using radiolabeled monoclonal antibodies and on EBV-transformed B-lymphocytes is performed using flow cytometry, immunoprecipitation, and immunoblot analysis. The decreased or undetectable binding of both GPIIb/IIIa ($a_{IIb} b_3$)- and $a_v b_3$-specific monoclonal antibodies indicates a defect in GPIIIa, whereas decreased or undetectable binding of GPIIb/IIIa ($a_{IIb} b_3$)-specific, but not $a_v b_3$-specific monoclonal antibody indicates a defect in GPIIb. This analysis has been utilized to identify defective GPIIb or GPIIIa subunits in numerous patients with Glanzmann thrombasthenia and subsequent identification of the mutation either in the GPIIb or GPIIIa gene has confirmed the reliability and usefulness of this assay [63,81-86].

Important information on the biosynthesis and assembly of GPIIb and GPIIIa subunits (reviewed in [6]) can be obtained by immunoblot analysis of whole platelet lysates using subunit specific monoclonal antibodies. The GPIIb polypeptide is synthesized as a single chain precursor (proGPIIb) that associates with GPIIIa within the endoplasmic reticulum (ER) of the cell. The two subunits must assemble into a complex to be transported to the Golgi and expressed on the cell surface. If the subunits do not form a proper complex, the subunits will be retained in the ER and degraded. Within the Golgi apparatus, complex oligosaccharides are added to the GPIIb subunit and processing or cleavage of this subunit into heavy and light chains occurs. The GPIIb heavy and light chains are covalently associated by a single disulfide bond[87]. When performing immunoblot analyses on whole platelet lysates, a great deal of information can be gained by analysis of the GPIIb subunit[88]. The presence of mature GPIIb heavy chain indicates appropriate assembly and processing of the GPIIb and GPIIIa subunits, while the presence of the proGPIIb subunit without mature GPIIb heavy chain indicates that the GPIIb protein was synthesized in the ER, but transport to the Golgi and/or processing of the GPIIb heavy chain was disrupted. Undetectable or trace levels of both proGPIIb and GPIIIa and the lack of GPIIb heavy chain suggests that GPIIb/IIIa subunit assembly was disrupted, resulting in expression of individual subunits which through mechanisms that are not completely understood, are degraded with a half-life of four to six hours[89].

Another informative parameter of GPIIb/IIIa expression and function is the determination of platelet fibrinogen levels[90]. Megakaryocytes do not synthesize fibrinogen[91,92] and the transport of plasma fibrinogen into the platelet a-granules occurs through the binding of fibrinogen to the GPIIb/IIIa receptor[90,93,94]. The majority of patients with Glanzmann thrombasthenia have low to undetectable levels of platelet fibrinogen[90,92], but some patients have been identified with levels of platelet fibrinogen that are higher than the 3-11% of normal levels that are frequently detected[3,63,95]. The presence of significant levels of platelet fibrinogen suggests that some receptors must be expressed on the platelet surface and that the receptors remain functional in protein trafficking of plasma fibrinogen. This parameter has been used to generate a new category of patients and will be discussed in the section decribing the identified mutations in patients with Glanzmann thrombasthenia.

18.4 Molecular identification of mutations

The field of platelet molecular biology was revolutionized by the PCR technology[96] and by the discovery that human platelets contained enough RNA to generate DNA fragments using reverse transcriptase (RT)-PCR[97]. The molecular biologic basis of Glanzmann thrombasthenia is now routinely determined by sequence analysis of genomic DNA and platelet RNA using GPIIb- and GPIIIa-specific oligonucleotide primers and PCR. A comprehensive review has been written on the molecular identification of mutations in inherited platelet disorders[8]. Briefly, the commonly used methodologies for the identification of mutations in Glanzmann thrombasthenia include RT-PCR of platelet RNA, PCR of genomic DNA, and single-stranded conformational polymorphism (SSCP) analysis of PCR fragments containing intron and exon boundaries. PCR fragments are sequenced either directly or from a cloned product, but the direct sequencing of PCR fragments amplified from genomic DNA provides the added advantage of immediately determining the homozygosity or heterozygosity of the mutation. The resulting defects from mutations affecting RNA splicing are identified by sequence analysis of the patients' platelet RNA.

18.4.1 Glanzmann Thrombasthenia Mutations

Glanzmann thrombasthenia is caused by homozygous or compound heterozygous mutations in the genes encoding GPIIb or GPIIIa. Of the 45 propositi kindreds with affected individuals described in this chapter, 17 of the affected individuals have been identified as compound heterozygotes and 28 have been identified as homozygotes, in which one of the latter may have arisen from a rare uniparental disomy event[98]. Since both GPIIb and GPIIIa subunits are required for normal biosynthesis, surface expression, and function of the GPIIb/IIIa receptor[15-17], a mutational defect in either subunit can disrupt the biogenesis and/or ligand-binding function of the heterodimeric complex and cause disease. The GPIIb and GPIIIa mutations that have been identified in patients with Glanzmann thrombasthenia are listed in Tables 1 and 2, respectively. The defects that result from each of these mutations will be discussed according to the biochemical characterization of GPIIb/IIIa expression in the patients' platelets[2]. The patients expressing detectable levels (>10% of normal) of GPIIb/IIIa receptors generally have mutations that affect the function of surface expressed receptor complexes. Within this group are also included mutations that result in both functional and biosynthetic defects. The mutations in this group are primarily missense mutations and one is a nonsense mutation. The patients expressing undetectable levels (^{2}10% of normal) of platelet GPIIb/IIIa receptors primarily have mutations that affect subunit expression and biosynthesis. These mutations include DNA deletions, insertions, and/or inversions; RNA splice site mutations, missense mutations, and nonsense mutations.

The GPIIb and GPIIIa mutations that affect functional aspects of the receptor will be described first and have been divided into those that affect 1) receptor ligand binding sites and complex stability, 2) aggregation but not internalization of fibrinogen, and 3) activation of the receptor complex. The GPIIb and GPIIIa mutations that result in biosynthetic defects will then be described and have been divided into those that affect 1) RNA accumulation, 2) subunit assembly, and 3) transport of assembled complexes from the ER to the Golgi.

18.4.2 Mutations resulting in functional defects

The biologic link between GPIIb/IIIa receptors and platelet aggregation is based on the highly regulated ligand-binding activity of this receptor complex[99]. In a resting state, this receptor is functional in binding ligand with low-affinity and in the trafficking of fibrinogen into the platelet a–granules[90,94]. In an activated state, signal transduction mechanisms[100] mediating platelet aggregation trigger a conformational change in the receptor to a high affinity ligand-binding state that is competent to bind soluble fibrinogen, von Willebrand factor, and other adhesive glycoproteins[101,102]. This "inside-out" signal transduction mechanism that mediates high-affinity ligand-binding is followed by an "outside-in" signal transduction mechanism[103] that is mediated by integrin-cytoskeleton interactions and required for postoccupancy events such as cell spreading and focal adhesion formation[104]. The sites on ligands that are recognized by integrin receptors have been identified from structural studies as short peptide sequences that are often presented on extended loops [105]. In the absence of a crystal structure, less precise information is available concerning the sites within integrins that recognize ligands, but the ligand-binding sites in the platelet GPIIb/IIIa receptor are most likely comprised of crucial residues that are brought together in the three-dimensional structure of the receptor complex[106,107]. The expression of truncated GPIIb and GPIIIa subunits have identified the minimal sequences that are necessary for functional ligand-binding regions[107,108] and structural models of the GPIIb subunit between residues 15-452[57] and the GPIIIa subunit between residues 110-294[67] have been generated by computer modeling. The locations of the Glanzmann thrombasthenia mutations, whether they result in functional or biosynthetic defects, that are within these putative ligand-binding regions are shown in schematic drawings of the structural models in Figures 1 (GPIIIa) and 3 (GPIIb) and within the primary sequences shown in Figures 2 (GPIIIa) and 4 (GPIIb).

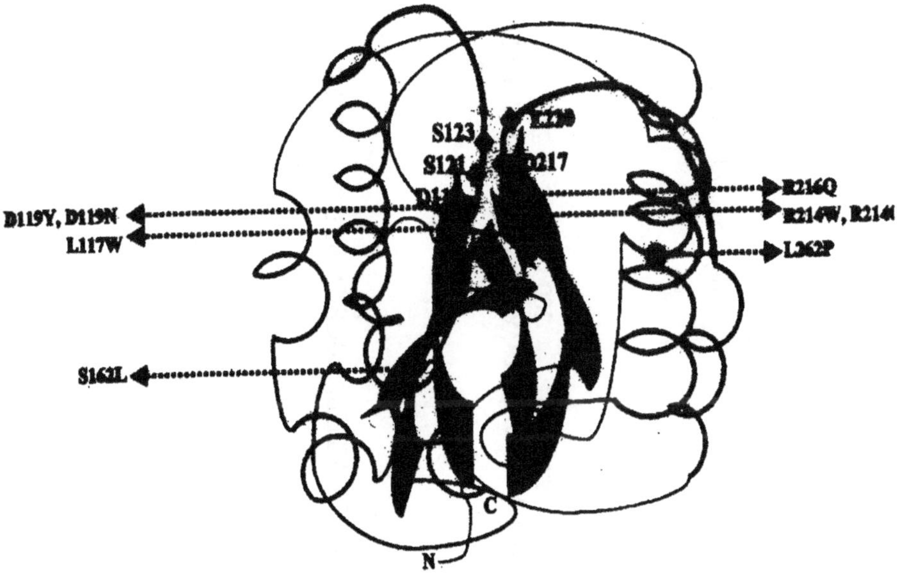

Fig.1. Glazmann thrombasthenia mutations within and surrounding the GPIIIa MIDAS-like motif. A schematic drawing of the GPIIIa structural model showing the β-strand (arrow), α-helix, and connecting strand arrangements between residues 110-294[67]. The $D^{119}XS^{121}XS^{123}$ and $D^{127}E^{220}$ residues that form the MIDAS-like motif are designated by black diamonds. The location of the Glanzmann thrombasthenia mutations within the structure are designated by black dots except those mutations that are located at D119.

18.4.3 Mutations Affecting Ligand Binding Sites and Complex Stability

The mutations that fall within this category have primarily been identified in GPIIIa and will be discussed first. These mutations were some of the first to be identified at the molecular level and their characterization identified important ligand and cation binding sites in GPIIIa. The majority of mutations that have been identified within GPIIb result in undetectable GPIIb/IIIa expression due to biosynthetic defects. Recently, some

```
                110       120       130       140       150       160
                YPVDIYYLMDLSMKDDLWSIQNLGTKLATQMRKLTSNLRIGFGAFVDKPVSPYMYISP
MK                     W
Cam                      Y
NR                       N
BL                                                                L

                170       180       190       200       210       220
                PEALENPCYDMKTTCLPMFGYKHVLTLTDQVTRFNEEVKKQSVSRNRDAPEGGFDAIMQA
Strasbourg I                                                 W
CM                                                           W
ET                                                           Q

                230       240       250       260       270       280
                TVCDEKIGWRNDASHLLVFTTDAKTHIALDGRLAGIVQPNDGQCHVGSDNHYSASTTMDY
LD                                              P
```

Fig.2. Glanzmann thrombasthenia mutations within the GPIIIa sequence including the MIDAS-like motif. The amino terminal sequence of GPIIIa from residues 110-289 showing the residues (in bold) that form the MIDAS-like motif[67]. The Glanzmann thombasthenia mutations are shown under the sequence and are listed by the patient designations as represented in Table 2.

mutations have been identified that result in surface expression levels of >10% of normal and may identify regions within GPIIb that affect ligand-binding sites. These mutations will be discussed following the GPIIIa mutations.

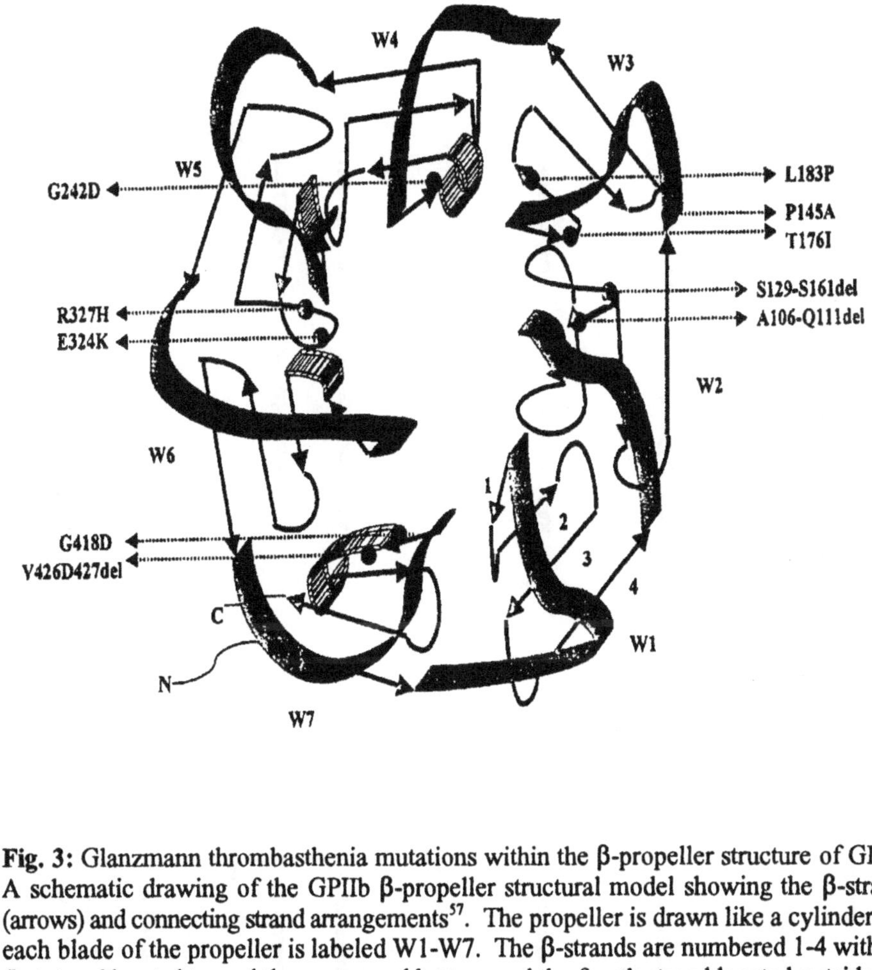

Fig. 3: Glanzmann thrombasthenia mutations within the β-propeller structure of GPIIb: A schematic drawing of the GPIIb β-propeller structural model showing the β-strands (arrows) and connecting strand arrangements[57]. The propeller is drawn like a cylinder and each blade of the propeller is labeled W1-W7. The β-strands are numbered 1-4 with the first strand located toward the center and bottom and the fourth strand located outside and top of the cylinder. The wide ribbons represent the connections between the fourth β-strand of one blade and the first β-strand of the next blade. The ladder-like connections represent the calcium-binding domains and are located on the bottom of the cylinder. The Glanzmann thrombasthenia mutations are represented by black dots and the black bar in W2 represents an intradisulfide bond.

Two regions within the GPIIIa subunit (D109-E171 and S211-G222) have been implicated in ligand and cation binding[106,109] and these findings were based on peptide cross-linking[110]

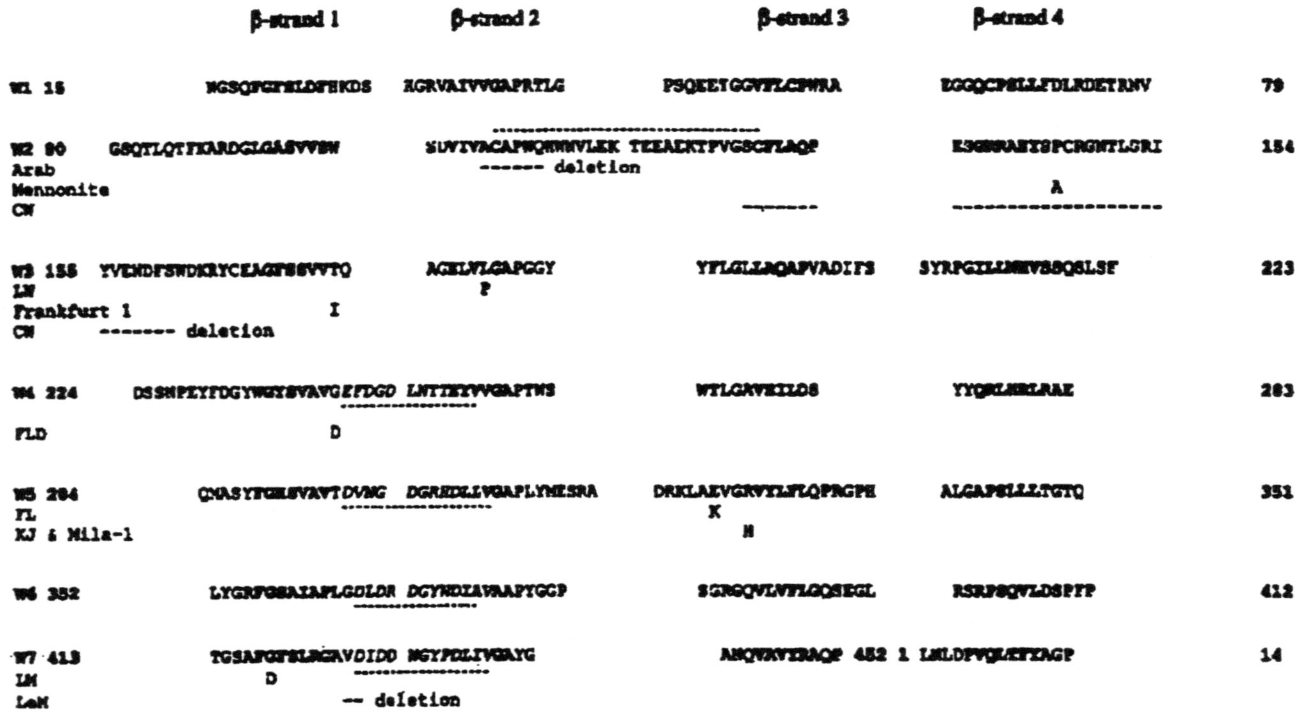

Fig. 4. **Glazmann thrombasthenia mutations within the β-propeller sequence for GPIIb:** The amino-terminal sequence of GPIIb from residues 1-452 that form the seven blades of the β-propeller.[57] The Glanzmann thrombosthenia mutations are shown under the GPIIb sequence and are listed by the patient designations as represented in Table 1. W1-W7 refer to the seven blades and the bold letters designate the amino acids that form the β-strands.[57] The dashed line above the sequence in the second and third β-strand sequence of W2 designates the disulfide bond formed by the two cysteine residues affected by the Arab and patient CW deletion mutations and the italic letter that are underlined by dashes in W4-W7 represent the calcium-binding domains.[48]

Table 1.Glanzmann Thrombasthenia GPIIb Mutations

Patient	Genotype	Exon[1]	Mutation[2]	Mutation Phenotype	RNA Splicing	Aminoacid Substitution[3]	Ref
KW	Homoztgote	1	IVS1-9del4.5kb	Del:Out of frame	Alter.	Premature termination	82
Arab	Homoztgote	4	IVS3(-3)-418del	Del.In frame	Alter.	(A(107)-Q(111)del	81
Mennonite	Homoztgote	4	526C ➔T	Missense	-	P176A(P145A)	122
Frankfurt I	Homoztgote	5	620C ➔ T	Missense	-	T2071(T1761)	121
LW	Homoztgote	6	641C ➔ T	Missense	-	L214P(L183P)	86
FLD	Homoztgote	8	818G ➔ A	Missense	-	G273D(G242D)	83
FL	Homoztgote	12	1063G ➔ A	Missense	-	E355K(E324K)	151
KJ	Homoztgote	12	1073G ➔ A	Missense	-	R358H(R327H)	153
Mila-1	Homoztgote	12	1073G ➔ A	Missense	-	R358H(R327H)	154
LM	Homoztgote	13	1346G ➔ A	Missense	-	G449D(G418D)	150
LeM	Comp, Heterozygote	13	1366-1371del,unk.	Del:in frame,unk.	-	V(425)D(426)del,unk.	152
Hispanic	Comp, Heterozygote	14	1413C ➔ G	Nonsense	-	Y471X(Y440X)	29
		29	3015insG	Ins:Out of frmae	-	Frameshift	
Gypsy	Homoztgote	15	IVS15(+1)G ➔ A	Del:Out of frame	Alter.	Premature termination	22
CW	Comp, Heterozygote	17	1750C ➔ T	Nonsense	-	R584X(R553X)	84
		4	480C ➔ G	Del:Out of frame	Alter.	S(129)-S(161)Del	
Chinese-10	Comp, Heterozygote	17	1750C ➔ T	Nonsense; Unk.	-	R584X(R553X), Unk.	84
SK	Comp, Heterozygote	17	1750C ➔T	Nonsense	-	R584X(R553X)	130
		26	IVS25(-3)C ➔G	Del.In frame	Alter.	V868-V909Del	
Family L	Comp, Heterozygote	17	1750C ➔T;Unk.	Nonsense; Unk	-	R584X(R553X), Unk.	131
Family II	Comp, Heterozygote	17	1750C ➔T;Unk	Nonsense; Unk	-	R584X(R553X), Unk.	132
MO	Comp, Heterozygote	17	1750C ➔T;Unk	Nonsense; Unk	-	R584X(R553X), Unk.	133
Iran-Jewish	Homoztgote	25	2473-247del/ins	Del/Ins:In frame	Alter.	L(786)-(795)del ➔Ins	143
SS	Homoztgote	26	2609C ➔ A	Nonsense	-	S(901)X(S870X)	141
NR	Comp, Heterozygote	28	2941C ➔ T	Nonsense; Unk	Alter.	Q(950)X ➔P917-50del;Unk	142
CG	Comp, Heterozygote	29	IVS29(+2)T ➔C	Del.In frame	Alter.	.V(951)-K(989)del	85
		18	1787T ➔C	Missense	-	1596T(1565T)	
MC	Comp, Heterozygote	29	IVS29(+2)T ➔C	Del.In frame	Alter.	V(951)-K(989)del	85
		30	3094insTG	Ins:Out of frame	-	Frameshift	
AP	Comp, Heterozygote	30	3077G ➔A;unk.	Missense; Unk.	-	R1026Q(R995Q); Unk.	95, 158

[1]Numbering according to references 45. [2]Nomenclature is based on recommendations by Beauder et. al. Hum Mutat 8: 197-202, 1996; Ibid. 203-206. The cDNA nucleotide numbering begins with the A nucleotide of the ATG start codon as +1[48]. Nucleotide substitutions: cDNA nucleotide number followed by nucleotide ➔ nucleotide substitution. Abbreviations: del-deletion, ins-insertion, inv-inversion, IVs-intervening sequence. [3]Amino acid numbering begins with methionine of the ATG start codon[48] and the amino acid codon number excluding the leader sequence is in parenthesis. Amino acid substitution are designated by amino acid-codon number-amino acid. Single letter amino acid code: A-alanine; D-aspartate; G-glycine; H-histidine; I-isoleucine; K-lysine; L-leucine; P-proline; Q-glutamine, R-arginine; S-serine; T-threonine; V-valine; Y-tyrosine; X-nonsense mutation; Comp: Compound; Unk: Unknown; Alter: Alternative.

Table 2. Glanzmann Thrombasthenia GPIIIa Mutations

Patient	Genotype	Exon[1]	Mutation[2]	Mutation Phenotype	RNA Splicing	Aminoacid Substitution[3]	Ref
Amsterdam I	Homozygote	2	IVS2(+1)G →T	Del:Out of frame	Alter.	Premature termination	147
Family I	Homozygote	3	262C →T	Missense	-	R88X(R62X)	132
MK	Homozygote	4	428T →G	Missense	-	L143W(L117W)	157
Cam	Homozygote	4	433G → T	Missense	-	D145Y(D119Y)	64
NR	Homozygote	4	433G → A	Missense	-	D145N(D119N)	116
BL	Homozygote	4	563 → T	Missense	-	S188L(S162L)	120
Strasbourg I	Homozygote	5	718C → T	Missense	-	R240W(R214W)	113
CM	Homozygote	5	718C → T	Missense	-	R240W(R214W)	117
ET	Homozygote	5	719G → A	Missense	-	R240Q(R214Q)	118
SH	Homoztgote	5	725G → A	Missense	-	R242Q(R216Q)	119
GT3	Compound	5	IVS5(+1)G →A	Del/Ins:Out of frame	Alter.	Premature termination	25,
	Heterozygote	1	IVS1-5Aluinv 15kb+IVS1del1kb	Inv/Del		No transcript detected	26
LD	Compound	6	847delGC	Del: Out of frame	-	Premature termination	28
	Heterozygote	6	863T →C	Missense	-	L288P(L262P)	
HS	Homozygote	8	1053-1058del	Del/Ins: In frame	-	351-353del →M351ins	146
Chinese-20	Homozygote	9	1199G →A	Missense	-	C374Y(C400Y)	63
CB	Homozygote	9	IVS9ins3-4kb	Ins: Out of frame	-	No transcript detected	139
RS	Homozygote	9	1260G →A	Del: Out of frame	Alter.	K(350)-S(396)del	98
	Disomy		+1143C →A	Ins: Restore frame	-	V(350)-S(351)ins	
RM	Compound	11	1791delT	Del: Out of frame	-	Premature termination	129
	Heterozygote	14	2248C →T	Missense		R750X(R724X)	
I-J_1	Homozygote	13	2031-2041del	Del: Out of frame	-	Premature termination	81
I-J_2	Compound	13	2031-2041del	Del: Out of frame	-	Premature termination	30
	Heterozygote	9	IVS9Alu-2163 or 2166del11.2kb	Del: Out of frame		Premature termination	
Paris I	Compound	15	2332T →C	Missense	-	S778P(S752P)	27
	Heterozygote		Unknown	Unknown		Unknown	

[1]Modified exon numbering: the leader sequence is encoded by exon 1[47] and exon numbers from reference 46 are increased by one. [2]Nomenclature is based on recommendations by Beaudet et al. Hum Mutat 8: 197-202, 1996; *Ibid*: 203-206. The cDNA nucleotide numbering begins with the A nucleotide of the ATG start codon as +1[49].Nucleotide substitutions: cDNA nucleotide number followed by nucleotide → nucleotide substitution. Abbreviations: del-deletion, ins-insertion, inv-inversion, IVS-intervening sequence. [3]Amino acid numbering begins with methionine of the ATG start codon[48] and the amino acid codon number excluding the leader sequence is in parenthesis. Amino acid substitution are designated by amino acid-codon number-amino acid. Single letter amino acid code: A-alanine; D-aspartate; G-glycine; H-histidine; I-isoleucine; K-lysine; L-leucine; P-proline; Q-glutamine, R-arginine; S-serine; T-threonine; V-valine; Y-tyrosine; X-nonsense mutation; Comp: Compound; Unk: Unknown; Alter: Alternative.

and monoclonal inhibition[111-113] studies. Two of the original Glanzmann thrombasthenia mutations identified at the molecular level are within these regions and result in D145Y(D119Y) (Cam variant)[64] and R240W(R214W) (Strasbourg I)[114] substitutions. The observation that the D145Y(D119Y) mutation in GPIIIa altered the conformation of the GPIIb/IIIa receptor in a manner consistent with loss of divalent cation binding led to the hypothesis that this site may constitute a cation ion binding region[36,64]. Mutational analyses of residues near D119 identified a DXSXS sequence that is highly conserved in all b-subunits[115]. This motif is highly conserved in the inserted (I) domain of six integrin a-subunits[65] and high resolution crystal structures have established this motif as part of a unique metal coordination site designated the metal ion-dependent adhesion site (MIDAS)[66]. The conservation of consensus motifs and the generation of similar hydropathy plots between GPIIIa and an I-domain suggested that this region in GPIIIa may adopt a similar I-domain secondary structure[66,116].

In addition to the MIDAS domain, molecular modeling and mutagenesis studies identified two additional residues, D217 and E220, that contribute to the ligand-binding function of the GPIIb/IIIa receptor[67]. The D217 residue is analogous to a metal-coordinating residue in the MIDAS domain suggesting that the D217 and E220 residues participate in the cation-binding sphere with residues D119, S121, and S123[67]. In addition to the original two mutations, four more mutations have been identified within these sites and are D145N(D119N) in patient NR[117], a second R240W(R214W) substitution in patient CM[118], an R240Q(R214Q) in patient ET[38,119], and an R242Q(R216Q) in patient SH[120]. The locations of these six Glanzmann thrombasthenia mutations are shown in a schematic drawing of the GPIIIa structural model (Figure 1) and in the primary sequence (Figure 2). These mutations all reside within the site where the two b-strands come together in the three-dimensional structure forming the metal-binding sphere of the MIDAS domain. All of these mutations result in normal to somewhat reduced levels of GPIIb/IIIa on the platelet surface, but the receptors fail to bind fibrinogen and some are easily dissociable with EDTA. The D145N(D119N) in patient NR supports clot retraction and binds fibrin while the D145Y(D119Y) mutation does not[117]. These mutations support the original observations that these sites within GPIIIa contribute to cation- and ligand-binding functions of the receptor complex.

Two additional missense mutations that result in ~30% of normal levels of GPIIb/IIIa receptors[28,121] are located within the vicinity of the MIDAS domain and are shown in the model (Figure 1) and sequence (Figure 2) of GPIIIa. The S162L mutation in patient BL results in the expression of GPIIb/IIIa receptors that are easily dissociated by calcium and undetectable by complex-dependent monoclonal antibodies[121]. Pulse-chase studies demonstrated that the mutant GPIIIa subunit formed an unstable complex, which rapidly disappeared, suggesting a trafficking defect. The L262P mutation in patient LD also results in the formation of GPIIb/IIIa complexes that are easily dissociated, but clot formation in this patient is normal[28]. Mammalian cell expression studies showed that complex maturation of the mutant receptor from the ER to the Golgi was delayed, but the expression of the mutant GPIIIa subunit did not affect clot formation.

Three mutations in GPIIb have recently been identified that result in surface expressed receptors of >10% of normal and affect ligand binding function of the receptor. These mutations result in L183P(L214P)[86], T207I(T176I)[122], and P145A(P176A)[123] substitutions. The analysis of these mutations has been greatly aided by the generation of a predicted fibrinogen in the patients' platelets. Other mutations included within this group are a GPIIb R1026Q(R995Q) mutation identified in patient AP[95] and a GPIIIa S778P(S752P) mutation identified in patient Paris I[27]. Patient AP was reported as having normal levels of platelet fibrinogen and the platelets from patient Paris I were reported to adhere normally to immobilized fibrinogen. Both defects result in surface expressed receptors that are apparently normal in mediating the trafficking of plasma fibrinogen. As will be discussed, the S778P(S752P) mutation affects the activation state of the receptor and the R1026Q(R995Q) mutation affects the transport of the receptor to the cell surface.

18.5.1 Mutations Affecting the Activation State of the Receptor Complex

Two GPIIIa mutations have been identified that affect the signaling mechanisms associated with GPIIb/IIIa receptor activation. A S752P mutation identified in patient Paris I[27,128,129] and an R724X mutation identified in patient RM[130] are located within the GPIIIa cytoplasmic domain. Resting platelets from patient Paris I expressed ~44% of normal levels of GPIIb/IIIa receptors and after activation expressed ~56% of normal levels, but platelet aggregation was absent in response to ADP, thrombin, and collagen[27]. Mammalian cell expression studies showed that the extracellular domain of the mutant receptor could obtain a high-affinity ligand-binding conformation and mutant receptors in an unactivated state could adhere to immobilized fibrinogen[128]. Mutagenesis studies showed reduced cell spreading and focal adhesion plaque formation suggesting that the S752P mutation disrupted the secondary structure of the cytoplasmic tail affecting the signaling events required for receptor activation[128,129]. The platelets from patient RM express a mutant receptor that contains only the first eight amino acids of the 47 amino acids that are normally present in the GPIIIa cytoplasmic domain. Resting platelets from patient RM expressed significant levels of a stable GPIIb/IIIa complex that was unresponsive to all physiologic agonists, but responsive to conformational activators[130]. Functional analyses of the mutant receptor showed normal adhesion to immobilized fibrinogen, but abnormal cell spreading and tyrosine phosphorylation of focal adhesion kinase pp125[FAK130]. This study demonstrates the importance of the membrane-distal portion of the GPIIIa cytoplasmic domain and the bidirectional signaling events that are required for the function of activated platelet GPIIb/IIIa receptors.

18.6 Mutations resulting in biosynthetic defects

Mutations Affecting RNA Accumulation

Decreased or undetectable RNA transcript levels can result from mutations affecting RNA half-life or cytoplasmic accumulation, as for example premature termination of protein translation, and from mutations affecting promoter regions. In Glanzmann thrombasthenia, a number of mutations including DNA deletions, insertions, inversions; splice mutations; and nonsense mutations have been identified that affect the accumulation of RNA. Of interest are six identical GPIIb nonsense mutations that result in R584X(R553X) substitutions and diminished RNA levels[84,131-134]. A proposed explanation for the reduced levels of RNA has been destabilization of RNA due to the presence of a termination codon in the middle of the GPIIb subunit. The relationship between abnormal protein translation and RNA accumulation has been observed in both prokaryotic and eukaryotic organisms[135,136]. In yeast, nonsense mutations can reduce RNA levels in a process designated nonsense-mediated mRNA decay[137]. One proposed explanation for this phenomenon is that premature termination of protein translation results in the lose of protection of mRNA by ribosomes, causing susceptibility to endogenous RNase digestion[138]. In addition, feed-back mechanisms from prematurely terminated proteins to intranuclear events have been proposed, in which specific mRNA transport from the nucleus may be inhibited[139].

Other mutations that have been shown to result in reduced RNA levels are a 2 base pair insertion in GPIIb exon 30 which encodes the cytoplasmic domain in patient MC[85], a 3-4 kb deletion in GPIIIa intron 9 in patient CB[140], a single base deletion resulting in a premature termination in GPIIIa exon 11 in patient RM[130], and a large inversion and deletion affecting GPIIIa exons 1 to 5 in one allele[25] and a donor splice mutation affecting GPIIIa exon 5 in the other allele[26] in patient GT3.

An interesting observation in prokaryotic operons is that the presence of nonsense mutations located at the amino-terminal end of the protein result in decreased levels of RNA accumulation while nonsense mutations located in the carboxy-terminal end of the protein have less of an effect[135]. Within eukaryotic systems, nonsense mutations identified in the b-globin gene have been shown to result in reduced accumulation of mRNA levels[139,141], but a relationship between the location of the mutation in the protein and RNA accumulation could not be determined. From the mutations identified in Glanzmann thrombasthenia, this location-dependent effect is suggestive but not proven. Within GPIIb, the nonsense mutation in exon 17 identified in six different patients resulted in decreased mRNA levels, whereas the nonsense mutations identified in exon 26 of patient SS[142] and exon 28 of patient NR[143] had little effect on RNA levels. Data regarding RNA levels in patients with GPIIIa defects are more difficult to interpret but perhaps also support a polar effect. The R88X(R62X) nonsense mutation in exon 3 of Family I was associated with undetectable levels of surface GPIIb/IIIa receptors[133], whereas the R750X(R724X) nonsense mutation in exon 14 of patient RM resulted in expression of surface GPIIb/IIIa receptors[130], suggesting the association of the latter mutation with the presence of normal amounts of RNA.

Mutations Affecting Subunit Assembly

The defects in the platelet GPIIb/IIIa receptors caused by a number of mutations identified in patients with Glanzman thrombasthenia can be explained by structural abnormalities affecting the biosynthetic pathway of the surface expressed receptor. An important and diagnostic indicator of a biosynthetic defect of GPIIb/IIIa receptors is the presence of proGPIIb in whole platelet lysates[88]. Mutations have been identified in which cleavage of proGPIIb into a mature GPIIb heavy chain subunit is inhibited and these mutations can be divided into those that prevent complex formation, which will be described in this section, and those that inhibit egress of the assembled GPIIb/IIIa complex from the ER to the Golgi, which will be described in the next section. Other mutations that can affect the conformational integrity of the GPIIb/IIIa receptor are defects affecting disulfide bonds. A conformational alteration in either subunit can disrupt subunit association and prevent transport and cleavage of the proGPIIb precursor subunit. The mutations that will be discussed all result in undetectable to reduced levels of mature GPIIb and GPIIIa in the patients platelet samples. Two of these mutations, identified in the Arab and Iraqi-Jewish populations in Israel[79,88], were among the original group of Glanzmann thrombasthenia mutations identified at the molecular level[81]. The Arab mutation is a 13 bp deletion in

GPIIb exon 4, which includes a splice acceptor site, leading to use of an alternative splice site. This alternative splice results in an inframe deletion of 6 amino acids (A106-Q111), including cysteine 107 in the GPIIb heavy chain. The location of this deletion is shown in the schematic drawing of the b-propeller model of GPIIb in Figure 3 and in the primary sequence in Figure 4. GPIIb was markedly reduced in these patients' platelets and the fraction of GPIIb in the uncleaved proGPIIb form was increased[88]. A similar mutation identified in patient CW[84] results in an inframe deletion of GPIIb amino acid residues S129-S161, which includes cysteine residues 130 and 146. This mutation is also shown in the schematic drawing in Figure 3 and in the primary sequence in Figure 4. This deletion includes b-strands 3 and 4 within the second blade of the propeller and the 4-1 strand that connects blades 2 and 3. These amino acid deletions and disulfide bond disruptions probably result in misfolded proteins that cannot form a stable complex.

The Iraqi-Jewish (I-J_1) mutation is an 11 bp deletion in GPIIIa exon 13 that results in a premature termination[81]. This deletion eliminates a cysteine residue that forms a long-range C406-C655 disulfide bond, but mutagenesis studies have shown that C655 can be deleted without any apparent effect on synthesis, surface expression, or function of GPIIb/IIIa receptors[61]. A second mutation (I-J_2) has been identified in the Iraqi-Jewish population which also disrupts the long-range disulfide bond in GPIIIa and results in a large deletion flanked by exons 10 and 14[30]. Both of these mutations result in premature terminations that delete the carboxy-terminal end of GPIIIa and probably result in defective complex formation.

Other mutations that appear to affect complex formation and prevent transport and cleavage of proGPIIb have been identified. In an Iranian-Jewish family in Israel, a 6 bp deletion and 31 bp insertion were identified in GPIIb exon 25[144]. This mutation creates an alternative splice site, resulting in an inframe deletion of ten amino acids, and addition of 8 new amino acids. This mutation is located within one of the proposed inter-subunit association sites for the GPIIb heavy chain[145], thus possibly disrupting stable heterodimer formation. This mutation is also located close to the proposed cleavage site of the GPIIb subunit[146] and may disrupt the conformation of the receptor complex for enzyme recognition. Another mutation, identified in patient SK[131], is located in the splice acceptor site of GPIIb exon 26 and results in an inframe deletion of this exon, which encodes the R858-Q859 enzyme cleavage site for the generation of the GPIIb heavy and light chains. The conformational change due to the 42 amino acid deletion may affect subunit association. Mammalian cell expression studies suggest that the cleavage of proGPIIb to heavy and light chains may not be essential for cell surface expression of the GPIIb/IIIa complex[146], thus mutations that affect the cleavage site may involve additional conformational defects that affect subunit association and/or processing. Within GPIIIa, an $Ile^{325}Pro^{326}Gly^{327}$ Æ Met deletion/insertion has been identified in patient HS and CHO cell expression studies show that this mutation prevents subunit association[147].

Other mutations affecting subunit assembly include transmembrane deletions of GPIIb and a deletion and insertion in GPIIIa. Two identical mutations identified in patients CG and MC were in the splice donor site of GPIIb exon 29 resulting in inframe deletions of the transmembrane domains[85]. The second mutation in patient CG was identified as a I596T(I565T) missense mutation. Immunoblot analyses showed reduced levels of mature GPIIb heavy chain suggesting that some complex formation and normal processing occurred. In addition, patient CG had platelet fibrinogen levels of 20% of normal suggesting that some surface receptors may be expressed and functional in trafficking of plasma fibrinogen. Another GPIIb transmembrane deletion has been identified in two Hispanic siblings in which a single base insertion results in a frameshift and premature termination affecting the carboxy-terminal end of the transmembrane domain and the cytoplasmic domain[29]. The other defect in the GPIIb subunit of these patients is a Y471X(Y440X) mutation located at the end of the fourth calcium-binding domain[29]. By immunoblot analyses, these mutations result in undetectable levels of GPIIb and trace levels of GPIIIa. A splice mutation identified in the gene encoding GPIIIa of patient RS required the presence of two different base changes and resulted in a deletion of GPIIIa exon 9 and a 5 base-pair insertion which restored the reading frame[98]. This deletion is located within a proposed inter-subunit interaction domain of GPIIIa[6] and may destablize complex formation.

A number of other mutations have been identified that result in frameshifts and premature terminations. Within GPIIb, a large DNA deletion resulting in a premature termination in intron 1 has been identified in patient KW[82] and a deletion and premature termination in exon 15 caused by a splice site mutation has been identified in kindreds of the Gypsy population in France[22]. Within GPIIIa, a 2 base-pair deletion in exon 6 was identified in patient LD[28] and a deletion and premature termination in exon 3 caused by a splice site mutation was identified in patient Amsterdam I[148].

Mutations Affecting Transport of Receptor Complexes to the Golgi

Calcium is required for stabilization and ligand binding of the GPIIb/IIIa complex[5,149,150]. The calcium binding sites within GPIIb are located between amino acid residues E243-V254, D297-V308, D365-V376, and D426-V437[48] and the occupation of these sites has been shown to result in optimal fibrinogen binding[56]. Glanzmann thrombasthenia mutations that are located within and surrounding the GPIIb calcium binding domains have indicated that these domains are not required for complex assembly, but are required for transport of the assembled complexes from the ER to the Golgi[83,151-155]. Six missense mutations that affect the calcium-binding motifs have been identified in Glanzmann thrombasthenia patients and are shown in the schematic drawing of GPIIb in Figure 3 and the primary sequence in Figure 4. These mutations include a G273D(G242D) substitution which precedes the first calcium binding domain in patient FLD[83], an E355K(E324K) substitution located between the second and third calcium-binding domains in patient FL[152], two R358H(R327H) mutations also located between the second and third calcium-binding domains in patients KJ[154] and Mila-1[155]; a G449D(G418D) substitution which precedes the fourth calcium binding domain in patient LM[151], and a V425D426 deletion at the start of the fourth calcium binding domain in patient LeM[153].

The hypothesis that the GPIIb calcium-binding domains are not required for subunit association, but are required for trafficking of the GPIIb/IIIa complex from the ER to the Golgi has been supported by a number of studies. Mammalian cell expression studies have shown that removal of the GPIIb calcium-binding domains does not affect assembly of the GPIIb/IIIa complex[153]. Association of the GPIIIa subunit with truncated GPIIb fragments containing the signal peptide and amino terminal residues ending before the first, second, or third calcium-binding domains has been shown to occur[154]. Studies using peptides containing Glanzmann thrombasthenia mutations and recombinant polypeptides containing GPIIb calcium binding domains showed that cation binding was affected by 1) the position of the mutation relative to the calcium-binding domain and 2) the amino acid substitution[156]. Putative mechanisms underlying the intracellular retention of GPIIb/IIIa complexes has been shown by in vitro mutagenesis and computer modeling studies to involve conformational changes in the GPIIb/IIIa complex[157]. The cumulative findings from all of these studies support the hypothesis that GPIIb/IIIa complex assembly can occur in the presence of disrupted calcium-binding motifs, but transport of these complexes from the ER requires conformationally intact and functional calcium-binding motifs.

An interesting mutation in GPIIIa that results in the expression of a phenotype identical to the phenotypes of the GPIIb mutations described above is a L143W(L117W) mutation[158]. This mutation identified in patient MK (Figures 1 and 3) has been shown by mammalian cell expression studies to result in the intracellular retention of misfolded GPIIb/IIIa heterodimeric complexes.

Another GPIIb mutation that appears to affect transport of GPIIb/IIIa receptor complexes to the cell surface is an R995Q mutation in patient AP[95,159]. This mutation is located within the highly conserved GFFKR sequence in GPIIb which is part of the proposed hinge region that keeps the receptors in a default low affinity ligand-binding conformation[160]. This mutation results in a slow and much reduced platelet aggregation response using agonists such as ADP and epinephrine. In addition, this mutation is associated with abnormal expression of GPIIb/IIIa complexes since total platelet GPIIb/IIIa levels were ~40-50% of normal, but platelet surface GPIIb/IIIa expression was only ~12-20% of normal. Platelet fibrinogen levels were normal and fibrinogen binding to activated platelets was reduced by ~80%.

18.7 Mutation hotspots

Analysis of the missense and nonsense mutations identified in patients with Glanzmann thrombasthenia show that 8/20 (40%) missense and 9/10 (90%) nonsense mutations occur in CpG dinucleotides and may be due to putative hotspot mutations[2]. One common cause of single base-pair mutations in human disease is due to the modification of DNA by cytosine methylation[161]. A cytosine to thymine mutation can occur by spontaneous deamination of the modified base, 5-methylcytosine, designating these sites as putative hotspots for point mutations[162]. A study analyzing the frequency of point mutations in human diseases showed that 35% of the mutations were found within CpG dinucleotides and over 90% of these mutations were due to CGÆTG or CGÆCA transitions[161]. Within the 8 Glanzmann thrombasthenia missense mutations, 2 result from CGÆTG transitions and 6 result from CGÆCA transitions. Within the 9 Glanzmann thrombasthenia nonsense mutations, 8 result from CGÆTG transitions and 1 results from a CGÆCA transition. Within the group of nonsense mutations, 6 of the CGÆTG transitions result in a GPIIb R584X (R553X) substitution in 6 different patients originating from China[84,132], Japan[131,134], and France[133]. A population study of 107 normal individuals from the Hunan Province in China failed to identify a polymorphism at this site, supporting the hypothesis that this site in GPIIb exon 17 may represent a mutational hotspot (unpublished data from Chen FP, Coller BS, French DL). Within the missense mutations, two different mutations that occur in CpG dinucleotides have been identified in separate kindreds and include two GPIIIa R240W(R214W) mutations resulting from CGÆTG transitions[114,118] and two GPIIb R358H(R327H) mutations resulting from CGÆCA transitions[154,155].

18.8 Prenatal diagnosis, carrier detection, and gene therapy

Carrier detection of Glanzmann thrombasthenia was originally performed by analysis of platelet surface or total GPIIb/IIIa receptors by biochemical or immunological methods, but unequivocal determinations always remained a concern[163]. DNA-based methodologies result in unequivocal carrier assessment and the enhanced sensitivity of PCR enables detection of DNA from cells present in a routine urine sample[164]. For more than a decade, prenatal diagnosis of Glanzmann thrombasthenia has been carried out by studying the GPIIb/IIIa receptor on platelets from fetal blood samples, but the identification of mutations at the molecular level in selected kindreds[22,81] has permitted the development of DNA-based diagnostic methods. Within the Glanzmann thrombasthenia families in Israel and one family in the United States, 23 prenatal diagnoses have been performed[2]. The first prenatal diagnoses were based on platelet antibody binding studies and performed on fetal blood samples obtained by cordocentesis[163,165]. In more recent prenatal diagnoses, mutations have been identified by PCR-based methods from DNA extracted from chorionic villus samples and/or cells in amniotic fluid. Fifteen prenatal diagnostic procedures in Israel have been performed in eight women who previously delivered an affected child. Seven of the fetuses were affected, six fetuses were diagnosed as carriers, and two fetuses had normal genotypes. None of the infants diagnosed as carriers or normals demonstrated bleeding manifestations. In the case of the family in the United States, prenatal diagnosis has been performed on 3 pregnancies by cordocentesis, PCR-based haplotype analysis of chromosomal polymorphic markers, and direct sequence analysis, respectively. The first prenatal diagnosis identified the fetus as affected and the fetus died immediately after the cordocentesis procedure. At the time of the second prenatal diagnosis, the mother's mutation had been identified but the father's mutation was still unknown. The identification of the mother's mutation was used with linkage analysis of polymorphisms within the father's chromosome 17 to determine haplotypes on DNA samples from the fetus, mother, father, and affect child[166]. The fetus was identified as a carrier, and unfortunately, due to an unrelated birth defect, this pregnancy was terminated. Both maternal and paternal mutations have been identified[1,166] and a third prenatal diagnosis was performed by direct sequence analysis. This fetus inherited both normal alleles from each parent, the pregnancy went to term, and the baby was born without any bleeding manifestations.

Since Glanzmann thrombasthenia is an autosomal recessive disorder, gene therapy could provide an alternative approach of treatment which may reverse the bleeding manifestations associated with the disease. A gene expression system has been established in which GPIIIa synthesis occurs under the control of the megakaryocyte-specific GPIIb promoter[167]. Gene constructs were expressed in $CD34^+$ peripheral blood cells isolated from two patients with Glanzmann thrombasthenia. These patients had two different GPIIIa mutations that resulted in a complete lack of surface expressed GPIIb/IIIa receptors. From *in vitro* analyses, surface and intracellular expression of GPIIb/IIIa receptors were identified in the transduced cells and these cells were shown to mediate normal clot retraction. These studies resulted in the *ex vivo* correction of the Glanzmann's thrombasthenia phenotype and demonstrated the potential utilization of the GPIIb promoter for megakaryocyte-targeted expression of proteins. Since naturally occuring polymorphic forms of GPIIb and GPIIIa subunits give rise to platelet alloantigenic responses[168-170], the induction of an alloantigenic response in patients that have never expressed GPIIb/IIIa receptors is a concern. The GPIIIa knock-out mouse[171] will be helpful in deciphering these responses and in the further development of gene therapy for Glanzmann thrombasthenia.

18.9 Conclusions

Our appreciation for the diversity of abnormalities that underlie Glanzmann thrombasthenia has been enriched by the clinical, biochemical, and molecular characterization of the increasing number of mutational defects identified in patients having this disorder. Due to the wealth of discoveries in the past decade, the mutations can be more precisely defined and distinct groups of mutational defects can be identified. The molecular characterization of patients with this disorder has provided the foundation for performing DNA-based carrier detection and prenatal diagnoses and the future of gene therapy is under development. The information generated from this work has enhanced our understanding of platelet physiology and has provided important insights into the biogenesis, structure, and function of the platelet GPIIb/IIIa and integrin family of adhesion receptors.

References

1. French DL, Coller BS: Hematologically important mutations: Glanzmann thrombasthenia. Blood Cells, Molecules, and Diseases 23:39-51, 1996.

2. French DL: The molecular genetics of Glanzmann's thrombasthenia. Platelets 9:5-20, 1998.

3. George JN, Caen JP, Nurden AT: Glanzmann's thrombasthenia: The spectrum of clinical disease. Blood 75:1383-1395, 1990.

4. Coller BS, Seligsohn U, Peretz H, Newman PJ: Glanzmann thrombasthenia: New insights from an historical perspective. Seminars in Hematol 31:301-311, 1994.

5. Phillips DR, Charo IF, Parise LV, Fitzgerald LA: The platelet membrane glycoprotein IIb-IIIa complex. Blood 71:831-843, 1988.

6. Calvete JJ: Clues for understanding the structure and function of a prototypic human integrin: The platelet glycoprotein IIb/IIIa complex. Thromb Haemost 72:1-15, 1994.

7. Calvete JJ: On the structure and function of platelet integrin aIIbb3, the fibrinogen receptor. Proc Soc Exp Biol Med 208:346-360, 1995.

8. Bray PF: Inherited diseases of platelet glycoproteins: Considerations for rapid molecular characterization. Thromb Haemost 72:492-502, 1994.

9. Glanzmann E: Hereditare hamorrhagische thrombasthenie: ein beitrag zur pathologie der blut plattchen. J Kinderke 88:113-141, 1918.

10. Braunsteiner H, Pakesch F: Thrombocytoasthenia and thrombocytopathia. Old name and new diseases. Blood 11:965-976, 1956.

11. Hardisty RM, Dormandy KM, Hutton RA: Thrombasthenia: Studies on three cases. Br J Haematol 10:371-387, 1964.

12. Zucker MB, Pert JH, Hilgartner MW: Platelet function in a patient with thrombasthenia. Blood 28:524-534, 1966.

13. Nurden AT, Caen JP: An abnormal platelet glycoprotein pattern in three cases of Glanzmann's thrombasthenia. Br J Haematol 28:253-260, 1974.

14. Phillips DR, Jenkins CSP, Luscher EF, Larrieu M-J: Molecular differences of exposed surface proteins on thrombasthenic platelet plasma membranes. Nature 257:599-600, 1975.

15. Duperray A, Troesch A, Berthier R, Chagnon E, Frachet P, Uzan G, Marguerie G: Biosynthesis and assembly of platelet GPIIb-IIIa in human megakaryocytes: Evidence that assembly between pro-GPIIb and GPIIIa is a prerequisite for expression of the complex on the cell surface. Blood 74:1603-1611, 1989.

16. O'Toole TE, Loftus JC, Plow EF, Glass AA, Harper JR, Ginsberg MH: Efficient surface expression of platelet GPIIb-IIIa requires both subunits. Blood 74:14-18, 1989.

17. Rosa J-P, McEver RP: Processing and assembly of the integrin, glycoprotein IIb-IIIa, in HEL cells. J Biol Chem 264:12596-12603, 1989.

18. Reichert N, Seligsohn U, Ramot B: Clinical and genetic aspects of Glanzmann's thrombasthenia in Israel. Thromb Diathesis Haemor 34:9806-9820, 1975.

19. Awidi AS: Increased incidence of Glanzmann's thrombasthenia in Jordan compared with Scandinavia. Scand J Haematol 30:218-222, 1983.

20. Seligsohn U, Rososhansky S: A Glanzmann's thrombasthenia cluster among Iraqi Jews in Israel. Thromb Haemost 52:230-231, 1984.

21. Khanduri U, Pulimood R, Sudarsanam A, Carman RH, Jadhav M, Pereira S: Glanzmann's thrombasthenia. A review and report of 42 cases from South India. Thromb Haemost 46:717-721, 1981.

22. Schlegel N, Gayet O, Morel-Kopp MC, Wyler B, Hurtaud-Roux MF, Kaplan C, McGregor J: Themolecular genetic basis of Glanzmann's thrombasthenia in a Gypsy population in France: Identification of a new mutation on the a_{IIb} gene. Blood 86:977-982, 1995.

23. Hall JM, Lee MK, Newman B, Morrow JE, Anderson LA, Huey B, King M-C: Linkage of early-onset familial breast cancer to chromosome 17q21. Science 250:1684-1689, 1990.

24. Bray PF, Shuman MA: Identification of an abnormal gene for the GPIIIa subunit of the platelet fibrinogen receptor resulting in Glanzmann thrombasthenia. Blood 75:881-888, 1990.

25. Li L, Bray PG: Homologous recombination among three intragene Alu sequences causes an inversion-deletion resulting in the hereditary bleeding disorder Glanzmann thrombasthenia. Am J Hum Genet 53:140-149, 1993.

26. Jin Y, Dietz HC, Nurden A, Bray PF: Single-strand conformation polymorphism analysis is a rapid and effective method for the identification of mutations and polymorphisms in the gene for glycoprotein IIIa. Blood 82:2281-2288, 1993.

27. Chen Y-P, Djaffar I, Pidard D, Steiner B, Cieutat A-M, Caen JP, Rosa J-P: Ser-752‡Pro mutation in the cytoplasmic domain of integrin b_3 subunit and defective activation of platelet integrin $a_{IIb}b_3$ (glycoprotein Iib-IIIa) in a variant of Glanzmann thrombasthenia. Proc Natl Acad Sci USA 89:10169-10173, 1992.

28. Ward CM, Newman PJ: A Leu262Pro mutation in the integrin b_3 subunit results in an $a_{IIb}b^3$ complex which binds fibrin but not fibrinogen. Blood 90:25a, 1997 (Abstract).

29. Ward CM, Scott JPI, Scott JPI, Newman PJ: A frameshift mutation at Gly975 in the transmembrane domain of a_{IIb} prevents $a_{IIb}b_3$ expression. Blood 90:25a, 1997 (Abstract).

30. Rosenberg N, Yatuv R, Orion Y, Zivelin A, Dardik R, Peretz H, Seligsohn U: Glanzmann thrombasthenia caused by an 11.2-kb deletion in the glycoprotein IIIa (b_3) is a second mutation in Iraqi Jews that stemmed from a distinct founder. Blood 89:3654-3662, 1997.

31. Caen JP: Glanzmann's thrombasthenia. Clin Haematol 1:383-392, 1972.

32. Phillips DR, Agin PP: Platelet membrane defects in Glanzmann's thrombasthenia. Evidence for decreased amounts of two major glycoproteins. J Clin Invest 60:535-545, 1977.

33. Hagen I, Nurden A, Bjerrum OJ, Solum NO, Caen J: Immunochemical evidence for protein abnormalities in platelets from patients with Glanzmann's thrombasthenia and Bernard Soulier syndrome. J Clin Invest 65:722-731, 1980.

34. Kunicki TJ, Nurden AT, Pidard D, Russell NR, Caen JP: Characterization of human platelet glycoprotein antigens giving rise to individual immunoprecipitates in crossed-immunoelectrophoresis. Blood 58:1190-1197, 1981.

35. Holahan JR, White GCI: Heterogeneity of membrane surface proteins in Glanzmann's thrombasthenia. Blood 57:174-181, 1981.

36. Ginsberg MH, Lightsey A, Kunicki TJ, Kaufmann A, Marguerie G, Plow EF: Divalent cation regulation of the surface orientation of platelet membrane glycoprotein IIb. Correlation with fibrinogen binding function and definition of a novel variant of Glanzmann's thrombasthenia. J Clin Invest 78:1103-1111, 1986.

37. Nurden AT, Rosa J-P, Fournier D, Legrand C, Didry D, Parquet A, Pidard D: A variant of Glanzmann's thrombasthenia with abnormal glycoprotein IIb-IIIa complexes in the platelet membrane. J Clin Invest 79:962-969, 1987.

38. Fournier DJ, Kabral A, Castaldi PA, Berndt MC: A variant of Glanzmann's thrombasthenia characterized by abnormal glycoprotein IIb/IIIa complex formation. Thromb Haemost 62:977-983, 1989.

39. Kato A: The biologic and clinical spectrum of Glanzmann's thrombasthenia: implications of integrin $a_{IIb}b_3$ for its pathogenesis. Crit Rev Oncol/Hematol 26:1-23, 1997.

40. Bray PF, Rosa JP, Lingappa VR, Kan YW, McEver RP, Shuman MA: Biogenesis of the platelet receptor for fibrinogen: Evidence for separate precursors for glycoproteins IIb and IIIa. Proc Natl Acad Sci USA 83:1480-1484, 1986.

41. Silver SM, McDonough MM, Vilaire G, Bennett JS: The in vitro synthesis of polypeptides for the platelet membrane glycoproteins IIb and IIIa. Blood 69:1031-1037, 1987.

42. Bray PF, Barsh G, Rosa J-P, Juo XY, Magenis E, Shuman MA: Physical linkage of the genes for platelet membrane glycoproteins IIb and IIIa. Proc Natl Acad Sci USA 85:8683-8687, 1988.

43. Sosnoski DM, Emanuel BS, Hawkins AL, van Tuilen P, Ledbetter DH, Nussbaum RL, Kaos F-T, Schwartz E, Phillips DR, Bennett JS, Fitzgerald LA, Poncz M: Chromosomal localization of the genes for the vitronectin and the fibronectin receptors a subunits and for platelet glycoprotein IIb and IIIa. J Clin Invest 81:1993-1998, 1988.

44. Peretz H, Yakobson E, Korostishevsky M, Usher S, Seligsohn U: Recombination distance between platelet glycoprotein IIB and glycoprotein IIIA genes. Thromb Haemost 73:1195, 1995 (Abstract).

45. Heidenreich R, Eisman R, Surrey S, Delgrosso K, Bennett JS, Schwartz E, Poncz M: Organization of the gene for platelet glycoprotein IIb. Biochemistry 29:1232-1244, 1990.

46. Zimrin AB, Gidwitz S, Lord S, Schwartz E, Bennett JS, White GCI, Poncz M: The genomic organization of platelet glycoprotein IIIa. J Biol Chem 265:8590-8595, 1990.

47. Villa-Garcia M, Li L, Riely G, Bray PF: Isolation and characterization of a TATA-less promoter for the human b3 integrin gene. Blood 83:668-676, 1994.

48. Poncz M, Risman R, Heidenreich R, Silver SM, Vilaire G, Surrey S, Schwartz E, Bennett JS: Structure of the platelet membrane glycoprotein IIb. J Biol Chem 262:8476-8482, 1987.

49. Fitzgerald LA, Steiner B, Rall RCJ, Lo S-S, Phillips DR: Protein sequence of endothelial glycoprotein IIIa derived from a cDNA clone. J Biol Chem 262:3936-3939, 1987.

50. Hynes RO: Integrins: a family of cell surface receptors. Cell 48:549-554, 1987.

51. Ginsberg MH, Loftus JC, Plow EF: Cytoadhesins, integrins, and platelets. Thromb Haemost 59:1-6, 1988.

52. Hynes RO: Integrins: versatility, modulation, and signaling in cell adhesion. Cell 69:11-25, 1992.

53. Calvete JJ, Henschen A, Gonzalez-Rodriguez J: Complete localization of the intrachain disulfide bonds and the N-glycosylation points in the a-subunit of human platelet glycoprotein IIb. Biochem J 261:561-568, 1989.

54. Poncz M, Newman PJ: Analysis of rodent platelet glycoprotein IIb: Evidence for evolutionarily conserved domains and alternative proteolytic processing. Blood 75:1281-89, 1990.

55. Tuckwell DS, Brass A, Humphries MJ: Homology modelling of integrin EF-hands. Evidence for widespread use of a conserved cation-binding site. Biochem J 285:325-331, 1992.

56. Gulino D, Boudignon C, Zhang L, Concord E, Rabiet M-J, Marguerie G: Ca^{2+}-binding properties of the platelet glycoprotein IIb ligand-interacting domain. J Biol Chem 267:1001-1007, 1992.

57. Springer TA: Folding of the N-terminal, ligand-binding region of integrin a-subunits into a b-propeller domain. Proc Natl Acad Sci USA 94:65-72, 1997.

58. Bogaert T, Brown N, Wilcox M: The Drosophila PS2 antigen is an invertebrate integrin that, like the fibronectin receptor, becomes localized to muscle attachments. Cell 51:929-940, 1987.

59. DeSimone DW, Hynes RO: *Xenopus laevis* integrins. Structural conservation and evolutionary divergence of integrin b subunits. J Biol Chem 263:5333-5340, 1988.

60. Calvete JJ, Henschen A, Gonzalez-Rodriguez J: Assignment of disulfide bonds in human platelet GPIIIa. A disulfide pattern for the b-subunits of the integrin family. Biochem J 274:63-71, 1991.

61. Wang R, Peterson J, Aster RH, Newman PJ: Disruption of a long-range disulfide bond between Cys_{406} and Cys_{655} in glycoprotein IIIa does not affect the function of platelet glycoprotein IIb-IIIa. Blood 90:1718-1719, 1997.

62. Santoso S, Kalb R, Kroll H, Walka M, Kiefel V, Mueller-Eckhardt C, Newman PJ: A point mutation leads to an unpaired cysteine residue and a molecular weight polymorphism of a functional platelet b_3 integrin subunit. J Biol Chem 269:8439-8444, 1994.

63. Grimaldi CM, Chen FP, Scudder LE, Coller BS, French DL: A Cys374Tyr homozygous mutation of platelet glycoprotein IIIa (b_3) in a Chinese patient with Glanzmann's thrombasthenia. Blood 88:1666-1675, 1996.

64. Loftus JC, O'Toole TE, Plow EF, Glass A, Frelinger AL, III, Ginsberg MH: A b_3 integrin mutation abolishes ligand binding and alters divalent cation-dependent conformation. Science 249:915-918, 1990.

65. Michishita M, Videm V, Arnaout M: A novel divalent cation-binding site in the A domain of the b2 integrin CR3 (CD11b/CD18) is essential for ligand binding. Cell 72:857-867, 1993.

66. Lee J-O, Rieu P, Arnaout MA, Liddington R: Crystal structure of the A domain from the a subunit of integrin CR3 (CD11b/CD18). Cell 80:631-638, 1995.

67. Tozer EC, Liddington RC, Sutcliffe MJ, Smeeton AH, Loftus JC: Ligand binding to integrin aIIbb3 is dependent on a MIDAS-like domain in the b3 subunit. J Biol Chem 271:21978-21984, 1996.

68. Uzan G, Prenant M, Prandini M-H, Martin F, Marguerie G: Tissue-specific expression of the platelet GPIIb gene. J Biol Chem 266:8932-8939, 1991.

69. Block KL, Poncz M: Platelet glycoprotein IIb gene expression as a model of megakaryocyte-specific expression. Stem Cells 13:135-145, 1995.

70. Prandini M-H, Uzan G, Martin F, Thevenon D, Marguerie G: Characterization of a specific erythro-megakaryocytic enhancer within the glycoprotein IIb promoter. J Biol Chem 267:10370-10374, 1992.

71. Denarier E, Martin F, Martineau S, Marguerie G: PCR cloning and sequence of the murine GPIIb promoter. Biochem Biophys Res Commun 195:1360-1364, 1993.

72. Block KL, Shou Y, Poncz m: An Ets/Sp1 interaction in the 5'-flanking region of the megakaryocyte-specific alpha Iib gene appears to stabilize Sp1 binding and is essential for expression of this TATA-less gene. Blood 88:2071-2080, 1996.

73. Cheresh DA, Berliner SA, Vincente V, Ruggeri ZM: Recognition of distinct adhesive sites on fibrinogen by related integrins on platelets and endothelial cells. Cell 58:945-953, 1989.

74. Wagner CL, Mascelli MA, Neblock DS, Weisman HF, Coller BS, Jordan RE: Analysis of GPIIb/IIIa receptor number by quantitation of 7E3 binding to human platelets. Blood 88:907-914, 1996.

75. Coller BS, Cheresh DA, Asch E, Seligsohn U: Platelet vitronectin receptor expression differentiates Iraqi-Jewish from Arab patients with Glanzmann thrombasthenia in Israel. Blood 77:75-83, 1991.

76. Reverter JC, Beguin S, Kessels H, Kumar R, Hemker HC, Coller BS: Inhibition of platelet-mediated, tissue factor-induced thrombin generation by the mouse/human chimeric 7E3 antibody. J Clin Invest 98:863-874, 1996.

77. Jennings LK, Ashmun RA, Wang WC, Dockter ME: Analysis of human platelet glycoproteins IIb-IIIa and Glanzmann's thrombasthenia in whole blood by flow cytometry. Blood 68:173-179, 1986.

78. Nurden AT, Didry D, Kieffer N, McEver RP: Residual amounts of glycoproteins IIb and IIIa may be present in the platelets of most patients with Glanzmann's thrombasthenia. Blood 65:1021-1024, 1985.

79. Coller BS, Seligsohn U, Little PA: Type I Glanzmann thrombasthenia patients from the Iraqi-Jewish and Arab populations in Israel can be differentiated by platelet glycoprotein IIIa immunoblot analysis. Blood 69:1969-1703, 1987.

80. Rosenberg N, Dardik R, Rosenthal E, Zivelin A, Seligsohn U: Mutations in the alphaIIb and beta3 genes that cause Glanzmann thrombasthenia can be distinguished by a simple procedure using transformed B-lymphocytes. Thromb Haemost 79:244-248, 1998.

81. Newman PJ, Seligsohn U, Lyman S, Coller BS: The molecular genetic basis of Glanzmann thrombasthenia in the Iraqi-Jewish and Arab populations in Israel. Proc Natl Acad Sci USA 88:3160-3164, 1991.

82. Burk CD, Newman PJ, Lyman S, Gill J, Coller BS, Poncz M: A deletion in the gene for glycoprotein IIb associated with Glanzmann's thrombasthenia. J Clin Invest 87:270-276, 1991.

83. Poncz M, Rifat S, Coller BS, Newman PJ, Shattil SJ, Parrella T, Fortina P, Bennett JS: Glanzmann thrombasthenia secondary to a Gly273Asp mutation adjacent to the first calcium-binding domain of platelet glycoprotein IIb. J Clin Invest 93:172-179, 1994.

84. French DL, Chen F, Xu L, Coller BS: A common Arg584Stop mutation in glycoprotein (GP) IIb in 2 additional patients with Glanzmann thrombasthenia. Blood 88:30a, 1996 (Abstract).

85. French DL, Chen FP, Xu L, Weiss HJ, Coller BS: Two unrelated doubly heterozygous patients with Glanzmann thrombasthenia have the same GPIIb mutation resulting in loss of the transmembrane domain. Thromb Haemost 77:360, 1997 (Abstract).

86. Grimaldi CM, Chen FP, Wu CH, Weiss HJ, Coller BS, French DL: Glycoprotein IIb Leu214Pro mutation produces Glanzmann thrombasthenia with both quantitative and qualitative abnormalities in GPIIb/IIIa. Blood 91:1562-1571, 1998.

87. Phillips DR, Agin PP: Platelet plasma membrane glycoproteins. Evidence for the presence of nonequivalent disulfide bonds using nonreduced-reduced two-dimensional gel electrophoresis. J Biol Chem 252:2121-2126, 1977.

88. Seligsohn U, Coller BS, Zivelin A, Plow EF, Ginsberg MH: Immunoblot analysis of platelet glycoprotein Iib in patients with Glanzmann thrombasthenia in Israel. Br J Haematol 72:415-423, 1989.

89. Kolodziej MA, Vilaire G, Rifat S, Poncz M, Bennett JS: Effect of deletion of glycoprotein IIb exon 28 on the expression of the platelet GPIIb/IIIa complex. Blood 74:14-18, 1991.

90. Coller BS, Seligsohn U, West SM, Scudder LE, Norton KJ: Platelet fibrinogen and vitronectin in Glanzmann thrombasthenia: Evidence consistent with specific roles for glycoprotein IIb/IIIa and a_vb_3 integrins in platelet protein trafficking. Blood 78:2603-2610, 1991.

91. Handagama P, Rappolee DA, Werb Z, Levin J, Bainton DF: Platelet a-granule fibrinogen, albumin, and immunoglobulin G are not synthesized by rat and mouse megakaryocytes. J Clin Invest 86:1364-1368, 1990.

92. Harrison P: Platelet a-granular fibrinogen. Platelets 3:1-11, 1992.

93. Harrison P, Wilbourn B, Debili N, Vainchenker W, Breton-Gorius J, Lawrie AS, Masse J-M, Savidge GF, Cramer EM: Uptake of plasma fibrinogen into the alpha granules of human megakaryocytes and platelets. J Clin Invest 84:1320-1324, 1989.

94. Handagama P, Scarborough RM, Shuman MA, Bainton DF: Endocytosis of fibrinogen into megakaryocytes and platelet a-granules is mediated by $a_{IIb}b_3$ (glycoprotein IIb-IIIa). Blood 82:135-138, 1993.

95. Hardisty R, Pidard D, Cox A, Nokes T, Legrand C, Bouillot C, Pannocchia A, Heilmann E, Hourdille P, Bellucci S, Nurden A: A defect of platelet aggregation with an abnormal distribution of glycoprotein IIb-IIIa complexes within the platelet. Blood 80:696-708, 1992.

96. Saiki RK, Gelfand DH, Stoffel S, Scharf SJ, Higuchi RG, Horn GT, Mullis KB, Erlich HA: Primer-directed enzymatic amplification of DNA with a thermostable DNA polymerase. Science 239:487, 1988.

97. Newman PJ, Gorski J, White GCI, Gidwitz S, Cretney CJ, Aster RH: Enzymatic amplification of platelet-specific messenger RNA using the polymerase chain reaction. J Clin Invest 82:739-743, 1988.

98. Jin Y, Dietz HC, Montgomery RA, Bell WR, McIntosh I, Coller BS, Bray PF: Glanzmann thrombasthenia. Cooperation between sequence variants in *cis* during splice site selection. J Clin Invest 98:1745-1754, 1996.

99. Phillips DR, Charo IF, Scarborough RM: GPIIb-IIIa: The responsive integrin. Cell 65:359-362, 1991.

100. Shattil SJ, Ginsberg MH, Brugge JS: Adhesive signaling in platelets. Curr Opin Cell Biol 6:695-704, 1994.

101. Savage B, Shattil SJ, Ruggeri ZM: Modulation of platelet function through adhesion receptors. J Biol Chem 267:11300-11308, 1992.

102. Weiss HJ: Flow-related platelet deposition on subendothelium. Throm Haemost 74:117-122, 1995.

103. Schwartz MA, Schaller MD, Ginsberg MH: Integrins - emerging paradigms of signal-transduction. Annu Rev Cell Biol 11:549-599, 1995.

104. Yamada KM, Geiger B: Molecular interactions in cell adhesion complexes. Current Biol 9:76-85, 1997.

105. Haas TA, Plow EF: Integrin-ligand interactions: a year in review. Curr Opin Cell Biol 6:656-662, 1994.

106. Ginsberg MH, Du X, O'Toole TE, Loftus JC: Platelet integrins. Thromb Haemost 74:352-359, 1995.

107. Loftus JC, Halloran CE, Ginsberg MH, Feigen LP, Zablocki JA, Smith JW: The amino-terminal one-third of a_{IIb} defines the ligand recognition specificity of integrin $a_{IIb}b_3$. J Biol Chem 271:2033-2039, 1996.

108. Wippler J, Kouns WC, Schlaeger E-J, Kuhn H, Hadvary P, Steiner B: The integrin aIIb-b3, platelet glycoprotein IIb-IIIa, can form a functionally active heterodimer complex without the cysteine-rich repeats of the b3 subunit. J Biol Chem 269:8754-8761, 1994.

109. D'Souza SE, Haas TA, Piotrowicz RS, Byers-Ward V, McGrath DE, Soule HR, Cierniewski C, Plow EF, Smith JW: Ligand and cation binding are dual functions of a discrete segment of the integrin b3 subunit: Cation displacement is involved in ligand bindin. Cell 79:659-667, 1994.

110. D'Souza S, Ginsberg MH, Burke T, Lam S, Plow E: Localization of an Arg-Gly-Asp recognition site within an integrin adhesion receptor. Science 242:91-93, 1988.

111. Andrieux A, Rabiet MJ, Chapel A, Concord E, Marguerie G: Highly conserved sequence of the Arg-Gly-Asp-binding domain of the integrin b3 subunit is sensitive to stimulation. J Biol Chem 266:14202-14207, 1991.

112. Calvete JJ, Arias J, Alvarez MV, Lopez MM, Henshen A, Gonzalez-Rodriguez J: Further studies on the topography of the N-terminal region of human platelet glycoprotein IIIa. Localization of monoclonal epitopes and the putative fibrinogen-binding sites. Biochem J 274:457-463, 1991.

113. Charo IF, Nannizzi L, Phillips DR, Hsu MA, Scarborough RM: Inhibition of fibrinogen binding to GP IIb-IIIa by a GP IIIa peptide. J Biol Chem 266:1415-1421, 1991.

114. Lanza F, Stierle A, Fournier D, Morales M, Andre G, Nurden AT, Cazenave J-P: A new variant of Glanzmann's thrombasthenia (Strasbourg I). Platelets with functionally defective glycoprotein IIb-IIIa complexes and a glycoprotein IIIa ^{214}Arg‡^{214}Trp mutation. J Clin Invest 89:1995-2004, 1992.

115. Bajt ML, Loftus JC: Mutation of a ligand binding domain of b3 integrin. Integral role of oxygenated residues in aIIbb3 (GPIIb/IIIa) receptor function. J Biol Chem 269:20913-20919, 1994.

116. Lee J-O, Bankston LA, Arnaout MA, Liddington RC: Two conformations of the integrin A-domain (I-domain): a pathway for activation? Structure 3:1333-1340, 1995.

117. Ward CM, Chao YL, Kato GJ, Casella J, Bray PF, Newman PJ: Substitution of Asn, but not Tyr, for Asp119 of the b_3 integrin subunit preserves fibrin binding and clot retraction. Blood 90:26a, 1997 (Abstract).

118. Djaffer I, Rosa J-P: A second case of variant of Glanzmann's thrombasthenia due to substitution of platelet GPIIIa (integrin b_3) Arg214 by Trp. Hum Mol Genet 2:2179-2180, 1993.

119. Bajt ML, Ginsberg MH, Frelinger AL, III, Berndt MC, Loftus JC: A spontaneous mutation of integrin $a_{IIb}b_3$ (platelet glycoprotein IIb-IIIa) helps define a ligand binding site. J Biol Chem 267:3789-3794, 1992.

120. Newman PJ, Weyerbusch-Bottum S, Visentin GP, Gidwitz S, White GCI: Type II Glanzmann thrombasthenia due to a destabilizing amino acid substitution in platelet membrane glycoprotein IIIa. Thromb Haemost 69:1017, 1993 (Abstract).

121. Jackson DE, Jennings LK, Newman PJ: A mutation within the ligand binding domain of glycoprotein (GP) IIIa (integrin b_3) results in an unstable GPIIb-IIIa complex that retains partial function in a novel form of Glanzmann thrombasthenia. Blood 88:280a, 1996 (Abstract).

122. Westrup D, Santoso S, Becker-Hagendorff K, Just M, Jablonka B, Siefried E, Kirchmaier CM: Transfection of GPIIb$_{Ile176}$/IIIa (Frankfurt I) in mammalian cells. Thromb Haemost 77:671, 1997 (Abstract).

123. Basani R, Bennett JS, Poncz M: A Glanzmann thrombasthenia variant due to an a_{IIb} mutation suggests that an additional N-terminal loop is involved in ligand-binding. Blood 90:26a, 1997 (Abstract).

124. Kirchmaier CM, Westrup D, Becker-Hagendorff K, Just M, Jablonka B, Seifried E: A new variant of Glanzmann thrombasthenia (Frankfurt I). Thromb Haemost 73:1058, 1995 (Abstract).

125. Tuckwell DS, Humphries MJ, Brass A: A secondary structure model of the integrin a subunit N-terminal domain based on analysis of multiple alignments. Cell Adhesion Commun 2:385-402, 1994.

126. Kamata T, Irie A, Tokuhira M, Takada Y: Critical residues of integrin aIIb subunit for binding of aIIbb3 (glycoprotein IIb-IIIa) to fibrinogen and ligand-mimetic antibodies (PAC-1, OP-G2, and LJ-CP3). J Biol Chem 271:18610-18615, 1996.

127. Irie A, Kamata T, Puzon-McLaughlin W, Takada Y: Critical amino acid residues for ligand binding are clustered in a predicted b-turn of the third N-terminal repeat in the integrin a4 and a5 subunits. EMBO J 14:5550-5556, 1995.

128. Chen Y-P, O'Toole TE, Ylanne J, Rosa J-P, Ginsberg MH: A point mutation in the integrin b_3 cytoplasmic domain (S^{752}‡P) impairs bidirectional signaling through $a_{IIb}b_3$ (platelet glycoprotein IIb-IIIa). Blood 84:1857-1865, 1994.

129. Ylanne J, Huuskonen J, O'Toole TE, Ginsberg MH, Virtanen I, Gahmberg CG: Mutation of the cytoplasmic domain of the integrin b_3 subunit. J Biol Chem 270:9550-9557, 1995.

130. Wang R, Shattil SJ, Ambruso DR, Newman PJ: Truncation of the cytoplasmic domain of b_3 in a variant form of Glanzmann thrombasthenia abrogates signaling through the integrin $a_{IIb}b_3$ complex. J Clin Invest 100:2393-2403, 1997.

131. Kato A, Yamamoto K, Miyazaki S, Jung SM, Moroi M, Aoki N: Molecular basis for Glanzmann's thrombasthenia (GT) in a compound heterozygote with glycoprotein IIb gene: A proposal for the classification of GT based on the biosynthetic pathway of glycoprotein IIb-IIIa complex. Blood 79:3212-3218, 1992.

132. Gu JM, Xu WF, Wang XD, Wu QY, Chi CW, Ruan CG: Identification of a nonsense mutation at amino acid 584-arginine of platelet glycoprotein IIb in patients with type I Glanzmann thrombasthenia. Br J Haematol 83:442-449, 1993.

133. Vinciguerra C, Trzeciak MC, Philippe N, Frappaz D, Reynaud J, Dechavanne M, Negrier C: Molecular study of Glanzmann thrombastheni in 3 patients issued from 2 different families. Thromb Haemost 74:822-827, 1994.

134. Tomiyama Y, Kashiwagi H, Kosugi S, Shiraga M, Kanayama Y, Kurata Y, Matsuzawa Y: Abnormal processing of the glycoprotein IIb transcript due to a nonsense mutation in exon 17 associated with Glanzmann's thrombasthenia. Thromb Haemost 73:756-762, 1995.

135. Morse DE, Yanofsky C: Polarity and the degradation of mRNA. Nature 224:329-331, 1969.

136. Peltz SW, Brown AH, Jacobson A: Regulation of mRNA turnover in eukaryotic cells. Crit Rev Eukaryot Gene Expression 1:99-126, 1993.

137. Peltz SW, Feng H, Welch E, Jacobson A: Nonsense-mediated mRNA decay in yeast. Prog Nucleic Acid Res Mol Biol 47:271-298, 1994.

138. Hagan KW, Ruiz-Echevarria MJ, Quan Y, Peltz SW: Characterization of cis-acting sequences and decay intermediates involved in nonsense-mediated mRNA turnover. Mol Cell Biol 15:809-823, 1995.

139. Baserga SJ, Benz EJJ: Nonsense mutations in the human b-globin gene affect mRNA metabolism. Proc Natl Acad Sci USA 85:2056-2060, 1988.

140. Djaffar I, Caen JP, Rosa J-P: A large alteration in the human platelet glycoprotein IIIa (integrin b_3) gene associated with Glanzmann's thrombasthenia. Hum Mol Genet 2:2183-2185, 1993.

141. Atweh GF, Brickner HE, Zhu X-X, Kazazian HHJ, Forget BG: New amber mutation in a b-thalassemic gene with nonmeasurable levels of mutant messenger RNA in vivo. J Clin Invest 82:557-561, 1988.

142. Vinciguerra C, Khelif A, Alemany M, Morle F, Grenier C, Uzan G, Gulino D, Dechavanne M, Negrier C: A nonsense mutation in the GPIIb heavy chain (Ser 870Æstop) impairs platelet GPIIb-IIIa expression. Br J Haematol 95:399-407, 1996.

143. Iwamoto S, Nishiumi E, Kajii E, Ikemoto S: An exon 28 mutation resulting in alternative splicing of the glycoprotein IIb transcript and Glanzmann's thrombasthenia. Blood 83:1017-1023, 1994.

144. Peretz H, Rosenberg N, Usher S, Graff E, Newman PJ, Coller BS, Seligsohn U: Glanzmann's thrombasthenia associated with deletion-insertion and alternative splicing in the glycoprotein IIb gene. Blood 85:414-420, 1995.

145. Calvete JJ, Mann K, Alvarez MV, Lopez MM, Gonzalez-Rodriguez J: Proteolytic dissection of the isolated platelet fibrinogen integrin GPIIb/IIIa. Localization of GPIIb and GPIIIa sequences putatively involved in the subunit interface and in intrasubunit and intrachain contacts. Biochem J 282:523-532, 1992.

146. Kolodziej MA, Vilaire G, Gonder D, Poncz M, Bennett JS: Study of the endoproteolytic cleavage of platelet glycoprotein IIb using oligonucleotide-mediated mutagenesis. J Biol Chem 266:23499-23504, 1991.

147. Morel-Kopp M-C, Kaplan C, Proulle V, Jallu V, Melchior C, Peyruchaud O, Aurousseau M-H, Kieffer N: A three amino acid deletion in glycoprotein IIIa is responsible for type I Glanzmann's thromasthenia: Importance of residues $Ile^{325}Pro^{326}Gly^{327}$ for b3 integrin subunit association. Blood 90:669-677, 1997.

148. Simsek S, Heyboer H, de Bruijne-Admiraal LG, Goldschmeding R, Cuijpers HTM, von dem Borne AEGK: Glanzmann's thrombasthenia caused by homozygosity for a splice defect that leads to deletion of the first coding exon of the glycoprotein IIIa mRNA. Blood 81:2044-2049, 1993.

149. Peerschke EJ, Grant RA, Zucker MB: Decreased association of ^{45}calcium with platelets unable to aggregate due to thrombasthenia or prolonged calcium deprivation. Br J Haematol 46:247-256, 1980.

150. Brass LF, Shattil SJ: Identification and function of the high affinity binding sites for Ca^{2+} on the surface of platelets. J Clin Invest 73:626-632, 1984.

151. Wilcox DA, Wautier JL, Pidard D, Newman PJ: A single amino acid substitution flanking the fourth calcium binding domain of a_{IIb} prevents maturation of the integrin $a_{IIb}b_3$ complex. J Biol Chem 269:4450-4457, 1994.

152. Bourre R, Peyruchaud O, Bray P, Combrie R, Nurden P, Nurden AT: A point mutation in the gene for platelet GPIIb leads to a substitution in a highly conserved amino acid located between the second and the third Ca^{++}-binding domain. Blood 86:452a, 1995 (Abstract).

153. Basani RB, Vilaire G, Shattil SJ, Kolodziej MA, Bennett JS, Poncz M: Glanzmann thrombasthenia due to a two amino acid deletion in the fourth calcium-binding domain of a_{IIb}: Demonstration of the importance of calcium-binding domains in the conformation of $a_{IIb}b_3$. Blood 88:167-173, 1996.

154. Wilcox DA, Paddock CM, Lyman S, Gill JC, Newman PJ: Glanzmann thrombasthenia resulting from a single amino acid substitution between the second and third calcium-binding domains of GPIIb. J Clin Invest 95:1553-1560, 1995.

155. Ferrer M, Fernandez-Pinel M, Gonzalez-Manchon C, Gonzalez J, Ayuso MS, Parrilla R: A mutant (Arg^{327}‡His) GPIIb associated to thrombasthenia exerts a dominant negative effect in stably transfected CHO cells. Thromb and Haemost 76:292-301, 1996.

156. Jackson DE, Poncz M, Holyst MT, Newman PJ: Inherited mutations within the calcium-binding sites of the integrin aIIb subunit (platelet glycoprotein IIb). Effects of the amino acid side chain and the amino acid position on cation binding. Eur J Biochem 240:280-287, 1996.

157. Kahn MJ, Kieber-Emmons T, Vilarie G, Murali R, Poncz M, Bennett JS: Effect of mutagenesis of GPIIb amino acid 273 on the expression and conformation of the platelet integrin GPIIb-IIIa. Biochemistry 35:14304-14311, 1996.

158. Basani RB, Brown DL, Vilaire G, Bennett JS, Poncz M: A Leu^{117}ÆTrp mutation within the RGD-peptide cross-linking region of b_3 results in Glanzmann thrombasthenia by preventing $a_{IIb}b_3$ export to the platelet surface. Blood 90:3082-3088, 1997.

159. Peyruchaud O, Nurden AT, Bourre F: Use of PCR-SSCP to screen the exons of the GP IIb and GP IIIa genes of a variant with Glanzmann thrombasthenia: A mutation in the nucleotide sequence for the GFFKR cytoplasmic domain of the integrin subunit a_{IIb} (GPIIb). Thromb Haemost 73:1189, 1995 (Abstract).

160. Hughes PE, Diaz-Gonzalez F, Leong L, Wu C, McDonald JA, Shattil SJ, Ginsberg MH: Breaking the integrin hinge. A defined structural constraint regulates integrin signaling. J Biol Chem 271:6571-6574, 1996.

161. Cooper DN, Youssoufian H: The CpG dinucleotide and human genetic disease. Hum Genet 78:151-155, 1988.

162. Ramsahoye BH, Davies CS, Mills KI: DNA methylation: biology and significance. Blood Rev 10:249-261, 1996.

163. Coller BS, Seligsohn U, Zivelin A, Zwang E, Lusky A, Modan M: Immunologic and biochemical characterization of homozygous and heterozygous Glanzmann thrombasthenia in the Iraqi-Jewish and Arab populations in Israel: comparison of techniques for carrier detection. Br J Haematol 62:723-735, 1986.

164. Peretz H, Seligsohn U, Zwang E, Coller BS, Newman PJ: Detection of the Glanzmann's thrombasthenia mutations in Arab and Iraqi-Jewish patients by polymerase chain reaction and restriction analysis of blood or urine samples. Thromb Haemost 66:500-504, 1991.

165. Seligsohn U, Mibashan RS, Rodeck CH, Nicolaides KH, Millar DS, Coller BS: Prenatal diagnosis of Glanzmann's thrombasthenia. Lancet II:1419, 1985 (Letter).

166. French DL, Coller BS, Usher S, Berkowitz R, Eng C, Seligsohn U, Peretz H: Prenatal diagnosis of Glanzmann thrombasthenia using the polymorphic markers BRCA1 and THRA1 on chromosome 17. submitted , 1998.

167. Wilcox DA, Olsen JC, Ishizawa L, Bray PF, French DL, Bell WR, Griffith M, White GC: Phenotypic correction of Glanzmann's thrombasthenia following megakaryocyte-targeted synthesis of the integrin b_3-subunit. Blood 90:281a, 1997 (Abstract).

168. Newman PJ: Platelet GPIIb-IIIa: molecular variations and alloantigens. Thromb Haemostat 66:111-118, 1991.

169. Kunicki TJ, Newman PJ: The molecular immunology of human platelet proteins. Blood 80:1386-1404, 1992.

170. Nurden AT: Polymorphisms of human platelet membrane glycoproteins: Structure and clinical significance. Thromb Haemostat 74:345-351, 1995.

171. Hodivala-Dilke KM, McHugh K, Tsakiris DA, Rayburn H, Ross FP, Coller BS, Teitelbaum S, Hynes RO: Beta 3 integrin knockout mice display a Glanzmann thrombasthenia phenotype. Blood 90:2550a, 1997 (Abstract).

19

Congenital Disorders of Platelet Signal Transduction and Secretion

A. Koneti Rao, M.D.
Departments of Medicine and Pathology
Sol Sherry Thrombosis Research Center
Temple University School of Medicine
Philadelphia, PA
19140, USA

19.1 INTRODUCTION

Platelets have been recognized as bona fide members of the hematologic community only since the middle of the 19th century (1). Their critical role in hemostasis and in the pathogenesis of thrombosis and atherosclerosis is irrefutably established. Reviewed here are selected aspects of the platelet role in hemostasis and the inherited defects that lead to impaired platelet function. Recognition and delineation of the abnormal mechanisms in patients with platelet dysfunction have provided invaluable insights into several key aspects of platelet physiology, and lay the foundation for developing newer antiplatelet agents.

Dr. A. Koneti Rao has made significant contributions to this area of research, specifically on congenital disorders of platelet signal transduction & secretion. Dr. Rao has published numerous papers in this area.

Hemostasis is a complex sequence of interrelated events involving the vessel wall, platelets, and the coagulation system. Following injury to the blood vessel, platelets adhere to exposed subendothelium by a process (adhesion) which involves the interaction of a plasma protein, von Willebrand factor (vWF), and a specific protein on the platelet surface, glycoprotein Ib (GPIb) (Fig 1). Adhesion is followed by recruitment of additional platelets which form clumps, a process called aggregation (cohesion). This platelet-platelet interaction involves binding of fibrinogen to specific platelet surface receptors- a complex comprised of glycoproteins IIb-IIIa (GPIIb-IIIa). In the resting state, platelets do not bind fibrinogen; platelet activation induces a conformational change in the GPIIb-IIIa complexes leading to fibrinogen binding and a sequence of events resulting in aggregation. On activation, GPIIb-IIIa complexes acquire also the capacity to bind other adhesive proteins, including vWF, fibronectin, and vitronectin. Activated platelets release contents of their granules (secretion or release reaction), such as adenosine diphosphate (ADP) and serotonin from the dense granules, which cause recruitment of additional platelets. There is also release of proteins present in the alpha granules and the acid hydrolase containing vesicles. In addition, platelets play a major role in hemostasis by contributing to coagulation mechanisms; several key enzymatic reactions in blood coagulation occur on the platelet membrane lipoprotein surface. Thrombin generation and formation of a clot composed of blood cells and fibrin strands leads to restoration of hemostasis.

Platelets are tiny anucleate but biochemically highly active cells that contain three types of storage granules: dense granules, a-granules and acid hydrolase-containing organelles (lysosomes). Platelet stimulation results in the extrusion by exocytosis of the granule contents by an energy dependent process. Platelets have an active machinery for production and use of ATP. The adenine nucleotides (ATP, ADP) in platelets exist in two compartments: the storage (secretable) pool within the dense granules and the metabolic (cytoplasmic) pool outside the granules (2). A number of physiological agonists interact with specific receptors on platelet surface to induce a series of responses including a change in platelet shape from discoid to spherical (shape change), aggregation, secretion, and thromboxane A2 (TxA2). The agonists include ADP, epinephrine, TxA2, thrombin, platelet activating factor (PAF), collagen, vasopressin and serotonin, which vary in their relative abilities to induce various responses.

19.2 Signal transduction mechanisms (Fig 1)

Extensive studies over the last 3 decades have unraveled some of the intricate biochemical events that occur on platelet activation and culminate in the end responses such as shape change, aggregation and secretion. Interaction of platelets with an agonist initiates the production or release of several intracellular messenger molecules including Ca2+ ions, products of phosphoinositide (PI) hydrolysis (diacylglycerol, DG, and inositol 1,4,5-triphosphate, InsP3), TxA2 and cyclic nucleotides (cAMP) (3-5) (Fig 1). These modulate the various discernible platelet responses of Ca2+ mobilization, protein phosphorylation, aggregation, secretion and liberation of arachidonic acid. The interaction between the agonist receptors on the platelet surface and the key intracellular effector enzymes (e.g. phospholipases A2 and C, adenylyl cyclase) are mediated by a group of GTP-binding proteins which are modulated by GTP (6). These G-proteins function as molecular off and on switches in regulating the transduction of signals from the surface receptors to intracellular effectors and enzymes. Phosphoinositides constitute a small fraction of the total platelet membrane lipids and, on platelet stimulation, they are hydrolyzed by phospholipase C (PLC) to DG and various inositol phosphates including 1,4,5 InsP3(4). Diacylglycerol is hydrolyzed by lipases to free arachidonic acid and glycerol, or is phosphorylated by a kinase to form phosphatidic acid, which then recycles to PI. As in most secretory cells, platelet activation results in a rise in cytoplasmic ionized calcium concentration which is a combination of release of Ca2+ from intracellular stores and influx of external Ca2+ (7). InsP3 functions as a messenger to mobilize Ca2+ from intracellular source, the dense tubular system in platelets (5). Functional consequences of intracellular Ca2+ mobilization include phosphorylation of myosin light chain by a specific Ca2+ dependent kinase. This has been considered to play a role in shape change and secretion. Another Ca2+ dependent process in platelets is the release of arachidonic acid from phospholipids by the action of the Ca2+ dependent phospholipase A2; this is the rate-limiting step in thromboxane A2 (TxA2) synthesis. The free arachidonic acid, liberated by the hydrolysis of phospholipids (predominantly phosphatidylcholine) by phospholipase A2, is converted by cyclooxygenase to prostaglandins G2 and H2, and subsequently by thromboxane synthetase to TxA2. In the presence of suitable phospholipids, diacylglycerol activates protein kinase C (PKC) at basal intracellular Ca2+ levels and this results in the phosphorylation of a protein of molecular weight approximately 47 kD (Pleckstrin) (8). PKC activation is considered to play a major role in platelet secretion (8) and in the expression of platelet surface fibrinogen binding sites (consisting of GP IIb-IIIa), which is a requisite for platelet aggregation.

19.3 Role of platelets in blood coagulation

An important role of platelets in hemostasis is their contribution to several of the key enzymatic events of the coagulation cascade which occur on the platelet surface membrane (9). Platelet a-granules are endowed with several coagulation factors, and platelets promote blood coagulation by various mechanisms including exposure of specific binding sites for the coagulation proteins, providing the surface for assembly of the involved coagulation proteins thereby inducing a profound acceleration of the enzymatic reactions, and by protecting the coagulation enzymes from inactivation by naturally occurring inhibitors. They play a remarkable role in specific coagulation steps such as in activation of prothrombin and factor X. For example, activated platelets express activated factor V (Va) on their surface which serves as a binding site for factor Xa; thus, platelets participate in the conversion of prothrombin to thrombin (Fig 1).

19.4 Congenital disorders of platelet function

Disorders of platelet function are characterized by bleeding manifestations that are highly variable among patients with different underlying defects. The manifestations are mucocutaneous in nature and excessive hemorrhage may follow surgical procedures or trauma. Platelet counts and morphology are normal in most patients. While the majority of patients have a prolonged bleeding time, it is not always the case. Platelet aggregation and secretion studies provide evidence for the defect, however, these parameters are not universally predictive of the severity of clinical manifestations. In the studies performed using platelet-rich plasma, as is the practice in most hospital clinical laboratories, platelet aggregation in response to increasing concentrations of agonists such as ADP and epinephrine consists of two phases - the initial wave of aggregation (primary aggregation) resulting from the interaction of the agonist with the platelet receptors, and the secondary irreversible aggregation related to the release of dense granule contents (10). Thus, patients with impaired release of dense granule contents or TxA2 production generally demonstrate a primary wave but not the second wave of aggregation. Patients with thrombasthenia or congenital afibrinogenemia have absent or markedly decreased primary wave of aggregation as well (10).

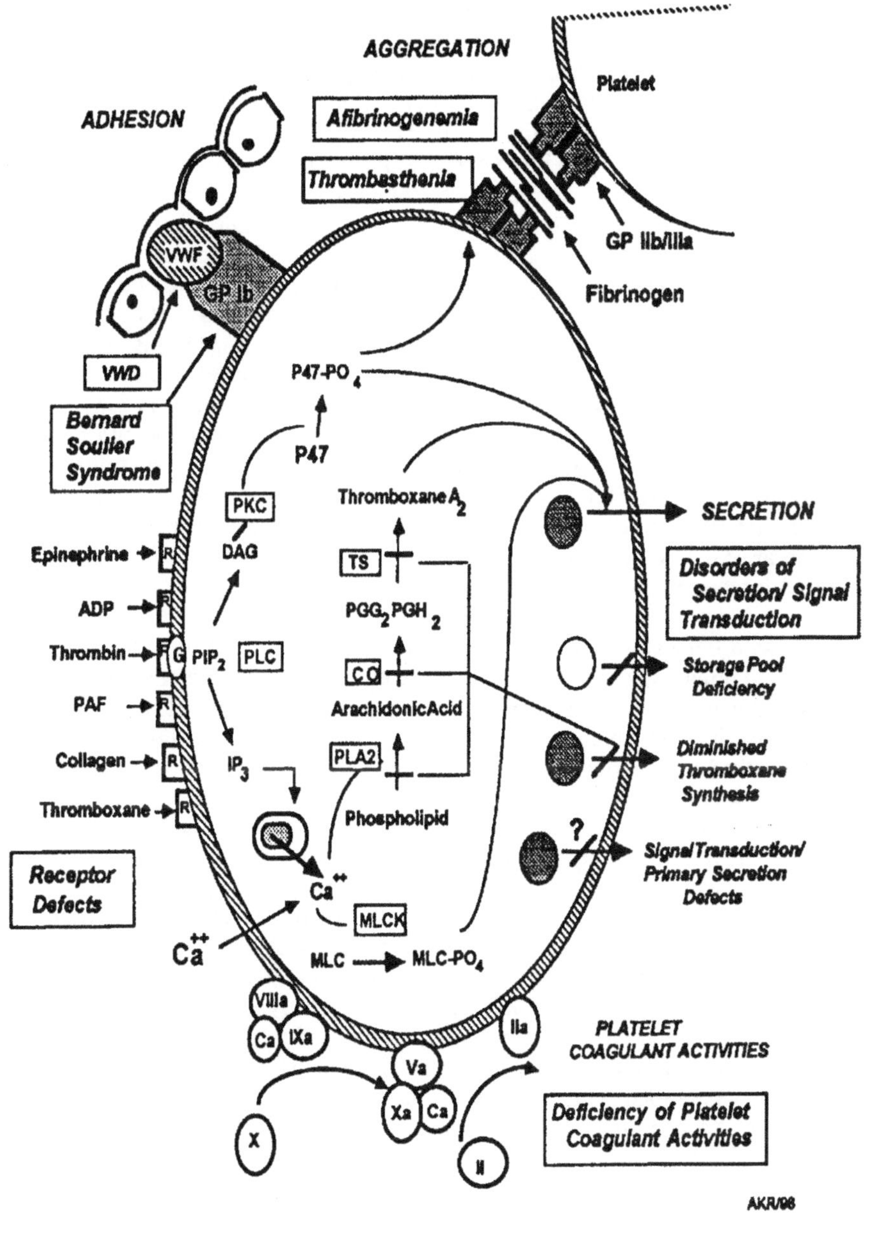

Fig 1. A schematic representation of the normal platelet responses and the congenital disorders of platelet function. Abbreviations: CO, cyclooxygenase; DAG, diacylglycerol; IP3, inositoltrisphosphate; MLC, myosin light chain; MLCK, myosin light chain kinase; PIP2, phosphatidylinositol bisphosphate; PKC, protein kinase C; PLC, phospholipase C; PLA2, phospholipase A2; TS, thromboxane synthase; vWF, von Willebrand factor; vWD, von Willebrand disease.

Although it is clear that platelet dysfunction may arise by diverse mechanisms (11-14), the specific molecular mechanisms underlying the altered function remain unknown in a large proportion of patients with congenital disorders of platelet function. Table 1 provides a classification based on the platelet functions or responses that are abnormal and these are depicted in Fig 1. In patients with defects in platelet-vessel wall interactions, adhesion of platelets to subendothelium is abnormal. The two disorders in this group are the von Willebrand disease (vWD), (15-16) due to a deficiency or abnormality in plasma vWF, and the Bernard-Soulier syndrome, where platelets are deficient in GPIb (and GPV and IX) and the binding of vWF to platelets is abnormal (17). Disorders of platelet-platelet interaction (aggregation) are characterized by severe deficiency of plasma fibrinogen (congenital afibrinogenemia) or by quantitative or qualitative abnormalities of the platelet membrane GPIIb-IIIa complex (Glanzmann's thrombasthenia) (18). The activated GPIIb-IIIa complexes constitute the platelet binding sites for fibrinogen; this binding is a pre-requisite for platelet aggregation. Platelets from these patients fail to aggregate in response to all physiological platelet agonists, and even the primary wave of platelet aggregation is absent. Patients with the Bernard Soulier syndrome and Glanzmann's thrombasthenia have defects in specific platelet membrane glycoproteins (GPIb and GPIIb-IIIa) that mediate binding of plasma protein, vWF and fibrinogen, respectively, to platelets. In contrast, in vWD and congenital afibrinogenemia the abnormalities in platelet function arise due to deficiencies of plasma proteins that interact with platelets and are essential for their role in hemostasis.

Patients with defects in platelet secretion and signal transduction are a heterogeneous population grouped together for convenience of classification rather than on the basis of an understanding of the specific underlying abnormality. The major common characteristics in these patients, as currently perceived, are blunted aggregation responses and diminished release of intracellular granule contents upon activation of platelet-rich plasma with agonists such as ADP, epinephrine, and collagen. In aggregation studies the second wave of aggregation is generally blunted or absent while the primary wave is present, although it too may also be impaired in some patients. A small proportion of these patients have a deficiency of dense granule stores (storage pool deficiency). Recent studies suggest that in some patients, the impaired secretion results from aberrations in the signal transduction events that govern end responses such as secretion and aggregation. This review will focus on these patients who are clearly encountered more often than thrombasthenia or the Bernard Soulier syndrome. Lastly, are the patients who have an abnormality in interactions of platelets with proteins of the coagulation system, so that the platelet contribution to the coagulation cascade (platelet coagulant activity) is impaired (19). The best described is the Scott syndrome where the platelets have impaired binding of factor Va-Xa and factor VIIIa-IXa resulting in impaired factor X and prothrombin activation, platelet-dependent fibrin-formation, and an abnormality in platelet factor 3 activity (19). These defects have been linked to an abnormality in calcium mediated vesiculation of the plasma membrane and decreased membrane expression of phosphatidylserine. In addition to the above groups, there are patients who have abnormalities in platelet function associated with systemic disorders such as Down's syndrome and the May-Heggelin anomaly where the specific aberrant mechanisms in platelets still needs to be delineated.

19.5 Disorders of platelet secretion and signal transduction

In patients lumped in this group, there is impairment in secretion of granule contents during platelet activation and they demonstrate absence of the second wave of aggregation during stimulation of platelet-rich plasma with ADP and epinephrine.

Table 1: Classification of Congenital Disorders of Platelet Function

1. Defects in platelet-vessel wall interaction (disorders of adhesion)

   a. von Willebrand disease (deficiency or defect in plasma vWF)
   b. Bernard-Soulier syndrome (deficiency or defect in GPIb)

2. Defects in platelet-platelet interaction (disorders of aggregation)

   a. Congenital Afibrinogenemia (deficiency of plasma fibrinogen)
   b. Glanzmann's thrombasthenia (deficiency or defect in GPIIb-IIIa)

3. Disorders of platelet secretion and signal transduction

   a. Abnormalities of Granules

      i. Storage pool deficiency
      ii. Quebec platelet disorder

   b. Signal transduction defects (primary secretion defects)
      i. Defects in platelet-agonist interaction (receptor defects)
      ii. Defects in G-protein activation
      iii. Defects in phosphatidylinositol metabolism
      iv. Defects in calcium mobilization
      v. Defects in protein phosphorylation (pleckstrin)

   c. Abnormalities in arachidonic acid pathways and thromboxane A2 synthesis

      i. Impaired liberation of arachidonic acid
      ii. Cyclooxygenase deficiency
      iii. Thromboxane synthase deficiency

4. Disorders of platelet coagulant-protein interaction

   Defect in factor Va-Xa interaction on platelets (Scott Syndrome)

5. Miscellaneous congenital disorders
   May-Heggelin anomaly
   Down's syndrome

Aggregation in response to collagen, arachidonic acid, and platelet-activating factor (PAF) may also be impaired. In these patients platelet function is abnormal either when the granules or their contents are diminished (storage pool deficiency, SPD) or when there is an aberration in the activation mechanisms governing secretion and other responses (Table 1).

Deficiency of Granule Stores

The term storage pool deficiency (SPD) refers to patients with deficiencies in platelet content of dense granules (d-SPD), alpha-granules (a-SPD) or both groups of granules (ad-SPD) (13,14). By electron microscopy, d-SPD platelets have a marked decrease in dense granules, and the total platelet and granule ATP and ADP contents are decreased. Other dense granule substance including serotonin, calcium, and pyrophosphate are also decreased. d-SPD has been reported in association with other inherited systemic disorders, such as the Hermansky-Pudlak syndrome (associated with oculocutaneous albinism), the Chediak-Higashi syndrome, and the Wiskottt-Aldrich syndrome. Abnormal granule morphogenesis is considered as the mechanism for d-SPD and these platelets are deficient in granulophysin, a 40 Kd granule protein. In patients with isolated deficiency of platelet a-granule contents (Gray Platelet syndrome) the megakaryocytes and platelets are selectively deficient in a-granule proteins such as platelet factor 4, b-thromboglobulin, fibrinogen, and the platelet-derived growth factor (PDGF) (13,14) The platelets appear gray on the peripheral smear. Impaired platelet aggregation and secretion have been reported consistently in response to thrombin but not other agonists. Gray platelet syndrome appears to result from a defect in the packaging of a-granule proteins into the granules in megakaryocytes with premature release of substance including PDGF. The Quebec platelet disorder is an autosomal dominant disorder associated with abnormal proteolysis of a-granule proteins (14,20), deficiency of platelet a-granule multimerin, and markedly impaired aggregation with epinephrine as a striking feature. Multimerin, a multimeric factor V binding protein, is decreased in the a-granule and, on immunoblot analyses, platelet factor V is degraded along with several other a-granule proteins including fibrinogen, vWF, thrombospondin, osteonectin, fibrinonectin and P-selectin.

Defects in Platelet Signal Transduction (Primary Secretion Defects)

Signal transduction events encompass processes that are triggered by the initial interaction of agonists with specific receptors on platelets and result in the activation of effectors, such as phospholipase C and phospholipase A2, leading ultimately to discernible end responses such as aggregation and secretion. The link between the surface receptors and the effector enzymes is provided in many instances by G-proteins. If the key components of platelet signal transduction mechanisms are the surface receptors, the G-proteins that modulate the intracellular effectors, and the effectors, evidence now exists for specific platelet abnormalities at each of these levels.

Defects in Platelet-Agonist Interaction (Receptor Defects)

Defects in platelet-agonist interaction are exemplified by patients with impaired responses because of an abnormality at the level of platelet surface receptors for a specific agonist. Such receptor defects have been documented for epinephrine, collagen, ADP, and thromboxane A2. Hirata et al (21) have described an Arg 60 to Leu mutation of the human thromboxane A2 receptor in a dominantly inherited bleeding disorder. Evidence has been presented by Cattaneo et al (22,23) and Nurden et al (24) for defects in the interaction of ADP and its receptors on platelets. Several patients have been described, (11,13) including by us, (25) in whom responses to epinephrine only are blunted; some of them have decreased numbers of platelet a2-adrenergic receptors (25). A few patients have been described in whom platelet responses to collagen only are blunted associated with deficiencies in membrane proteins including GPIa (26,27), GPVI (28) and others (29).

Defects in GTP-Binding Proteins

G-proteins are a group of heterogeneous proteins that form the link between surface receptors and intracellular enzymes although the specific Ga subunits mediating different agonist-responses may be different (6). Because of their modulating role, G-proteins constitute an important locus for aberrations leading to platelet dysfunction. Convincing evidence for the existence of such a defect has been provided by Gabbeta et al (30) in a patient with a mild bleeding disorder, abnormal aggregation and secretion responses to a number of agonists, and diminished GTPase activity (a reflection of GTP a-subunit function) on activation. In addition, the binding of 35S-GTPgS to platelet membranes was diminished. This patient was found to have a selective decrease in platelet membrane Gaq subunit with normal levels of Gai, Ga12, Ga13 and Gaz. This patient has been reported previously to have abnormalities in other downstream events: impaired Ca2+ mobilization (31), diminished release of free arachidonic acid from phospholipids on platelet activation (32) despite presence of normal platelet cytosolic and membrane levels of phospholipase A2 (30). These findings have been corrborated by essentially identical abnormal platelet responses in the Gaq deficient mice (33).

Defects in Other Signal Transduction Events: Phospholipase C Activation, Calcium Mobilization, Pleckstrin Phosphorylation

Several patients have been identified who have a relatively mild bleeding diathesis and impaired dense granule secretion, although their platelets have normal granule stores and, in general, appear to synthesize substantial amounts of thromboxane A2 (34-36). These patients demonstrate abnormalities in aggregation and secretion particularly in response to weaker agonists (ADP, epinephrine, PAF); the response to relatively stronger agonists such as arachidonate and high concentrations of collagen may be normal. In our experience, this group of patients is far more common than those with SPD or defects in TxA2 synthesis. Lages and Weiss (34) have described eight such patients who had decreased initial rates and extents of aggregation in response to ADP, epinephrine, and U44069. Defects in early platelet activation events were postulated in these patients. They subsequently demonstrated a defect in phosphatidylinositol hydrolysis and phosphatidic acid formation (37), and in pleckstrin phosphorylation (38) in one patient.

Koike et al (36) described platelet dysfunction in 12 patients with the behavioral disorder attention deficit disorder (ADD). These patients had a mild bleeding disorder characterized predominantly by easy bruising. The bleeding times were normal in eleven of these patients. Platelet aggregation and 14C-serotonin secretion responses during stimulation of platelet-rich plasma with ADP, epinephrine, and collagen revealed only minor and sporadic abnormalities. In contrast, the aggregation responses of gel-filtered platelets to a divalent cationophore A23187 were markedly impaired in ADD patients compared with a control group. In response to stimulation with A23187 and thrombin (at low concentrations), dense granule and acid hydrolase (b-glucuronidase, b-hexosaminidase, a-mannosidase) secretion was impaired whereas a-granule secretion was normal. The platelet contents of all of these granule constituents were normal, thereby excluding a storage pool deficiency. Additional studies revealed that liberation of arachidonic acid from phospholipids and its conversion to TxA2 were normal in these platelets. The ADD patients and those referred to above attest to the existence of a group of patients who have platelet dysfunction despite the presence of normal granule stores and normal TxA2 synthesis.

An early response to platelet stimulation is the rise in cytoplasmic ionized Ca2+ concentration. Therefore, attention was focused on this process to explain the impaired secretion. In several patients, defects in calcium mobilization have been proposed based on impaired platelet responses to the calcium ionophore (A23187) (11,39); however, this evidence is indirect, at best. We have provided direct evidence that some of these patients have impaired Ca2+ mobilization upon platelet activation (31,40). Detailed studies in two patients with impaired aggregation and secretion responses revealed that resting cytoplasmic Ca2+ concentration was normal, but the peak Ca2+ concentrations following activation with ADP, collagen, PAF or thrombin were diminished (40). Subsequent studies revealed that their platelets had abnormalities in the release of Ca2+ from intracellular stores as well as in the influx of extracellular Ca2+ (31). Our recent studies indicate that these platelets have a defect in formation of InsP3, the key intracellular mediator of Ca2+ release, as well as in diacylglycerol formation and pleckstrin phosphorylation (41), indicating a defect in PLC activation. Our studies indicate that human platelets contain at least seven PLC isozymes in the quantitative order PLC-g2> PLC-b2> PLC-b3> PLC-b1> PLCgb1> PLC-d1> PLC-b4> (42). Studies in one of these patients with impaired PLC activation revealed a selective decrease in PLC-b2 with normal levels of other PLC isoforms (42). To our knowledge, such a deficiency of a PLC isozyme has hitherto not been reported in human tissues. PLC- b isozymes are activated by G-protein mediated mechanisms while PLC-g isozymes are activated by tyrosine kinase-dependent mechanisms (43,44); the relative importance of the various isozymes in cellular responses remains unknown. Our finding in the patient indicate that PLC-b2, the predominant G-protein linked PLC isozyme in platelets, plays a major physiological role in platelet responses to activation. PLC- b2 is present predominantly in hematopoietic tissues (43,44) and interestingly, agonist-induced calcium mobilization has been reported to be diminished in neutrophils of knockout mice deficient in PLC- b2 (45).

Defects in phosphatidylinositol metabolism and protein phosphorylation have also been described by other investigators (37,38,46-49). Holmsen et al (46) described a patient with abnormalities in platelet aggregation and dense granule secretion who had impaired release of free arachidonic acid from phospholipids and phosphoinositide hydrolysis on thrombin activation. This patient had significant reduction in membrane GPIIb and IIIa as well; however, no studies were performed on Ca2+ mobilization or I(1,4,5)P3 production. Another patient has been described with impaired platelet response and diminished phosphoinositide metabolism in whom the altered stimulus-response coupling has been attributed to abnormal membrane phospholipid composition (47). Fuse et al (48) have reported a patient with a mild bleeding disorder whose platelets had impaired aggregation, secretion, InsP3 formation, and Ca2+ mobilization in response to a TxA2 mimetic (STA2) associated with normal TxA2 formation. Interestingly, GTPase activity upon activation with STA2 was also impaired leading to the conclusion that the platelets had an abnormality in coupling between TxA2 receptor and PLC. In the recently described patient by Mitsui et al (49), the abnormal platelet aggregation responses were associated with decreased InsP3 formation but with normal GPTase activity on stimulation with TxA2 analog U46619 and normal platelet TxA2 receptors, suggesting the possibility of an abnormality in PLC activity downstream of the receptor. Together, these studies provide evidence for abnormalities in signal transduction pathways in patients with diminished platelet aggregation and secretion responses.

We have recently summarized (35) our detailed studies on signaling mechanisms in eight patients with abnormal aggregation and secretion in response to several different surface receptor-mediated agonists despite presence of normal dense granule contents. Both protein kinase C induced pleckstrin phosphorylation and cytoplasmic Ca2+ mobilization play a major role in secretion on activation (8). Receptor-mediated Ca2+ mobilization and/or pleckstrin phosphorylation were abnormal in seven of the patients. We postulated that combined platelet activation with a cell permeable direct PKC activator DiC8 and ionophore A23187, which possibly bypass two major intracellular mediators (inositol trisphosphate, diacylglycerol), may induce normal dense granule secretion in patients with impaired receptor mediated secretion. Platelet activation with a combination of ADP with DiC8 or A23187 improved secretion in four patients. However, combination of DiC8 and A23187 induced normal secretion in platelet-rich plasma in all patients. These studies suggest that in these patients the ultimate process of exocytosis or secretion per se is intact and impaired secretion results from abnormalities in early signal transduction events.

19.6 Signal Transduction Defects and Activation of GPIIb-IIIa

Activation of GPIIb-IIIa and fibrinogen-binding to platelets is a prerequisite for aggregation, is a signal transduction dependent process, and has been linked to pleckstrin phosphorylation (50,51). Therefore, it is likely that abnormalities in signaling mechanisms would impair activation of otherwise normal GPIIb-IIIa on platelets. Evidence that this is indeed the case is provided by the report (52) of a patient with markedly abnormal platelet aggregation associated with impaired receptor activated pleckstrin phosphorylation and decreased activation of the platelet GPIIb-IIIa complexes despite the presence of adequate number of receptors with intact ligand (fibrinogen) binding capacity on the platelets. A similar abnormality in the activation of GPIIb-IIIa has also been observed in the patient with the Gaq deficiency (30) attesting to the role of the G-protein mediated signaling mechanisms in modulating the conformational change in GPIIb-IIIa on activation. More importantly, such a defect in GPIIb-IIIa activation, secondary to abnormalities in upstream signaling events, may provide a cogent explanation for abnormalities in initial aggregation responses noted by Lages and Weiss (34) in a number of their patients. Abnormalities in activation of GPIIb-IIIa secondary to impaired signal transduction may be a more common mechanism for blunted aggregation (particularly primary wave) than specific defects in the GPIIb-IIIa complex per se (53).

19.7 Abnormalities in Thromboxane Production and Arachidonic Acid Pathways

A major platelet response during activation is liberation of arachidonic acid from phospholipids and its subsequent oxygenation to TxA2. Patients have been described with impaired liberation of arachidonic acid from membrane phospholipids during platelet stimulation (32,54,46). In the patients described by Rao et al (32) platelet TxA2 production was diminished following stimulation with ADP and thrombin but was normal with arachidonic acid. Detailed studies in one patient revealed that platelet phospholipase A2 levels, the major enzyme involved in the release of arachidonic acid, were normal but Ca2+ mobilization was abnormal suggesting that the impaired arachidonate liberation may be secondary to reduced expression of the Ca2+ -dependent phospholipase A2 activity (30,31). Several patients have been described with platelet dysfunction associated with congenital deficiencies of cyclooxygenase (55-60) and thromboxane synthase (61,62).

Rao et al (63) have described a patient with bleeding manifestations whose platelets had impaired dense granuel secretion, normal granule stores, and normal liberation of arachidonic acid from phospholipids but had markedly impaired TXA2 synthesis during stimulation of platelet-rich plasma with thrombin. However, the platelets synthesized considerable amounts of TXA2 on activation of platelets suspended in a buffer containing no albumin. These findings suggest that the platelets had diminished levels of enzyme activity that could express itself adequately only in the absence of albumin, a protein known to bind free arachidonic acid avidly. Although the exact site of the enzyme defect was not elucidated, the studies in this case reflect the modulating role of albumin in platelet arachidonate metabolism.

19.8 Relative frequency of various platelet abnormalities

Thrombasthenia, the Bernard-Soulier syndrome and afibrinogenemia are rare disorders. It is generally accepted that vWD is the most common congenital platelet function disorder although the severe forms are rare. Patients currently classified in the heterogeneous category of defects in platelet secretion and signal transduction are probably the most frequently encountered inherited platelet function abnormalities excluding vWD. In our experience, the SPD is present in less than 10-15% of patients with congenital platelet defects. Abnormalities in thromboxane production occur in about 20% of patients. A large proportion of the remaining patients with abnormal aggregation and secretion responses demonstrate adequate dense granule stores and produce substantial amounts of thromboxane A2. They may have defects in the signaling mechanisms. In this heterogeneous group, the underlying mechanisms still need to be established.

19.8.1 Treatment of platelet function disorders

In patients with severe vWD and afibrinogenemia the treatment for bleeding episodes and management during surgical procedures is replacement of the deficient plasma factor. Milder cases of vWD are treated with 1-desamino-D-arginine vasopressin (DDAVP) which induces a rise in plasma vWF and factor VIII. However, not all of the subtypes of vWD respond to DDAVP (15,64). Platelet transfusions have been the major therapeutic modality to manage bleeding in patients with intrinsic platelet defects and this approach needs to be individualized. A viable alternative is intravenous administration of DDAVP which shortens the bleeding time in a substantial group of these patients, particularly those with normal dense granule stores (64,65). This response appears to be dependent on the underlying platelet mechanism leading to the platelet dysfunction (64,65). Patients with thrombasthenia appear not to respond to DDAVP infusion with a shortening of the bleeding time (65).

19.9 Conclusions

Rapid strides continue to be made in delineating the molecular and biochemical aspects of platelet function. A part of this understanding has come from characterization of experiments of nature, such as the platelet defects in patients with abnormal membrane glycoproteins. A better delineation of the involved platelet mechanisms will undoubtedly translate into development of novel therapeutic strategies for a diverse group of disorders - hemorrhagic, thrombotic, atherosclerotic and proliferative. It is indeed impressive to reflect on all of the biochemical machinery that exists in platelets and on the roles played by these tiny circulating anucleated cytoplasmic fragments of the megakaryocyte.

Acknowledgments

Supported by RO1 HL 056724 from NIH-NHLBI, and a Grant-in-Aid award from the American Heart Association.

References

1. Spaet TH. Platelets: The blood dust. In: Wintrobe MM ed. Blood, pure and eloquent. New York: McGraw-Hill; 1980:549-71.

2. Holmsen H. Platelet secretion and energy metabolism. In: Colman RW, Hirsh J, Marder VJ, Salzman EW eds. Hemostasis and thrombosis. Basic principles and clinical practice. Philadelphia: J.B. Lippincott; 1994:524-45.

3. Schafer AI. Biochemical mechanisms of platelet activation. Blood 1989; 74:1181-1195.

4. Majerus PW. Inositol phosphate biochemistry. Ann Rev Biochem. 1992; 61:225-250.

5. Berridge MJ. Inositol triphosphate and calcium signaling. Nature 1993; 361:315-325.

6. Brass LF, Hoxie JA, Manning DR. Signaling through G-proteins and G-protein-coupled receptors during platelet activation. Thromb Haemostas. 1993; 70:217-223.

7. Rink TJ, Sage SO. Calcium signaling in human platelets. Ann Rev Physiol. 1990; 52:431-446.

8. Nishizuka Y. The role of protein kinase C in cell surface signal transduction and tumour promotion. Science 1986; 233:305-311.

9. Walsh PN. Platelet-coagulant protein interactions. In: Colman RW, Hirsh J, Marder VJ, Salzman EW eds. Hemostasis and Thrombosis. Basic principles and clinical practice. Philadelphia: J.B. Lippincott; 1994:629-51.

10. Day HJ, Rao AK. Platelet function testing. Semin Hematol. 1986; 23:89-101.

11. Rao AK. Congenital disorders of platelet function. Hematol Oncol Clin N Amer. 1990; 4:65-86.

12. Rao AK. Case studies in hemostasis and thrombosis at Temple University School of Medicine. Platelet bleeding disorder in a 30-year-old female. Mechanisms of congenital platelet function defects. Am J Hemat. 1993; 44:139-144.

13. Weiss HJ. Inherited abnormalities of platelet granules and signal transduction. In: Hemostasis and thrombosis. Basic principles and clinical practice. Colman RW, Hirsh J, Marder VJ, Salzman EW, eds. Philadelphia: J.B. Lippincott; 1994:673-84.

14. Hayward CPM. Inherited disorders of platelet (-granules. Platelets 1997; 8:197-209.

15. Nichols WC, Ginsburg D. von Willebrand Disease. Medicine 1997; 76:1-20.

16. Triplett DA. Laboratory diagnosis of von Willebrand's disease. Mayo Clin Proc. 1991; 66:832-840.

17. George JN, Nurden AT. Inherited disorders of the platelet membrane: Glanzmann thrombasthenia, Bernard-Soulier syndrome, and other disorders. In: Hemostasis and thrombosis. Basic principles and clinical practice. Colman RW, Hirsh J, Marder VJ, Salzman EW, eds. Philadelphia: J.B. Lippincott; 1994: 652-71.

18. Coller BS, Seligsohn U, Peretz H, Newman PJ. Glanzmann thrombasthenia: new insights from an historical perspective. Sem Hemat. 1994; 31:301-311.

19. Weiss HJ. Scott Syndrome. A disorder of platelet coagulant activity. Sem Hemat. 1994; 31:312-319.

20. Hayward CPM, Rivard GE, Kane WH et al. An autosomal dominant, qualitative platelet disorder associated with multimerin deficiency, abnormalities in platelet factor V, thrombospondin, von Willebrand factor, and fibrinogen, and an epinephrine aggregation defect. Blood 1996; 87:4967-4978.

21. Hirata T, Kakizuka A, Ushikubi F, Fuse I, Okuma M, Narumiya S. Arg60 to leu mutation of the human thromboxane A2 receptor in a dominantly inherited bleeding disorder. J Clin Invest. 1994; 94:1662-1667.

22. Cattaneo M, Lecchi A, Randi AM, McGregor JL, Mannucci PM. Identification of a new congenital defect of platelet function characterized by severe impairment of platelet responses to adenosine diphosphate. Blood 1992; 80:2787-2796.

23. Cattaneo M, Lombardi R, Zighetti ML, et al. Deficiency of (33P)2MeS-ADP binding sites on platelets with secretion defect, normal granule stores and normal thromboxane A2 production. Thromb Haemostas. 1997; 77:986-90.

24. Nurden P, Savi P, Heilmann E, et al. An inherited bleeding disorder linked to a defective interaction between ADP and its receptor on platelets. J Clin Invest. 1995; 95:1612-1622.

25. Rao AK, Willis J, Kowalska MA, Wachtfogel Y, Colman RW. Differential requirements for epinephrine induced platelet aggregation and inhibition of adenylate cyclase. Studies in familial alpha2-adrenergic receptor defect. Blood 1988; 71:494-501.

26. Nieuwenhuis HK, Akkerman JWN, Houdijk WPM, Sixma JJ. Human blood platelets showing no response to collagen fail to express surface glycoprotein Ia. Nature 1985; 318:470-472.

27. Kehrel B, Balleisen L, Kokott R et al. Deficiency of intact thrombospondin and membrane glycoprotein 1a in platelets with defective collagen-induced aggregation and spontaneous loss of disorder. Blood 1988; 71:1074-1078.

28. Moroi M, Jung SM, Okuma M, Shinmyozu K. A patient with platelets deficient in glycoprotein VI that lack both collagen-induced aggregation and adhesion. J Clin Invest 1989; 84:1440-1445.

29. Ryo R, Yoshida A, Sugano W, et al. Deficiency of P62, a putative collagen receptor, in platelets from a patient with defective collagen-induced platelet aggregation. Am J Hematol. 1992; 39:25-31.

30. Gabbeta J, Yang X, Kowalska MA, Sun L, Dhanasekaran N, Rao AK. Platelet signal transduction defect with G(subunit dysfunction and diminished G(q in a patient with abnormal platelet responses. Proc Natl Acad Sci, USA. 1997; 94:8750-8755.

31. Rao AK, Disa J, Yang X. Concomitant defect in internal release and influx of calcium in congenital platelet dysfunction and impaired agonist-induced calcium mobilization. J Lab Clin Med. 1993; 121:52-63.

32. Rao AK, Kioke K, Willis J, et al. Platelet secretion defect associated with impaired liberation of arachidonic acid and normal myosin light chain phosphorylation. Blood 1984; 64:914-922.

33. Offermanns S, Toombs CF, Hu YH, Simon MI. Defective platelet activation in Gaq-deficient mice. Nature 1997; 389:183-186.

34. Lages B, Weiss HJ. Heterogeneous defects of platelet secretion and responses to weak agonists in patients with bleeding disorders. Br J Haematol. 1988; 68:53-60.

35. Yang X, Sun L, Gabbeta J, Rao AK. Platelet activation with combination of protein kinase C activator and ionophore A23187 induces normal secretion in patients with impaired receptor mediated secretion and congenital signal transduction defects. Thromb Res 1997;88:317-328.

36. Koike K, Rao AK, Holmsen H, Mueller PM. Platelet secretion defect in patients with the attention deficit disorder and easy bruising. Blood 1984;63:427-433.

37. Lages B, Weiss HJ. Impairment of phosphatidylinositol metabolism in a patient with a bleeding disorder associated with defects of initial platelet responses. Thrombos Haemostas. 1988; 59:175-179.

38. Speiser-Ellerton S, Weiss HJ. Studies on platelet protein phosphorylation in patients with impaired responses to platelet agonists. J Lab Clin Med. 1990; 115:104-111.

39. Lages B, Malmstein C, Weiss HJ, et al. Impaired platelet response to thromboxane-A2 and defective calcium mobilization in a patient with a bleeding disorder. Blood 1981;57:545-552.

40. Rao AK, Kowalska MA, Disa J. Impaired cytoplasmic ionized calcium mobilization in inherited platelet secretion defects. Blood 1989; 74:664-672.

41. Yang X, Sun L, Ghosh S, Rao AK. Human platelet signaling defect characterized by impaired production of inositol 1,4,5-triphosphate and phosphatidic acid and diminished pleckstrin phosphorylation. Evidence for defective phospholipase C activation. Blood 1996; 88:1676-1683.

42. Lee SB, Rao AK, Lee K-H, Yang Y, Bae YS, Rhee SG. Decreased expression of phospholipase C-(2 isozyme in human platelets with impaired function. Blood 1996; 88:1684-1691.

43. Rhee SG, Choi K. Regulation of inositol phospholipid-specific phospholipse C isozymes. J Biol Chem 1992;267:12393-12396.

44. Lee SB, Rhee SG. Significance of PIP2 hydrolysis and regulation of phosholipase C isozymes. Curr Opin Cell Biol 1995;7:183-189.

45. Jiang H, Kuang Y, Wu Y, et al. Role of phospholipase C b2 in chemoattractant-elicited responses. Proc Natl Aca Sci (USA) 1997;94:7971-7975.

46. Holmsen H, Walsh PN, Koike K, et al. Familial bleeding disorder associated with deficiencies in platelet signal processing and glycoproteins. Br J Haem. 1987; 67:335-344.

47. Cartwright J, Hampton KK, Macneil S, Colvin BT, Preston FE. A haemorrhagic platelet disorder associated with altered stimulus-response coupling and abnormal membrane phospholipid composition. Br J Haem. 1994; 88:129-136.

48. Fuse I, Mito M, Hattori A, et al. Defective signal transduction induced by thromboxane A2 in a patient with a mild bleeding disorder: impaired phospholipase C activation despite normal phospholipase A2 activation. Blood 1993; 81:994-1000.

49. Mitsui T, Yokoyama S, Shimizu Y, Katsuura M, Akiba K, Hayasaka K. Defective signal transduction through the thromboxane A2 receptor in a patient with a mild bleeding disorder. Deficiency of the inositol 1,4,5-triphosphate formation despite normal G-protein activation. Thromb Haemostas. 1997; 77:991-995.

50. Shattill SJ, Brass LF. Induction of the fibrinogen receptor on human platelets by inracellular mediators. J Biol Chem 1987;262:992-1000.

51. van Willigen G, Akkerman J-W. Protein kinase C and cyclic AMP regulate reversible exposure of binding sites for fibrinogen on the glycoprotein IIb-IIIa complex of human platelets. Biochem J 1991;273:115-120.

52. Gabbeta J, Yang X, Sun L, McLane M, Niewiarowski S, Rao AK. Abnormal inside-out signal transduction-dependent activation of glycoprotein IIb-IIIa in a patient with impaired pleckstrin phosphorylation. Blood 1996; 87:1368-1376.

53. Gabbeta J, Rao AK. Impaired platelet aggregation and abnormal activation of GPIIb-IIIa. Thromb Haemostas 1997; 78:1302.

54. Deykin K, Rittenhouse-Simmons, Russell PA. Impaired Ca++ mobilization: a possible cause of platelet syndrome. Clin Res 1978;21:503A.

55. Malmsten C, Hambert M, Svensson J, et al. Physiological role of an endoperoxide in human platelets: hemostatic defect due to platelet cyclooxygenase deficiency. Proc Natl Acad Sci USA 1975;72:1446-1450.

56. Horellou MH, Lecompte T, Lecrubier C, et al. Familial and constitutional bleeding disorder due to platelet cyclooxygenase deficiency. Am J Hematol 1983;114:1-9.

57. Lagarde M, Byron PA, Vargaftig BB, et al. Impairment of platelet thromboxane A2 generation and of the platelet release reaction in two patients with congenital deficiency of platelet cyclooxygenase. Br J Haematol 1978;38:251-266.

58. Pareti FI, Manucci PM, D'Angelo A, et al. Congenital deficiency of thromboxane and prostacyclin. Lancet 1980;1:890-898.

59. Rak K, Boda Z. Hemostatic balance in congential deficiency of platelet cyclooxygenase. Lancet 1980;2:44.

60. Roth GJ, Machuga R. Radioimmune assay of human platelet prostaglandin synthetase. J Lab Clin Med 1982;99:187-196.

61. Defiryn G, Machin SJ, Carreras LO, et al. Familial bleeding tendency with partial platelet thromboxane synthetase deficiency: reorientation of cyclic endoperoxide metabolism. Br J Haematol 1981;49:29-41.

62. Mestel F, Oetliker O, Beck E, et al. Severe bleeding associated with defective thromboxane synthetase [letter]. Lancet 1980;1:157.

63. Rao AK, Koike K, Day HJ, Smith JB, Holmsen H. Bleeding disorder associated albumin-dependent partial deficiency in platelet thromboxane production. Effect of albumin on arachidonate metabolism in platelets. Amer J Clin. Path 1985;83:687-696.

64. Mannucci PM. Desmopressin (DDAVP) in the treatment of bleeding disorders; the first 20 years. Blood 1997; 90:2515-2521.

65. Rao AK, Gosh S, Sun L, et al. Mechanisms of platelet dysfunction and response to DDAVP in patients with congenital platelet function defects. Thromb Haemostas 1995; 74:1071-1078.

20

BIOCHEMISTRY OF ALTERED PLATELET REACTIVITY IN HYPERTENSION

Huzoor-Akbar, Ph.D.
Department of Biomedical Sciences
Ohio University
College of Osteopathic Medicine
Athens, OH
45701, USA

20.1 INTRODUCTION

Over the years it has become evident that blood platelets play a very important role in health and disease. Normal circulatory function and supply of blood is essential for optimal performance of body as a whole. Platelets on one hand, aggregate at the site of injury and prevent the loss of blood. On the other hand, platelets may excessively aggregate in blood vessels and thus interfere with the normal supply of blood to various parts of the body. Occlusion of blood vessels and consequently decreased or lack of blood supply severely impairs the performance of vital organs such as heart, kidneys and brain and may even result in death. A number of researchers have shown that platelets from hypertensive humans and spontaneously hypertensive rats (SHR) exhibit increased sensitivity to aggregation agonists (1-25) Thus abnormal platelet aggregation in hypertension constitutes an additional risk factor for cardiovascular and

Dr. Huzoor-Akbar, Ph. D., is an Associate Professor of Pharmacology at Ohio University. He has published numerous research articles and received several research grants over the years. He is a member of the American Society of Pharmacology & Experimental Therapeutics. His current research is focused on: a) the biochemistry of altered blood platelet sensitivity in hypertension, b) the role of methy-lation & phosphorylation of G-proteins in platelet function.

cerebrovascular diseases. Prevention and/or treatment of abnormal platelet aggregation in hypertension requires a better understanding of the mechanisms of platelet aggregation as well as characterization of the factors responsible for increased aggregation response in hypertension.

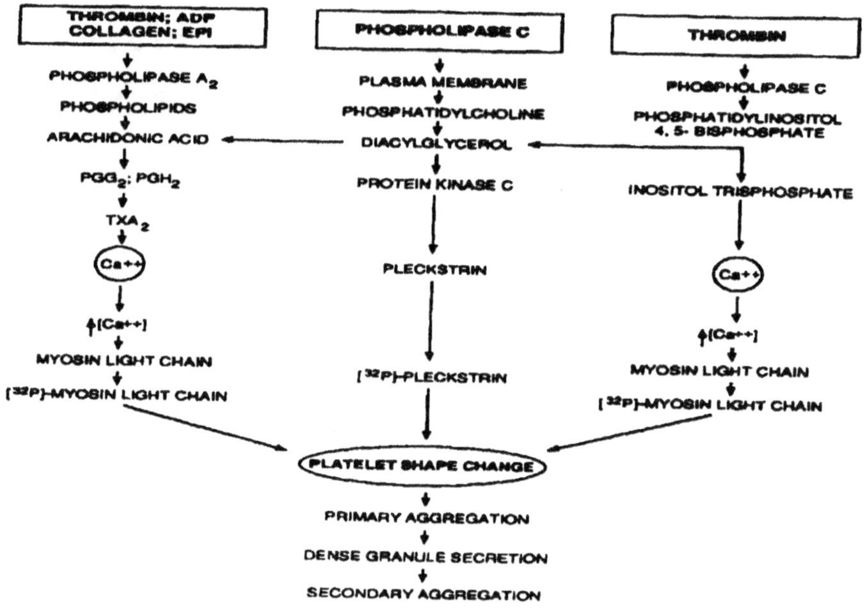

Fig. 1. Agonist-receptor mediated signal transduction leads to activation of phosphoinositide specific phospholipase C (PI-PLC) and phospholipase A_2 (PLA_2). Inositol tris phosphate (IP_3) and thromboxane A_2 (TXA_2) induce a rise in cytosolic calcium level. Elevated calcium leads to activation of myosin light chain (a 20 kDa protein, P20) kinase and thereby phosphorylation of P20. Diacylglycerol, produced by agonist-mediated reactions or by exogenously added bacterial phospholipase C (PLC) activates protein kinase C. Protein kinase C phosphorylates a 47 kDa protein (P47), named pleckstrin. Phosphorylation of pleckstrin and myosin light chain has been linked with platelet shape change. Binding of fibrinogen to its receptors (glycoprotein IIb/IIIa) on platelet surface, exposed due to shape change, results in primary platelet aggregation. Secretion of fibrinogen from α-granules is essential for *in vitro* washed platelet aggregation in the absence of exogenously added fibrinogen in the medium. (EPI = epinephrine).

Interaction of platelets with a variety of agonists leads to platelet activation, which consists of one or more of the following phenomena: platelet shape change, secretion from α-granules, primary aggregation, secretion from dense granules and secondary aggregation. A number of biochemical reactions initiate, promote, sustain and terminate physiological responses associated with platelet activation. The biochemical reactions involved in platelet activation include: agonist-receptor mediated signal transduction, phosphoinositide turnover, generation of diacylglycerol and inositol trisphosphate, mobilization of intra- and extracellular calcium, release of arachidonic acid from platelet phospholipids and its conversion into prostaglandins and thromboxanes, and phosphorylation of pleckstrin and myosin light chain (see Fig. 1). However, the rapid speed of these reactions and the quick onset of associated physiological phenomena makes it somewhat difficult to establish cause-effect relationships. This difficulty has been further compounded by the fact that multiple biochemical pathways, some proaggregation and some antiaggregation, simultaneously influence platelet activation. Flaws in experimental designs, such as: (a) use of excessive amounts of agonists and inhibitors, (b) determination of biochemical change at a time point significantly distant from the onset of the physiological response such as shape change, and (c) the use of non-specific inhibitors to ascertain the role of a given biochemical reaction, have also contributed to the confusion regarding possible cause-effect relationships between biochemical reactions and the physiological responses in platelets.

In this chapter, literature dealing with altered platelet function in hypertension will be reviewed to develop a better understanding of the role of agonist-receptor mediated signal transduction and the ensuing biochemical reactions that may be responsible for altered platelet reactivity in hypertension.

20.2 Platelet adhesion and aggregation responses in hypertension

Platelet adhesion, as an indicator of platelet responsiveness, has been measured for over fifty years (26). In 1968 Poplawski et al. (1) were the first to report that platelet

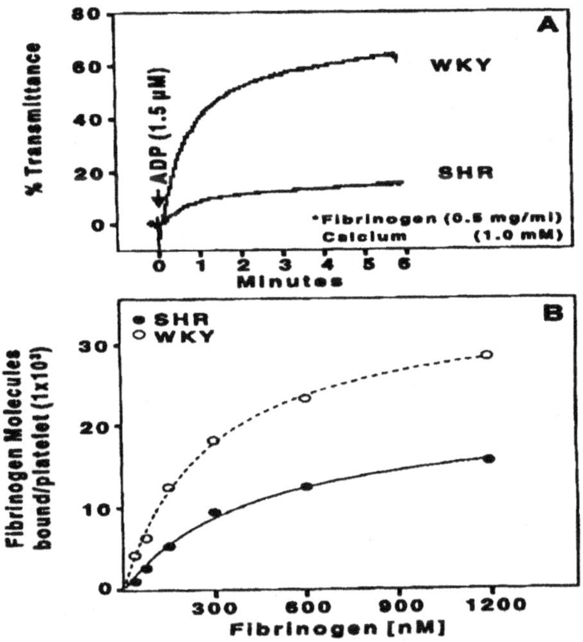

Fig. 2. Thrombin-induced platelet aggregation (Panel A) and [^{125}I]-fibrinogen binding (Panel B) was examined in washed platelets WKY and SHR rats. This Fig. has been taken from Life. Sci. 53, 1967-1974, 1993.

adhesiveness is increased in hypertensive patients by about two-fold. Recently Andrioli et al. (10), using the modern flow cytometry technology, have confirmed that thrombin induces greater platelet adhesion in patients with uncomplicated hypertension. Evidence of *in vivo* platelet activation in hypertension has also been reported by Blann et al. (11).

These authors have demonstrated that plasma levels of β-thromboglobulin and soluble P-selectin, two markers of platelet activation, are elevated in hypertensive patients.

A number of researchers have reported increased platelet aggregation in hypertensive humans (1-12) as well as in SHR (13-25). Platelet aggregate formation involves binding of fibrinogen to its receptors on platelet membrane (27-29). These fibrinogen receptors are exposed during platelet activation by agonists such as thrombin (27). We have shown that thrombin induces a significantly increased binding of fibrinogen to fibrinogen receptors on SHR than on normotensive (Wistar Kyoto, WKY) platelets (Fig. 2). Furthermore, thrombin-induced greater fibrinogen binding on SHR platelets appears to lead to a greater platelet aggregation response in SHR than in WKY rats (Fig. 2).

However, increased platelet aggregation response in SHR has not been found to be universal. In our laboratory, under identical experimental conditions utilizing the same washed platelet samples at the same time, we have observed that while thrombin induces a greater platelet aggregation response in SHR, ADP induces a lesser fibrinogen binding

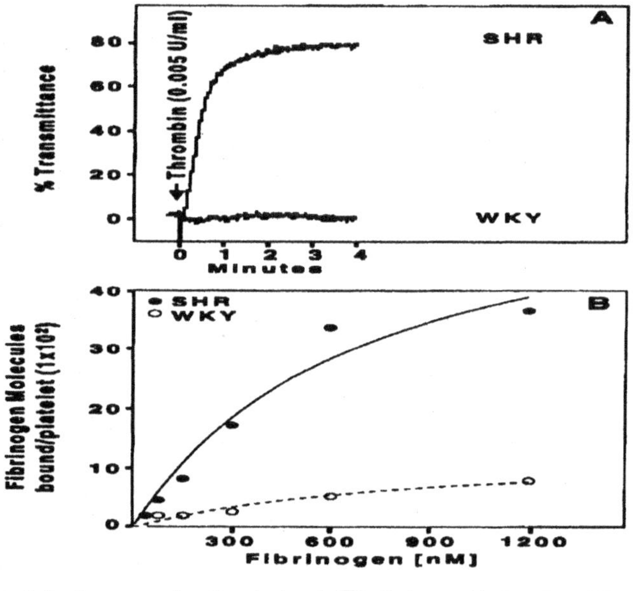

Fig. 3. ADP-induced platelet aggregation (Panel A) and [^{125}I]-fibrinogen binding (Panel B) was examined in washed platelets from WKY and SHR rats. Fibrinogen (0.5 mg/ml) and calcium (1.0 mM) was added to platelets prior to addition of ADP (1.5 μM).

and platelet aggregation response in washed SHR than WKY platelets (Fig. 3). Exogenously added bacterial phospholipase C (PLC) has also been shown to induce a lesser aggregation response in SHR than in WKY platelets. Other researchers have also reported decreased platelet aggregation responses in hypertensive humans and animals. In 1971 Nagaoka et al. (13) reported that thrombin as well as ADP induces less aggregation in SHR than in WKY platelets. Moderate hypertensive patients have been shown to exhibit a decreased aggregation response to all agonists (30). Tomita et al. (31-37) have published several reports describing decreased platelet aggregation response in stroke-prone spontaneously hypertensive rats (SHRSP).

Possible explanations for the conflicting reports on aggregation responses in hypertension range from differences in methodologies used by different researchers to the differences in the underlying pathological changes that may be occurring due to hypertension and possibly atherosclerosis. All of these explanations are plausible and there is certainly merit to the notion that SHRSP platelets, due to prolonged pathological conditions, may become refractory to aggregation agents. However, based on the data presented in Fig. 2, and Fig. 3, we suggest that platelets from SHR and WKY respond differently to different agonists even under identical experimental conditions. Therefore, the explanation for increased aggregation response to thrombin and a decreased aggregation response to ADP or PLC in SHR than in WKY may reside in the differences in the biochemical signaling mechanisms utilized by various agonists. The biochemical mechanisms that may be responsible for increased platelet reactivity to thrombin are discussed in the following sections.

20.3 Role of phosphoinositide in platelet reactivity inhypertension

Agonist-receptor mediated activation of phosphoinositide-specific phospholipase C (PI-PLC) has been shown to be involved in the regulation of a variety of cellular functions (38-41). There is sufficient evidence to show that thrombin, by interacting with specific receptors on the platelet membrane (38), stimulates PI-PLC and thereby induces hydrolysis of phosphoinositide namely, phosphatidylinositol (PI), phosphatidylinositol-4-phosphate (PIP), and phosphatidylinositol-4,5-bisphosphate (PIP_2). Hydrolysis of phosphoinositide generates diacylglycerol (DAG) and, depending on the parent compound, either inositol phosphate (IP), inositol bisphosphate (IP_2) or inositol trisphosphate (IP_3). Both DAG and IP_3 have been shown to induce platelet aggregation by activating protein kinase C (PKC)

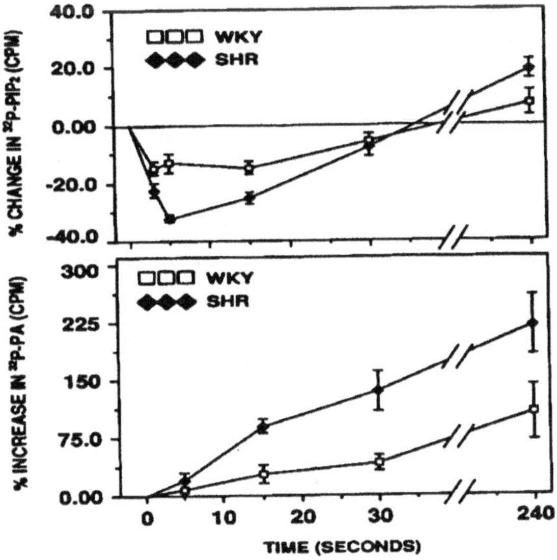

Fig. 4. Thrombin-induced loss in phosphatidylinositol-4,5-bisphosphate (PIP_2) and formation of phosphatidic acid (PA) was quantified in washed WKY and SHR platelets. This Fig. has been adapted from Thromb. Res. 49, 5-21, 1988.

and by mobilizing intracellular calcium respectively (29).

The role of phosphoinositide turnover in thrombin-induced greater aggregation in SHR than WKY platelets has been examined by us and others (18-23). In these experiments, platelets, prelabelled with ^{32}P-orthophosphate, are challenged with varying concentrations of thrombin for specified periods of time and the loss of ^{32}P-orthophosphate from phosphoinositide is quantified as an indicator of phosphoinositide hydrolysis (18-21). Data from these experiments have shown that SHR and WKY platelets do not differ in their ability to incorporate ^{32}P-orthophosphate into phosphoinositide (22). Moreover, the rate of turnover of phosphoinositide has been found to be the same in unstimulated SHR and WKY platelets (20). These findings have demonstrated that there is no inherent difference in either the amount or the rate of metabolism of phosphoinositide in WKY and SHR platelets. However, thrombin, as low as 0.05 U/ml, has been shown to induce a greater hydrolysis of PIP_2 in SHR (23$\pm$4%) than in WKY (15$\pm$3%) platelets within three seconds (21). A maximal decrease in ^{32}P-labelled PIP_2 in SHR (32$\pm$ 2%) and WKY (13$\pm$5%) platelets has been observed within five seconds of stimulation with thrombin (Fig. 4). The decrease in ^{32}P-labelled PIP_2 has been shown to coincide with a greater accumulation of ^{32}P-labelled phosphatidic acid in SHR than in WKY platelets (Fig. 4). Thrombin has also been shown to generate greater amounts of inositol-1,4,5-tris phosphate (IP_3) in SHR than in WKY platelets (42,43). Based on these data it has been suggested that thrombin-induced greater aggregation in SHR than in WKY platelets is linked with a greater hydrolysis of phosphoinositide in SHR platelets (21,42). This possibility has been further supported by the reports that DAG and IP_3, both products of phosphoinositide hydrolysis, expose fibrinogen receptors on permeabilized platelets (29).

20.3.1 Cytosolic calcium levels and platelet reactivity in hypertension

Free cytosolic calcium has been shown to play an important role in the regulation of cellular processes in a variety of cells including platelets (44). Free cytosolic calcium concentration in unstimulated platelets is maintained by a number of mechanisms which control the influx of extracellular calcium ions, the release of calcium ions from storage sites within platelets as well as the removal of ionized calcium from cytosol (44). Platelets

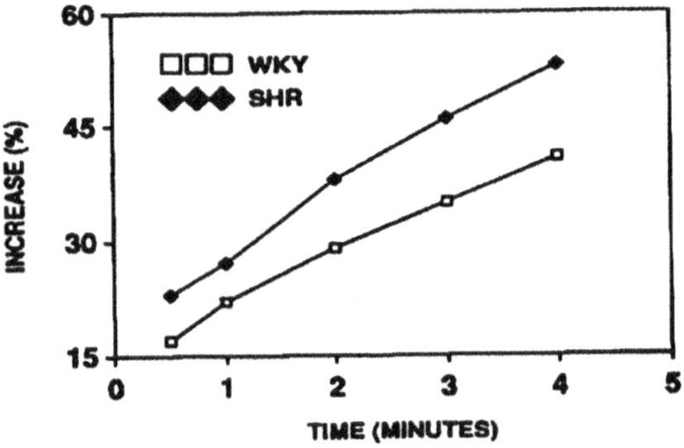

Fig. 5. Thrombin (5 mU/ml) induced increase in cytosolic calcium ($[Ca^{2+}]_i$) in WKY and SHR platelets was measured in Fura 2/AM loaded platelets. This Fig. has been taken from Life. Sci. 53, 1967-1974, 1993.

from hypertensive humans as well as SHR have been shown to have elevated basal cytosolic calcium levels as compared to normotensive humans or rats (8,16,44-51). Based on these reports it has been suggested that elevated basal cytosolic calcium level is responsible for agonist-mediated greater aggregation response in hypertensive than in normotensive platelets (44-52). However, the data on basal cytosolic calcium levels, particularly in WKY and SHR platelets, is controversial (53). The basal cytosolic calcium levels have been reported to range from 62 nM to 308 nM in SHR and from 54 nM to 246 nM in WKY platelets (16,47,53). Most investigators have not taken into account the leakage of fura-2, the dye used for quantifying cytosolic calcium, from rat platelets. Moreover, it is always possible that a slight activation of platelets during preparation of gel-filtered or washed platelets may differentially induce phosphoinositide hydrolysis leading to an apparently higher basal cytosolic calcium level in hypertensive than normotensive platelets. In our laboratory the basal cytosolic calcium levels, without correction for the leakage of fura-2, were found to be similar in WKY (96$\pm$5 nM) and SHR (100$\pm$5 nM) platelets (24). Lower but similar cytosolic calcium levels were observed in WKY (61$\pm$2 nM) and SHR (63$\pm$2 nM) even after correction for the leakage of fura-2 (24). Obviously these data do not support the view that greater aggregation response in SHR than in WKY is some how linked with elevated basal cytosolic calcium levels. Furthermore, if elevated calcium levels are responsible for a greater aggregation response in SHR then all agonists including ADP and PLC should produce a greater aggregation response. But ADP and PLC induce a lesser aggregation response in SHR than in WKY platelets.

Thrombin (Fig. 5) and ionomycin, which induces a rise in cytosolic calcium by a receptor-independent mechanism, have been shown to induce a greater rise in SHR than in WKY platelets (42,43).
Thrombin-induced rise in cytosolic calcium, particularly in a calcium poor medium, is mediated, at least in part, by IP_3. Therefore, it is likely that thrombin-induces a greater platelet aggregation response in hypertensive than in normotensive platelets by inducing a greater hydrolysis of PIP_2, generation of IP_3 and consequently a greater rise in cytosolic calcium (Fig. 5).

20.3.2 Role of phosphorylation of myosin light chain and pleckstrin in platelet reactivity in hypertension

Thrombin and other agonists have been shown to induce phosphorylation of a number of platelet proteins (21,40,54,55). Phosphorylation of two of these proteins namely, myosin light chain, a 20 kDa protein (P20), and pleckstrin, a 47 kDa protein (P47), has been shown to be involved in the regulation of platelet function (40,54,55). The possibility that the greater platelet aggregation response in hypertensive than in normotensive platelets may be linked with phosphorylation of P47 and P20 has been examined by us and others (19,21,56-58). The basal phosphorylation profiles of P47 and P20 have been found to be the same in unstimulated WKY and SHR platelets (57) as well as in unstimulated platelets from normotensive and essential hypertensive humans (58). These findings imply that in unstimulated normotensive and hypertensive platelets the levels of cytosolic calcium and DAG, two agents responsible for phosphorylation of P20 and P47, should be within normal range.

We have shown that thrombin, within five seconds of stimulation, induces significantly greater phosphorylation of P47 and P20 in SHR than in WKY platelets (Fig. 6). These findings have been confirmed by others using platelets from normotensive and hypertensive humans and animals (57,58). Therefore, it appears that thrombin, within 3-5 seconds, induces: platelet shape change (which exposes fibrinogen receptors); hydrolysis of phosphoinositide; and phosphorylation of P47 and P20 (Fig. 2, 4, 6). Chronological proximity of these events leads us to suggest that thrombin-induces a greater aggregation response in hypertensive than in normotensive platelets by inducing a greater hydrolysis of phosphoinositide which in turn generates greater amounts of IP_3 and DAG in hypertensive platelets. Higher levels of DAG and IP_3 induce greater fibrinogen binding, possibly due to increased phosphorylation of P47 and P20, by activating PKC and elevating cytosolic calcium levels respectively (21,29).

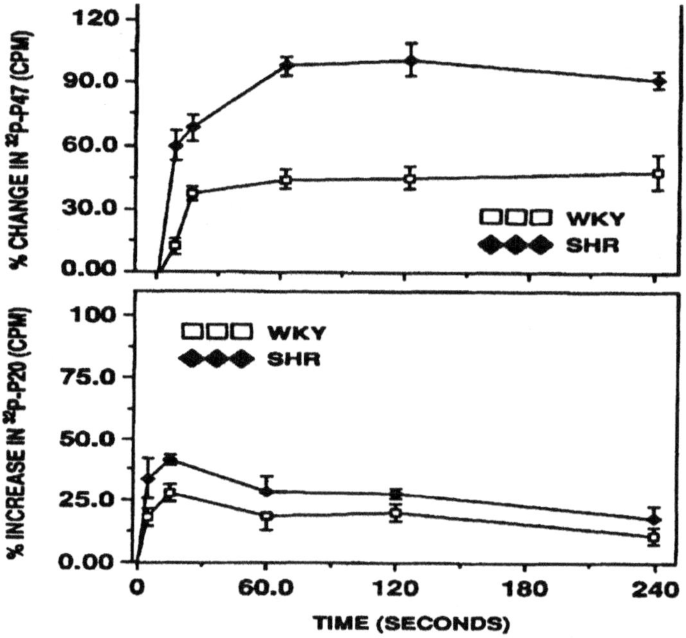

Fig. 6. Thrombin-induced phosphorylation of a 47 kDa (P47) and a 20 kDa (P20) protein was examined in washed WKY and SHR platelets. This Fig. has been taken from Thromb. Res. 49, 5-21, 1988.

Tomita and co-workers (31-37) have shown that platelets from stroke-prone spontaneously hypertensive rats (SHRSP) exhibit not only hypoaggregation but also a decrease in biochemical parameters such as calcium mobilization, phosphoinositide hydrolysis, and phosphorylation of P47 and P20. These reports clearly demonstrate that platelets from SHRSP respond differently to stimulation by thrombin than SHR platelets. Nevertheless, these findings do support a link between thrombin-induced platelet activation and thrombin-induced biochemical changes. Thrombin-induces a greater aggregation response in SHR and a lesser aggregation response in SHRSP platelets possibly because it induces greater biochemical changes in SHR and lesser biochemical changes in SHRSP.

20.3.3 Increased platelet sensitivity to Prostaglandin E_1 in hypertension

Prostaglandin E_1 (PGE_1) has been shown to induce a greater accumulation of cyclic AMP in platelets from SHR and essential hypertensive humans than in platelets from WKY or normotensive humans (14,23,30,59). In other reports, PGE_1 has been shown to generate less cyclic AMP in platelets from essential hypertensive than normotensive humans (60,61). PGE_1 generates cyclic AMP by stimulating adenylyl cyclase. A greater accumulation of cyclic AMP in PGE_1 stimulated platelets may be due to an increase in generation of cyclic AMP or due to a decrease in metabolism of cyclic AMP by cyclic nucleotide phosphodiesterase (23). We have shown that PGE_1 induces a greater

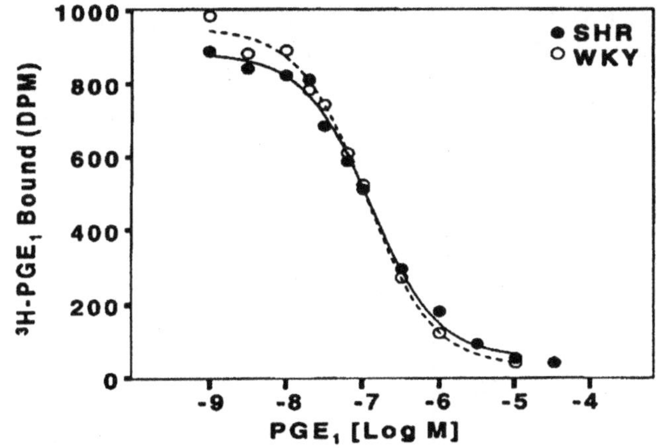

Fig. 7. Binding of [^{3}H]-PGE_1 (Top panel) and [^{125}I]-thrombin (Bottom panel) was determined in the presence of increasing concentrations of unlabeled PGE_1 and thrombin respectively. The displacement curves, shown here, for [^{3}H]-PGE_1 binding and [^{125}I]-thrombin binding to WKY and SHR platelets have been taken from Am. J. Hyperten. 6, 857-862, 1993 and Life. Sci. 53, 1967-1974, 1993

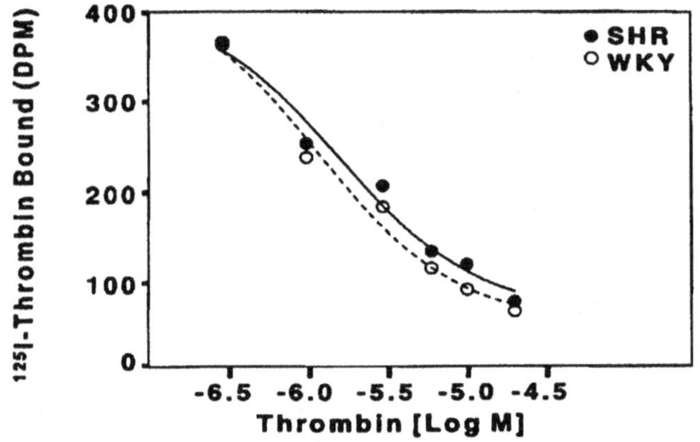

accumulation of cyclic AMP in SHR than in WKY platelets (21,23) and this greater accumulation of cyclic AMP does not appear to be due to the differences in the ability of platelets to metabolize cyclic AMP or due to differences in the catalytic activity of adenylyl cyclase (23).

These findings demonstrate that platelets from hypertensive humans and animals exhibit a greater sensitivity not only to thrombin, which is an inducer of aggregation, but also to PGE_1 which is an inhibitor of platelet aggregation. Possible mechanisms that may be responsible for the increased sensitivity to both PGE1 and thrombin are discussed below.

20.4 Thrombin- and PGE_1-receptor mediated signal transduction mechanisms in hypertension

The cellular interaction of agonists depends on the physico-chemical nature and on the topographical distribution of the membrane receptors (62). Thus thrombin and PGE_1 mediated responses would depend on the number and the type (high affinity/low affinity) of receptors available for interaction with thrombin or PGE_1 respectively. The possibility that platelets from SHR exhibit greater sensitivity to thrombin and PGE_1 because of differences in number or affinity of their respective receptors has been examined by us (23,24). Thrombin as well as PGE_1 binding sites/platelets and dissociation constants have been found to be similar in WKY and SHR platelets (Fig.7, ref. 23,24). Therefore, increased platelet reactivity to thrombin or PGE_1 is not attributable to changes in the number or the affinity of their respective receptors.

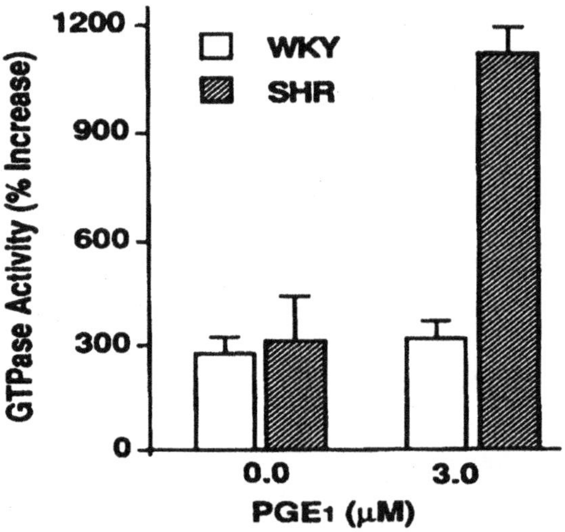

Fig. 8. GTPase activity in WKY and SHR platelet membranes was determined in the presence or absence of PGE_1 (3 μM). This Fig. has been taken from Am. J. Hyperten. 6, 857-862, 1993.

It has been suggested that differences in the biochemical signal transduction pathway, which is initiated by the agonist-receptor interaction, may be responsible for altered platelet reactivity in certain pathological conditions including hypertension (63-65). A closer look at the mechanism of action of thrombin, i.e. activation of PI-PLC, and PGE_1, i.e. activation of adenylyl cyclase, reveals that the two seemingly different agonists utilize a common biochemical signal transduction pathway. Following interaction with their specific receptors both thrombin and PGE_1 transduce their signal via specific guanine nucleotide binding proteins (G proteins) to the appropriate effectors. Thrombin activates PI-PLC via a G protein named G_q and PGE_1 stimulates adenylyl cyclase via a G protein called G_s (66).

The possibility that either the expression of G proteins or the GTPase activity associated with the G proteins may be altered in platelets from hypertensive than normotensive humans or animals has been examined (23,65,67). McLellan et al. (67) and Marcil et al. (65) have shown that platelets from normotensive and essential hypertensive humans express similar amounts of G_s. The data for expression of G_i appear to be controversial. Mclellan et al. (67) have reported that there is no difference in the expression of $G_{i\alpha 2}$ in platelets from normotensive and hypertensive humans. But Marcil et al. (65) have shown a 70 % decrease in $G_{i\alpha 2}$ levels in platelets from hypertensive than from normotensive humans. Expression of G_q in platelets from hypertensive humans or rats has not been examined so far.

If the expression of G_s, responsible for stimulation of adenylyl cyclase, is the same in the platelets from normotensive and hypertensive humans and animals then the increased platelet reactivity to PGE_1 in hypertension may be due to differences in the G protein associated GTPase activity. We have examined this possibility and shown that PGE_1 induces a greater stimulation of GTPase activity in platelet membranes from SHR (806%) than from WKY (46%) platelets (Fig. 8). These data clearly suggest that increased platelet reactivity to PGE_1, and possibly thrombin, in hypertension may be due to altered signal transduction.

Differences between hypertensive and normotensive platelet membrane structure may be responsible for altered signal transduction and thereby increased platelet reactivity in hypertension. A number of researchers have examined this possibility (68-74). Platelet membrane microviscosity has been examined using diphenylhexatriene (DPH), which probes the hydrocarbon backbone of the lipid core, and trimethylamino-diphenylhexatriene (TMA-DPH), which anchors the polar heads of phospholipids and labels glycerol and fatty acyl chain region of the membrane (75). However, the data on membrane fluidity is controversial. Platelet membrane fluidity, measured with DPH, has been shown to be decreased (75), increased (73) or not changed (70) in hypertension. Similarly, platelet membrane fluidity, measured with TMA-DPH, has been reported to be increased (70), or not changed (73) in hypertension. Attempts have been made to link changes in platelet membrane microviscosity with changes in platelet cholesterol content, cholesterol/phospholipid (C/P) ratio, phospholipid composition and fatty acid composition (72-74,75). Benjamin et al. (76) and Caimi et al. (72) have shown that C/P ratio is increased in platelet plasma membrane of hypertensive than normotensive humans. Increase in C/P ratio has been shown to decrease membrane fluidity, increase number of thrombin receptors, and increase platelet responsiveness to thrombin (77). However, Naftilan et al. (75) have shown that platelet cholesterol content is the same in normotensive and hypertensive humans but the DPH-probed membrane fluidity is decreased. Moreover, we have shown that cholesterol content, the C/P ratio as well as the number of thrombin receptors is the same in the platelets from WKY and SHR platelets (24,78).

While an increase in the membrane cholesterol content may be linked with an increase in the number of thrombin receptors, the changes in platelet fatty acid composition, which also affects membrane fluidity, do not affect binding of thrombin or yohimbin to platelets (79). It has been suggested that changes in fatty acid composition in SHR platelet membranes leading to changes in microenvironment fluidity may be a possible cause of altered signal transduction (79). A 5% increase in platelet membrane linoleate has been shown to increase DPH-probed membrane fluidity by 10% and inhibit platelet aggregation by 80% (79). The possibility that differences in membrane fatty acid composition or phospholipid composition in normotensive and hypertensive platelets may be responsible for altered signal transduction and therefore increased platelet reactivity in hypertension has been examined (72,78). We have reported that percent composition of phosphatidylcholine in SHR (46%) platelets is significantly higher than in WKY (34%) platelets (78). Percent composition of arachidonic acid in SHR (24%) platelets has also been found to be greater than WKY (15%) platelets (78). It is possible, although at present there is no evidence to suggest it, that increased phosphatidylcholine and arachidonic acid composition in SHR may be linked with altered platelet signal transduction and consequently with increased platelet sensitivity to agonists such as thrombin and PGE_1.

20.5 Role of nitric oxide and cyclic GMP in platelet reactivity in hypertension

Nitric oxide, also known as the endothelium derived relaxing factor or EDRF, has been shown to inhibit platelet adhesion as well as aggregation (80-84) by elevating cyclic GMP levels (81,84). We have shown that cyclic GMP inhibits platelet activation by inducing phosphorylation of rap1b, a low molecular weight G protein (84). Agonist-induced rise in cytosolic calcium not only leads to platelet aggregation but also activates nitric oxide synthase which produces nitric oxide (85). This suggests that thrombin-induced rise in cytosolic calcium participates not only in platelet activation but also in inhibition of platelet activation via nitric oxide-cyclic GMP pathway. It is possible that differences in the agonist induced activation of nitric oxide synthase, generation of nitric oxide or nitric oxide-induced formation of cyclic GMP may render SHR platelets more sensitive to aggregation agents. It has been suggested that NG-monomethyl-L-arginine (L-NMMA), an inhibitor of nitric oxide synthesis, promotes platelet aggregation by preventing the synthesis of nitric oxide in normotensive but not in hypertensive platelets (82). Woods et al. (83) have shown that platelets from hypertensive patients are at least three-fold less sensitive to the inhibitory effect of sodium nitroprusside, a nitric oxide donor, than platelets from normotensive humans. We have shown that nitric oxide induces a four-fold lesser increase in cyclic GMP accumulation in SHR than in WKY platelets (Fig. 9, ref. 86). It appears, based on these findings, that increased platelet reactivity to thrombin in SHR than in WKY platelets may be, at least in part, due to a decrease in the nitric oxide-induced production of cyclic GMP.

20.6 Role of thromboxane A_2 and lipoxygenase metabolites in hypertension

Agonist-receptor mediated generation of thromboxane A_2 (TXA_2) has been shown to be involved in regulation of human platelet activation. Increased prostaglandins/TXA_2 production has been reported in platelets spontaneously hypertensive rats (87,88). We have shown that thrombin induces production of greater amounts of TXA_2 in SHR than in WKY platelets in a concentration-dependent manner (88). However, prostaglandin H_2 (PGH_2) does not induce aggregation in rat platelets (89). Moreover, imidazole and aspirin, inhibitors of prostaglandin synthesis, inhibit ADP-induced aggregation in human but not in rat platelets (88,90). These findings suggest that a greater generation of TXA_2 in SHR than in WKY platelets may not be the cause of increased platelet sensitivity to thrombin (88). Tomita et al. (91) have shown that thrombin induces production of similar levels of

TXA_2 in SHRSP and WKY platelets but induces less aggregation in platelets from SHRSP than from WKY. Taken together, these observations imply that rat platelets are capable of producing TXA_2, but TXA_2 formation is not involved in either hypo-aggregation in SHRSP or hyper-aggregation in SHR than in WKY platelets.

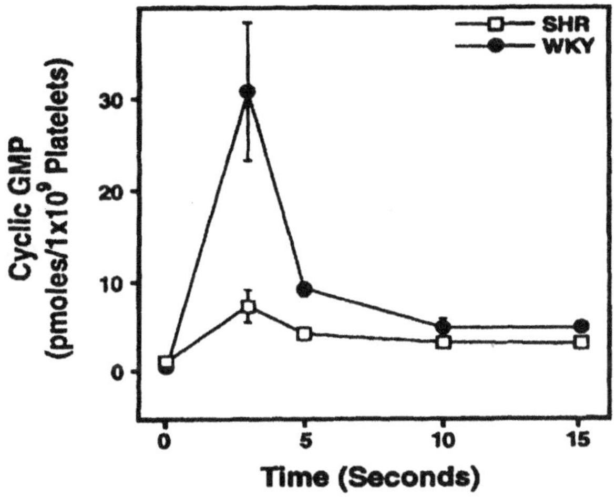

Fig. 9. Nitric oxide-induced increase in cyclic GMP was quantified in washed WKY and SHR platelets. A 3 seconds incubation with nitric oxide (10 µM) induced a 4-fold lesser increase in cyclic GMP in SHR than in WKY platelets.

Lipoxygenase metabolites of arachidonic acid such as 12-hydroxyeicosa- tetraenoic acid (12-HETE) have been shown to be involved in the release of neurotransmitters and other cellular functions (92-94). Basal generation of 12-HETE has been reported to be about 3.7-fold greater in SHR than in WKY platelets (95,96). Arachidonic acid has been shown to stimulate platelets by activating protein kinase C, by a prostaglandin-independent mechanism (97). It is possible that the lipoxygenase metabolites of arachidonic acid may also be involved in altered reactivity in hypertension.

20.7 Summary

Platelets from SHR exhibit increased sensitivity to thrombin and PGE_1 than platelets from WKY rats. Both of these agonists produce their effects by a G protein mediated signal transduction mechanism leading to activation of PI-PLC and adenylyl cyclase respectively. Enhanced signal transduction at the G protein level may be responsible for a greater adenylyl cyclase, and possibly PI-PLC, activity in SHR than in WKY platelets.

Thrombin induces a greater turnover of phosphoinositide, due to an increased signal transduction leading to an enhanced PI-PLC activity, in SHR than in WKY platelets. Increased production of IP_3 and DAG, due to a greater hydrolysis of phosphoinositide leads to exposure of a greater number of fibrinogen receptors and therefore a greater aggregation response in SHR than in WKY platelets.

PGE_1 induces a greater accumulation of cyclic AMP in SHR than in WKY platelets. This increased accumulation of cyclic AMP in SHR does not appear to be due to a decrease in the metabolism of cyclic AMP or due to an increase in the catalytic activity of adenylyl cyclase. A greater increase in the GTPase activity of the Gs, G protein responsible for activation of adenylyl cyclase, in SHR than in WKY platelets appears to be the biochemical mechanism responsible for an enhanced sensitivity to PGE_1 in SHR than in WKY platelets.

Finally, it appears that at least one of the negative feed back mechanisms, involved in controlling the extent of agonist-induced platelet aggregation, may also be altered in SHR platelets. Agonist-induced rise in cytosolic calcium leads to generation of nitric oxide which inhibits platelet aggregation by elevating cyclic GMP levels. Platelets from SHR appear to have a diminished capacity to generate nitric oxide. Moreover, nitric oxide produces a significantly lesser amount of cyclic GMP in SHR than in WKY platelets. The diminished ability to produce cyclic GMP, one of the antiplatelet signals, may be responsible, at least in part, for the increased platelet reactivity to thrombin in SHR than in WKY platelets.

20.8 Acknowledgements

Dr. Huzoor-Akbar's research has been supported, in part, by the American Heart Association (Ohio Affiliate), the American Health Assistance Foundation, the Ohio University Baker Fund, and the Ohio University College of Osteopathic Medicine.

References

1. Poplawski, A., Skorulska, M. and Niewiarowski, S. Increased platelet adhesiveness in hypertensive vascular disease. J. Atheroscler. Res. 8, 721-723, 1968.

2. Coccheri, S. and Florentinin, P. Platelet adhesiveness and aggregation in hypertensive patients. Acta Med. Scand. Suppl. 525, 273-275, 1971.

3. Mehta, J. and Mehta, P. Platelet function in hypertension and effect of therapy. Am. J. Cardiol. 47, 331-334, 1981.

4. Kjeldsen, S. E., Gjesdal, K., Eide, I., Aakesson, I., Amundsen, R., Foss, O. P. and Leren, P. Increased beta-thromboglobulin in essential hypertension: Interaction between arterial plasma adrenaline, platelet function and blood lipids. Acta Med. Scand. 213(5), 369-373, 1983.

5. Lechi, A., Lechi, C., Bonadonna, G., Sinigaglia, D., Corradini, P., Polignano, R., Arosio, E., Covi, G. and DeTogni, P. Increased basal and thrombin-induced free calcium in platelets from hypertensive patients. Hyperten. 9, 230-235, 1987.

6. Nyrop, M. and Zweifler, A. J. Platelet aggregation in hypertension and the effect of antihypertensive treatment. J. Hyperten. 6, 263-269, 1988.

7. Nityanand, S., Pande, I., Bajpai, V. K., Singh, L., Chandra, M. and Singh, B. N. Platelets in essential hypertension. Thromb. Res. 72, 447-454, 1993.

8. Touyz, R. M. and Schiffrin, E. L. Effects of angiotensin II and endothelin-1 on platelet aggregation, cytosolic pH and free Ca++ concentrations in essential hypertension. Hyperten. 22, 853-862, 1993.

9. Gleerup, G., Vind, J. and Winther, K. Platelet function and fibrinolytic activity during rest and exercise in borderline hypertensive patients. Eur. J. Clin. Invest. 25, 266-270, 1995.

10. Andrioli, G., Ortolani, R., Fontana, L., Gaino, S., Bellavite, P., Lechi, C., Minuz, P., Manzato, F., Tridente, G. and Lechi, A. Study of platelet adhesion in patients with uncomplicated hypertension. J. Hyperten. 14, 1215-1221, 1996.

11. Blann, A. D., Lip, G. Y., Islim, I.F. and Beevers, D. G. Evidence of platelet activation in hypertension. J. Human Hyperten. 11, 607-609, 1997.

12. Spalding, A., Vaitkevicius, H., Dill, S., MacKenzie, S., Schmaier, A. and Lockette, W. Mechanism of epinephrine-induced platelet aggregation. Hyperten. 31, 603-607, 1998.

13. Nagaoka, A., Sudo, K., Orita, S., Kikuchi, K. and Aramaki, Y. Hematological studies on the spontaneously hypertensive rats with special reference to the development of thrombosis. Jpn. Circ. J. 35(11), 1379-1390, 1971.

14. Hamet, P., Fraysse, J. and Fronks, D. J. Cyclic nucleotides and aggregation in platelets of spontaneously hypertensive rats. Circ. Res. 43, 583-591, 1978.

15. McGregor, L., Morazain, R. and Renaud, S. Platelet function and fatty acid composition of platelet phospholipids in spontaneously hypertensive rats fed saturated or polyunsaturated fats. Atheroscler. 38, 129-136, 1981.

16. Bruschi, G., Bruschi, M. E., Caroppo, M., Orlandini, G. Spaggiari, M. and Cavatorta, A. Cytoplasmic free Ca++ is increased in the platelets of spontaneously hypertensive rats and essential hypertensive patients. Clin. Sci. 68, 179-184, 1985.

17. Huzoor-Akbar and Anwer, K. Evidence that the rat is not an appropriate model to study the role of prostaglandins in normal or abnormal platelet aggregation. Thromb. Res. 41, 555-566, 1986.

18. Anwer, K. and Huzoor-Akbar. Abnormal platelet aggregation in spontaneously hypertensive rats is linked with phosphoinositide turnover but not prostaglandin synthesis. Fed. Proc. 45, 226, 1986.

19. Huzoor-Akbar and Anwer, K. Phosphorylation of 47kDa protein and phosphoinositide turnover are involved in abnormal platelet activation in spontaneously hypertensive rats. Proc. VIth International Conference on Cyclic Nucleotides, Calcium and Protein Phosphorylation, 241, 1986.

20. Koutouzov, S., Remmal, A., Marche, P. and Meyer, P. Hypersensitivity of phospholipase C in platelets of spontaneously hypertensive rats. Hyperten. 10(5), 497-504, 1987.

21. Huzoor-Akbar and Anwer, K. Thrombin-induced abnormal platelet activation in spontaneously hypertensive rats is linked to phosphoinositide turnover and phosphorylation of 47,000 and 20,000 Dalton proteins. Thromb. Res. 49, 5-21, 1988.

22. Koutouzov, S., Remmal, A., Meyer, P. and Marche, P. Relationship between enhanced phosphoinositide turnover and cellular responses in platelets from spontaneously hypertensive rats. J. Cardiovasc. Pharmacol. 11 (Suppl 1):S6-9, 1988.

23. Huzoor-Akbar, Dean, T. and Kornhauser, R. Increased platelet reactivity to prostaglandin E1 in hypertension is linked with altered signal transduction. Am. J. Hyperten. 6, 857-862, 1993.

24. Huzoor-Akbar, Kundu, N. and Kornhauser, R. Normal thrombin binding leads to greater fibrinogen binding and increased platelet aggregation in spontaneously hypertensive rats. Life. Sci. 53, 1967-1974, 1993.

25. Lominadze, D., Joshua, I. G. and Schuschke, D. A. In vivo platelet thrombus formation in microvessels of spontaneously hypertensive rats. Am. J. Hyperten. 10, 1140-1146, 1997.

26. Wright, H. P. Changes in adhesiveness of blood platelets following parturitions and surgical conditions. J. Path. Bact. 54, 461, 1942.

27. Fitzgerald, L. and Phillips, D. R. Calcium regulation of platelet membrane glycoprotein Iib-IIIa complex. J. Biol. Chem. 260, 11366-11374, 1985.

28. Brass, L. F., Shattil, S. J., Kunicki, T. J. and Bennett, J. S. Effect of calcium on the stability of the platelet membrane glycoprotein IIb-IIIa complex. J. Biol. Chem. 260, 7875-7881, 1985.

29. Shattil, S. J. and Brass, L. F. Induction of fibrinogen receptors on human platelets by intracellular mediators. J. Biol. Chem. 262, 992-1000, 1987.

30. Mikhail, N., Fukuda, N., Tremblay, J. and Hamet, P. Platelets, growth factors, and vascular smooth-muscle cells in hypertension and diabetes. J. Cardiovasc. Pharmacol. 22 (suppl. 6), S64-S74, 1993.

31. Tomita, T., Umegaki, K., Nakamura, K. and Hayashi, E. Impaired function of TXA2 and/or calcium in the platelets from SHRSP at hypertensive ages. Thromb. Res. 35, 219-230, 1984.

32. Umegaki, K., Ikeda, M., Nakamura, K., Inoue, Y. and Tomita, T. Platelet dysfunction in stroke-prone spontaneously hypertensive rats: Defective Ca++ functions in protein phosphorylation and in coupling of protein phosphorylation to physiological responses. J. Hyperten. 4 (Suppl. 3), S279-S281, 1986.

33. Umegaki, K., Ikeda, M., Nakamura, K. and Tomita, T. Defects of thrombin-induced protein phosphorylation in platelets from stroke-prone spontaneously hypertensive rats. FEBS Lett. 196, 139-144, 1986.

34. Umegaki, K., Nakamura, K. and Tomita, T. Primary dysfunction in aggregation and secretion of SHRSP platelets: Not secondary to the circulation of "exhausted" platelets. Blut 52, 17-27, 1986.

35. Ikeda, M., Onda, T., Tomita, I. and Tomita, T. Reduced functions of intracellular Ca++ in aggregation, secretion and protein phosphorylation of permeabilized platelets from stroke-prone spontaneously hypertensive rats. Thromb. Res. 67, 655-663, 1992.

36. Ikeda, M., Onda, T., Mitsubori, T., Umegaki, K., Tomita, I. and Tomita, T. Phospholipid metabolism in platelets from stroke-prone spontaneously hypertensive rats and Wistar Kyoto rats. J. Pharmacobio-Dyn. 15, 49-57, 1992.

37. Ikeda, M., Onda, T., Tomita, I. and Tomita, T. The differences in Ca++-sensitivity of protein kinase C in platelets from Wistar Kyoto rat and stroke-prone spontaneously hypertensive rat. Thromb. Res. 82, 417-427, 1996.

38. Brass, L., Ahuja, M., Belmonte, E., Pizarro, S., Tarver, A. and Hoxie, J. The human platelet thrombin receptor: turning it on and turning it off. Ann. N. Y. Acad. Sci. 714, 1-12, 1994.

39. Berridge, M. J. Inositol trisphosphate and diacylglycerol as second messengers. Biochem. J. 220, 345-360, 1984.

40. Imaoka, T., Lynham, J. A. and Haslam, R. J. Purification and characterization of the 47,000 Dalton protein phosphorylated during degranulation of human platelets. J. Biol. Chem. 258, 11404-11414, 1983.

41. Haslam, R. J. and Davidson, M. M. L. Receptor-induced diacylglycerol formation in permeabilized platelets: possible role for a GTP-binding protein. J. Recept. Res. 4, 605-629, 1984.

42. Huzoor-Akbar and Kundu, N.: Mobilization of intra- and extracellular calcium, but not the basal calcium level, is increased in platelets from spontaneously hypertensive rats. Proc. 20th FEBS Meeting p. 150 (1990).

43. Kundu, N. and Huzoor-Akbar: Inositol trisphosphate generation and mobilization of calcium is increased in platelets from spontaneously hypertensive rats. FASEB J. 5, A899 (1991).

44. Buhler, F. R. and Resink, T. J. Platelet membrane and calcium control abnormalities in essential hypertension. Am. J. Hyperten. 1, 42-46, 1988.

45. Erne, P., Bolli, P., Burgiesser, E. and Buhler, F. R. Correlation of platelet calcium with blood pressure: Effect of antihypertensive therapy. N. Engl. J. Med. 310, 1084-1088, 1984.

46. Ahn, Y. S., Jy, W., Harrington, W. J., Shanbaky, N., Fernandez, L. F. and Haynes, D. H. Increased platelet calcium in thrombosis and related disorders and its correction by nifedipine. Thromb. Res. 45, 135-143, 1987.

47. Oshima, T., Young, E. W., Bukoski, R. D., and McCarron, D. A. Abnormal calcium handling by Platelets of spontaneously hypertensive rats. Hyperten. 15, 606-611, 1990.

48. Oshima, T., Young, E. W. and McCarron, D. A. Abnormal platelet and lymphocyte calcium handling in prehypertensive rats. Hyperten. 18, 111-115, 1991.

49. Gulati, S., Khullar, M., Sharma, B. K. and Ganguly, N. K. Platelet calcium pump activity in essential hypertensives and their first-degree relatives. Moll. Cell. Biochem. 156, 37-42, 1996.

50. Eberhard, M., Evequoz, D. and Erne, P. Hypertension does not affect intracellular calcium uptake in human platelets. Hyperten. 9, 136-142, 1996.

51. Ohno, Y., Matsuo, K., Suzuki, H., Tanase, H., Takano, T. and Saruta, T. Increased intracellular Ca++ is not coinherited with an inferred major gene locus for hypertension (ht) in the spontaneously hypertensive rats. Am. J. Hyperten. 10, 282-288, 1997.

52. Zicha, J., Kunes, J. and Devynck, M-A. Platelet calcium handling is different in rats with salt-dependent and spontaneous forms of genetic hypertension. Am. J. Hyperten. 9, 812-818, 1996.

53. Murakawa, K., Kanayama, Y., Kohno, M., Kawarabayashi, T., Yasunari, K., Takeda, T. and Hyono, A. Cytoplasmic free [Ca2+] is not increased in the platelets of deoxycorticosterone--salt and spontaneously hypertensive rats. Clin. Sci. 71, 121-123, 1986.

54. Yamanishi, J., Takai, Y., Kaibuchi, K., Sano, K., Castagna, M. and Nishizuka, Y. Synergistic function of phorbol ester and calcium in serotonin release from human platelets. Biochem. Biophys. Res. Commun. 112, 778-786, 1983.

55. Daniel, J. L., Holmsen, H. and Adelstein, R. S. Thrombin-stimulated myosin phosphorylation in intact platelets and its possible involvement in secretion. Thromb. Haemost. 38, 984-989, 1977.

56. Takaori, K., Itoh, S., Kanayama, K. and Takeda, T. Protein kinase C activity in platelets from spontaneously hypertensive rats (SHR) and normotensive Wistar Kyoto rats (WKY). Biochem. Biophys. Res. Commun. 141, 769-773, 1986.

57. Koutouzov, S., Limon, I., and Marche, P. Receptor-dependent and -independent protein phosphorylation in platelets of spontaneously hypertensive rats. Thromb. Res. 59, 475-487, 1990.

58. Haller, H., Lindschau, C., Quass, P. and Distler, A. Protein phosphorylation and intracellular free calcium in platelets of patients with essential hypertension. Am. J. Hyperten. 5, 117-124, 1992.

59. Resink, T. J., Burgisser, E. and Buhler, F. R. Enhanced platelet cyclic AMP response to prostaglandin E1 in essential hypertension. Hyperten. 8, 662-668, 1986.

60. Mazeaud, M. M., Le Quan Sang, K.-H. and Devynck, M-A. Platelet cyclic AMP in essential hypertensive and normotensive offspring. J. Hyperten. 7 (Suppl. 6), S148-S149, 1989.

61. Hoffmann, G., Gobel, B. O., Harbercht, U., Vetter, H. and Dusing, R. Platelet cyclic AMP and cyclic GMP in essential hypertension. Am. J. Hyperten. 5, 847-850, 1992.

62. Donner, M., Bouchy, M., Viriot, M-L. and Andre, J.-C. Spectrofluorometric methods for the measurement of the so called Amicroviscosity@ of lipid structures. Biochimie, 63, 961-965, 1981.

63. Spiegel, A. M. G proteins in cellular control. Curr. Opin. Cell. Biol. 4, 203-211, 1992.

64. Gawler, D., Milligan, G., Spiegel, A. M., Unson, C. G. and Houslay, M. D. Abolition of the expression of inhibitory guanine nucleotide regulatory protein G_i activity in diabetes. Nature 327, 229-232, 1987.

65. Marcil, J., Schiffrin, E. L. and Anand-Srivastava, M. B. Aberrant adenylyl cyclase/cyclic AMP signal transduction and G protein levels in platelets from hypertensive patients improve with antihypertensive drug therapy. Hyperten. 28, 83-90, 1996.

66. Shenker, A., Goldsmith, P., Unson, C. G. and Spiegel, A. M. The G protein coupled to the thromboxane A2 receptor in human platelets is a member of the novel Gq family. J. Biol. Chem. 266, 9309-9313, 1991.

67. McLellan, A. R., Milligan, G., Houslay, M. D. and Connell, J. M. C. G-proteins in essential hypertension: a study of human platelet plasma membranes. J. Hyperten. 11, 543-549, 1993.

68. Aragon-Birlouez, J., Montenay-Garestier, T. and Devynck, M-A. Further analysis of cell membrane changes in genetic hypertension in rats by diphenylhexatriene fluorescence polarization. Clin. Sci. 66, 717-723, 1984.

69. Dominiczak, A. F., Lazar, D. F., Das, A. K. and Bohr, D. F. Lipid bilayer in genetic hypertension. Hyperten. 18, 748-757, 1991.

70. Le Quan Sang, K.-H., Montenay-Garestier, T. and Devynck, M-A. Alterations of platelet membrane microviscosity in essential hypertension. Clin. Sci. 80, 205-211, 1991.

71. Le Quan Sang, K.-H., Mazeaud, M., Astarie, C., Montenay-Garestier, T. and Devynck, M-A. Plasma lipids and platelet membrane fluidity in essential hypertension. Thromb. Haemost. 69, 70-76, 1993.

72. Caimi, G., Presti, R. L., Montana, M., Contorno, A., Canino, B., Catania, A., Sarno, A. and Cerasola, G. Platelet membrane fluidity and platelet membrane lipid pattern in essential hypertension. Am. J. Hyperten. 8, 82-86, 1995.

73. Caimi, G. Erythrocyte, platelet and polymorphonuclear leukocyte membrane dynamic properties in essential hypertension. Clin. Hemorheol. Microcirc. 17, 199-208, 1997.

74. Devynck, M-A., Kunes, J., Le Quan Sang, K.-H. and Zicha, J. Membrane microviscosity and plasma triacylglycerols in the rat. Clin. Sci. 94, 79-85, 1997.

75. Naftilan, A. J., Dzau, V. and Loscalzo, J. Preliminary observations on abnormalities of membrane structure and function in essential hypertension. Hyperten. 8 (Suppl. II), II-174- II-179, 1986.

76. Benjamin, N., Robinson, B. F., Graham, J. G. and Wilson, R. W. Cholesterol:phospholipid ratio is elevated in platelet plasma membrane in patients with hypertension. J. Human Hyperten. 4, 273-276, 1990.

77. Tandon, N., Harmon, J. T., Rodbard, D. and Jamieson, G. A. Thrombin receptors define responsiveness of cholesterol-modified platelets. J. Biol. Chem. 258, 11840-11845, 1983.

78. Huzoor-Akbar, Anwer, K., Wince, L. and Kundu, N. Platelet phospholipid metabolism and fatty acid composition, but not the cholesterol/phospholipid ratio, is altered in spontaneously hypertensive rats. FASEB J. 3, A 311 (1989).

79. MacIntyre, D. E., Hoover, R. L., Smith, M., Steer, M., Lynch, C., Karnovsky, M. and Salzman, E.W. Inhibition of platelet function by cis-unsaturated fatty acids. Blood 63, 848-857 1984.

80. Radomski, M. W., Palmer, R. M. J. and Moncada, S. Comparative pharmacology of endothelium-derived relaxing factor, nitric oxide and prostacyclin in platelets. Br. J. Pharmacol. 92, 181-187, 1987.

81. Radomski, M. W., Palmer, R. M. J. and Moncada, S. The role of nitric oxide and cyclic GMP in platelet adhesion to vascular endothelium. Biochem. Biophys. Res. Commun.148, 1482-1489, 1987.

82. Cadwgan, T. M. and Benjamin, N. Evidence for altered platelet nitric oxide synthesis in essential hypertension. J. Hyperten. 11, 417-420, 1993.

83. Woods, J. D., Edwards, J. S. and Ritter, J. M. Inhibition by nitroprusside of platelet calcium mobilization: Evidence for reduced sensitivity to the nitric oxide in essential hypertension. J. Hyperten. 11, 369-373, 1993.

84. Huzoor-Akbar, Calub, F., Jensen, BO, Holmsen, H.: Nitric oxide-induced phosphorylation of platelet rap 1b serves as one of the signals for inhibition of platelet aggregation. Canadian J. Physiol. Pharmacol. 72, 479 (1994).

85. Radomski, M. W., Palmer, R. M. J. and Moncada, S. Modulation of platelet aggregation by an L-arginine-nitric oxide pathway. TiPS 12, 87,88, 1991

86. Huzoor-Akbar: Role of nitric oxide/cyclic GMP signaling mechanisms in increased platelet sensitivity to thrombin in hypertension. FASEB J. 9, A922 (1995).

87. Clerk, F. D., Gorp, L. V., Xhonneux, B., Somers, Y. and Wouters, L. Enhanced platelet turnover and prostaglandin production in spontaneously hypertensive rats. Thromb. Res. 27, 243-249, 1982.

88. Huzoor-Akbar and Anwer, K. Evidence that the rat is not an appropriate model to study the role of prostaglandins in normal or abnormal platelet aggregation. Thromb. Res. 41, 555-566, 1986.

89. Nishizawa, E. E., Williams, D. J. and Connell, C. L. Arachidonate induced aggregation of rat platelets may not require prostaglandin endoperoxide or thromboxane A_2. Thromb. Res. 30, 289-296, 1983.

90. Hwang, D. H. Aggregation and inhibition of rat platelets and the formation of endoperoxide metabolites. Prostaglandins and Med. 5, 163-173, 1980.

91. Tomita, T., Umegaki, K., Nakamura, K. and Hayashi, E. Impaired function of TXA2 and/or calcium in the platelets from SHRSP at hypertensive ages. Thromb. Res. 35, 219-230, 1984.

92. Wu, X. C., Richards, N. T., Michael, J. and Johns, E. Relative roles of nitric oxide and cyclo-oxygenase and lipoxygenase products of arachidonic acid in the contractile responses of rat renal arcuate arteries. Br. J. Pharmacol. 112, 369-376, 1994.

93. Piomelli, D. and Greengard, P. Lipoxygenase metabolites of arachidonic acid in neuronal transmembrane signalling. TiPS 11, 367-373, 1990.

94. Piomelli, D., Volterra, A., Dale, N., Siegelbaum, S. A., Kandel, E. R., Schwartz, J. H. and Belardetti, F. Lipoxygenase metabolites of arachidonic acid as second messengers for presynaptic inhibition of Aplysia sensory cells. Nature, 328, 38-43, 1987.

95. Stern, N., Kisch, E. S. and knoli, E. Platelet lipoxygenase in spontaneously hypertensive rats. Hyperten. 27, 1149-1152, 1996.

96. Mezei, Z., Kis, B., Gecse, A., Telegdy, G., Abraham, G. and Sonkodi, S. Platelet eicosanoids and the effect of captopril in blood pressure regulation. Eur. J. Pharmacol. 340, 67-73, 1997.

97. Nishikawa, M., Hidaka, H. and Shirakawa, S. Possible involvement of direct stimulation of protein kinase C by unsaturated fatty acids in platelet activation. Biochem. Pharmacol. 37, 3079-3089, 1988.

21

NITRIC OXIDE-MEDIATED REGULATION OF PLATELET FUNCTION

Marek W. Radomski and
Anna S. Radomski
Division of Research and Development Lacer, S.A.,
08025 Barcelona, Spain and Department of Pharmacology
University of Alberta
Edmonton, AB
Canada

21.1 INTRODUCTION

Marek W. Radomski, Ph. D., has served on the Faculty of Pharmacology at the University of Alberta, Canada. Now he is back in Spain as Director of R & D. His work was supported by grants from the Medical Research Council of Canada. MWR is a Scholar of the Heritage Foundation for Medical Research. Dr. Radomski's contribution has been significant.

21.1 History of the discovery of nitric oxide as an endogenous regulator of vascular homeostasis

The discovery of the biological significance of nitric oxide (NO) in the mammalian systems stemmed from two lines of investigation that were conducted apart from each other. The first line was focused on the mechanism of acetylcholine-induced vascular relaxation in vitro and in vivo by the endothelium-derived relaxing factor, EDRF (Furchgott and Zawadzki, 1980), while the second examined the mechanism of macrophage cytotoxicity (Stuehr and Marletta, 1985; Hibbs et al, 1987).

Both lines of research converged with the discovery that NO gas can account for the biological activity of EDRF (Palmer et al., 1987) and to mediate the some of the cytotoxic reactions of cytokine-activated macrophages.

The appreciation of the role of NO as an endogenous regulator of platelet function arose from the research aimed at elucidation of non-prostaglandin mediators involved in inhibition of platelet activation. These investigations came to fruition by the findings that EDRF is also a potent inhibitor of platelet aggregation (Azuma et al, 1986), and that the release of NO from the vascular endothelial cells inhibits platelet adhesion (Radomski et al, 1987a,b), aggregation (Radomski et al, 1987c) and stimulates disaggregation of preformed platelet aggregates (Radomski et al, 1987d).

Other milestones in research on the biological significance of NO in vascular homeostasis were the description of the molecular pathway of NO biosynthesis by NO synthase from a semi-essential amino acid, L-arginine (Palmer et al, 1988), and the identification of this enzyme and its role as modulator of platelet function in human platelets (Radomski et al, 1990a).

At the time when this contribution has been written (spring 1998) combined data bank search (1988-1998) generated more than 18,000 references on NO, out of which approximately 1000 dealt with the effects of NO on platelets. Therefore, the presentation and discussion of all the developments in this large area of research falls beyond the scope of the chapter. This contribution is rather intended to give a selected outlook on the current status of NO as regulator of platelet hemostasis and thrombosis.

21.2 Enzymology

Gene cloning and other molecular and biochemical techniques identified three distinct isoforms of NOS, coded by genes residing in different chromosomes. These are endothelial-type (eNOS), neuronal-type (nNOS) and an isoform expressed during various reactions associated with the release of the pro-inflammatory cytokines, iNOS (Radomski, 1995). Although almost every mammalian cell has the capacity to express NOS, in the hemostatic system eNOS and iNOS appear to play important roles in the cellular biology. Both in the endothelial cells (Marsden et al, 1992) and platelets (Radomski et al, 1990a, Muruganandam and Mutus, 1994) eNOS is constitutively expressed. The expression of iNOS in the endothelium appears to be cytokine- and protein synthesis-dependent (Radomski et al., 1990b). A controversy exists, as to the expression of iNOS in platelets. Some researchers found the presence of the iNOS mRNA and/or protein in normal platelets (Mehta et al, 1995; Wallerath et al, 1997), whereas other failed to confirm these findings (Sase and Michel, 1995). Platelets have a limited capacity to synthesize proteins de novo and acquire most of their proteins from megakaryocytes. We and others have shown that cytokine-stimulated megakaryoblasts and bone marrow megakaryocytes express iNOS (Lelchuk et al 1992; Wallerath et al 1997). As megakaryocytopoiesis and thrombopoiesis are cytokine-dependent and regulated by the stimulatory and inhibitory cytokines (Brown amd Martin, 1994) the presence or absence of iNOS in platelets may reflect the interactions between these hematopoietic factors.

The eNOS and iNOS are homodimer hemproteins with the molecular weight of 130-135 kDa per subunit. The subcellular localization of eNOS and iNOS appears to be different as the presence of the myristoylation site targets eNOS to the membrane fractions of the cell, while the distribution of iNOS is largely cytosolic and asscoayed with cell caveolae (Garcia-Cardena et al, 1996). Interestingly, similar to some other enzymes eNOS can be translocated during activation with lysophosphatidylcholne (Dudek et al, 1995). Calmodulin-binding domain is an important part of NOS that mediates calcium binding to the enzyme. The enzymes require L-arginine and cofactors including NADPH, terahydrobiopterin and flavin nucleotides (Marsden et al, 1992; Radomski et al, 1990a; Muruganandam and Mutus, 1994). Both isoforms convert L-arginine substrate via a two-step hydroxylation-oxidation reaction to NO and L-citrulline, a co-product of the reaction (Palmer et al, 1988).

21.3 Molecular targets and metabolism of NO

Although NO can be classified as a radical its reactivity with the molecular targets is not indiscriminate. The soluble guanylate cyclase (GC-S) is a major molecular, intracellular target of NO (Mellion et al, 1981). This is due to the affinity of NO to iron in protoporphyrin IX. The interaction of NO with GC-S leads to its activation and conversion of GTP to cyclic GMP (cGMP). Nitric oxide can interact also with other heme-containing proteins such as oxyhemoglobin and various enzymes of the respiratory chain (Martin et al, 1985; Hibbs et al, 1987). The reaction of NO with oxyhemoglobin may yield methemoglobin (the reaction with heme, Martin et al, 1985) and nitrosylated hemoglobin (the reaction with thiol groups of hemoglobin, Stamler et al, 1997). Some thiols such as glutathione (Radomski et al, 1992) and albumin (Stamler et al, 1992) may also be S-nitrosylated by NO.

Superoxide is also an affinity target of NO and non-enzymatic, very efficient (the rate of the reaction approaches the diffusion rate) reaction of these two molecules yields peroxynitrite ($ONOO^-$) (Beckman and Tsai, 1994).

21.4 Nitric oxide as physiological regulator of platelet function

There is now convincing evidence that NO is a major physiological factor modulating platelet function.

21.4.1 Adhesion

Both endogenous and exogenous NO has been now shown to inhibit platelet adhesion to the endothelium, the extracellular matrix, collagen fiber, collagen-coated beads, collagen-coated microtiter plates and fibrinogen-coated cover slips (Radomski et al 1987a,b; Sneddon and Vane 1988; Venturini et al, 1989; Pohl and Busse, 1989; Polanowska-Grabowska and Gear, 1994; Wu et al, 1997) both in static and flow systems.

21.4.2 Aggregation and thrombus formation

In vitro NO, released from fresh or cultured endothelial cells, inhibits aggregation induced by a variety of agonists as well as by shear stress (Radomski et al, 1987c,d; Furlong et al, 1987; Busse et al, 1987; Macdonald et al, 1988; Alheid et al, 1989; Houston et al, 1990; Broekman et al, 1991). Nitric oxide also causes disaggregation of preformed platelet aggregates (Radomski et al, 1987d) and inhibits platelet recruitment (Freedman et al, 1997).

In vivo, basal or agonist-stimulated release of NO result in inhibition of platelet aggregation induced by some aggregating agents or endothelial injury and increase bleeding time (Rosenblum et al, 1987; Bhardwaj et al, 1988; Hogan et al, 1988; Humphries et al, 1990; Herbaczynska-Cedro et al, 1991; May et al 1991; Golino et al, 1992; Yao et al, 1992; Houston and Buchanan, 1994). In addition, there is luminal release of NO from human vasculature causing increases in intraplatelet cGMP levels (Andrews et al, 1994). Finally, the administration of NOS inhibitor L-NMMA into healthy volunteers increased platelet aggregation granule release (Bodzenta-Lukaszyk et al, 1994) and shortens bleeding time (Simon et al, 1995) whereas L-arginine, the substrate for NO synthesis, leads to the inhibition of platelet activation (Caren and Corbo, 1973; Adams et al, 1995; Bode-Boger et al, 1998). Both the vasodilator (Houston and Buchanan, 1994) and platelet-inhibitory (Yao et al, 1992; Golino et al, 1992; Bodzenta-Lukaszyk et al, 1994; Adams et al, 1995; Bode-Boger et al, 1998) components contribute to the haemostatic action of NO.

21.5 Mechanisms of NO action on platelets

Nitric oxide inhibits the release of ADP and thromboxane from platelets (Radomski et al, 1990c, Bode-Boger et al, 1998). In addition, NO inhibits the release of MMP-2 that mediates non-thromboxane, non-ADP-dependent pathway of aggregation (Sawicki et al, 1997). Mechanisms of platelet-inhibitory effects of NO are still subject of investigation. Nitric oxide is relatively lipophilic and freely diffusible molecule that potentially could interact with many molecular targets. However, it has been known from some time now that NO has a high affinity to heme in heme proteins (Craven and DeRubertis, 1978). Since the soluble guanylyl cyclase (GC-S) is a major platelet heme protein it is not surprising that platelet fractions containing GC-S readily sequester NO gas (Liu et al, 1993). This study has demonstrated that NO is sequestered preferentially by subcellular fractions of these cells that contain GC-S activity and that the sequestration of NO in these fractions stimulates the catalytic activity of GC-S. The interaction of NO with GC-S results in enzyme activation and leads to the conversion of magnesium guanosine 5'-triphosphate to guanosine 3',5'-monophosphate (cGMP) (Mellion et al, 1981).

21.5.1. Cyclic GMP-dependent and independent effects of NO

Three different proteins appear to mediate the cellular actions of cGMP: cGMP-dependent protein kinase, cGMP-binding cAMP phosphodiesterase and cGMP-regulated ion channels (Walter, 1989)

Stimulation of cGMP-dependent protein kinase results in phosphorylation of various target proteins (Walter, 1989). Among proteins phosphorylated in response to cGMP best characterised is 46/50 kDa vasodilator-stimulated phosphoprotein (VASP; Haffner et al, 1995). In adhering platelets VASP is associated with actin filaments and focal contact areas, i.e. transmembrane junctions between microfilaments and the extracellular matrix (Reinhard et al, 1992). In particular, the association of VASP with the platelet cytoskeleton may be of importance for its inhibitory effect on the fibrinogen receptor (Horstrup et al, 1994). Cyclic GMP-induced protein phosphorylation may be also involved in the uptake of serotonin by platelets (Launay et al, 1994)

Cyclic GMP decreases basal and stimulated concentrations of intracellular Ca^{2+} (Nakashima et al, 1986; Johansson and Haynes, 1992). A number of Ca^{2+} handling systems have been identified in platelets including receptor operated channels, passive leak, Ca^{2+}-ATPase extrusion pump, the Na^{+}/ Ca^{2+} exchanger, Ca^{2+}-accumulating ATPase pump of the dense tubular membrane (an intraplatelet membrane Ca^{2+} store) and passive leakage and receptor operated Ca^{2+} channels in the dense tubular membrane. In principle, all these processes could be affected by cGMP. It has been shown that cGMP increases the activity of Ca^{2+} -ATPase extrusion pump and leakage across the plasma membrane (Johansson and Haynes, 1992). In addition, cGMP causes inhibition of Ca^{2+} mobilisation from intraplatelet stores including the dense tubular membrane (Nakashima et al, 1986). In contrast to cAMP, cGMP does not stimulate the dense tubular Ca^{2+} pump. Consequently, cGMP cannot result in increased dense tubular sequestration of Ca^{2+} whereas cAMP does. Thus, cGMP may be "a better Ca^{2+} antagonist" (Johansson and Haynes 1992).

Phosphorylation of a cGMP-inhibited cAMP phosphodiesterase by the cAMP-dependent protein kinase increase phosphodiesterase activity and this may represent a negative feedback mechanism of cellular cAMP levels (Macphee et al). It has been shown that cGMP by inhibiting cGMP-inhibited cAMP phosphodiesterase may delay the hydrolysis of cAMP and enhance the biological effects of the latter nucleotide (Maurice and Haslam, 1990). However, the physiological and pharmacological relevance of this "cross-talk" between cAMP and cGMP pathways is unclear (Radomski et al, 1992).

Some authors suggested that in addition to cGMP-dependent effects some actions of NO on platelets may be independent from the generation of this cyclic nucleotide. These could be due to direct actions of NO on calcium flux (Menshikov et al, 1993), metabolism via inhibition of ADP ribosylation (Brune and Lapetina, 1989) and inhibition of 12-lpoxygenase (Nakatsuka and Osawa, 1994). We have recently characterised 1H-[1,2,4]oxadiazolo[4,3-a]quinoxalin-1-one (ODQ) as a potent and selective inhibitor of GC-S (Moro et al 1996, Martinez-Cuesta and Radomski, 1998). Using this compound we have shown that both the antiaggregatory and adhesion-inhibitory effects of NO in vitro could by completely reversed by ODQ showing their dependence on cGMP.

21.5.2 Cyclic GMP, membrane phospholipids and receptors

Metabolism of membrane phospholipids may be also a target for the action of cGMP. Indeed, the inhibition of both phospholipase C and A_2 has been implicated in the mechanism of this action on platelets (Nakashima et al, 1986; Sane et al, 1989).

Cyclic GMP down-regulates the function of some platelet receptors including the fibrinogen receptor IIb/IIIa and protein kinase C-induced expression of P-selectin and the release of lysosomal protein CD63 (Salas et al, 1994; Murohara et al, 1995; Mendelsohn et al, 1990; Michelson et al, 1996). Interestingly, von Willebrand and fibronectin receptors appear no to be regulated by cGMP (Michelson et al, 1996; Shahbazi et al, 1994).

The biological actions of cGMP are terminated by cGMP phosphodiesterase, by its efflux from platelets and may also depend on the activity of protein phosphatases (Walter 1989).

21.6 Peroxynitrite

In addition to its platelet- and vessel wall-regulatory effects NO was also shown to exert some detrimental effects such as inhibition of cellular respiration (Hibbs et al, 1987). These effects were originally ascribed to the actions of very large amounts of NO. Beckman and colleagues (1994) suggested that some of detrimental effects of excessive amounts of NO could be the result of its reaction with superoxide to form a potent and aggressive radical, peroxynitrite ($ONOO^{-}$). We have found that ONOO- can be considered as a biological opponent of NO in its actions on platelets and the vascular wall. Indeed, the oxidant can induce platelet aggregation and counteract the inhibitory effects of NO and prostacyclin on platelets (Moro et al, 1994). In addition, ONOO- may render the vasculature non-reactive to the vasodilator effects of NO (Villa et al, 1994). There is now indirect evidence that ONOO- may be generated during platelet aggregation (Naseem et al, 1997). Moreover, inhibition of superoxide action may inhibit thrombin-induced platelet aggregation (Salvemini et al, 1989). Al these data suggest that, indeed, the release of ONOO- is likely to attenuate the physiological effects of NO on platelets and the vasculature. The mechanism of platelet-activator effects of ONOO- may depend on its ability to cause nitration and phosphorylation of platelets (Mondoro et al, 1997).

21.7 Nitric oxide and vascular disorders

The vasodilator and platelet-regulatory functions of endothelium are impaired during the course of vascular disorders including atherosclerosis, the coronary artery disease, essential hypertension, diabetes mellitus and preeclampsia (DeBelder and Radomski, 1994), however the pathomechanism of these changes remains unclear.

Since oxidative modification of low-density lipoproteins (LDL) plays a key role in atherogenesis, a number of studies (reviewed by Radomski and Salas, 1995a) have examined the effects of native and oxidised LDL on NO-mediated vascular functions. In most of these studies lipoproteins decreased the bioactivity of NO. Several hypotheses have been proposed to explain these effects of LDL including inhibition of NOS activity and direct inactivation of NO (Luscher et al, 1993; Flavahan, 1992; Cooke and Tsao, 1992 and refs. therein). The decreased bioactivity of NO in atherosclerosis could also result from changes in the metabolism of this molecule and generation of $ONOO^-$ from superoxide and inducible NO (Beckman and Tsao, 1994). These data indicate that the generation of secondary oxidants such as $ONOO^-$ may play an important role in the pathogenesis of atherosclerosis (Radomski and Salas, 1995a). In addition, LDL may inhibit arginine uptake into platelets and through this mechanism decrease NOS activity and promote thrombosis (Chen and Mehta, 1994), an effect reversible by the administration of L-arginine in diet (Tsao et al, 1994). In contrast to LDL, high-density lipoproteins (HDL) decreased platelet function by increasing NOS activity in platelets (Chen and Mehta, 1994). Moreover, human apolipoprotein E that mediates hepatic clearance of lipoproteins exerts a significant inhibitory effect on platelets through stimulation of the intraplatelet levels of NO and cGMP (Riddell et al, 1997). The ischemic heart disorder and myocardial infarction are common manifestations of coronary atherosclerosis. The endogenous NO inhibited microthromboembolism in the ischemic heart, protected myocardium against intracoronary thrombosis and decreased platelet deposition due to the carotid endarterectomy (Komamura et al, 1994; Olsen et al, 1996). Thus, the development and complications of atherosclerosis may be linked to altered metabolism and actions of NO in platelet aand the vascular wall.

An impaired NO generation or action may also underlie the pathomechanism of vasospastic and thrombotic changes of essential hypertension (Cadwgan and Benjamin, 1993; Calver et al, 1992).

There are indications that the changes in the bioactivity and metabolism of NO are involved in the pathogenesis of diabetes mellitus. Insulin, at physiological concentrations, exerts a platelet anti-aggregating effect that is mediated through stimulation of platelet NOS and the resultant increase in cGMP and, possibly, also cAMP levels (Trovati et al 1996; 1997). This suggests that a reduced generation of NO in insulin-deficient states could contribute to platelet hyperactivity and diabetic angiopathy. Recently, we have examined the formation of NO in the vasculature and platelets of JCR-LA-cp rat, a model of insulin-resistant states associated with obesity and complicated by atherosclerosis. We have found that generation of NO is crucial for the preservation of vascular homeostasis under these conditions (McKendrick et al 1998).

Vasoconstriction and increased platelet activation are also characteristic for pre-eclampsia, a severe disease that may complicate normal pregnancy. We have examined the activity of NOS and the soluble guanylate cyclase in non-pregnant, healthy pregnant and pre-eclamptic women. Increased generation of NO in platelets of pre-eclamptic women was associated with increased platelet activation (Salas et al 1998). However, despite increased formation of NO the activity of the soluble gunaylate cyclase was reduced in pre-eclamptic women suggesting that this NO was not bioactive i.e. failed to protect platelets from activation. It is possible that enhanced generation of ONOO- could be held repsonsible for platelet dysfunction in preeclampsia.

21.8 Nitric oxide, platelets and septicemia

The expression of iNOS has been also implicated in the pathogenesis of septicemia and septic shock. The iNOS is induced following the invasion of gram-negative bacteria and exposure of cells to endotoxin (LPS) and cytokines (for refs. see DeBelder and Radomski 1994; Radomski 1995). LPS- and cytokine-mediated expression of iNOS is likely to have complex repercussions for vascular hemostasis. On one hand, inducible NO acts to promote hemostasis and inhibit thrombosis since the inhibition of its generation by NOS inhibitors greatly potentiated cytokine-stimulated platelet adhesion to cultured human endothelial cells (Radomski et al 1993), precipitated renal glomerular thrombosis (Schultz and Raij 1992) and exacerbated sepsis-induced renal hypoperfusion (Spain et al 1994). On the other hand, the exposure of endothelial cells to cytokine-induced NO may result in cell toxicity and destruction (Palmer et al 1992) and it has been reported that the inhibition of NOS may be beneficial in the treatment of septic shock (Kilbourn et al 1990; Wright et al 1992). A partial explanation for this discrepancy may be that the currently available inhibitors of iNOS are not selective and inhibit the activities of other NOS isoenzymes. Indeed, an intact generation of constitutive NO may be important to maintain the integrity of the microvasculature during sepsis (Radomski and Salas 1995b). Clinical investigations are in progress to address the efficacy and safety of NOS inhibition treatment in this condition.

21.8.1 Nitric oxide, platelets and uremia

Bleeding is a well-known complication of uraemia that is attributed to the suppression of platelet function by the disease process. Interestingly, L-arginine and some other platelet-inhibitory metabolites of the urea cycle are accumulated in uraemia (Horowitz et al 1970) and this is associated with an increase in TNF? levels (Noris et al 1993). It has been shown that platelets obtained from uraemic patients generate more NO than controls so that increased expression and/or activity of NOS may play a role in platelet dysfunction observed in uraemia (Noris et al 1993).

21.8.2 Nitric oxide, platelets, tumor growth and metastasis

Recent studies have shown that platelets are one of the cytotoxic effector cells against neoplasia (for review Okada et al 1996). It has been shown that the cytotoxic effects of activated platelets against lung cancer cells in culture may depend on the activity of platelet NOS (Okada et al 1996).

Platelets also play a role in the pathogenesis of tumour metastasis by increasing the formation of tumour cell-platelet aggregates thus facilitating cancer cell arrest in the microvasculature. We have demonstrated that that tumour cell-induced platelet aggregation in vitro is modulated by the ability of tumour cells to generate NO and this correlated with their propensity for metastasis (Radomski et al 1991). Indeed, human colon carcinoma cells isolated from metastases exhibited lower NO activity than cells isolated from the primary tumour. Moreover, the expression of iNOS by murine melanoma cells inversely correlated with their ability to form metastases in vivo (Dong et al 1994). These data suggest that differential synthesis of NO may distinguish between cells of low and high metastatic potential. Interestingly, NOS has been found in some human gynaecological malignancies and the highest NOS activities detected in poorly differentiated tumours (Thomson et al 1994). Thus, further work is needed to unravel the biological significance of NO for the growth of tumor, tumor metastasis and platelet-tumor cell interactions.

21.9 Pharmacology of NO generation and action in the platelet microenviro-nment

21.9.1 L-Arginine

L-Arginine is the endogenous substrate from which NO is generated in various cells and tissues (Radomski 1995). Under physiological conditions the amounts of endogenous L-arginine in the platelet microenvironment are high (mM) suggesting that the availability of substrate is unlikely to constitute a rate-limiting factor for activation of eNOS. Indeed, stimulation of resting platelets with L-arginine does not result in generation of NO (Radomski et al 1990a). In contrast, when eNOS is stimulated L-arginine is metabolised to NO. Platelet activation is a potent stimulus for stimulation of NOS and under these conditions L-arginine is converted by platelet eNOS to NO (Radomski et al 1990a, Malinski et al 1993, Bode-Boger et al 1998). The presence of the extracellular calcium appears to be crucial for the activation of platelet eNOS by L-arginine, as the platelet-inhibitory activity of L-arginine in the citrated platelet-rich plasma is lower than in hirudinised whole blood (Radomski et al 1990a, Bode-Boger et al 1998). The expression of iNOS appears to provide an appropriate stimulus for the conversion of L-arginine to NO. This may occur during acute or chronic inflammatory states exemplified by sepsis, preeclampsia and atherosclerosis. Under physiological conditions platelets have been shown to express both the message and protein for iNOS (Wallerath et al 1997), although evidence for the expression of enzyme activity under these conditions is lacking. In contrast, atherosclerosis and preeclampsia lead to a clear-cut increase in the activity of Ca^{2+}-independent NOS implying that iNOS is activated during these pathologies (Dube et al 1998, Salas et al 1998). Interestingly, Cooke and colleagues (1992) provided a convincing evidence for platelet-inhibitory actions of L-arginine when administered to animals and humans with atherosclerosis. Thus, L-arginine may be a useful pharmacological tool in vascular disorders associated with platelet activation.

21.9.2 Stimulators, inhibitors of NOS and NO scavengers

Some endogenous and exogenous compounds have been reported to inhibit platelet function via stimulation of the activity of platelet NOS. Relaxin is a uterine hormone responsible for the relaxation of this organ. It has been shown that cGMP-dependent inhibition of platelet aggregation by relaxin is associated with stimulation of platelet NO and inhibited by selective inhibitors of NOS. Trilinolein is triacylglycerol containing linoleic acid that can be purified from *Panax pseudoginseng*, a medcinal herb widely used from cardiovascular disorders in traditional Chinese medicine (Shen and Hong 1995). It has been shown that the platelet-inhibitory effect of trilinolein is dependent of the generation of NO in platelets (Shen and Hong 1995).

Some analogues of L-arginine such as N^G-monometyl-L-arginine act as potent and highly selective inhibitors of NOS (Radomski 1995). These compounds counteract the platelet-inhibitory effects of L-arginine (Radomski et al 1990a) shorten bleeding time in humans (Simon et al 1995) and may precipitate thrombosis under conditions of vascular stress (Schulz and Raij 1992). Interestingly, long-term smoking impairs the activity of platelet NOS (Ichiki et al 1996), although the mechanism responsible for this impairment remains to be elucidated.

Several different preparations of purified human hemoglobin, cross-linked to prevent renal damage, are currently undergoing clinical trials as an oxygen-carrying agent. In addition to binding oxygen, hemoglobin has a very high affinity to NO. Olsen and colleagues (1996) showed that cross-linked hemoglobin enhances platelet deposition in a rat carotid endarterectomy model, an effect prevented by administration of L-arginine. Thus, cross-linked hemoglobins may stimulate platelet activation through the mechanism involving NO inhibition.

21.9.3 Nitric oxide gas

Early studies using NO gas showed that the molecule was a potent, but short-acting (the biological half-life< 4min) inhibitor of platelet adhesion, aggregation and stimulator of platelet disaggregation (Radomski et al 1987a-d). Because of its short-lasting pharmacological effects inhaled NO is used increasingly as a selective pulmonary vasodilator to treat critically ill adults and infants (for review see Cheung et al 1997). Some investigators studied the effects of inhaled NO on platelet function in health and disease. Interestingly, there are very limited effects of inhaled NO on platelets of healthy subjects (Albert et al 1996). However, both in adults with adult respiratory distress syndrome (Samama et al 1995) and critically ill neonates (Cheung et al 1998) NO treatment causes inhibition of platelet function. Under some conditions this effect is unexpectedly long lasting and may enhance the risk of the intracranial bleeding. Therefore, caution should be taken during administration of NO gas to critically ill patients. Interestingly, animal experiments showed that inhaled NO may inhibit the development of coronary thrombosis (Adrie et al 1996). The pharmacological significance of these findings remains to be investigated.

21.9.4 Nitric oxide donors and their clinical relevance as inhibitors of platelet activation

Mechanism of pharmacological activity of NO donors is most probably related to the release of NO (Feelisch and Noack 1987). As these compound release NO in a spontaneous, catalyst- or enzyme-dependent manner the biological effects of compounds are usually longer lasting when compared with NO gas.

Organic nitrates are poor spontaneous releasers of NO and require the presence of a thiol co-factor (e.g. N-acetylcysteine) for acceleration of this liberation (Feelisch 1991). However, in vivo the release of NO from organic nitrates is greatly enhanced by thiols and enzyme(s). The enzymes whose actions may be relevant to inhibition of platelet aggregation by glyceryl trinitrate are plasma glutathione-s-transferases (Chen et al 1996). Whether or not these or similar enzymes are present in platelets remains controversial (Gerzer et al 1988, Weber et al 1996). However, following stimulation of platelet aggregation by ADP in vitro glyceryl trinitrate has been shown to exert a potent disaggregating effect which may be due to a temporary elevation in the sensitivity of GC-S to NO (Chirkov et al 1991). Nitrate-induced inhibition of platelet aggregation in vitro can be greatly potentiated in the presence of thiols or cultured vascular cells (Loscalzo 1985, Feelisch 1991, Benjamin et al 1991). This indicates that the conversion of organic nitrates by the vascular tissue in vivo can result in the release of sufficient amounts of NO for inhibition of platelet function. Indeed, in healthy volunteers, oral and intravenous administrations of glyceryl trinitrate and isosorbide mononitrates resulted in inhibition of platelet aggregation ex vivo (for review see Radomski et al 1997). In addition, glyceryl trinitrate and isosorbide dinitrate inhibited experimental thrombosis and reocclusion after thrombolysis in dogs and rats (Plotkine et al 1991, Werns et al 1994). The effectiveness of organic nitrates as antithrombotics increases with the extent of vascular injury. In normal pigs, the deposition of platelets on arterial segments following injury using balloon angioplasty is inhibited by intravenous infusion of nitroglycerin when arterial injury is deep (extending through the internal elastic lamina) rather than mild (deendothelialization only) (Lam et al 1988). Furthermore, short- and long-lasting administration of nitroglycerin and isosorbide dinitrate to patients suffering from coronary artery disease and acute myocardial infarction resulted in a significant inhibition of platelet adhesion and aggregation (Gebalska 1990, Diodati et al 1990, Sinzinger et al 1992).

What is a position of organic nitrates among "classical" inhibitors of platelet function? Actelylsalicylic acid (aspirin) is by far the most widely used anti-platelet drug in clinical practice and its benefits in terms of decreasing mortality due to re-infarction have been unequivocally demonstrated (ISIS-2 1988, ISIS-3 1992, for review see Patrono 1989) while those of organic nitrates have not yet been established. A meta-analysis found significant reduction in mortality when intravenous glyceryl trinitrate or nitroprusside were used during acute course of myocardial infarction (Yusuf et al 1988). Moreover, when combined with N-acetylcysteine, glyceryl trinitrate substantially reduced myocardial infarction in unstable angina, an effect compatible with an anti-platelet effect of glyceryl trinitrate (Horowitz et al 1988). Surprisingly, GISSI III (1994) and ISIS-4 (1993) studies failed to show clinically beneficial effect of organic nitrates on mortality after myocardial infarction. However, further analysis of GISSI III suggests that the apparent additive effect of glyceryl trinitrate and lisinopril could be attributed to anti-platelet effects of this NO donor (Andrews, May et al 1994). In addition, it is possible that nitrates may act by reducing the infarct size in small rather than large infarcts so that the neutral results of GISSI-3 and ISIS-4 may be explained by the heterogeneity of effect. Interestingly aspirin, a cyclooxygenase inhibitor, blocks only thromboxane-mediated platelet aggregation (for refs. see Patrono 1989) leaving the remaining pathways of adhesion and aggregation unopposed. In contrast, NO inhibits the activation cascade of mediators generated by all known pathways of platelet aggregation (Radomski et al 1997) and some pathways of platelet adhesion to sub-endothelium (Shahbazi et al 1994). Inhibition of platelet adhesion may be of particular importance to decrease platelet accumulation due to ischemia-reperfusion injury following organ ischemia and drugs such as NO donors that increase cGMP levels may afford effective anti-platelet action (Chintala et al 1994). Thus, there is a clear need for further clinical studies to determine the place and the effectiveness of traditional organic nitrates and new NO donors for the treatment of vascular thrombotic and ischemic disorders.

Whether or not platelets similar to the vessel wall become tolerant to the platelet-inhibitory effects of organic nitrates is again controversial (Weber et al 1996; Booth et al 1996). There have been many attempts to synthesise tolerance-free NO donors. One of more promising groups of drugs are cysteine-containing nitrates. The incorporation of a cellular thiol cysteine to the structure of organic nitrate resulted in a high effectiveness of these compounds as inhibitors of platelet and leukocyte functions both in vitro and in vivo (Lefer et al 1993).

The phenomenon of tolerance is of lesser pharmacological significance for other NO donors including sodium nitroprusside, molsidomine and SIN-1.

Because of its powerful vasodilator action sodium nitroprusside is often used to treat vascular emergencies associated with hypertensive crisis. Since this compound shows some anti-platelet activity both in vitro and in vivo (Levin et al 1982, Hines and Barash 1989) its acute clinical effects may also be mediated, in part, through inhibition of platelet function. Recently, sodium nitroprusside was administered intrapericardially to treat experimentally induced coronary thrombosis in dogs (Willerson et al 1996). As this route of administration of sodium nitroprusside produced less vasodilatation than systemic one, localized administration of this drug may offer new therapeutic possibilities for the treatment of the coronary thrombosis.

Molsidomine and its active metabolite SIN-1 inhibit experimental thrombosis and platelet aggregation in healthy volunteers and in patients suffering from acute myocardial infarction (Wautier et al 1989). Interestingly, SIN-1 in addition to NO generates superoxide and $ONOO^-$ (Hogg et al 1993). Since $ONOO^-$ causes platelet aggregation and counteracts the platelet inhibitory activity of NO (Moro et al 1994), the formation of this radical may offset the anti-platelet activity of NO released from SIN-1.

21.9.5 Novel NO donors

The platelet-inhibitory actions of organic nitrates cannot be separated from their effects on vascular wall. The concept of platelet-selective NO donors has arisen from our experiments with S-nitrosoglutathione (GSNO) (Radomski et al 1992). S-Nitrosoglutathione is a tripeptide S-nitrosothiol that is formed by S-nitrosylation of glutathione, the most abundant intracellular thiol. We have found that the intravenous administration of GSNO into conscious rat inhibits platelet aggregation at doses that have only small effect on the blood pressure (Radomski et al 1992). Moreover, similar platelet/vascular differentiation is detected following intaarterial administration of GSNO into the circulation of human forearm (DeBelder et al 1994). Finally, we have infused GSNO into patients undergoing balloon angioplasty and found that this NO donor effectively protected platelets from activation at the site of angioplastic injury without altering blood pressure (Langford et al 1994). Interestingly, the exposure of human neutrophils to NO led to depletion of glutathione stores, activation of hexose monophosphate shunt, synthesis of endogenous GSNO and inhibition of superoxide generation by neutrophils. Synthetic GSNO resulted in similar effects (Clancy et al 1994). Moreover, the administration of GSNO inhibited leukocyte activation, expression of iNOS and bypass-induced myocardial lesion in dogs (Mayers et al 1998). These observations show that GSNO is a potent regulator of platelet and neutrophil functions and it may be a prototype for the development of blood cell-selective NO donors.

We have previously shown that NO readily synergizes in its platelet-inhibitory action with other inhibitors including prostacyclin (an agonist of adenylate cyclase) and zaprinast and isobutylmetylxanthine (inhibitors of cyclic nucleotide phosphodiesterases) (Radomski et al 1987a).

Recently, nitroderivatives of aspirin and RGDS peptide have been synthesised in order to capitalise on the synergy of NO with cyclooxygenase inhibitors (aspirin) and the antagonists of the fibrinogen receptor (RGDS) (Lechi et al 1996; Gurevich et al 1997)). The pharmacological and clinical potential of such hybrid compounds remains to be investigated.

In summary, the generation of endogenous NO from L-arginine represents a major bioregulatory mechanism for preservation of platelet hemostasis and prevention of thrombosis. Pharmacological exploration of these findings offers a new strategy for treatment of vascular disorders associated with platelet sctivation.

Acknowledgements

This work was supported by a grant from the Medical Research Council of Canada. MWR is a Scholar of the Heritage Foundation for Medical Research.

REFERENCES

Adams MR, Forsyth CJ, Jessup W, Robinson J, Celermajer DS (1995) Oral L-arginine inhibits platelet aggregation but does not enhance endothelium-dependent dilation in healthy young men. J Am Coll Cardiol 26:1054-1061.

Adrie C. Bloch KD. Moreno PR et al (1996) Inhaled nitric oxide increase coronary artery patency after thrombolysis Circulation 94: 1919-1926.

Albert J,Wallen NH, Broijersen A, Frostell C, Hjemdahl P (1996) Effects of inhaled nitriic oxide compared with aspirin on platelet function in vivo inhealthy volunteers Clin Sci 91: 225-231.

Alheid U, Reichwehr I, Forstermann U (1989) Human endothelial cells inhibit platelet aggregation by separately stimulating platelet cyclic AMP and cyclic GMP. Eur J Pharmacol 164:103-110.

Andrews NP, Dakak N, Schenke WH, Quyyumi AA (1994) Platelet-endothelium interactions in humans: changes in platelet cyclic guanosine monophosphate content in patients with endothelial dysfunction.Circulation 90:I-397.

Andrews R, May JA, Vickers J, Heptinstall S (1994) Inhibition of platelet aggregation by transdermal glyceryl trinitrate. Br Heart J 1994; 72:575-579

Azuma H, Ishikawa M, Sekizaki S (1986) Endothelium-dependent inhibition of platelet aggregation. Br J Pharmacol 88; 411-415.

Bani D, Bigazzi M, Masini E, Bani G, Sacchi TB (1995) relaxin depresses platelet aggregation: invitro studies on isolated human and rabbit platelets Lab Invest 73: 709-716.

Beckman J, Tsai JH (1994) Reactions and diffusion of nitric oxide and peroxynitrite. The Biochemist 16:8-10.

Benjamin N, Dutton JAE, Ritter JM (1991) Human vascular smooth muscle cells inhibit platelet aggregation when incubated with glyceryl trinitrate: evidence for generation of nitric oxide. Br J Pharmacol 102:847-850.

Bhardwaj R, Page CP, May GR, Moore PK (1988) Endothelium-derived relaxing factor inhibits platelet aggregation in human whole blood in vitro and in the rat in vivo Eur J Pharmacol 157:83-91.

Bode-Boger SM, Boger RH, Galland A, Frolich JC (1998) Differential inhibition of human platelet aggregation and thromboxane A_2 formation by L-arginine in vivo and in vitro Naunyn-Schmiedeberg's Arch Pharmacol 357; 143-150.

Bodzenta-Lukaszyk A, Gabryelewicz A, Lukaszyk A et al (1994) Nitric oxide synthase inhibition and platelet function. Thromb Res 75:667-672.

Booth BP, Jacob S, Bauer JA, Fung HL (1996) Sustained antiplatelet properties of nitroglycerin during hemodynamic tolerance in rats J Cardiovasc Pharmacol 28: 432-438.

Broekman MJ, Eiroa AM, Marcus AJ (1991) Inhibition of human platelet reactivity by endothelium-derived relaxing factor from human umbilical vein endothelial cells in suspension: blockade of aggregation and secretion by an aspirin-insensitive mechanism. Blood 78:1033-1040.

Brown AS, Martin JF (1994) The megakaryocyte platelet system and vascular disease Eur J Clin Invest 24(S1); 9-15.

Brune B, Lapetina EG (1989) Activation of a cytosolic ADP-ribosyltransferase by nitric oxide-generating agents. J Biol Chem 264:8455-8458.

Busse R, Luckhoff A, Bassenge E (1987) Endothelium-derived relaxing factor inhibits platelet activation. Naunyn Schmiedeberg's Arch Pharmacol 336:566-571.

Cadwgan TM, Benjamin N (1993) Evidence for altered platelet nitric oxide synthesis in essential hypertension J Hypertens 11:417-420.

Calver A, Collier J, Moncada S, Vallance P (1992) Effect of local infusion of N^G-monomethyl-L-arginine in patients with hypertension. The nitric oxide dilator mechanism appears abnormal J Hypertens 10:1025-1031.

Caren R, Corbo L (1973) Response of plasma lipids and platelet aggregation to intravenous arginine. Proc Soc Exp Biol Med 143:1067-1071.

Chen LY, Mehta JL (1994) Inhibitory effect of high-density lipoprotein on platelet function is mediated by increase in nitric oxide synthase activity in platelets. Life Sci 23:1815-1821.

Chen LY, Mehta P, Mehta JL (1996) platelet inhibitory effect of nitroglycerin in platelet-rich plasma: relevance og lutathione-ss- transferase in plasma J Investig Med 44: 5611-565.

Cheung PY, Salas E, Schulz R, Radomski MW (1997) Nitric oxide and platelet function: implications for neonatology. Sem. Perinatol. 21:409-417.

Cheung P-Y, Salas E, Etches PC, Phillipos E, Schulz R, Radomski MW (1998) Inhaled nitric oxide and inhibition of platelet aggregation in critically ill neonates. The Lancet 351: 1181-1182.

Chintala MS, Bernardino V, Chiu PJ (1994) Cyclic GMP but not cyclic AMP prevents renal platelet accumulation after ischemia-reperfusion in anesthetized rats. J Pharmacol Exp Ther 271:1203-1208.

Chirkov YY, Belushkina NN, Tyshchuk IS, Severina IS, Horowitz JD (1991) Increase in reactivity of human platelet guanylyl cyclase during aggregation potentiates the disaggregating capacity of sodium nitroprusside. Clin Exper Pharmacol Physiol 18:517-524

Clancy RM, Levartovsky D, Leszczynska-Piziak J, Yegudin J, Abramson SB (1994) Nitric oxide reacts with intracellular glutathione and activates the hexose monophosphate shunt in human neutrophils: evidence for S-nitrosoglutathione as a bioactive intermediary. Proc Natl Acad Sci USA 91:3680-3684.

Cooke JP, Tsao P (1992) Cellular mechanisms of atherogenesis and the effects od nitric oxide. Curr Opin Cardiol 7:799-804.

Craven PA, DeRubertis FR (1978) Restoration of the reponsiveness of purified guanylate cyclase to nitrosoguanidine, nitric oxide, and related activators by heme and heme proteins. Evidence for involvement of the paramagnetic nitrosyl-heme complex in enzyme activation. J Biol Chem 253:8433-8443.

DeBelder AJ, Radomski MW. (1994) Nitric oxide in the clinical arena. J Hypertens 12:617-624.

Diodati J, Theroux P, Latour JG et al (1990) Effects of nitroglycerin at therapeutic doses on platelet aggregation in unstable angina pectoris and acute myocardial infarction. Am J Cardiol 66:683-688.

Dong Z, Staroselsky AH, Qi X, Xie K, Fiedler I (1994) Inverse correlation between expression of inducible nitric oxide synthase activity and production of metastasis in K-1735 murine melanoma cells. Cancer Res 54:789-793.

Dube G, Salas E, Kurtz W, Christie R, Radomski M. Effects of estrogen and an estrogen receptor modulator on atherogenesis in rabbits: role of NO. The biology of nitric oxide. Part 6. Eds. Moncada, S., Toda, N., Maeda, H., Higgs, E.A. Portland Press London and Miami 1998, p. 146.

Dudek R, Wildhirt S, Suzuki H, Winder S, Bing RJ (1995) Intracellular translocation of endothelial nitric oxide synthase by lysophosphatidylcholine Pharmacol 50; 257-260.

Feelisch M (1991) The action and metabolism of organic nitrates and their similarity with endothelium-derived relaxing factor (EDRF). In: Moncada S, Higgs EA, Berrazueta JR eds. Clinical relevance of nitric oxide in the cardiovascular system, Edicomplet Madrid, pp. 29-43

Feelisch M, Noack EA (1987) Correlation between nitric oxide formation during degradation of organic nitrates and activation of guanylyl cyclase. Eur J Pharmacol 139:19-30.

Flavahan NA. Atherosclerosis or lipoprotein-induced endothelial dysfunction. Circulation (1992) 85:1927-1938.

Freedman JE, Loscalzo J, Barnard MR, Alpert C, Keaney JF, Jr, Michelson AD (1997) Nitric oxide released from activated platelets inhibits platelet recruitment J Clin Invest 100; 350-356.

Furchgott RF, Zawadzki JV (1980) The obligatory role of endothelial cells in the relaxation of arterial smooth muscle by acetylcholine. Nature 288:373-376.

Furlong B, Henderson AH, Lewis MJ, Smith JA (1987) Endothelium-derived relaxing factor inhibits in vitro platelet aggregation Br J Pharmacol 90:687-692.

Garcia-Cardena G, Oh P, Liu J, Schnitzer JE, Sessa WC (1996) Targeting of nitric oxide synthase to endothelial cell caveolae via palmytoylation: implications for nitric oxide signalling Proc Natl Acada Sci USA 93; 6448-6453.

Gebalska J (1990) Platelet adhesion and aggregation in relation to clinical course of acute myocardial infarction. M.D. thesis Warsaw, in Polish.

Gerzer R, Karrenbrock B, Siess W, Heim JM (1988) Direct comparison of the effects of nitroprusside, SIN-1 and various nitrates on platelet aggregation and soluble guanylyl cyclase activity. Thromb Res 52:11-21.

GISSI III study group (1994). GISSI-3: effects of lisinopril and transdermal glyceryl trinitrate singly and together on 6-week mortality and ventricular function after acute myocardial infarction. Lancet 343:1115-1122.

Golino P, Capelli-Bigazzi M, Ambrosio G et al (1992) Endothelium-derived relaxing factor modulates platelet aggregation in an in vivo model of reccurent platelet activation Circ Res 71:1447-1456.

Gurevich VS, Lominadze DG, Adeagbo ASO et al (1997) Antithrombotic and vasorelaxant properties of a novel synthetic RGD peptide containing nitric oxide Pharmacol 55: 1-9.

Haffner C, Jarchau T, Reinhard M et al (1995) Molecular cloning, structural analysis and functional expression of the proline-rich focal adhesion and microfilament-associated protein VASP. EMBO J 14:19-27.

Herbaczynska-Cedro K, Lembowicz K, Pytel B (1991) N^G-monomethyl-L-arginine increases platelet deposition on damaged endothelium in vivo. A scanning electron microscopy study Thromb Res 64:1-9.

Hibbs JB Jr, Taintor RR, Vavrin Z (1987) Macrophage cytotoxicity: role for L-arginine deiminase and imino nitrogen oxidation to nitrite. Science 235; 473-476.

Hines R, Barash PG (1989) Infusion of sodium nitroprusside induces platelet dysfunction in vitro. Anesthesiology 71:805-806.

Hogan JC, Lewis MJ, Henderson AH (1988) In vivo EDRF activity influences platelet function Br J Pharmacol 94:1020-1022.

Hogg N, Darley-Usmar VM, Wilson MT, Moncada S (1993) Oxidation of alpha-tocopherol in human low density lipoprotein by the simultaneous generation of superoxide and nitric oxide. FEBS Lett 326:199-203.

Horowitz HI, Stein IM, Cohen BD, White JG (1970) Further studies on the platelet-inhibitory effect of guanidinosuccinic acid and its role in uremic bleeding. Am J Med 49:336-345.

Horowitz JD, Henry CA, Syrjanen ML et al (1988) Combined use of nitroglycerin and N-acetylcysteine in the management of unstable angina pectoris. Circulation 77:787-794.

Horstrup K, Jablonka B, Hönig-Liedl P et al (1994) Phosphorylation of focal adhesion vasodilator-stimulated phosphoprotein at Ser157 in intact human platelets correlates with fibrinogen receptor inhibition. Eur J Biochem 225:21-27.

Houston DS, Buchanan MR (1994) Influence of endothelium-derived relaxing factor on platelet function and hemostasis in vivo. Thromb Res 74:25-37.

Houston DS, Robinson P, Gerrard JM (1990) Inhibition of intravascular platelet aggregation by endothelium-derived relaxing factor: reversal by red blood cells. Blood 76:953-958.

Humphries RG, Tomlinson W, O'Connor SE, Leff P (1990) Inhibition of collagen- and ADP-induced platelet aggregation by substance P in vivo: Involvement of endothelium-derived relaxing factor J Cardiovasc Pharmacol 16:292-297.

Ichiki K, Ikeda H, Haramaki N, Ueno T, Imaizumi T (1996) Long-term smoking impairs platelet-derived nitric oxide release Circulation 94: 3109-3114.

ISIS-2 (1988) (Second International Study of Infarct Survival) Collaborative Group: Randomized trial of intravenous streptokinase, oral aspirin, both or neither among 17, 187 cases of suspected acute myocardial infarction: ISIS-2. Lancet 2:349-360.

Johansson JS, Haynes DH (1992) Cyclic GMP increases the rate of the calcium extrusion pump in intact platelets but has no direct effect on the dense tubular calcium accumulation system. Biochim Biophys Acta 1105:40-50.

Kilbourn RG, Gross SS, Adams J et al (1990) N^G-methyl-L-arginine inhibits tumor necrosis factor-induced hypotension: implications for the involvement of nitric oxide. Proc Natl Acad Sci USA 87:3029-3032.

Komamura K, Node K, Kosaka H, Inoue M (1994) Endogenous nitric oxide inhibits microthromboembolism in the ischemic heart. Circulation 90:I-345.

Lam JYT, Chesebro JH, Fuster V (1988) Platelets, vasoconstriction, and nitroglycerin during arterial wall injury. A new antithrombotic role for an old drug. Circulation 78:712-716.

Langford EJ, Brown AS, Wainwright RJ et al (1994) Inhibition of platelet activity by S-nitrosoglutathione during coronary angioplasty. Lancet 344:1458-1460.

Launay JM, Bondoux D, Oset-Gasque MJ et al (1994) Increase of human platelet serotonin uptake by atypical histamine receptors. Am J Physiol 266:R526-R536.

Lechi C, Gaino S, Tommasoli V et al (1996) In vitro study of the anti-aggregating activity of two nitroderivatives of acetylsaicylic acid Blood Coagulation Thromb 7: 206-209.

Lefer DJ, Nakanishi K, Vinten-Johansen J (1993) Endothelial and myocardial cell protection by a cysteine-containing nitric oxide donor after myocardial ischemia and reperfusion. J Cardiovasc Pharmacol 22(Suppl. 7):S34-S43.

Lelchuk R, Radomski MW, Martin JF, Moncada S (1992) Constitutive and inducible nitric oxide synthases in human megakaryoblastic cells J Pharmacol Exp Ther 262:1220-1224.

Levin RL, Weksler BB, Jaffe EA (1982) The interaction of sodium nitriprusside with human endothelial cells and platelets: nitroprusside and prostacyclin synergistically inhibit platelet function. Circulation 66:1299-1307.

Liu Z, Nakatsu K, Brien JF et al (1993) Selective sequestration of nitric oxide by subcellular components of vascular smooth muscle and platelets: relationship to nitric oxide stimulation of the soluble guanylyl cyclase. Can J Physiol Pharmacol 71:938-945.

Loscalzo J (1985) N-Acetylcysteine potentiates inhibition of platelet aggregation nitroglycerin. J Clin Invest 76:703-8.

Luscher TF, Tanner FC, Tschudi MR, Noll G (1993) Endothelial dysfunction in coronary artery disease. Ann Rev Med 44:395-418.

Macdonald PS, Read MA, Dusting GJ (1988) Synergistic inhibition of platelet aggregation by endothelium-derived relaxing factor and prostacyclin Thromb Res 49:437-449.

Malinski T, Radomski MW, Taha Z, Moncada S (1993) Direct electrochemical measurement of nitric oxide released from human platelets Biochem Biophys Res Commun 194:960-965.

Marsden PA, Shappert KT, Chen HS et al (1992) Molecular cloning and characterization of human endothelial nitric oxide synthase FEBS Lett 307:287-293.

Martin W, Villani GM, Jothianandan D, Furchgott RF (1985) Selective blockade of endothelium-dependent and glyceryl trinitrate-induced relaxation by hemoglobin and by methylene blue in the rabbit aorta J Pharmacol Exp Ther 232; 708-716.

Martinez-Cuesta M, Radomski M (1998) Cyclic GMP mediates the inhibitory effect of nitric oxide on platelet adhesion. The biology of nitric oxide. Part 6. Eds. Moncada, S., Toda, N., Maeda, H., Higgs, E.A. Portland Press London and Miami p. 141.

Maurice DH, Haslam RJ (1990) Molecular basis of the synergistic inhibition of platelet function by nitrovasodilators and activators of adenylate cyclase: inhibition of cyclic AMP breakdown by cyclic GMP. Mol Pharmacol 37:671-681.

May GR, Crook P, Moore PK, Page CP (1991) The role of nitric oxide as an endogenous regulator of platelet and neutrophil activation within the pulmonary circulation Br J Pharmacol 102:759-763.

Mayers I, Jonhson, D, Salas, E, Hurst ., Radomski, M (1998) Myocardial activity is increased following canine cardiopulmonary bypass. The biology of nitric oxide. Part 6. Eds. Moncada, S., Toda, N., Maeda, H., Higgs, E.A. Portland Press London and Miami, p. 181.

McKendrick JD, Salas E, Dube GP, Murat J, Russell JC, Radomski MW (1998) Inhibition of nitric oxide generation unmasks vascular dysfunction in insulin-resistant, obese JCR:LA-cp rats. Br J Pharmacol. 124:361-369.

Mehta JL, Chen LY, Kone CB, Mehta P, Turner P (1995). Identification of consitutive and inducible forms of nitric oxide synthase in human platelets J Lab Clin Med 125; 370-377.

Mellion BT, Ignarro LJ, Ohlstein EH, Pontecorvo EG, Hyman AL, Kadowitz PJ (1981) Evidence for the inhibitory role of huanosine 3′, 5′-monophospahte in ADP-induced human platelet aggregation in the presence of nitric oxide and related vasodilators Blood 57; 946-955.

Mendelsohn ME, O'Neill S, George D, Loscalzo J (1990) Inhibition of fibrinogen binding to human platelets by S-nitroso-N-acetylcysteine. J Biol Chem 265:19028-19034.

Menshikov MY, Ivanova K, Schaefer M, Drummer C, Gerzer R (1993) Influence of the cGMP analogue 8-PCPT-cGMP on agonist-induced increases in ionized calcium and on aggregation of human platelets Eur J Pharmacol 245; 281-284.

Michelson AD, Benoit SE, Furman MI et al (1996) Effects of nitric oxide/EDRF on platelet surface glycoproteins Am J Physiol 39; H1640-H1648.

Mondoro TH, Shafer C, Vostal JG (1997) Peroxynitrite-induced tyrosine nitration and phosphorylation in human platelets Free Rad Biol Med 22; 1055-1063.

Moro MA, Darley-Usmar VM, Goodwin DA et al (1994) Paradoxical fate and biological action of peroxynitrite on human platelets. Proc Natl Acad Sci USA 91:6702-6706.

Murohara T, Parkinson SJ, Waldman SA, Lefer AM. Inhibition of nitric oxide biosynthesis provides P-selectin expression in platelets. Role of protein kinase C (1995) Art Thromb Vasc Biol 15:2068-2075.

Muruganandam A, Mutus B (1994) Isolation of nitric oxide synthase from human platelets Biochim Biophys Acta 1200:1-6.

Nakashima S, Tohmatsu T, Hattori H, Okano Y, Nozawa Y (1986) Inhibitory action of cyclic GMP on secretion, phosphoinositide hydrolysis and calcium mobilization in thrombin-stimulated human platelets. Biochem Biophys Res Commun 135: 1099-1104.

Nakatsuka M, Osawa Y (1994) Selective inhibition of the 12-lipoxygenase pathway of arachidonic acid metabolism by L-arginine or sodium nitroprusside in intact human platelets. Biochem Biophys Res Commun 200:1630-1634.

Naseem KM, Khan J, Jacobs M, Bruckdorfer KR (1997) Nitration of platelet cytosolic proteins by peroxynitrite Biochem Soc Trans 25; 397S.

Noris M, Benigni A, Boccardo P et al (1993) Enhanced nitric oxide synthesis in uremia: implications for platelet dysfunction and dialysis hypotension. Kidney Int 44:445-450.

Okada M, Sagawa T, Tominaga A, Kodama T, Hitsumoto Y (1996) Two mechanisms of platelet-mediated killing of tumour cells: one cyclo-oxygenase dependent and the other nitric oxide dependent Immunology 89: 158-164.

Olsen SB, Tang DB, Jackson MR et al (1996) Enhancement of platelet deposition by cross-linked hemoglobin in a rat carotid endarterectomy model. Circulation. 93; 327-32.

Palmer RMJ, Ashton DS, Moncada S (1988) Vascular endothelial cells synthesize nitric oxide from L-arginine. Nature 333:664-666.

Palmer RMJ, Bridge L, Foxwell NA, Moncada S (1992) The role of nitric oxide in endothelial cell damage and its inhibition by glucocorticoids Br J Pharmacol 105:11-12.

Palmer RMJ, Ferrige AG, Moncada S (1987) Nitric oxide release accounts for the biological activity of endothelium-derived relaxing factor. Nature 327:524-526.

Patrono C (1989) Aspirin and human platelets: from clinical trials to acetylation of cyclooxygenase and back. Trends Pharmacol Sci 10:453-458.

Plotkine M. Allix M, Guillou J, Boulu R (1991) Oral administration of isosorbide dinitrate inhibits arterial thrombosis in rats. Eur J Pharmacol 201:115-116.

Pohl U, Busse R (1989) EDRF increases cyclic GMP in platelets during passage through the coronary vascular bed Circ Res 65:1798-1803.

Polanowska-Grabowska R, Gear ARL (1994) Role of cyclic nucleotides in rapid platelet adhesion to collagen. Blood 83:2508-2515.

Radomski MW (1995) Nitric oxide-biological mediator, modulator and effector molecule. Ann Med 27:321-330.

Radomski MW, Jenkins DC, Holmes L, Moncada S (1991) Human colorectal adenocarcinoma cells: differential nitric oxide synthesis determines their ability to aggregate platelets. Cancer Res 51:6073-6078.

Radomski MW, Palmer RMJ, Moncada S (1987a) Endogenous nitric oxide inhibits human platelet adhesion to vascular endothelium. Lancet 2:1057-1058.

Radomski MW, Palmer RMJ, Moncada S (1987b) The role of nitric oxide and cGMP in platelet adhesion to vascular endothelium Biochem Biophys Res Commun 148:1482-1489.

Radomski MW, Palmer RMJ, Moncada S (1987c) Comparative pharmacology of endothelium-derived relaxing factor, nitric oxide and prostacyclin in platelets Br J Pharmacol 92:181-187.

Radomski MW, Palmer RMJ, Moncada S (1987d) The anti-aggregating properties of vascular endothelium: interactions between prostacyclin and nitric oxide Br J Pharmacol 92:639-646.

Radomski MW, Palmer RMJ, Moncada S (1990a). An L-arginine/nitric oxide pathway present in human platelets regulates aggregation Proc Nat Acad Sci USA 87:5193-5197.

Radomski MW, Palmer RMJ, Moncada S. (1990b) Glucocorticoids inhibit the expression of an inducible but not the constitutive, nitric oxide synthase in vascular endothelial cells Proc Natl Acad Sci USA 87:10043-10047.

Radomski MW, Palmer RMJ, Moncada S (1990c). Characterization of the L-arginine: nitric oxide pathway in human platelets Br J Pharmacol 101: 325-328.

Radomski MW, Rees DD, Dutra A, Moncada S (1992) S-Nitrosoglutathione inhibits platelet activation in vitro and in vivo. Br J Pharmacol 107:745-749.

Radomski MW, Salas E. (1995a). Nitric oxide-biological mediator, modulator and factor of injury:its role in the pathogenesis of atherosclerosis. Atherosclerosis 118(Suppl.) S69-S80.

Radomski MW, Salas E. (1995b). Platelet regulation and damage in vascular thrombotic and septic disorders. In: Role of nitric oxide in sepsis and ARDS. Fink MP, Payen D eds. Springer- Verlag Berlin Heidelberg New York, pp. 138-154.

Salas E, Miszta-Lane H Radomski MW (1997) Regulation of platelet function by nitric oxide and other nitrogen- and -oxygen-derived species. In: Handbook of Experimental Pharmacology vol. 126 Platelets and their Factors, Eds. Von Bruchhausen F., Walter U. Springer-Verlag Berlin Heidelberg New York pp. 371-397

Radomski MW, Vallance P, Whitley G, Foxwell N, Moncada S (1993) Platelet adhesion to human vascular endothelium is modulated by constitutive and cytokine induced nitric oxide Cardiovasc Res 27:1380-1382.

Reinhard M, Halbrügge M, Scheer U et al (1992) The 46/50 kDa phosphoprotein VASP purified from human platelets is a novel protein associated with actin filaments and focal contacts. EMBO J 11:2063-2070.

Riddell DR, Graham A, Owen JS (1997) Apolipoprotein E inhibits platelet aggregation through the L-arginine:nitric oxide pathway. Implications for vascular disease J Biol Chem 272; 89-95.

Rosenblum WI, Nelson GH, Povlishock JT (1987) Laser-induced endothelial damage inhibits endothelium-dependent relaxation in the cerebral microcirculation of the mouse Circ Res 60:169-176.

Salas, E, Busletta A, Radomski MW (1998) Generation of NO, cGMP and platelet function in normal pregnancy and preeclampsia. The biology of nitric oxide. Part 6. Eds. Moncada, S., Toda, N., Maeda, H., Higgs, E.A. Portland Press London and Miami p. 44.

Salas E, Moro MA, Askew S et al (1994) Comparative pharmacology of analogues of S-nitroso-N-acetyl-DL-penicillamine in platelets. Br J Pharmacol 112:1071-1076.

Salvemini D, de Nucci G, Sneddon JM, Vane JR (1989) Superoxide anions enhance platelet adhesion and aggregation Br J Pharmacol 97; 1145-1150.

Samama CM, Diaby M, Fellahi JL et al (1995) Inhibition of platelet aggregation by inhaled nitric oxide in patients with acute adult respiratory distress syndrome Anesthesiology 83: 56-65.

Sane DC, Bielawska A, Greenberg CS, Hannun YA (1989) Cyclic GMP analogs inhibit gamma thrombin-induced arachidonic acid release in human platelets. Biochem Biophys Res Commun 165:708-714.

Sase K, Michel T (1995) Expression of constitutive, endothelial nitric oxide synthase in human blood platelets Life Sci 57; 2049-2055.

Sawicki G, Salas E, Murat J, Miszta-Lane H, Radomski MW (1997) Release of gelatinase A from human platelets mediates aggregation Nature 386; 616-619.

Schultz PJ, Raij L (1992) Endogenously synthesized nitric oxide prevents-endotoxin-induced glomerular thrombosis J Clin Invest 90:1718-1725.

Shahbazi T, Jones N, Radomski MW, Moro MA, Gingell D (1994) Nitric oxide donors inhibit platelet spreading on surfaces coated with fibrinogen but not fibronectin Thromb Res 75:631-642.

Shen YC, Hing CY (1995) Effect of trilinolein on cyclic nucleotide formation in human platelets: relationship with its antiplatelet effect and nitric oxide synthesis Br J Pharmacol 116: 1644-1648.

Simon DI, Stamler JS, Loh E, Loscalzo J, Francis SA, Craeger MA (1995) Effect of nitric oxide synthase inhibition on bleding time in humans J cariovasc Pharmacol 26; 339-342.

Sinzinger H, Virgolini I, O'Grady J, Rauscha F, Fitscha P (1992) Modification of platelet function by isosorbide dinitrate in patients with coronary artery disease. Thromb Res 65:323-335.

Sneddon JM, Vane JR (1988) Endothelium-derived relaxing factor reduces platelet adhesion to bovine endothelial cells Proc Natl Acad Sci USA 85:2800-2804.

Spain DA, Wilson MA, Garrison RN (1994) Nitric oxide synthase inhibition exacerbates sepsis-induced renal hypoperfusion. Surgery 116:322-331.

Stamler JS, Jia L, Eu JP, McMahon TJ et al (1997) Blood flow regulation by S-nitrosohemoglobin in the physiological oxygen gradient Science 276; 2034-2037.

Stamler JS, Simon DJ, Osborne JA et al (1992) S-Nitrosylation of proteins with nitric oxide: synthesis and characterization of biologically active compounds. Proc Natl Acad Sci USA 89:444-448.

Stuehr DJ, Marletta MA (1985) Mammalian nitrate synthesis: Mouse macrophages produce nitrite and nitrates in response to Escherichia coli lipopolysaccharide. Proc Natl Acad Sci USA 82; 7738-7742.

Thomson L, Lawton FG, Knowles RG et al (1994) Nitric oxide synthase activity in human gynecological cancer. Cancer Res 54:1352-1354.

Trovati M, Massucco P, Mattiello L et al (1996) The insulin-induced increase of guanosine-3′,5′-cyclic monophosphate in human platelets is mediated by nitric oxide Diabetes 45: 768-770.

Trovati M Anfossi G, Massucco P et al (1997) Insulin stimulates nitric oxide synthesis in human platelets and, through nitric oxide, increases platelet concentrations of both guanosine-3′,5′-cyclic monophosphate and adenosine-3′,5′-cyclic monophospahte Diabetes 46: 742-749.

Tsao PS, Theilmeier G, Singer AH, Leung LLK, Cooke JP (1994) L-Arginine attenuates platelet reactivity in hypercholesterolemic rabbits. Arterioscler Thromb 14:1529-1533.

Venturini CM, Del Vecchio PJ, Kaplan JE (1989) Thrombin-induced platelet adhesion to endothelium is modified by endothelial derived relaxing factor (EDRF). Biochem Biophys Res Commun 159:349-354.

Villa LM, Salas E, Darley-Usmar VM, Radomski MW, Moncada S (1994) Peroxynitrite induces both vasodilatation and impaired vascular relaxation in the rat isolated perfused heart. Proc Natl Acad Sci USA 91; 12383-12387.

Wallerath T, Gath I, Aulitzky WE, Pollock JS, Kleinert H, Forstermann U (1997) Identification of NO synthase isoforms expressed in human neutrophil granulocytes, megakaryocytes and platelets. Thromb Haemostas 77; 163-167.

Wautier JL, Weill D, Kadeva H, Maclouf J, Soria C (1989) Modulation of platelet function by SIN-1A. J Cardiovasc Pharmacol 14:S111-S114.

Weber AA, Neuhaus T, Seul C et al (1996) Biotransformation of glyceryl trinitrate by blood platelets as compared vascular smooth muscle cells Eur J Pharmacol 309: 209-213.

Werns SW, Rote WE, Davis JH, Guevara T, Lucchesi BR (1994) Nitroglycerin inhibits experimental thrombosis and reocclusion after thrombolysis. Am Heart J 127:727-737.

Willerson JT, Igo SR, Yao SK, Ober JC, Macris MP, Ferguson JJ (1996) Localized administration od sodium nitroprusside enhances its protection against platelet aggregation in stenosed and injured coronary arteries Texas Heart Inst J 23: 1-8.

Wright CE, Rees DD, Moncada S (1992) Protective and pathological role of nitric oxide in endotoxin shock. Cardiovasc Res 26:48-57.

Wu CC, Ko FN, Teng CM (1997) Inhibition of platelet adhesion to collagen by cGMP-elevating agents. Biochem Biophys Res Commun 231; 412-416.

Yao SK, Ober JC, Krishnaswami A et al (1992) Endogenous nitric oxide protects against platelet aggregation and cyclic flow variations in stenosed and endothelium-injured arteries. Circulation 86:1302-1309.

Yusuf S, MacMahon S, Collins R, Peto R (1988) Effect of intravenous nitrates on mortality in acute myocardial infarction: an overview of the randomised trials. Lancet i:1088-1092.

22

ASPIRIN, PROSTAGLANDINS AND PLATELET FUNCTION: PHARMACOLOGY AND THROMBOSIS PREVENTION

David C. Calverley, MD
Division of Hematology, Department of Medicine
University of Southern California
Los Angeles, CA 90033

Gerald J. Roth, MD
Division of Hematology, Department of Medicine
University of Washington
Hematology Section, Medical and Research Services
VA Puget Sound Health Care System
Seattle, WA 98108

22.1 INTRODUCTION

> Dr. David Calverley is a hematologist with a research interest in the molecular mechanisms of platelet adhesion and activation. His clinical interest lies in the area of platelet and hemostatic disorders. In addition to research, clinical, and teaching activities, he is Associate Director of the University of Southern California/Norris Cancer Hospital Clinical Coagulation Laboratory. Dr. Gerald Roth is a hematologist with interests in both research and clinical aspects of platelets. He has studied molecular aspects of aspirin's effect on platelets and observed the acetylation reaction between aspirin and platelet cyclo-oxygenase.

Since the early 1970's, there has been significant progress made in defining the mechanism of action and pharmacology of aspirin, one of the most significant anti-thrombotic agents known. Today, a wealth of studies, extending over the past three decades and ranging from molecular biology to clinical epidemiology, all point to the conclusion that aspirin exerts it's anti-thrombotic effect via a single specific reaction with cyclo-oxygenase, an enzyme found in platelets (1-3). The purpose of this chapter is to elucidate mechanisms of platelet function by outlining how aspirin acts on platelets to inhibit its prothrombotic properties; this physiologic information will then be integrated with information from randomized clinical trials that address aspirin's important role in the prevention of thromboembolic disease.

Aspirin is widely accepted as a treatment for disorders at three different dosage levels, with each dose reflecting the relative aspirin sensitivity of different target cells. As shown in Table 1, aspirin can act as an *anti-thrombotic* (60-325mg dosage), an *analgesic/ antipyretic* (650mg dosage) or as an *anti-rheumatic* agent (6,000mg dosage)(4). This chapter focusses on the first dosage level; namely, the clinical use of aspirin as an anti-platelet and anti-thrombotic agent.

Table 1: Three Levels of Aspirin Therapy: 10-Fold Dose Increments

Indication	Dosage	Comment
Antithrombotic (25,28)	"Platelet"-- 60mg	Anucleate target, exquisite sensitivity, permanent/ limited effect
Analgesic/ antipyretic (89)	"Headache"-- 600mg	"Take two aspirin and call me in the morning"
Antirheumatic (90)	"Salicylate"-- 6,000 mg	Does not work through prostaglandins alone

22.1.1 Historical Perspective

From a historical perspective, as an acetylated salicylate, aspirin was initially con-sidered to be a prodrug or precursor of salicylic acid. Investigators in the 1940's reported aspirin to be a more potent anti-hemostatic agent than salicylic acid (5) while a decade later, it was suggested that aspirin may be useful as an anti-thrombotic agent since it was able to induce significant clinical bleeding (6). This was followed in 1968 by the observation that aspirin's anti-thrombotic actions could be ascribed to sup-pression of platelet function (7,8).

Prostaglandins (PG's) were discovered as unique biologic activities in human semen, and this led to their structural and substrate characterization (9,10). Biosynthesis of PG's depends on exogenous stimuli that release arachidonate from phospholipid stores (11). Later, it was reported that aspirin works through inhibition of prostaglandin synthesis (12). This critical work joined together two separate areas of investigation, namely prostaglandin physiology and the pharmacology of aspirin-like agents, and over the next two decades, the basic tenets of the "PG hypothesis" were confirmed; namely, 1) PG's mediate responses such as inflammation that are blocked by NSAIDs (non-steroidal anti-inflammatory drugs) and 2) NSAID's act by blocking PG synthesis (12,13). Examples of compounds in different classes of NSAIDs are shown in Figure 1.

22.2 Prostaglandin structure and function

PG synthesis involves four steps (Figure 2). The first two steps are common to all cells involved in prostaglandin synthesis while the final two steps are cell-specific (14-16). Release of the substrate, arachidonic acid, from membrane phospholipid stores by phospholipase is the initial event in prostaglandin synthesis, and this is followed by formation of the common PG intermediate, PGH$_2$, catalyzed by PGH synthase. At this point, rearrangement of PGH$_2$ to form either *stable* (PGD$_2$/ E$_2$/ F$_{2\alpha}$) or *unstable* (platelet thromboxane - TxA$_2$, endothelial prostacyclin - PGI$_2$) products takes place. The final step, also cell-specific, involves breakdown of the active compounds to inactive metabolites.

Stable prostaglandin products are not readily hydrolyzed when solubilized in water, and these compounds mediate a number of physiologic processes (inflammatory, reproductive, renal, neurologic) in a variety of tissues and organs (15). It is the unstable prostanoids, however, that are of greatest relevance to thrombosis and it's prevention by aspirin. These compounds (platelet TxA$_2$ and endothelial PGI$_2$) undergo rapid hydrolysis in water and exert opposing local hemostatic and anti-hemostatic effects. Thromboxane promotes the formation of hemostatic plugs by aggregating platelets and constricting blood vessels (17) while prostacyclin counteracts this by suppressing platelet aggregation and dilating blood vessels (18).

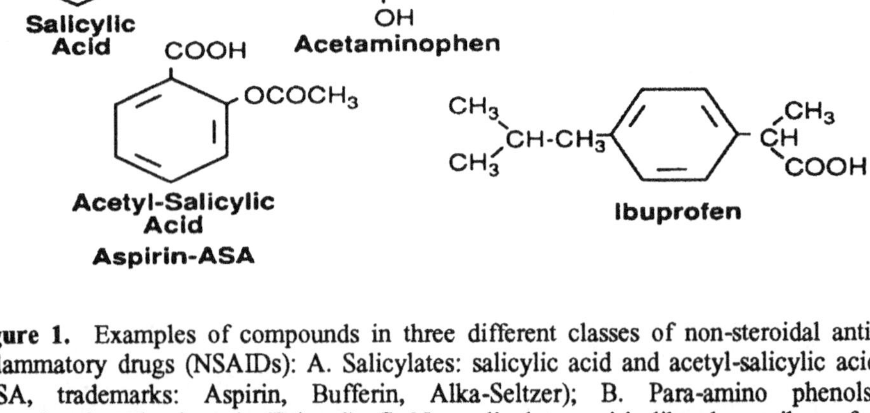

Figure 1. Examples of compounds in three different classes of non-steroidal anti-inflammatory drugs (NSAIDs): A. Salicylates: salicylic acid and acetyl-salicylic acid (ASA, trademarks: Aspirin, Bufferin, Alka-Seltzer); B. Para-amino phenols: acetaminophen (trademark: Tylenol); C. Non-salicylate aspirin-like drugs: ibuprofen (trademarks: Advil, Motrin, Nuprin)

22.3 Effect of aspirin on prostaglandin synthesis

PGH synthase is the key initial enzyme of PG synthesis and consists of two components (Figure 2). It initially both oxygenates and cyclizes arachidonic acid to PGG_2 through it's *cyclo-oxygenase* activity and then reduces PGG_2 to PGH_2 through it's *peroxidase* component. Aspirin affects PG synthesis in a highly specific manner by inhibiting only the cyclo-oxygenase function of PGH synthase while leaving other PG elements untouched (Figure 3)(19). PGH synthase-2, a second recently dis-covered PGH synthase, is not found in platelets and does not appear to relate to the anti-hemostatic and anti-thrombotic effects of aspirin (20).

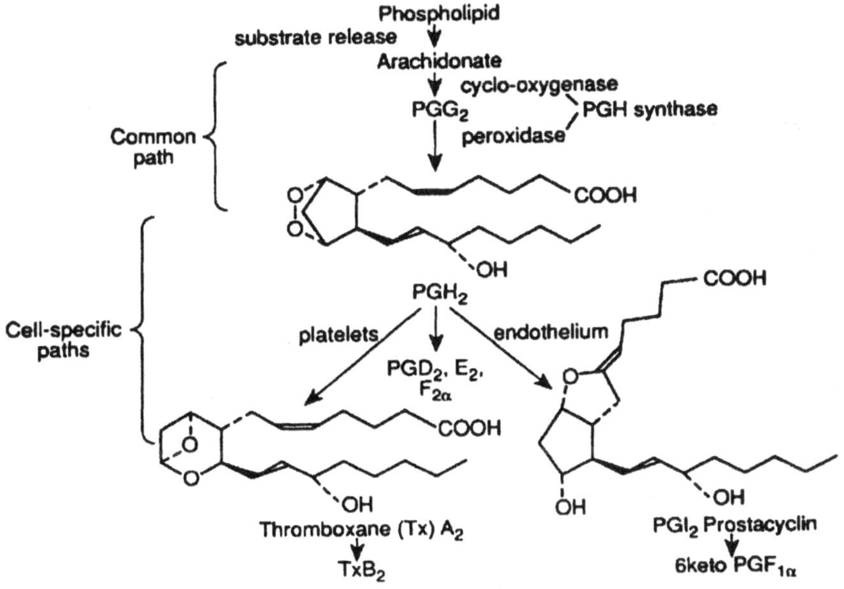

2. Steps in prostaglandin (PG) biosynthesis: PG synthesis involves four steps: 1) Release of the substrate, arachidonic acid (20:4), from membrane phospholipids by a phospholipase. 2) Formation of the common intermediate, PGH_2, from arachi-donate by PGH synthase through sequential cyclo-oxygenase and peroxidase activities, both of which are present within the same enzyme. 3) Formation of cell-specific PG products, either "stable" (D_2, E_2, F_{2a}) or "unstable" (thromboxane, prostacyclin), by distinct enzymes found in different cells and tissues. 4) Breakdown of active compounds to inactive metabolites, using TxA_2 -> TxB_2 and PGI_2 -> 6keto PGF_{1a} as examples.

Aspirin blocks cyclooxygenase by acetylating the protein. The reaction depends on the affinity of the drug for the enzyme's active site (2,3,21). The $-COCH_3$ acetyl moiety of aspirin (Figs 1 and 3) is labile and capable of being transferred to many different biological substrates (acetylation). For example, aspirin *non-specifically* acetylates a variety of proteins, lipids, and nucleic acids at *milli-molar* concentrations (22). In contrast, aspirin acetylates cyclo-oxygenase in a highly *specific* fashion with the reaction going to completion within minutes at aspirin concentrations in the *micro-molar* range (2,3). These reaction conditions are due to aspirin's affinity for the active site of cyclo-oxygenase that includes the serine-529 residue which undergoes acetylation (Figure 3) (21). Recent work has demonstrated that this single serine residue is not required for catalytic function but that the addition of a bulky acetyl moiety at this site interferes with the enzyme's activity (23,24).

Acetylation by aspirin of serine 529 within cyclo-oxygenase produces a covalent O-acetyl bond that resists hydrolysis under intra-cellular conditions (Figure 3). This results in the irreversible inactivation of cyclooxygenase by aspirin that persists for the rest of the platelet's circulating lifespan. This phenomonon accounts for the pro-gressive daily increase in the aspirin-induced functional defect of circulating platelets observed when the drug is given in low doses (5 to 75mg per day) on successive days (25).

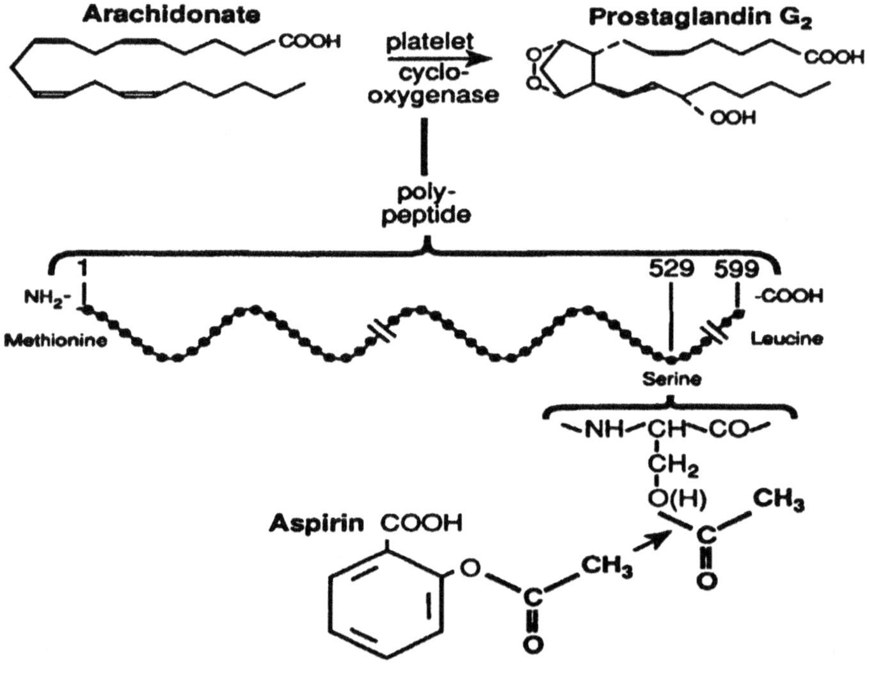

Figure 3. The mechanism of aspirin's effect on platelets: The 599 amino acid polypeptide chain of PGH synthase (center wavy line: NH_2-terminal methionine-1, COOH-terminal leucine-599) exerts cyclo-oxygenase activity as shown above (oxygenation/cycli-zation of arachidonate to PGG_2) and interacts with aspirin as shown below. The serine residue located at position 529 of the polypeptide chain of cyclo-oxygenase is acetylated through transfer of aspirin's acetyl group as indicated in bold face. Covalently-modified, acetylated PGH synthase carries a single acetyl group in its active site and lacks all cyclo-oxygenase activity.

Both platelet and endothelial cyclooxygenase appear to be equally sensitive to the effects of aspirin since they are identical products derived from the same gene. This point has been demonstrated by direct experimentation in intact cells (26, 27). However, probably attributable to the inability of platelets to synthesize new protein, patient studies show that inhibition of TXA_2 formation occurs with lower doses of aspirin than does inhibition of PGI_2 synthesis (25, 28). Regardless, considerable over-lap exists between the effects, and no dose of aspirin will block TXA_2 formation completely without affecting PGI_2 production. Since platelet and endothelial cyclo-oxygenase are equally sensitive to aspirin (27), the exact physiological basis for, and the clinical significance of, the differential sensitivity of prostacyclin and thromboxane A_2 to the drug remains somewhat unclear and provides a topic for continuing debate (4).

22.4 Aspirin's unique effect on platelet physiology

Because aspirin blocks prostanoid synthesis, the ability of aspirin to impair platelet

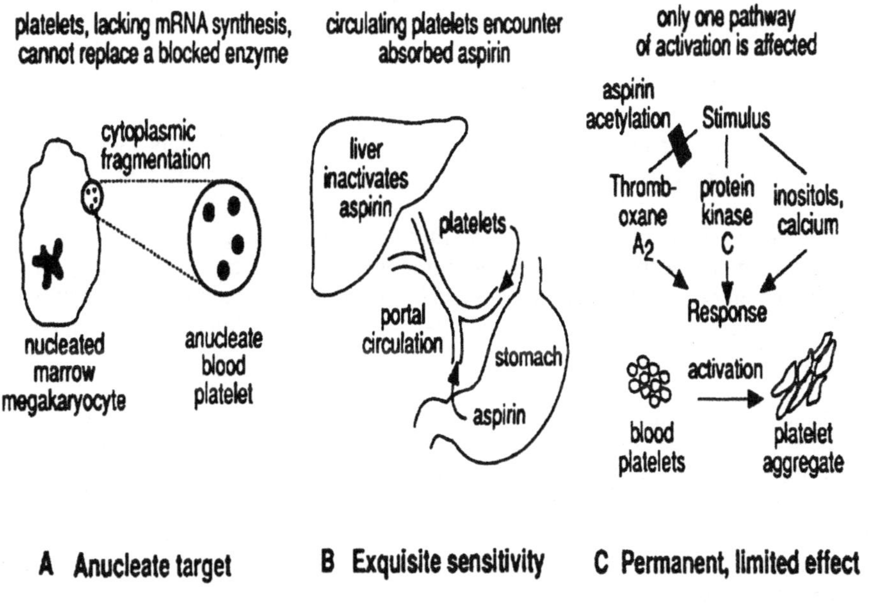

Figure 4. The features of the unique relationship between aspirin and platelets: A. Anucleate target. Nucleated megakaryocytes in the bone marrow undergo cytoplasmic fragmentation and shed platelets into the circulating blood. Lacking a nucleus (anucleate) and being incapable of mRNA synthesis, blood platelets cannot synthesize new protein and are unable to replace an inactive enzyme such as aspirin-modified cyclo-oxygenase; B. Exquisite sensitivity. Following oral ingestion, aspirin is absorbed from the upper GI tract with about the same time course as that of platelet entry into the portal circulation from pre-portal blood. During the time of aspirin absorption, the entire mass of platelets in a patient ingesting 100 mg or more of the drug will be exposed to a sufficient aspirin concentration to inhibit PG synthesis nearly completely. Following transit through the portal circulation, aspirin is partially deacetylated/inactivated in the liver; C. Permanent, limited effect. Platelets activate and aggregate in response to exogenous stimuli such as

thrombin or collagen. Several intracellular pathways of activation mediate the response of a platelet to a stimulus, but only one, thromboxane A_2 synthesis, is blocked by aspirin. Other pathways, such as protein kinase C, inositol release and calcium flux, are unaffected by aspirin and preserve a degree of platelet function despite aspirin inhibition of TxA_2. Therefore, aspirin's inhibitory effect on platelet function is partial because it is limited to only one of several intra-cellular signalling pathways. The aspirin effect is permanent since it involves covalent acetylation of cyclo-oxygenase.

activation implies that a prostanoid is involved in the process (7,8). In this respect, one small dose of aspirin irreversibly impairs the function of all available platelets for several days (2,25,28). The gradual recovery of platelet function following aspirin administration reflects the appearance in the circulation of new, unaffected platelets formed in the bone marrow following ingestion of the drug (2,25). Aspirin's specific effect on platelet function can be ascribed to three key features discussed below and entitled: 1) anucleate target, 2) exquisite sensitivity, 3) limited effect (Figure 4).

(a) Anucleate target

As cytoplasmic fragments of bone marrow megakaryocytes, platelets lack the bio-synthetic machinery necessary to make new messenger RNA or protein (Fig. 4A). Therefore, the enzyme defect induced by aspirin cannot be repaired by the platelet during its lifetime (approximately 10 days). Upon discontinuation of aspirin, cyclo-oxygenase activity will recover as a function of platelet turnover (29). This *anucleate* quality, along with the process of mixing aspirin and platelets in the portal circulation, is responsible for the ability of this drug, with it's 20 minute half life, to exert a long-lasting therapeutic effect despite being given only once a day (30).

Aspirin's permanent inactivation of the anucleate platelet is also responsible for the cumulative effect seen with repeated daily dosing below 100 mg (25). Daily administration of 30 to 50 mg of aspirin has been shown to lead to complete suppression of platelet thromboxane synthesis in 7 to 10 days (25, 30, 31). This, in turn, leads to inhibition of platelet aggregation and prolongation of the bleeding time, a clinical test often useful in identifying patients with platelet functional defects for further study and characterization (28, 32).

Nucleated cells such as endothelium can replace their cyclooxygenase supply following aspirin treatment by synthesizing new enzyme. This ability to recover from aspirin treatment contributes to the reduced aspirin sensitivity of these cells as compared with platelets (25, 28).

(b) Exquisite Sensitivity

The admixture of aspirin and platelets in the portal circulation accounts in part for the remarkable sensitivity of platelets to relatively small and / or infrequently administered amounts of the drug. The evidence in this respect comes from a study comparing intravenous to orally administered aspirin that demonstrated cyclo-oxygenase acetyl-ation was present before the drug passed through the liver (33). During the 30 minute period after aspirin is ingested orally, the entire circulating platelet mass passes through the portal venous system and is exposed to active aspirin, as the drug is absorbed from the GI tract (Fig. 4B) (28, 33-35). Access to platelets in the portal circulation is thus an important component of aspirin's anti-platelet effect since the active drug is partially deacylated during its subsequent passage through the liver (33-35). This deacylation of active aspirin is incomplete, however, and so other prostanoid-forming cells are affected beyond the portal circulation and liver such as marrow megakaryocytes (36).

As noted above, a single daily dose of more than 100 mg of aspirin will completely inhibit platelet PG synthesis. Implicit in this observation is the fact that aspirin affects preformed platelets in marrow megakaryocytes before they enter the circulating blood over the next day (36). If this were not the case, one would expect platelet PG synthesis would increase by approximately 10% during the day after exposure to aspirin reflecting the platelet's circulating lifespan of approximateley 10 days. Consistent with this observation is the fact higher aspirin doses affect an even earlier cohort of developing platelets in the marrow, accounting for the clinical efficacy of every-other-day treatment schedules in thrombosis prevention.

(c) Permanent, Limited Effect

Aspirin has a limited effect on platelet function *in vivo* causing minimal prolongation of the bleeding time, a modest inhibition of platelet aggregation, and only a slight decline in normal hemostatic function at a clinical level (7, 8). The highly selective effect of aspirin on platelet PG's underpins its limited effect on overall platelet function. This is because thromboxane synthesis is only one of several signaling pathways that participate in platelet activation (Fig 4C). Other pathways involving key mediators such as phospholipase C, phosphotidylinositol 3-kinase, and intracellular calcium flux mediated by inositol triphosphate are unaffected by cyclo-oxygenase acetylation (37, 38). Thus, selective elimination of the TXA_2 pathway by aspirin causes a clear-cut, though modest decrease in platelet function and provides a limited but definite anti-thrombotic effect.

22.5 Molecular and Clinical Correlates: How Aspririn-Mediated Cyclooxygenase Inhibition Prevents Arterial Thrombosis

The possibility exists that aspirin's anti-thrombotic effect may arise from the drug's actions on pathways that lie outside those of prostaglandin synthesis such as plasma coagulation and fibrinolysis. However, these prostaglandin-independent effects occur at aspirin concentrations that are irrelevant to the anti-platelet activity of the drug, and this fact makes it unlikely that aspirin prevents thrombosis by affecting a non-prostaglandin pathway. In this respect, it is important to stress that a close correlation exists between the micromolar plasma concentrations of aspirin associated with taking 30 to 300 mg orally per day and four critically important aspects of the effect of aspirin on platelets; namely, 1) acetylation of platelet cyclooxygenase, 2) the blocking of enzyme activity, 3) inhibition of PG synthesis, and 4) inhibition of platelet function. These four effects of low-dose aspirin administration occur under similar reaction conditions and with the same time course (2,3,7,8,25,28,29). The only observable physiological effect of aspirin in this critically-important micromolar concentration is exerted on platelet cyclooxygenase, and so there exists a close correlation between the molecular data noted in the preceding paragraphs and the *in vivo* clinical effects of aspirin in thrombosis prevention noted below.

22.6 Prevention of Clinical Thrombosis by Aspirin

Two large studies have provided evidence that aspirin is effective in the primary prevention of occlusive cardiovascular disease in men (that is, it helps prevent a first heart attack or stroke in otherwise healthy patients) (39,40). An overview analysis of the studies indicated a significant benefit due to aspirin in preventing both nonfatal heart attacks (myocardial infarction, MI) and total vascular occlusive events (41). The absolute benefit of aspirin from these trials was significantly lower, though, compared with the effect shown in secondary prevention trials of patients who had already had a cardiovascular event and were taking aspirin to prevent a second event. In this respect only four primary events were prevented in the two studies for every 1,000 healthy men taking aspirin for five years. Additional evidence will be needed to determine whether the beneficial effects of aspirin administration outweigh the hazards in primary prevention of cardiovascular events.

Aspirin has been shown to reduce the risk of a first myocardial infarction in patients with chronic stable angina in two studies. In the first (a subgroup of the Physicians' Health Study referenced above) there was a risk reduction at least as significant as that for the healthy men enrolled in the same trial (42). In the second randomized trial of 2,035 patients, there was a 34% reduction in the incidence of first myocardial infarction and sudden cardiac death associated with aspirin (43).

With respect to secondary MI prevention of recurrent events in patients following an initial MI, there are eleven randomized double-blind studies of aspirin involving 40,000 patients (reviewed as part of a larger meta-analysis of 151 randomized antiplatelet trials, ref 44) that collectively suggest a reduction of approximately 31% in incidence of nonfatal recurrent MI among those taking aspirin following an initial heart attack. Antiplatelet agents reduced the vascular mortality rate by 13% and the nonfatal stroke rate by 42% in this group (44).

Based on four randomized, double-blind trials aspirin is clearly effective in unstable angina (45-48). It reduces the risk of MI and cardiac death by 50 to 70% either alone or in concert with the anticoagulant heparin. Aspirin also has a role in reducing the incidence of early closure of saphenous vein bypass grafts of the coronary arteries and following coronary artery angioplasty (49-58).

Aspirin is effective in the prevention of cerebrovascular thrombosis (stroke). In this respect, a meta-analysis of 17 trials of patients with cerebrovascular disorders has demonstrated that among the larger trials in which aspirin was given following a "transient ischemic attack" (in which patients at risk for stroke can have stroke-related symptoms that resolve within 24 hours), there was a statistically significant risk reduc-tion in death, nonfatal stroke, or nonfatal MI (59). A follow up meta-analysis suggested that the administration of anti-platelet therapy for about three years can reduce the risk of nonfatal stroke, nonfatal MI, and death from cardiovascular causes during this interval by approximately 25% (44). Whereas aspirin is recommended after a TIA or minor ischemic stroke (60,61), the same cannot be said after a completed major stroke since the data is unclear in terms of aspirin benefit in this setting (62).

Several studies have also demonstrated a role for aspirin, with or without the antiplatelet agent dipyridamole, in preventing thrombotic occlusion of prosthetic grafts placed in peripheral arteries that are significantly narrowed due to atherosclerosis (63-67). These patients are at high risk for heart attacks and strokes which aspirin also helps to prevent in this setting.

In summary, aspirin is beneficial in a large number of patients at high risk of occlusive cardiovascular disease. The recently reported large meta-analysis of antiplatelet therapy clinical trials suggests risk reduction associated with long term aspirin administration is independent of age, sex, and cardiovascular risk factors (44).

22.7 The Pharmacology of Aspirin Dose and Its Antithrombotic Effect: How Lower Doses May Minimize Prostacyclin Synthesis Inhibition

Based on current data, 80 to 160 mg of aspirin per day (1-2 baby aspirin tablets) provides the maximal antithrombotic effect of the drug. There is considerable in vitro and clinical evidence supporting the choice of a daily dose of 80 to 160 mg for the prevention of arterial thromboembolism in all high risk situations (4).

Because aspirin can block endothelial prostacyclin (PGI_2) synthesis, the drug is potentially harmful if a decrease in PGI_2 synthesis predisposes patients to thrombosis (18,27). To what extent aspirin influences prostacyclin synthesis, and at what threshhold dose below which its effect on prostacyclin is spared, is debatable. Direct studies of the issue have demonstrated that aspirin has an equal effect on both platelet and endothelial cyclooxygenase (27). However, earlier work suggested, in fact, that the platelet enzyme might be more susceptible (68). Aspirin was also considered to spare prostacyclin production if urinary metabolites were the measure of the aspirin effect, but not to spare the same parameter when quantitated in vascular tissue or as PGI_2 levels in blood from bleeding time wounds (28,69-71). Further evidence suggests prostacyclin synthesis in humans may be miniscule and hence incapable of significantly influencing hemostasis (72).

In light of the above conflicting findings regarding aspirin and prostacyclin synthesis inhibition under *in vitro* and controlled *in vivo* conditions, the potential clinical importance of "PGI_2 sparing" can potentially be elucidated through clinical trials employing aspirin doses and formulations postulated to spare prostacyclin synthesis. A recent example of this is the Dutch TIA trial comparing 30 mg and 283 mg of daily aspirin in 3,131 patients (60). The lower dose that would be expected to "spare PGI_2" was slightly more effective with fewer side effects than the higher dose, but the difference was not statistically significant. Thus the limited amount of clinical data on the subject does not conclusively demonstrate a role for a dose-related PGI_2 sparing effect in prevention of thrombosis and so the importance of aspirin's anti-PGI_2 effect in antithrombotic therapy remains unclear more than 20 years after its discovery.

While the data above does not resolve the question of the importance of prostacyclin sparing by low dose aspirin, the results of the Second Antiplatelet Trialist overview support (nonsignificant trend, P=0.06) the use of lower doses of aspirin (44). In this respect, trials using less than 160 mg per day of aspirin showed a 28% odds reduction in vascular events, compared with 26% and 20% odds reduction in trials employing 160 to 325 mg per day and 500 to 1,500 mg per day respectively (4,44).

Higher aspirin doses of 900 to1,500 mg per day have not been shown to have increased efficacy in arterial thromboembolism prevention when compared with 300 mg per day or lower doses. In one metaanalysis of 19 antiplatelet trials in several thrombotic disorders, the issue of aspirin dosage was indirectly analyzed (44). It was found that trials employing 900 to 1,500 mg of daily aspirin versus placebo had similar outcomes to trials employing doses of 300 to 325 mg per day versus placebo. Patients treated with 900 to 1,500 mg experienced a 23% reduction in new vascular events, whereas those taking 320 mg showed a decrease of 24%. Thus a dose-response plateau appears to exist in aspirin's antithrombotic effect. Clinical and laboratory observations are complimentary in this respect since aspirin inhibition of in vitro platelet PG synthesis shows a similar dose-response plateau with maximal blockade at micromolar concentrations (2,25,28).

Among the clinical indications for aspirin, the dose for prevention of stroke in patients following a transient ischemic attack or initial minor stroke is presently disputed. In this respect, direct comparisons reveal no differences in efficacy between 300 mg and 1200 mg daily or between doses of 30 and 283 mg daily, though a small sample size was involved (60,61). It should also be noted that studies in which patients on larger aspirin doses were seen to do better also received another antiplatelet drug (62).

In summary, all observers do not agree on any single effective dose of aspirin for preventing thrombosis and they do not agree that the same dose of aspirin will work for every indication (62). Since aspirin blocks platelet PG synthesis in every thrombotic disorder in which it is effective and platelets in thrombotic disorders do not differ from normal platelets in their susceptibility to aspirin, then the least amount of aspirin that blocks platelet PG synthesis to a maximal extent is likely to be the most effective and least toxic dose. Such a dose would be 80 mg to 160 mg orally per day.

22.8 Side Effects of Aspirin

Aspirin induces gastrointestinal irritation and bleeding by a mechanism that may involve both inhibition of PG synthesis and direct damage to gastric and intestinal mucosa (73,74). Results from additional studies have suggested the existence of dose dependency of these GI side effects (75-77). Subjective GI symptoms and patient noncompliance with taking aspirin is higher when doses of 900-1,500 mg are taken compared with 300 mg or placebo (78-80). Enteric-coated and buffered preparations of aspirin may ameliorate GI toxicity (73). Trials incorporating aspirin doses less than 320 mg have suggested GI toxicity may be either low or absent in this group compared with higher doses (44,80 - 82).

Aspirin is capable of reducing the renal synthesis of vasodilator prostaglandins which can impair blood pressure control in those with essential hypertension and increase the risk of developing chronic renal disease (83,84). These effects are generally present at antiinflamatory level doses (2 grams - 6 grams per day), though, and clinically relevant suppression of renal prostaglandins is unlikely to be present in the anti-thrombotic dosing range (80 mg - 320 mg per day).

The bleeding and antithrombotic effects of aspirin are inseperable since both result from platelet inhibition. The bleeding-related side effects associated with aspirin are usually subtle and not recognized in otherwise hemostatically normal individuals (7,8,85,86). In the two aspirin primary prevention trials noted earlier among 22,000 men, 10 severe strokes related to intracerebral bleeding were noted in those taking aspirin over five years compared with two in the placebo group (40). Another primary prevention trial did not demonstrate similar toxicity (39) while in secondary prevention trials intracerebral bleeding may be countered by a concommitant decrease in thrombotic strokes (87,88). Thus it is apparent from the above that the incidence of intracranial bleeding caused by long term aspirin administration is not easily quantitated. On the other hand, an aspirin dose-related increase in extracranial bleeding events (e.g. GI bleeding) has been noted in several clinical trials, two of which employed doses less than 100 mg (60,88).

References

1. Hirsh J, Dalen JE, Fuster V, Harker LA, Patrono C, Roth GJ: Aspirin and other platelet-active drugs The relationship among dose, effectiveness, and side effects. Chest 108 (No.4, Suppl.): 247S, 1995

2. Roth GJ, Majerus PW: The mechanism of the effect of aspirin on human platelets I. Acetylation of a particulate fraction protein. J. Clin. Invest. 56: 624, 1975

3. Roth GJ, Stanford N, Majerus PW: Acetylation of prostaglandin synthase by aspirin. Proc. Natl. Acad. Sci. USA 72: 3073, 1975

4. Roth GJ, Calverley DC: Aspirin, platelets and thrombosis. Theory and practice. Blood 83: 885, 1994.

5. Link KP, Overman RS, Sullivan WR, Nuebner CF, Scheel LD: Studies on hemorrhagic sweet clover disease. J. Biol. Chem. 147: 463, 1943

6. Craven LL: Experiences with aspirin (acetylsalicylic acid) in the nonspecific prophylaxis of coronary thrombosis. Mississippi Valley Med. J. 75: 38, 1953

7. Weiss HJ, Aledort LM, Kochwa S: The effect of salicylates on the hemostatic properties of platelets in man. J. Clin. Invest. 47: 2169, 1968

8. Mustard JF, Packham MA: Factors influencing platelet function: Adhesion, release and aggregation. Pharmacol. Rev. 22: 97, 1970

9. Bergstrom S, Sjovall J: The isolation of prostaglandin. Acta Chem. Scand. 11: 1086, 1957

10. Van Dorp DA, Beerthuis RK, Nugteren DH, Vonkeman H: The biosynthesis of prostaglandins. Biochim. Biophys. Acta 90: 204, 1964

11. Lands WEM, Samuelsson B: Phospholipid precursors of prostaglandins. Biochim. Biophys. Acta 164: 426, 1968

12. Vane JR: Inhibition of prostaglandin synthesis as a mechanism of action for aspirin-like drugs. Nature New Biology 231: 232, 1971

13. Vane JR, Flower RJ, Botting RM: History of aspirin and its mechanism of action. Stroke 21 (suppl.IV): IV-12, 1990

14. Samuelsson B, Granstrom E, Green K, Hamberg M, Hammerstrom S: Prostaglandins. Ann. Rev. Biochem. 44: 669, 1975

15. Oliw E, Granstrom E, Anggard E: The prostaglandins and essential fatty acids. in Pace-Asciak C, Granstrom E (eds): Prostaglandins and Related Substances. Amsterdam, Elsevier, 1983, p 1

16. Roth GJ: Platelet arachidonate metabolism and platelet-activating factor, in Phillips DR, Shuman MA (eds): Biochemistry of Platelets. Orlando, FL, Academic, 1986, p 69

17. Hamberg M, Svensson J, Samuelsson B: Thromboxanes: A new group of biologically active compounds derived from prostaglandin endoperoxides. Proc. Natl. Acad. Sci. USA. 72: 2994, 1975

18. Moncada S, Gryglewski R, Bunting S, Vane JR: An enzyme isolated from arteries transforms prostaglandin endoperoxides to an unstable substance that inhibits platelet aggregation. Nature 263: 663, 1976

19. Smith WL, Lands WEM: Stimulation and blockade of prostaglandin biosyn-thesis. J. Biol. Chem. 246: 6700, 1971

20. Kujubu DA, Fletcher BS, Varnum BC, Lim RW, Herschman HR: TIS10, a phorbol ester tumor promoter-inducible mRNA from Swiss 3T3 cells, encodes a novel prostaglandin synthase/cyclooxygenase homologue. J. Biol. Chem. 266: 12866, 1991

21. Roth GJ, Machuga ET, Ozols J: Isolation and covalent structure of the aspirin-modified, active-site region of prostaglandin synthetase. Biochemistry 22: 4672, 1983

22. Pinckard RN, Hawkins D, Farr RS: *In vitro* acetylation of plasma proteins, en-zymes and DNA by aspirin. Nature 219: 68, 1968

23. DeWitt DL, Smith WL: Primary structure of prostaglandin G/H synthase from sheep vesicular gland determined from the complementary DNA sequence. Proc. Natl. Acad. Sci. USA 85: 1412, 1988

24. Shimokawa T, Smith WL: Prostaglandin endoperoxide synthase The aspirin acetylation region. J. Biol. Chem. 267: 12387, 1992

25. Patrignani P, Filabozzi P, Patrono C: Selective, cumulative inhibition of plate-let thromboxane production by low-dose aspirin in healthy subjects. J. Clin. Invest. 69: 1366, 1982

26. Funk CD, Funk LB, Kennedy M, Pong A, FitzGerald GA: Human platelet / erythroleukemia cell PGG / H synthase: cDNA cloning, expression, mutagenesis and gene chromosomal assignment. FASEB J. 5: 2306, 1991

27. Jaffe EA, Weksler BB: Recovery of endothelial cell prostacyclin production after inhibition by low doses of aspirin. J. Clin. Invest. 63: 532, 1979

28. FitzGerald GA, Oates JA, Hawiger J, Maas RL, Roberts LJ II, Lawson JA, Brash AR: Endogenous biosynthesis of prostacyclin and thromboxane and platelet function during chronic administration of aspirin in man. J. Clin. Invest. 71: 676, 1983

29. Burch JW, Stanford N, Majerus PW: Inhibition of platelet prostaglandin syn-thetase by oral aspirin. J. Clin. Invest. 61: 314, 1978

30. Patrono C, Ciabattoni G, Patrignani P: Clinical pharmacology of platelet cyclo oxygenase inhibition. Circulation 72: 1177, 1985.

31. De Caterina R, Giannessi D, Bernini W: Low-dose aspirin in patients recov-ering from myocardial infarction: Evidence for a selective inhibition of thromboxane-related platelet function. Eur. Heart J. 6: 409, 1985.

32. De Caterina R, Giannessi D, Boem A: Equal antiplatelet effects of aspirin 50 or 324 mg / day in patients after acute myocardial infarction. Thromb. Haemost. 54: 528, 1985.

33. Pedersen AK, FitzGerald GA: Dose-related kinetics of aspirin. Presystemic acetylation of platelet cyclo-oxygenase. N. Eng. J. Med. 311: 1206, 1984.

34. Rowland M, Riegelman S, Harris PA, Sholkoff SB: Absorption kinetics of aspirin in man following oral administration of an aqueous solution. J. Pharm. Sci. 61: 379, 1972

35. Levy G: Clinical pharmacokinetics of salicylates: A re-assessment. Brit. J. Clin. Pharmacol. 10 (suppl.1): 285S, 1980

36. Di Minno G, Silver MJ, Murphy S: Monitoring the entry of new platelets into the circulation after ingestion of aspirin. Blood 61: 1081, 1983

37. Asaoka Y, Nakamura S, Yosida K, Nishizuka Y: Protein kinase C, calcium, and phospholipid degradation. Trends Biochem. Sci. 17: 414, 1992

38. Majerus PW: Inositol phosphate biochemistry. Ann. Rev. Biochem. 61: 225, 1992

39. Peto R, Gray R, Collins R, Wheatley K, Hennekens C, Jamrozik K, Warlow C, Hafner B, Thompson E, Norton S, Gilliland J, Doll R: Randomized trial of prophylactic daily aspirin in British male doctors. Br Med J 296: 313, 1988

40. Steering Committee of the Physicians' Health Study Research Group: Final report on the aspirin component of the ongoing Physicians Health Study. N. Eng. J. Med. 321: 129, 1989

41. Hennekens CH, Buring JE, Sandercock P, Collins R, Peto R: Aspirin and other anti-platelet agents in the secondary and primary prevention of cardiovascular disease. Circulation 80: 749, 1989

42. Ridker PM, Manson JE, Gaziano JM, Buring JE, Hennekens CH: Low-dose aspirin therapy for chronic stable angina: A randomized, placebo-controlled clinical trial. Ann. Intern. Med. 114: 835, 1991

43. Juul-Moller S, Edvardsson N, Jahnmatz B: Double-blind trial of aspirin in primary prevention of myocardial infarction in patients with stable chronic angina pectoris. Lancet 340: 1421, 1992

44. Antiplatelet Trialists' Collaboration: Collaborative overview of randomized trials of antiplatelet treatment. Part I: Prevention of death, myocardial infarction and stroke by prolonged antiplatelet therapy in various categories of patients. Brit. Med. J. 308: 81, 1994.

45. The RISC Group: Risk of myocardial infarction and death during treatment with low dose aspirin and intravenous heparin in men with unstable coronary artery disease. Lancet 336: 827, 1990

46. Lewis HD Jr., Davis JW, Archibald DG, Steinke WE, Smitherman TC, Doherty JE 3d, Schnaper HW, LeWinter MM, Linares E, Pouget JM, Sabharwal SC, Chesler E, DeMots H: Protective effects of aspirin against acute myocardial infarction and death in men with unstable angina. Results of a Veterans Administration Cooperative Study. New Eng. J. Med. 309: 396, 1983

47. Cairns JA, Gent M, Singer J, Finnie KJ, Froggatt GM, Holder DA, Jablonsky G, Kostuk WJ, Melendez LJ, Myers MG: Aspirin, sulfinpyrazone, or both in unstable angina. Results of a Canadian multicenter trial. New Eng. J. Med. 313: 1369, 1985

48. Theroux P, Ouimet H, McCans J, Latour JG, Joly P, Levy G, Pelletier E, Juneau M, Stasiak J, deGuise P: Aspirin, heparin, or both to treat acute unstable angina. New Eng. J. Med. 319: 1105, 1988

49. Lorenz RL, Weber M, Kotzur J, Theisen K, Schacky CV, Meister W, Reichardt B, Weber PC: Improved aortocoronary bypass patency by low-dose aspirin (100 mg daily): Effects on platelet aggregation and thromboxane formation. Lancet 1: 1261, 1984

50. Mayer JE Jr, Lindsay WG, Castaneda W, Nicholoff DM: Influence of aspirin and dipyridamole on patency of coronary artery bypass grafts. Ann. Thor. Surg. 31: 204, 1981

51. Chesebro JH, Clements IP, Fuster V, Elveback LR, Smith HC, Bardsley WT, Frye RL, Holmes DR Jr., Vlietstra RE, Pluth JR: A platelet-inhibitor drug trial in coronary-artery bypass operations: Benefit of perioperative dipyridamole and aspirin therapy on early postoperative vein-graft patency. New Eng. J. Med. 307: 73, 1982

52. Chesebro JH, Fuster V, Elveback LR, Clements IP, Smith HC, Holmes DR Jr, Bardsley WT, Pluth JR, Wallace RB, Puga FJ: Effect of dipyridamole and aspirin on late vein-graft patency after coronary bypass operations. New Eng. J. Med. 310: 209, 1984

53. Brown BG, Cukingnan RA, DeRouen T, Goede LV, Wong M, Fee HJ, Roth JA, Carey JS: Improved graft patency in patients treated with platelet-inhibiting therapy after coronary bypass surgery. Circulation 72: 138, 1985

54. Rajah SM, Salter MC, Donaldson DR, Subba Rao R, Boyle RM, Partridge JB, Watson DA: Acetylsalicylic acid and dipyridamole improve the early patency of aorto-coronary bypass grafts. A double-blind, placebo-controlled, randomized trial. J. Thor. Cardiovasc. Surg. 90: 373, 1985

55. Goldman S, Copeland J, Moritz T, Henderson W, Zadina K, Ovitt T, Doherty J, Read R, Chesler E. Sako Y: Improvement in early saphenous vein graft patency after coronary artery bypass surgery with antiplatelet therapy: Results of a Veterans Admin-istration Cooperative Study. Circulation 77: 1324, 1988

56. Schwartz L, Bourassa MG, Lesperance J, Aldridge HE, Kazim F, Salvatori VA, Henderson M, Bonan R, David PR: Aspirin and dipyridamole in the prevention of restenosis after percutaneous transluminal coronary angioplasty. N Engl J Med 318: 1714, 1988

57. Lembo NJ, Black AJ, Roubin GS, Wilentz JR, Mufson LH, Douglas JS Jr, King SB III: Effect of pretreatment with aspirin versus aspirin plus dipyridamole on frequency and type of acute complications of percutaneous transluminal coronary angioplasty. Amer. J. Cardiol. 65: 422, 1990

58. White CW, Chaitman B, Lassar TA, Marcus ML, Chisholm RJ, Knudson M, Morton B, Roy L, Khaja F: Antiplatelet agents are effective in reducing the immediate complications of PTCA: Results from the ticlopidine multicenter trial. Circulation : 76: 400, 1987

59. Antiplatelet Trialists' Collaboration: Secondary prevention of vascular disease by prolonged antiplatelet treatment. Brit. Med. J. 296: 320, 1988

60. The Dutch TIA Trial Study Group: A comparison of two doses of aspirin (30 mg vs. 283 mg a day) in patients after a transient ischemic attack or minor ischemic stroke. New Eng. J. Med. 325: 1261, 1991

61. UK-TIA Study Group: The United Kingdom transient ischaemic attack (UK-TIA) aspirin trial: final results. J. Neurol. Neurosurg. Psychiatry 54: 1044, 1991

62. Dyken ML, Barnett HJM, Easton JD, Fields WS, Fuster V, Hachinshi V, Norris JW, Sherman DG: Low-dose aspirin and stroke. Stroke 23: 1395, 1992

63. Harter HR, Burch JW, Majerus PW, Stanford N, Delmez JA, Anderson CB, Weerts CA: Prevention of thrombosis in patients on hemodialysis by low-dose aspirin. New Eng. J. Med. 301: 577, 1979

64. Harajola PT, Meurala H, Frick MH: Prevention of early reocclusion by dipyri-damole and ASA in arterial reconstructive surgery. J. Cardiovasc. Surg. 22: 141, 1981

65. Goldman M, Hall C, Dykes J, Hawker RJ, McCollum CN: Does 111-indium platelet deposition predict patency in prosthetic arterial grafts? Brit. J. Surg. 70: 635, 1983

66. Hess H, Mietaschk A, Deichsel G: Drug-induced inhibition of platelet function delays progression of peripheral occlusive arterial disease. A prospective double-blind arteriographically controlled trial. Lancet 1: 415, 1985

67. Clyne CAC, Archer TJ, Atuhaire LK, Chant ADB, Webster JHH: Random control trial of a short course of aspirin and dipyridamole (persantin) for femorodistal grafts. Brit. J. Surg. 74: 246, 1987

68. Burch JW, Baenziger NL, Stanford N, Majerus PW: Sensitivity of fatty acid cyclooxygenase from human aorta to acetylation by aspirin. Proc. Natl. Acad. Sci. USA 75: 5181, 1978

69. Clarke RJ, Mayo G, Price P, FitzGerald GA: Suppression of thromboxane A_2 but not of systemic prostacyclin by controlled-release aspirin. New Eng. J. Med. 325: 1137, 1991

70. Preston FE, Whipps S, Jackson CA, French AJ, Wyld PJ, Stoddard CJ: Inhibition of prostacyclin and platelet thromboxane A_2 after low-dose aspirin. New Eng. J. Med. 304: 76, 1981

71. Kyrle PA, Eichler HG, Jager U, Lechner K: Inhibition of prostacyclin and thromboxane A_2 generation by low-dose aspirin at the site of plug formation in man *in vivo*. Circulation 75: 1025, 1987

72. FitzGerald GA, Brash AR, Falardeau P, Oates JA: Estimated rate of prosta-cyclin secretion into the circulation of normal man. J. Clin. Invest. 68: 1272, 1981

73. Graham DY: The relationship between nonsteroidal anti-inflammatory drug use and peptic ulcer disease. Gastroenterol. Clin. North Amer. 19: 171, 1990

74. Gabriel SE, Jaakkimainen L, Bombardier C: Risk for serious gastrointestinal complications related to use of nonsteroidal anti-inflammatory drugs A meta-analysis. Ann. Int. Med. 115: 787, 1991

75. Levy M: Aspirin use in patients with major upper gastrointestinal bleeding and peptic ulcer disease. New Eng. J. Med. 290: 1158, 1974

76. Silvoso GR, Ivey KJ, Butt JH, Lockard OO, Holt SD, Sisk C, Baskin WN, Mackercher PA, Hewett J: Incidence of gastric lesions in patients with rheumatic disease on chronic aspirin therapy. Ann. Int. Med. 91: 517, 1979

77. Prichard PJ, Kitchingman GK, Walt RP Daneshmend TK, Hawkey CJ: Human gastric mucosal bleeding induced by low dose aspirin, but not warfarin. Brit. Med. J. 298: 493, 1989

78. ISIS-2 Collaborative Group: Randomized trial of intravenous streptokinase, oral aspirin, both or neither amoung 17,187 cases of suspected acute myocardial infarction: ISIS-2. Lancet 2: 349, 1988

79. Breddin K, Loew D, Lechner K, Uberla K, Walter E: Secondary prevention of myocardial infarction: A comparison of acetylsalicylic acid, placebo and phenpro-coumon. Haemostasis 9: 325, 1980

80. Manson JE, Stampfer J, Colditz GA Willet WC, Rosner B, Speizer ME, Hennekens CH: A prospective study of aspirin use and primary prevention in cardio-vascular disease in women. J. Amer. Med. Assoc. 266: 521, 1991

81. ISIS-3 Collaborative Group: A randomized comparison of streptokinase vs tissue plasminogen activator vs anistreplase and of aspirin plus heparin vs aspirin alone among 41,299 cases of suspected myocardial infarction. Lancet 339: 753, 1992

82. Roderick PJ, Wilkes HC, Meade TW: The gastrointestinal toxicity of aspirin: An overview of randomized trials. Brit. J. Clin. Pharmacol. 35: 219, 1993.

83. Oates JA, FitzGerald GA, Branch RA, Jackson EK, Knapp HR, Roberts LJ: Clinical implications of prostaglandin and thromboxane A2 formation. N. Eng. J. Med. 319: 761, 1988.

84. Sandler DP, Burr FR, Weinberg CR: Nonsteroidal anti-inflammatory drugs and the risk for chronic renal disease. Ann. Intern. Med. 115: 165, 1991

85. O'Brien JR: Effects of salicylates on human platelets. Lancet 1: 779, 1968

86. Mielke CH, Kaneshiro MM, Maher IA, Weiner JM, Rapaport SI: The stand-ardized normal Ivy bleeding time and its prolongation by aspirin. Blood 35: 204, 1969

87. Sherman DG, Dyken ML Jr, Gent M, Harrison MJG, Hart RG, Mohr JP: Anti-thrombotic therapy for cerebrovascular disorders: An update. Chest 108 (No.4, suppl.):444S, 1995

88. The SALT Collaborative Group: Swedish Aspirin Low-dose Trial (SALT) of 75 mg aspirin as secondary prophylaxis after cerebrovascular ischaemic events. Lancet 338: 1345, 1991

89. Insel PA: Analgesic-antipyretics and antiinflammatory agents; Drugs employed in the treatment of rheumatoid arthritis and gout, in Gilman AG, Rall TW, Nies AS, Taylor P (eds): Goodman and Gilman's The Pharmacological Basis of Therapeutics, 8th ed. New York, NY, Pergamon, 1990, p 638

90. Abramson SB, Korchak H, Ludewig R, Edelson H, Haines K, Levin RI, Herman R, Rider L, Kimmel S, Weissman G: Modes of action of aspirin-like drugs. Proc. Natl. Acad. Sci USA 82: 7227, 1983

23

CURRENT DEVELOPMENTS IN ANTICOAGULANT AND ANTITHROMBOTIC AGENTS

Jawed Fareed, Ph.D,FACB.
Debra Hoppensteadt, Ph.D., DIC.
Daniel Fareed, B. Sc.
Jeanine M. Walenga, Ph.D.
Helmut Wolf, M.D., Ph.D.
Rodger L. Bick, M.D., Ph.D.
Departments of Pathology and,
Thoracic and Cardiovascular Surgery
Loyola University Health Systems

23.1 OVERVIEW

> Over the years, Jawed Fareed, Ph.D, FACB, and his colleagues have made highly significant contributions to the development of newer therapeutic anticoagulant and antithrombotic agents. Dr. Fareed's group has published extensively in this area.

Thrombosis is clearly the most common cause of death in the United States. About two million individuals die each year from an arterial or venous thrombosis or the consequences of these disorders. About 80% to 90% of all causes of thrombosis can now be defined with respect to cause. Of these, over 50% of all patients harbor a congenital or acquired blood coagulation protein or platelet defect which caused the thrombotic event. It is obviously of major importance to define those individuals harboring such a defect, as this allows (1) appropriate antithrombotic therapy to decrease risks of recurrence (2) determine length of time the patient must remain on therapy for secondary prevention and (3) allow for testing of family members in those harboring a blood coagulation protein or platelet defect which is hereditary (about 50% of all coagulation and platelet defects mentioned above). Aside from mortality, significant additional morbidity occurs from both arterial or venous thrombotic events, including, but not limited to paralysis (non-fatal thrombotic stroke), cardiac disability (repeated coronary events), loss of vision (retinal vascular thrombosis) and fetal waste syndrome (placental vascular thrombosis), stasis ulcers and other manifestations of post-phlebitic syndrome, etc.

The incidence of DVT in the USA is about 159 per 1000,000 or about 400,000 per year. A definable etiology can be found in 80% to 90% of these patients which allows for effective therapy to be delivered and allows for the other advantages of defining the blood coagulation protein or platelet defects mentioned above, to be instituted.

The overall incidence of PE in the USA is about 139 per 100,000 or about 347,000 cases per year based on clinical data; the incidence of fatal PE in the USA based on autopsy data is 94 per 100,000 or 235,000 deaths. The same scenario as that for DVT prevails for PE. If the PE is not fatal, every attempt should be made to define the blood coagulation or platelet defect.

About 1.5 million individuals in the USA will have an acute myocardial infarction per year; 50% will be fatal and 50 will be a premature event. Thus, there are about 750,000 deaths from coronary artery thrombosis per year. Of these coronary thrombotic events, 67% of patients harbor a coagulation blood protein or platelet defect leading to thrombosis. Fifty percent of these coagulation protein or platelet defects will be hereditary, thus Figure

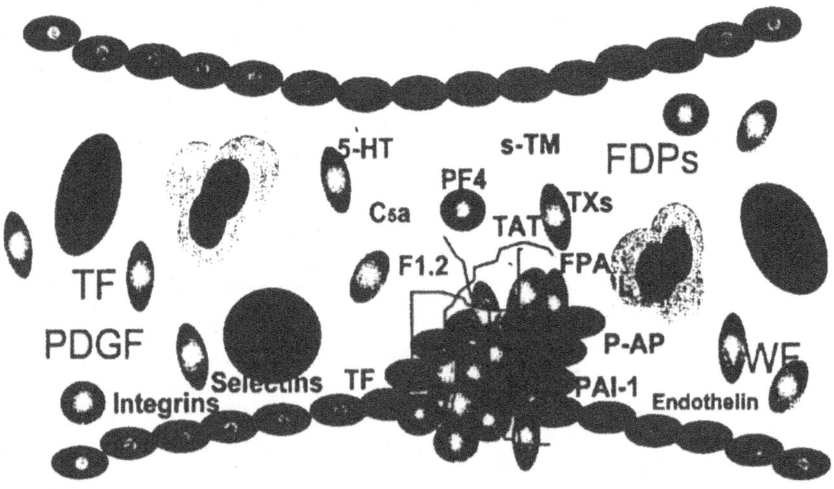

1. Blood and vascular activation processes resulting in thrombogenesis. Vascular damage, cell activation and related products contribute to the development of thrombus and vascular constriction. Antithrombotic drugs are capable of targeting single or multiple sites in the control of thrombogenesis.

emphasizing the importance of defining the presence and type of defect in survivors of acute myocardial infarction. Defining the defect will also allow one to optimize antithrombotic therapy for secondary prevention.

Cerebrovascular thrombosis (CVT) occurs in over 1.5 million individuals yearly in the USA; of these 66% suffer death or severe permanent paralysis. In those suffering CVT, including transient cerebral ischemic attacks (TIA's), small stroke syndrome (SSS) and frank thrombotic stroke, at least 30% harbor a blood coagulation protein or platelet defect causing thrombosis. Like the disorders discussed above, the need for defining the presence or absence and type of defect is of obvious importance.

Although the incidence of retinal arterial or venous thrombosis is unclear and death does not occur, significant visual morbidity is a major problem. Like CVT, about 30% of individuals sustaining retinal vascular thrombosis harbor a blood coagulation protein or platelet defect; reasons for defining these are obvious as for the other disorders discussed. Thrombosis therefore represents the leading cause of death and disability throughout the world. Antithrombotic and anticoagulant drugs therefore represent an important class of drugs with a major impact on healthcare.

The pathophysiology of thrombotic events is multicomponent and involves blood, vascular system and target sites. The process of thrombogenesis is depicted in Figure 1. Initially, vascular injury results in the localized alterations of the vessels, generation of tissue factor, and subsequent activation of platelets. Activated cells mediate several direct or signal transduction induced processes resulting in the activation of platelets. Cellular activation also results in the release of various mediators which amplify vascular spasm and the coagulation process. Drugs that target various sites of the activation process can be developed to control thrombotic events. Because of the coupled pathophysiology, a single drug may not be able to target these sites to produce therapeutic actions. Furthermore, many of these mediators produce localized actions at cellular and subcellular levels. The

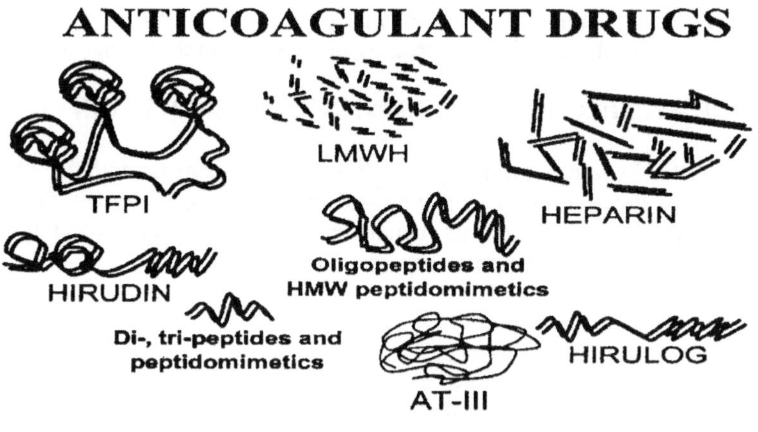

Figure 2. A diagramtic illustration of the structural diversity in anticoagulant drugs. These drugs represent protein, peptides, nucleic acids, carbohydrates and synthetic heterocyclic agents.

feedback amplification process also plays an important role in the pathology of these disorders. This understanding has led to the concept of polytherapy in the management of thrombotic disorders.

The newer developments in antithrombotic drugs are rather significant. Many advanced techniques to develop antithrombotic drugs are used at the present time. Advances in biotechnology and separation techniques have also contributed to the development of newer antithrombotic drugs (1-3). These drugs may prove to have a better efficacy in the control of thrombogenesis and its treatment. Drugs and devices which have been or are being developed based on newer concepts.

It can be appreciated that antithrombotic drugs represent a wide spectrum of natural, synthetic, semi-synthetic and biotechnology produced agents with marked differences in chemical composition, physicochemical properties, biochemical actions and pharmacologic effects. Figure 2 depicts the molecular diversities in the structure of various classes of new anticoagulant drugs. These represent proteins, peptides, carbohydrates and synthetic oligosaccharides. Beside the structural heterogeneity, each of these agents also show distinct functional properties. Many of these agents produce their antithrombotic effects via indirect mechanisms such as the modulation of endogenous processes (TFPI release).

The endogenous actions of the antithrombotic drugs are remarkably complex. It is no longer valid to assume that an antithrombotic drug must produce an anticoagulant action in blood as the conventional heparin and oral anticoagulants. Many of the drugs do not produce any alteration of blood clotting parameters, yet they are effective therapeutic agents because of their interactions with the various elements of the vasculature and other blood components. Another perspective is that several of these agents require endogenous transformation to become active products. Therefore, it becomes important to rely on the pharmacodynamic actions of these agents rather than on their in vitro characteristics to assess potency or efficacy of the product. Hematologic modulation plays a key role in the mediation of the antithrombotic actions of these drugs involving red cells, white cells, platelets and blood proteins. This is particularly true for the case of trauma induced thrombotic disorders where multiple processes are involved in thrombogenesis.

23.2 NEWER APPLICATIONS OF UNFRACTIONATED HEPARIN

The unfractionated heparin is primarily used as an anticoagulant for both the therapeutic and surgical indications. Usually beef lung and porcine mucosal derived products are available for these indications. The unfractionated heparins have been recently used for the following additional therapeutic uses:

1. management of unstable angina
2. adjunct to chemotherapeutic agents
3. adjunct to anti-inflammatory drugs
4. modulatory agent for growth factors
5. treatment of hemodynamic disorders

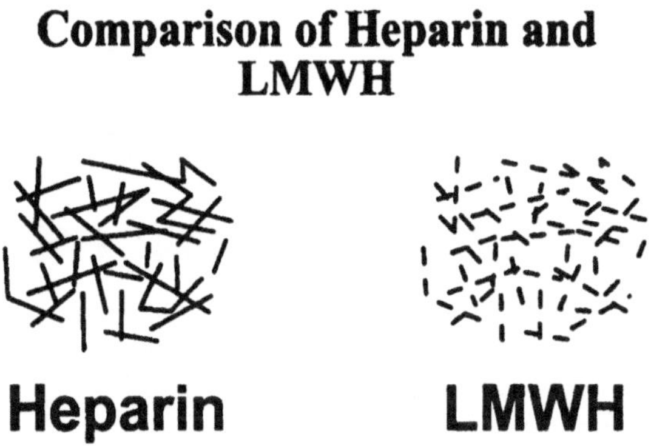

Figure 3. A diagrammatic illustration of the molecular heterogeneity in unfractionated heparin and a related depolymerized agent. Beside compositional variation, heparin and low molecular weight heparins exhibit functional heterogeneity in terms of its interaction with endogenous proteins and cells. The antithrombotic and anticoagulant effects of heparin are based on its polypharmacologic actions.

The heparins are heterogeneous in nature, as depicted in Figure 3. Based on the origin, the molecular weight of these anticoagulant drugs varies. The beef lung heparin preparations are usually of higher molecular weight, in contrast to the porcine mucosal heparin. However, the charge density and other characteristics, such as the serpin affinity is similar for both preparations. The low molecular weight heparins are derived from the unfractionated heparin controlled depolymerization process and are also depicted in this figure.

Several newer pharmacological effects, independent of the anticoagulant effects, have been described. These include glycosylation mediated functional modulation of macromolecules, growth factor modulation, release of endothelial products and interactions with various endogenous sites.

Chemical modification of the unfractionated heparin, such as desulfation, deamination and coupling with various agents have resulted in products of non-anticoagulant nature with selective actions on enzymes and cellular receptors. Thioxyloside derivatives have also been reported to produce oral antithrombotic actions in animal models. However, relatively larger dosages are needed to produce these effects. These heparin derivatives are currently tested for such indications as sepsis, viral infections and the treatment of proliferative disorders.

23.3 THE DEVELOPMENT OF LOW MOLECULAR WEIGHT HEPARINS

The development of low molecular weight heparin (LMWH) has added a new dimension to the clinical management of thrombotic disorders. These agents have revolutionized the prophylaxis of post-surgical thrombosis (4-5). More recently these drugs have been used for the treatment of trauma related thrombosis. In particular, their relative effects on platelets are minimal in comparison to heparin. Thus, these agents are of value in platelet compromised patients.

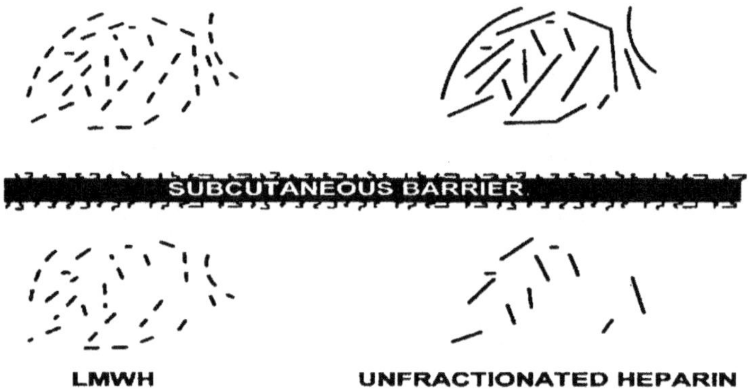

Figure 4. Bioavailability of heparin and LMWH after subcutaneous administration. The subcutaneous barrier only allows the absorption of lower molecular weight components of heparin. This observation was one of the major factors in the development of LMWHs.

For the past few decades, heparin has been widely used for the prevention of post-operative thromboembolism (6,7). However, there are several adverse side-effects associated with the use of heparin such as bleeding, heparin induced thrombocytopenia, heparin induced thrombosis (8,9) and osteoporosis (10). In addition, the regimen of prophylactic heparin used in the prevention of deep venous thrombosis (DVT) is tedious, requiring 2 to 3 daily injections because of the limited bioavailability and short half-life of heparin when administered subcutaneously.

The many clinical problems associated with heparin led several investigators to study the structure of heparin and to identify the active component(s) of this agent. The observation that only low molecular weight heparins are absorbed after subcutaneous administration has led to the development of LMWHs. As shown in Figure 4, the bioavailability of LMWH component is much greater than the unfractionated heparin. Experimental studies revealed the first LMWH had a subcutaneous bioavailability of about 90% compared to 15 to 25% for unfractionated heparin (UFH) (11). Furthermore, these agents exhibit a much longer biologic half-life in contrast to heparin (3). Thus, LMWH preparations could be administered with a single daily injection making them easy to administer as a prophylactic agent. Eight LMWHs have already been approved for the prophylaxis of thrombosis in the European community. Several of these LMWHs are under clinical trials in the US. Three of these agents have been approved for use prophylaxis of DVT. One product, namely Enoxaparin is approved for the management of acute coronary syndrome.

Low molecular weight components in unfractionated heparins have been identified long before the development of clinically effective agents. However, because of technologic problems, they were not made available for clinical use. The very first LMWH was obtained by a fractionation method (3). However, this method only yielded a limited amount of the active product and was cost prohibitive. During the past decade, different chemical and enzymatic processes have been developed to obtain LMWH from the parent unfractionated heparin (UFH). The depolymerization process for the production of LMWH is shown in Figure 5. Such methods as chemical degradation, enzymatic depolymerization and (γ-radiation cleavage can be used to produce the LMWHs.

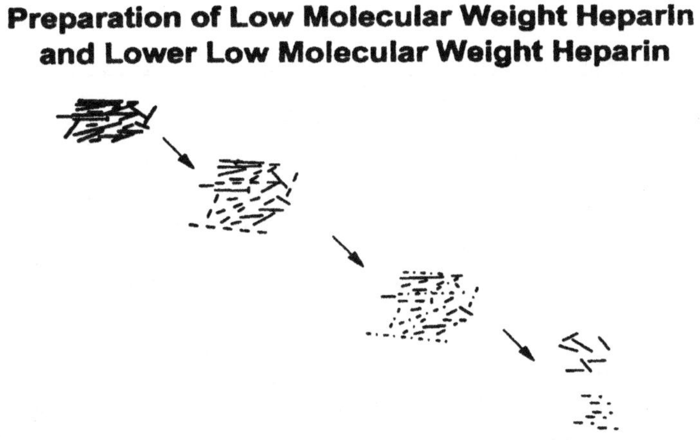

Figure 5. A diagrammatic illustration of the manufacturing processes for LMWHs. Unfractionated heparin of porcine or bovine origin is depolymerized by chemical or enzymatic digestion in controlled conditions. The degree of depolymerization is adjusted to obtain LMWH of varying weight.

Recent studies have shown that the individual LMWHs obtained from each process exhibit chemical and pharmacological differences (3). Such differences may influence the clinical outcome especially in the case of higher dosage use for treatment of cardiovascular indications. Limited clinical data on the differential pharmacologic efficacy of these agents is available at this time. For example, several of the LMWHs have been elucidated for the management of acute coronary syndromes in various clinical trials, however only one of these agents have exhibited a favorable profile leading to its approval for this indication.

23.4 LOW MOLECULAR WEIGHT HEPARINS IN THE MANAGEMENT OF THROMBOSIS

In the past, LMWHs have been used for the post-surgical prophylaxis of DVT. However, these agents are now also used in the treatment of pre-existing events utilizing both the subcutaneous and intravenous (Breddin, personal communication) routes of administration.

In some of the randomized studies in which LMWHs have been compared with UFH in patients undergoing abdominal surgery a significant decrease in DVT was observed in patients receiving LMWHs compared with UFH (12-15). There was no statistical difference in efficacy observed between the two treatments in the other studies (16-18). In most studies there was no significant difference between the observed bleeding effects of the LMWHs and UFH. In all of these studies, LMWH was given as a single daily dose while UFH was administered 2 or 3 times daily. The results of these studies suggest that LMWHs are effective and well tolerated in patients undergoing general surgery. The advantage of LMWHs over heparin was primarily in the reduced number of injections per day.

The most extensively studied method of prophylaxis of DVT in orthopedic patients in recent years has been with LMWH. Evidence that this treatment is both effective and safe is quickly accumulating.

Several randomized trials using venography or ^{125}I fibrinogen uptake as the endpoint have also been analyzed. Four studies compared LMWH with dextran. In two of these studies there was no significant difference between the LMWH and the dextran groups. However, in the other two studies, LMWH was more effective than dextran. The different results obtained in these studies is probably due to the different dosages selected for dextran as well as the difference in the dosages and composition of the different LMWHs. Two other studies compared LMWH with placebo. In one study, a difference in the incidence of DVT was found, but in the other study there was no significant difference.

Many surgeons and physicians around the world now appreciate that a large number of cases of venous thromboembolism, particularly those following surgery, can be avoided by correct use of prophylaxis. With additional education and with the introduction of newer and more efficacious agents, it should become routine for all patients undergoing surgery to be assessed for their risk of venous thrombosis and protected accordingly.

Of the many drugs used as prophylactic antithrombotic agents, heparin has a long history as therapy for both DVT and pulmonary embolism (PE). Many studies have shown that in moderate and high risk patients, heparin can prevent postoperative DVT and PE (4,19-21). Now, with the introduction of LMWHs, these benefits can be had together with easier dosing and potentially less risk of bleeding.

Many surgeons still, however, harbor fears and doubts about using thromboprophylaxis. One of the most common fears is that of bleeding. There are still some surgeons who are not convinced that the benefits of prophylaxis outweigh the risks. Optimization of the individual LMWHs in valid clinical trials is therefore important for the safe use of these agents. Recently, the US FDA has issued a warning against the use of LMWHs in patients undergoing spinal (epidural) anesthesia.

Available clinical data suggests that LMWHs can be safely substituted for low-dose heparin. To validate this, several comparative trials throughout the world on low dose heparin and LMWHs are in progress. In several European clinical trials, the efficacy of LMWHs in the prevention of post-surgical DVT has now been proven, and these agents are considered to be the drug of choice for this indication (12,22,23). When used for prophylactic treatment (subcutaneous), most LMWHs mediate their actions in a similar manner; however, their efficacy and tolerability profiles differ markedly as discussed above, and the recommended dosages for the various products differ. Because of the differences between products, such practices as standardization by a single in vitro assay and assignment of a single INN designation are deemed invalid for LMWHs as a group. The individualized agent approach to all LMWHs has recently been adopted by the World Health Organization (WHO), US Food and Drug Administration (FDA) and the Scientific and Standardization Subcommittee of the International Society of Thrombosis and Hemostasis (ISTH). The recognition of the individuality of each of the LMWHs is extremely important to avoid excessively high or low doses of a product. Dose finding studies are essential and will have a major impact on the prophylactic and therapeutic acceptance of LMWHs.

Having satisfactorily passed their first step in clinical use, LMWHs are now being applied to other clinical situations. LMWHs are moving into the area of established DVT and therapeutic treatment of thrombosis. LMWHs may be therapeutic alternatives to heparin in some, but not all patients who develop a sensitivity to heparin or who develop heparin-induced thrombocytopenia. At optimal dosages, these agents produce their antithrombotic effects, but whether they have fewer adverse effects than standard heparin may vary with the product and is unproven for many of them. LMWHs have proven to be equally effective as heparin in general surgery and orthopedic surgery (14,18,24).

More recently, LMWHs are being used for specific indications such as percutaneous transluminal coronary angioplasty (PTCA), as adjuncts to thrombolytic agents in disseminated intravascular coagulation (DIC) and the hypercoagulable state. A recent study has reported on the inhibition of cellular proliferation after experimental balloon angioplasty by LMWH in rabbits (25). There are several other indications such as the treatment of cardiovascular disorders where these agents may also prove to be useful.

LMWHs have been used in isolated areas as anticoagulants during cardiovascular bypass surgery procedures. However, a relatively larger amount of LMWHs is needed to produce comparable anticoagulant responses for anticoagulation during the surgical procedure. Furthermore, it is rather difficult to antagonize the effect of these agents with conventional protamine usage. Thus, their use in cardiovascular surgical indications requires additional clinical studies. Currently, LMWHs are undergoing clinical trials in the management of thrombotic and ischemic stroke. It has been suggested that the use of LMWHs reduces the risk of thrombosis and improves the quality of life in patients who suffered thrombotic stroke. However, the results of these clinical trials have provided mixed results. For example, in one of the initial trials a LMWH (Fraxiparin) was found to be effective, however in subsequent trials it failed to provide similar results.

23.5 GLYCOSAMINOGLYCAN RELATED ANTITHROMBOTIC AGENTS

Both the synthetic and natural analogues of various glycosaminoglycans have been developed for antithrombotic use. These agents are able to act through their interactions with serine protease inhibitors and by releasing TFPI from the endogeneous sites.

23.5.1 OTHER HEPARIN RELATED ANTITHROMBOTIC AGENTS

Beside the above clinically used approaches, several newer drugs to prevent thromboembolic disorders have been or are being developed. Most of these drugs are in their early phases of development and it will take some time before their clinical efficacy is proven.

A very-LMWH fraction (CY 222) has been developed for clinical trials. This agent produces its effects by multiple mechanisms. Pentasaccharide is a synthetic material whose structure is based on the critical binding region of heparin to AT. It functions by inhibiting activated factor X. These agents do not impair primary hemostasis and they exhibit a high bioavailability. They do not exhibit any anticoagulant effect at antithrombotic dosages. Attempts have also been made to produce ultra-LMWH preparations (MW <3000 daltons). One such preparation, CY 231, was developed by Sanofi and exhibited a MW of 2400 daltons.

Currently several dermatan sulfates are under development for the prophylaxis of venous thromboembolism. While similar in structure to heparin, these agents do not produce any effect on platelets. Furthermore, these agents are poorly absorbed after subcutaneous administration. More recently some low molecular weight dermatans have been produced which are absorbed subcutaneously, unlike dermatan sulfate.

Heparan sulfates have been developed as prophylactic antithrombotic agents. These agents are not homogeneous and contain other chondroitin sulfates. They bind to AT (antithrombin III is officially being designated as antithrombin; AT) and heparin cofactor II but to a lesser degree than heparin, and they are weakly anticoagulant. Thus, large dosages of heparan are needed for effective antithrombotic treatment. Depolymerized heparans have better bioavailability than the native heparans.

A synthetic hypersulfated lactobionic acid amide (Aprosulate) has been developed for prophylactic antithrombotic use. This agent produces its action via heparin cofactor II and by inhibiting protease generation. The bioavailability of this agent is better than that of dermatan and heparan sulfates. However, this product exhibits heparin-induced thrombocytopenic effects and some teratogenetic potential. Thus, its use in clinical trials has been stopped.

Many other glycosaminoglycans are being developed for the prophylaxis of thromboembolism. Some of these represent mixtures of glycosaminoglycans with varying molecular weight profiles. Noteworthy are Lomoparan and Suleparoide which are depolymerized heparan preparations. These agents exert their antithrombotic actions via unknown mechanisms; however, these are clinically very effective drugs. Additional synthetic heparinomimetics include synthetic oligosaccharides with high affinity to antithrombin III. More recently mixed inhibitors of factor Xa and IIa have also been designed.

23.6 NON-HEPARIN GLYCOSAMINOGLYCANS

In addition to the development of LMWH, many significant developments in the area of non-heparin glycosaminoglycan derived products as antithrombotic drugs have also taken place. Many drugs are developed as byproducts of heparin. It is no longer believed that a sulfomucopolysaccharide of natural origin must exhibit some interaction with antithrombin to have effective antithrombotic properties. Several agents without this interaction have been found to produce therapeutic effects on the blood and vascular system (26,27). Several mammalian glycosaminoglycan derived drugs are currently being used in European countries as antithrombotic, antilipemic and anti-atherosclerotic agents (28). These agents represent mixtures of native sulfomucopolysaccharides or its derivatives obtained by depolymerization and/or fractionation. At present, the chemistry and pharmacology of these drugs is not fully understood.

Due to the compositional variations, the mechanisms of pharmacologic and clinical actions of each these agents differ markedly. As with LMWHs, each product must be individually characterized. The pharmacologic action of these agents are mediated by several endogenous interactions in which both humoral and cellular receptors are involved. In addition to plasmatic effects, these agents are capable of producing several cellular actions. The measure of a single plasma parameter, therefore, cannot be a true reflection of the overall pharmacologic action.

When first developed, each of these agents was accepted as a derivative of heparin. However, with an increase in our knowledge of their structure and functional activity, preliminary pharmacologic studies were carried out to determine the proper indications for individual drugs. The glycosaminoglycan derived drugs are generally used as antithrombotic agents; however, several other indications such as atherosclerosis, stroke, hyperlipemia and senile dementia are now being considered.

23.6.1 CURRENTLY AVAILABLE NON-HEPARIN GLYCOSAMINOGLYCANS

A partial list of some of the available glycosaminoglycan derived drugs is given in Table I. Most of the agents listed in this table are in clinical development or actual usage at this time. The described agents are mainly of mammalian origin with the exception of SP 54 (Hemoclar) which is obtained from a plant. However, SP 54 is a hypersulfated pentosan polysulfate with structural and functional characteristics similar to the other sulfated glycosaminoglycans.

ORG 10172 is a depolymerized mixture of heparans, dermatans, heparin and other chondroitin sulfates. It has been currently developed under the commercial name of Lomoparan by Organon, Inc., Oss, The Netherlands. Currently this agent is undergoing clinical trials for the prophylaxis of D.T. after general and orthopedic surgery. This agent is also being used in the prevention of ischemic complications associated with stroke. It is claimed to have a better safety/efficacy ratio than heparin such that it produces minimal antihemostatic effects at antithrombotic doses (29).

MF 701 is the code name of a dermatan preparation which is currently being developed by Mediolanum Laboratories, Milan, Italy. This heterogeneous mixture of dermatan sulfate is claimed to be of mammalian mucosal origin. Currently it is being developed for prophylaxis against thrombosis after general and orthopedic surgery. Since the bioavailability of this agent via subcutaneous administration is rather limited, it is being administered intramuscularly. Several clinical trials are currently on-going with this agent.

Suleparoide is a widely used semi-synthetic glycosaminoglycan which has been used for prophylaxis of both arterial and venous thrombosis.

Another dermatan preparation is currently under development at Opocrin Laboratories, Corlo, Italy. This dermatan preparation (OP 435) is extracted from bovine mucosa. It is being developed for prophylactic antithrombotic usage in patients undergoing general surgery.

There have been several questions on the safety and efficacy of the higher molecular weight dermatans such as the MF 701 and OP 435. Both of these agents exhibit rather limited bioavailability, and have thus been administered in rather large dosages. Considering these issues, low molecular weight dermatan preparations have been recently introduced. Once such preparation is OP 370, or Desmin, currently developed by Alfa-Wasserman Company, Modena, Italy. This agent exhibits a much better bioavailability than high molecular weight dermatan sulfate and may also exhibit a longer duration of action. Currently this agent is being developed for prophylactic antithrombotic usage.

TABLE I

DEVELOPMENT OF GLYCOSAMINOGLCYANS DERIVED DRUGS

Drug	Composition	Status
ORG 10172	Depolymerized mixture of GAGs mainly containing	Prophylactic antithrombotic drug - ongoing clinical trials.
MF 701	Mixture of native and depolymerized dermatans.	Prophylactic antithrombotic drug - ongoing clinical trials.
Suleparoide	Semisynthetic GAG	Available for various indications.
OP 435	Mixture of dermatans.	Prophylactic antithrombotic - preclinical.
OP 370	Low molecular weight dermatans	Prophylactic antithrombotic - preclinical.
SP 54	Hypersulfated pentosan polysulfate.	Prophylactic antithrombotic - clinically used.
MPS	Depolymerized hypersulfated mixture of GAGs.	Antithrombotic agent. Developed for animal use.
Sulfomucopolysaccharide mixtures	Mixture of GAGs.	Clinically used.
Vasoflux	Semisynthetic GAG	Phase II Trial

GAGs: glycosaminoglycans

Hemoclar (SP 54) or pentosan polysulfate is a linear cationic polymer of beechwood tree origin which has been used for the prophylaxis of deep venous thrombosis. This agent is developed by Bene Chemical, Munich, Germany. Because of it origin, this agent can be obtained in relatively large quantities and may prove to be useful in various indications.

MPS represents a mixture of mucopolysaccharides obtained from mammalian trachea. This agent is manufactured by Luitpold-Werk, Munich, Germany for both human and equine usage for the treatment of joint diseases. Only limited data is available on the structure-activity relationship of this agent. This agent may have various applications as an antithrombotic agent. Additional pharmacologic studies may be needed for other indications.

The sulfated mucopolysaccharides or SMPS represent a large class of drugs which are currently sold under various commercial names. SMPS are non-characterized glycosaminoglycans which contain dermatan, heparan, chondroitin sulfate and other non-characterized glycosaminoglycans. While the composition of these agents does not differ markedly, each product is marketed for a specific indication. The mechanism of action and pharmacodynamics are not understood and the pharmacology of these mucopolysaccharides is poorly understood.

23.6.2 CLINICAL APPLICATIONS OF NON-HEPARIN GLYCOSAMINOGLYCANS

Despite the unresolved basic pharmacologic issues as discussed above, the developments in the area of glycosaminoglycans are progressing. While the thrust is to develop an antithrombotic agent, these agents may also be used as an alternate drug in conditions where heparin may not be safe, especially in the heparin induced thrombocytopenia and heparin-induced thrombosis related syndromes. Some of the newer indications for these agents include anti-inflammatory and anti-atherosclerotic applications, treatment of vascular disorders, as topical agents for wound healing, and as a treatment for AIDS. Other indications may include senile dementia and as a cytoprotective agent in surgery or transplant. Several small scale clinical trials have been done to study these non-heparin glycosaminoglycans for these different indications (30- 34).

Heparan sulfate has been studied in DVT, chronic venous insufficiency and intermittent claudication (30,33). Although small numbers of patients were included in these trials, there appears to be positive results obtained with this agent.

Dermatan sulfate has also been studied in several small scale clinical trials (34) Most recently a pilot study has been completed using dermatan sulfate in acute leukemia to control DIC (35). Other clinical trials on some of the previously mentioned agents are also underway to prove the effectiveness of these non-heparin glycosaminoglycans.

23.7 RECOMBINANT AND SYNTHETIC ANTITHROMBIN DRUGS

Thrombin is known to play a crucial role in the overall thrombotic events leading to both arterial and venous thrombosis (36). Beside the transformation of fibrinogen to fibrin, this enzyme is claimed to mediate the activation of platelets and macrophages and produces

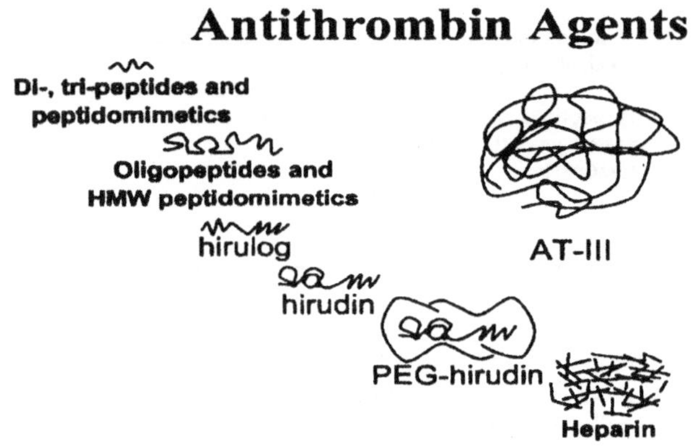

Figure 6. Structural heterogeneity in antithrombin agents. Besides the synthetic antithrombin agents, plasma derived antithrombin III and its recombinant equivalent and various recombinant equivalents of naturally occurring antithrombin agents and conjugates have also become available

on-site vascular effects leading to ischemia and vascular contraction. Furthermore, this enzyme is also linked with cellular proliferation and related events leading to restenosis. Thus, the development of agents which can solely target this enzyme is considered to be an important approach in providing newer drugs for the treatment of venous and arterial thrombosis. A list of the currently available thrombin inhibitors and their clinical development status is summarized in Table II. These agents are diagramatically shown on Figure 6. These agents also exhibit a marked heterogeneity in their structure which ranges from peptides to proteins and synthetic peptidomimetics. Antithrombin III preparations have also become available for therapeutic use. More recently, recombinant antithrombin III has also become available.

TABLE II

DEVELOPMENTAL STATUS OF SITE DIRECTED THROMBIN INHIBITORS

AGENTS	CHEMICAL NATURE	DEVELOPMENTAL STATUS
Hirudin and its analogues	Recombinant protein	Developed for both arterial and venous thrombosis
Hirulogs	Synthetic bifunctional oligopeptides	Several clinical studies are completed; additional studies are planned in various indications.
Peptidomimetics	Synthetic heterocyclic derivatives (argatroban, inogatran, napsagatran)	Phase II clinical development in the USA; used in Japan
Peptides and their derivatives	Peptide arginals (efegatran)	Phase II clinical development
Aptamers	DNA and RNA derived oligonucleotides with thrombin-binding domains (defibrotide)	Preclinical stage; limited animal data available
Plasma derived antithrombin	Protein and their recombinant equivalent products	Antithrombin)AT-III) is currently used. HC-II is still in developmental status
Transition state peptide analogues	Oligopeptides and synthetic organic agents (boronic acid derivatives)	Preclinical screening is being completed

One of the building blocks used to develop synthetic inhibitors is arginine with optimal C and/or N-terminal modifications. Many of these agents have been found to be highly toxic due to inhibition of butyl cholinesterase (37). After further modifications an isomer called argatroban (MD805 or MCI9038) has been generated as a selective reversible

inhibitor of thrombin (K_i=0.019 μM) (37). It acts by directly inhibiting thrombin and preventing it from acting in the coagulation and fibrinolytic system (38,39). Argatroban is effective in preventing thrombus formation in various animal models at low concentrations (>1 µ:M) (37,39). This compound is being tested clinically for several indications (40,41). Another promising low molecular weight (439 Da) reversible peptidomimetic thrombin inhibitor, Inogatran, is currently being developed by Astra Hassle Ab (Sweden). Phase I clinical studies on this compound have been completed on this potent thrombin inhibitor (ki=15nM) and have shown the half life of this agent to be about 1 hour (42).

TABLE III

COMPARISON OF R-HIRUDIN AND HEPARIN

R-Hirudin	Heparin
Monocomponent protein with single target (thrombin).	Polycomponent drug with multiple sites of action.
Thrombin-mediated amplification of coagulation is effected only under certain conditions.	Thrombin and factor Xa feedback amplification of clotting is affected. Fibrinolysis and platelet function is affected
No known interactions with endothelium other than blocking thrombin-thrombomodulation mediated activation of protein C.	Significant interactions with endothelium. Both physical and biochemical modulation of endothelial function.
Shorter half-life via i.v. route.	Short half-life via i.v. route.
Functional bioavailability is variable and dependent on the structure of r-hirudin.	Functional bioavailability is 20-30%. LMWHs are better absorbed.
Endogenous factors (PF4, FVIII) do not alter its antithrombotic action.	Marked modulation by the endogenous factors. Several factors may alter the anticoagulant actions.
Relatively inert proteins not altered by metabolic processes.	Transformed by several enzyme systems and reduces its anticoagulant actions.
Information on cellular uptake and depo formation is not presently known.	Significant cellular uptake and depo formation.

A series of tripeptide aldehydes containing arginine have been developed as the first reversible peptide thrombin inhibitors. The prototype compound to be synthesized is D-Phe-Pro-Arg-H (GYKI-14166) (43), which although is a very selective and potent inhibitor of thrombin, is very unstable in neutral aqueous solution where it cyclizes and is inactivated. In order to achieve compounds that are both stable and specific for thrombin, a series of N-alkyl derivatives have been synthesized (a basic amino terminus promotes thrombin specificity) and from this series the methyl derivative D-MePhe-Pro-Arg-H (GYKI 14766) (44) has been found to be as potent and selective reversible inhibitor of thrombin as the prototype aldehyde. One of these agents, Efegatran7 (D-MePhe-Pro-Arg-H), was developed by Eli Lily and studied briefly in clinical trials for prevention of reocclusion during interventional cardiovascular procedures.

With the availability of molecular biology techniques, it has become possible to produce pharmaceutical quantities of recombinant equivalent of hirudin, a potent antithrombin agents which as originally isolated from the medicinal leech, Hirudo medicinalis (45). This anticoagulant is a 65 amino acid protein which is much stronger in producing its anticoagulant effects than heparin. Furthermore, it does not require any endogenous factors for producing its effects. A comparison of this new anticoagulant and heparin is given on Table III.

Although r-hirudin is a stronger antithrombin agent than heparin, the thrombin generation pathways in the coagulation cascade appear to be inhibited only under certain conditions. Thus, a very high dose of r-hirudin as compared to heparin may be needed for effective antithrombotic activity as only one target site can be inhibited. By inhibiting thrombin, the bioregulatory actions of thrombin such as protein C activation, the release of t-PA and cellular function may also be inhibited.

A synthetic analogue of hirudin, namely Hirulog (Biogen), has also been developed and tested in various clinical trials (46). This agent represents a completely synthetic anticoagulant whose anticoagulant actions are comparable to heparin. However, it does not require any plasmatic factors for its anticoagulant actions. Currently, this anticoagulant is being evaluated in various indications such as in the management of unstable angina and interventional cardiology related occlusive phenomenon.

23.7.1 DEVELOPMENTAL PERSPECTIVE FOR NEW ANTITHROMBIN AGENTS

The current developmental perspectives on various thrombin inhibitors are summarized in Table II. Site-directed antithrombin agents such as recombinant hirudin, hirulog and synthetic tripeptides are currently undergoing various phases of clinical trials. These agents were initially developed with the premise that sole targeting of thrombin may be the most important step in the development of anticoagulants and antithrombotic drugs to treat thrombotic and cardiovascular disorders (47,48). It is has been over five years that these agents have undergone clinical trials. The primary focus of these trials has been in cardiovascular areas such as pretreatment of ischemic heart disorders, anticoagulation during coronary angioplasty, treatment of post-interventional abrupt closure and late reocclusion (restenosis) and adjunct usage during thrombolysis. In heparin compromised patients, these new anticoagulants may be useful on a compassionate basis.

Several peptidomimetic antithrombin agents are also currently in clinical development for both cardiovascular and therapeutic indications. These include argatroban (Mitsubishi, Japan), inogatran (Astra, Sweden) and napsagatran (Roche, Switzerland). The results of some of these clinical trials are expected later this year. In addition, many other antithrombin agents are in active preclinical development at this time. Of the various peptide derived antithrombin agents, efegatran (Eli Lilly, USA) was initially developed for cardiovascular indications. However, its development was stopped for several years. Currently, hirulog, PEG-hirudin, hirudin and argatroban are in active clinical trials.

Some of the thrombin inhibitors also possess direct vasodilatory effects, independent of their activity on thrombin. These effects are believed to be mediated through metabolic products containing guanidino groups which result in the increase of nitric oxide generation. Depending on the specific indication for which the thrombin inhibitor is developed, this vascular effect may be desirable. Nevertheless, this type of effect should be investigated and considered.

Although several reports on the experimental use of these antithrombin agents have been made available on their use in cardiovascular surgery in animal models, only isolated reports in human studies are available (49). Since a known antagonist to neutralize the effects of antithrombin agents is not available, concerns over the use of these agents have been expressed (50). Currently, no pharmacologic antagonists exist for hirudin or any of the other direct thrombin inhibitors, making the managing of precipitated hemorrhagic effects problematic. Current hirudin reversal strategies employed include prothrombin complex concentrate to neutralize the circulating antithrombin agents or dialysis to remove an antithrombin agent from the blood. Strategies that are under development of effective neutralizing agents are production of mutant thrombin molecules that are devoid of clotting activity but for which antithrombin agents retain a high affinity, thus saturating circulating levels of these agents.

Current clinical trends also point to polypharmacologic approaches for the treatment and management of cardiovascular disorders. Combination modalities of anticoagulants with other drug classes such as antiplatelet and fibrinolytic drugs has been employed. While clinically developed thrombin inhibitors are reportedly specific for thrombin inhibition, these agents exhibit interactions with other drugs. Antiplatelet drugs such as aspirin and ticlopidine have additive or synergistic actions when used in conjunction with thrombin inhibitors. Similar interactions would be expected when thrombin inhibitors would be used with heparin and LMWHs or antithrombin III concentrates. On the other hand, adjunct usage of thrombin inhibitors with thrombolytic agents, depending on the dose, could result in facilitation of thrombolysis or, if the thrombin inhibitor dose is high in thrombolytic compromise. Some of these agents may also directly inhibit activated protein C, or inhibit activated protein C formation by neutralizing thrombomodulin-bound thrombin (47). Furthermore, thrombin inhibitors may neutralize the coagulant actions of factor VIIa and prothrombin complex concentrates. Therefore, it is imperative that the interactions of thrombin inhibitors are well studied and considered when designing drug combination protocols for clinical applications, as such interactions may have both safety and efficacy implications.

23.7.2 PLASMA DERIVED INHIBITORS

Several inhibitors of plasma origin have been identified as important inhibitors of the coagulation process including antithrombin III (ATIII), protein C, TFPI and heparin cofactor II. Of these inhibitors, ATIII is the only agent which has been developed for commercial purposes. Patients with hereditary thrombophilia due to a classic deficiency of ATIII can be treated with ATIII concentrate. In addition patients with disseminated intravascular coagulation (DIC), renal insufficiency, post-surgical thrombosis, thermal injury and trauma which are associated with a decrease of ATIII in proportion to the degree of illness/injury, can also be treated (51).

Recently ATIII concentrates have been used successfully for prophylaxis using continuous infusion for the bone marrow transplant (BMT) associated toxicity (52). Organ dysfunction during BMT has also been treated with ATIII concentrate (53). In the same

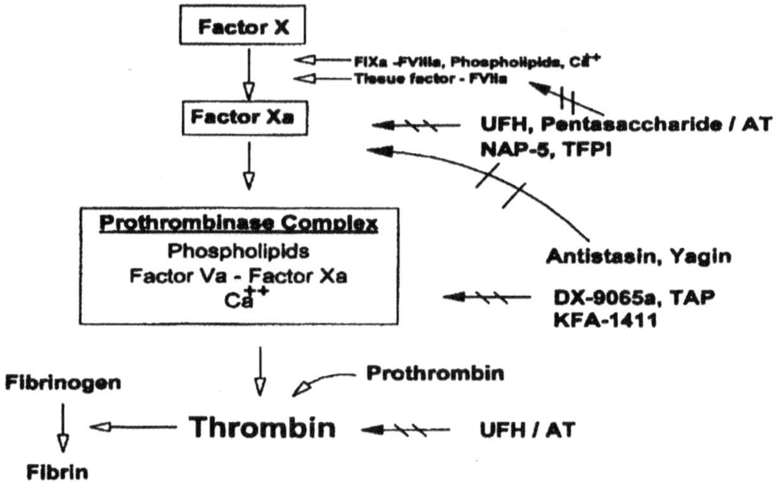

Figure 7. Factor Xa targeting in the development of new antithrombotic agents. Both the direct and indirect inhibitors of factor Xa is explored. These agents are also potent inhibitors of thrombin generation at both the plasmatic and cellular sites.

report ATIII was stated to have a major impact on both medical and economic outcomes of BMT (53). In patients with ATIII treatment hospital duration and hospital cost were significantly decreased. Protein C concentrates have only been used in pre-clinical settings. TFPI is also currently under development for clinical use.

23.7.3 FACTOR XA INHIBITORS

Currently, factor Xa targeting is one of the major focuses of drug development. The factor Xa inhibitor strategy was actually derived, in part, from heparin (Figure 7). The LMWHs have higher anti-Xa activity than anti-thrombin activity, whereas heparin has equal factor Xa and thrombin inhibitory activities. An indirect evidence of the validity of the hypothesis that factor Xa inhibition is important for the control of thrombogenesis is given by the efficacy of LMWHs which contain a large proportion of molecules with high anti-factor Xa activity (54). Pharmacologic development has now separated these two properties so that solely anti-factor Xa or anti-thrombin agents are available.

The first development of factor Xa inhibitors was met with less interest than thrombin inhibitors. These early inhibitors had low affinity to factor Xa, low selectivity and low potency (55,56). Because of the increase in enzymatic activity of the coagulation cascade once the prothrombinase complex is formed, a potent factor Xa inhibitor is required to have an extremely high affinity for the enzyme. The first factor Xa inhibitors did not fulfill this requirement.

Potent factor Xa inhibitors have several potential advantages. Factor Xa is in the common pathway of both the intrinsic and extrinsic systems, playing a central role in the coagulation pathway, so it is a logical focus of drug development for the control of thrombosis. Factor Xa is formed at an earlier stage than thrombin, and the procoagulant effect of factor Xa is strongly amplified by the prothrombinase complex. Factor Xa has no known activity other than as a procoagulant, as opposed to thrombin which has multiple activation roles at various plasmatic and cellular levels (platelets, endothelial cells, other cells), not the least of which is activation of the protein C inhibitor pathway. Inhibition of protein C by a thrombin inhibitor would lead to a reduced formation of activated protein C, an important natural anticoagulant.

Since factor Xa has relatively slow activation kinetics, as opposed to thrombin, effecting its function should result in easier management of the balance between the therapeutic and bleeding effects of a drug. From recent clinical trials thrombin inhibitors have shown to have a relatively narrow safety/efficacy margin which could lead to an overdose of the therapeutic dose with a resultant bleeding complication (57,58). Because of their different mechanism of action, factor Xa inhibitors are expected to have a better efficacy/safety profile than thrombin inhibitors.

TABLE IV

Direct Factor Xa Inhibitors

Agent	*Chemical Nature*	*Source*	*Developmental Status*
Yagin	leech protein (85 a.a.)	Animal derived	Terminated
Antistasin	eech protein (119 a.a.)	Recombinant	Terminated
TAP	Tick protein (60 a.a.)	Recombinant	Pre-clinical
NAP-5	Hookworm protein	Recombinant	Pre-clinical
TFPI	Human protein	Recombinant	Pre-clinical
DX-9065a	Propanoic acid derivative	Synthetic	Phase II clinical trials
SEL 2711	Pentapeptide	Synthetic	Pre-clinical
YM 60828	Peptidomietic	Synthetic	Pre-clinical
BX 807834	Peptidomimetic	Synthetic	Pre-clinical
KFA1411	Peptidomimetic	Synthetic	Pre-clinical

Factor Xa inhibitors are structurally diverse ranging from peptides to proteins to heparin saccharidic sequences (59,60). They can be either naturally derived, recombinant or synthetic in origin. Molecular size differs between the inhibitors, as does specificity and kinetics of factor Xa inhibition. The targeted binding site on factor Xa can differ between the inhibitors; they can be direct binding to factor Xa or indirect via a cofactor such as antithrombin (AT) III. Binding to the enzyme can be either reversible or irreversible. The protein inhibitors have limitations due to their size which limits access to factor Xa bound within clots, and, furthermore, they tend to be immunogenic, can carry viral or animal contaminants and can become limited in supply.

Because of the structural differences, the primary mechanism of action, i.e., direct or indirect acting on factor Xa as well as other attributes, differs with the various agents. The factor Xa inhibitors that are in development are shown in Table IV. Only a few of these agents are currently undergoing clinical trials.

The factor Xa inhibitors presently under clinical development are a diverse class of new antithrombotic agents with direct and indirect mechanisms of inhibition. The pentasaccharide represents a synthetic oligosaccharide which requires the endogenous cofactor ATIII for its activity. TAP is a low molecular weight peptide with direct inhibitory activity, whereas DX-9065a is a synthetic organic compound of lower molecular weight and direct inhibitory actions. DX-9065a is less specific than TAP, NAP-5, TFPI and several new agents also seem promising.

Despite differences in their mechanisms of action and in vitro activities, pentasaccharide, DX-9065a and TAP have been shown to be effective antithrombotic agents in experimental models of venous thrombosis, coronary artery occlusion, arterial thrombolysis and acute reocclusion, restenosis after angioplasty, dialysis, and DIC. Pentasaccharide has also demonstrated measurable antithrombotic effects in human trials. Both TAP and DX-9065a produce measurable in vitro anticoagulant effects. In contrast, pentasaccharide does not produce an anticoagulant effect by the typical clot based assays. Thus, with factor Xa inhibitors there is not necessarily a correlation between current lab assays and antithrombotic efficacy as there is with heparin.

Factor Xa inhibitors vary in their efficacy to inhibit factor Xa depending on molecular size, access to clot bound factor Xa, access to prothrombinase bound factor Xa, specificity and kinetics of inhibition. It is probably the inhibition of clot bound factor Xa that relates to the prolongation of the aPTT as observed with some of the factor Xa inhibitors and, therefore, the inability to inhibit clot bound factor Xa that relates to a lack of aPTT prolongation by other agents. DX-9065a with a relatively lower molecular weight may be more effective in the inhibition of clot bound factor Xa than TAP or pentasaccharide.

In contrast to thrombin inhibitors, the factor Xa inhibitors are devoid of bleeding effects even at dosages much higher than supra-therapeutic levels, which in the case of heparin are responsible for bleeding. Thus, a dissociation between the bleeding and antithrombotic responses is clearly evident for both DX-9065a and pentasaccharide.

There are potential advantages for factor Xa inhibitors over thrombin inhibitors, particularly a higher safety margin in prophylactic regimens and less frequent dosing requirements. With the development of factor Xa inhibitors it needs to be assured that the maximum benefit of factor Xa inhibitors has been achieved, be it in prophylactic or therapeutic use (i.e., in the choice of the clinical pathology, dose of drug, dosing regimen, etc. when establishing clinical trials). It is likely that each drug will have a role in specific clinical indications and that one drug will not be optimal for all thrombotic situations. How and where they are used clinically, and how they will compete with LMW heparins and direct thrombin inhibitors remains to be determined.

The limited data available on factor Xa inhibitors is favorable and thus warrants additional investigations to demonstrate the relative inhibitory profile of these agents in thrombotic and cardiovascular indications. The future holds promise for finding effective factor Xa inhibitors and their challenge to the widely accepted thrombin hypothesis.

23.8 ANTIPLATELET DRUGS IN DEVELOPMENT

Antiplatelet drugs such as aspirin and propionic acid derivatives have been conventionally used to manage platelet mediated thrombotic disorders such as peripheral arterial occlusive disorders thrombotic stroke and coronary artery diseases. Newer generation antiplatelet drugs such as ticlopidine and clopidogrel have been introduced for the management of stroke (61). More recently, these two antiplatelet drugs have been tested for other indications. While these drugs are useful, they only inhibit certain activation processes. None of these agents are capable of producing the inhibition of tissue factor and thrombin mediated activation of platelets. Direct thrombin inhibitors, such as hirudin (36), may be useful in targeting tissue factor and coagulation activation processes where platelet

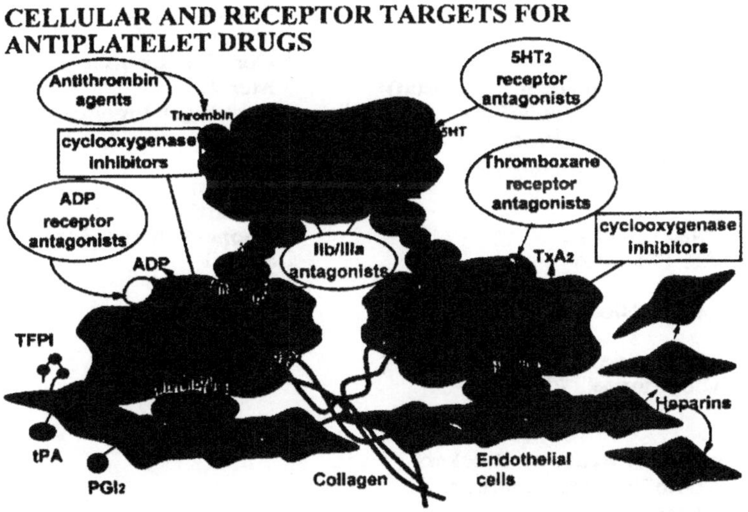

Figure 8. Pharmacologic target sites of the antiplatelet actions of various anticoagulant and antithrombotic drugs. Receptors enzymes and other membrane sites are targeted by these drugs.

activation is mediated by thrombin. It is therefore clear that different sites of activation are independently responsible for activating platelets in the mediation of thrombotic events related to platelets. Knowing this, glycoprotein IIb/IIIa (GpIIb/IIIa) antagonists, thromboxane A_2 receptor blockers, thromboxane synthase inhibitors, synthetic cyclooxygenase inhibitors, prostacyclin analogues and 5-HT_2 receptor blockers have been developed. Ticlopidine and clopidogrel represent ADP receptor antagonists and have been approved in the US for aspirin compromised patients. However, their use in interventional cardiology has expended.

Figure 8 shows a diagrammatic illustration of the sites of action of various anticoagulant and antithrombotic drugs on platelets. As can be seen, different drugs have different sites of action on the platelet activation process. Platelet receptor modulators, (GPIIb/IIIa, TxB2, ADP), cyclooxygenase inhibition and inhibition of the thrombin mediated function of platelets are some of these sites.

The GPIIb/IIIa receptor targeting drugs are claimed to act on the final common pathways for platelet activation by various activating processes (62). The interaction of fibrinogen with GPIIb/IIIa sites is partly mediated by the arginine-glycine-aspartic acid (RGD) recognition site (63). Several synthetic peptides and peptidomimetic agents have been developed to target this site in order to prevent the activation of platelets (64). Monoclonal antibodies, such as 7E3 (ReoPro7) (65) have been clinically tested. ReoPro7 has recently

TABLE V

ANTIPLATELET DRUGS IN DEVELOPMENT

A. Glycoprotein IIb/IIIa Receptor Antagonists

1. 7E3 (abcixima, ReoPro)	Centocor/Lilly
2. YM 337	Yamanouchi
3. Integrelin	Cor Therapeutics
4. MK-383 (tirofiban, Aggrastat)	Merck
5. IBU-104XX (fradafian)	Boehrnger Ingelheim
6. SC-54684A (xemlofian)	Searle
7. Ro-44-9883 (lamofiban)	Roche
8. DMP-757	DuPont
9. GPI-562	Novartis
10. RPR-110173	Rhone-Poulenc Rorer
11. SR 121566A	Sanofi

B. ADP Receptor Blockers

1. Ticlopidine, Ticlid	Sanofi
2. Clopidogrel, Plavix	Sanofi

C. Thromboxane A Receptor Blockers
D. ThromboxaneSynthase Inhibitors
E. Cyclooxygenase Inhibitors
F. Prostaglandin Analogues
G. Serotonin Receptor Blockers

been approved by the US FDA for the prevention of post-PTCA restenosis. More recently, a fragment of a humanized monoclonal antibody against the GpIIb/IIIa site has also been identified (YM-337). The use of monoclonal antibodies as therapeutic agents may have some limitations such as the immunogenicity, lack of reversibility and non-availability by subcutaneous or oral route. Thus, the development of analogues of RGD or modification of the RGD sequence is considered more practical, safe and cost-effective.

Many of the drugs listed on Table V are currently undergoing clinical trials for various indications. Abciximab®, Integrelin® and Aggrastat® have been approved for various indications in cardiovascular and the pharmacological data on each of these drugs is primarily obtained in experimental settings where the effect of these agents are solely investigated in platelet function. Non-platelet mediated effects on cells have not been taken into account in the clinical development of these agents. It is conceivable that some of these effects may also contribute to their overall clinical effects. Furthermore, very little is known on the drug interactions of these agents. For proper development of these agents, valid preclinical pharmacologic data and information on drug interactions may be crucial.

23.8.1 DEFIBROTIDE AND RELATED AGENTS

Aptamers are oligonucleotides (double or single stranded DNA, or single stranded RNA) which act directly on proteins to inhibit disease processes. Thirty two such aptamers have

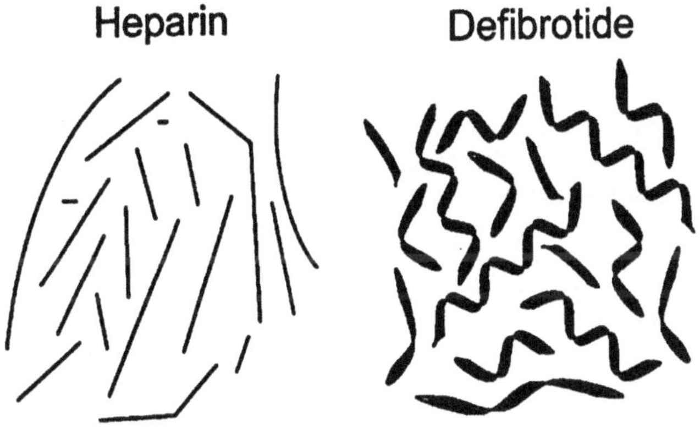

Figure 9. A comparison of the molecular heterogeneity in unfractionated heparin and defibrotide. Both drugs are polycomponent and exhibit an apparent mean molecular weight of 15 KDA. While heparin is composed of sulfated mucopolysaccharide chains, defibrotide is composed of single stranded nucleic acid (DNA fragments). Both drugs produce polypharmacological effects and are capable of targeting multiple sites in blood and vasculature.

been recently isolated as inhibitors of thrombin with binding affinities in the range of 20-200 nM (66). One of the most potent thrombin aptamers has been found to interact with thrombin's anion binding exosite, so that it competes with substrates that interact with that specific site, such as fibrinogen and thrombin platelet receptors (67,68). Another recent development in the area of oligonucleotide inhibitors of thrombin has been the isolation of two RNAs that bind thrombin with high affinity (Kd in the nM range). These oligonucleotides have been shown to inhibit fibrinogen clotting in an *in vitro* test (69).

Defibrotide is a polydeoxyribonucleotide derived drug of mammalian origin. Like heparin, this agent is also heterogeneous and is composed of nucleic acid strands ranging in molecular weight from 2,000 to 30,000 (70). The mean molecular weight of this agent is around 17,000. This product is obtained from mammalian tissues such as lung and spleen. Figure 9 shows a comparison of the heterogeneous nature of defibrotide and heparin. Like heparin, this agent is composed of several molecular components, whose molecular weight ranges from 2-50 KDA.

Despite the fact that defibrotide does not produce any systemic anticoagulant effects after intravenous injection, this agent is found to produce dose and time dependent antithrombotic effects in a stasis thrombosis model (70). Defibrotide has been found to contain nucleotide sequences capable of producing direct antithrombin actions (71). The antithrombotic mechanism by which defibrotide mediates its actions include endothelial modulation, increase in cAMP and TFPI release. Most interestingly, this agent produces its antithrombotic action without producing any systemic anticoagulant effects. It has therefore been used for the management of peripheral vascular disease and different microangiopathic disorders.

23.9 SYNOPSIS

The current management of thrombotic and vascular disorders has been strongly influenced by the introduction of several new drugs and the optimal use of physical devices (72-78). One of the major impacts in the management of thrombotic disorders has been the LMWHs. Not only are these agents being used for prophylaxis, but recent studies have shown their efficacy in the treatment of established thromboembolic disorders. Other developments in the clinical use of LMWHs include the management of coronary syndromes, in pregnancy, pediatrics, transplant associated vasculopathy and in cancer. More recently these drugs are compared with antithrombin agents for their efficacy in various indications. A recent report compared a recombinant hirudin preparation (Revasc®) with LMWH, Enoxaparin, for the prevention of DVT in post orthopedic surgical patients (72). The authors concluded that when administered 30 minutes before total hip replacement, Revasc was more efficacious than Enoxaparin.

Dermatan sulfate and heparan sulfate have provided disappointing results to date. The synthetic heparin pentasaccharide, which acts strictly via the inhibition of factor Xa, has been successfully used in angioplasty and is presently undergoing clinical trial for the prophylaxis of venous thrombosis in orthopedic surgical patients. Various large scale clinical trials have been organized. In addition, several derivatives of oligosaccharides are also developed with multiple protease inhibitory action.

Although synthetic and recombinant thrombin and factor Xa inhibitors have been widely studied as adjuncts to thrombolytic therapy, in the management of unstable angina and other cardiovascular indications, for safety reasons their development has been slowed.
Despite this, thrombin inhibitors have shown promise in the prevention of post-surgical venous thrombosis in orthopedic patients and they offer a useful anticoagulant for heparin compromised patients. Thus, these agents are being evaluated for the anticoagulant management of patients with heparin induced thrombocytopenia. Recombinant hirudin and a synthetic peptidomimetic drug, argatroban are in an extensive developmental phase for HIT patients for anticoagulation of thrombosis, for anticoagulation during cardiologic

interventional procedures and as the anticoagulant in cardiac surgery. The initial clinical trials with other synthetic thrombin inhibitors such as efegatran and inogatran have been rather disappointing. Similarly the development of oral thrombin inhibitors has been rather unsuccessful.

An aggressive developmental approach is in the antiplatelet drug area. In particular, numerous glycoprotein (GP) IIb/IIIa inhibitors are currently under investigation. The GP IIb/IIIa targeting antibody, ReoPro, has been approved for limited indications such as the management of high risk patients for angioplasty. Numerous synthetic GP IIb/IIIa inhibitors are being developed for both intravenous and oral use. However, marked variations in absorption and safety considerations represent important developmental issues which require further investigations. While these drugs may be useful in arterial thrombosis, their value in venous thrombosis is not yet proven. In addition, to the GP IIb/IIIa inhibitors, ADP receptor blockers such as ticlopidine and clopidogrel have provided impressive results in the management of thrombotic events during interventional cardiovascular procedures. The CAPRIE trial for the management of stroke and other arterial events has shown a superior effect of clopidogrel in comparison to aspirin in the overall outcome of patients where combined events are considered.

Combination modalities for the mangement of arterial and venous thrombosis are currently being explored. While these modalities provide an increase in the efficacy, safety considerations are of crucial importance. Aspirin is commonly used with various anticoagulants. Many other antiplatelet drugs, such as ticlopidine, are also being combined with heparin, hirudin, GP IIb/IIIa inhibitors and aspirin itself.

Several other approaches to antithrombotic drug development are in both pre-clinical and clinical investigations with only limited data available at this time. These include factor Xa and factor VII inhibitors, thrombomodulin, activated protein C, tissue factor pathway inhibitor (TFPI), PAI inhibitors and protease annexin.

The development of these newer drugs and the advanced understanding of the pathogenesis of cadiovascular disorders have provided some major challenges at both the clinical and basic science levels, not only to monitor but to understand the site of action of these drugs. There is a need for tests to study the endogenous level and their relevance to the modulatory effects of the pathophysiologic process. A recent review to address this issue in the pathogenesis of acute coronary syndrome has discussed the use of various markers of pathophysiologic activation and vascular modulation in this catastrophic condition (79). Such drugs as LMWH, hirudin and GP IIb/IIIa inhibitors have been found to exhibit various clinical effects in this syndrome. However, on the basis of their circulating levels or the pharmacodynamic effects not all of the clinical and toxic effects of these agents can be explained. Thus, each of these drugs is projected to produce multiple effects at plasmatic, cellular and target organ sites. The availability of highly sensitive methods to determine the multiple sites of action provides a unique opportunity to not only understand the mechanism of action by which these drugs work, but additional knowledge of the pathogenesis of various diseases can be readily gained.

The European community and Canada has been rather fast in expediting the regulatory process for the clinical use of new anticoagulant and antithrombotic drugs. Table VI shows a list of the newer drugs which are available for clinical use in the USA. Currently, their are three LMWHs approved for DVT prophylaxis. On of these LMWHs, namely Enoxaparin, is also approved for unstable angina. The use of these agents for the treatment of established thrombosis is approved in the European countries, Canada and is currently under consideration by the US FDA. One antithrombin agent, namely Refludan® (hirudin) has been approved for therapeutic anticoagulation in patients who are not

TABLE VI

NEWER ANTICOAGULANTS AND ANTITHROMBOTIC DRUGS FOR CLINICAL USE IN THE USA

Agents	Use
1. Low Molecular Weight Heparins	
a. Ardeparin	DVT Prophylaxis
b. Dalteparin*	DVT Prophylaxis
c. Enoxaparin*	DVT Prophylaxis, Acute Coronary Syndrome
2. Antithrombin Agents	
a. Refludan	Anticoagulation in HIT Syndrome
3. Antiplatelet Drugs	
a. Clopidogrel	Aspirin Substitute
b. ReoPro	High Risk Angioplasty
c. Integrelin	Unstable Angina
d. Aggrastat	Unstable Angina

* These agents are approved for therapeutic use in Canada and the European community.

treatable with heparin. A new ADP antagonist namely, clopidogrel (Plavix®) has been approved by the US FDA and the EMEA. Presently, two additional IIb/IIIa antagonists Integrelin and Aggrastat have been approved. It is expected that the forthcoming recommendations of the consensus conference will point to the use of some of these agents in specific indications. Newer indications for these agents have been continually developed and an expanded use of many of the newly approved drugs will be seen in the coming years.

Curricular updates at both the medical school and post graduate level, inclusion of substudies and training of allied health personnel in the development of these drugs will be of vital importance in the overall exceptance of the newer drugs for the management of thrombotic and cardiovascular disorders.

References

1. Fareed J, Walenga JM, Pifarre R: Newer approaches to the pharmacologic management of acute myocardial infarction. *Cardiac Surgery: State of the Art Reviews* (1992) 6(1):101-111.

2. Fareed J, Bacher P, Messmore HL, Walenga JM, Hoppensteadt DA, Strano A, Pifarre R: Pharmacological modulation of fibrinolysis by antithrombotic and cardiovascular drugs. *Prog Cardiovas Dis* (1992) **6**:379-398.

3. Fareed J, Walenga JM, Hoppensteadt D, Racanelli A, Coyne E: Chemical and biochemical heterogeneity in low molecular weight heparins: Implications for clinical use and standardization. *Semin Thromb Hemost* (1989) **15**:440-463.

4. Hull R, Raskob G, Pineo G: Subcutaneous low molecular weight heparin compared with continuous intravenous heparin in the treatment of proximal vein thrombosis. *N Eng J Med* (1992) **326**(15):975-982.

5. Prandoni P, Lensing A, Buller H, Carta M, Cogo A, Vigo M, Casara D, Ruol A, ten Cate JW: Comparison of subcutaneous standard low-molecular weight heparin with intravenous standard heparin in proximal deep-vein thrombosis. *Lancet* (1992) **339**:441-445.

6. Hirsh J, Levine M: The development of low molecular weight heparins for clinical use. In Verstraete M, et al. (Eds) *Thrombosis and Haemostasis*, pp. 425-448, Leuven University Press, Leuven, 1987

7. Hull R, Delmore T, Carter C, Hirsh J, Genton E, Gent M, Turpie G, McLaughlin D: Adjusted subcutaneous heparin versus warfarin sodium in the long-term treatment of venous thrombosis. *N Eng J Med* (1983) **306**:954-958.

8. Kelton JG: Heparin induced thrombocytopenia. *Haemostasis* (1986) **16**:173-186.

9. Silver D, Kapsch D, Tosi E: Heparin induced thrombocytopenia, thrombosis and haemorrhage. *Ann Surg* (1983) **198**:301-305.

10. Griffith GC, Nichols G Jr, Asher JD, Flanagan B: Heparin osteoporosis. *J Am Med Assoc* (1965) **193**:85-88.

11. Bara L, Billaud E, Gramond G, Kher A, Samama M: Comparative pharmacokinetics of low molecular weight heparin (PK 10169) and unfractionated heparin after intravenous and subcutaneous administration. *Thromb Res* (1985) **39**:631-636.

12. Bergqvist D, M@tzsch T, Burmark US, Frisell J, Guilbaud O, Hallb``k J, Horn A, Lindhagen A, LjungNr H, Ljungstr`m KG, Onarheim H, Risberg B, T`rngren S, _rtenwall P: Low molecular weight heparin given the evening before surgery compared with conventional low-dose heparin in prevention of thrombosis. *Br J Surg* (1988) **75**(9):888-891.

13. Borris LC, Hauch O, Jorgensen LN, Lassen MR: Enoxaparin versus dextran 70 in the prevention of post-operative deep vein thrombosis after total hip replacement. A Danish multicenter study. *Proceeding of the Danish Enoxaparin Symposium*, Feb 3, 1990

14. European Fraxiparin Study Group: Comparison of low molecular weight heparin and unfractionated heparin for the prevention of deep vein thrombosis is patients undergoing abdominal surgery. *Br J Surg* (1988) **75**:1058-1063.

15. Haas S, Stemberger A, Fritsche HM, Welzel D, Wolf H, Lechner F, Blumel G: Prophylaxis of deep vein thrombosis in high risk patients undergoing total hip replacement with low molecular weight heparin plus dihydroergotamine. *Arzneimittel-Forschung* (1987) **37**(7):839-843.

16. Baumgartner A, Jacot N, Moser G, Krahenbuhl B: Prevention of postoperative deep vein thrombosis by one daily injection of low molecular weight heparin and dihydroergotamine. *Vasa* (1989) **18**(2):152-156.

17. Caen JP: A randomized double-blind study between a low molecular weight heparin Kabi 2165 and standard heparin in the prevention of deep vein thrombosis in general surgery. A French multicenter trial. *Thromb Haemost* (1988) **59**(2):216-220.

18. Hartl P, Brucke P, Dienstl E, Vinazzer H: Prophylaxis of thromboembolism in general surgery Comparison between standard heparin and Fragmin. *Thromb Res* (1990) **57**(4):577-584.

19. Eriksson BI, Zachrisson BE, Teger-Nilsson AC, Risherg B: Thrombosis prophylaxis with low molecular weight heparin in total hip replacement. *Br J Surg* (1988) **75**(11):1053-1057.

20. Leyvraz PF, Richard J, Bachmann F, Van Melle G, Treyvaud JM, Livio JJ, Candardjis G: Adjusted versus fixed subcutaneous heparin in the prevention of deep vein thrombosis after total hip replacement. *N Eng J Med* (1983) **309**:954-958.

21. Matzsch T, Bergqvist D, Fredin H, Hedner U: Safety and efficacy of a low molecular weight heparin (Logiparin) versus dextran as prophylaxis against thrombosis after total hip replacement. *Acta Chirur Scand* (1988) **543**:80-84.

22. Kakkar VV, Murray WJG: Efficacy and safety of low molecular weight heparin (CY 216) in preventing postoperative venous thrombo-embolism A cooperative study. *Br J Surg* (1985) **72**:786-791.

23. Levine MN, Hirsh J: An overview of clinical trials with low molecular weight heparin fractions. *Acta Chirur Scand* (1988) **154**(543):73-39.

24. Koppenhagen K, Adolf J, Matthes M, Troster E, Roder JD, Haas S, Fritsche HM, Wolf H: Low molecular weight heparin and prevention of postoperative thrombosis in abdominal surgery. *Thromb Haemost* (1992) **67**(6):627-630.

25. Hanke H, Oberhoff M, Hanke S, Hassenstein S, Kamenz J, Schmid KM, Betz E, Karsch KR: Inhibition of cellular proliferation after experimental balloon angioplasty by low molecular weight heparin. *Circulation* (1992) **85**:1548-1556.

26. Hoppensteadt D, Racanelli A, Walenga JM, Fareed J: Comparative antithrombotic and hemorrhagic effects of dermatan sulfate, heparan sulfate and heparin. *Semin Thromb Hemost* (1989) **15**:378-385.

27. Marcum JA, Rosenberg RD: Role of endothelial cell surface heparin-like polysaccharides. In Heparin and related polysaccharides. Structure and activities. Eds. Ofosu FA, Danishefsky I, Hirsh J. *The New York Academy of Sciences*, NY 1989, pp. 81-94

28. Sirtori CR: Pharmacology of sulfomucopolysaccharides in atherosclerosis prevention and treatment. In Ricci G, et al, Eds. Selectivity and Risk-Benefit Assessment of Hyperlipidemia Drugs. Raven Press, New York, pp 189-94, 1982

29. Fareed J, Hoppensteadt D, Jeske W, Walenga JM: An overview of non-heparin glycosaminoglycans as antithrombotic agents. In *Blood Coagulation* (1993). Poller L (ed), Churchill Livingstone, New York.

30. Agrati AM, DeBartolo G, Palmieri G: Heparan sulfate Efficacy and safety in patients with chronic venous insufficiency. *Minerva Cardioangiologica* (1991) **39**(10):395-400.

31. Caramelli L, Mirchioni R, Carini: Effectiveness of short-term sulodexide treatment on peripheral vascular disease clinical manifestations. *Rivista Europea per le Scienze Mediche e Farmacologiche* (1988) **10**(1):55-58.

32. Bergqvist D, Kettunen K, Fredin H, Fauno P, Suomalainen O, Soimaakallio S, Karjalainen P, Cederholm C, Jensen LJ, Justesen T, Stiekema JCJ: Thromboprophylaxis in patients with hip fractures A prospective, randomized, comparative study between Org 10172 and dextran 70. *Surgery* (1991) **109**(5):617-622.

33. Romeo S, Grasso A, Costanzo C: A controlled clinical experiment "within subjects" with heparan sulfate in intermittent claudication. *Minerva Cardioangiol* (1991) **39**(9):345-352. 992) **339**(8789):3340335.

34. Lane DA, Ryan K, Ireland H, Curtis JR, Nurmohamed MT, Krediet RT, Roggekamp MC. Stevens P, ten Cate JW: Dermatan sulphate in haemodialysis. *Lancet* (1thrombocytopenia and thrombosis. *Australian & New Zealand J Med* (1991) **21**(1):52-54.

35. Cofrancesco E, Boschetti C, Leonardi P, Cortellaro M: Dermatan sulphate in acute leukaemia. *Lancet* (1992) **339**(8802):1177-1178.

36. Fareed J, Walenga JM, Iyer L, Hoppensteadt D, Pifarre R: An objective perspective on recombinant hirudin A new anticoagulant and antithrombotic agent. *Blood Coag Fibrino* (1991) **2**:135-147.

37. Hijikata-Okunomiya A, Okamoto S: A strategy for a rational approach to designing synthetic selective inhibitors. *Sem Thromb Hemost* (1992) **18**(1):135-149.

38. Tamao Y, Yamamoto T, Hirata T, Kinugasa M, Kimumoto M: Effect of argipidine (MD-805) on blood coagulation. *Jpn Pharmacol Ther* (1986) **14**:869-874.

39. Maruyama I: Synthetic anticoagulants. *Jpn J Clin Hematol* (1990) **31**:776-781.

40. Kumon K, Tanaka K, Nakajima N, Naito Y, Fijuta T: Anticoagulation with a synthetic thrombin inhibitor after cardiovascular surgery and for treatment of disseminated intravascular coagulation. *Crit Care Med* (1984) **12**:1039-1043.

41. Matsuo T, Kario K, Kodama K, Okamoto S: Clinical applications of the synthetic thrombin inhibitor, Argatroban (MD-805). *Sem Thromb Hemost* (1992) **18**(2):155-160.

42. Teger-Nilsson A, Eriksson U, Gustafsson D, Bylund R, Fager G, Held P: Phase I studies on Inogatran, a new selective thrombin inhibitor. *J Am Col Cardiol* (1995) 117A-118A.

43. Bajusz S, Barab<s E, SzJll E, Bagdy D: Peptide aldehyde inhibitors of the fibrinogen-thrombin reaction. InPeptides - Chemistry, Structure and Biology. Meienhofer J, ed. Ann Arbor Sci. Publ. Inc.m Ann Arbor, MI. 1975 pp603-8.

44. Bajusz S, SzJll E, Bagdy D, Barabas E, Horvath E, Dioszegi M, Fittler Z, Szabo G, Juhasz A, Tomori E, Szilagyi G: Highly active and selective anticoagulants D-Phe-Pro-Arg-H, a free tripeptide aldehyde prone to spontaneous inactivation, and its stable N-methyl derivative, D-MePhe-Pro-Arg-H. *J Med Chem* (1990) **33**:1729-35.

45. Markwardt F, Fink G, Kaiser B, Klocking HP, Nowak G, Richter M, Sturzebecher J: Pharmacological survey of recombinant hirudin. *Pharmazie* (1988) **43**:202-207.

46. Maraganore JM, Bourdon P, Jablonski J, Ramachandran KL, Fenton JW 2nd: Design and characterization of hirulogs A novel class of bivalent peptide inhibitors of thrombin. *Biochemistry* (1990) **29**:7095-7101.

47. Iyer L, Fareed J: Recombinant hirudin: a perspective. *Exp Opin Invest Drugs* (1996) **5**(5):469-494.

48. Topol EJ: Novel antithrombotic approaches to coronary artery disease. *Am J Cardiol* (1995) **76**(6):27B-33B.

49. Riess FC, Lower C, Seelig C, Bleese N, Kormann J, Muller-Berghaus G, Potzsch B: Recombinant hirudin as a new anticoagulant during cardiac operations instead of heparin: successful for aortic valve replacement in man. *J Thorac Cardiovasc Surg* (1995) **110**:265-267.

50. Edmunds LH: HIT, HITT and desulfatohirudin: look before you leap. *J Thorac Cardiovasc Surg* (1995) **110**(1):1-3.

51. Kowal-Vern A, Gamelli RL, Walenga JM, Hoppensteadt D, Sharp-Pucci M, Schumacher HR: The effect of burn wound size on hemostasis: a correlation of the hemostatic changes to the clinical state. *Journal of Trauma* (1992) **33**(1):50--57.

52. Morris, CL, Lutes R, Gruppo RA, Hashmi R, Harris R, Sambrano J, Morris JD: Prophylactic continuous infusion (CI) antithrombin III (AT-III) for prevention of regimen related toxicity (RRT) following bone marrow transplantation (BMT). *Blood* (1996) **88**(10)(Suppl 1):117a.

53. Haire WD, Stephens LC, Ruby EI: Antithrombin III (AT3) treatment of organ dysfunction during bone marrow transplantation (BMT) - results of a pilot study. *Blood* (1996) **88**(10)(Suppl 1):456a.

54. Clagett GP, Anderson Jr FA, Heit J, Levine MN, Wheeler HB. Prevention of venous thromboembolism. *Chest* (1995) **108**(4)(Suppl):312S-334S.

55. Sthrzebecher J, Sthrzebecher U, Vieweg H, Wagner G, Hauptmann J, Markwardt F. Synthetic inhibitors of bovine factor Xa and thrombin. Comparison of their anticoagulant efficiency. *Thromb Res* (1989) **54**:245-252.

56. Hauptmann J, Kaiser B, Nowak G, Sthrzebecher J, Markwardt F. Comparison of the anticoagulant and antithrombotic effects of synthetic thrombin and factor Xa inhibitors. *Thromb Haemost* (1990) **63**:220-223.

57. The Global Use of Strategies to Open Occluded Coronary Arteries (GUSTO) IIb Investigators. A comparison of recombinant hirudin with heparin for the treatment of acute coronary syndromes. *N Engl J Med* (1996) **335**(11):775-782.

58. Antman EM for the TIMI 9B Investigators. Hirudin in acute myocardial infarction. Thrombolysis and thrombin inhibition in myocardial infarction (TIMI) 9B trial. *Circulation* (1996) **94**:911-921.

59. Kaiser B, Hauptmann J. Factor Xa inhibitors as novel antithrombotic agents: Facts and perspectives. *Cardiovascular Drug Reviews* (1994) **12**(3):225-236.

60. Kaiser B. Factor Xa versus factor IIa inhibitors. *Clin Appl Thrombosis/Hemostasis* (1997) **3**(1):16-24.

61. Saltiel E, Ward A: Ticlopidinea review of its pharmacodynamic and pharmacokinetic properties and therapeutic efficacy in platelet-dependent disease states. *Drugs* (1987) **34**:222-262.

62. Pytele R, Pierschbacher MS, Ginsberg MH, Plow EF, Ruoslathi E: Platelet membrane glycoprotein Iib/IIIamember of a family od RGD specific adhesion receptors. *Science* (1986) **231**:1559-1562.

63. Philips DR, Charo IF, Scarborough RM: GPIIb-IIIathe responsive integrin. *Cell* (1991) **65**:359-362.

64. Ruggeri ZM, Houghton RA, Russel SR, Zimmerman TS: Inhibition of platelet function with synthetic peptide designed to be high affinity antagonist of fibrinogen binding to platelets. *Proc Natl Acad Sci USA* (1986) **83**:5708-5712.

65. Mickelson JK, Simpson PJ, Cronin M, Homeister JW, Laywell E, Kitzen J, Lucchesi BR: Antiplatelet antibody [7E3 $f(ab')_2$] prevents rethrombosis after recombinant tissue-type plasminogen activator-induced coronary artery thrombolysis in a canine model. *Circulation* (1990) **81**:617-627.

66. Bock LC, Griffin LC, Latham JA, Vermaas EH, Toole JJ: Selection of single-stranded DNA molecules that bind and inhibit human thrombin. *Nature* (1992) **355**(6360):564-6.

67. Macaya RF, Schultze P, Smith FW, Roe JA, Feigon J: Thrombin-binding DNA aptamer forms a unimolecular quadruplex structure in solution. *Proc Natl Acad Sci USA* (1993) **90**:3745-3749.

68. Paborsky LR, McCurdy SN, Griffin LC, Toole JJ, Leung LLK: The single-stranded DNA aptamer binding-site of human thrombin. *J Biol Chem* (1993) **268**:20808.

69. Kubik MF, Stephens AW, Schneider D, Marlar R, Tasset D: High-affinity RNA ligands to human "-thrombin. *Nucleic Acids Research* (1994) **22**(13):2619-2626.

70. Fareed J, Walenga JM, Hoppensteadt DA, Kumar A, Ulutin O, Cornelli U: Pharmacologic profiling of defibrotide in experimental models. *Semin Thromb Hemost* (1988) **14**:27-37.

71. Bracht F, Schror K: Isolation and identification of aptamers from defibrotide that act as thrombin antagonists in vitro. *Biochem Biophys Res Com* (1994) 200(2):933-937.

72. Eriksson BI, Wille-Jorgensen P, Kalebo P, Mouret P, Rosencher N, Bosch P, Baur M, Ekman S, Bach D, Lindratt S, Close P: A comparison of recombinant hirudin with a low molecular weight heparin to prevent thromboembolic complications after total hip replacement. *N Engl J Med* (1997) **337**:1329-1335.

73. Kaiser B, Callas D, Walenga JM, Fareed J: Synthetic and recombinant antithrombin drugs. *Exp Opin Invest Drugs* (1998) 7(6):963-985.

74. Kaiser . Hoppensteadt D, Fareed J: Recombinant TFPI and variants: potential implications in the treatment of cardiovascular disorders. *Exp Opin Invest Drugs* (1998) 7(7):1-20.

75. Gewirtz AM, Sokol DL, Ratajczak MZ: Nucleic acid therapeutics: State of the art and future prospects. *Blood* (1998) **92**:712-736.

76. Richardson P, Elias AD, Krishnan A, Wheeler C, Nath R, Hoppensteadt D, Kinchla NM, Neuberg D,Waller EK, Antin JH, Soiffer R, Vredenburgh J, Lill M, Woolfrey AE, earman SI, Iacobelli M, Fareed J,Guinan EC: Treatment of severe veno-occlusive disease with defibrotide: Compassionate use results in response without significant toxicity in a high-risk population. *Blood* (1998) **92**:737-744.

77. Zachaarski LR, Ornstein DL: Heparin and cancer. *Hemostasis Thromb* (1998) **80**:10-23.

78. Fareed J, Callas D, Hoppensteadt DA, Walenga JM, Bick RL: Antithrombin agents as anticoagulants and antithrombotics. Implications in drug development. *Medical Clinics of North America* (1998) **82**(3):569-586.

79. Fareed J, Hoppensteadt DA, Leya F, Iqbal O, Wolf H, Bick RL: Useful laboratory tests for studying thrombogenesis in acute cardiac syndromes. *Clinical Chemistry* (1998) **44**:8(B):1845-1853.

24

RANDOMIZED TRIALS OF ANTIPLATELET THERAPY

Cathie LM Sudlow, Colin N Baigent
Clinical Trial Service Unit
and Epidemiological Studies Unit
University of Oxford
Radcliffe Infirmary
Oxford OX2 6HE
United Kingdom

24.1. INTRODUCTION

Vascular disease is of major importance worldwide as a cause both of death and disability, mainly in middle-aged and elderly individuals. Ischemic heart disease and stroke are the two leading causes of death in the developing world, as well as in the developed world, and accounted for around one fifth of the 50 million deaths worldwide in 1990 (1). Cardiovascular disorders caused almost 10% of all disability worldwide in 1990 (2), and projections suggest that by the year 2020 this will have risen to 15%, with almost all of the increase occurring in developing countries (3).

The main global priorities for reducing the increasing burden of vascular disease are primary prevention

Cathie Sudlow qualified in medicine from Oxford University in the UK in 1991, and is training in epidemiology and neurology. She recently completed a masters degree in epidemiology. Her main research interest is in cardiovascular epidemiology, particularly stroke, and she is currently a wellcome Research Fellow in the vascular overview group in the Clinical Trial Service Unit in Oxford.

Colin Baigent is a UK Medical Research Council Scientist and coordinator of the Antithrombotic Trialists' Collaboration. His interests include the treatment and prevention of cardiovascular disease and the causes of such disease in chronic renal failure.

strategies, such as reducing cigarette consumption, reducing blood pressure, and the promotion of healthy diets (4,5). The role of pharmacological agents, however, is to treat or prevent vascular disease among those patients at increased risk of a serious vascular event. Because experimental studies show that platelet activation, adhesion and aggregation are fundamental components of thrombus formation during vascular occlusion, antiplatelet agents might be expected to be effective in the treatment and prevention of arterial and venous occlusive disease.

THE NEED FOR LARGE-SCALE RANDOMIZED EVIDENCE

The overwhelming majority of recent improvements in outcome in cardiovascular diseases have been only *moderate* in size. Relative risk reductions of only about 10 or 20% (typically generating absolute differences in a major outcome of a few percent at most) are therefore the most that should generally be anticipated. To detect such moderate-sized differences reliably, both moderate biases and moderate random errors must be excluded. This requires large-scale randomized evidence, which may take the form of a single mega-trial, or, as for antiplatelet therapy, a meta-analysis of all the available randomized trials (6). In the latter, data from similar trials are combined using standard statistical methods (7) to produce unbiased estimates of the effects of treatment on major vascular outcomes. Whilst some trials may be too small on their own to demonstrate a moderate effect of treatment, the increased statistical power attained by considering all similar trials together means that a moderate effect is more likely to be detected. This helps to ensure that effective interventions are identified and that those that are ineffective or hazardous are discarded.

DATA SOURCES

This chapter reviews all of the currently available randomized evidence for the effects of antiplatelet therapy in the treatment and prevention of major vascular outcomes, as summarised by the Antiplatelet Trialists' (APT) Collaboration in 1994 (8-10). Where appropriate, this meta-analysis is supplemented by data from more recent trials. For the purposes of this chapter, antiplatelet agents are those whose primary effects on the vascular system arise from inhibition of platelet adhesion, platelet aggregation, or both. They therefore include cyclo-oxygenase inhibitors (such as aspirin), inhibitors of thromboxane receptors or of thromboxane synthetase (such as ridogrel), platelet cyclic AMP phosphodiesterase inhibitors (such as dipyridamole), ADP receptor blockers (such as clopidogrel), and glycoprotein IIb/IIIa receptor antagonists (such as abciximab), but exclude agents with a major vasodilating action (such as epoprostenol, oxpentyfilline and ketanserin) and agents with a major anticoagulant action (8). Our focus in this chapter will be on the evidence from trials of any antiplatelet regimen versus control, or of one antiplatelet regimen versus another. Table 27.1 summarises the currently available randomized evidence, and is divided into (a) those trials available in March 1990 and contributing to the 1994 APT overview, and (b) the main trials subsequently becoming available, which are to be included in an updated meta-analysis by the Antithrombotic Trialists' (ATT) Collaboration.

The main outcome in the APT overview was the occurrence of a serious vascular event (i.e. non-fatal myocardial infarction (MI), non-fatal stroke or vascular death). Where appropriate, information on major bleeding, deep venous thrombosis, pulmonary embolism and arterial or graft occlusion was also sought.

STATISTICAL METHODS

A fuller description of the methods used in the APT overview can be found elsewhere (8), but Figures 27.1 and 27.2 respectively summarise the derivation of a proportional odds reduction and corresponding absolute risk reduction. In Figure 27.1 the overall 'typical odds ratio' for patients with a prior MI is the ratio of the odds of a serious vascular event among patients allocated antiplatelet treatment to the corresponding odds among control patients. This 'typical odds ratio' of 0.75 corresponds to a 'typical odds reduction' of 25%.

The corresponding *absolute* effects quantify the impact of treatment among different groups of patients. For a specific outcome, the proportional effects of a particular treatment are generally similar among different categories of patients, but the corresponding absolute risk reductions depend on baseline risk. For example, among a group of patients with a 4% risk of experiencing a vascular event, a 25% odds reduction would reduce the overall risk of a vascular event to around 3%. The absolute reduction in risk would be about 1%, or 10 vascular events avoided for every 1000 patients treated. In a group of patients with a 40% risk of experiencing a vascular event, the same proportional reduction would translate to an absolute reduction of about 10%, or 100 vascular events avoided for every 1000 patients treated.

24.2 Antiplatelet therapy in the prevention of vascular events among patients at high risk of occlusive arterial disease

ANTIPLATELET THERAPY VERSUS CONTROL AMONG HIGH RISK PATIENTS

Proportional effects

The 1994 APT overview assessed 142 randomized trials that had compared a prolonged course (i.e. a month or more) of antiplatelet therapy versus control among around 70,000 high risk patients with overt clinical manifestations of occlusive arterial disease. Overall, antiplatelet therapy produced a reduction of about a quarter in the odds of a serious vascular event. There were no statistically significant differences - formally, no statistical heterogeneity - between the odds reductions in each of the four categories of high risk patients: those with a prior MI; those with an acute MI; those with a prior stroke or transient ischemic attack (TIA); and a group of other patients at high risk, including, among others, patients with stable or unstable angina, patients undergoing revascularisation procedures, and those with peripheral arterial disease (Figure 24.1). This one quarter odds reduction arose from about a one third reduction in the odds of a non-fatal MI, about a one third reduction in the odds of a non-fatal stroke, and about a one sixth reduction in the odds of vascular death. There was also about a one sixth reduction in all-cause mortality, reflecting the fact that antiplatelet therapy does not result in an excess of non-vascular mortality.

The APT 1994 overview thus established conclusively the value of antiplatelet therapy for a wide range of patients at high risk of occlusive arterial disease. The proportional reductions in vascular events were shown to be similar in men and women, in middle age and old age, in hypertensive and normotensive patients, and in diabetic and non-diabetic patients (8).

Table 1. Summary of sources of randomized trials data used in the chapter. a) Data available by March 1990 and included in analysis for the 1994 APT overview.

Subject area	Comparison	Number of trials	Approximate number of patients
Prevention of vascular events among patients at high risk of occlusive arterial disease.	Any antiplatelet regimen versus control	142	70000
	One antiplatelet regimen versus another*	22	10000
	One aspirin dose versus another	3	2500
Primary prevention of vascular events among patients at low risk of occlusive arterial disease.	Any antiplatelet regimen versus control	3	30000
Maintenance of vascular graft or arterial patency among patients at high risk of occlusive arterial disease.	Any antiplatelet regimen versus control	46	8500
	One antiplatelet regimen versus another*	16	3500
	One aspirin dose versus another	2	450
Reduction in venous thrombosis and pulmonary embolism among surgical and medical patients.	Any antiplatelet regimen versus control	62	9000
	One antiplatelet regimen versus another*	14	1000
	One aspirin dose versus another	2	160

* Excluding comparisons of different doses of aspirin

(b): Main sources of additional data available since March 1990

Subject area	Comparisons	Main trials	Approximate number of individuals
Prevention of vascular events among individuals at high risk of occlusive arterial disease.	One antiplatelet regimen versus another*	ESPS-2[15], CAPRIE[16]	23500
	Addition of GP IIb/IIIa inhibitor to aspirin vs same aspirin regimen • Patients undergoing PTCA‡	EPIC[19-21], EPILOG[22], CAPTURE[23], IMPACT-II[24], RESTORE[25]	11000
	• Patients with unstable angina / non- Q-wave MI	PURSUIT[26], Canadian Lamifiban Study[27], PARAGON[28,29], PRISM-PLUS[30]	15000
	• Patients with acute ST elevation MI receiving reperfusion therapy	TAMI-8[31], IMPACT-AMI[32], RAPPORT[33], PARADIGM[34]	1000
	One aspirin dose versus another	Dutch-TIA[42]	3000
	Low-dose† aspirin versus control	SALT[43], SAPAT[44], ESPS-2[15]	10000
Management of acute ischemic stroke	Aspirin versus control	International Stroke Trial[47], Chinese Acute Stroke Trial[48], MAST-Italy[49]	40000
Primary prevention of vascular events among individuals at low risk of occlusive arterial disease.	Aspirin versus control	Thrombosis Prevention Trial[52], Hypertension Optimal Treatment Study[53]	24000

* Excluding comparisons of different doses of aspirin
† 75mg daily or less
‡ Percutaneous transluminal coronary angioplasty

Absolute effects

The absolute effects of antiplatelet therapy on vascular events among the various categories of patients at high risk of occlusive arterial disease are shown in Figure 24.2. Among about 20,000 patients with acute myocardial infarction there was a one month benefit of about 40 vascular events avoided per 1000 patients treated, whilst among about 20,000 patients with a prior MI a similar number of events were avoided over about two years. Among about 10,000 patients with a prior stroke or TIA there was a three year benefit of about 40 per 1000, and among about 20,000 patients with some other high risk condition there was a six month benefit of about 20 per 1000. Patients with a prior MI had a higher untreated risk of non-fatal reinfarction (6.5%) than of non-fatal stroke (1.5%), so that antiplatelet therapy prevented around three times as many non-fatal MIs (about 18 per 1000) as non-fatal strokes (about 6 per 1000). By contrast, patients with a prior stroke or

Category of trial	No of trials with data	MI, STROKE OR VASCULAR DEATH: Anti-platelet	MI, STROKE OR VASCULAR DEATH: Adjusted controls*	STRATIFIED STATISTICS: O–E	STRATIFIED STATISTICS: Variance	% odds reduction (SD)
Prior MI	11	1331/9877 (13.5%)	1693/9914 (17.1%)	−158·5	561·6	25% (4)
Acute MI	9	992/9388 (10.6%)	1348/9385 (14.4%)	−177·9	510·3	29% (4)
Prior stroke/TIA	18	1076/5837 (18.4%)	1301/5870 (22.2%)	−98·5	386·5	22% (4)
Other high risk	104	784/11 434 (6.9%)	1058/11 542 (9.2%)	−134·0	352·5	32% (4)
ALL HIGH RISK (four main categories)	142	4183/36 536 (11.4%)	5400/36 711 (14.7%)	−568·8	1810·9	27% (2)
ALL LOW RISK (primary prevention)	3	652/14 606 (4.46%)	708/14 604 (4.85%)	−28·5	273·5	10% (6)
ALL TRIALS* (high or low risk)	145	4835/51 144 (9.5%)	6108/51 315 (11.9%)	−597·3	2084·4	25% (2)

Heterogeneity of odds reductions:
– between four high risk categories $\chi^2_3 = 4.1$; NS
– between high risk and low risk $\chi^2_1 = 10.5$; P=0.001

*Crude, unadjusted control total = 5274/45 172.

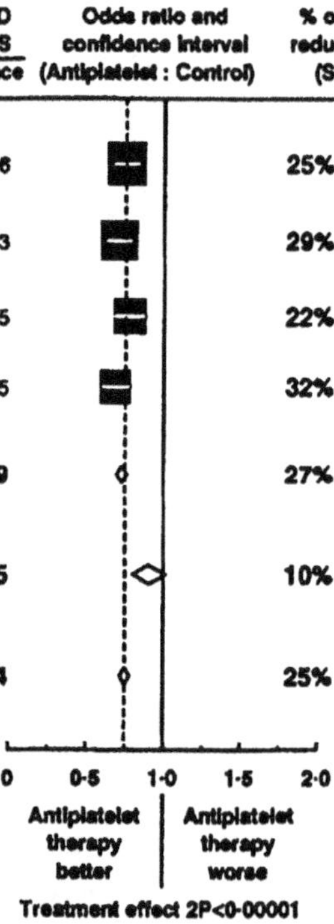

Figure 24.1. Proportional effects of antiplatelet therapy on vascular events (myocardial infarction, stroke, or vascular death) in four main high risk categories of trial and in low risk (primary prevention).

(Stratified ratio of odds of an event in treatment groups to that in control groups is plotted for each group of trials (black square, area proportional to amount of statistical information in contributing trials), along with its 99% confidence interval (horizontal line). Overviews of results for certain subtotals (and 95% confidence intervals) are represented by diamonds. Odds reductions observed in similar groups of trials are given to right of solid vertical line.)

MI = myocardial infarction. TIA = transient ischemic attack.

TIA were at much higher untreated risk of a subsequent non-fatal stroke (10.2%) than of a non-fatal MI (2.9%), so that antiplatelet therapy prevented around twice as many non-fatal strokes (about 20 per 1000) as non-fatal MIs (about 9 per 1000) among these patients.

How long should antiplatelet therapy continue?

There are currently no randomized trials comparing different durations of treatment with antiplatelet therapy. Although the 1994 APT data appeared to suggest that the additional effects of antiplatelet therapy among high risk patients became smaller over the years that treatment was continued, there are difficulties in interpreting such patterns (8). Patients at high risk of occlusive arterial disease tend to remain so, and therefore the benefits of antiplatelet therapy are likely to continue to outweigh any bleeding risks unless a strong contraindication develops, or the risks of treatment become unacceptably high. In the absence of clear evidence to the contrary, therefore, it is generally appropriate to consider indefinite treatment with antiplatelet therapy in high risk patients.

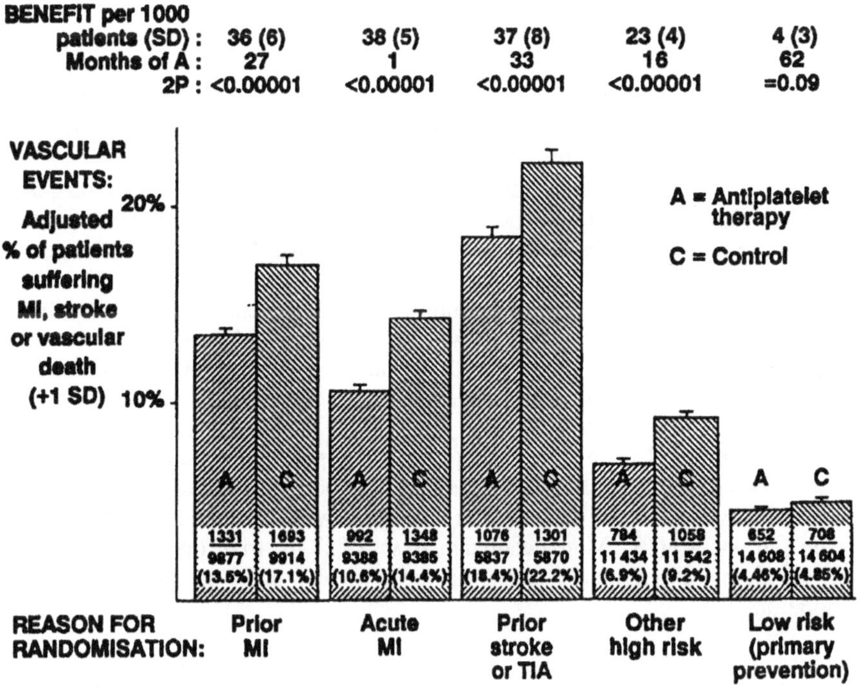

Figure 24.2. Absolute effects of antplatelet therapy on vascular events (myocardial infarction, stroke or vascular death) in four main high risk categories of trial and in low risk (primary prevention).

(Adjusted totals calculated after converting any unevenly randomized trials to even ones by counting control groups more than once, for calculating adjusted percentages and events prevented per 1000 patients allocated antiplatelet therapy. Statistical significance (2p) based on stratified analyses of original unadjusted numbers in each trial.)

Months of A = Means of scheduled durations of antiplatelet therapy. No trial lasted under one month.

DIFFERENT ANTIPLATELET REGIMENS AMONG HIGH RISK PATIENTS

Medium-dose aspirin (75-325mg daily) was the most widely tested antiplatelet regimen among the trials included in the 1994 APT overview, and around two-thirds of the randomized evidence involved an aspirin regimen. In order to assess whether some other antiplatelet regimen was more effective than aspirin alone, the relative effects of different regimens were assessed both through making indirect comparisons between randomized trials comparing one antiplatelet regimen versus control (Figure 24.3), and through direct comparisons of one antiplatelet regimen versus another (Figure 24.4). There was no clear evidence that any antiplatelet regimen was more effective than aspirin alone, although small differences could not be excluded (8). In terms of their effects on major outcomes, proportional differences between active antiplatelet regimens are likely to be small (say, around 10% or less) compared with proportional differences between antiplatelet therapy and control (for example, the 25% odds reduction in vascular events among high risk patients). Trials involving direct randomized comparisons between such regimens therefore need to be particularly large. Several such large-scale trials have been completed since the previous APT report.

The combination of aspirin and dipyridamole

Aspirin inhibits platelet cyclo-oxygenase and, thereby, the platelet aggregating agent thromboxane-A_2 (11). One approach to improving on the antiplatelet effects of aspirin might be through the use of drugs that inhibit pathways other than those mediated by thromboxane-A_2. Dipyridamole has been shown to have several effects on platelet function, including inhibition of platelet phosphodiesterase, direct stimulation of vascular endothelial prostacyclin release, and inhibition of cellular uptake and metabolism of adenosine (12). In the 1994 APT overview the addition of dipyridamole to aspirin did not appear to produce any worthwhile reduction in vascular events (Figure 24.4), but the possibility of a small additional benefit was not excluded. More recently, the second European Stroke Prevention Study (ESPS-2), randomized over 6000 patients with a previous TIA or stroke in a 2x2 factorial design to one of four treatment arms: (i) aspirin (50mg daily); (ii) dipyridamole (400mg daily); (iii) the combination of aspirin and dipyridamole; or (iv) placebo (11). The addition of dipyridamole to aspirin produced a significant 23% reduction in the risk of stroke (2p=0.006) compared with aspirin alone, but the 13% reduction in the risk of 'stroke or death' was more modest and statistically uncertain (2p = 0.06) (13). More detailed analyses of all the trials that have assessed the effects of adding dipyridamole to aspirin are currently being prepared by the ATT Collaboration (14).

Clopidogrel and the thienopyridines

Clopidogrel is a thienopyridine derivative similar to ticlopidine, and produces significant inhibition of in-vivo ADP-induced platelet aggregation (15). The Clopidogrel versus Aspirin in Patients at Risk of Ischemic Events (CAPRIE) study was a randomized trial comparing clopidogrel (75 mg daily) versus aspirin (325 mg daily) among around 20,000 high risk patients with a recent ischemic stroke, a recent MI, or peripheral arterial disease (16). Clopidogrel produced an 8.7% (95% CI 0.3 to 16.5%) proportional reduction in the risk of a vascular event (defined, in this trial, as a non-fatal MI, non-fatal stroke, or non-hemorrhagic vascular death), which was conventionally statistically significant (p=0.04). A similar proportional reduction in vascular events was observed among around 3500 high risk patients included in previous trials comparing ticlopidine and aspirin (Figure 24.4) (8).

Antiplatelet regimen	No of trials with data	MI, STROKE OR VASCULAR DEATH: Anti-platelet	Adjusted controls*	STRATIFIED STATISTICS: O–E	Variance	Odds ratio and confidence interval (Antiplatelet : Control)	% odds reduction (SD)
Aspirin alone (mg daily):							
Aspirin (500–1500)	30	1243/9223	1514/9248	−127·8	549·3		21% (4)
Aspirin (160–325)	12	1303/11 808	1740/11 862	−215·7	657·6		28% (3)
Aspirin (<160) [mostly 75–150]	7	129/1440	168/1438	−19·6	64·2		26% (11)
Subtotal: Aspirin*	46	2675/22 471 (11.9%)	3422/22 548 (15.2%)	−357·5	1217·4		25% (2)
Six other regimens:							
Aspirin + Dip	34	752/6842	989/6876	−112·4	342·9		28% (5)
Aspirin + Sulph	2	38/283	50/278	−6·5	18·5		30% (20)
Dipyridamole	10	87/719	113/755	−10·5	40·5		23% (14)
Sulphinpyrazone	17	299/2231	358/2242	−30·1	136·2		20% (8)
Ticlopidine	39	264/3250	371/3278	−50·6	125·5		33% (7)
Suloctidil	6	47/364	59/367	−5·6	20·5		24% (19)
Subtotal: six other regimens*	105	1487/13 689 (10.9%)	1940/13 796 (14.1%)	−214·0	662·0		28% (3)
ALL TRIALS*	142	4183/36 536 (11.4%)	5400/36 711 (14.7%)	−568·8	1810·9		27% (2)

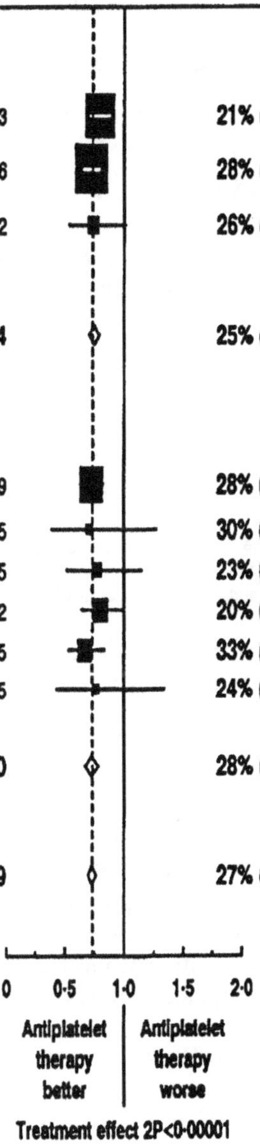

Heterogeneity of odds reductions:
- between three aspirin categories $\chi^2_2 = 2.7$; NS
- between seven regimens (any aspirin & six others) $\chi^2_6 = 2.6$; NS

*Crude, unadjusted control totals 3130/21 147 (aspirin), 1773/12 737 (six other regimens).
NB: Some trials contributed to more than one comparison.

Figure 24.3. Indirect comparisons of proportional effects on vascular events of different antiplatelet regimens in high risk patients.
Symbols and conventions as in Figure 24.1.
Dip = dipyridamole. Sulph = sulphinpyrazone.

Regimens compared (A1 v A2)	No of trials	MI, STROKE OR VASCULAR DEATH		STRATIFIED STATISTICS		Odds ratio and confidence interval (A1 : A2)	% odds reduction (SD)
		A1	A2	O–E	Variance		
High* v medium Asp	3	195/1212	201/1213	−3·8	79·8		5% (11)
Asp + Dip v same Asp	14	316/2661	312/2656	1·0	134·1		−1% (9)
Sulphinpyrazone v Asp	5	98/674	88/673	5·5	34·1		−18% (19)
Ticlopidine v Aspirin	3	378/1730	414/1741	−16·6	149·9		10% (8)

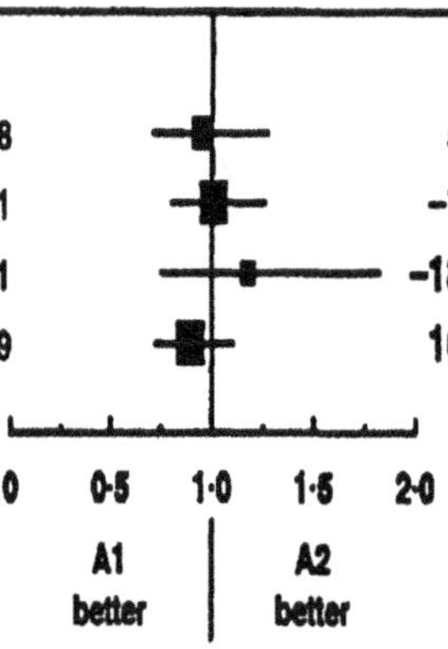

*Aspirin doses(mg/day): high=500–1500, medium=75–325
**Crude, unadjusted sulphinpyrazone total = 85/528

Figure 24.4. Direct comparisons of proportional effects on vascular events of different antiplatelet regimens in high risk patients.
Symbols and conventions as in Figure 24.1.

Hence, if the trials of thienopyridine antiplatelet agents are considered together, they suggest that such agents are at least as effective as medium-dose aspirin, and may offer somewhat greater protection against vascular events. Clopidogrel was also shown to be safe in CAPRIE, with a much lower incidence of neutropenia than had previously been observed in trials of ticlopidine. However, the size of any absolute benefits that may arise from replacing aspirin with clopidogrel remains statistically uncertain. Since aspirin is substantially less expensive than clopidogrel, for the time being the main clinical value of clopidogrel is likely to be among those patients for whom aspirin is contraindicated or poorly tolerated. However, the *addition* of clopidogrel to aspirin seems a more promising approach than substituting clopidogrel for aspirin; the two drugs have different mechanisms of action and so might well act synergistically, and trials of clopidogrel plus aspirin versus aspirin alone are now being planned.

Adding a IIb/IIIa receptor antagonist to aspirin

Glycoprotein (GP) IIb/IIIa receptor antagonists prevent a conformational change in the GP IIb/IIIa receptor which occurs prior to binding of fibrinogen, and thereby block the final step in platelet aggregation (17,18). Several randomized trials have now assessed the effects of a short (12 hours to 5 days) intravenous infusion of a GP IIb/IIIa blocker in the prevention of thrombotic complications associated with: (i) percutaneous transluminal coronary angioplasty and other forms of coronary intervention (19-25); (ii) unstable angina and non-Q-wave MI in non-interventional settings (26-30); and (iii) reperfusion therapy in acute MI with ST segment elevation (31-34). In all of these trials aspirin (and sometimes heparin) was given routinely to all patients, so that the treatment strategy being tested was that of adding a GP IIb/IIIa inhibitor to aspirin. Overall, the trials in the first two of these clinical categories have suggested that the combined endpoint of death or MI at 30 days might be reduced by about one-fifth and that this benefit may persist long-term (21,35,36). The trials of GP IIb/IIIa inhibitors given with reperfusion therapy (mainly fibrinolytic therapy) for acute MI have also produced some promising results (31-34), but far too few patients have yet been randomized to be certain of the size of the benefits.

Any benefits of treatment with GP IIb/IIIa inhibitors must be weighed against the risks. Several studies have found an increased risk of major bleeding, which is likely to be of particular importance in patients whose absolute risk of bleeding is high, such as those receiving fibrinolytic therapy (see below for further discussion). A small excess risk of thrombocytopenia has also been reported in some trials (37).

Oral GP IIb/IIIa inhibitors have now been developed, and several large-scale randomized trials are currently assessing the effects of longer term use of these agents on major vascular outcomes among patients with acute coronary syndromes (18,38).

DIFFERENT DOSES OF ASPIRIN AMONG HIGH RISK PATIENTS

The 1994 APT overview showed that daily aspirin doses of <160mg (mostly 75-150mg), 160-325mg, and 500-1500mg prevented similar proportions of vascular events, although limited numbers had been studied in trials using daily doses of <160mg (Figures 27.3 and 27.4). It has been suggested that doses of >1000mg daily may be preferable for patients with a prior ischemic stroke or TIA (39,40), but higher doses of aspirin increase the risk of gastrointestinal bleeding (41,42), and the APT overview did not provide any clear evidence to support their use (8).

Considerably more evidence about the effects of low-dose aspirin regimens is now available. Two recent trials have investigated the effects of a daily aspirin dose of exactly 75mg. The Swedish Aspirin Low-dose Trial (SALT) included around 1400 patients with a recent TIA, minor ischemic stroke or retinal artery occlusion (43), whilst the Swedish Angina Pectoris Aspirin Trial (SAPAT) included about 2000 patients with stable angina (44). Although not included in the formal analyses in the APT overview, summary data from these two trials were made available to the APT Collaboration in 1994 and, following their inclusion, the evidence from all trials assessing 75mg daily showed that this dose produced a definite reduction in vascular events (odds reduction 29%, 95% CI 17 to 39%, $2p<0.0001$) (8).

Virtually complete suppression of platelet TxA_2 synthesis after 7-10 days occurs with daily aspirin doses as low as 30-50mg (45), raising the possibility that such low doses might prevent serious vascular events as effectively as medium dose aspirin (75-325mg daily). In ESPS-2, low-dose (50mg daily) aspirin produced only a 13% (95% CI 1 to 25%) reduction in the risk of 'stroke or death' compared with placebo (13). In the only directly randomized comparison of medium versus low-dose aspirin, the effects of daily doses of 30mg and of 283mg were found to be similar among around 3000 patients with a recent TIA or minor ischemic stroke, but this trial was too small to exclude the possibility of moderate (but clinically important) differences between these two regimens (42). Hence, although aspirin 75mg daily appears to be about as effective as higher aspirin doses in preventing serious vascular events, at present there is no reliable evidence that doses lower than 75mg daily are adequate.

ANTIPLATELET THERAPY IN ACUTE ISCHEMIC STROKE

At the time of the 1994 APT overview no large-scale randomized trial had assessed the effects of antiplatelet therapy during the first few weeks after an acute ischemic stroke. Antiplatelet therapy given early after an ischemic stroke might decrease the extent of infarction, reduce the risk of a recurrent stroke, and prevent venous thromboembolism. However, the risks of hemorrhage into infarcted brain tissue ('hemorrhagic transformation') and of extracranial bleeding might also be particularly high (46). Two large-scale randomized trials of antiplatelet therapy in acute stroke were reported in 1997. Between them the International Stroke Trial (IST) and the Chinese Acute Stroke Trial (CAST) randomized almost 40,000 patients within 48 hours of symptom onset to receive either aspirin (300mg daily in IST, 160mg daily in CAST) or no aspirin for the first few weeks following an acute ischemic stroke (47,48). Overall, among patients randomized into IST, CAST and one smaller trial, MAST-Italy (49), an average of about 3 weeks of aspirin reduced the odds of death or non-fatal stroke by 11% (95% CI 4 to 17%; 2p = 0.001), and the odds of death or dependency by 5% (95% CI 1 to 9%; 2p=0.007) (48). This corresponded to about 9 per 1000 fewer deaths or non-fatal strokes in the first few weeks, and to around 13 per 1000 fewer patients dead or dependent after a few months. Although there was evidence of an increased risk of hemorrhagic stroke (including symptomatic hemorrhagic transformation of the initial infarct), this risk was small (overall about 2 extra events per 1000 patients treated) and more than offset by the benefits.

There were no significant differences in the effects of aspirin among the various subgroups studied in IST and CAST. Thus the proportional benefits were similar in the presence and absence of atrial fibrillation, among older and younger patients, among the various subtypes of ischemic stroke, and regardless of whether patients were randomized early or later in the 48 hour period following symptom onset (48).

When antiplatelet therapy is commenced some months after an acute stroke, and continued for several years, the randomized evidence shows that for every 1000 patients treated around 1 serious vascular event is avoided during each month of treatment (8). The results of IST and CAST show that early treatment prevents around 9 recurrent strokes or deaths per 1000 patients during the first month (48). Moreover, if aspirin is started early then it is more likely to be continued after hospital discharge. Hence aspirin 160-300mg should now be considered as early as possible for almost all patients presenting with acute ischemic stroke, provided that there are no strong contraindications and that hemorrhagic stroke can be excluded reliably.

24.3 Antiplatelet therapy in the prevention of vascular events among patients at low risk of occlusive arterial disease

In the APT 1994 overview there were almost 30,000 low risk individuals with no clinically apparent arterial disease randomized between antiplatelet therapy and control, most of them in two large trials of aspirin among male doctors in the United Kingdom (50) and the United States (51). Antiplatelet therapy produced a small (10%; 95% CI –1 to 20%) and non-significant reduction in vascular events among these patients, in contrast to the highly significant one quarter odds reduction among high risk patients (Figure 24.1) (8). There was a highly significant reduction of about one third in the odds of non-fatal MI among the low risk patients, but this effect was small in absolute terms: around 5 years of antiplatelet therapy led to the avoidance of only about 5 non-fatal MIs per 1000 patients treated. This small beneficial effect on non-fatal MI was offset by a non-significant 21% (95% CI –5 to 53%) proportional increase in non-fatal strokes, equivalent to 2 extra such events per 1000, with no apparent effect of antiplatelet therapy on vascular deaths. When the etiology of all strokes (fatal and not) observed among low risk individuals was considered, there was not only a small, marginally significant ($2p<0.05$) excess of strokes attributed to hemorrhagic causes but also a non-significant excess of 'other' strokes (8). This may have been because other strokes included those of unknown cause, a few of which may have been hemorrhagic and caused by antiplatelet therapy. But as a large proportion of these other strokes were confirmed as ischemic, the adverse effects on stroke among these low risk individuals may well have been inflated by the play of chance. Hemorrhagic strokes are more likely to be fatal, and a small excess of these with antiplatelet therapy may account, at least in part, for the lack of any significant effect on vascular mortality.

The APT 1994 overview concluded that uncertainty about the overall balance of benefit and hazard among individuals at low or intermediate risk of arterial occlusive disease meant that more randomized evidence was required. Two recently published trials among asymptomatic individuals with identifiable risk factors for vascular events have addressed this issue. The Thrombosis Prevention Trial (TPT) was a 2x2 factorial randomized trial of aspirin 75mg daily versus placebo, and of low-intensity oral anticoagulation with warfarin (target INR 1.5) versus placebo, among around 5000 men aged 45-69 years. Participants recruited from the general population were considered to be in the top 20 to 25% of a coronary heart disease risk score distribution (52). The Hypertension Optimal Treatment (HOT) trial was also a factorial randomized trial, with comparisons of aspirin 75 mg daily versus placebo, and of three different intensities of blood pressure reduction, in around 19000 hypertensive individuals (53). Table 24.2 (a) provides the published results for vascular events (as defined by the APT) from these trials in combination with the results from trials among low risk individuals included in the 1994 APT overview. Although both TPT and HOT aimed to recruit individuals at intermediate risk of vascular events, the risk among those allocated no aspirin treatment in both trials was very similar to that in the previously published low risk trials.

Overall, antiplatelet therapy produced a 13% (95% CI 5-20%) proportional reduction in the odds of a vascular event, corresponding to an absolute risk reduction of only about one vascular event avoided per 1000 patients treated per year (Table 24.2 (a)). As in the other low risk trials, the risk of MI was reduced in both TPT and HOT: in TPT the rate of all (i.e. fatal or non-fatal) ischemic heart disease was reduced by 20% (95% CI 1-35%); and in HOT the rate of all MI was reduced by 36% (95% CI 15-51%). But, although antiplatelet therapy appeared to reduce the risk of ischemic stroke, it also appeared to increase the risk of hemorrhagic stroke: overall in TPT there was a non-significant 3% (95% CI –45-35%) reduction in the rate of all stroke; and in HOT there was a non-significant 2% (95% CI –24-22%) reduction in the rate of all stroke. As in the other low risk trials, there was no overall effect on vascular death or on all-cause mortality (52,53).

Table 24. 2. Effects of antiplatelet therapy in low risk individuals

	Approx. average duration (years)	Approx. control group risk (per year)	Antiplatelet	Adjusted controls	Odds ratio and confidence interval†	Approx. vascular events avoided per 1000 patients treated per year
APT 1994[8] (3 trials)	5.2*	0.9%	652 / 14608 (4.5%)	708 / 14604 (4.8%)	0.90 (0.80 to 1.01)	0.7
TPT[52¶]	6.4	1.5%	211 / 2545 (8.3%)	250 / 2540 (9.8%)	0.82 (0.68 to 1.00)	2.4
HOT[53]	3.8	1.0%	315 / 9399 (3.4%)	368 / 9391 3.9%)	0.85 (0.70 to 1.04)	1.5
All trials	5.0*	1.0%	1178 / 26552 (4.4%)	1326 / 26535 (5.0%)	0.87 (0.80 to 0.95)	1.1

* weighted by study size

† 95% CI for TPT, HOT and APT 1994; 99% CI for 'All trials'

¶ The effects of aspirin were similar in the absence or presence of warfarin, so the data presented are not stratified by warfarin allocation. Heterogeneity of odds ratios between 5 trials: χ^2 (4 df) = 4.5; NS

(b): Effects of antiplatelet therapy on intracranial and on major extracranial bleeding

	Approx. average duration (years)	Antiplatelet	Adjusted controls	Approx. excess bleeds per 1000 patients treated per year
INTRACRANIAL BLEEDS				
APT 1994 (2 trials)[8]	5.2	44/14466 (0.3%)	24/14454 (0.2%)	
TPT[52]	6.4	9/2545 (0.4%)	1/2540 (0.04%)	
HOT[53]	3.8	14/9399 (0.1%)	15/9391 (0.2%)	
All trials	5.0	67/26410 (0.3%)	40/26385 (0.2%)	0.2 (2p=0.01)
MAJOR EXTRACRANIAL BLEEDS				
APT 1994 (2 trials)[8]	5.2	69/14466 (0.5%)	49/14454 0.3%)	
TPT[52]	6.4	20/2545 (0.8%)	13/2540 (0.5%)	
HOT[53]	3.8	122/9399 (1.3%)	63/9391 (0.7%)	
All trials	5.0	211/26410 (0.8%)	125/26385 (0.5%)	0.7 (2p<0.00001)

Individuals in primary prevention trials were at much lower risk of vascular events (average 1% per year in the control group) than the high risk patients included in the APT overview (average 9% per year in the control group) (8), and so there is a finer balance between the benefits and risks of antiplatelet therapy. The reliable identification of apparently healthy individuals who are likely to benefit from aspirin (and of those most likely to be harmed) is, therefore, likely to require very large-scale randomized evidence. A meta-analysis of detailed individual patient data from the available randomized trials is currently being conducted by the Antithrombotic Trialists' Collaboration. At present, however, there is no good evidence for the widespread use of aspirin among apparently healthy individuals.

24.4 Antiplatelet therapy in the maintenance of vascular graft or arterial patency among patients at high risk of occlusive arterial disease

After coronary revascularisation, whether by coronary bypass grafting or by percutaneous transluminal angioplasty, angiographic studies show substantial rates of reocclusion (54). For example, about one fifth of coronary artery bypass grafts occlude during the first postoperative year (55) and a few percent per year occlude thereafter (56,57). These occlusions are often subclinical, though some may produce clinical signs of MI (8). Occlusion or reocclusion is also seen after peripheral artery revascularisation, although many such occlusions are subclinical. Finally, in patients with renal disease, occlusion of arteriovenous fistulas or shunts established to allow hemodialysis is common (and such patients are also at particularly high risk of other serious thrombotic complications (58)). Experimental (59,60) and clinical (8,61,62) evidence suggests that antiplatelet therapy may help prevent vascular graft or arterial occlusion, particularly during the period after vascular procedures, before any intimal damage has healed.

To assess the effects of antiplatelet therapy on vascular occlusion, the APT Collaboration conducted a meta-analysis of all randomized trials of antiplatelet therapy in which vascular graft or arterial patency was studied systematically (9). Vascular patency was assessed by a variety of methods, including angiography, Doppler ultrasonography, limb plethysmography, and clinical examination. Data were sought on the number of patients in each treatment group who had suffered at least one main proximal graft or vessel occlusion during follow-up.

Antiplatelet therapy produced a highly significant 44% (95% CI 37 to 50%, $2p<0.00001$) reduction in the odds of vascular occlusion. The odds reductions were separately significant among each of the three specific categories of patients (i.e. those undergoing coronary artery revascularisation, those undergoing peripheral artery revascularisation, and those having hemodialysis access established). Typically, therefore, antiplatelet therapy produces worthwhile absolute reductions in the odds of vascular occlusion, as well as preventing a quarter of serious vascular events among patients undergoing vascular procedures. For example, among 1000 patients who have coronary artery revascularisation procedures, about six months of antiplatelet therapy should protect several dozen patients from coronary arterial occlusion as well as preventing a few dozen major vascular events (Figure 24.5).

The proportional effects on vascular occlusion of different antiplatelet regimens appeared similar, and there was no clear evidence that high daily doses of aspirin (i.e. 500-1500mg) were any more effective than medium-dose aspirin (75-325mg). At present medium-dose aspirin is the most widely tested, most convenient, and least expensive antiplatelet treatment. The choice of which antiplatelet regimen to use is, however, of secondary importance: what chiefly matters is that some such treatment should be considered routinely for the vast majority of patients undergoing vascular procedures.

It had been suggested that starting antiplatelet therapy *after* invasive vascular procedures may

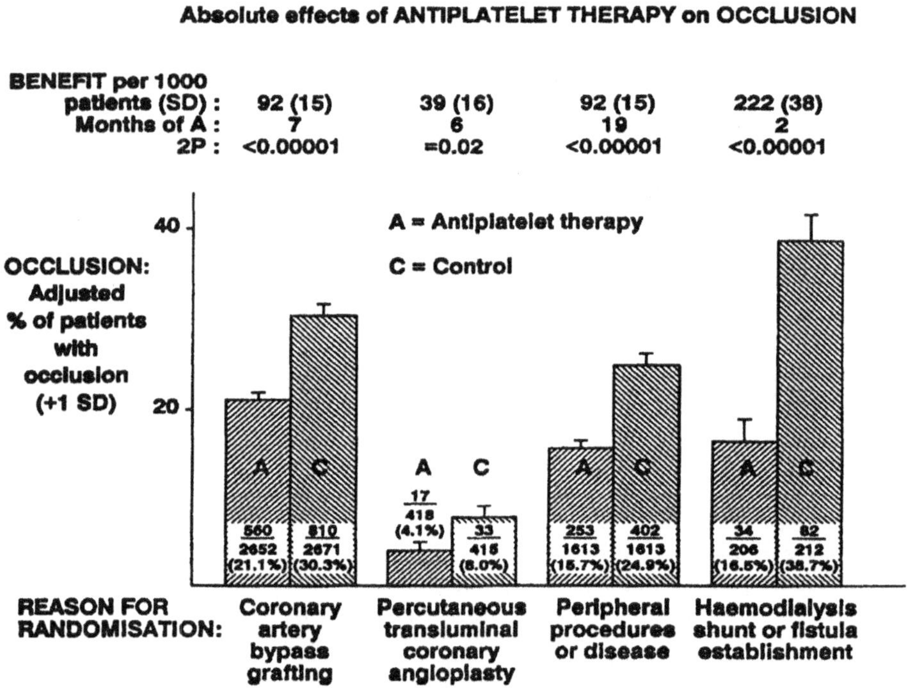

Figure 24.5. Absolute effects of antiplatelet therapy on occlusion. Conventions as in Figure 24.2.

reduce the risk of bleeding while still preventing occlusion (62). Data on bleeding were incomplete, but no large excess with antiplatelet therapy started either before or after the procedure was apparent. Indirect comparisons suggested that similar proportional reductions in occlusion were produced by starting antiplatelet therapy before and just after the procedure (9), but absolute benefits are likely to be greater if antiplatelet therapy is started sooner rather than later.

24.5 Antiplatelet therapy in patients at risk of venous thromboembolism

Deep venous thrombosis (DVT) and pulmonary embolism (PE) are important causes of morbidity and mortality in surgical patients or immobilised medical patients (63). Randomized trials have shown that perioperative subcutaneous unfractionated or low molecular weight heparin approximately halves the risk of DVT and PE among patients undergoing surgery (64,65). However, at the time of the APT 1994 overview, although many small trials had examined the effects of antiplatelet therapy in the prevention of venous thromboembolism, none had provided convincing evidence of benefit. The APT Collaboration conducted a meta-analysis of all the randomized trials of antiplatelet therapy with systematic assessment of the effects on venous thromboembolism among patients undergoing orthopedic or general surgery, and among patients immobilised for some medical reason, such as acute stroke or myocardial infarction (10). Among 53 trials, antiplatelet therapy produced a 39% (95% CI 30 to 47%, 2p<0.00001) reduction in the odds of DVT, with similar proportional benefits observed among the different categories of medical and surgical patients studied. The absolute benefits were therefore greatest

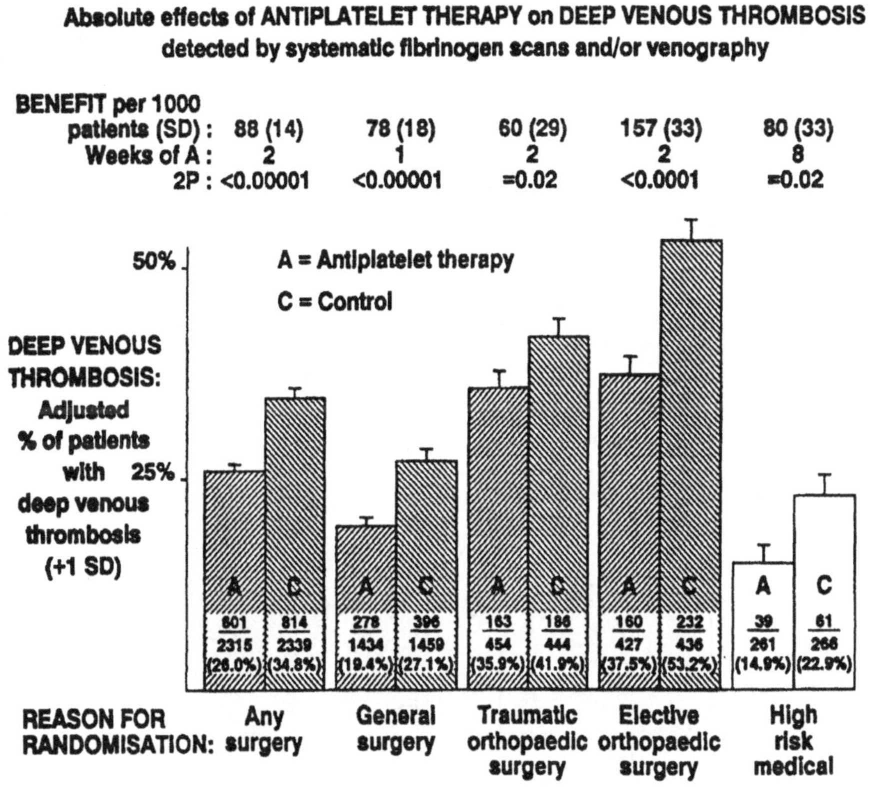

Figure 24.6. Absolute effects of antiplatelet therapy on deep venous thrombosis detected by systematic fibrinogen scans and / or venography. Conventions as in Figure 24.2. Weeks of A = Means of scheduled durations of antiplatelet therapy.

among those at highest risk, in particular the orthopedic surgery patients studied, among whom DVT was avoided in around 160 patients per 1000 treated (Figure 24.6). There was also a highly significant 64% (95% CI 51 to 74%, 2p<0.00001) reduction in the odds of PE, with no significant heterogeneity between the different categories of patients studied. This corresponded to an absolute benefit of around 20 fewer patients with PE per 1000 treated (Figure 24.7).

The proportional reductions in DVT or PE produced by different antiplatelet regimens and by different daily aspirin doses were similar. The risks of bleeding with antiplatelet therapy are discussed further in the next section, but they were relatively small, and were clearly outweighed by the benefits of reduced thromboembolic risk.

Before this meta-analysis was performed, it had generally been concluded that antiplatelet therapy had little or no effect on DVT or PE (63,66). But when all the randomized

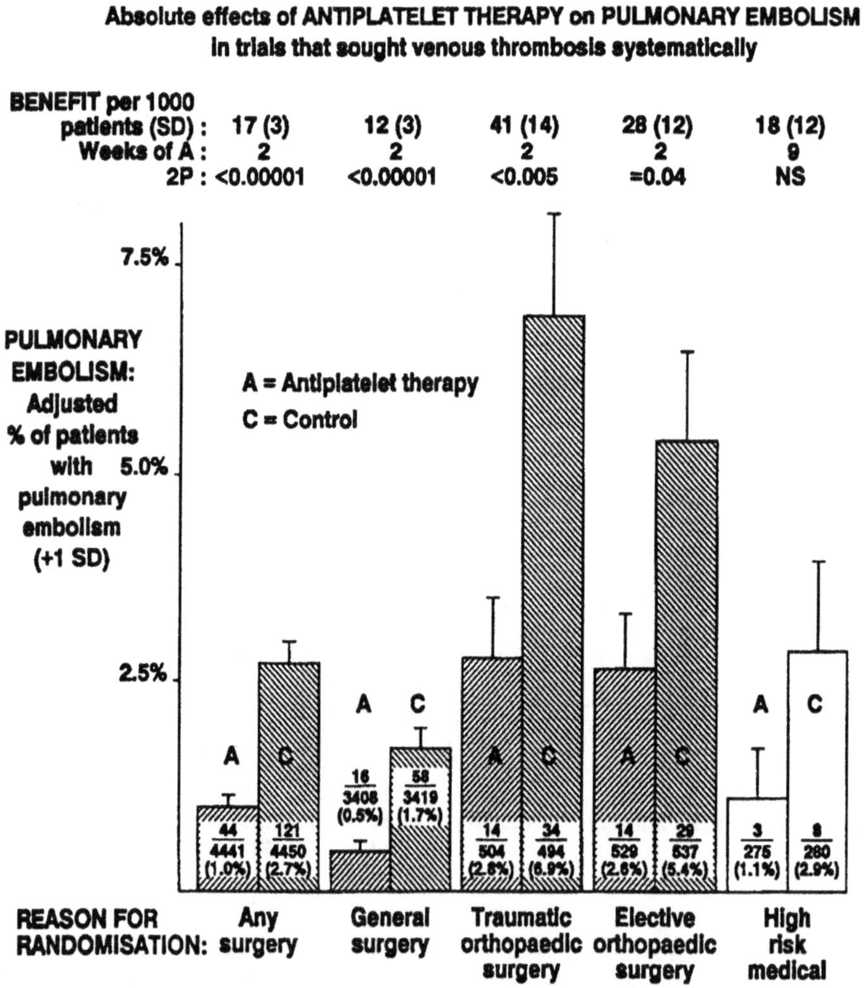

Figure 24.7. Absolute effects of antiplatelet therapy on pulmonary embolism in trials that sought venous thrombosis systematically. Conventions as in Figure 24.2. Weeks of A = Means of scheduled durations of antiplatelet therapy.

evidence is considered it is clear that antiplatelet therapy reduces the odds of DVT and of PE by about a half – an effect similar in size to that of low-dose subcutaneous heparin

(67). The risk of thromboembolism can remain high for some months after surgery (68), and antiplatelet therapy has the advantage that it does not require monitoring and can be continued after discharge for as long as the risk of thromboembolism remains substantial.

24.6 The risks of serious bleeding with antiplatelet therapy

Minor bleeding that does not require hospitalisation or blood transfusion (and is not life threatening) may reduce compliance with treatment. However, it is serious, potentially life threatening bleeding that has to be weighed against the benefits of antiplatelet therapy when deciding whether or not to treat an individual patient. There are two main types of serious bleeding: extracranial bleeds occur mainly in the gastrointestinal tract, may require hospital admission or blood transfusion, but do not generally result in permanent disability and are rarely fatal (69); by contrast, intracranial bleeds are uncommon, but they are often fatal and usually cause significant disability in survivors (70). For patients at high risk of a serious vascular event, therefore, the main hazard offsetting treatment benefit is intracranial hemorrhage.

INTRACRANIAL BLEEDS

The APT 1994 overview showed that any excess risk of intracranial bleeding with antiplatelet therapy was small, at most one or two per 1000 patients per year in trials of long-term (more than one month) treatment (8). Among patients at high risk of arterial occlusive disease, therefore, the large absolute reductions in serious vascular events produced by antiplatelet therapy (Figure 24.2) far outweigh any absolute hazards.

Among the more recently developed antiplatelet agents, GP IIb/IIIa receptor antagonists have the potential to produce substantial inhibition of platelet function. A small excess of intracranial bleeds has been observed in randomized trials of these agents, but the number of events was small, and as yet there are too few data to draw reliable conclusions about the size of any hazard, particularly among patients also receiving other antithrombotic treatments or fibrinolytic therapy (37).

Patients with an acute ischemic stroke have a higher baseline risk of intracranial bleeding, and medium-dose aspirin produced an excess of about two intracranial bleeds per 1000 during a treatment period of just a few weeks in two recent, large, randomized trials (47,48). However, it is clear from these trials that any increased risk of intracranial bleeding is still substantially outweighed by the associated reductions in recurrent stroke, death and disability (particularly in the absence of concomitant heparin use).

MAJOR EXTRACRANIAL BLEEDS

In the 1994 APT overview fatal extracranial bleeding was rare, and there was no clear excess with prolonged antiplatelet therapy. Most major extracranial bleeds are gastrointestinal, and a meta-analysis of the placebo-controlled trials demonstrated that prolonged aspirin treatment increased the risk of gastrointestinal bleeding by around 1.5 to 2-fold (70). However, unlike serious vascular events, such bleeds rarely result in permanent disability or death.

In short-term trials among patients having invasive vascular procedures (including coronary artery bypass grafting and coronary or peripheral arterial angioplasty), the APT overview showed that there was a slight excess of both fatal and non-fatal major bleeds, but the substantial benefits of treatment clearly outweighed the risks of bleeding (9). Similar results were found among surgical patients who had received antiplatelet therapy as prophylaxis against venous thromboembolism (10). Trials of antiplatelet therapy tended to exclude patients who were considered to be at high risk of bleeding (such as those with a recent gastrointestinal bleed, active gastrointestinal ulceration, or a previous hemorrhagic stroke). But, given that the ratio of benefit to hazard was so extreme among most patients entered into randomized trials, those with only modest absolute increases in bleeding risk could still be expected to benefit substantially from antiplatelet therapy. Thus, a decision to commence prolonged aspirin treatment in a patient with an elevated risk of bleeding may often be appropriate.

The GP IIb/IIIa receptor antagonists have been associated with an excess of major extracranial bleeds, mainly from vascular access sites in the groin (37). To date these agents have been used mainly among patients with acute coronary syndromes. Such patients usually receive aspirin, heparin, and occasionally fibrinolytic therapy, and many undergo invasive vascular procedures. Consequently, the risk of procedure-related bleeding is substantial. In the Evaluation of 7E3 for the Prevention of Ischemic Complications (EPIC) trial, for example, about 2000 patients scheduled for percutaneous transluminal coronary angioplasty (PTCA) or atherectomy were randomized to receive a bolus and 12 hour infusion of either abciximab or placebo. At 30 days the risk of major bleeding was 7% in the placebo group and 14% in the abciximab group, a substantial absolute excess of about 70 per 1000 patients treated (19). Modification of heparin dosing and early arterial sheath removal from the groin may reduce the baseline bleeding risk and therefore the absolute excess hazard (37). Adopting such a policy in the c7E3 Fab Antiplatelet Therapy in Unstable Refractory Angina (CAPTURE) trial, in which about 1200 unstable angina patients scheduled for PTCA were randomized to a similar abciximab regimen or placebo, led to 30 day risks of 1.9% in the placebo group and 3.8% in the abciximab group, an absolute excess of about 20 per 1000 patients treated (23). Hence it may be possible to limit the absolute excess of major bleeding associated with such invasive procedures, but very little is yet known about either the excess risk of major bleeding in patients also receiving fibrinolytic therapy, or the risk associated with oral GP IIb/IIIa receptor antagonists when used for longer periods of months or years.

24.7 Conclusions and recommendations for practice

Overviews of randomized trials show that prolonged antiplatelet therapy is clearly beneficial for patients at high risk of occlusive arterial disease, and should be given to those with a prior MI, an acute MI, a prior stroke or TIA, and to those with other clinical manifestations of arterial occlusive disease. There is still no evidence that any other antiplatelet regimen is clearly superior to medium-dose aspirin (i.e. 75-325mg daily). However, both clopidogrel and the combination of aspirin and dipyridamole appear to be safe and effective alternatives to medium-dose aspirin. The addition of a GP IIb/IIIa receptor antagonist to aspirin in the early treatment of some patients with acute coronary syndromes appears to reduce the risk of occlusive vascular events, but increases the risk of major bleeds. There is now clear evidence that, in the absence of a definite contraindication, aspirin (160-300mg daily) should be started as early as possible after an acute stroke of presumed ischemic origin. The role of antiplatelet therapy in the primary prevention of vascular events among asymptomatic individuals remains unclear, and at present cannot be recommended. Antiplatelet therapy also prevents vascular occlusion among patients undergoing arterial revascularisation procedures, and venous thromboembolism among patients undergoing surgery or immobilised for medical reasons, and should therefore be considered for all such patients.

REFERENCES

1. Murray CJL, Lopez AD. Mortality by cause for eight regions of the world: Global Burden of Disease Study. Lancet 1997; 349: 1269-1276.

2 Murray CJL, Lopez AD. Global mortality, disability, and the contribution of risk factors: Global Burden of Disease Study. Lancet 1997; 349:1436-1442.

3. Murray CJL, Lopez AD. Alternative projections of mortality and disability by cause 1990-2020: Global Burden of Disease Study. Lancet 1997; 349:1498-1504.

4. Peto R, Lopez AD, Boreham J, Thun M, Heath C Jr. *Mortality from smoking in developed countries 1950-2000. Indirect estimates from National Vital Statistics.* Oxford: Oxford University Press, 1994.

5. Collins R, Peto R. Antihypertensive Drug Therapy: Effects on Stroke and Coronary Heart Disease. In: Swales JD, ed. *Textbook of Hypertension.* Oxford: Blackwell Scientific Publications, 1994: 1156-1164.

6. Collins R, Peto R, Gray R, Parish S. Large-scale randomized evidence: trials and overviews. In: Weatherall DJ, Ledingham JGG, Warrell DA, eds. *Oxford Textbook of Medicine.* Oxford: Oxford University Press, 1996: 21-32.

7. Early Breast Cancer Trialists' Collaborative Group. *Treatment of early breast cancer.* Vol 1. *Worldwide evidence 1985-1990.* Oxford: Oxford University Press, 1990: 12-18.

8. Antiplatelet Trialists' Collaboration. Collaborative overview of randomised trials of antiplatelet therapy – I: Prevention of death, myocardial infarction, and stroke by prolonged antiplatelet therapy in various categories of patients. BMJ 1994; 308: 81-106.

9. Antiplatelet Trialists' Collaboration. Collaborative overview of randomised trials of antiplatelet therapy – II: Maintenance of vascular graft or arterial patency by antiplatelet therapy. BMJ 1994; 308: 159-168.

10. Antiplatelet Trialists' Collaboration. Collaborative overview of randomised trials of antiplatelet therapy – III: Reduction in venous thrombosis and pulmonary embolism by antiplatelet prophylaxis among surgical and medical patients. BMJ 1994; 308: 235-246.

11. Patrono C. Aspirin as an antiplatelet drug. N Engl J Med 1994; 330: 1287-1294.

12. Fitzgerald GA. Dipyridamole. N Engl J Med 1987; 316: 1247-57.

13. Diener HC, Cunha L, Forbes C, Sivenius J, Smuts P, Lowenthal A. European Secondary Prevention Study 2. Dipyridamole and acetylsalicylic acid in the secondary prevention of stroke. J Neurol Sci 1996; 143: 1-13.

14. Sudlow C, Baigent C, on behalf of the Antithrombotic Trialists' Collaboration. Different antiplatelet regimens in the prevention of vascular events among patients at high risk of stroke: new evidence from the Antithrombotic Trialists' Collaboration. (Seventh European Stroke Conference, Edinburgh, May, 1998). *Cerebrovascular Diseases* 1998; 8(suppl 4): 68.

15. Savi P, Heilmann E, Nurden P, et al. Clopidogrel: an antithrombotic drug acting on the ADP-dependent activation pathway of human platelets. Clin Appl Thromb Hemost 1996; 2: 35-42.

16. CAPRIE Steering Committee. A randomised, blinded, trial of clopidogrel versus aspirin in patients at risk of ischemic events. Lancet 1996; 348: 1329-1339.

17. Ferguson JJ, Waly HM, Wilson JM. Fundamentals of coagulation and glycoprotein IIb/IIIa receptor inhibition. Eur Heart J 1998; 19 (Suppl D): D3-D9.

18. Braunwald E, Maseri A, Armstrong PW, et al. Rationale and clinical evidence for the use of GP IIb/IIIa inhibitors in acute coronary syndromes. Eur Heart J 1998; 19 (Suppl D): D22-D30.

19. The EPIC Investigators. Use of a monoclonal antibody directed against the platelet glycoprotein IIb/IIIa receptor in high risk coronary angioplasty. N Engl J Med 1994; 330: 956-961.

20. Lincoff AM, Califf RM, Anderson K, et al. Evidence for prevention of death and myocardial infarction with platelet membrane glycoprotein IIb/IIIa receptor blockade by abciximab (c7E3 Fab) among patients with unstable angina undergoing percutaneous coronary revascularization. J Am Coll Cardiol 1997; 30: 149-156.

21. Topol EJ, Ferguson JJ, Weisman HF, et al. Long-term protection from myocardial ischemic events in a randomized trial of integrin β_3 blockade with percutaneous coronary intervention. JAMA 1997; 278: 479-484.

22. The EPILOG Investigators. Platelet glycoprotein IIb/IIIa receptor blockade and low-dose heparin during percutaneous coronary revascularization. N Engl J Med 1997; 336: 1689-1696.

23. The CAPTURE Investigators. Randomised placebo-controlled trial of abciximab before and during coronary intervention in refractory unstable angina: the CAPTURE Study. Lancet 1997; 349: 1429-1435.

24. The IMPACT-II Investigators. Randomised placebo-controlled trial of effect of eptifibatide on complications of percutaneous coronary intervention: IMPACT-II. Lancet 1997; 349: 1422-1428.

25. The RESTORE Investigators. Effects of platelet glycoprotein IIb/IIIa blockade with tirofiban on adverse cardiac events in patients with unstable angina or acute myocardial infarction undergoing coronary angioplasty. Circulation 1997; 96: 1445-1453.

26. Califf RM. PURSUIT. Presented at the 70th Scientific Sessions of the American Heart Association; November, 1997, Orlando, Florida, USA.

27. Theroux P, Kouz S, Roy L, et al. Platelet membrane receptor glycoprotein IIb/IIIa antagonism in unstable angina: the Canadian Lamifiban Study. Circulation 1996; 94: 899-905.

28. Alexander JH, Newby LK, Moliterno DJ, et al. Relationship of outcomes to treatment with lamifiban in patients undergoing PTCA: analysis of PARAGON A (abstract). J Am Coll Cardiol 1997; 29: 409A.

29. Van de Werf F. PARAGON. Presented at the 19th Congress of the European Society of Cardiology; August, 1997, Stockholm, Sweden.

30. The PRISM-PLUS Study Investigators. Inhibition of the platelet glycoprotein IIb/IIIa receptor with tirofiban in unstable angina and non-Q-wave myocardial infarction. New Engl J Med 1998; 338: 1488-1497.

31. Kleiman NS, Ohman EM, Califf RM, et al. Profound inhibition of platelet aggregation with monoclonal antibody 7E3 Fab after thrombolytic therapy: results of the Thrombolysis and Angioplasty in Myocardial Infarction (TAMI) 8 pilot study. J Am Coll Cardiol 1993; 22: 381-389.

32. Ohman EM, Kleiman NS, Gacioch G, et al. Combined accelerated tissue-plasminogen activator and platelet glycoprotein IIb/IIIa integrin receptor blockade with integrilin in acute myocardial infarction: results of a randomized, placebo-controlled, dose-ranging trial. Circulation 1997; 95: 846-854.

33. Brener SJ, Barr LA, Burchenal J, et al. A randomized, placebo-controlled trial of abciximab with primary angioplasty for acute MI: the RAPPORT trial (abstract). Circulation 1997; 96 (suppl I) : I-473.

34. Harrington RA. PARADIGM. Presented at the 69th Scientific Sessions of the American Heart Association; November 1996, New Orleans, La, USA.

35. Moliterno DJ, Topol EJ. Meta-analysis of platelet GP IIb/IIIa antagonist randomized clinical trials in ischemic heart disease: consistent, durable, salutary effects. Circulation 1997; 96 (suppl I) : I-475.

36. Narins CR, Topol EJ. Attention shifts to the white clot. Lancet 1997; 350 (suppl III): 2.

37. Ferguson JJ, Kereiakes DJ, Adgey AAJ, et al. Safe use of GP IIb/IIIa inhibitors. Eur Heart J 1998; 19 (Suppl D): D40-D51.

38. Cannon CP, McCabe CH, Borzak S, et al. Randomized trial of an oral platelet glycoprotein IIb/IIIa antagonist, sibrofiban, in patients after an acute coronary syndrome: results of the TIMI 12 Trial. Circulation 1998; 97: 340-349.

39. Dyken ML, Barnett HJM, Easton JD, Fields WS, Fuster V, Hachinski V et al. Low-dose aspirin and stroke. "It ain't necessarily so". Stroke 1992; 23: 1395-1399.

40. Barnett HJM, Kaste M, Meldrum H, Eliasziw M. Aspirin dose in stroke prevention. Beautiful hypotheses slain by ugly facts. Stroke 1996; 27: 588-592.

41. UK-TIA Study Group. The United Kingdom transient ischemic attack (UK-TIA) aspirin trial: final results. J Neurol Neurosurg Psychiatry 1991; 54: 1044-54.

42. The Dutch TIA Study Group. A comparison of two doses of aspirin (30mg vs 283mg a day) in patients after a transient ischemic attack or minor ischemic stroke. N Engl J Med 1991; 325: 1261-1266.

43. SALT Collaborative Group. Swedish aspirin low-dose trial (SALT) of 75mg aspirin as secondary prophylaxis after cerebrovascular ischemic events. Lancet 1991; 338: 1345-1349.

44. Juul-Moller S, Edvardsson N, Jahnmatz B, Rosen A, Soreneson S, Omblus R, for the Swedish Angina Pectoris Aspirin Trial (SAPAT) Group. Double-blind trial of aspirin in primary prevention of myocardial infarction in patients with stable chronic angina pectoris. Lancet 1992; 340: 1421-1425.

45. Patrono C, Ciabattoni G, Patrigiani P, et al. Clinical pharmacology of platelet cyclooxygenase inhibition. Circulation 1985; 72: 1177-1184.

46. Sandercock PAG, van den Belt AGM, Lindley RI, Slattery J. Antithrombotic therapy in acute ischemic stroke: an overview of the completed randomised trials. J Neurol Neurosurg Psychiatry 1993; 56: 17-25.

47. International Stroke Trial Collaborative Group. The International Stroke Trial (IST): a randomised trial of aspirin, subcutaneous heparin, both or neither among 19,435 patients with acute ischemic stroke. Lancet 1997; 349: 1569-1581.

48. CAST (Chinese Acute Stroke Trial) Collaborative Group. CAST: randomised placebo-controlled trial of early aspirin use in 20,000 patients with acute ischemic stroke. Lancet 1997; 349: 1641-1649.

49. Multicentre Acute Stroke Trial – Italy (MAST-I) Group. Randomised controlled trial of streptokinase, aspirin, and combination of both in treatment of acute ischemic stroke. Lancet 1995; 346: 1509-1514.

50. Peto R, Gray R, Collins R, et al. Randomized trial of prophylactic daily aspirin in British male doctors. BMJ 1988; 296: 313-316.

51. Steering Committee of the Physicians' Health Study Research Group. Final report on the aspirin component of the ongoing physicians' health study. N Engl J Med 1989; 321: 129-135.

52. The Medical Research Council's General Practice Research Framework. Thrombosis prevention trial: randomised trial of low-intensity anticoagulation with warfarin and low-dose aspirin in the primary prevention of ischemic heart disease in men at increased risk. Lancet 1998; 351: 233-241.

53. Hansson L, Zanchetti A, Carruthers SG, et al. Effects of intensive blood-pressure lowering and low-dose aspirin in patients with hypertension: principal results of the Hypertension Optimal Treatment (HOT) randomised trial. Lancet 1998; 351: 1755-1762.

54. Gillum RF. Coronary artery bypass surgery and coronary angiography in the United States, 1979-1985. Am Heart J 1987; 113: 1255-1260.

55. Fuster V, Chesebro JH. Role of platelets and platelet inhibitors in aorto-coronary vein graft disease. Circulation 1986; 73: 227-232.

56. Cainpean L, Enjalbert M, Leoperance J. The relation of risk factors to the development of atherosclerosis in saphenous vein bypass grafts. N Engl J Med 1984; 311: 1329-1332.

57. Detre KM, Holubkov R, Kelsey SF, Cowley M, Kent KM, Williams DK, et al. Percutaneous transluminal coronary angioplasty in 1985-1986 and 1977-1981. N Engl J Med 1988; 318: 265-270.

58. MacMahon S, Raine A. Mortality from myocardial infarction among dialysis and renal transplant patients in Britain. Circulation 1990; 82 (Suppl 3): 615.

59. Pirk J, Ruzbarsky V, Konig J. The effect of antiaggregating drugs on the patency of grafts in the arterial system. World J Surg 1980; 4: 615-620.

60. Bonchek LI, Boerboom LE, Olinger GN. Prevention of lipid accumulation in experimental vein bypass by antiplatelet therapy. Circulation 1982; 66: 338-341.

61. Antiplatelet Trialists' Collaboration. Secondary prevention of vascular disease by prolonged antiplatelet treatment. BMJ 1988; 296: 320-331.

62. Israel DH, Adams PC, Stein B, Chesebro JH, Fuster V. Antithrombotic therapy in the coronary vein graft patient. Clin Cardiol 1991; 14: 283-295.

63. Thromboembolic Risk Factors (THRIFT) Consensus Group. Risk of and prophylaxis for venous thromboembolism in hospital patients. BMJ 1992; 305: 567-574.

64. Collins R, Scrimgeour A, Yusuf S, Peto R. Reduction in fatal pulmonary embolism and venous thrombosis by perioperative administration of subcutaneous heparin. N Engl J Med 1988; 318: 1162-1173.

65. Leizorovicz A, Haugh MC, Chapuis F-R, Boissel J-P. Low molecular weight heparin in the prevention of perioperative thrombosis. BMJ 1992; 305: 567-574.

66. Consensus Development Conference Report. Prevention of venous thrombosis and pulmonary embolism. JAMA 1986; 256: 744-749.

67. Collins R, Baigent C, Sandercock P, Peto R, for the Antiplatelet Trialists' Collaboration. Antiplatelet therapy for thromboprophylaxis: the need for careful consideration of the evidence from randomised trials. BMJ 1994; 309: 1215-1217.

68. Huber O, Bounameaux H, Borst F, Rohner A. Postoperative pulmonary embolism after hospital discharge. An underestimated risk. Arch Surg 1992; 127: 310-313.

69. Roderick PJ, Wilkes HC, Meade TW. The gastrointestinal toxicity of aspirin: an overview of randomised controlled trials. Br J Clin Pharmacol 1993; 35: 219-226.

70. Bamford J, Sandercock P, Dennis M, Burn J, Warlow C. A prospective study of acute cerebrovascular disease in the community: the Oxfordshire Community Stroke Project 1981-1986. 2. Incidence, case fatality and overall outcome at one year of cerebral infarction, primary intracerebral hemorrhage and subarachnoid hemorrhage. J Neurol Neurosurg Psychiatry 1990; 53: 16-22.

SUBJECT INDEX

GPSR Compliance

The European Union's (EU) General Product Safety Regulation (GPSR) is a set of rules that requires consumer products to be safe and our obligations to ensure this.

If you have any concerns about our products, you can contact us on ProductSafety@springernature.com

In case Publisher is established outside the EU, the EU authorized representative is:

Springer Nature Customer Service Center GmbH
Europaplatz 3
69115 Heidelberg, Germany

Batch number: 10761599

Printed by Printforce, the Netherlands